AF411431

Energy Storage Systems for Electric Vehicles

Energy Storage Systems for Electric Vehicles

Editor

Erik Schaltz

MDPI • Basel • Beijing • Wuhan • Barcelona • Belgrade • Manchester • Tokyo • Cluj • Tianjin

Editor
Erik Schaltz
Aalborg University
Denmark

Editorial Office
MDPI
St. Alban-Anlage 66
4052 Basel, Switzerland

This is a reprint of articles from the Special Issue published online in the open access journal *Energies* (ISSN 1996-1073) (available at: https://www.mdpi.com/journal/energies/special_issues/storage_EV).

For citation purposes, cite each article independently as indicated on the article page online and as indicated below:

LastName, A.A.; LastName, B.B.; LastName, C.C. Article Title. *Journal Name* **Year**, *Article Number*, Page Range.

ISBN 978-3-03936-962-1 (Hbk)
ISBN 978-3-03936-963-8 (PDF)

Contents

About the Editor

Erik Schaltz (Associate Professor) obtained his MSc and PhD degrees in electrical engineering from the Department of Energy Technology, Aalborg University, Aalborg, Denmark, in 2005 and 2010, respectively. From 2009 to 2012, he was an assistant professor, and since 2012, he has been an associate professor. Both positions are at the Department of Energy Technology, Aalborg University. In this Department, he is the program leader of the research program 'E-mobility and Industrial Drives', and the vice program leader of 'Battery Storage Systems'. He has been the main supervisor in four completed PhD projects, and has participated in more than 15 national and international research projects. He has been a guest and associate editor in several journals related to batteries and e-mobility, and he has authored or co-authored more than 100 publications. His research interests include usage of power electronics, electric machines, fuel cells, batteries, ultracapacitors, etc., in electric and hybrid electric vehicles. In addition, he is also focused on battery state-estimation, management (electric and thermal), and the modelling (electric, thermal and lifetime) of battery cells and packs.

Preface to "Energy Storage Systems for Electric Vehicles"

This book is based on the papers published in the Special Issue of *Energies* on 'Energy Storage Systems for Electric Vehicles'. Energy storage systems for electric vehicles is a multidisciplinary field, and this book therefore covers a wide range of topics, e.g. electric engineering, mechanical engineering, control engineering, environmental engineering, material science, etc. Energy storage systems is a key technology for future transportation systems and it is my hope that this book will give a good overview of the relevant research disciplines applied for energy storage systems for electric vehicles.

I would like to thank all the authors for their contributions, and the *Energies* Editorial Office for the professional management of the Special Issue. I am very grateful for the support of the authors and the Editorial Office, as it would not have been possible to create this book without them.

Erik Schaltz
Editor

Article

An Energy Consumption Model for Designing an AGV Energy Storage System with a PEMFC Stack

Roman Niestrój [1], Tomasz Rogala [2] and Wojciech Skarka [2,*]

[1]   Department of Electrical Engineering and Computer Science, Silesian University of Technology, Akademicka 2A, 44-100 Gliwice, Poland; roman.niestroj@polsl.pl

[2]   Deparment of Fundamentals of Machinery Design, Silesian University of Technology, Akademicka 2A, 44-100 Gliwice, Poland; tomasz.rogala@polsl.pl

*   Correspondence: wojciech.skarka@polsl.pl

Received: 2 May 2020; Accepted: 1 July 2020; Published: 3 July 2020

Abstract: This article presents a methodology for building an AGV (automated guided vehicle) power supply system simulation model with a polymer electrolyte membrane fuel cell stack (PEMFC). The model focuses on selecting the correct parameters for the hybrid energy buffering system to ensure proper operating parameters of the vehicle, i.e., minimizing vehicle downtime. The AGV uses 2×1.18 kW electric motors and is a development version of a battery-powered vehicle in which the battery has been replaced with a hybrid power system using a 300 W PEMFC. The research and development of the new power system were initiated by the AGV manufacturer. The model-based design (MBD) methodology is used in the design and construction of a complete simulation model for the system, which consists of the fuel cell system, energy processing, a storage system, and an energy demand models. The energy demand model has been developed based on measurements from the existing AGV, and the remaining parts of the model are based on simulation models tuned to the characteristics obtained for the individual subsystems or from commonly available data. A parametric model is created with the possibility for development and determination by simulation of either the final system or from the parameters of the individual models' elements (components of the designed system). The presented methodology can be used to develop alternative versions of the system, in particular the selection of the correct size of supercapacitors and batteries which depend on the energy demand profile and the development of the DC/DC converter and controllers. Additionally, the varying topology of the whole system was also analyzed. Minimization of downtime has been presented as one of many possible uses of the presented model.

Keywords: fuel cell; automated guided vehicle; hybrid energy storage system; model-based design; waveforms modeling; autoregressive models of nonstationary signals

---

1. Introduction

The use of electric drives in various types of vehicles is becoming increasingly popular. The growing use of such drives is due to the many advantages of electric motors compared to internal combustion engines. This is particularly evident in closed areas in internal transport where automated guided vehicles (AGVs) are heavily utilized. High torque, quiet operation, and zero-emissions are just some of the advantages over other primitive solutions. However, these vehicles have operational problems such as insufficient work duration and limited ranges. This is caused by the relatively low energy density of the energy sources used in these vehicles. The development dynamics of the basic energy sources used in AGVs, such as lithium-ion batteries, does not indicate that this problem will be solved in the short-term (within the next decade). For this reason, designers are searching for other energy sources that provide significantly higher levels of energy density while having the same advantages as modern batteries. One of the proposed solutions is to use hydrogen fuel cell stacks to power

AGVs. However, the power supply system itself, based on a fuel cell (FC), is much more complex than that of battery power. Usually, the fuel cell is supplemented with a hybrid energy storage system constituting an energy buffer that eliminates the imperfections associated with using hydrogen fuel cell stacks. This is due to the operational characteristics of the fuel cell that must produce electricity after commissioning; thus, the efficiency of electricity generation varies significantly depending on the load on the fuel cell. It is particularly unfavorable to operate the fuel cell at a very low/high load or with high dynamics of load change, which significantly reduces the efficiency of this device. Typically, an energy buffer comprises a battery and a set of supercapacitors with appropriately selected parameters. Control of the operation of the hydrogen fuel cell, integration of the appropriate battery size, buffering problems, multidirectional energy conversion, adaptation of electrical energy to various parameters, and hydrogen storage and supply all have specific characteristics and require appropriate adjustment of the power supply system to meet these energy demand characteristics. This means designing an optimal power supply system using a hydrogen fuel cell is a complex task.

As part of the work, the design of a hydrogen fuel cell-based power supply system for an existing AGV (Formica-1, AIUT Ltd., Gliwice, Poland) powered with a lithium-ion battery was undertaken, with an effort to preserve the vehicle's operational characteristics, minimize any structural changes, and significantly increase the vehicle's operating time.

Due to the limited number of commercially available FC's capacities, the fuel cell selection is usually based on average demand power. The authors note that the main problem in designing the entire power system based on FC is the correct selection of the energy buffer. Therefore, particular attention is paid to the selection of a hybrid energy storage system because the correct choice of this system allows one to adjust the characteristics of the entire system to one's needs, whilst minimizing the fuel cell's power.

The justification for using an energy buffer with an FC is to temper large fluctuations in power demand and to accumulate energy from regenerative braking. The energy buffer, in this case, corrects FC deficiencies as the FC is not able to rapidly increase its power output, has a limited peak power, and is not able to absorb braking energy. The nature of FC's work dedicates them rather to independent work in stationary applications. For traction applications, an energy buffer is needed that is tailored exactly to the nature of the energy demand.

To solve the selection problem for hybrid energy storage system in an AGV powered by a polymer electrolyte membrane fuel cell stack (PEMFC) outlined in this section, we urge the reader to refer to the literature review regarding the model-based design for methodologies used, FC modeling, for discussion on the components utilized, and for an overview of FC-based power systems (Section 2). Section 3 describes the assumptions of the general methodology for designing the entire system, and the assumptions, modeling methods, and model bases for the individual subsystems of the entire AGV, in particular, the energy demand at various operational states, the hybrid vehicle power systems model such as the FC, DC/DC converter model, as well as supercapacitors and batteries. For the system's application (Section 4), details of AGV development research involving the change of the power supply system from a battery system to a system based on FC is described. Section 4 highlights the identification of the vehicle's energy demand at various operating states, the model for this demand, the use of a power system model with the energy demand model for optimizating the newly developed power system based on a previously selected FC, and the selection of the structure and parameters of the buffering system energy. Optimizations were carried out through simulation experiments using developed models. The last section provides a detailed discussion of the results from earlier studies.

2. State-of-the-Art

Model-based design (MBD) methodology is often used to design complex mechatronic systems [1,2]. The methodology consists of building computer simulation models of the designed system and simulating its operation. The use of such a methodology is particularly beneficial in the design of systems requiring the cooperation of specialists from various fields and systems. In our case, the system crosses fields that include mechanical, electrical, and chemical sciences, which address difficult to describe phenomena and require personalized and very specific system control. A typical example of such a system is a drive and power supply system for vehicles, such as a power supply system consisting of a hydrogen fuel cell, battery, supercapacitors, and the respective control systems and energy flow processes. The purpose of using MBD methodology is to initially plan how to design a system, its operation, and control its parameters, all whilst meeting the criteria and fulfilling its desired functions. It is also possible to determine appropriate or optimal technical parameters of the individual subsystems, such as the technical parameters and the control method. Usually, the complete model includes not only the designed system, the power supply system, and the drive system, but also the entire facility on which the designed system is built, as well as external conditions affecting the operation of the whole. For vehicles, this is usually a power system model, the propulsion system model, and the entire vehicle model, and often includes the route model and the conditions they encounter. Depending on the situation, the complexity of the model should be adjusted to obtain satisfactory results [2,3]. If we have a prototype or a copy of the system available, we can determine an appropriate model using experimental tests, but if the system is in the concept phase we must build a model, e.g., a model based on the sum of the general theoretical phenomena occurring in the system. Likewise, the model for the power supply or propulsion system itself is much more complex, whilst the model of the routes and the whole vehicle is simplified or considers the relevant data to enable simulation. For hydrogen-powered electric vehicles, the most important and substantive input model is the hydrogen fuel cell itself, which forms the entire power supply system as the whole vehicle is driven and powered by such a system. Usually, choosing the correct system parameters makes the most sense when the vehicle is traveling along a fixed route or a finite and known set of routes where the load and driving conditions are set or predictable. With this knowledge, one can accurately determine the features of the power system. This is the case with certain types of vehicles such as AGVs or racing vehicles. However, if the load conditions and route conditions are unknown or difficult to predict, the task is much more difficult, and the results will not be as definitive as expected.

This section describes the problems associated with modeling vehicle system components and is was divided into two parts: The first concerns the hydrogen cells themselves and the second deals with the remaining elements of the energy conversion system. In these subsections, the authors focus primarily on the energy buffer, but FCs are also analyzed because it is the operational features of the FCs that have a significant impact on the selection of the energy buffer. Another element that affects the form and characteristics of the energy buffer is the fluctuating nature of energy demand and the need to recover energy under vehicle braking. Correct and detailed modeling of these sections of the whole system (and not only the energy buffering system) is, therefore, a condition for completing simulation experiments from which the energy buffering system will be selected.

2.1. Modeling of the Fuel Cell Stack

A hydrogen fuel cell is an electrochemical device that converts chemical energy via an electrolytic reaction directly into electricity. For modeling purposes, it is not necessary to know all physicochemical conditions related to energy production in the FC.

There are several classes of simulation models in the literature, which can be divided into three sub-groups; electrical, chemical, and experimental. Electric fuel cell models are used to compute power systems. This model treats the fuel cell as an element of an electrical circuit and does not include phenomena underlying electrical production. Phenomena such as particle diffusion, mass transport, and thermodynamic transformations are addressed in a chemical model.

The commonly used generic simulation model using a MATLAB/Simulink system includes two types of models: Simplified and detailed. Such models include the calculation of the irreversibles that affect the voltage drop of a cell during operation relative to the theoretical voltage resulting from a chemical reaction, which in turn results in changes in the energy characteristics. This is influenced by the following types of irreversibles: Activation losses, fuel crossover and internal currents, ohmic losses, mass transport, and concentration losses. The origins and descriptions of these irreversibles, as well as the modeling method, have been previously discussed [4,5]. When creating a simplified model, two points from the activation region and two points from the ohmic region are utilized from the polarization curve. However, for the construction of a detailed model, further data are required, such as the number of cells, nominal Low Heating Value (LHV) stack efficiency, nominal operating temperature, nominal air flow rate, absolute supply pressure, and the nominal composition of fuel and air; these are typically provided in the fuel cell manufacturer's documentation. When it comes to modeling fuel cell dynamics, current step and interrupt tests must be completed for a given cell. The necessary parameters to construct this part of the model are then determined from these tests or can be obtained directly from the manufacturer's data, because they depend on the fuel cell itself. If such tests cannot be performed, the data can be assumed from a recommended range [4]. Occasionally, the fuel cell manufacturer does not provide basic technical data for the FC, and in this case more tests on the system are required to determine a full range of parameters. Other parameters obtained in tests depend on the whole system in which the FC works and its load, and they are specific for a particular configuration of the system.

Other fuel cell equivalent circuit models for passive mode testing and dynamic mode design have been compared in [6]. This comparison includes the following dynamic models: Larmie [7], Dicks–Larminie [5], Yu-Yuvarajan [8], Choi [9], and shows that complex models are not always effective for practical applications. These four dynamic models are used to simulate the power-generating cell, whilst the passive equivalent circuit model represents the fuel cell which is not producing electric power. These models represent the response to an external electric stimulus to determine the condition of the fuel cell. Additionally, in [6], Page [10], and Garnier [11], passive models are compared. The work [6] does not present any relationship between passive mode test responses and dynamic mode performance.

Not all of the fuel cell's irreversibles are relevant under normal operating conditions. While commissioning and rated conditions are the most common conditions, overloading is not a common condition. Some systems do not function under FC operation with such overloading conditions at all. Therefore, irreversibles that affect work under such unusual conditions are not considered or modeled at all. However, sometimes this is needed, and irreversible mass transport and concentration losses must be modeled. A model for mass transport losses in the form of a theoretical model is presented in [12] and in the form of an empirical model in [4]. This model was developed to simulate transport phenomena in a proton exchange membrane fuel cell (PEMFC).

The hydrogen fuel cell is complex and expensive, and in systems with high dynamics of power demand where it is required to supplement such a cell with additional elements such as startup batteries, buffers, inverters/converters, then the whole system needs to be modified to handle a specific load. Testing such a system can be completed using a computer simulation model presented later in the article, but it is also possible to create a physical simulator which is a cheap alternative to testing. Such a solution built based on a programmable DC power supply, control interface, and software written in LabVIEW has been proposed for testing the entire system and acts as a guide in the development of power conditioning equipment [13] with the ability to work in steady-state and transient modes.

2.2. Modeling of the System Using a Fuel Cell Stack

To generate energy in FCs, it is necessary to use a hydrogen tank together with a hydrogen pressure reduction system and a control valve mediated by a controller to regulate the amount of hydrogen supplied on an ongoing basis. Oxygen is usually supplied from the air through a fan system to the fuel cell. It is also possible to supply oxygen from a high-pressure tank similar to the whole hydrogen supply system.

For large FCs (larger than 10 kW), the installation of the FC itself becomes very complicated and maintaining balanced operational parameters becomes a problem. These issues are the subject of separate research, and balance of the plant (BOP) [14] and incorrect configuration and selection of inappropriate operating parameters of individual elements of the FC system can lead to insufficient cell performance and rapid degradation of the cell. A simple solution to the complexity of the installation on the FC preparation side can be made by using an FC configuration based on a dead-end anode (DEA) structure. This type of installation, unlike the flow-through anode (FTA) configuration, significantly simplifies the need to prepare hydrogen and guarantees the appropriate humidity of the cell, ensuring close to 100% hydrogen use by controlling the (normally closed) purge valve [15,16]. This configuration is popular for low power FCs, but also developed for higher power applications. DEA installations operating differently to FTA are not managed by a regulated control valve and have to be purged periodically by the purge valve [17].

Since the fuel cell itself is an energy generator operating under specific parameters, usually this produced energy must be adapted to the purposes of the energy demand characteristics. If the power take-off is not variable, this system may be simpler, but with high variability of energy demand, it is necessary to consider the electric converter/inverter and energy buffer supplied via batteries or supercapacitors. Supervising the work of these devices can take place at various levels, most often at the basic level through ongoing control of the parameters of individual devices, and frequently at the strategic level by adapting the operating parameters not only to the ongoing demand but also to future demand.

Modeling the power supply load is a separate problem. A power supply load model takes the form of a specific load profile based on the behavior of the powered system and optimized with measurements taken during the experiment or by considering physical phenomena, e.g., a model outlining the dynamics of a moving object. The choice to develop this model depends on the design phase. If one has a prototype or physical copy of the system required to be powered, one can choose the first solution, but if one only has the concept or accurate documentation, the second solution is needed. Interesting solutions can be found in various works for modeling system fragments or the entire system oriented at determining specific parameters. An example model for a complete power supply system for hybrid vehicles is described in [18]. The modeled system consists of a hydrogen power supply, DC/DC converter, battery, inverter, electric motor, and the vehicle body. A complete model of the system was developed based on the experimental data. The model was then used to develop a power management control algorithm for fuel cell hybrid vehicles using a stochastic programing technique. This approach requires a complete system which can be subjected to a series of experiments. Another approach is to use model-based design (MBD), where a model is created at the design and concept stage, and numerical simulation experiments outline various potential solutions and determine the impact of various parameters on the system's performance (sensitivity analysis), or to formally optimize the system or its components [1,2].

Improved modeling of the Proton Exchange Membrane (PEM) fuel cell power stack for electric vehicles in which a separate oxygen tank supply system was used to improve performance is presented in [19]. Simulation calculations were oriented towards finding an optimal control strategy for the pressure that facilitates the output power according to the power demand of the load.

In addition to the holistic approach to modeling the hydrogen fuel cell system, researchers are interested in individual elements of the system. Furthermore, the hydrogen cell itself, with a series of tanks, controllers, control valves, and fans supplying air, may include power electronics which process and adapt energy to meet demand from energy buffering units, including various types of batteries and supercapacitors. Regulators and controllers are indispensable to these units and operate at various levels, and often operate with a complex strategy for a given application.

Selecting the power electronics for the FC's energy conversion system is quite a difficult task. The situation is additionally complicated by the fact that the energy-receiving system requires the conversion of energy to different voltages, types of currents, and their power simultaneously. We chose to only focus on work completed on general modeling of energy flow and power losses, not energy-electronic phenomena or their modeling. Therefore, only models for average value converters were assessed, and those cooperating with basic energy buffers and thus implementing alternative strategies of constant current and constant output voltage were analyzed. There are no universal solutions to control of energy flow because the characteristics of the energy demand received from FC are application dependent. The selection of elements of the entire FC system is of interest to many scientists. An essential element in the system is the boost converter. A simple model of the cell as an electrical circuit has been previously described [20]. The model, taking into account a portion of the irreversibles appropriate to the nature of the application, is used to select the suitable type of DC/DC boost converter and to select the parameters of the energy storage constituting the energy buffer which compensates for rapid changes in energy demand. Various connection options (behind or before the converter) of the supercapacitors are also discussed.

For energy buffers, there has been significant progress in the development of the latest types of batteries. The multitude of solutions is not conducive when making optimal decisions, especially at the development stage of the system. Therefore, simple battery models using the most popular battery types are used. The basic battery models are lithium-ion, lead-acid, nickel-cadmium, and nickel–metal hydride [21], and their various parameters are also defined, including charge and discharge, temperature effects, and ageing. This enables the modeling of various connections cells in series and/or in parallel [22–25].

A "Theoretical Modeling Methods for Thermal Management of Batteries" review has been previously completed [26]. In addition to typical models, various new approaches are presented, e.g., in [27,28].

In [27] a novel lumped electrothermal circuit of a single battery cell was presented, including the extraction procedure of the parameters of the single-cell from experimental evidence and a simulation environment, given in SystemC-WMS for the simulation of a battery pack.

In [28] a new open-circuit voltage (OCV) model is proposed. The new model can simulate the OCV curves of a lithium iron magnesium phosphate (LiFeMgPO$_4$) battery at different temperatures. It also considers both charging and discharging. The most remarkable feature from the different models, in addition to the proposed OCV model, is their integration into a single hybrid electrical model. A lumped thermal model is implemented to simulate the temperature development in the battery cell. The synthesized electro-thermal battery cell model is extended to model a battery pack of an actual electric vehicle.

Typically, the problem of choosing a buffer system includes what type of energy buffers will be selected, the size of the buffer, and the features of individual parts (batteries, supercapacitors). Buffer hybridization is a common solution which involves a combination of a supercapacitor with a battery and is outlined in [29]. Various configurations and sets of supercapacitors and batteries together with DC/DC converters are discussed in several papers [30–32]. The correct selection of the buffer parameters and the topology of this system allows one to overcome most of the FC's weaknesses. Selecting the optimal parameters and topology for these subsystems in the FC is important, as the FC is strongly dependent on the energy demand characteristics in the system [30,32].

Modeling supercapacitors (SC) requires consideration of the electrical, self-discharge, and thermal behavior. A comprehensive review of the modeling techniques is described [33–35]

The equivalent mathematical model derived from the electrical model, which was used to simulate the voltage response of the supercapacitor, is presented in [33].

The review presented in [33] discusses SC modeling, state estimation, and their industrial applications, intending to summarize recent research progress and stimulate innovative thoughts for SC control/management. For the SC modeling, state-of-the-art models for electrical, self-discharge, and thermal behavior are systematically reviewed, where the electrochemical, equivalent circuit, intelligent, and fractional-order models describing the electrical behavior simulation are highlighted. For SC state estimation, methods for state-of-charge (SOC) estimation and state-of-health (SOH) monitoring are covered, together with an underlying analysis of the ageing mechanism and its influencing factors.

The models which are described in the literature have various advantages and disadvantages, ranging from the ease of use down to the complexity of characterization and parameter identification. Work presented in [35] presents a comprehensive review and compares these models, specifically focusing on the models that predict the electrical characteristics of double-layer capacitors (DLC), showing the strengths and weaknesses of different available models and their various areas for improvement.

Experience in implementing the various applications of the hydrogen fuel cell system is very helpful when designing a complete system. One can find many interesting descriptions of applications with different degrees of maturity and covering both stationary and mobile applications in ground, water, and aerial vehicles. Research has described the various aspects of the whole system and its hybridization [36–39], current energy management and energy management strategy [40–44], energy control and processing [45,46], optimization of power systems based on fuel cells for matching operational parameters [47,48], power transmission in hydrogen cell-powered propulsion systems [49], and general aspects of the development of hydrogen cell-based systems [50,51].

3. Model of Energy Transfer in the System

A general methodology for building an energy transfer model enabling simulation experiments when designing a hybrid power supply (HPS) system based on a hydrogen cell stack for an AGV is shown in Figure 1.

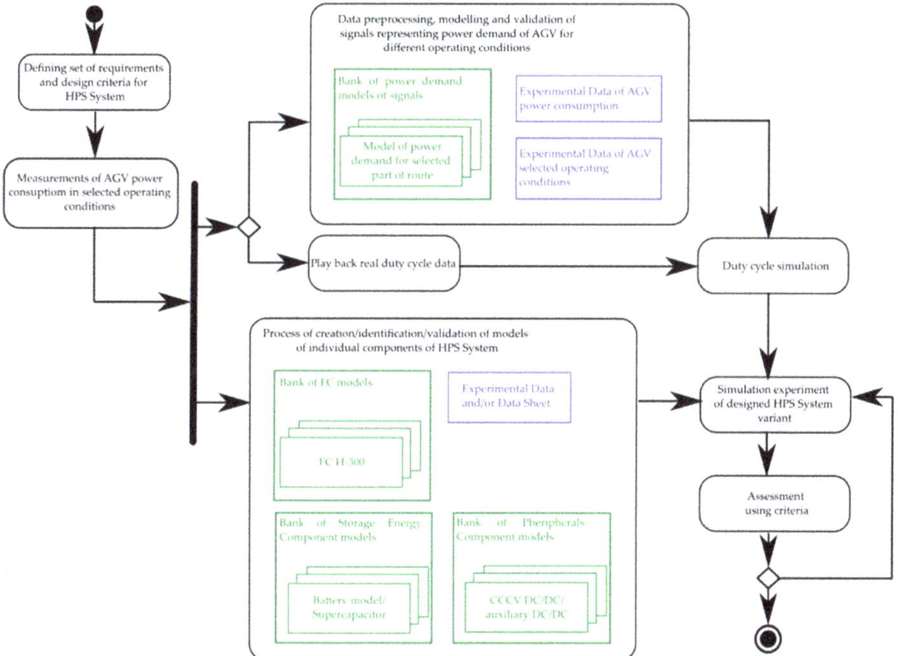

Figure 1. A general methodology for building a model enabling simulation experiments for designing a hybrid energy storage system with a hydrogen cell stack for an automated guided vehicle (AGV).

Conducting simulation experiments requires the definition of the HPS system in the AGV. Since these vehicles are designed for close repetitive transport operations over long periods and have known operating conditions, i.e., speed and load, one can adapt the HPS system to individual needs, such as the demand for instantaneous power during a specific operating condition. At this stage, the criterion for assessing the designed HPS system should also be determined. For the next step, it is necessary to measure and identify the instantaneous power demand by the AGV at different operating conditions. These measurements should include the power demand for expected operating conditions over the planned route. From this, work can be completed on the data preprocessing, modeling, and validation of the models representing the power demand. These models are identified based on data from instantaneous power measurements at various operating conditions. Based on a set of such models, it is possible to simulate the power demand for the new AGV route and other operating conditions. A detailed discussion on this subject is presented in Section 3.1.

Simultaneously, by the defining power demand models, it is possible to create component models of the HPS system. It should be noted that these models of HPS can be identified based on additional measurements or characteristics provided by the manufacturers. More information about creating and identifying component models of an HPS including a hydrogen cell, models of storage components, and other auxiliary components, are described in Section 3.2. To validate the hybrid power supply system model, it is possible to provide the load in the form of played-back real values of instantaneous power demand and creating comparisons, e.g., concerning the current power supply system installed in AGV.

The proposed methodology described herein and in further sections of this manuscript allowed us to design a customized HPS system for operating conditions over a preplanned route. This can be achieved by conducting simulation experiments to find the optimal solution or a set of possible solutions which satisfy defined criteria. The optimal criteria can refer to finding the optimal battery or supercapacitor capacity for the HPS or other objectives. More information on this subject is discussed in Sections 4.2 and 4.3.

3.1. A Generic Model for Instantaneous Power Demand

This section describes a generic procedure for building a model to compute instantaneous power demand. This generic model is used to estimate the instantaneous power demand under the AGV's different operating conditions during working duty cycles. The model results are used as a load for the hybrid hydrogen power supply system model discussed in Section 3.2. The use of both models makes it possible to perform different simulation experiments, which allows one to examine different configurations of a power supply system with varying parameters. The generic model for instantaneous power demand is the first part of this model.

The presented methodology for building an instantaneous power demand model, ultimately to develop a new hybrid vehicle power supply system, depends on the available data sources. Two possible options defining the data source availability can be distinguished:

- Variant I: Data which describe the full dynamic model of the AGV are available. In this case, the developed model allows one to implement any scenario of AGV operation and estimate the instantaneous power demand. The data includes all the dynamic parameters of the vehicle including the mechanical system of the vehicle transmission system, the model of the control system, as well as the electric power supply system. It should be noted this is a seldom case and is a time-consuming modeling activity that requires a lot of information about the considered object, i.e., access to information about the dynamic parameters of the vehicle, information about how the vehicle is controlled, including the operation of supervised control system, etc. Unfortunately, some sections of this information are often unavailable due to companies protecting their intellectual property.

- Variant II: Only data with selected operating conditions are available, such as the speed of individual main drives that accompany the measurements of the instantaneous power demand of the vehicle. It should be noted that the use of this variant is purposeful, especially for AGV which has a limited number of possible settings of selected operating conditions, e.g., rotation speed of drives as well as acceleration and braking ramps. In such a case, it is not necessary to identify the entire domain defined by the space of possible values under the parameters of the operating conditions but only selected characteristic parameters.

In the further sections of this script, only variant II is considered. This approach requires one to define the following operating conditions:

- A speed parameter v of a vehicle or rotational speed of drive or drives. Under stationary conditions, this parameter should be measured at typical velocities for the type vehicle. For instance, 0.3 or 1.2 m/s are used as standard velocities [52] and some values are set by the manufacturer, e.g., 0.5 m/s (according to safety requirements [52]), and the maximum speed adjusted to the maximum permissible load. In this work, the safe velocity value for the maximum load of 1.2 t is 0.8 m/s. Measurements of velocities under transient conditions also allow identification of the acceleration and deceleration ramps;
- A carried load L with respect to the maximum limit load;
- Description of the characteristic route and driving direction, e.g., straight route ahead, straight route reverse, right turn, left turn, rotation around the AGV normal axis;
- Information about the inclination of the route (maximum 3% for AGVs according to the standard [52] on a technical floor); in this study this value has been omitted.

For the aforementioned values, under a combination of operating conditions, a bank of autoregressive models has been applied. These models are representations of signals which, for selected operating conditions, represent instantaneous power demand for the selected type of vehicle. The main task of the models, in detail, is to:

- Represent expected values and variance of the instantaneous power demand under selected operating conditions;
- Reflect the dynamics of changes in the instantaneous power demand and their frequency amplitude characteristics.

3.1.1. Models for Stationary Conditions

The autoregressive model of the signal [53–55] of instantaneous power demand is given by the formula:

$$M(k): \ y(k) = E\{y\} + \frac{\epsilon(k) \times Var\{y\}}{A\left(q_p^{-1}\right)} \tag{1}$$

where y is the instantaneous power demand, ϵ is the noise which follows a Gaussian distribution, $A\left(q_n^{-1}\right)$ is a polynomial of the n order represented by $A\left(q_n^{-1}\right) = 1 + a_1 \times q^{-1} + a_2 \times q^{-2} + a_n \times q^{-n}$, and $E\{y\}$, $Var\{y\}$ are the expected value and variance of the instantaneous power demand. The expected value and variance can be additionally represented by other linear or quadratic functions of $f(V, L)$. To account for dynamic changes in the instantaneous power demand, the model can be represented in the frequency domain [55] using the following formula:

$$M(j\omega): P_y\left(e^{j\omega}\right) = \frac{\epsilon(k) \times Var\{Y\}}{\left|A\left(e^{j\omega}\right)\right|^2} = E\{Y\}^2 + \frac{\epsilon(k) \times Var\{Y\}}{|1 + \sum_{k=1}^{p} a(k) \times e^{-j\omega k}|^2} \tag{2}$$

where $P_y\left(e^{j\omega}\right)$ represents the power spectral density of the modeled signal. The above model can be applied under stationary operating conditions.

3.1.2. Models for Nonstationary Conditions

Similarly to stationary conditions, a signal model can be built for the nonstationary conditions [55,56]. This applies to parameters such as acceleration, braking, and emergency braking, etc. The model to apply for this case has the following formula:

$$M(k): \ y(k) = E\{y_d\} + \frac{\epsilon(k) \times Var\{y_d\} \times En\{y_d\}}{A\left(q_p^{-1}\right)} + Tr\{y\} \tag{3}$$

where $Tr\{y\}$ is the linear model of the acceleration, deceleration ramp, etc. This part of the model can be determined using a least-squares criterion. $En\{y_d\}$ is the envelope model established for the signal after removing the trend from the nonstationary signal y_d. The model of the envelope can be evaluated for the following signal:

$$En\{y\} = \frac{|z[k]| + |z[k-1]| + \cdots + |z[n-N]|}{N-1} \tag{4}$$

where $|z[k]|$ is a module of an analytical signal obtained using a Hilbert transform [57] and N is a length of the moving average filter. The envelope signal can be represented by a regressive model given by Equation (1). If the envelope is flat and monotonical then a linear model can be used.

After removing the trend and by eliminating the second-order nonstationarity resulting from the variable variance, the frequency assessment of the model presented in the Equation (2) can then also be used.

3.1.3. Model Validation

The validation of the model describing the route section under selected operating conditions can be calculated by using the following measures:

- Using a training data set to develop the model and validation data $y_{val}(k)$, the following measures of model compliance can be determined:

$$Er_{Energy} = \frac{\left| \int_0^t M(k)dt - \int_0^t y_{val}(k)dt \right|}{\int_0^t y_{val}(k)dt} 100\% \tag{5}$$

where $\int_0^t M(k)dt$ is the energy computed for the signal model, and $\int_0^t y_{val}(k)dt$ is the energy of the validation signal.

- The second measures (as a functional feature) of model compliance are executed with the use of relative error of power in the frequency domain:

$$Er_{FreqStruct} = \left| P_y\!\left(e^{j\omega}\right) - P_{y_{val}}\!\left(e^{j\omega}\right) \right| \tag{6}$$

where $P_y\!\left(e^{j\omega}\right), P_{y_{val}}\!\left(e^{j\omega}\right)$ are the power spectral densities of the model obtained as an output of the model and the power spectral density of validation data, respectively. The measure determined here is a functional assessment in the frequency domain and it determines the difference in signal power for the frequency components. The selection of the model order is determined, based on the similarity of the power spectral density characteristics, to reflect the dynamics of the signal changes by the signal model.

3.1.4. Combining Models

After validating the individual models representing the signal from instantaneous power demand, a selected scenario can be built which represents the AGV route. Usually, this route is planned and the AGV moves along the route under established operating condition parameters such as speed, load, etc.

Before creating a power demand model for a selected vehicle scenario, it was necessary to divide the scenario into appropriate route sections for which appropriate models would be assigned to generate the instantaneous power demand signals.

An important element when building full waveforms for the entire scenario was the points where the signals of the partial models would be combined. To combine waveforms of the individual models, it is possible to use the following window (Equation (7)), which is a modified version of the window previously shown in [58], the length of which can correspond to the length of the modeled waveforms:

$$
w(y) = \begin{cases}
\left(-\frac{1}{2} \times d_l + \frac{1}{2}\right) \times \left\{1 + \cos\left(\frac{2\Pi}{r} \times \left[x - \frac{r}{2}\right]\right)\right\} + d_l, & 0 \leq y < \frac{r}{2} \\
1, & \frac{r}{2} \leq y < 1 - \frac{r}{2} \\
\left(-\frac{1}{2} \times d_r + \frac{1}{2}\right) \times \left\{1 + \cos\left(\frac{2\Pi}{r} \times \left[x - 1 + \frac{r}{2}\right]\right)\right\} + d_r, & 1 - \frac{r}{2} \leq y < 1
\end{cases}
\tag{7}
$$

3.2. Hybrid Power Supply System Model for the AGV

The model of the hybrid power supply system for the AGV was developed in the MATLAB/Simulink environment partly using the Simscape Electrical library components. This model is a numerical tool supporting the selection of elements for the hybrid power supply system. The block diagram of a hybrid power supply system is shown in Figure 2. The numerical model was built based on this block diagram (Supplementary Materials). This model could be used to optimize the parameters of the power supply system after a specific operation scenario for the AGV is chosen (length and diversity of the route, load, driving dynamics) and after assuming the optimization criteria (for example, minimizing the capacity of the main energy store).

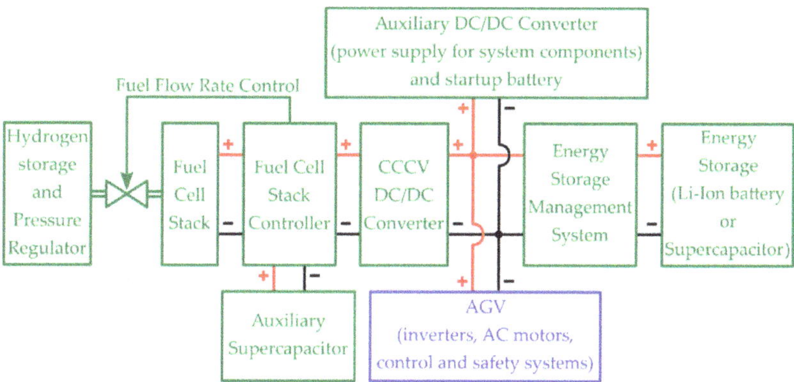

Figure 2. A block diagram describing the hybrid power supply system for the AGV.

The electrical energy source in the hybrid power supply system was the fuel cell stack fueled by hydrogen. It was assumed that hydrogen was stored in a metal hydrides tank equipped with a pressure regulator [59]. The flow of hydrogen through the fuel cell stack was regulated by a proportional control valve. The control signal for this valve was generated using a hydrogen flow regulator. This regulator was a component of the fuel cell stack controller. The controller additionally protected the stack against operation from moving outside the safe operating range of the electrical and thermal parameters. In addition, the fuel cell stack controller contained the SCU (short circuit unit), which periodically short-circuited the stack and improved its performance [60]. Due to the operation of the SCU, it was necessary to install an auxiliary supercapacitor in the system, which maintained the supply voltage for the duration of the stack short-circuit, and additionally provided an energy buffer for rapid changes in the load current of the stack when the stack was not able to impulsively provide adequate power due to limitations imposed by its own dynamics and the hydrogen fueling system dynamics.

Electrical energy from the fuel cell stack was supplied to the main AGV power busbars through a DC/DC Constant Current - Constant Voltage (CCCV) converter working at a Constant Current (CC) or Constant Voltage (CV) output, where the output current setpoint for CC mode could be invariable or could be set by the stack load power regulator, which was part of the converter control system. The method for determining the output current setpoint depended on the configuration of the hybrid power supply system used and the method of its optimization.

A lithium-ion battery or supercapacitor could act as the main electrical energy storage. The reasons for using electrical energy storage (as energy buffer) together with the fuel cell stack, in traction applications, were the large fluctuations in the power demand and the need to accumulate energy from regenerative braking. The energy storage, in this case, complemented the deficiencies of the fuel cell stack, meaning the stack was not able to increase the output power impulsively, had limited peak power, and was not able to absorb braking energy. The nature of the fuel cell stack was rather dedicated to independent work in stationary applications. In traction applications, an additional energy storage device was necessary [30–32].

There was a management system between the main power busbars and the main energy storage, the primary role of which was to protect the energy storage against operation outside the safe range of electrical and thermal parameters. The management system also allowed for pre-charging of the main energy storage with energy from the fuel cell stack after starting the hybrid power supply system, which was needed when the supercapacitor acted as the energy storage. It was assumed that the energy storage could also be charged from an external energy source, depending on the adopted configuration of the hybrid power supply system and the scenario of the AGV operation.

While developing the numerical model for the hybrid power supply system, assumptions were considered from the practical conditions or were the result of previous preliminary analyses. The initial selection of the fuel cell stack was guided by the average power demand of the AGV and from economic criteria. The cheapest fuel cell stack was selected that would meet the AGV requirements according to preliminary estimates. It was assumed that a horizon fuel cell stack, type H-300, with 300 W power, a rated voltage 36 V, and rated current of 8.3 A would be used [61]. This stack consisted of 60 PEM fuel cells connected in series, low-temperature operation, powered by hydrogen from the pressure tank and oxygen obtained from atmospheric air. The nominal efficiency of the H-300 stack was 40%. This was a low power fuel cell stack that had a very simple "balance of plant" structure. The stack was equipped with three fans that provided cooling to the stack with a suitable amount of the air. The fuel cell stack was equipped with a factory controller that regulated the rotation speed of the fans by supplying them using the Pulse Width Modulation (PWM) method, and which controlled the hydrogen two-state valves. This fuel cell stack with factory controller functioned as a dead-end anode stack [15,16] without external humidification and hydrogen recirculation. It was assumed that the functionality of this controller could be extended to meet the needs of the power supply system under development by controlling the proportional hydrogen valve for flow-through anode operation [17], which was included in the numerical model.

The presented numerical model was primarily used to determine the flow of electrical energy in a hybrid power supply system, so several simplifications were assumed when developing this model. It was assumed that the fuel cell operated at a constant temperature and the airflow from which oxygen was extracted was always sufficient, regardless of the power load of the cell. The assumption regarding airflow was also fulfilled for the modeled fuel cell in the absence of external restrictions, which has been previously determined [62], where it was stated that even with the smallest used fan efficiency the cell worked with an air excess coefficient of ~20. Both thermal phenomena occurring in the hydrogen tank and the hydrogen release dynamics from the metal hydride storage were not taken into consideration. It was assumed that the hydrogen in the fueling system always had sufficient pressure to achieve the required hydrogen flow. Additionally, thermal phenomena in other elements of the power supply system were deemed to be negligible, assuming that they worked in optimal and constant thermal conditions. The phenomena related to the pulse operation of power electronic devices in the DC/DC converter were also not taken into account together with any ageing of the lithium-ion battery.

It was assumed that an external energy source was required to start the hybrid power supply system, ensuring the power needed to start the fuel cell stack and the stack controller, especially when the main energy storage was discharged. A low-capacity start-up battery could be used as an auxiliary energy source, which, if necessary, could be charged from an external source and, after starting the power supply system, could be recharged from the main power busbars. The energy needed to start the power supply system was small; however, the starter battery model was omitted for simplicity.

Modeling of the AGV drive system had been simplified to just model the instantaneous power demand, while the demand for the power of the components of the drive system (inverters, motors) during vehicle movement and related to the operation of the vehicle's control, safety, and signaling systems also had to be taken into account. The instantaneous power demand model for a selected AGV operation scenario was created by submitting multiple data samples obtained during measurements made by a real AGV with different load states and with different operating states, both during steady driving and in dynamic states (acceleration, braking). The data samples were recorded for an AGV powered by a standard (factory) lithium-ion battery that was charged from an external source at the end of the operation. Then the data were subjected to filtering and processing as described in Section 3.1. It was assumed that the instantaneous power demand for a vehicle powered by a standard battery and in a vehicle powered by a hybrid power supply system with a fuel cell stack under the same operating conditions and load conditions was the same. In connection with the adopted method of modeling the AGV drive system, the phenomena associated with switching power electronic devices in the inverters of the vehicle's drive nodes were excluded from the research.

Optimization of the structure and parameters of the hybrid power supply system could be carried out considering various criteria by setting selected parameters for the numerical model and analyzing the obtained waveforms, both utilizing experiments performed by trial and error and by automatic optimization algorithms. Usually, the parameters of the fuel cell stack were assumed at the beginning of the optimization process because the choice of the stack was not very flexible and the rated powers of the available stacks were highly graduated. The choice of energy storage was more flexible, so the parameters of this storage device could be optimized. During the simulation, the ongoing analysis of the selected waveforms of electrical quantities were carried out in terms of exceeding the defined criteria (critical values). This analysis is conducted regardless of the applied optimization method in the numerical model. If such an exceedance occurred during the simulation, then the simulation would be stopped and the model would return an error code that determined which criterion had been violated. A total of fifteen different criteria were defined in the numerical model for the various components of the hybrid power supply system. These criteria are:

- For the fuel cell stack: A minimum voltage, maximum load current, maximum load power, and the conditions of long-term power overload;
- For the auxiliary supercapacitor: The maximum charging or discharging current;
- For the DC/DC converter: A minimum supply voltage, maximum load power, and the conditions of long-term power overload;
- For the main energy storage: The maximum charging and discharging current, and the conditions of long-term overload during charging and discharging;
- For the main power busbars load model (i.e., the AGV power demand model): A minimum voltage, maximum voltage, and the maximum difference between the achieved power and the required power.

These criteria resulted from the catalogue of real element parameters of the hybrid power supply system and the conditions imposed by the elements of the AGV drive system (e.g., for inverters: The minimum and maximum supply voltage). Not all the criteria needed to be active at the same time. The selection of active criteria depended on which power supply parameters were unknown in the design aid process and which were imposed as project assumptions. For example, if the required minimum DC/DC converter power rating was unknown, then the criteria related to the power overload of the converter was turned off. If a specific DC/DC converter type needed to be used in the design, then in this situation the parameters of this converter should have been treated as project assumptions and the appropriate criteria values in the model were to be set, following the datasheet of the converter. In addition, the model for the hydrogen fueling system analyzed the hydrogen consumption during the simulation and returned the appropriate error code if the hydrogen tank was emptied. In this situation, the simulation was also stopped.

The numerical model of the hybrid power supply system defined the allowable voltage range and allowable state-of-charge (SOC) range of the main energy storage. Exceeding the voltage or state of charge for energy storage was not treated as a critical error and did not stop the simulation. However, it affected the way the energy storage worked, which was signaled in the model by the appropriate status signals. If the minimum voltage or the minimum state of charge was exceeded during discharge, the energy storage could only be charged. If with such limited use of energy storage, there was an increased power demand from the AGV model, the voltage of the main power busbars would fall below the criterion value. Similarly, if the maximum voltage or maximum charge was exceeded during charging, the energy storage could only be discharged. If under this condition, the AGV model attempted to achieve a return of braking energy to the energy storage, then the voltage of the main power busbars would rise above the criterion value. Exceeding the criterion values of the main power busbars voltage was treated as a critical error and stopped the simulation by returning an appropriate error code. In this situation, the error code had to be analyzed together with the main energy storage status to detect the reason for stopping the simulation.

The simulation model developed in the MATLAB/Simulink environment was built according to the block diagram shown in Figure 2. In addition to the blocks outlined in Figure 2, it also contained elements that allowed one to record the simulation results in the MATLAB for automatic optimization, and it also contained elements that allowed an ongoing view of waveforms, important parameters, error and status signals for the trial and error experiments.

To model the fuel cell stack, a block from the Simscape Electrical library was used, which is described in detail in [4]; the addition of concentration or mass transport losses in accordance with the method presented in [5] was applied. The losses of concentration or mass transport ΔV_{trans} are described by the equation:

$$\Delta V_{trans} = m \times \exp(n \times I_{FC}) \tag{8}$$

where the coefficients m and n are selected experimentally and I_{FC} is the stack load current. To tune the model for the fuel cell stack's activation area and load losses (ohmic losses), the results of measurements completed on the real H-300 stack and the genetic algorithm were used. During measurements this stack operated as a dead-end anode with the factory controller. In addition, the concentration losses model was experimentally tuned to obtain the appropriate stack voltage drop when overloaded. The thresholds for stack voltage and current were taken into account, and when they reached the stack were disconnected from the load by the factory stack controller.

The power of the fuel cell stack's own needs ("balance of plant") was modeled as being linearly dependent on the stack load power. The H-300 stack balance of plant was very simple (containing only fans, a controller, and hydrogen valves). However, it would be possible to model the balance of plant for a more sophisticated system, if the power demand characteristics of the components were available.

The fuel cell stack controller model included a hydrogen flow regulator that generated the $FFR_{(ref)}$ control signal for the hydrogen proportional control valve, which determined the flow through the anode of the stack. The principle of proportional control for this regulator was derived from the equations of the fuel cell stack model used in MATLAB presented in publications [4] and [5]. This regulator calculated the hydrogen flow needed to meet the hydrogen needs of the fuel cell stack at a given load current and a given hydrogen utilization. With a set number of cells in the stack, stack temperature, pressure and purity of hydrogen, the control principle is described as follows:

$$FFR_{(ref)} = C_{FFR} \times \frac{I_{FC(avg)}}{U_{H2\%(ref)}} \tag{9}$$

where the value of the coefficient C_{FFR} can be determined using the relationships given in [4] or [5]. $U_{H2\%(ref)}$ is the percentage setpoint of hydrogen utilization and the input quantity is the average current $I_{FC(avg)}$ of the fuel cell stack. This regulator ensured the hydrogen flow when the load current in the stack increased dynamically, which in turn ensured the rapid opening of the hydrogen control valve and prevented a voltage drop in the stack. Due to the strong averaging of the stack load current at the regulator input and the dynamics of the control valve (which was modeled by first-order inertia), the setpoint of hydrogen utilization by the stack should have been slightly less than the nominal hydrogen utilization to ensure proper fueling of hydrogen in fast transient states. The nominal hydrogen utilization could be calculated using the stack's rated parameters and relationships, as given in [4]. For an H-300 stack, it was 83%. When starting the hybrid power supply system and its associated transient states, the flow regulator ensured a sufficiently high initial hydrogen flow. The stack controller model contained a stack power demand model (power of its own needs), implemented as an array of values with interpolation that models "balance of plant". This power demand was included in the load model of the main power busbars.

The characteristics of an H-300 stack for nominal hydrogen utilization, obtained by the numerical model and tuned based on the results of the measurements are presented in Figure 3.

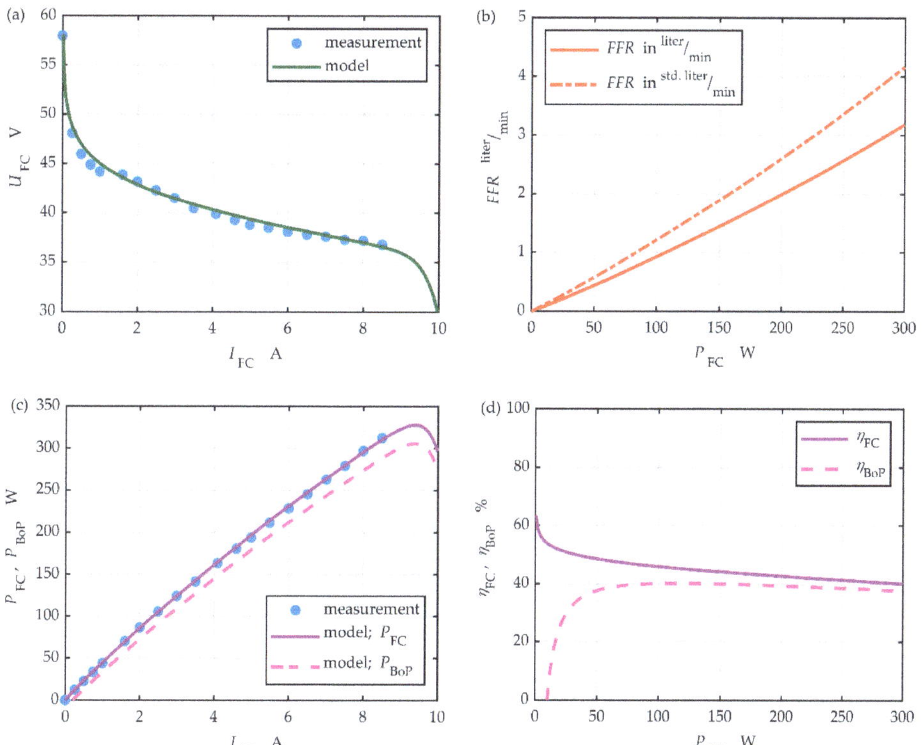

Figure 3. The characteristics of the H-300 fuel cell stack obtained from a numerical model at a temperature of 40 degrees Celsius, absolute hydrogen pressure of 1.5 bar, nominal hydrogen utilization of 83%. The stack's rated parameters are 36 V, 8.3 A, 300 W, with nominal efficiency of 40%. (**a**) The current-voltage characteristic, (**b**) the fuel flow rate vs. stack load power, (**c**) the stack load (gross) power and available (net) power vs. the stack load current, and (**d**) stack efficiency and system efficiency (stack efficiency taking into account "balance of plant").

The "auxiliary supercapacitor" block in Figure 2 also contains the controller that charges the auxiliary supercapacitor in a precise manner during the power supply system start-up to the required minimum voltage and then connects it to the output busbars of the fuel cell stack.

The DC/DC converter model was an average value model that considered the efficiency characteristics implemented as an array of values with interpolation and a no-load current. Additionally, it included the characteristics of the output power limitation as a function of the converter supply voltage. The output power limitation could be used interchangeably with the power threshold detection (and error code) depending on the purpose of the simulation test. The setpoint of the output current in CC mode could be constant or it could come from the regulation of the fuel cell stack load power. It was a Proportional Integral (PI) type, anti-windup regulator.

The model of the main energy storage management system, depending on the state of charge and voltage of the energy storage, allowed for its normal operation (such as charging and discharging) or to operate with restrictions (only discharging or only charging). This allowed a pre-charge of the energy storage after starting the power supply system if this function was needed.

The main energy storage model contained models of supercapacitor or lithium-ion battery, alternatively selected.

The model of the main power busbar loading system is included in the "AGV" block shown in Figure 2, which loads the power supply system with the power required by the AGV. Additionally, the power for the fuel cell's own need is represented by the "auxiliary DC/DC converter" block in the same diagram. The power required by an AGV is shown in the value tables, containing samples of the power demand while driving and samples of the vehicle's own needs.

The hybrid power supply system model included control signals that enforced the appropriate order of switching on its elements during start-up, thus mapping the operation of the real system.

4. Optimization Process Use Case

4.1. Automated Guided Vehicle (AGV)

An automated guided vehicle is designed for the transport of goods, materials, and semi-finished products as part of internal transport carried out in closed production or warehouse halls. The vehicle is designed to travel at ground level and can transport goods directly by itself by placing a loaded pallet on the upper loading surface of the vehicle or by pulling an attached transport trolley. The vehicle moves independently throughout the hall, performing tasks independently without human assistance in accordance with its pre-planned action and along a planned route. Usually, the vehicle travels along fixed routes according to a fixed schedule adapted in conjunction with the production cycle. The reproducible nature of the travel route and loads is important for matching the planned hydrogen fuel cell stack-based power supply system to the application. The vehicle monitors the surroundings via a sensor system to avoid collisions with them. The vehicle is powered by a lithium-ion battery placed in an easily accessible and replaceable cassette and the drive consists of two electric motors. A low-power AGV (Formica-1, AIUT Ltd., Gliwice, Poland) was used in this research.

4.2. Instantaneous Power Demand Model—Route Scenario

4.2.1. Identification Experiment

Identification of the instantaneous power demand model whose output is the input of the hybrid power supply system model requires proper planning of the identification experiment. The first step of these activities was to develop a common test plan for different operating conditions that take into account various stationary and nonstationary operations carried out on the real AGV.

The experiment was completed for different operating conditions at different route sections. The experiments are listed in Table 1. Due to the autonomous operation of the AGV control and the stochastic nature of the interaction between the vehicle surface and the AGV, the selected experiments were repeated several times and the average results obtained in this way were used for testing the signal models.

Table 1. List of conducted experiments using AGV.

No.	Operating Conditions Related to Routes	Other Operating Conditions
Ex 1–4	Straight route ahead (start, driving with constant speed, stop)	V = 0.3m/s, L = 100% V = 0.5m/s, L = 100% V = 0.8m/s, L = 100% V = 1.0m/s, L = 0%
Ex 5–8	Straight route reverse (start, driving with constant speed, stop)	V = 0.3m/s, L = 100% V = 0.5m/s, L = 100% V = 0.8m/s, L = 100% V = 1.0m/s, L = 0%
Ex 9–12	Slalom route (making three turns by 180 deg)	V = 1.0m/s, L = 0% CW V = 1.0m/s, L = 0%, CCW V = 1.0m/s, L = 100% CW V = 1.0m/s, L = 100% CCW
Ex 13–14	Rotation around its axis	V = 0.2m/s, L = 100% CCW, CW
Ex 14	Emergency stop	not applicable

During the conducted experiments, the following values were recorded: The voltage and the current returned by the batteries, the current values recorded on the main drive, and the current value on the stabilizing converter. Additionally, measurements of the resistance of the drive that was not directly measured were made. Due to these measurements, it was possible to record the instantaneous power demand. A schematic of the measuring system is shown in Figure 4. The data were recorded using an oscilloscope and with a sampling frequency of 100 or 50 kHz, depending on the duration of the selected route section.

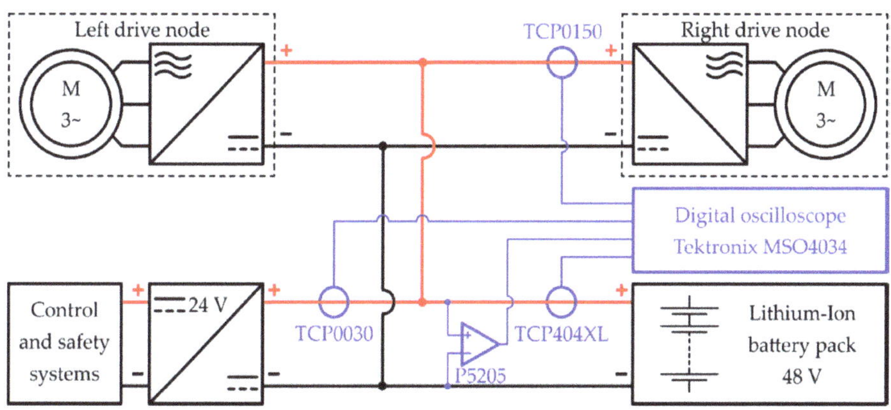

Figure 4. Block diagram of the low-power AGV drive system including the oscilloscope probes used to measure the power demands.

Restrictions on the safety and control of AGV are specified in the standard [52], including various responsibilities imposed on manufacturers and users. Due to the above reasons, the AGV was equipped with a logger system to record or monitor selected parameters during operating conditions around the route.

Selected logger data was used to observe the operating conditions. The data gathered concerned the rotational speeds of integrated Tekno TO-62 drives (left and right drive nodes according to Figure 4) equipped with an induction motor (nominal power 1.18 kW), the mechanical transmission with gear ratio 8.12 with a maximum continuous wheel torque of 25 Nm, and the power with a nominal voltage of 33 V. This element was also equipped with a 48 VDC nominal brake and a 5000 pulses speed encoder. The data were recorded using the AGV's inbuilt logger with a sampling frequency of ~2.5 Hz and were not synchronized with the instantaneous power demand signals recorded with the use of an oscilloscope. The recorded speed data were used to identify the operating conditions associated with the route section covered and its identification. This is a necessary part of the proposed approach, in particular, which is forced by conducting measurements in situ conditions where synchronization of measurements with the logger data (operating condition) was not possible. Figure 5 shows selected waveforms, speed signals from the logger, and the auxiliary computed signals.

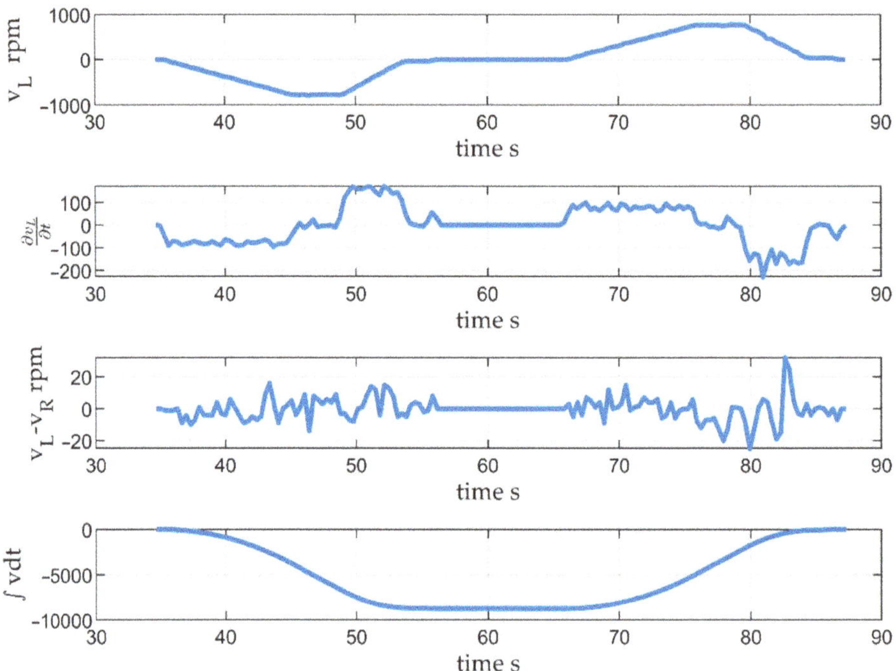

Figure 5. Example waveforms achieved from the logger and the additionally computed auxiliary signals.

To synchronize the measurements and thereby identify individual sections of the route for which signal models can be developed, the data were preprocessed by determining the auxiliary signals which were used to enhance the recognition of different operating conditions (some examples are shown in Figure 5), using resampling methods and identifying common starting points for both sources of data. For the obtained segments of the labeled measurement data related to route sections and selected operating conditions, models for stationary and nonstationary conditions were identified defining the banks of models.

Figure 6 shows the selected labeled measurement data based on previously determined data labels from the logger and auxiliary data. Based on the data labels, it was possible to segment the data and create a bank of signal models representing the instantaneous power demand for selected operating conditions.

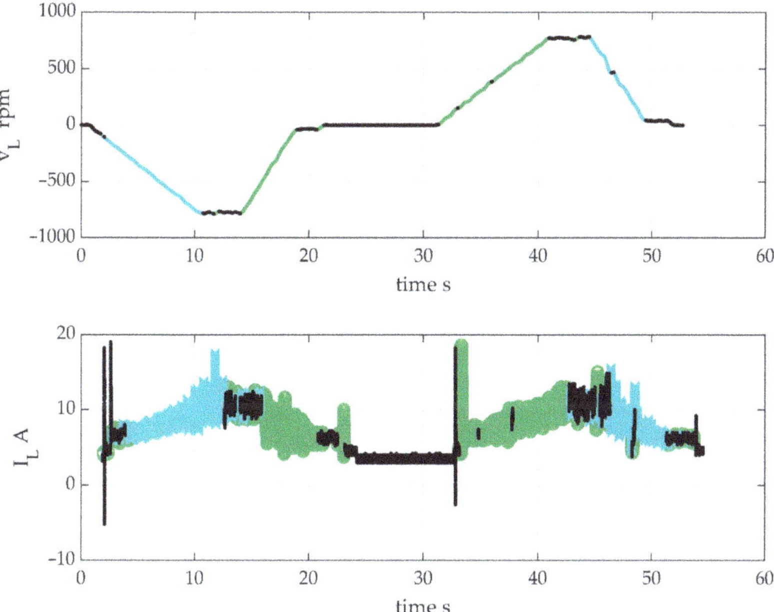

Figure 6. An example of labeling of the measurement results (in this case, the measurement of the current on the left inverter) based on operating conditions (in this case, based on the speed of straight route both backwards and forwards).

4.2.2. Instantaneous Power Demand Model for a Selected Scenario

The route scenario presented in Figure 7, developed for the AGV, consists of a section of the slalom route with the load (marked in red) and the rest of route unloaded.

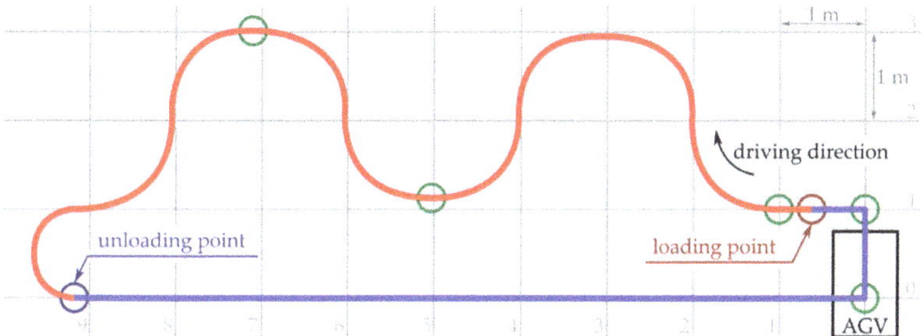

Figure 7. An example scenario of the AGV route for which the instantaneous power demand model is being built.

For the presented scenario, a signal of instantaneous power demand for a section of the route without load, shown in Figure 7, was modeled with the use of a set of models. For this section of the route the following models were prepared:

- Increasing speed models from the stationary vehicle to 1 m/s velocity;
- Models for a constant speed of 1 m/s for where the expected value of instantaneous power demand was read from the average power demand for the assumed speed;
- Models for decreasing speed from 1 m/s to vehicle stop;
- Models for 90 degrees left turns.

The needed lengths (number of samples) of the individual waveforms computed by the models was determined based on information about the time necessary to achieve the required speed (in the case of braking and accelerating). The model output did not compute any velocity, as this value could be read from the inverse of the average power demand versus the average velocity which had been identified based on the collected data sets presented in Table 1. This was determined using the linear approximation $P_{inst-ave} = C_1 \times v_{ave} + C_2$, where C_1 is 351.4 $\frac{Ws}{m}$ and C_2 is 279.3 W (valid for the average velocities v_{ave} between 0.3 and 0.8 m/s). The required number of samples for a constant speed period could be determined from the required length of the route and sampling frequency.

The waveforms were generated for the considered route section shown in Figure 7 by using previously listed models. The errors of the individual models are presented in Table 2. An example assessment of the selected model (with constant speed) for the instantaneous power signal distribution in the frequency domain is presented in Figure 8. The calculated errors were obtained from the test measurement data. Next, the individual waveforms generated with signal models were combined using the window indicated in Equation (7). An example of the joined data from two models is shown in Figure 9.

Table 2. List of models and their relative errors for the considered scenario.

Model Name	Er_{Pwr}
Model for increasing speed	3.3%
Model with constant speed	0.51%
Model for decreasing speed	3.1%
Model for turning left	15.2%

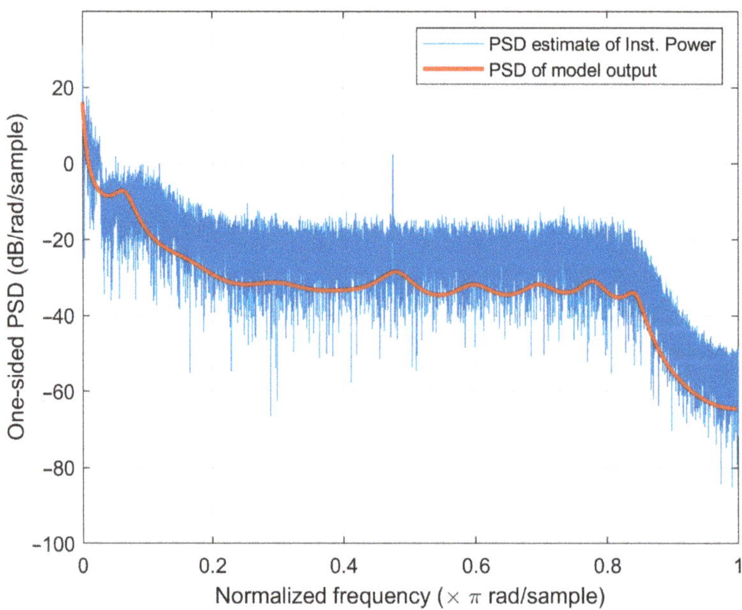

Figure 8. A periodogram estimate of Power Spectral Density (PSD) of instantaneous power for a model with twentieth order and the PSD measurement data for a straight profile at a constant speed.

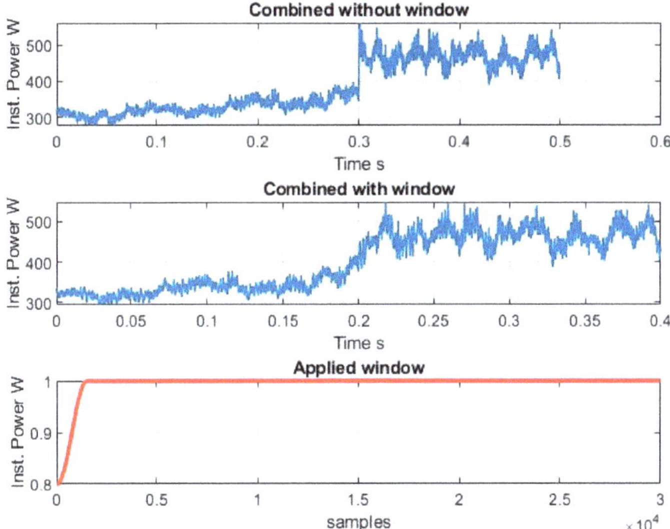

Figure 9. The combination of the two waveforms generated from the signal models for nonstationary and stationary signals (**a**) without using a prepared window; (**b**) with the use of window; and (**c**) with a used window to combine the two signals from the models.

The relative energy error for the modeled route was mainly (excluding the influence of windowing) a weighted average of errors for the individual models used to determine the instantaneous power demand, and the weights of this average resulted from the fraction used of the individual signals in the whole combined waveform.

The example presented in this section shows the possibility of modeling the instantaneous power demand using a well-known class of autoregressive models of signals. The proposed approach requires a simulation experiment by recording the instantaneous power demand for various operating conditions. The advantage of the presented method is the lack of interference from the AGV software, including its control system where this information is often unavailable due to company intellectual property issues, and there is no need to create a dynamic vehicle model.

4.3. An Example of Using the Model to Optimize the Hybrid Power Supply System

To demonstrate the practical use of the numerical model for an AGV hybrid power supply system, a short model route was designed, as outlined in Figure 7. The AGV moved with a load of 1.2 tons along a model route (marked in red) and then moved along a route without a load (marked in blue). The loading and unloading points and control points are marked in green, where the vehicle stopped for a maximum of a few seconds. When driving without a load, the vehicle accelerated and braked more rapidly than when driving with a load.

The demand for power (P_{AGV}) for the AGV during the model route was determined by the measurement results obtained for a real AGV, using the processing methods described in Section 3. An example of the waveform of the power demand while driving is shown in Figure 10. The results of measurements for the power demand when the vehicle was stopped were used to model the AGV stoppage.

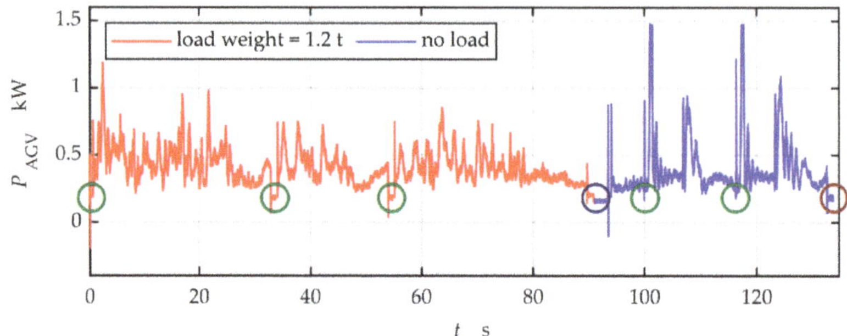

Figure 10. The power demand for an AGV driving on a model route; brown circle—loading point, blue circle—unloading point, green circles—control points (the points marked with circles coincide with points in Figure 7).

It was assumed that the model cycle for AGV operation included: Waiting time for the first drive after starting the power supply system of 30 s, five drives along the model route, a standstill after each drive, and waiting time for switching off after the driving cycles of 10 s. An example of the AGV's power demand waveform during the operation cycle is shown in Figure 11. The standstill time after driving was one of the parameters that changed during the optimization process and, in this case, was 255 s. The total electric energy consumption during the entire operation cycle was 131.2 Wh, with an average power demand of 236 W.

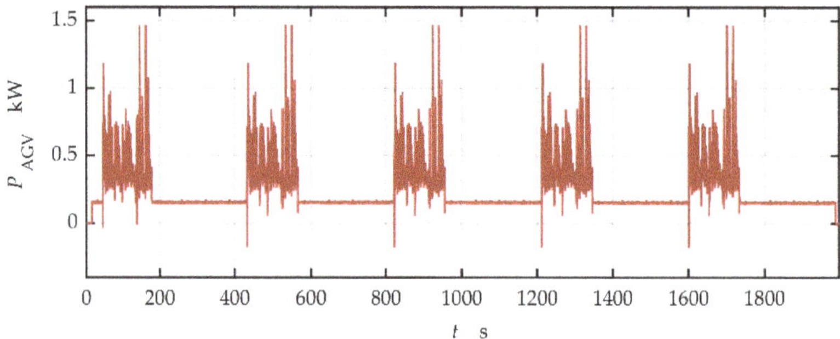

Figure 11. The AGV's demand for power during the model operation cycle.

The optimization aimed to minimize the energy storage capacity and the duration of stops between drives, assuming that all the energy needed to power the AGV came from the fuel cell stack, i.e., the state of charge of the energy storage should have been the same after an entire operation cycle as at its beginning. In the model hybrid power supply system, none of the fifteen electrical parameter criteria could be violated. It was essential that, during the simulated vehicle operation between the main energy storage voltage and the threshold (criterion) values of the supply voltage of the load system, a safety margin of ~3 V was maintained. These threshold values were 30 and 60 V, respectively. Additional parameters that were tuned in the optimization process and which had an essential impact on the results obtained were the allowable range of the energy storage voltage (in particular the energy pre-charge storage voltage), the output voltage of the DC/DC converter in CV mode, and the output current of the DC/DC converter in CC mode. An important result obtained from the model was the hydrogen consumption for the assumed operation cycle of the AGV, which allowed one to choose the required capacity of the hydrogen tank.

The preliminary simulation tests were carried out for the assumed operation cycle, assuming that the main energy storage was a LiFePO$_4$ battery with a capacity of 10 Ah and a rated voltage of 48 V, achieved from a real test bench. This is a low-cost battery that can provide a large enough impulse discharge current, with an expected value of ~3 C without degradation. This battery is built of sixteen 10 Ah prismatic cells connected in series. The energy storage model was tuned using optimization methods and the datasheet from the battery cells. It was assumed that the battery was pre-charged before starting the AGV power supply system (the initial state of charge was 50%). The selected results of preliminary simulation tests are shown in Figure 12. The results for the turned off SCU are presented so that transients that are associated with the operation of the SCU do not impair their readability.

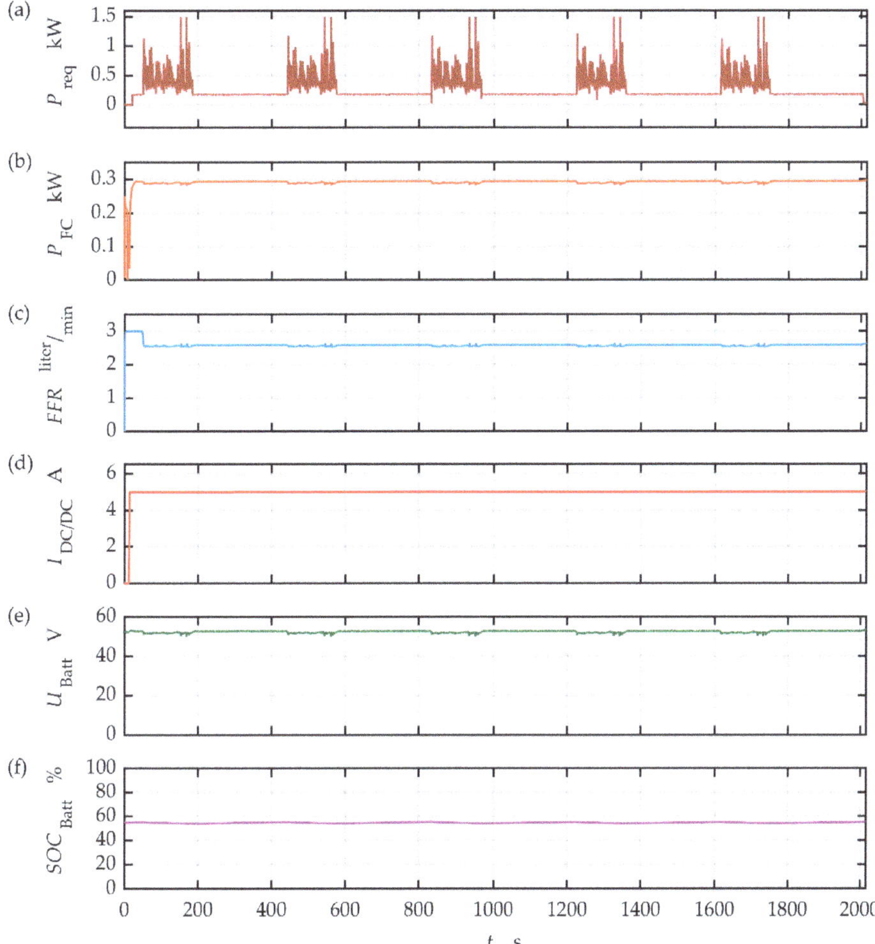

Figure 12. The results of preliminary simulation tests obtained assuming that the main energy storage was a LiFePO$_4$ battery with a capacity of 10 Ah; (**a**) the total demand for power; (**b**) the load power of the fuel cell stack; (**c**) the hydrogen flow rate; (**d**) the output current of the DC/DC converter; (**e**) the voltage of the battery and (**f**) the state of charge of the battery.

Figure 12a shows the total demand for power P_{req}, including P_{AGV} power for the AGV and P_{Aux} power for the hybrid power supply system's own needs. Figure 12b shows the load power of the fuel cell stack P_{FC}. It can be seen that the stack was utilized optimally and correctly (without overloading) throughout the entire operating cycle of the AGV stack and was loaded with power close to the rated power.

Figure 12c shows the hydrogen flow rate *FFR* obtained at the control valve output. Integrating this waveform, after converting it to the standard liters, it can be determined that ~141 L of hydrogen were consumed during the entire AGV cycle of operation, which corresponds to 423 Wh hydrogen energy, assuming that the change in enthalpy of formation was equal to the lower heat value [5]. When considering the energy consumption of the AGV from Figure 12a (~144 Wh), the efficiency of the hybrid power supply system was 34%. Figure 12d shows the current $I_{DC/DC}$ output waveform of the DC/DC converter. It can be seen that this converter works permanently in CC mode and the setpoint of the output current is constant and equal to 5 A. Figure 12e shows the U_{Batt} voltage waveform of the 10 Ah battery. This voltage was practically constant, which resulted from the relatively rigid discharge characteristics and the slight changes in the state of charge $SOC_{Batt\%}$ for this battery, shown in Figure 12f. A practically constant voltage of the battery at a constant output current of the DC/DC converter caused the fuel cell stack to be loaded with constant power throughout the entire operation cycle. The standstill time needed to restore the battery state of charge to the condition before driving was 257 s.

The battery used in the preliminary simulation tests had too large a capacity for the energy demand for the selected scenario of AGV operation, which was uneconomical. In subsequent tests, the battery capacity was reduced to a value of 0.2 Ah; this still ensured the correct operation of the AGV. The capacity of the battery was chosen so that its SOC varied from 20% to 80% during the operation of the AGV. It was assumed that the battery was pre-charged to an SOC of 20%, with an SOC of at least 70% required to start the vehicle. Therefore, when the power supply system was turned on, the battery was pre-charged by the fuel cell stack. The results did not change significantly. A slightly poorer utilization of the fuel cell stack was obtained while the AGV was driving. This was due to greater voltage drops in the smaller capacity battery, which, with a constant output current of the DC/DC converter, resulted in a decrease in the stack load power. The consumption of hydrogen increased to ~148 standard liters due to the initial charging of the battery, which absorbed 7.3 standard liters of hydrogen and lasted ~90 s. After omitting the supercapacitor pre-charge energy, the system efficiency was similar to previously, at ~34%. The used battery had a capacity of only 0.2 Ah, which was practically impossible due to too low current values for batteries with such a small capacity. Simulation tests using a selected scenario for vehicle operation and a 0.2 Ah battery were not of practical importance but were used to present the issue and how to use the model as a design aid. Therefore, during these tests, no criterion values for battery current were determined. A similar battery operation regime with real, higher capacity could be obtained for the real scenario AGV operation with higher energy demand. The use of battery capacity in such a work regime seems optimal, but it should also be noted that a battery working continuously in such a regime can quickly degrade.

Due to the possibility of quick battery degradation, further simulation tests were completed using a supercapacitor as the main energy storage. The most important advantage of supercapacitors, outlined in [30,32], is their use as an energy buffer for a fuel cell stack in traction applications due to their higher power density compared to batteries. Supercapacitors have higher efficiency and a higher number of charge and discharge cycles without degradation compared to the battery. By optimization, the criterion for the minimum capacity of the supercapacitor was chosen. The capacity of the selected SC maintained the voltage of the main power busbars over their required operating range whilst preserving a safety margin from the criterion values. The results of these simulation tests are given in Figure 13. It was assumed that the supercapacitor was not pre-charged, so its pre-charging was implemented by the fuel cell stack on start-up of the power supply system.

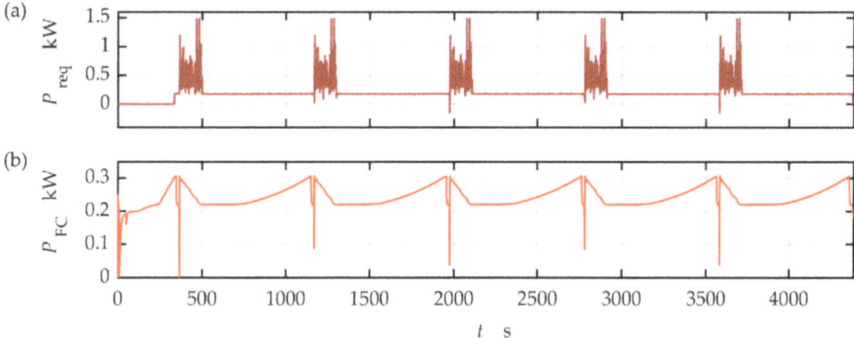

Figure 13. Selected results from the simulation tests obtained using a supercapacitor with a capacity of 35 F as the main energy storage; (**a**) the total demand for power and (**b**) the load power of the fuel cell stack.

The main difference between the operation of the power supply system with a LiFePO$_4$ battery and a supercapacitor is due to the different discharge characteristics of these energy storage devices. The voltage U_{SC} of the supercapacitor changes significantly more when discharged than the voltage U_{Batt} of the LiFePO$_4$ batteries (at least in the SOC range between 20% and 80%). There is a constant output current of the DC/DC converter in CC mode, resulting in a much worse utilization of the fuel cell stack due to significantly lower stack load power at a low supercapacitor voltage (Figure 13b). The presented results were obtained using a real supercapacitor model which consisted of 44 component supercapacitors with a capacity of 385 F each in a 2P22S connection system (two connected in parallel, 22 in series), which gave a resultant capacity of 35 F and a rated voltage of 61.6 V.

Using a supercapacitor, the hydrogen consumption was now 251 standard liters, of which 19.4 standard liters were required for pre-charging of the supercapacitor. The pre-charge time was approximately 340 s. The energy consumption of the AGV was 243.7 Wh (Figure 13a) and the hydrogen energy used for the AGV operation was ~695 Wh. The obtained power supply system efficiency was similar to that previously shown in the battery case, but the vehicle's standstill time after driving required to charge the supercapacitor to its pre-driving condition was 670 s. Such a long standstill time was due to the low power utilization of the fuel cell stack at low supercapacitor voltage when the stack provided slightly more power than that of the AGV's own needs.

Further simulation tests were completed with the fuel cell load power regulator turned on, which affected the setpoint of the DC/DC converter output current under the CC mode. As a result of the re-optimization, the capacity of the supercapacitor in the main energy storage was reduced to 24.5 F, obtained by connecting the 2P22S component supercapacitors with a capacity of 270 F each. The results of the simulation tests are shown in Figure 14. It can be seen that the utilization of the fuel cell stack was again optimal and correct (Figure 14b). At the same time, a much shorter standstill time (240 s) after driving is needed to charge the supercapacitor to its pre-driving condition. The hydrogen consumption was 150 standard liters, of which ~14.5 standard liters are required for the initial charge of the supercapacitor, which lasts ~200 s. The energy consumption of the AGV is 139.6 Wh (Figure 14a) and hydrogen energy used for AGV operation was ~406.5 Wh. Again, a system efficiency of ~34% was obtained, but the required standstill was much shorter than for a 35 F supercapacitor. Improvement in the use of the fuel cell stack compared to the previous simulation was obtained as the stack load power regulator increased the setpoint of the DC/DC converter output current in the CC mode at low supercapacitor voltage, just enough not to overload the stack. The fast transients shown in Figure 14b–d result from when the supercapacitor was charged to a certain maximum voltage, and the DC/DC converter then went into CV mode. In this situation, the load power of the fuel cell stack dropped sharply, and after the converter returned to CC mode, it increased again rapidly.

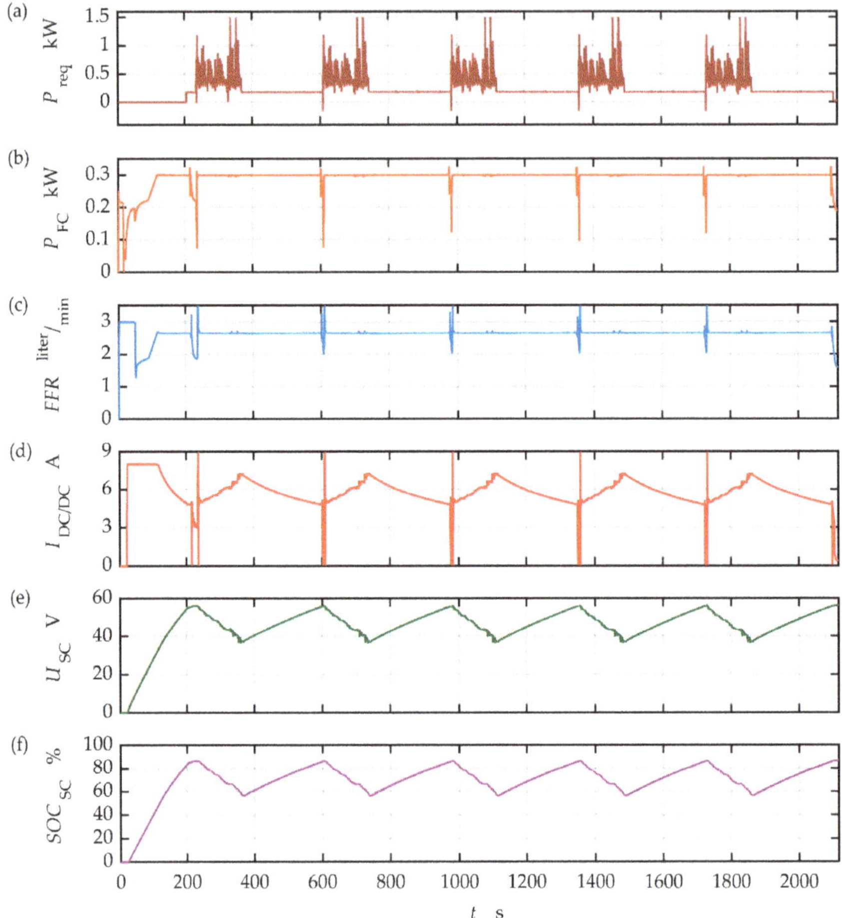

Figure 14. The results of simulation tests obtained assuming that the main energy storage is a supercapacitor with a capacity of 24.5 F and the fuel cell stack load power regulator is turned on; (**a**) the total demand for power; (**b**) the load power of the fuel cell stack; (**c**) the hydrogen flow rate; (**d**) the output current of the DC/DC converter; (**e**) the voltage of the supercapacitor and (**f**) the state of charge of the supercapacitor.

It can be seen that the presented simulation studies achieved the optimization goals and aid the design of the hybrid power supply system. The given results were valid for the assumed scenario of the AGV operation. In subsequent simulation tests, how the SCU operation affects the obtained optimization results was checked.

The SCU short-circuited the fuel cell stack for 100 ms every 10 s. However, transients lasted longer than 100 ms and were associated, among other things, with the need to recharge the auxiliary supercapacitor which was partially discharged when supporting the DC/DC converter supply voltage during stack short-circuit. The obtained results were accurate and similar to those presented in Figure 14, the main difference was that the standstill time after driving was extended to 250 s to charge the supercapacitor back to its condition before driving. Hydrogen consumption increased slightly to 155 standard liters and the power supply system efficiency decreased by 0.7%.

5. Discussion

The energy transfer numeric model for hybrid power supply system was the main objective of the article. The model consisted of the elements of supply system such as: Power converter, energy buffer, FC, and control devices. For the load of the supply system, the instantaneous power demand model was used. This model represents a generic instantaneous power demand for a given route for stationary and nonstationary conditions for an AGV under experimentally determined selected operating conditions. The main reason for developing the generic instantaneous power demand model is its simplicity and the fact that it does not require any additional information about the subsystems of AGVs and any supplementary information from the AGV manufacturers. It should be emphasized that the method proposed here could be improved by using further models that consider nonstationary conditions, i.e., that include nonstationarity of frequency components.

Section 4.3 describes the optimization process of the hybrid power supply system parameters for a model route and an AGV operation scenario. A similar process of optimization and design-aid can be completed using the real vehicle operation cycle by obtaining the route parameters and driving scenario from the vehicle manufacturer or vehicle user. In a situation where the AGV cannot make stops after driving, which allow for recharging of the main energy storage as described in Section 4.3, it is possible to minimize the capacity of the energy storage using a numerical model with the appropriate utilization of a fuel cell stack, under the assumption that the energy storage will be recharged using an external source after the entire operation cycle of the vehicle. In this situation, a compromise can be made between the capacity of the main energy storage (lithium-ion battery) and the hydrogen consumption, to consequently determine the capacity of the hydrogen tank. It is possible to use a stack load power regulator to intentionally reduce stack utilization and not exceed the assumed hydrogen consumption.

Further development of the presented numerical model, in the part related to the production of electrical energy, may include issues such as the dynamics of hydrogen release from the metal hydride storage under various operating conditions, and the dynamics of the cell response to a change in the hydrogen flow at the control valve output by considering the dynamics of the hydrogen distribution inside the cell. In the section of the model in which the power demand for the AGV is modeled, a dynamic model of the vehicle can be used which requires the drive torque for the specific route conditions, vehicle load, and the traction parameters (speed and acceleration). The hybrid power supply system power demand can be calculated based on the required drive torque in the dynamic models for the vehicle's drive nodes. Further development of the numerical model requires conducting additional tests on the fuel cell stack together with the hydrogen tank and examination of the real AGV to collect additional data and verify the extended numerical model.

The overall assessment of the proposed solution was carried out quantitatively for selected model elements (models of instantaneous power demand and the tuned hydrogen cell model). For other elements of the model, the assessment is qualitative as it is dependent on the specific instance of the AGV.

6. Conclusions

The research aim was to develop a model of a hybrid power supply system with a fuel cell stack for designing an energy storage system. The power supply system model is a numerical tool supporting the design and optimization of the power supply system following the MBD methodology. The article presents an example of the process of energy storage optimization for the AGV hybrid power supply system, which implements an example cycle of operation. Data for the AGV power demand model were obtained from measurements carried out on a real factory battery-powered AGV. These data were processed, and allowed the extract models for standard route fragments that can be interpolated under various load conditions. Based on these fragmentary models, it is possible to develop a power demand model for any route and optimize the hybrid power supply system for this route.

The conclusions from the generic instantaneous power demand model are:

- The proposed generic model allows for the determination of the instantaneous electric power for any route without the need to identify the dynamic drive system parameters;
- The model enables the determination of both stationary and nonstationary operating conditions using a simple approach with autoregressive models from signals with additional elements used for modeling the first-order and second-order nonstationarity with the application of additional linear, quadratic, or autoregressive models;
- Building a generic model for instantaneous power demand is possible since the AGV object is a system with constant control settings and operating conditions, and the AGV usually moves along an unchanged route for a long period. For more complex objects, the proposed approach may not be cost-effective as it would require more identification experiments.

Conclusions related to the model of the hybrid power supply system:

- The model seems to correctly imitate the energy transfer in the hybrid power system. The waveforms calculated by the model are reliable and all the phenomena visible are correct and explainable. The effectiveness of the model, however, must be confirmed by measurements of real cases with the design and optimization of the hybrid power supply system, which will be the subject of future research;
- The methodology used to model the components of the hybrid power supply system, using a few original ideas, means that the results of computer simulations are calculated relatively quickly, even for long routes taken by the AGV.

Supplementary Materials: The following are available online at http://www.mdpi.com/1996-1073/13/13/3435/s1: Preview of simulation model, graphical abstract, figures.

Author Contributions: Conceptualization, W.S.; data curation, T.R.; formal analysis, R.N., T.R., and W.S.; funding acquisition, W.S.; investigation, R.N., T.R., and W.S.; methodology, R.N., T.R., and W.S.; software, R.N. and T.R.; supervision, W.S.; validation, R.N. and T.R.; writing—original draft, R.N., T.R., and W.S.; writing—review and editing, R.N., T.R., and W.S. All authors have read and agreed to the published version of the manuscript.

Funding: Publication supported by the Rector's professor's grant. Silesian University of Technology grant number 10/060/RGP18/0102.

Acknowledgments: The authors would like to express their gratitude to AIUT Ltd., Gliwice, Poland for providing the research objects and materials for the construction of this research and to Barbara Skarka for support in developing the English version of the article.

Conflicts of Interest: The authors declare no conflict of interest.

References

1. Tyczka, M.; Skarka, W. *Optimisation of Operational Parameters Based on Simulation Numerical Model of Hydrogen Fuel Cell Stack Used for Electric Car Drive, Proceedings of the ISPE Inc. International Conference on Transdisciplinary Engineering Location: Fed Univ Technol, Curitiba, Brazil, 3–7 October 2016*, 23rd ed.; Transdisciplinary Engineering: Crossing boundaries Book Series: Advances in Transdisciplinary Engineering; IOS Press: Amsterdam, The Netherlands, 2016; pp. 622–631.
2. Skarka, W. *Model-Based Design and Optimization of Electric Vehicles, Proceedings of the 25th ISPE Inc International Conference on Transdisciplinary Engineering Location, Univ Modena & Reggio Emilia, Modena, Italy, 3–6 July 2018*; Transdisciplinary Engineering Methods for social innovation of industry 4.0 Book Series: Advances in Transdisciplinary Engineering; IOS Press: Amsterdam, The Netherlands, 2018; pp. 566–575.
3. Targosz, M.; Skarka, W.; Przystałka, P. Model-based optimization of velocity strategy for lightweight electric racing cars. *J. Adv. Transp.* **2018**, *2018*. [CrossRef]
4. Njoya, S.M.; Tremblay, O.; Dessaint, L.-A. A generic fuel cell model for the simulation of fuel cell vehicles. *IEEE Veh. Power Propuls. Conf. VPPC* **2009**, *1*, 1722–1729. [CrossRef]
5. Larminie, J.; Dicks, A. *Fuel Cell Systems Explained*, 2nd ed.; John Wiley & Sons Ltd.: Chichester, UK, 2003; p. 418.
6. Runtz, K.J.; Lyster, M.D. Fuel cell equivalent circuit models for passive mode testing and dynamic mode design. In Proceedings of the Canadian Conference on Electrical and Computer Engineering, Saskatoon, SK, Canada, 1–4 May 2005; pp. 794–797. [CrossRef]

7. Larminie, J.R.J. Current interrupt techniques for circuit modelling. In Proceedings of the IEE Colloquium on Electrochemical Measurement, London, UK, 17 March 1994; pp. 12/1–12/6.

8. Yu, D.; Yuvarajan, S. A novel circuit model for PEM fuel cells. *APEC '04* **2004**, *1*, 362–366.

9. Choi, W.; Enjeti, P.N.; Howze, J.W. Development of an equivalent circuit model of a feul cell to evaluate the effects of inverter ripple current. *APEC '04* **2004**, *1*, 355–361.

10. Page, S.C.; Krumdieck, S.P.; Anbuky, A. Testing Procedure for Passive Fuel Cell State of Health. In Proceedings of the Australian Universities Power Engineering Conference, Hobart, TAS, Australia, 25–28 September 2004. Available online: https://www.researchgate.net/publication/255578836_TESTING_PROCEDURE_FOR_PASSIVE_FUEL_CELL_STATE_OF_HEALTH (accessed on 15 May 2020).

11. Garnier, A.J.; Pera, M.; Hissel, D.; Harel, F.; Candusso, D.; Glandut, N.; Diard, J.; de Bernardinis, A.; Kauffmann, J.; Coquery, G. Dynamic PEM fuel cell modeling for automotive applications. *IEEE VTC 2003-Fall* **2003**, *5*, 3284–3288.

12. Singh, D.; Lu, D.M.; Djilali, N. A two-dimensional analysis of mass transport in proton exchange membrane fuel cells. *Int. J. Eng. Sci.* **1999**, *37*, 431–452. [CrossRef]

13. Acharya, P.; Enjeti, P.; Pitel, I.J. An advanced fuel cell simulator. In Proceedings of the Nineteenth Annual IEEE Applied Power Electronics Conference and Exposition, Anaheim, CA, USA, 22–26 February 2004; pp. 1554–1558. [CrossRef]

14. Rojas, A.C.; Lopez, G.L.; Gomez-Aguilar, J.F.; Alvarado, V.M.; Torres, C.L.S. Control of the air supply subsystem in a PEMFC with balance of plant simulation. *Sustainability* **2017**, *9*, 73. [CrossRef]

15. Kurnia, J.C.; Sasmito, A.P.; Shamim, T. Advances in proton exchange membrane fuel cell with dead-end anode operation: A review. *Appl. Energy* **2019**, *252*, 113416. [CrossRef]

16. Chen, J.; Siegel, J.; Stefanopoulou, A.; Waldecker, J. Optimization of Purge Cycle for Dead-Ended Anode Fuel Cell Operation. *Int. J. Hydrog. Energy* **2013**, *38*, 5092–5105. [CrossRef]

17. Woo, C.; Benziger, J. PEM fuel cell current regulation by fuel feed control. *Chem. Eng. Sci.* **2007**, *62*, 957–968. [CrossRef]

18. Kim, M.J.; Peng, H.; Lin, C.C.; Stamos, E.; Tran, D. Testing, modeling, and control of a fuel cell hybrid vehicle. In Proceedings of the 2005, American Control Conference, Portland, OR, USA, 8–10 June 2005; pp. 3859–3864. [CrossRef]

19. Yu, Q.; Srivastava, A.K.; Choe, S.Y.; Gao, W. Improved Modeling and Control of a PEM Fuel Cell Power System for Vehicles. In Proceedings of the IEEE SoutheastCon 2006, Memphis, TN, USA, 31 March–2 April 2006; pp. 331–336. [CrossRef]

20. Kong, X.; Khambadkone, A.M. Dynamic modelling of fuel cell with power electronic current and performance analysis. In Proceedings of the The Fifth International Conference on Power Electronics and Drive Systems, Singapore, 17–20 November 2003; pp. 607–612. [CrossRef]

21. Tremblay, O.; Dessaint, L.-A.; Dekkiche, A.-I. A Generic Battery Model for the Dynamic Simulation of Hybrid Electric Vehicles. In Proceedings of the 2007 IEEE® Vehicle Power and Propulsion Conference, Arlington, TX, USA, 9–13 September 2007.

22. Omar, N.; Monem, M.A.; Firouz, Y.; Salminen, J.; Smekens, J.; Hegazy, O.; Gaulous, H.; Mulder, G.; van den Bossche, P.; Coosemans, T.; et al. Lithium iron phosphate based battery—Assessment of the aging parameters and development of cycle life model. *Appl. Energy* **2014**, *113*, 1575–1585. [CrossRef]

23. Saw, L.H.; Somasundaram, K.; Ye, Y.; Tay, A.A.O. Electro-thermal analysis of Lithium Iron Phosphate battery for electric vehicles. *J. Power Sources* **2014**, *249*, 231–238. [CrossRef]

24. Tremblay, O.; Dessaint, L.A. Experimental Validation of a Battery Dynamic Model for EV Applications. *World Electr. Veh. J.* **2009**, *3*, 13–16. [CrossRef]

25. Zhu, C.; Li, X.; Song, L.; Xiang, L. Development of a theoretically based thermal model for lithium ion battery pack. *J. Power Sources* **2013**, *223*, 155–164. [CrossRef]

26. Shabani, B.; Biju, M. Theoretical Modelling Methods for Thermal Management of Batteries. *Energies* **2015**, *8*, 10153–10177. [CrossRef]

27. Orcioni, S.; Buccolini, L.; Ricci, A.; Conti, M. Lithium-ion Battery Electrothermal Model, Parameter Estimation, and Simulation Environment. *Energies* **2017**, *10*, 375. [CrossRef]

28. Alhanouti, M.; Gießler, M.; Blank, T.; Gauterin, F. New Electro-Thermal Battery Pack Model of an Electric Vehicle. *Energies* **2016**, *9*, 563. [CrossRef]

29. Supercapacitor Model. Available online: https://www.mathworks.com/help/physmod/sps/examples/supercapacitor-model.html (accessed on 15 May 2020).

30. Zhao, H.; Burke, A. *Fuel Cell Powered Vehicles Using Supercapacitors: Device Characteristics, Control Strategies, and Simulation Results*; Working Paper Series; UC Davis Institute of Transportation Studies: Davis, CA, USA, 2010; p. 10. Available online: https://escholarship.org/uc/item/23w1m5bb (accessed on 15 May 2020). [CrossRef]

31. Erdinc, O.; Uzunoglu, M. Recent trends in PEM fuel cell-powered hybrid systems: Investigation of application areas, design architectures and energy management approaches. *Renew. Sustain. Energy Rev.* **2010**, *14*, 2874–2884. [CrossRef]

32. Xun, Q.; Liu, Y.; Holmberg, E. A comparative study of fuel cell electric vehicles hybridization with battery or supercapacitor. In Proceedings of the 2018 International Symposium on Power Electronics, Electrical Drives, Automation and Motion (SPEEDAM), Amalfi, Italy, 20–22 June 2018. [CrossRef]

33. Cultura, A.B.; Salameh, Z. International Conference on Computer Information Systems and Industrial Applications (CISIA 2015). *Modeling, Evaluation and Simulation of a Supercapacitor Module for Energy Storage Application*; Atlantis Press Publisher, 2015; pp. 876–882. Available online: https://www.atlantis-press.com/proceedings/cisia-15/22669 (accessed on 15 May 2020). [CrossRef]

34. Zhang, L.; Hu, X.; Wang, Z.; Sun, F.; Dorrell, D.G. A Review of Supercapacitor Modeling, Estimation, and Applications: A Control/Management Perspective. *Renew. Sustain. Energy Rev.* **2017**. [CrossRef]

35. Jiya, I.N.; Gurusinghe, N.; Gouws, R. Electrical Circuit Modelling of Double Layer Capacitors for Power Electronics and Energy Storage Applications: A Review. *Electronics* **2018**, *7*, 268. [CrossRef]

36. Morin, B.; Van Laethem, D.; Turpin, C.; Rallières, O.; Astier, S.; Jaafar, A.; Verdu, O.; Plantevin, M.; Chaudron, V. Direct Hybridization Fuel Cell–Ultracapacitors. *Fuel Cells* **2014**, *14*. [CrossRef]

37. Kadyk, T. Integrated direct hybridization of fuel cell and supercapacitor by materials design. *Electrochim. Acta* **2019**, *319*. [CrossRef]

38. Arora, D.; Hinaje, M.; Bonnet, C.; Rael, S.; Lapicque, F. Sizing Supercapacitor for Direct Hybridization with Polymer Electrolyte Membrane Fuel Cell. In Proceedings of the 2018 IEEE Vehicle Power and Propulsion Conference (VPPC), Chicago, IL, USA, 27–30 August 2018. [CrossRef]

39. Ayad, M.Y.; Becherif, M.; Henni, A. Vehicle hybridization with fuel cell, supercapacitors and batteries by sliding mode control. *Renew. Energy* **2011**, *36*, 2627–2634. [CrossRef]

40. Souleman, N.; Motapon, L.-A.; Dessaint, K.A.-H. A Comparative Study of Energy Management Schemes for a Fuel-Cell Hybrid Emergency Power System of More-Electric Aircraft. *IEEE Trans. Ind. Electron.* **2014**, *61*, 1320–1334.

41. Han, J.; Charpentier, J.-F.; Tang, T. An Energy Management System of a Fuel Cell/Battery Hybrid Boat. *Energies* **2014**, *7*, 2799–2820. [CrossRef]

42. Gerardin, K.; Raël, S.; Bonnet, C.; Arora, D.; Lapicque, F. Direct Coupling of PEM Fuel Cell to Supercapacitors for Higher Durability and Better Energy Management. *Fuel Cells* **2018**, *18*, 315–325. Available online: https://onlinelibrary.wiley.com/doi/abs/10.1002/fuce.201700041 (accessed on 15 May 2020). [CrossRef]

43. Thounthong, P.; Raël, S.; Davat, B. Energy management of fuel cell/battery/supercapacitor hybrid power source for vehicle applications. *J. Power Sources* **2009**, *193*, 376–385. [CrossRef]

44. Hu, J.; Jiang, X.; Jia, M.; Zheng, Y. Energy Management Strategy for the Hybrid Energy Storage System of Pure Electric Vehicle Considering Traffic Information. *Appl. Sci.* **2018**, *8*, 1266. [CrossRef]

45. Suh, K.-W.; Stefanopoulou, A. Coordination of Converter and Fuel Cell Controllers. *Int. J. Energy Res.* **2005**, 563–568. [CrossRef]

46. Gauchia, L.; Bouscayrol, A.; Sanz, J.; Trigui, R.; Barrade, P. Fuel Cell, Battery and Supercapacitor Hybrid System for Electric Vehicle: Modeling and Control via Energetic Macroscopic Representation. In Proceedings of the 2011 IEEE Vehicle Power and Propulsion Conference, Chicago, IL, USA, 6–9 September 2011. [CrossRef]

47. Zhao, H.; Burke, A. Optimization of fuel cell system operating conditions for fuel cell vehicles. *J. Power Sources* **2008**, *186*, 408–416. [CrossRef]

48. Aschilean, I.; Varlam, M.; Culcer, M.; Iliescu, M.; Raceanu, M.; Enache, A.; Raboaca, M.S.; Rasoi, G.; Filote, C. Hybrid Electric Powertrain with Fuel Cells for a Series Vehicle. *Energies* **2018**, *11*, 1294. [CrossRef]

49. Matsunaga, M.; Fukushima, T.; Ojima, K. Powertrain System of Honda FCX Clarity Fuel Cell Vehicle. *World Electr. Veh. J.* **2009**, *3*, 820–829. [CrossRef]

50. Wee, J.-H. Applications of proton exchange membrane fuel cell systems. *Renew. Sustain. Energy Rev.* **2007**, *11*, 1720–1738. [CrossRef]

51. Alejandro, M.; Teresa, J.L.; Miguel, A. Herreros: Current State of Technology of Fuel Cell Power Systems for Autonomous Underwater Vehicles. *Energies* **2014**, *7*, 4676–4693. [CrossRef]

52. ANSI/ITSDF B56.5-2019 (Revision of ANSI/ITSDF B56.5-2012). Safety Standard for Driverless, Automatic Guided Industrial Vehicles and Automated Functions of Manned Industrial Vehicles. 2019. Available online: http://www.itsdf.org/cue/b56-standards.html (accessed on 15 May 2020).

53. Ljung, L. *System Identification: Theory for the User*, 2nd ed.; Prentice Hall: Upper Saddle River, NJ, USA, 1999.

54. Marple, S.L., Jr. *Digital Spectral Analysis with Applications*; Chapter 8; Prentice Hall: Englewood Cliffs, NJ, USA, 1987.

55. Soderstroem, T.; Stoica, P. System Identification; Publisher Prentice Hall International Series in Systems and Control Engineering. 1989. Available online: https://www.diva-portal.org/smash/record.jsf?pid=diva2%3A52642&dswid=-3745 (accessed on 15 May 2020).

56. Niedźwiecki, M.; Kołek, M. Identification of nonstationary multivariate autoregressive processes–Comparison of competitive and collaborative strategies for joint selection of estimation bandwidth and model order. *Digit. Signal Process.* **2018**, *78*, 72–81. [CrossRef]

57. Oppenheim, A.V.; Ronald, W.S.; John, R.B. *Discrete-Time Signal Processing*, 2nd ed.; Upper Prentice Hall: Saddle River, NJ, USA, 1999.

58. Bloomfield, P. *Fourier Analysis of Time Series: An Introduction*; Wiley-Interscience: New York, NY, USA, 2000.

59. Billur, S.; Farida, L.-D.; Michael, H. Metal hydride materials for solid hydrogen storage: A review. *Int. J. Hydrog. Energy* **2007**, *32*, 1121–1140.

60. Gupta, G.; Wu, B.; Mylius, S.; Offer, G.J. OfferA systematic study on the use of short circuiting for the improvement of proton exchange membrane fuel cell performance. *Int. J. Hydrog. Energy* **2017**, *42*, 4320–4327. [CrossRef]

61. *Horizon Fuel Cell Technologies: H-300 Fuel Cell Stack User Manual*. 2013. Available online: https://www.fuelcellstore.com/manuals/horizon-pem-fuel-cell-h-300-manual.pdf (accessed on 20 April 2020).

62. Strahl, S.; Husar, A.; Puleston, P.; Riera, J. Performance Improvement by Temperature Control of an OpenCathode PEM Fuel Cell System. *Fuel Cells* **2014**, *14*, 466–478. [CrossRef]

Article

Battery-Aware Electric Truck Delivery Route Exploration [†]

Donkyu Baek [1], Yukai Chen [2,*], Naehyuck Chang [3], Enrico Macii [4] and Massimo Poncino [2]

[1] School of Electronics Engineering, Chungbuk National University, Cheongju 28644, Korea; donkyu@cbnu.ac.kr

[2] Department of Control and Computer Engineering (DAUIN), Politecnico di Torino, 10129 Torino, Italy; massimo.poncino@polito.it

[3] School of Electrical Engineering, Korea Advanced Institute of Science and Technology, lDaejeon 34141, Korea; naehyuck@cad4x.kaist.ac.kr

[4] Interuniversity Department of Regional and Urban Studies and Planning (DIST), Politecnico di Torino, 10129 Torino, Italy; enrico.macii@polito.it

* Correspondence: yukai.chen@polito.it

† This paper is an extended version of our paper published in 2019 IEEE/ACM International Symposium on Low Power Electronics and Design (ISLPED), 29–31 July 2019, Lausanne, Switzerland.

Received: 13 March 2020; Accepted: 16 April 2020; Published: 24 April 2020

Abstract: The energy-optimal routing of Electric Vehicles (EVs) in the context of parcel delivery is more complicated than for conventional Internal Combustion Engine (ICE) vehicles, in which the total travel distance is the most critical metric. The total energy consumption of EV delivery strongly depends on the order of delivery because of transported parcel weight changing over time, which directly affects the battery efficiency. Therefore, it is not suitable to find an optimal routing solution with traditional routing algorithms such as the Traveling Salesman Problem (TSP), which use a static quantity (e.g., distance) as a metric. In this paper, we explore appropriate metrics considering the varying transported parcel total weight and achieve a solution for the least-energy delivery problem using EVs. We implement an electric truck simulator based on EV powertrain model and nonlinear battery model. We evaluate different metrics to assess their quality on small size instances for which the optimal solution can be computed exhaustively. A greedy algorithm using the empirically best metric (namely, distance × residual weight) provides significant reductions (up to 33%) with respect to a common-sense heaviest first package delivery route determined using a metric suggested by the battery properties. This algorithm also outperforms the state-of-the-art TSP heuristic algorithms, which consumes up to 12.46% more energy and 8.6 times more runtime. We also estimate how the proposed algorithms work well on real roads interconnecting cities located at different altitudes as a case study.

Keywords: Electric Truck Simulator; Electric Vehicle (EV); Vehicle Routing Problem (VRP); Traveling Salesman Problem (TSP); least-energy routing algorithm; energy efficiency; EV batteries; metric evaluation

---

1. Introduction

Electric Vehicles (EVs) are currently a tiny fraction of the car market, which is dominated by Internal Combustion Engine Vehicles (ICEVs); however, the growth of the EV market over last ten years is remarkable, and they are expected to progressively replace ICEVs in the next 20 years.

EVs have high energy efficiency and are sustainable transportation, since their electric motor has high dynamic performance and they are more environmentally-friendly than ICEVs on the market today; even when evaluating the emissions generated during electricity production for the charge of EVs, their overall Greenhouse Gas (GHG) emission is up to 58% that lower than the emissions of average mid-size passenger ICEVs [1]. Moreover, the impact on climate change by the production of electricity and operation of EVs is up to 30% less compared with ICEVs when considering the average generation of electricity in Europe. In recent years, the landscape of EVs is widening and expands to domains such as electric racing cars, electric buses, and electric trucks.

In particular, Tesla announced electric trucks will replace existing ICE trucks in the future [2]. The electric truck can accelerate more quickly than conventional diesel trucks because of the characteristic of the electric motor: high torque at low Rotations Per Minute (RPM) with high efficient. Furthermore, 98% of the kinetic energy can be recovered with electric energy during regenerative braking, which makes the electric truck more energy efficient. According to the announcement by Tesla, electric truck owners can save more than $200,000 over a million miles compared to fuel cost of a conventional truck.

The optimal energy-efficient delivery route can further improve the energy efficient of electric trucks. The typical delivery scenario is defined as an electric truck loads all packages for customers at a depot, visits each customer to deliver their package, and then returns to the depot without payload. For a conventional ICEV, the "cost" of a path is strongly driven by the distance (even if weight also matters) and the problem nicely fits into the well-known Traveling Salesman Problem (TSP) using distance as a metric. However, when considering EVs, the solution is not as straightforward; although distance obviously matters, the total energy consumption also strongly depends on the order of delivery as the efficiency of the EV is affected by the total (vehicle + payload) weight. Since the impact of the package weight on the battery SOC of electric trucks is much higher than that of the ICE trucks due to the characteristics of battery, the energy-optimal delivery method for electric trucks should be newly considered. As a matter of fact, one key characteristic of a battery is that it is progressively less efficient in delivering its energy as its State Of Charge (SOC) decreases [3,4]. A fully charged lithium-ion battery in EVs has better performance to deliver a high power demand than when it is partially discharged [5]. As the power consumption of the electrical motor strongly depends on the total weight, apparently, if we deliver the heaviest package first, the overall vehicle weight is reduced the most after unloading this package and following such order would be optimal [5]. On the other hand, it is obvious that the delivery distance should be considered; if we deliver the heaviest package first and this package has a very long distance from the depot, the battery will be discharged by carrying the heaviest weight for a long time, which might lead to a non-optimal delivery route if following the rule of deliver heaviest package first.

One first difference with respect to a plain ICEVs delivery is therefore in terms of metrics: for EVs, some combination of weight and distance should be considered. However, the most significant difference (and complication) lies in the fact that the calculation of the optimal energy path cannot be done incrementally, as the energy cost of a path is "dynamic", i.e., it depends on the previous choices as a consequence of the dependence on the residual weight, which means the weight and distance are the variables during the delivery process.

In this paper, we propose an overall electric trucks delivery simulation framework implemented by SystemC and SystemC-AMS for the least-energy electric truck delivery routing problem. We first implement an electric truck simulator with a powertrain model and a non-linear battery model of the Tesla Semi [6] that adopts the methodology introduced in [3] to trace the SOC evolution during the package delivery. From the simulation results, we show that a conventional metric for TSP, total delivery distance, as with any other "static" metric, does not minimize total energy consumption for an EV delivery. Since only an exhaustive exploration of all path guarantees to find the optimal path, we evaluate different static metrics (functions of weight and distance) on small graph instances to assess their quality; then, using the best metric derived in this calibration phase, we show how a greedy algorithm using that metric can provide significant reductions (up to 33%) with respect to the common-sense heaviest first package delivery.

The work is an extension of our previous conference paper [7], the main contribution of this work is to devise a heuristic algorithm to determine the energy-optimal routing of an electric truck. This main results encompasses a number of elements that we summarize as follows:

- Implement an electric truck simulator with a powertrain model and a non-linear battery model.
- Explore all delivery paths and evaluate the the correlation between energy consumption and various delivery metrics.
- Introduce heuristic algorithms using the best metrics, and show simulation results providing significant reductions of energy consumption.
- Analyze the effects of package weight distribution and number of packages on energy saving.
- Perform the proposed routing methods on the real routing condition, which consider average road speed, road distance between cities and corresponding road slope.

The paper is organized as follows. Section 2 introduces the motivation for the least-energy electric truck delivery routing problem. A comparison between the traditional shortest distance route and heaviest first route presented in this section to illustrate the limitation of distance-based routing method for the electric truck delivery routing problem. Section 3 describes the EV powertrain model and battery model used in our simulation framework and the typical vehicle routing problem. Section 4 shows our analysis of the routing problem; we discuss new metric candidates considering both of delivery distance and package weight. Then, new heuristic methods are proposed including greedy algorithm and TSP methods. Section 5 firstly shows how to implement the powertrain model of an electric truck and the related battery model in our simulation; then, the model validation follows. After that, we compare the energy consumption for the electric truck delivery problem by the metrics and approximate algorithms. To evaluate our proposal in the real delivery environment, we performed a case study for the delivery in a province located in Italy. Finally, Section 6 gives the conclusion of our work.

2. Motivation

In order to clearly illustrate the motivation of this work, we built an example to indicate how it is not possible to derive an energy-optimal delivery policy simply using a single "static" metric. Figure 1a shows a simple three-destinational delivery task from a depot (D) with a rough mapping on the plane, and the distance matrix between any pair of destinations. We assumed symmetry between node pairs to guarantee the generality.

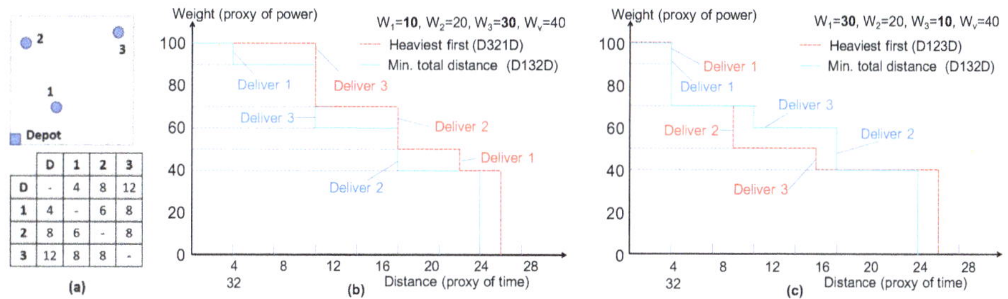

	D	1	2	3
D	-	4	8	12
1	4	-	6	8
2	8	6	-	8
3	12	8	8	-

Figure 1. Example to illustrate the motivation of energy-optimal delivery route exploration cannot only depend on single static metric: a delivery task (**a**) and delivery routes for two different delivery weights (**b**,**c**).

To evaluate the energy cost of a delivery path, we use a time diagram that plots the evolution of the total transported weight over delivery distance. We use weight as a proxy of electric truck power consumption, since it is proportional to the weight of the vehicle plus the total payload. Notice that it is clearly a simplification and does not take into account non-linearities of a battery, but even such approximation can help reveal the point we are making.

Distance is used as a proxy of time; we assume a constant speed for the deliveries. Again, this is an approximation of the real setting, where speed can be extremely varying due to the driving habits and the road traffic. When using a real battery model in the loop, however, the real speed profile of the vehicle can be accounted for. Therefore, evolution of weight over distance is a proxy of power over time, and the area of one such curve is then an estimate of the energy consumed for that delivery route.

Figure 1b shows two such delivery routes for a case in which the weights are $W_1 = 10$, $W_2 = 20$, $W_3 = 30$ and vehicle weight is $W_v = 40$. The dotted red curve represents a route for which packages are delivered in heaviest-first order ($D \rightarrow 3 \rightarrow 2 \rightarrow 1 \rightarrow D$), whereas the solid blue line denotes a route with the minimum total distance ($D \rightarrow 1 \rightarrow 3 \rightarrow 2 \rightarrow D$). In this specific case, the "min. total distance" policy works best (smaller area under the curve) and, by exhaustive exploration of the 3!=6 combinations of deliveries, it can be shown to yield the best value of the metric. Notice that the ($D \rightarrow 3 \rightarrow 2 \rightarrow 1 \rightarrow D$) route is not just slightly sub-optimal, and it ranks fourth out of the six routes, $D \rightarrow 3 \rightarrow 1 \rightarrow 2 \rightarrow D$ being the worst one.

Figure 1c shows two other waveforms for the same delivery task in which the weights change to $W_1 = 30$, $W_2 = 20$, $W_3 = 10$, that is, in which the heaviest package corresponds to the closest destination (node 1). In this case, the red dotted profile of the "heaviest-first" yields the best value of the metric, while the "min. total distance" yields a slightly worse value. Notice that the blue solid line corresponds to the same order of delivery (yet with a different cost) as in Figure 1b as the distance has not changed in the two examples.

Notice also that (due the symmetric distances) there are two paths (i.e., $D \rightarrow 1 \rightarrow 3 \rightarrow 2 \rightarrow D$ and $D \rightarrow 2 \rightarrow 3 \rightarrow 1 \rightarrow D$) with the same distance but with different "energy" cost, the cost of the path $D \rightarrow 2 \rightarrow 3 \rightarrow 1 \rightarrow D$ being larger than the other one shown in Figure 1c. Therefore, an algorithm that picks edge simply based on distance could even get farther from the optimal solution.

In this straightforward example, although there are several approximations during the power consumption estimation, it shows the main two critical points raised by our work. Firstly, no simple single static metric can solve ideally the problem of searching the energy-optimal delivery route. Secondly, due to the state-dependent characteristic of the cost function, only the brutal exhaustive exploration can find the optimal solution, but, since this is only feasible for very small instances, we need to find a provably good static metric that can be used in a heuristic algorithm. According to the above finding from the motivating example, this static metric should combine both *weight and distance* as twinned factors affecting the energy consumption of the EV.

3. Background and Related Work

3.1. Powertrain Model in EV

As is common, the vehicle powertrain model is from the vehicle dynamics. There are four forces acting on a vehicle driving on a road with θ road slope, as shown in Figure 2: rolling resistance F_R, gradient resistance F_G, inertia resistance F_I, and aerodynamic resistance F_A.

The powertrain model by vehicle dynamics (P_{dyna}) is described as

$$P_{dyna} = T\omega = F\frac{ds}{dt} = (F_R + F_G + F_I + F_A)v$$

$$F_R \propto C_{rr}W, F_G \propto W\sin\theta, F_I \propto ma, \text{ and } F_A \propto \frac{1}{2}\rho C_d A v^2 \tag{1}$$

$$P_{dyna} \approx (\alpha + \beta\sin\theta + \gamma a + \delta v^2)mv$$

where C_{rr} is the rolling coefficient, W is vehicle weight, θ is road slope, m is vehicle mass, v is vehicle speed, a is vehicle acceleration, C_d is drag coefficient, and A is the vehicle facial area [8]. The coefficients

α, β, γ, and δ of the dynamic power P_{dyna} are coefficients of rolling resistance, gradient, inertia, and aerodynamic, respectively.

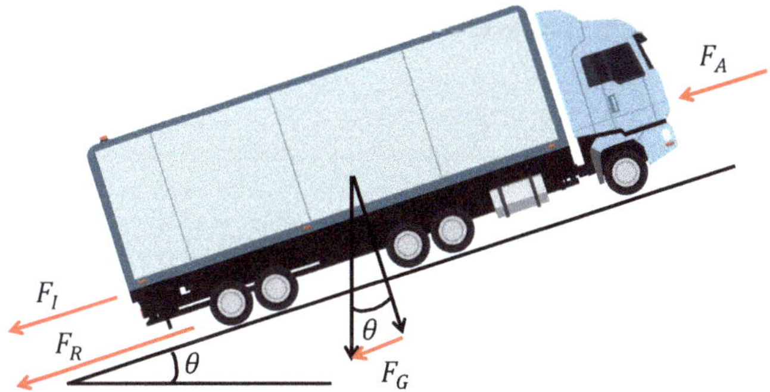

Figure 2. Forces acting on a truck.

In addition to the forces, there are several losses on a rotating motor: a copper loss is proportional to the square of motor torque and iron and friction loss are related to motor RPM. In addition, there is a constant loss while the vehicle is operating. The powertrain model of EV (P_{EV}) includes the motor losses in addition to P_{dyna}:

$$P_{EV} = P_{dyna} + C_0 + C_1 v + C_2 v^2 + C_3 T^2. \tag{2}$$

where C_0, C_1, C_2, and C_3 are the coefficients for constant loss, iron and friction loss, copper loss, and drivetrain loss, respectively.

EVs and hybrid vehicles mostly use regenerative braking during deceleration. The regenerative braking converts kinetic energy to electric energy from generation process. The harvested energy is related to the electromagnetic flux inside of the motor, which is proportional to the motor RPM. Therefore, the regenerative braking model can be simplified as

$$P_{regen} = \epsilon T v + \zeta. \tag{3}$$

where ϵ and ζ are regenerative braking coefficients to model the regenerative power as a function of the current velocity.

3.2. Battery Model in EV

The EV is normally powered by a battery pack that includes a large number of Lithium battery cells connected in parallel and series. The battery pack of Tesla Model 3 with long range version comprises 4416 2170-size lithium-ion cells of 4800 mAh nominal capacity with 46p96s arrangement, and 800 km range. Tesla Semi truck has a 750 kWh battery pack and weighs about 5.1 tons. To build the battery pack model directly is a non-trivial task, therefore, the battery pack can be built by composing the individual battery cell model, and the model must be able to accurately account for the varied load current and SOC variations of the usable battery capacity to capture the non-ideal discharge characteristics of battery. We select a circuit-equivalent battery model that can model the effects of load current magnitude and dynamics on real-time battery usable capacity [9], and then use this model to compose the whole battery pack model.

Figure 3 depicts the circuit-equivalent model of one battery cell adopted in this work. The left-hand part for modeling the battery lifetime (usable capacity) and the the right-hand part represents the transient battery voltage. Notice that the left-hand part also account for current magnitude and load frequency dependency on actual battery capacity. The left-hand part comprises a capacitor C representing the nominal storage capacity of the battery in *Ah* and a current generator representing the battery current I_{batt} requested by the load. As the available capacity of the battery is affected by the load current variations distribution, there are two voltage generators on the left part: one represents the dependency of the battery capacity on the current values, while the other generator models the dependency on load current frequency. Both decrease the voltage at node SOC, representing the SOC, for larger current magnitudes and frequencies. The right-part has a variable voltage generator affected by the SOC of battery, the internal resistance is also influenced by the current SOC of battery, and two pairs of RC express the instantaneous battery voltage.

The methodology to extract the relationship between SOC and internal resistance, capacitance, and open circuit voltage is presented in [10], and the implementation of adding two different dependencies of load current on the left-hand side of model is introduced in [9].

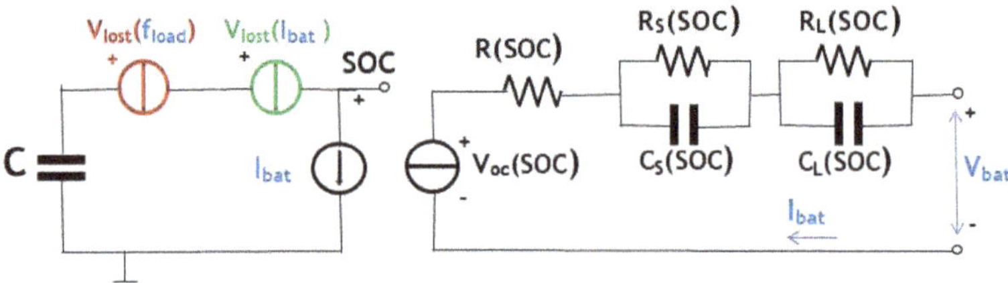

Figure 3. Adopted circuit-equivalent model for one single battery cell.

Based on this single cell model, we derive the battery pack model by ideally scaling all electrical parameters according to the series/parallel configuration in the pack, which leads to faster simulation runs and a higher flexibility in the modeling of large battery pack, so that not all cells of a large pack have to be modeled and simulated individually. Notice that the implementation of battery pack model is somehow ideal (e.g., cell mismatches are not considered), while this is still more accurate than a linear battery model that neglects state-dependent battery characteristic.

Given this model, we can track the energy consumed by the EV by applying to the model the drawn power (as current and voltage waveforms) corresponding to the electrical motor consumption on a given leg of the route. In the most general case, there is a non-ideal power conversion step between the electrical motor and the battery. In this case, it suffices to scale the motor current and voltage according to the converter efficiency $\eta < 1$, which can be any complex function of the motor parameters, i.e., $P_{batt} = P_{motor} \cdot \eta$. We assume the converter efficiency is a fixed valued in this work as in [3].

3.3. Work Flow of Proposed Methodology

Figure 4 shows the conceptual flow of our methodology for the estimation of the operation range of EVs. Three descriptive datasets are required as inputs.

Vehicle data includes (1) motor information: motor efficiency by motor torque and RPM, operation range of motor torque and RPM, (2) vehicle information: weight, drive train and body shape, and (3) other electrical systems. Facial area of the vehicle affects aerodynamic resistance. Route information include road distance between cities, road slope, and traffic on each road. Also payload by a delivery task is given. Route information and vehicle data is used in the vehicle model, which generates instantaneous power demand (V(t) and I(t)) during delivery.

We implement a battery model from a given battery specification: nominal capacity, voltage-to-SOC curve, impedance, and structure of the battery pack. Then we combine the vehicle model and battery model together to conduct simulation, the power demand of given delivery tasks over time derived from vehicle model is pass into battery model, finally the battery model computes the residual SOC of battery pack.

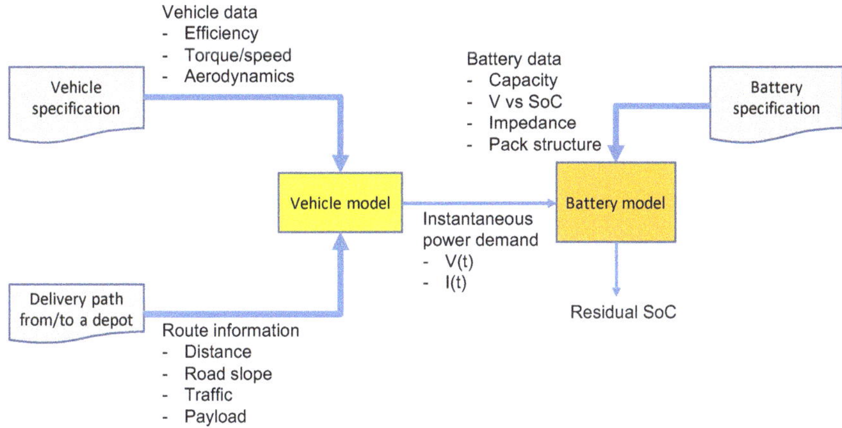

Figure 4. Overall concept of the proposed methodology.

3.4. Vehicle Routing Problem

The vehicle routing problem is formulated as a graph $G(V, E, C)$ where $V = \{v_0, \cdots, v_N\}$ is the set of vertices including N destinations and a depot, $E = \{e_{ij}|_{i,j \in V}\}$ is the set of edges between two vertices v_i and v_j, and $C = \{c_{ij}|_{i,j \in V}\}$ is the cost related to each edge e_{ij}. Vertex v_0 is the depot, while the remaining vertices in V represent customers that need to be served. The TSP consists in finding a route based at the depot, such that each of the vertices is visited exactly once while minimizing the overall routing cost.

The formulation of the vehicle routing problem is generalized as the TSP. Because TSP is known as NP-hard, several approximation algorithms are proposed during last several decades. Christofides designed an approximation algorithm for TSP using the Minimum Spanning Tree (MST) algorithm, which obtains approximated results less than 1.5 times the optimal solution [11]. From the general TSP, there are several variants of TSP to consider various constraints and delivery requirements. There is a variant of TSP considering a set of potential customers living near secured customers [12]. The salesman finds the shortest path to cover all potential customers within a certain distance from the path. A fleet of delivery vehicles characterized by different capacities and costs is an important variant of TSP [13,14]. There is a set of customers and a set of different types of vehicles. Each vehicle has different capacities in terms of the number of customers and operation cost; the goal is to find a set of routes for each vehicle minimizing total delivery cost. Some studies consider the number of customers that each vehicle should be responsible for, but they do not consider the vehicle weight changing with each delivery.

Recently, the vehicle routing problem with pick-up and delivery considers the situation in which packages have to be picked-up from one of customers and delivered to another location [15,16]. During the pick-up and delivery process, visiting each pickup and delivery places occurs exactly once and total package weight during the delivery should not exceed the capacity. This problem considers the weight of each package; however, it does not consider the energy consumption that changes after unloading each package.

There is a paper minimizing energy in the vehicle routing problem [17]. This paper solves the vehicle routing problem using integer linear programming to minimize the product of distance and weight of each arc. However, there is no result validation using energy simulation. Therefore, there is no analysis of energy consumption in the points of view of package weight distribution and routing methods in the real road condition.

4. Analysis of Routing Algorithms

The examples in Figure 1 suggest that a reasonable metric to track the energy spent on a delivery path could be to use a quantity that is correlated to the product of *distance* × *weight*. More precisely, let i and j be two vertices of the delivery graph, d_{ij} the distance between them, and W_i and W_j the weights to be delivered, respectively, at i and j. This metric should be proportional to $d_{ij} \times (W_{current} - W_i)$, assuming the edge is traveled in the direction $i \rightarrow j$. $W_{current}$ is the current weight of the vehicle when reaching node i (Figure 5a); this metric, which we call DxW_r hereafter (i.e., distance times residual weight), uses the information contained in the solid oval circle.

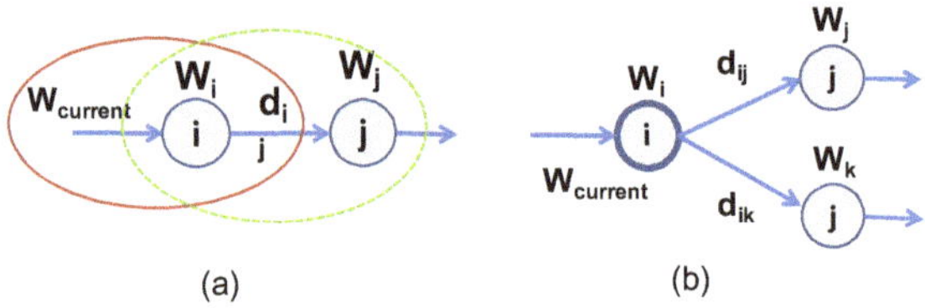

Figure 5. Features of the Target Metric: Generic instance between two nodes i and j (**a**); and a greedy decision based on this metric (**b**).

By accruing this quantity over the delivery sequence, we would therefore be able to measure the area below the curves in the (weight, distance) space of Figure 1 as a proxy of the total energy spent.

Therefore, as the delivery problem is an instance of TSP, intuitively one could be tempted to run some TSP heuristic algorithm using the above cost in place of distance as in traditional TSP instances. Although approximate, (besides also neglecting battery non-idealities), this strategy will leverage well-consolidated heuristics for the solving the TSP and could be relatively efficient. Unfortunately, there is a subtlety in this argument. TSP formulations do assume the use of a "static" metric, i.e., *whose value does not depend on the currently built solution*, such as distance. As a matter of fact, state-of-the-art TSP heuristics rely on the calculation of the MST algorithm as a pre-processing. This is because *the cost of a MST is the simplest lower bound for the TSP*: it can be shown that the removal of one edge from any Hamiltonian cycle (i.e., a solution of the TSP) yields a spanning tree [11].

Algorithms to compute the MST systematically grow the tree by *greedily picking edges* in increasing order of the cost function, which clearly implies the need of a "static" metric, otherwise it would not be possible to guarantee global optimality by using local optimal choices (e.g., shortest edge) at each step. It is therefore immediately clear that the above cost function $d_{ij} \times (W_{current} - W_i)$ cannot be used in a conventional TSP algorithm and in particular based on MST, for two reasons. First, MST runs on an undirected graph and there is no intuition about in what direction the edge is traversed. Secondly, and most importantly, MST algorithms take a "local" greedy decision independent of the specific previous decisions; this is at the basis of greedy algorithms, in which the global optimum is a sequence of locally optimal choices. Therefore, even replacing the traditional distance-based metric in the MST with DxW_r, it would simply not work, as shown in Figure 5b) (notice that for simplicity we assume the graph is directed to emphasize the direction of the delivery path).

Consider the decision to be taken at node i, at which we arrive with a given value of $W_{current}$. Without loss of generality, let us assume that there are only two possible choices at i, i.e., nodes j and k. A MST algorithm based on DxW_r, since $W_{current}$ represents a "state" information of the current path at i and as such is a fixed value, will then greedily pick the edge with smallest value $\min(d_{ij}, d_{ik})$ to decide about whether to grow the MST along j or k. In other words, as $W_{current}$ does not change, $\min((W_{current} - W_i) * d_{ij}, (W_{current} - W_i) * d_{ik})$ and $\min(d_{ij}, d_{ik})$ result in the same choice.

Therefore, as already shown by the example of Section 2, the energy-optimal (or more precisely, DxW_r-optimal) solution can only be obtained by enumeration of all the possible paths. A greedy, local metric used for a MST-based TSP algorithm would then result in strongly sub-optimal solutions even if we should adapt an MST algorithm to use DxW_r.

Algorithms Used in Our Analysis

Based on the previous discussion, an exhaustive exploration of all paths to collect the one with the smallest cumulative DxW_r is the only way to achieve the optimal solution. This algorithm has obviously factorial complexity ($O(n!)$ for n vertices), which is clearly applicable only to small instances and can be used for evaluating the quality of different approximate algorithms.

Concerning TSP-based algorithms, the computational complexity of the TSP heuristic with the best approximation, i.e., Christofides' algorithm, is $O(n^3)$, assuming that the graph is fully connected [11]. Although polynomial, cubic complexity can already be significant as the number of instances grow in the order to a few tens. Therefore, given the approximations of a TSP-based solution (the intrinsic approximation of the algorithm plus that of the metric described above), it makes sense to devise a simpler and *greedy* algorithm that *builds up the cycle as a path, one edge at a time, starting from the depot*, possibly using different greedy metrics. The greedy heuristic would clearly be linear in the number of nodes. Should some of the greedy heuristics be roughly as approximate as the TSP heuristic, it would at least guarantee that it can handle larger problem instances.

This choice would allow one to use the DxW_r metric that more closely tracks the energy value; by forming a path, in fact, we can calculate the equivalent of $W_{current}$ for the path being built. Moreover, since we start from the depot node, edges have an implicit direction and DxW_r can be calculated correctly. The approximation lies obviously in the fact that the greedy solution is not optimal, and, unlike the TSP heuristic, the approximation cannot be bounded. Christofides' algorithm, for example, can be shown to yield a solution that is no more than $3/2$ of the optimal cost.

In our analysis, we therefore compare three classes of algorithms to solve the optimal routing problem:

1. a set of algorithms based on the exhaustive enumeration of all paths, which are applicable only to small instances and used to evaluate the quality of the approximations;
2. heuristic greedy algorithms using different metrics;
3. heuristic TSP algorithms using different metrics.

They are listed in Table 1 where algorithms belonging to the three above categories are separated by a double line in the table. Each algorithm is labeled with an abbreviation for ease of reference in the simulation results.

Table 1. Exhaustive exploration of paths: list of algorithms.

Name	Description
Min$D$	Paths are sorted in order of total length, and the shortest path is selected.
MinDxW_r	Paths are sorted in order of total $D \times W_r$ and the path with the smallest aggregate value is selected.
Heaviest first	Paths are built by greedily choosing vertices in decreasing order of weights.
Shortest first	Paths are built by greedily picking edges **starting from the depot node** in increasing order of distance.
Smallest DxW_r first	paths are built by picking edges **starting from the depot node** in increasing order of $D \times W_r$.
TSPD	TSP heuristic algorithm using distance as a metric.
TSPDxW	TSP heuristic algorithm using DxW as a metric.

The *TSPDxW* heuristic of the last line requires a further explanation. Given that, as shown above, using DxW_r in a TSP based algorithm would be immaterial as it will coincide with a distance-based metric, we devised an alternative metric that mimics DxW_r but that is suitable for a TSP-based heuristic. This metric, which we call DxW, assumes: (i) a 50% chance of traveling the edge in each direction; and (ii) approximates $W_{current}$ with the total weight (vehicle + payload). This results in a quantity that does not depend on the currently built solution, and uses also the information about the destination node (as shown in the dashed oval of Figure 5a), i.e.,

$$d_{ij} \times (W_{total} - (W_i + W_j)/2).$$

Notice that a plain *"Shortest first"* and *"Smallest DxW_r"* would be the same algorithm. Therefore, to resolve this problem, for *"Smallest DxW_r first"* algorithm, we used $d_{ij} * (W_{current} - W_j)$, which considers both: (1) shortest distance first with d_{ij}; and (2) heaviest package first by abstracting weight of the destinations W_j from $W_{current}$.

5. Simulation Results

5.1. Simulation Setup

5.1.1. Powertrain Model

We implemented a powertrain model of a Tesla Semi truck from the vehicle specification based on the presentation by Elon Musk; this is currently the only source of information for the specs as Tesla is preparing to release the Semi in 2020 or later [6,18]. The powertrain consists of four Model 3 electric motors; each motor is a three-phase AC permanent magnet electric motor with maximum power of 192 kW from 4700 to 9000 RPM, and maximum torque is 410 Nm below 4500 RPM [19,20]. We estimated curb weight of Semi as the sum of typical weight of class 8 truck and battery pack weight [21].

We first implemented a vehicle model in ADVISOR (ADvanced VehIcle SimulatOR) [22] by using the above vehicle specification. Then, we extracted the coefficients of EV powertrain model with a number of ADVISOR simulations, as described in [23]. Table 2 summarizes the model coefficients of Tesla Semi. We compared the results computed between the derived power model and ADVISOR to validate the model we used. Figure 6 shows the difference between the estimation of power consumption by the ADVISOR vehicle simulator and the powertrain models we derived; the normalized root-mean-square error is 4.93%.

Table 2. Powertrain model coefficients for Tesla Semi truck.

α	0.098	β	10.1522	γ	1.006	δ	2.5×10^{-5}	C_0	10000
C_1	0.03	C_2	0.02598	C_3	1.54×10^{-5}	ϵ	0.5912	ζ	0.0

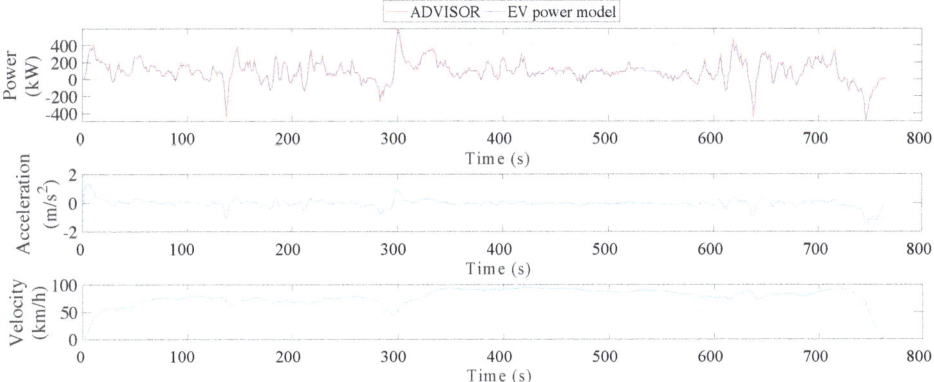

Figure 6. Powertrain model validation results compared with ADVISOR.

5.1.2. Battery Pack Model

There is no exact specification of battery pack in Tesla Semi truck until now; therefore, we assumed that each electric motor is connected to one battery pack of Model 3 in our experiments. Each battery pack is composed of four modules that are connected in series; each module consists of Panasonic NCR18650B 3400 mAh Lithium battery cells arranged in a 46p24s configuration [24].

Table 3 summarizes the physical electrical parameters of each cell, each module, and the whole battery pack.

Table 3. Electrical parameters of the battery pack.

Parameters	Cell	Module	Whole Pack
Nominal Capacity	3400 mAh	156.4 Ah	156.4 Ah
Nominal Voltage	3.6 V	86.4 V	345.6 V
Cut-off voltage	2.75 V	66.0 V	264.0 V

We built our battery single cell model based on the measurement data by adopting the method described in [9]. We assumed such 7104 battery cells in the pack to be ideally balanced in the following experiments, and then built battery pack model, as indicated in Section 3.2. Concerning the regenerative braking phase, we assumed that regenerative charging efficiency is 20% in our simulation, i.e., 20% of the kinetic energy is converted to electric energy and transferred into the battery pack.

5.1.3. Simulation Framework

We adopted the simulation framework proposed in [3], which targets the modeling and simulation of energy flow in the EV. The simulation framework is built by SystemC and SystemC-AMS, which are the extension of C/C++ with libraries to describe HW constructs and analog/mixed-signal subsystems. SystemC-AMS provides different abstraction levels to cover a wide variety of domains, three different Model of Computations (MoCs) supported by SystemC-AMS that allow the simulation framework to integrate circuit equivalent battery model and empirical powertrain model simultaneously. Another main advantage of the SystemC-AMS simulation framework is the fast simulation speed while keeping the same accuracy with regard to state-of-the-art tools such as Matlab/Simulink, with speedups up to two orders of magnitude and a high level of accuracy. Such quick estimation of energy consumption of EV and the battery lifetime give the opportunity to conduct exhaustive exploration in a short time for different delivery routes in our following experiments.

5.2. Simulation Results

5.2.1. Comparison Against Exact Results

In this section, we compare the energy consumption of various delivery strategies based on different policies for a set of small-sized (4–7) instances for which an exhaustive exploration of all the possible delivery paths is feasible.

For each number of destinations, we randomly generated 50 instances with different distributions of locations of the depot and of the destinations by uniformly distributing them in a 30 km × 30 km area. We selected the area for the delivery so that all the delivery sequences can be completed without exhausting the battery energy before returning to a depot. For each of the 50 instances, package weights for each destination were chosen as uniformly distributed from 0.1 to 3 ton.

For each problem instance (destination and weight distribution), we calculated by exhaustive exploration the route yielding the smallest value of energy for a number of metrics. Energy was calculated by conducting simulation described in Section 5.1. In this test example, a constant speed of 76 km/h was assumed, which was the average truck speed in metropolitan area interstates in the US in 2015.

Figure 7a shows the energy consumption of the optimal route, averaged over the 50 instances, for problems with 4–7 destinations and for the set of algorithms described in Table 1.

The leftmost blue bars represent the optimal routes yielding the minimum energy consumption among all routes, obtained by computing the actual energy consumption per each segment using the battery model. This is the reference value against which the other results were compared. All the other bars refer to solutions (i.e., routes) returned by the above algorithms and evaluated using the battery model. The objective of the simulation was to check how the greedy algorithms (TSP-based or not) differ from the optimal solution and how the error increases with increased problem sizes. Bars in the plot are in the same order as in Table 1. For ease of reading, bars referring to path-based algorithms (first part of the table) are shown as solid bars, whereas those referring to greedy or TSP algorithms are shown as patterned bars.

Concerning path-based algorithms, we can notice how the *MinDxW$_r$* metric (third bar from left) tracks very well the true energy value, much better than distance alone (second bar from left). Concerning approximation algorithms, as a first general comment, we can see that all algorithms overestimate the actual energy consumption. Then, we immediately observe that weight alone (Heaviest-first) tracks quite poorly the actual consumed energy, somehow contradicting the intuition suggested by the battery property; the distance from the reference is already $> 20\%$ even for the four-destination instance. Although the actual error may differ depending on the weight distribution (as shown in Figure 1), the results are the average 50 different runs, thus we can safely assume this is not a good metric. Notice also that the tracking error increases with larger instances.

Another observation is that a traditional TSP with distance metric (second bar from the right) performs reasonably only for the smallest instance; that average error increases quickly and is already around 18% for seven destinations. Therefore, we can also rule out this algorithm from the list.

The remaining ones (*Smallest DxW$_r$ first*, *Shortest first*, and *TSPDxW*) have errors below 10%, with the greedy algorithms being below 5% and scaling better with problem size than *TSPDxW*. Especially, *Smallest DxW$_r$ first* algorithm saves energy consumption from 6.58% to 12.46% compared with traditional *TSPD* algorithm. The gap between the best one and *TSPD* increases by the number of destinations.

Figure 7b shows the worst case error among the 50 instances for the same set of algorithms. Results are consistent with average error, with the maximum error being significantly larger than the average one. The greedy algorithms have show again the best results, in terms of both error and scalability. The *Smallest DxW$_r$* is the only algorithm with worst-case error around 20% (as opposed to about 30–35% of the others) for the seven-node case.

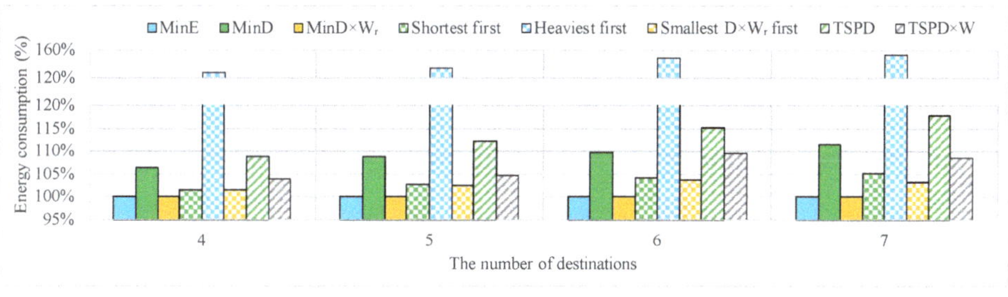

(**a**) Average energy consumption comparison.

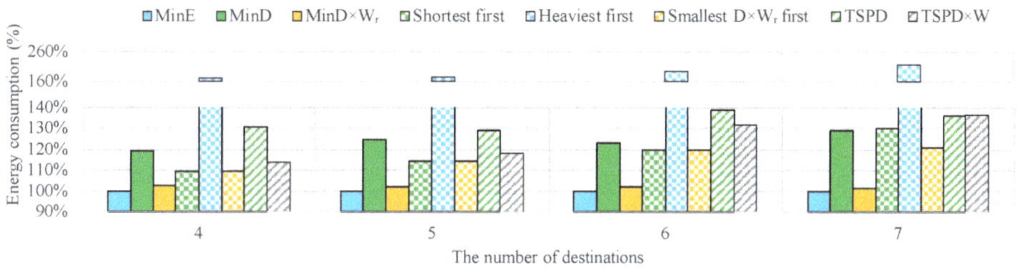

(**b**) Worst energy consumption comparison.

Figure 7. Energy consumption for different metrics (exhaustive exploration).

5.2.2. Comparison by Package Weight Distribution

In this section, we discuss how the error increases with the increased package weight. Figure 8 shows the energy consumption of different metrics by package weight distribution. Each column means different weight distribution of delivery package: 0.1–0.5, 0.5–1.0, 1.0–1.5, 1.5–2.0, and 2.5–3.0 ton, respectively. In this comparison, one depot and seven destinations were uniformly distributed in a 30 km × 30 km area, and we extracted average energy consumption of 50 instances by different metrics. In order not to provide unnecessary information, we do not consider $MinDxW_r$ and *Heaviest first* metrics here because these metrics show either sufficiently accurate or irrelevant results from the previous section.

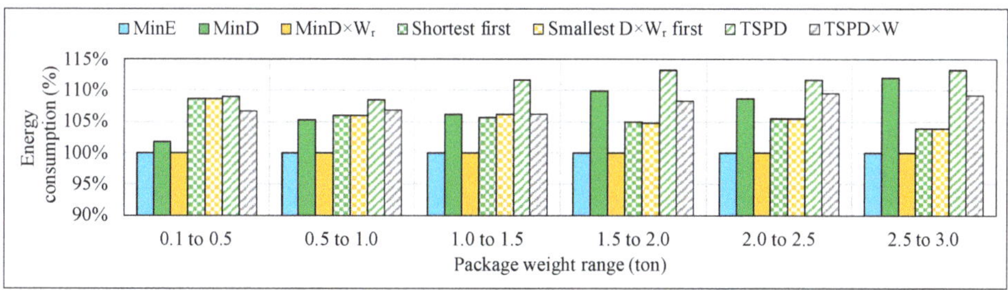

Figure 8. Energy consumption by package weight distribution.

When the weight of the package is less than 0.5 ton, the sum of all package weight can be ignored in most instances compared with total weight of the electric truck. Thus, all results by different metrics show less than 10% error. As the range of the package weight increases, however, the effect of the package weight on the overall energy consumption increases. Therefore, metrics considering distance only show worse results, namely the exhaustive exploration with distance (*minD*) and the Traditional heuristic TSP (*TSPD*). On the other hand, heuristic TSP with *DxW* (*TSPDxW*) and two other greedy heuristics show less than 10% error for all weight ranges.

5.2.3. Application to Larger-Scale Instances

We generated a number of instances with 10, 20, 30, 50, and 100 destinations; for each problem size, we generated 20 random instances and collected the average value of energy and execution time. In all cases, weights were scaled so that the delivery task could be completed.

Figure 9 compares the absolute energy values for the three competitive algorithms resulting from the previous section: one TSP with the proposed metric (*TSPDxW*) and the the two greedy heuristics (*Shortest first* and *Smallest DxW$_r$ first*). From the results in Figure 7a we know that all approximations are overestimations, thus we can assume that lower values of energy imply higher accuracy. In Figure 9, the *Smallest DxW$_r$ first* shows the best results: its energy consumption is 10% smaller than the TSP heuristic and 4% smaller than the shortest first one, for the larger 100-node instance.

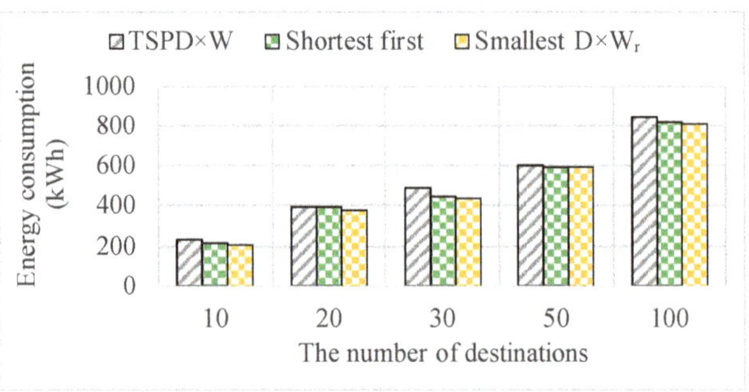

Figure 9. Greedy and approximate TSP algorithms on large problem instances.

Figure 10 shows the slowdown of the TSP heuristic with respect to the *Smallest DxW$_r$ first* algorithm. The TSP execution time is obviously independent of the metric used (*D* vs. *DxW*). The TSP heuristic is significantly slower than the greedy method; the slowdown increases for increasing problem sizes, reaching 8.6 times for the 100 destination case.

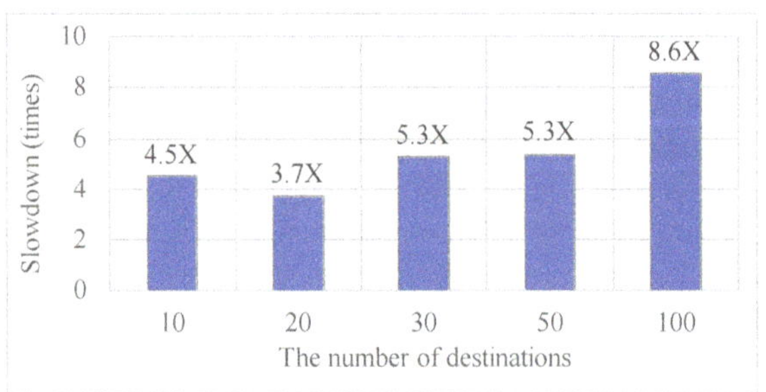

Figure 10. Slowdown of TSP heuristics vs. greedy algorithm.

5.3. Case Study: Routing Problem in Real Roads

In the experiments presented in Section 5.2, the distribution of the locations were synthetically generated on a plane. In this section, we show the routing algorithms in a real case, consisting of a set of locations taken from a map and for which actual distances, road slope, and road traffic between destinations are taken into account. We generated 50 instances, in which package weights for each destination were chosen as uniformly distributed from 0.1 to 3 ton.

5.3.1. Extraction of Road and Elevation Information

We used Piedmont region in Italy as the delivery destinations. There are 10 destinations in Figure 11 including a depot. The destinations of the delivery are limited to towns or cities in the province.

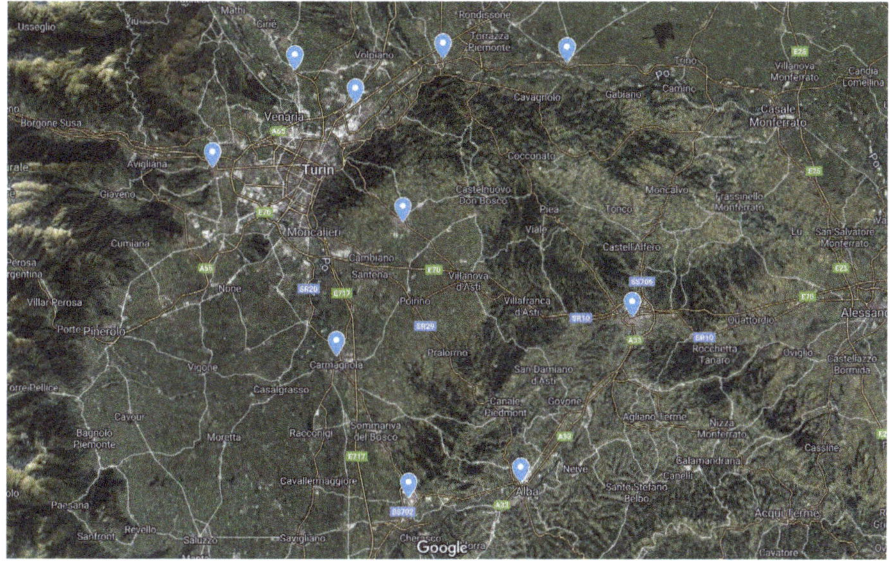

Figure 11. Destinations for deliveries on real roads.

To travel between destinations, we cannot drive on a straight path, but we must use given roads. There are several route options connecting destinations to each other. Among them, the most recommended one by Google Maps is picked. We extracted the distance of each route and related average driving time.

Table 4 shows the altitude information of each city. We extracted average road slope from road distance and altitude difference between two cities. The average altitude difference among 10 cities is 84 m.

The practical distance of the route, road slope, and driving time are different by direction. Therefore, we implemented several matrices containing road distance, road slope, and driving time. The average distance of the roads is 44 km, and the average time for driving the roads is 51 min. The driving time was used to calculate average electric truck velocity. Therefore, the energy consumption for the delivery was obtained from the practical road distance, velocity, and road slope.

Table 4. Destination information.

City	Torino	Chivaso	Crescentino	Asti	Chieri
Altitude (m)	216	186	155	126	283
City	**Alba**	**Bra**	**Carmagnola**	**Torino Airport**	**Rivoli**
Altitude (m)	167	277	233	282	400

5.3.2. Simulation Results

Figure 12a shows the average energy consumption of the optimal route in the case study described in Figure 11 on the 50 instances with routing algorithms described in Table 1. When the number of destinations was four, we randomly picked one depot and four destinations among the 10 cities in Figure 11.

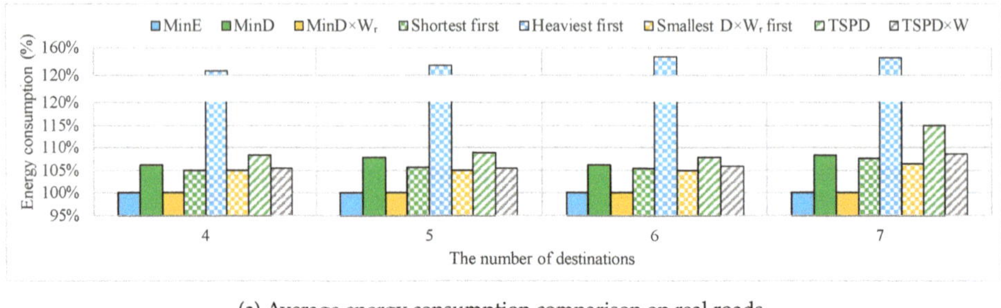

(**a**) Average energy consumption comparison on real roads.

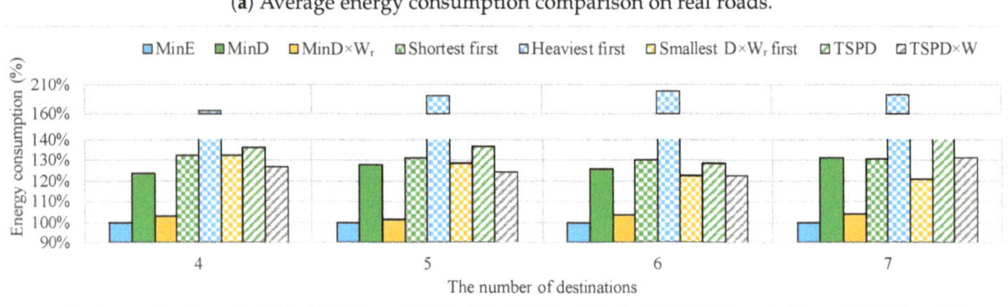

(**b**) Worst energy consumption comparison on real roads.

Figure 12. Energy consumption for different metrics (exhaustive exploration) on real roads.

Similar to Figure 7a, the leftmost blue bars (MinE) represent energy consumption by the energy-optimal routes among all routes. This is the reference value to compare with the other results. Bars in the plot are in the same order as in Table 1.

As confirmed by the results in Figure 7a, the path-based algorithms using the metric *MinDxW$_r$* (third bar from the left) track very well the true energy value, much better than distance alone (second bar from left), in all simulation results with different number of destinations. Concerning approximation algorithms, weight alone (Heaviest-first) tracks quite poorly the actual consumed energy as in Section 5.2. The average error is up to 47.9%.

On the other hand, shortest first and *Smallest DxW$_r$ first* metrics track the actual consumed energy very well. The average error is less than 10% for all numbers of destinations. Especially, *Smallest DxW$_r$ first* shows the best results for all different number of destinations. A traditional TSP with a distance metric (second bar from right) tracks well only for the smallest instance as in Figure 7a; average error increases up to 15.0%. TSP with a metric *DxW* shows better results than distance metric.

Figure 12b shows the worst case error among the 50 instances for the same set of algorithms in Figure 12a. The results are consistent with average error, but significantly larger. The best algorithms in Figure 12a (*Smallest DxW$_r$ first*, *Shortest first*, and *TSPDxW*) again show the best results, in terms of both error and scalability. *TSPD* and *TSPDxW* methods are not better in some cases (e.g., six destinations) because triangular inequality is not applicable, and the distance between cities is asymmetric.

6. Conclusions

The total energy consumption for parcel delivery with an electric truck strongly depends on the order of delivery because battery efficiency is affected by how the transported weight changes over time. However, it is impossible to consider the transported weight changes with the traditional routing algorithms, which use "static" metrics such as distance. In this paper, we demonstrate that the functions of weight and distance as metrics provide significant energy reductions with respect to the traditional routing algorithms. A traditional TSP with distance metric performs reasonably only for the smallest instance; average error increases quickly. On the other hand, a greedy algorithm minimizing driving distance by residual weight (DxW_r) shows better results with almost $10\times$ faster runtime. We also performed a comparison in the real delivery case, where curved and sloped roads connect cities. In this case study, the greedy algorithm also shows better results than traditional TSP. In addition, the package weight affects the result of routing algorithm. As we increase the package weight distribution, the gain by the proposed method increases.

Author Contributions: Conceptualization, N.C. and M.P.; Writing—original draft, D.B., Y.C. and M.P.; and Writing—review and editing, N.C. and E.M. All authors have read and agreed to the published version of the manuscript.

Funding: This work was supported by the National Research Foundation of Korea (NRF) grant funded by the Korea government (MSIP) (No. NRF- 2018R1A2B3007894).

Conflicts of Interest: The authors declare no conflict of interest.

References

1. *Electric Vehicles from Life Cycle and Circular Economy Perspectives*; Technical Report; European Environment Agency (EEA): København, Denmark, 2018.
2. Tesla Press Information. Available online: https://www.tesla.com/presskit#semi (accessed on 13 March 2020).
3. Chen, Y.; Baek, D.; Kim, J.; Di Cataldo, S.; Chang, N.; Macii, E.; Vinco, S.; Poncino, M. A systemc-ams framework for the design and simulation of energy management in electric vehicles. *IEEE Access* **2019**, *7*, 25779–25791. [CrossRef]
4. Baek, D.; Chen, Y.; Chang, N.; Macii, E.; Poncino, M. Optimal Battery Sizing for Electric Truck Delivery. *Energies* **2020**, *13*, 709. [CrossRef]
5. Chen, Y.; Baek, D.; Bocca, A.; Macii, A.; Macii, E.; Poncino, M. A Case for a Battery-Aware Model of Drone Energy Consumption. In Proceedings of the 2018 IEEE International Telecommunications Energy Conference (INTELEC), Turin, Italy, 7–11 October 2018; pp. 1–8.
6. Tesla Semi Official Website. Available online: https://www.tesla.com/semi (accessed on 13 March 2020).
7. Baek, D.; Chen, Y.; Macii, E.; Poncino, M.; Chang, N. Battery-Aware Electric Truck Delivery Route Planner. In Proceedings of the International Symposium on Low Power Electronics and Design (ISLPED), Lausanne, Switzerland, 29–31 July 2019; pp. 1–6.
8. Baek, D.; Chen, Y.; Bocca, A.; Bottaccioli, L.; Di Cataldo, S.; Gatteschi, V.; Pagliari, D.J.; Patti, E.; Urgese, G.; Chang, N.; et al. Battery-aware operation range estimation for terrestrial and aerial electric vehicles. *IEEE Trans. Veh. Technol.* **2019**, *68*, 5471–5482. [CrossRef]
9. Chen, Y.; Macii, E.; Poncino, M. A circuit-equivalent battery model accounting for the dependency on load frequency. In Proceedings of the Design, Automation & Test in Europe Conference & Exhibition (DATE), Lausanne, Switzerland, 27–31 March 2017; pp. 1177–1182.
10. Petricca, M.; Shin, D.; Bocca, A.; Macii, A.; Macii, E.; Poncino, M. An automated framework for generating variable-accuracy battery models from datasheet information. In Proceedings of the International Symposium on Low Power Electronics and Design (ISLPED), Beijing, China, 4–6 September 2013; pp. 365–370.
11. Christofides, N. *Worst-Case Analysis of a New Heuristic for the Traveling Salesman Problem*; Report 388; Carnegie Mellon University: Pittsburgh, PA, USA, 1976.
12. Dumitrescu, A.; Mitchell, J.S. Approximation algorithms for TSP with neighborhoods in the plane. *J. Algorithms* **2003**, *48*, 135–159. [CrossRef]
13. Baldacci, R.; Battarra, M.; Vigo, D. Routing a heterogeneous fleet of vehicles. In *The Vehicle Routing Problem: Latest Advances and New Challenges*; Springer: New York, NY, USA, 2008; pp. 3–27.
14. Li, F.; Golden, B.; Wasil, E. A record-to-record travel algorithm for solving the heterogeneous fleet vehicle routing problem. *Comput. Oper. Res.* **2007**, *34*, 2734–2742. [CrossRef]
15. Desaulniers, G.; Desrosiers, J.; Erdmann, A.; Solomon, M.M.; Soumis, F. The VRP with Pickup and Delivery. In *The Vehicle Routing Problem*; SIAM: Philadelphia, PA, USA, 2002; pp. 225–242.
16. Gansterer, M.; Küçüktepe, M.; Hartl, R.F. The multi-vehicle profitable pickup and delivery problem. *OR Spectr.* **2017**, *39*, 303–319. [CrossRef]

17. Kara, İ.; Kara, B.Y.; Yetis, M.K. Energy Minimizing Vehicle Routing Problem. In *Combinatorial Optimization and Applications*; Springer: Berlin/Heidelberg, Germany, 2007; pp. 62–71.

18. Meet the Tesla Semitruck, Elon Musk's Most Electrifying Gamble Yet. 2017. Available online: https://www.tesla.com/presskit#model3 (accessed on 13 March 2020).

19. Tesla Model3 Technical Specifications and Performance Figures. Available online: http://www.zeperfs.com/en/fiche7083-tesla-model-3-75.htm (accessed on 13 March 2020).

20. Tesla Model 3: 2018 Motor Trend Car of The Year Finalist. Available online: https://www.motortrend.com/news/tesla-model-3-2018-car-of-the-year-finalist/ (accessed on 13 March 2020).

21. *Technologies and Approaches to Reducing the Fuel Consumption of Medium and Heavy-Duty Vehicles*; Technical Report; National Academy of Sciences: Washington, DC, USA, 2010.

22. Markel, T.; Brooker, A.; Hendricks, T.; Johnson, V.; Kelly, K.; Kramer, B.; O'Keefe, M.; Sprik, S.; Wipke, K. ADVISOR: A systems analysis tool for advanced vehicle modeling. *J. Power Sources* **2002**, *110*, 255–266. [CrossRef]

23. Baek, D.; Chang, N. Runtime Power Management of Battery Electric Vehicles for Extended Range With Consideration of Driving Time. *IEEE Trans. Very Large Scale Integr. Syst.* **2019**, *27*, 549–559. [CrossRef]

24. Tesla Model 3 & Chevy Bolt Battery Packs Examined. Available online: https://cleantechnica.com/2018/07/08/tesla-model-3-chevy-bolt-battery-packs-examined/ (accessed on 13 March 2020).

Article

Torque Coordination Control of an Electro-Hydraulic Composite Brake System During Mode Switching Based on Braking Intention

Yang Yang [1,2,*], Yundong He [2], Zhong Yang [3], Chunyun Fu [1,2] and Zhipeng Cong [2]

1   State Key Laboratory of Mechanical Transmissions, Chongqing University, Chongqing 400044, China; fuchunyun@cqu.edu.cn
2   School of Automotive Engineering, Chongqing University, Chongqing 400044, China; heyundong@cqu.edu.cn (Y.H.); cong@cqu.edu.cn (Z.C.)
3   Chongqing Changan Automobile Co., Ltd., Chongqing 400023, China; yangzhong1@changan.com.cn
*   Correspondence: yangyang@cqu.edu.cn; Tel.: +86-136-0831-1819

Received: 3 March 2020; Accepted: 17 April 2020; Published: 19 April 2020

Abstract: The electro-hydraulic composite braking system of a pure electric vehicle can select different braking modes according to braking conditions. However, the differences in dynamic response characteristics between the motor braking system (MBS) and hydraulic braking system (HBS) cause total braking torque to fluctuate significantly during mode switching, resulting in jerking of the vehicle and affecting ride comfort. In this paper, torque coordination control during mode switching is studied for a four-wheel-drive pure electric vehicle with a dual motor. After the dynamic analysis of braking, a braking force distribution control strategy is developed based on the I-curve, and the boundary conditions of mode switching are determined. A novel combined pressure control algorithm, which contains a PID (proportional-integral-derivative) and fuzzy controller, is used to control the brake pressure of each wheel cylinder, to realize precise control of the hydraulic brake torque. Then, a novel torque coordination control strategy is proposed based on brake pedal stroke and its change rate, to modify the target hydraulic braking torque and reflect the driver's braking intention. Meanwhile, motor braking torque is used to compensate for the insufficient braking torque caused by HBS, so as to realize a smooth transition between the braking modes. Simulation results show that the proposed coordination control strategy can effectively reduce torque fluctuation and vehicle jerk during mode switching.

Keywords: electric vehicles; electro-hydraulic braking; braking intention; mode switching; torque coordinated control

---

1. Introduction

The electro-hydraulic composite braking system of an electric vehicle (EV) consists of the motor braking system (MBS) and the hydraulic braking system (HBS), which realize the pure electric, pure hydraulic, and hybrid braking modes. The composite braking system converts the kinetic energy of the vehicle into electric energy and ensures braking stability and braking efficiency during braking [1–4]. The braking modes of the electro-hydraulic composite system switch between each other as braking conditions vary. However, the MBS and HBS dynamic response characteristics are not consistent, which leads to total braking torque fluctuations during mode switching, thus affecting braking safety and ride comfort. Therefore, it is of great significance to study the braking torque coordination control during braking mode switching.

Current research on the electro-hydraulic composite braking system mainly focuses on the distribution of braking forces and recovery of braking energy. For example, for the problem of braking force distribution in different modes, Sun et al. [5] established the optimal distribution coefficient response surface by optimizing the distribution coefficient of hydraulic braking torque and regenerative braking torque offline, which improved braking stability and energy recovery efficiency during braking. Shi et al. [6] designed a regenerative braking system that can achieve braking energy recovery during emergency braking. Considering the tire, hydraulic, and motor losses, Sun et al. [7] proposed an on-line control strategy for electro-hydraulic composite braking force, which improved the regenerative braking power. For the problem of braking torque coordination control during mode switching, Okano et al. [8] adopted a filtering algorithm to assign the MBS response to high frequency braking torque and the HBS response to low frequency braking torque, making full use of the dynamic characteristics of both systems. He et al. [9] designed a combined controller for torque disturbance in mode switching, incorporating linear quadratic optimal and sliding mode controllers—the former controller is used for anti-interference, and the latter is used to compensate the performance index offset of the nonlinear part—and achieved a good match of the target speed during mode switching. Considering the influence of the half axle elasticity and backlash nonlinearity of the transmission system on the control performance and dynamic characteristics of the MBS, Lv et al. [10,11] proposed an active control algorithm based on a hierarchical structure to realize the clearance compensation for the transmission system, and Zhang et al. [12] proposed a method of backlash sliding mode compensation and an elastic double closed-loop PID compensation for the control of a permanent magnet synchronous motor; the approaches of both research groups effectively compensated for the influence of the transmission system on the control performance of a permanent magnet synchronous motor. According to whether the HBS provides braking torque, Yang et al. [13] proposed to reduce the torque fluctuation by controlling the change rate of the clutch engagement torque and motor braking torque, and by modifying the target braking torque at different stages during mode switching. Yu et al. [14,15] proposed a double closed-loop feedback control and motor braking torque modifying method, based on the differences in the characteristics of the MBS and the HBS, using the motor braking torque to modify the hydraulic braking torque, thus reducing vehicle jerk during mode switching. Although the aforementioned research has improved braking torque coordination control during mode switching, leading to better control of the electro-hydraulic composite braking system and reduced vehicle jerking, they did not reflect the driver's braking intention during mode switching; that is, the motor and hydraulic braking torque cannot be adjusted reasonably according to whether the driver pays more attention to brake safety or ride comfort.

In this paper, the problem of the total braking torque fluctuation and the jerk of the complete vehicle is addressed for the electro-hydraulic braking system of a four-wheel-drive pure electric vehicle with a dual motor. Firstly, the dynamic characteristics of the vehicle during braking are analyzed. Braking force distribution control strategies that take into account braking safety and regenerative braking energy recovery are established based on the vehicle state parameter constraints. Then, a combined control method, which contains a PID controller and a fuzzy controller, is used to control the brake pressure in each wheel cylinder. Finally, the fuzzy control rules based on the brake pedal stroke and its change rate are designed to modify the target hydraulic braking torque and to reflect the driver's braking intention; that is, the target hydraulic braking torque is modified according to whether the driver pays more attention to brake safety or ride comfort. At the same time, the rapid response of the motor braking torque is used to compensate for the insufficient braking torque caused by the slow response of the HBS, so as to realize the smooth transition of the braking mode, which enhances braking safety and ride comfort during mode switching.

2. Vehicle Dynamics Model and Braking Force Distribution Control Strategies

2.1. The Electric Vehicle Structure

The pure electric vehicle analyzed in this paper is a front–rear centralized dual-motor driving system with two three-phase permanent magnet synchronous motors (PMSM), as shown in Figure 1.

The driving forces are transmitted to the wheels from the motors through final drives I and II. The vehicle-state variables, such as vehicle speed, braking pedal displacement, and battery state of charge (SOC), are collected by an electronic control system and transmitted to the vehicle control unit (VCU) via a controller area network (CAN) bus, and the braking torques required of the MBS and the HBS are determined by the vehicle control unit. The motor braking torque and hydraulic braking torque are controlled by the motor control unit (MCU) and hydraulic control unit (HCU), respectively. The basic parameters of the vehicle are shown in Table 1.

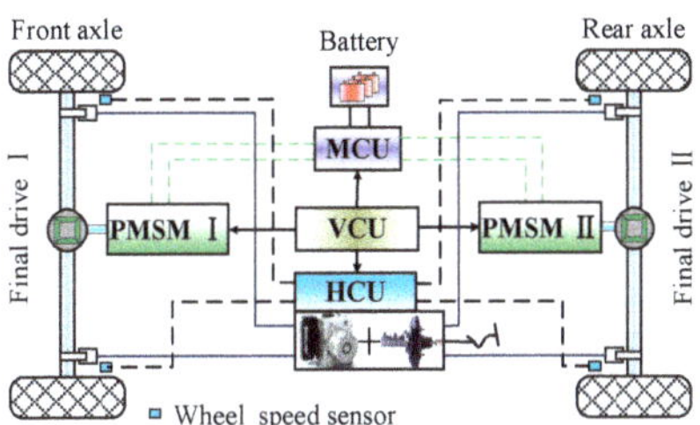

Figure 1. Structure of a four-wheel-drive pure electric vehicle.

Table 1. Parameters of the vehicle.

Parameters	Value
Vehicle mass (kg)	1800
Rolling radius of tire (mm)	362
Height of the center of mass of vehicle (mm)	560
The distance between the center of mass and front axle (mm)	1600
The distance between the center of mass and rear axle (mm)	1100

2.2. Dynamics Analysis of Braking

When a vehicle is braking, and the air resistance moment, rolling resistance moment, and moment of inertial generated by the rotating mass are ignored, the normal acting force of the ground on the front wheel [7] is

$$F_{Z1} = \frac{G\left(L_r + zH_g\right)}{L} \tag{1}$$

The normal force of the ground acting on the rear wheel is

$$F_{Z2} = \frac{G\left(L_f - zH_g\right)}{L} \tag{2}$$

where F_{Z1} and F_{Z2} are the normal acting force of the ground on the front and rear wheels; L is wheelbase; H_g denote the height of the center of mass of vehicle; L_f and L_r are the distance from the center of mass to front and rear axles; G is the weight of vehicle; and z represents the braking strength, and it is the ratio of the vehicle deceleration to the gravitational acceleration.

The dynamic equation of wheel rotation during braking is

$$J_t\dot{\omega}_w = F_{Xb}r - (T_m + T_h) \tag{3}$$

where J_t represent the moment of inertia of transmission equivalent to the wheel; F_{Xb} denote brake force of ground; r is the tire radius; T_m is the braking torque acting on the wheel by PMSM I or PMSM II; T_h is hydraulic braking torque; and ω_w is angular velocity of the wheel.

Dynamics equation of the vehicle transmission system during braking is

$$k_1 T_{m1} i_{O1} + k_2 T_{m2} i_{O2} + k_3 T_h = \frac{G}{g}\frac{dv}{dt} r \tag{4}$$

where v denote vehicle speed; g is acceleration of gravity; according to the working state of PMSM I, PMSM II, and HBS, k_1, k_2, and k_3 is 0 or 1; and i_{O1} and i_{O2} are the gear ratio of the final drive I.

2.3. The Braking Force Distribution Control Strategies

The distribution of braking force is determined by the braking state of the vehicle and must meet the requirements of the brake regulations. Due to deceleration during braking of the vehicle, the stable braking strength provided by the motors varies between 0.086 and 0.2. According to the theory of braking stability, the braking force distribution curve under the I-Curve can ensure the stability of the vehicle; that is, to maintain the ability of straight-line driving of the vehicle during braking [16–18]. Therefore, the maximum braking strength provided by the motors in the pure electric braking mode is determined to be 0.17. The dynamic distribution control strategies based on the I-Curve are shown in Figure 2.

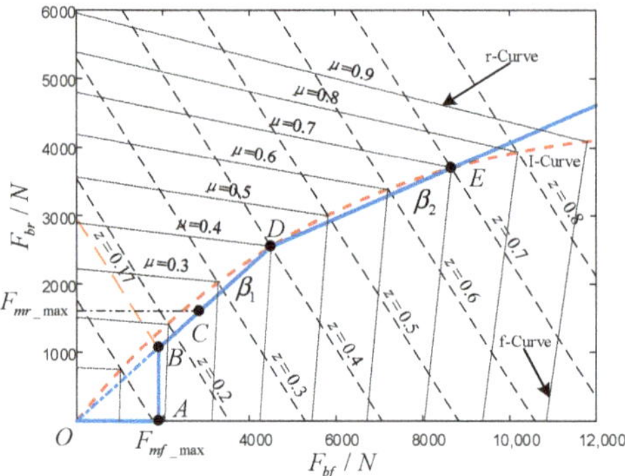

Figure 2. Braking force distribution curve of the front and rear axles.

The braking strength at points A, B, C, D, and E in Figure 2 are $z(A)$, $z(B)$, $z(C)$, $z(D)$, and $z(E)$. I-Curve is the ideal braking force distribution over the front and rear axles. F_{bf} and F_{br} denote the braking force of the front and rear axles. F_{mf_max} and F_{mr_max} denote the maximum braking forces provided by the PMSM I and the PMSM II. The fixed braking force distribution coefficients of the front and rear axles are β_1 and β_2. μ is the road adhesion coefficient.

The minimum speed of the vehicle at which the motors maintain a stable braking torque is v_{min}; the maximum speed of the vehicle at which the motors can perform regenerative braking is v_{max}; and the maximum state of charge in which the battery can be charged is SOC_h. During braking, the speed of the vehicle and the SOC of the battery should obey the restriction that $v_{min} \leq v \leq v_{max}$ and $SOC \leq SOC_h$, respectively. The motor and hydraulic braking force distribution based on the braking strength required by driver is as follows:

When $SOC < SOC_h$ and $v_{min} \leq v \leq v_{max}$:

(1) $0 < z \leq z(A)$, in order to guarantee the braking stability of the vehicle during braking, the PMSM I is given priority to provide braking force, the braking force is provided by PMSM I individually in this condition.

(2) $z(A) < z \leq z(B)$, the braking force is provided by PMSM I and PMSM II simultaneously, and PMSM I provides the maximum braking force, the remaining force is provided by PMSM II.

(3) $z(B) < z \le z(C)$, the HBS starts to provide braking force in this condition, and the braking force of the front and rear axles are distributed according to the fixed braking force distribution coefficient β_1. PMSM I maintains the maximum braking force and the insufficient braking force of the front axle is provided by the HBS. The braking force of PMSM II continues to increase with braking strength, until the maximum braking force of PMSM II is reached.

(4) $z(C) < z \le z(E)$, the required braking force is provided by the motor braking system and hydraulic braking system simultaneously in this condition. Both PMSM I and PMSM II are working at the maximum braking force that can be supplied, and the insufficient braking force of the front and rear axles are provided by the hydraulic braking system.

When $SOC \ge SOC_h$ or $v < v_{\min}$ or $v > v_{\max}$, or $z \ge z(E)$, the braking force is provided only by the HBS, and the braking force at the front and rear axles are distributed according to the fixed braking force distribution coefficient β_1 and β_2.

3. Modeling and Characteristics Analysis of Braking Systems

In order to obtain the dynamic response characteristics of the MBS and the HBS, a simulation model was established in MATLAB/Simulink (2016b, MathWorks, MA, USA) based on the mathematical models of the motor and the hydraulic brake system. In order to accurately control the hydraulic braking torque, a combined control method, which contains a PID controller and a fuzzy controller, was designed to control the brake pressure in each wheel cylinder.

3.1. The Modeling of PMSM

The main parameters of the PMSM used in this paper are shown in Table 2. The PMSM is a strong complex-coupling, high-order, and multivariable nonlinear system [19,20]. In order to realize the vector control of the motor, the mathematical model of the PMSM in the two-phase rotating reference frame (d-q axis) was used to establish the simulation model. The three-phase PMSM in the d-q axis can be described as [21]

$$\begin{pmatrix} u_d \\ u_q \end{pmatrix} = \begin{pmatrix} R_s + L_d \frac{d}{dt} & -\omega_m L_q \\ \omega_m L_d & R_s + L_q \frac{d}{dt} \end{pmatrix} \begin{pmatrix} i_d \\ i_q \end{pmatrix} + \begin{pmatrix} 0 \\ \omega_m \psi_f \end{pmatrix} \tag{5}$$

The electromagnetic torque equation of the PMSM in d-q axis is

$$T_m = 1.5 p_n \left[\psi_f i_q + \left(L_d - L_q \right) i_d i_q \right] \tag{6}$$

where u_d and u_q are the armature voltage components in the d-q axis, respectively; i_d and i_q denote the armature current in the d-q axis, respectively; L_d and L_q represent the equivalent armature inductance in the d-q axis respectively; ψ_f is the rotor flux corresponding to the permanent magnet; R_s represent the stator resistance; ω_m denote the rotational angular velocity of d-q axis; and p_n represents the pole pairs of motor.

Table 2. Key parameters of the motor.

Parameters	PMSM I	PMSM II
Rated/peak power (kw)	24.5/49	13.5/27
Peak torque (Nm)	155.1	171.9
Rated/peak speed (rpm)	3000/6000	1500/6000

3.2. The Modeling of Hydraulic Components

The hydraulic braking system mainly includes system components, such as the brake master cylinder, the wheel cylinder, and the brake pedal simulator, as well as the control components, such as the high-speed on–off valve [22,23]. The structure of the hydraulic braking system used in this article is shown in Figure 3. The pressure of the wheel cylinder is adjusted by the high-speed on–off valve to meet the requirement of the hydraulic braking torque, while the hydraulic braking torque is directly determined by the pressure in the wheel cylinder; i.e., these components reflect the dynamic characteristics of the HBS. Therefore, the mathematical models of the high-speed on–off valves and the brake wheel cylinders are mainly described.

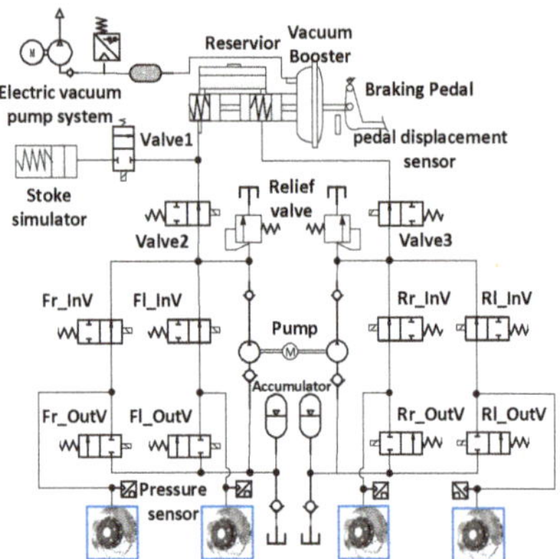

Figure 3. The hydraulic braking system (HBS) structure diagram.

3.2.1. The Modeling of the High-Speed On–Off Valve

The pressure of a brake wheel cylinder is controlled by a pair of high-speed on–off valves: the inlet valve that is normally opened and the outlet valve that is normally closed. The structure of the outlet valve and the force analysis of the valve core are shown in the Figure 4. The key parameters of the high-speed on–off valves selected in this paper are summarized in Table 3.

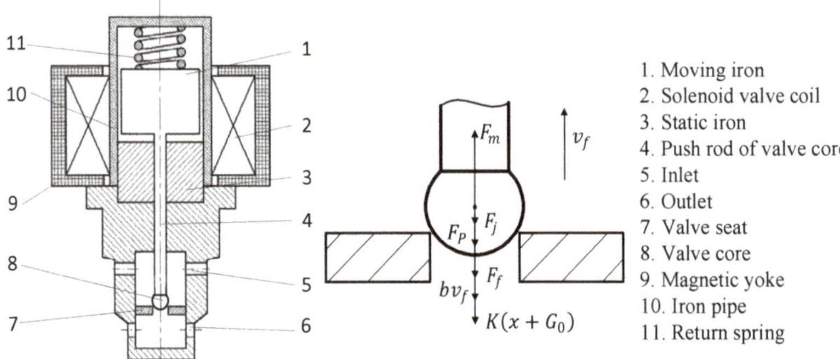

Figure 4. The structure diagram of the outlet valve and the force analysis of the core.

According to the control signals of the vehicle controller, the pressure in the wheel cylinder is adjusted by the hydraulic controller through the combined control of the inlet and outlet valves. Through the force analysis of the valve core in Figure 4, the kinetic equation of the valve core can be obtained [24]:

$$\begin{cases} \dfrac{dv}{dt} = \dfrac{1}{m_f}\left[F_m(x,i) - K(x + G_0) - F_P(x) - bv_f - F_f - F_j\right] \\ v_f = \dfrac{dx}{dt} \end{cases} \tag{7}$$

where F_m is the electro-magnetic force; F_f denotes the frictional force, it takes an estimated value of 0.01 N in this study; F_j represents the jet force; F_P is the flow force of the core assembly; m_f is the mass of the core assembly; K is the stiffness of the return spring; b and x are the velocity viscosity coefficient and the displacement of the core, respectively; and G_0 represents the return spring's pre-compression force.

Table 3. Main parameters of the high-speed switch valve.

Parameters	Value
Return Spring Stiffness (N/mm)	1.6
Moving-iron Mass (g)	15
Initial Air Gap (mm)	0.3
Coil Turns	380
Core Displacement (mm)	0.22
Spring Pre-tightening Force (N)	7

3.2.2. The Modeling of the Wheel Cylinder

During hydraulic braking, the dynamic characteristics of the wheel cylinder piston can be expressed as a spring–mass–damper system, and its dynamic equation [25] is

$$p_w A_p - F_{k0} = m_p \ddot{x}_p + C_p \dot{x}_p + k_p x_p \tag{8}$$

where F_{k0} represents pre-tightening force; p_w denotes the wheel cylinder pressure; A_p is the effective action area of the piston; m_p denotes the mass of piston; x_p represents the displacement of piston; C_p is the damping of the brake; and k_p is the equivalent stiffness.

The relationship between the pressure in the wheel cylinder and the hydraulic braking torque in front and rear axles can be expressed as [13]

$$\begin{cases} T_{hf} = 2p_w \dfrac{\pi D_f^2}{4} R_f K_f \\[2ex] T_{hr} = 2p_w \dfrac{\pi D_r^2}{4} R_r K_r \end{cases} \tag{9}$$

where T_{hf} and T_{hr} represent the hydraulic braking torque of the front and rear axles, respectively; K_f and K_r are the brake factors of the front and rear axles; D_f and D_r denote the diameter of the front and rear wheel cylinders; and R_f and R_r represent the effective radius of the front and rear wheel brake discs.

The key parameters of the front and rear brakes is shown in Table 4.

Table 4. Key parameters of front and rear wheel brakes.

Parameters	Value
Front/rear brake cylinder diameter (mm)	49/21
Front/rear brake effective factor	0.8
Front/rear brake disc radius (mm)	120
Regulating valve pressure (bar)	150

3.2.3. Design of a Combined Controller for Hydraulic Braking Torque

The HBS should have good control performance to achieve a fast and accurate response to the hydraulic braking force. A cooperative control strategy is used to improve the response speed and control the accuracy of the hydraulic braking torque. When the pressure error $|\Delta p(t)|$ between the target pressure and the tracking pressure is larger than the value of the threshold, the PID controller controls the wheel cylinder pressure to achieve a rapid adjustment. Conversely, when the pressure error $|\Delta p(t)|$ is smaller than the value of the threshold, the fuzzy controller is used to stabilize the wheel cylinder pressure near the target pressure and reduce the hydraulic pressure fluctuation [26,27]. The control schematic is shown in Figure 5. Wheel cylinder control is determined by the logical threshold controller according to the pressure error $\left|\Delta p(t)\right|$, and then either the PID or fuzzy controller determines the duty ratio of the high-speed on–off valve. The inlet and outlet valves of the wheel cylinder are directly controlled by PWM (pulse width modulation). Thus, the tracking pressure of the wheel cylinder follows the target pressure. A threshold value of 0.5 bar was set for $\left|\Delta p(t)\right|$ as simulation results indicated that both the response speed and the control accuracy of the pressure in the wheel cylinder were optimal at this level.

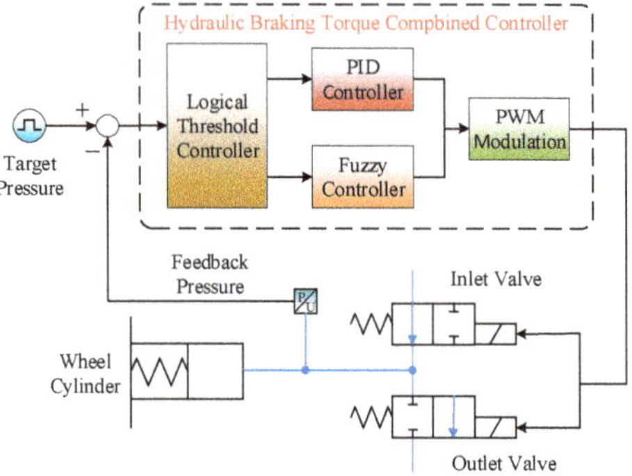

Figure 5. The pressure control algorithm of the wheel cylinder.

When the pressure error $\left|\Delta p(t)\right|$ is larger than the value of the threshold, the PID controller is used to adjust the pressure in the wheel cylinder, so that the tracking pressure in the wheel cylinder rapidly responds to the target pressure. The mathematical model of the PID controller [24] is

$$D(t) = k_p\Delta p(t) + k_i \int_{t_0}^{t_1} \Delta p(t)dt + k_d\frac{d\Delta p(t)}{dt} \tag{10}$$

where $D(t)$ is the duty ratio of the PWM signal; k_p, k_i, and k_d are the proportional, integral, and differential coefficients of the PID controller, respectively; t_0 and t_1 are the time when the HBS starts and ends to work.

When the pressure error $\left|\Delta p(t)\right|$ is smaller than the value of the threshold, the fuzzy controller is used to adjust the pressure of the wheel cylinder. The opening degree of the inlet and outlet valves are determined by the target pressure $p(t)$ and the pressure error $\Delta p(t)$. When the target pressure $p(t)$ is small (S′) and the pressure error $\Delta p(t)$ is negative (N), the tracking pressure needs to be reduced, so that the duty ratio of the inlet valve D_{in} takes a smaller value (S′) and the duty ratio of the outlet valve D_{out} takes a larger value (VL′). When the target pressure $p(t)$ is large (VL′) and the pressure error $\Delta p(t)$ is positive (P), the tracking pressure needs to be increased, so that the duty ratio of the inlet valve D_{in} takes a larger value (VL′) and the duty ratio of the outlet valve D_{out} takes a smaller value (S′). The fuzzy controller rules for the degree of opening of the high-speed on–off valve are shown in Tables 5 and 6.

In Table 5, N, Z, and P denote less than zero, equal to zero, and greater than zero, respectively; whereas S′, MS′, M′, ML′, L′, and VL′ represent small, small medium, medium, medium large, large, and very large, respectively.

The pressure variation of a wheel cylinder controlled by the combined controller under the input of a sinusoidal signal is shown in Figure 6. The tracking pressure in the wheel cylinder is precisely controlled, and the pressure error between the target pressure and tracking pressure is very small.

Table 5. The fuzzy control rules of the inlet valve.

D_{in}		$\Delta p(t)$		
		N	**Z**	**P**
$p(t)$	S′	S′	MS′	MS′
	MS′	MS′	M′	M′
	M′	M′	ML′	ML′
	ML′	ML′	L′	L′
	L′	L′	VL′	VL′
	VL′	L′	VL′	VL′

Table 6. The fuzzy control rules of the outlet valve.

D_{out}		$\Delta p(t)$		
		N	**Z**	**P**
$p(t)$	S′	VL′	VL′	L′
	MS′	L′	L′	ML′
	M′	ML′	ML′	M′
	ML′	M′	M′	MS′
	L′	MS′	MS′	S′
	VL′	MS′	MS′	S′

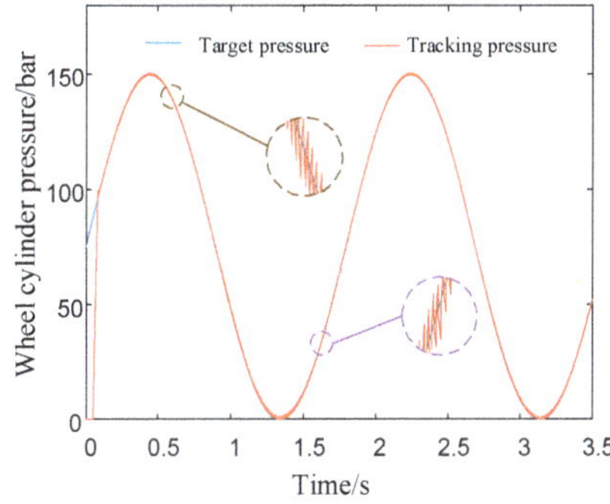

Figure 6. The response of the wheel cylinder with a sinusoidal input.

3.3. Dynamic Characteristics Analysis of the MBS and the HBS

Using the mathematical models described above, a dynamic model of the electro-hydraulic composite braking system was built in MATLAB/Simulink, and a separate HBS physical model was built in the Simulink sub-module, Simscape. The dynamic responses of the MBS and HBS are shown in Figure 7. Under the input of the same demand braking torque, the response time of the MBS is t_m, the response time of the HBS is t_h, and the response time difference is Δt_{mh}. Compared with the HBS, the dynamic response of the MBS is fast, and the braking torque rise time is short, but there is a certain amount of overshoot. There are two main reasons for the differences in dynamic characteristics: In the initial stage of the hydraulic braking torque response, the HBS needs high-pressure brake fluid to fill the circuit and liquid chamber, and during the rising period of the hydraulic braking torque, there are viscosity resistance, flow force, and orifice compensation of the hydraulic braking system.

Because of the differences in dynamic response characteristics between the MBS and the HBS, the total braking torque fluctuates significantly during mode switching, which cannot meet the braking torque required by the driver, and may also lead to an increase in the jerk of the complete vehicle and the false trigger of the ABS braking, as shown in the simulation part of this study. So it is necessary to coordinate the motor braking torque and hydraulic braking torque in the process of mode switching, to ensure the stability of the braking torque during braking.

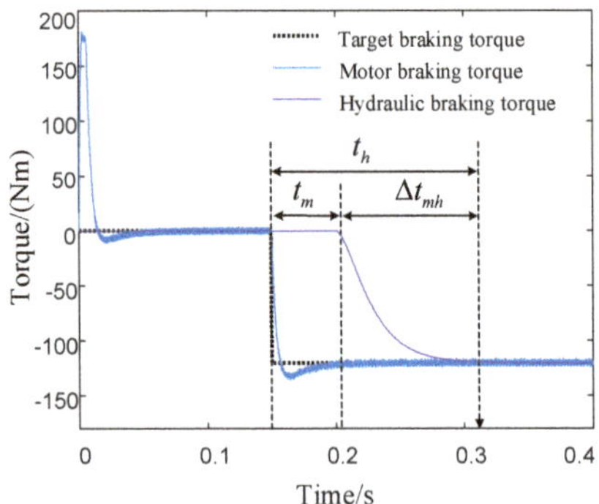

Figure 7. Dynamic response characteristics of the HBS and the motor braking system (MBS).

4. The Torque Coordinated Control Strategy of Mode Switching

4.1. The Condition Analysis of Mode Switching

By controlling the working state of the motor and switching of the HBS coupling valve, the electro-hydraulic composite braking system of the electric vehicle can realize various braking modes. According to the braking force distribution control strategies of this paper, the working state of each component of the braking system of each braking mode are shown in Table 7. Based on the response differences between the MBS and HBS, and setting aside the discontinuous change due to brake pedal action, the processes responsible for the torque fluctuation during mode switching mainly exist in the working conditions that the braking torque step changed. Therefore, the coordinated control strategy is mainly applied to the braking conditions in which the braking torque of the MBS or HBS step changes.

Table 7. The working state of each component in the different braking modes.

Mode	PMSM I	PMSM II	HBS of Front Axle	HBS of Rear Axle
Pure electric braking	●	○	○	○
	●	●	○	○
Hybrid braking	●	●	●	○
	●	●	●	●
Pure hydraulic braking	○	○	●	●

In Table 7, "●" represents MBS working or HBS working, and "○" represents MBS not in operation or HBS not in operation.

4.2. The Design of the Coordination Controller

The dynamic coordination control strategy of brake mode switching developed in this paper is shown in the Figure 8. Firstly, the target hydraulic braking torque, T_{h_req}, and the target motor braking torque, T_{m_req}, are preliminarily distributed based on the vehicle state parameters by the braking force distribution controller. Secondly, the target hydraulic braking torque is modified through fuzzy control rules based on the pedal opening and its change rate, to reflect the driver's braking intention. Then, the target motor braking torque is corrected by the actual hydraulic braking torque (T_h) output by the HBS, and the rapid response of the MBS is used to compensate for the hydraulic braking torque, to achieve a smooth transition of the braking mode. Finally, the HBS and the MBS are controlled by the hydraulic and motor braking controller, respectively, to respond to the modified target braking torque T'_{h_req} and T'_{m_req}, and then output the actual hydraulic braking torque, T_h, and actual motor braking torque, T_m, to decelerate the vehicle.

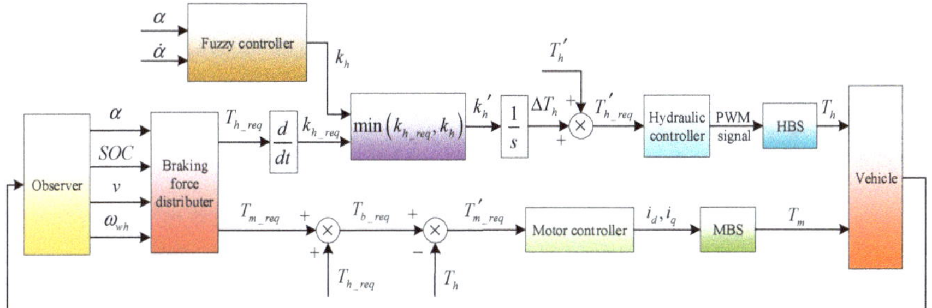

Figure 8. Dynamic coordination control structure.

4.2.1. The Modification of Target Hydraulic Braking Torque

The target change rates of target hydraulic braking torque during braking are

$$k_{h\_req} = \frac{d}{dt}T_{h\_req} \tag{11}$$

where k_{h_req} represents the target change rates of hydraulic braking torque.

During the mode switching, the upper limit of the change rate of the hydraulic braking torque, k_h, is determined by the fuzzy controller, which is designed based on the brake pedal stroke, α, and its change rate, $\dot{\alpha}$. Then the upper limit of the change rate, k_h, is compared with the target change rate k_{h_req}, to determine the modified target change rate.

$$k'_h = \min\left(k_{h\_req}, k_h\right) \tag{12}$$

The increment of the modified target hydraulic braking torque, ΔT_h, can be obtained by the integration of the modified target change rate. Therefore, the modified target braking torque of the HBS is

$$T'_{h\_req} = T'_h + \int \min\left(k_{h\_req}, k_h\right)dt \tag{13}$$

where T'_{h_req} represents the target braking torque of the HBS modified by the coordination controller; and T'_h is the initial braking torque at the moment when the mode is switched.

During mode switching, the driver's braking intention is reflected by the brake pedal stroke and the brake pedal stroke change rate. Then the fuzzy controller outputs the upper limit of the change rate of the target hydraulic braking torque, so as to realize the modification of the target hydraulic braking torque. The fuzzy subsets of brake pedal stroke, brake pedal stroke change rate, and the upper limit of the change rate of the hydraulic braking torque are {VS, S, MS, M, ML, L, VL}; therefore, the input and output of the fuzzy controller can be described as

$$\{\alpha\} = \{VS, S, MS, M, ML, L, VL\}$$
$$\{\dot{\alpha}\} = \{VS, S, MS, M, ML, L, VL\} \tag{14}$$
$$\{k_h\} = \{VS, S, MS, M, ML, L, VL\}$$

where VS, S, MS, M, ML, L, and VL represent very small, small medium, medium, medium large, large, and very large, respectively.

The fuzzy control rules are shown in Table 8, and the membership function of input and output variables of fuzzy controller are shown in Figures 9–11. The fuzzy control rules are formulated based on the following experiences:

Criterion 1: If α and $\dot{\alpha}$ are S, then k_h is MS. In this case, the brake pedal stroke and its change rate are small; it can be considered that the driver pays more attention to the ride comfort during mode switching, and the upper limit of the change rate of the hydraulic braking torque takes a medium-small value.

Criterion 2: If α is S and $\dot{\alpha}$ is L, then k_h is ML. In this case, the braking pedal opening is small and its change rate is large, which indicates that the driver pays more attention to the braking safety during mode switching. Therefore, the upper limit of the change rate of the hydraulic braking torque takes a medium-large value.

Criterion 3: If α is VL and $\dot{\alpha}$ is S, then k_h is L. In this case, the braking pedal opening is very large and its change rate is small, indicating that the driver pays attention to both braking safety and ride comfort during mode switching, so the upper limit of the change rate of the hydraulic braking torque takes a large value.

Table 8. Fuzzy control rules of target hydraulic braking torque change rate.

k_h		$\dot{\alpha}$						
		VS	**S**	**MS**	**M**	**ML**	**L**	**VL**
	VS	VS	S	MS	M	M	ML	ML
	S	S	MS	M	M	ML	ML	L
	MS	MS	M	M	ML	ML	L	L
α	M	M	M	ML	ML	L	L	VL
	ML	M	ML	ML	L	L	VL	VL
	L	ML	ML	L	L	VL	VL	VL
	VL	ML	L	L	VL	VL	VL	VL

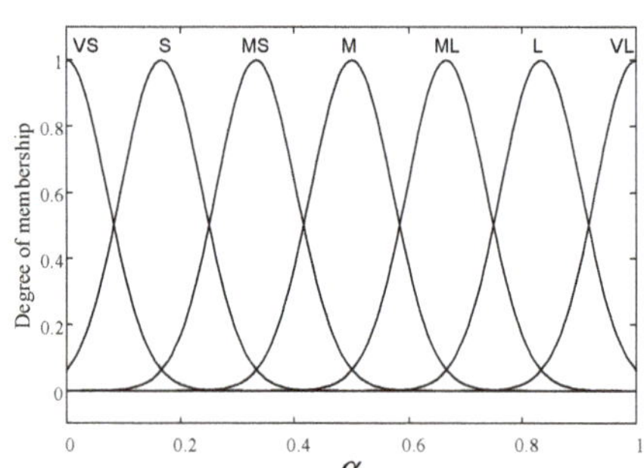

Figure 9. Pedal opening degree membership function.

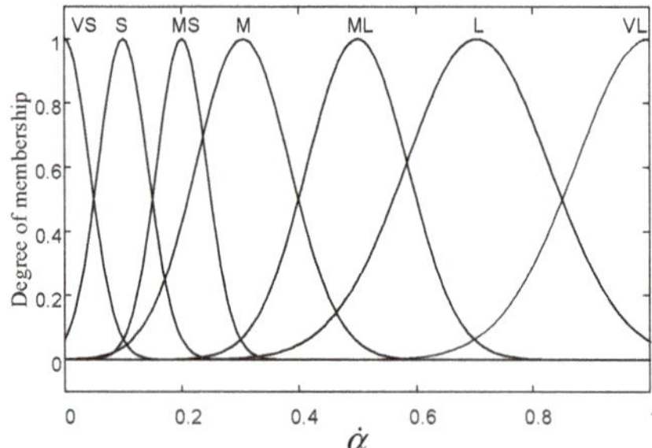

Figure 10. Pedal opening change rate membership function.

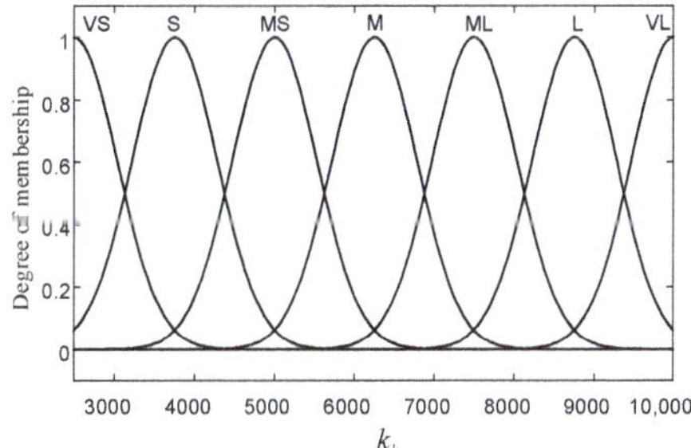

Figure 11. The upper limit of the change rate of the hydraulic braking torque membership function.

According to the modified target hydraulic braking torque T'_{h_req}, the hydraulic controller determines the duty ratio of high-speed on–off valve, and directly controls the inlet valve and outlet valve of wheel cylinder through PWM modulation, so as to make the tracking pressure of wheel cylinders follow the modified target pressure changes.

4.2.2. The Modification of Target Motor Braking Torque

According to the analysis in Section 3.3 of this paper, the response time of the MBS is shorter than that of the HBS, thus the rapid response of MBS can be used to compensate for the insufficient braking torque caused by the slow response of HBS, so as to solve the fluctuation of total braking torque and the jerk of the complete vehicle during brake mode switching. Therefore, during mode switching, the MBS need to provide the target motor brake torque T_{m_req}, which is determined by the braking force distribution controller and additionally provide the difference between the target hydraulic braking torque T_{h_req} and the current actual hydraulic braking torque T_h; that is,

$$T'_{m\_req} = T_{m\_req} + T_{h\_req} - T_h \tag{15}$$

The sum of target hydraulic braking torque T_{h_req} and target motor braking torque T_{m_req} is the total braking torque T_{b_req} required by the driver, so Equation (15) can be rewritten as

$$T'_{m\_req} = T_{b\_req} - T_h \tag{16}$$

The motor control parameters i_d and i_q are output by the motor controller according to the modified target motor braking torque T'_{m_req}, so that the MBS outputs the actual motor braking torque T_m to act on the vehicle.

According to the braking force distribution control strategy of this paper, the HBS starts to provide braking torque when the braking strength required by driver is greater than $z(B)$. If the braking torque required by driver changes, in order to maintain the coordinated compensation ability of the MBS to the hydraulic braking torque during mode switching, the target braking torque T_{m_req} allocated by the braking force distribution controller to the MBS should be smaller than the maximum braking torque T_{m_max} that can be provided by the MBS. When the braking torque required by the driver continues to increase, and the required braking strength satisfy $z(B) < z \leq z(C)$, the target braking torque T_{m_req} determined by braking force distribution controller should increase to T_{m_max} gradually. If the braking torque required by driver remains unchanged, the maximum braking torque that can be provided by the MBS remains at T_{m_max} at the braking torque distribution stage.

5. Simulation and Analysis

The forward simulation model of a pure electric vehicle was established in MATLAB/Simulink in this study, as shown in Figure 12. This simulation model includes the PMSM I and PMSM II models, the HBS model, the battery model, the vehicle longitudinal resistance model, and the controller model. Because the switching between different braking modes is similar, a specific mode switching can be selected for verification and analysis. The same conclusion can be obtained by the simulation of other switching processes with the proposed torque coordinated control method.

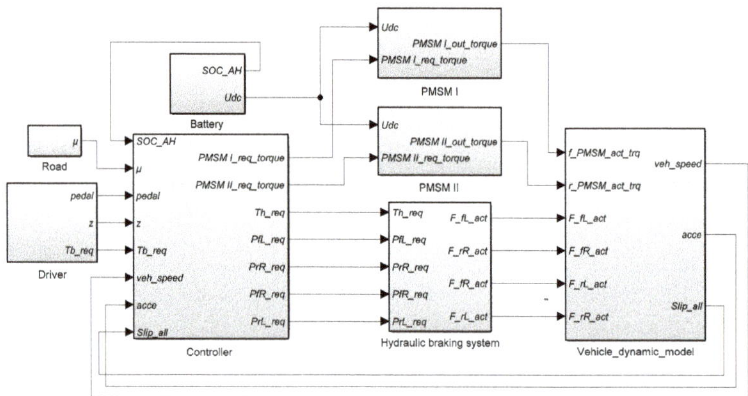

Figure 12. The simulation model of the electric vehicle.

The conditions of constant and variable braking strength are simulated to verify the effectiveness of the dynamic coordination control strategy. For the switching from a constant braking strength, the mode switching between the pure electric and pure hydraulic braking modes were selected for the simulation test, and the motor braking switching to a hybrid braking mode was selected for simulation verification in the mode switching of variable braking strength. In the switching from a constant braking strength, the motor and hydraulic braking torque with coordination will respond to the modified target braking torque output by the coordination controller, while the motor and hydraulic braking torque without coordination will respond to the target braking torque that is assigned by the braking force distributor. Under the condition of a variable braking strength, the braking torque and the jerk of the complete vehicle, focusing on safety and ride comfort, were compared, to verify whether the coordinated control strategy reflects the driver's braking intention.

5.1. Simulation and Verification of the Constant Braking Strength of Mode Switching

5.1.1. Switch from Pure Hydraulic to Pure Electric Braking Mode

The speed variation condition shown in Figure 13a is designed to verify the effectiveness of the coordinated control algorithm during pure hydraulic switching to the pure electric braking mode. In this braking condition, the initial speed of the vehicle is 110 km/h, the initial *SOC* of the battery is 0.6, and the road adhesion coefficient is 0.8. The braking strength required by the driver is increased from 0 to 0.15 within 0 to 0.5 s, and then remains constant.

At the start of braking, the speed of the vehicle is too high for the motors to perform regenerative braking; hence, braking torque is provided by the HBS. After 1.5 s, when the speed of the vehicle has decreased sufficiently, the braking mode switches to pure electric, and braking torque is provided only by the MBS. Because of the compensation of the motor braking torque to the hydraulic braking torque, the *SOC* with coordination is slightly higher than the *SOC* without coordination at the start of braking; moreover, the slow increase of the motor braking torque with coordination during mode switching resulted in a lower *SOC* with coordination than that without coordination, as shown in Figure 13a. As shown in Figure 13b, during the increase of braking strength, the response speed of the HBS is slow due to the orifice compensation and viscosity resistance within the HBS, hence the total braking torque without coordination cannot quickly respond to the target total braking torque required by the driver.

The coordinated total braking torque follows the target total braking torque well because the MBS can compensate for the insufficient braking torque caused by the slow response of the HBS.

During mode switching, Figure 13c demonstrates that the motor braking torque without coordination increased rapidly, while the slower responding HBS was still providing a high braking torque. According to Figure 13d, the rapid increase in total braking torque without coordination resulted in a 31.29 m/s^3 jerk of the complete vehicle. In contrast, the motor braking torque with coordination increases as the hydraulic braking torque decreases, and the total braking torque is maintained at a constant level as far as possible. The maximum jerk of the vehicle was 5.91 m/s^3; thus, the ride comfort of the vehicle is improved during mode switching.

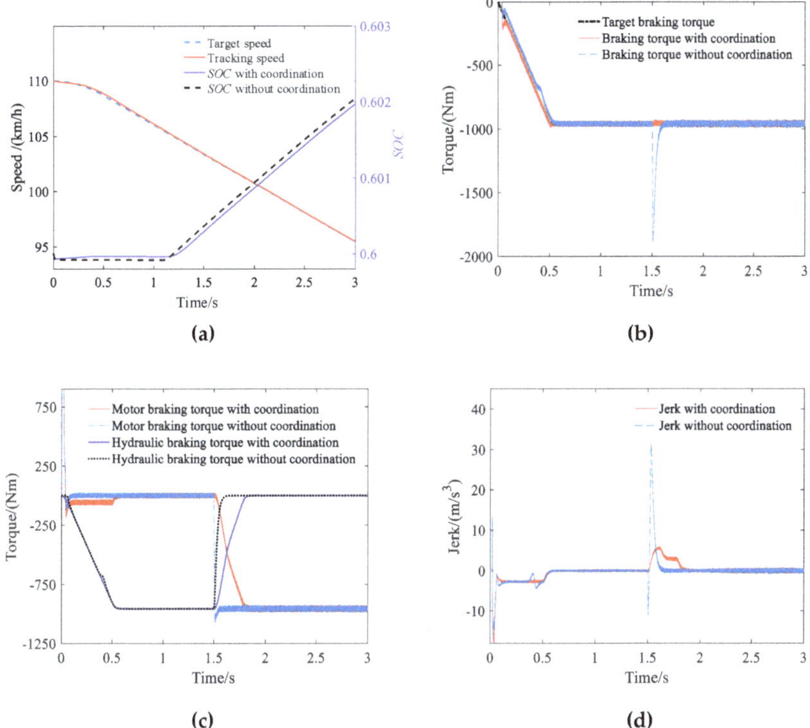

Figure 13. Simulation results of the pure hydraulic switch to a pure motor braking mode under a constant braking strength: (**a**) The variation in vehicle speed and *SOC*, (**b**) the variation in total braking torque, (**c**) the variation in motor and hydraulic braking torque, and (**d**) the variation in vehicle jerk.

5.1.2. Switch from Pure Electric to Pure Hydraulic Braking Mode

The speed variation condition shown in Figure 14a is designed to verify the effectiveness of the coordinated control algorithm during pure electric switching to the pure hydraulic braking mode. In this braking condition, the initial speed of the vehicle is 25 km/h, the initial *SOC* of the battery is 0.6, and the road adhesion coefficient is 0.8. The braking strength required by driver increased from 0 to 0.1 within 0 to 0.5 s, and then remains unchanged.

With the deceleration of the vehicle, the vehicle speed was reduced to 20 km/h at 1.67s, which is too slow for the MBS to maintain a stable regenerative braking torque; hence, the braking mode is switched from pure electric to pure hydraulic. Due to the coordination of motor braking torque to hydraulic braking torque during mode switching, the *SOC* with coordination is higher than that without coordination, as shown in Figure 14a. The simulation results of Figure 14b,c show that the mode switching without coordination cannot provide the braking torque required by the driver because of the rapid withdrawal of the motor braking torque, giving a wrong braking felling to the driver, resulting in a mis-operation by the driver. With coordination, the motor braking torque decreases as the hydraulic braking torque increases, creating a smooth transition, so that the total braking torque changes gradually and the speed of vehicle is steadily reduced. According to Figure 14d, the maximum jerk of the complete vehicle during mode switching with and without coordination was -3.14 m/s3 and -21.18 m/s3, respectively. Thus, the coordination control strategy improves the vehicle's safety and ride comfort during braking mode switching.

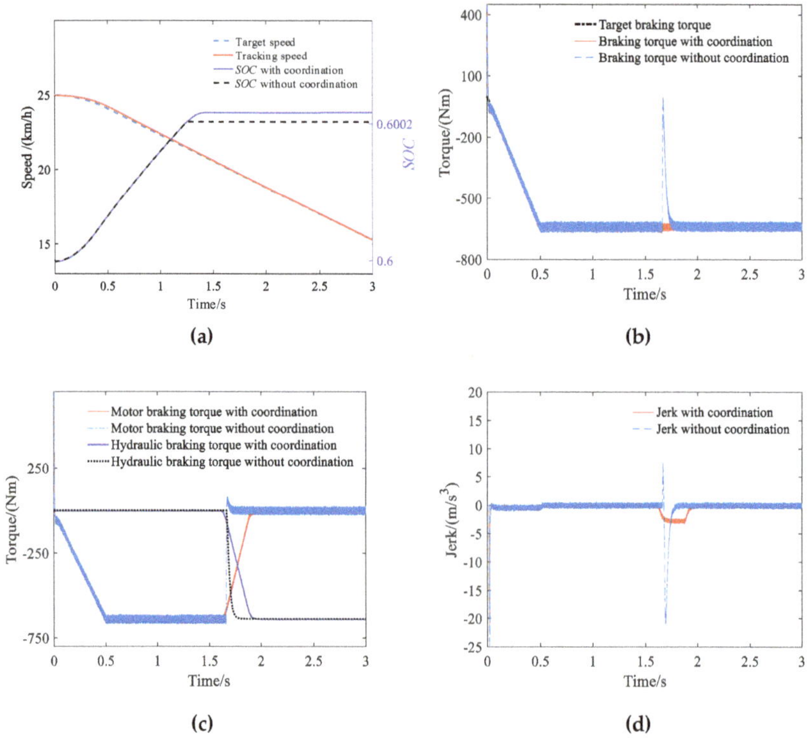

Figure 14. Simulation results of the pure hydraulic switch to a pure motor braking mode under a constant braking strength: (**a**) The variation in vehicle speed and *SOC*, (**b**) the variation in total braking torque, (**c**) the variation in motor and hydraulic braking torque, and (**d**) the variation in vehicle jerk.

5.2. Simulation and Verification of Variable Braking Strength of Mode Switching

Under the condition of a variable braking strength, the initial speed of vehicle is 60 km/h, the initial SOC of battery is 0.6, and the road adhesion coefficient is 0.8. The braking strength required by the driver is increased from 0 to 0.1 within 0 to 0.5 s, and then remains unchanged. At 1.5 s, the driver depressed the brake pedal with two different brake pedal stroke change rates, and the braking strength gradually increases to 0.3. The variation in the braking strength is shown in Figure 15a.

The driver depressed the brake pedal to the same opening with two different brake pedal stroke change rates; therefore, both the switching processes are switched from pure electric to the hybrid braking mode. Because the change rate in brake pedal stroke is different, one switching process focuses on brake safety while the other focuses on ride comfort. The simulation results in Figure 15 are all obtained by the coordinated control strategy proposed in this paper. It can be seen from Figure 15a that since the variation in each braking torque is almost the same for both the safety-focused and comfort-focused intentions, the *SOC* changes of the two braking intention are almost the same, but the SOC with a safety-focused intention is slightly higher than that of the comfort-focused intention.

As shown in Figure 15b, c, since the motor and hydraulic braking torque is modified by the torque coordination controller according to the different braking intentions of the driver, the coordinated total braking torque, motor, and hydraulic braking torque that focus on ride comfort are changed more gently than that focusing on braking safety. As shown in Figure 15d, the jerk of the vehicle during mode switching that focuses on ride comfort is -4.84 m/s^3, while the mode switching that focuses on braking safety is -9.97m/s^3. The simulation results show that the coordinated control strategy of this paper can modify the motor and hydraulic braking torque reasonably according to the driver's braking intention, so as to take into account the braking safety and ride comfort during braking mode switching.

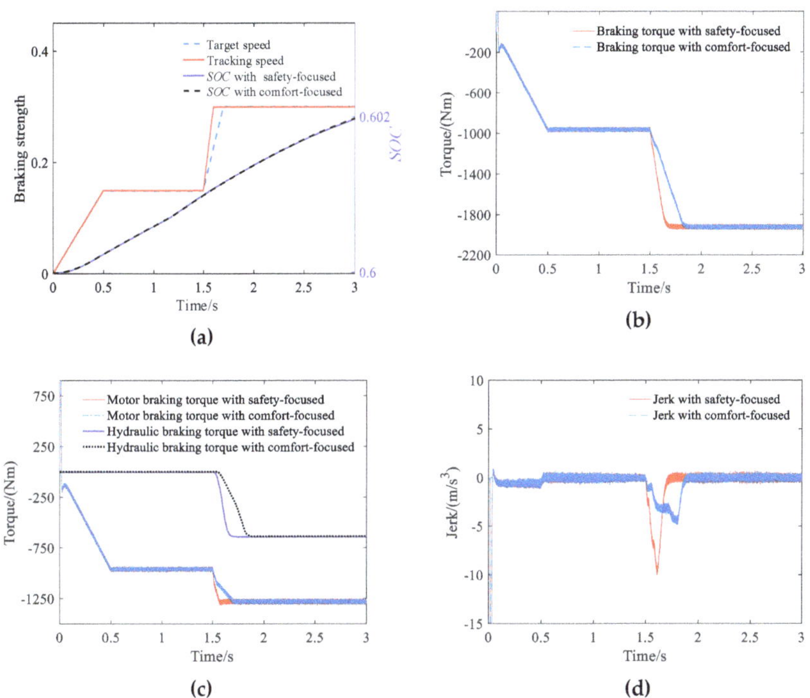

Figure 15. Simulation results of the pure motor switch to a hybrid braking mode under variable braking strengths: (**a**) The variation in the braking strength and *SOC*, (**b**) the variation in total braking torque, (**c**) the variation in motor and hydraulic braking torque, and (**d**) the variation in vehicle jerk.

6. Conclusions

The configuration of a four-wheel-drive pure electric vehicle with a dual motor is considered in this paper, and a braking torque dynamic coordinated control strategy, based on the braking intention of the driver, is proposed. The dynamic coordinated control strategy effectively reduces the torque fluctuation and the jerk of the complete vehicle during mode switching.

A controller combining a PID controller and a fuzzy controller is used to adjust the pressure in the wheel cylinder of the HBS to achieve precise control of the hydraulic braking torque. The brake pedal stroke and its change rate are used to reflect the braking intention of the driver, and to modify the target hydraulic braking torque based on a fuzzy control algorithm according to whether the driver pays more attention to brake safety or ride comfort. At the same time, the rapid response of the MBS is used to compensate for the insufficient braking torque caused by the slow response of the HBS, so as to ensure the ride comfort and stability of the braking during mode switching.

Based on the mathematical model of the electro-hydraulic composite braking system, a simulation verification platform is constructed. The typical mode switching of the constant and variable braking strength conditions are simulated to verify the effectiveness of the dynamic coordinated control strategy. The simulation results show that the torque coordination control strategy described in this paper can not only modify the motor and hydraulic braking torque according to the driver's braking intention, but also significantly reduce the braking torque fluctuation and the jerk of the complete vehicle, thereby improving ride comfort and safety. The influence of the proposed control strategy on energy recovery mainly depends on the type of mode switching condition.

Intelligent transportation system (ITS) and vehicle-to-vehicle communication (V2V) are future development trends; therefore, in future research, the authors will combine ITS to make active predictions of braking mode. In addition, energy consumption will be considered by reducing the frequency of mode switching.

Author Contributions: Conceptualization, Y.Y. and Y.H.; methodology, Y.Y. and Y.H.; software, Y.H. and Z.C.; validation, Y.H., Z.C. and C.F.; formal analysis, Y.Y.; investigation, Y.H.; resources, Y.Y. and Y.H.; data curation, Z.Y.; writing—original draft preparation, Y.H.; writing—review and editing, Y.Y., Y.H. and C.F.; visualization, Y.H.; supervision, Y.Y., C.F. and Z.Y.; project administration, Y.Y., C.F. and Z.Y.; funding acquisition, Y.Y. and C.F. All authors have read and agreed to the published version of the manuscript.

Funding: This research was funded by the National Key R&D Program of China (grant No. 2018YFB0106100), and the National Natural Science Foundation of China (grant No.51575063).

Conflicts of Interest: The authors declare no conflict of interest.

References

1. Niu, G.; Arribas, A.P.; Salameh, M.; Krishnamurthy, M.; Garcia, J.M. Hybrid energy storage systems in electric vehicle. In Proceedings of the Transportation Electrification Conference and Expo (ITEC), Dearborn, MI, USA, 14–17 June 2015; pp. 1–6.

2. Wu, J.; Wang, X.; Li, L.; Qin, C.; Du, Y. Hierarchical control strategy with battery aging consideration for hybrid electric vehicle regenerative braking control. *Energy* **2018**, *145*, 301–312. [CrossRef]

3. Dizqah, A.M.; Lenzo, B.; Sorniotti, A.; Gruber, P.; Fallah, S.; De Smet, J. A Fast and Parametric Torque Distribution Strategy for Four-Wheel-Drive Energy-Efficient Electric Vehicles. *IEEE Trans. Ind. Electron.* **2016**, *63*, 4367–4376. [CrossRef]

4. Niu, G.; Shang, F.; Krishnamurthy, M.; Garcia, J.M. Design and Analysis of an Electric Hydraulic Hybrid Powertrain in Electric Vehicles. *IEEE Trans. Transp. Electrif.* **2017**, *3*, 48–57. [CrossRef]

5. Sun, F.; Liu, W.; He, H.; Guo, H. An integrated control strategy for the composite braking system of an electric vehicle with independently driven axles. *Veh. Syst. Dyn.* **2016**, *54*, 1031–1052. [CrossRef]

6. Shi, J.; Wu, J.; Zhu, B.; Zhao, Y.; Deng, W.; Chen, X. Design of Anti-lock Braking System Based on Regenerative Braking for Distributed Drive Electric Vehicle. *SAE Int. J. Passeng. Cars-Electron. Electr. Syst.* **2018**, *11*, 205–218. [CrossRef]

7. Sun, H.; Wang, H.; Zhao, X. Line Braking Torque Allocation Scheme for Minimal Braking Loss of Four-Wheel-Drive Electric Vehicles. *IEEE Trans. Veh. Technol.* **2019**, *68*, 180–192. [CrossRef]

8. Okano, T.; Sakai, S.; Uchida, T. Braking Performance Improvement for Hybrid Electric Vehicle Based on Electric Motor's Quick Torque Response. In Proceedings of the 19th International Electric Vehicle Symposium and Exhibition, Busan, Korea, 19–23 October 2002; pp. 1285–1296.

9. He, C.; Zhang, J.; Wang, L.; Gou, J.; Li, Y. Dynamic Load Emulation of Regenerative Braking System during Electrified Vehicle Braking States Transition. In Proceedings of the Vehicle Power & Propulsion Conference, Beijing, China, 15–18 October 2013; pp. 1–5.

10. Lv, C. Dynamical Blending Control of Regenerative Braking and Frictional Braking for Electrified Vehicles. Ph.D. Thesis, Tsinghua University, Beijing, China, 2015.

11. Lv, C.; Zhang, J.; Li, Y.; Yuan, Y. Synthesis of a Hybrid-Observer-Based Active Controller for Compensating Powetrain Backlash Nonlinearity of an Electric Vehicle during Regenerative Braking. *SAE Int. J. Altern. Powertrains* **2015**, *4*, 190–198. [CrossRef]

12. Zhang, Z.; Ma, R.; Wang, L.; Zhang, J. Novel PMSM Control for Anti-Lock Braking Considering Transmission Properties of the Electric Vehicle. *IEEE Trans. Veh. Technol.* **2018**, *67*, 10378–10386. [CrossRef]

13. Yang, Y.; Wang, C.; Zhang, Q.; He, X. Torque Coordination Control during Braking Mode Switch for a Plug-in Hybrid Electric Vehicle. *Energies* **2017**, *10*, 1684. [CrossRef]

14. Yu, Z.; Shi, B.; Xiong, L.; Han, W.; Shu, Q. Coordinated Control of Hybrid Braking Based on Integrated-Electro-hydraulic brake system. *J. Tongji Univ. Nat. Sci. Ed.* **2019**, *47*, 851–856. [CrossRef]

15. Yu, Z.; Shi, B.; Xiong, L.; Han, W. Coordinated Control under Transitional Conditions in Hybrid Braking of Electric Vehicle. In Proceedings of the Brake Colloquium & Exhibition-36th Annual, Palm Desert, CA, USA, 14–17 October 2018; pp. 1–7.

16. Xu, W.; Chen, H.; Zhao, H.; Ren, B. Torque optimization control for electric vehicles with four in-wheel motors equipped with regenerative braking system. *Mechatronics* **2019**, *57*, 95–108. [CrossRef]

17. Fujimoto, H.; Harada, S. Model-Based Range Extension Control System for Electric Vehicles With Front and Rear Driving–Braking Force Distributions. *IEEE Trans. Ind. Electron.* **2015**, *62*, 3245–3254. [CrossRef]

18. Qiu, C.; Wang, G.; Meng, M.; Shen, Y. A novel control strategy of regenerative braking system for electric vehicles under safety critical driving situations. *Energy* **2018**, *149*, 329–340. [CrossRef]

19. Tehrani, M.G.; Kelkka, J.; Sopanen, J.; Mikkola, A.; Kerkkänen, K. Electric Vehicle Energy Consumption Simulation by Modeling the Efficiency of Driveline Components. *SAE Int. J. Commer. Veh.* **2016**, *9*, 31–39. [CrossRef]

20. Zhong, Z.; Li, J.; Zhou, S.; Zhou, Y.; Jiang, S. Torque Ripple Description and Its Suppression through Flux Linkage Reconstruction. *SAE Int. J. Altern. Powertrains* **2017**, *6*, 175–182. [CrossRef]

21. Shen, J.-Q.; Yuan, L.; Chen, M.-L.; Xie, Z. Flux Sliding-mode Observer Design for Sensorless Control of Dual Three-phase Interior Permanent Magnet Synchronous Motor. *J. Electr. Eng. Technol.* **2014**, *9*, 1614–1622. [CrossRef]

22. Wang, C.; Zhao, W.; Li, W. Braking sense consistency strategy of electro-hydraulic composite braking system. *Mech. Syst. Signal Process.* **2018**, *109*, 196–219. [CrossRef]

23. Han, W.; Xiong, L.; Yu, Z. A novel pressure control strategy of an electro-hydraulic brake system via fusion of control signals. *Proc. Inst. Mech. Eng. Part D: J. Automob. Eng.* **2019**, *233*, 3342–3357. [CrossRef]

24. Yang, Y.; Li, G.; Zhang, Q. A Pressure-Coordinated Control for Vehicle Electro-Hydraulic Braking Systems. *Energies* **2018**, *11*, 2336. [CrossRef]

25. Jiang, G.; Miao, X.; Wang, Y.; Chen, J.; Li, D.; Liu, L.; Muhammad, F. Real-time estimation of the pressure in the wheel cylinder with a hydraulic control unit in the vehicle braking control system based on the extended Kalman filter. *Proc. Inst. Mech. Eng. Part D: J. Automob. Eng.* **2016**, *231*, 1340–1352. [CrossRef]

26. Kumar, Y.S.R.R.; Sonawane, D.B.; Subramanian, S.C. Application of PID control to an electro-pneumatic brake system. *Int. J. Adv. Eng. Sci. Appl. Math.* **2012**, *4*, 260–268. [CrossRef]

27. Pi, D.; Cheng, Q.; Xie, B.; Wang, H.; Wang, X. A Novel Pneumatic Brake Pressure Control Algorithm for Regenerative Braking System of Electric Commercial Trucks. *IEEE Access* **2019**, *7*, 83372–83383. [CrossRef]

Article

Li-Ion Battery Performance Degradation Modeling for the Optimal Design and Energy Management of Electrified Propulsion Systems

Li Chen [1,*], Yuqi Tong [1,2] and Zuomin Dong [1]

[1] Department of Mechanical Engineering, Institute for Integrated Energy Systems, University of Victoria, Victoria, BC V8W 2Y2, Canada; yuqitong@uvic.ca (Y.T.); zdong@uvic.ca (Z.D.)
[2] State-assigned Electric Vehicle Power Battery Center, Beijing 100072, China
* Correspondence: chenli@uvic.ca

Received: 1 March 2020; Accepted: 20 March 2020; Published: 2 April 2020

Abstract: Heavy-duty hybrid electric vehicles and marine vessels need a sizeable electric energy storage system (ESS). The size and energy management strategy (EMS) of the ESS affects the system performance, cost, emissions, and safety. Traditional power-demand-based and fuel-economy-driven ESS sizing and energy management has often led to shortened battery cycle life and higher replacement costs. To consider minimizing the total lifecycle cost (LCC) of hybrid electric propulsion systems, the battery performance degradation and the life prediction model is a critical element in the optimal design process. In this work, a new Li-ion battery (LIB) performance degradation model is introduced based on a large set of cycling experiment data on $LiFePO_4$ (LFP) batteries to predict their capacity decay, resistance increase and the remaining cycle life under various use patterns. Critical parameters of the semi-empirical, amended equivalent circuit model were identified using least-square fitting. The model is used to calculate the investment, operation, replacement and recycling costs of the battery ESS over its lifetime. Validation of the model is made using battery cycling experimental data. The new LFP battery performance degradation model is used in optimizing the sizes of the key hybrid electric powertrain component of an electrified ferry ship with the minimum overall LCC. The optimization result presents a 12 percent improvement over the traditional power demand-driven hybrid powertrain design method. The research supports optimal sizing and EMS development of hybrid electric vehicles and vessels to achieve minimum lifecycle costs.

Keywords: li-ion battery; performance degradation modelling; electrified propulsion; battery sizing; powertrain optimization; optimal energy management

1. Introduction

With the increasing concerns on the emissions of greenhouse gases (GHG) and other air pollutants, the automotive and marine industry are adopting hybrid electric or pure electric propulsion systems for vehicles and marine vessels with large onboard battery energy storage system (ESS) at an increasing pace. Today lithium-ion batteries (LIBs) become the primary type of batteries used in various electric ESS due to their significant longer life and much higher energy density. Among different kinds of LIBs, the $LiFePO_4$ (LFP) battery has been widely used in heavy-duty transportation applications, due to its lower cost and non-toxicity, well-defined performance, better long-term stability, and capability to fit for more extensive variations in temperature. The service life of battery ESSs is a critical issue for various types of electrified vehicles (EV), as well as their marine counterparts. Considerable efforts have been devoted to capturing the performance degradation and extending the operating life of batteries. However, there are limited efforts on the quantitative analysis of how battery capacity loss would influence the optimization on the sizing of key powertrain components and the powertrain energy management system (EMS), and how improperly sized powertrain components and an inflexible EMS would accelerate the battery capacity loss.

Large vehicles, ships and smart local power grids share a common feature that is entirely different from conventional passenger cars and light trucks, having case-dependent driving and load cycles and an enormous energy demand. The high power and energy requirements of those applications can significantly aggravate the total cost of hybrid electric propulsion systems with a large battery ESS. Though the price of LIBs has been considerably reduced in recent years, it is still one of the most expensive components in the hybrid electric and pure electric propulsion systems, compared to the internal combustion engines (ICE) and electrical machines (EM), etc. [1]. The battery ESS is also the component with the shortest life in the powertrain system, likely needing replacements. The lifecycle cost of battery ESSs includes investment, operation, replacements, and recycling costs. The cost and performance of battery ESSs remain the main concerns of adopting hybrid electric propulsion for heavy-duty vehicular and marine applications. Optimal sizing and energy management of battery ESSs can reduce the overall lifecycle costs (LCC) of the electrified propulsion system of the vehicle or vessel, and these tasks cannot be accomplished without an accurate battery performance degradation model to predict its deterioration rate and the remaining useful life (RUL) of the batteries.

The performance degradation of LIBs are relevant to battery materials, manufacturing quality, use patterns, and many other factors. The anode and cathode materials used in the LIB are the main factors that determine battery performance, and hence the degradation during the use of the battery [2–4]. With lower cost and the ability to resist thermal runaway at elevated temperatures, the LFP battery is popular for light vehicles and the dominant choice for heavy-duty transportation applications. The use pattern of a LIB, determined by the operating temperature, charge and discharge currents, stage of charge (SOC) variation in a cycle, working time, etc., can prominently affect the degradation rate and life of the LIB. Specifically, battery life can be categorized into cycle life and calendar life, in which the battery will be charged/discharged, and stored, respectively. Battery cycle life performance degradation is an inevitable process that happens right after the first charge/discharge process. The capacity of a battery will be reduced due to many reasons. The formation of a solid electrolyte layer (SEI) on carbon-based anodes, though it can help the active carbon material resist further corrosion from the electrolyte, will consume available cyclic ions and increase cell impedance [5]. Ramadass, *et al.* [6] showed that battery film resistance rises with the cycling numbers. Furthermore, it has been concluded that battery cyclic capacity fading has a direct link with the thickness of the SEI layer [7]. The loss of available lithium is the main reason for cyclic capacity decay. Moreover, microscopic electrochemical side reactions happen all the time inside the battery, leading to reduced calendar life [5].

Temperature, SOC, and operating current rate are key elements causing battery performance degradation. Operations under high temperatures not only aggravate side reactions and material exfoliations but also cause battery deformation and affect the battery's overall performance [8]. Storage of a battery at an elevated temperature can also accelerate the decay of its calendar life [9]. Battery charging at a low temperature below −10 °C decreases its reaction rate, causing Li plating along the carbon surface. Dendritic Li plate not only consumes available capacity but also can penetrate the separator and cause inner short circuit [10]. The charge and discharge current rate (C_{rate}) is another factor that profoundly affects battery life. If a battery is cycled at a higher current rate, a large amount of Li ions will accumulate on the surface of active materials over a short time. If the diffusion process of ion is restricted, dendrite Li might be generated [11]. Low C_{rate}, on the other hand, is more favorable for safer performance and longer life. The SOC indicates the percentage of remaining energy that the battery can release compared to the rated capacity. The range of SOC variation in a cycle, referred to as the depth of discharge (DOD), can also affect battery life. Higher DOD means harsher usage of the battery, thus accelerating its degradation [12]. Battery performance degradation, might be caused by the loss of active material (LAM) of electrodes, the loss of lithium inventory (LLI), and the increase of the internal resistance of the cell [13]. The observed degradation involves complicated electro-chemical processes, making its mathematical quantification very difficult. All of the stated factors must be considered if an accurate performance degradation model is to be introduced to predict the remaining lifetime of the battery under given operating conditions.

The models for capturing battery performance degradation and predicting its RUL can be classified into the chemistry-physics-based models, or the data-driven, machine-learning-based models. In the physics-based models, mechanical fatigue and chemical degradation are mathematically quantified individually [12] or jointly [14]. The thermal analysis [15] and electro-thermal coupled modelling [16] of LIBs can reveal how temperature affects battery capacity fading. The commonly-used semi-empirical modelling method is based on the Arrhenius kinetics equations, such as those reported by Bloom, Cole, Sohn, Jones, Polzin, Battaglia, Henriksen, Motloch, Richardson and Unkelhaeuser [12] and Wang, et al. [17]. More detailed molecular or atomist models were introduced to describe the electrochemical reactions and represent battery degradation at a fine-grained level [18,19]. However, the complexity of an electrochemical model usually led to intensive computation, making their applications in real-time energy management difficult. The data-driven modelling method was introduced recently with the advance of machine-learning techniques [13,20]. However, the complete ignorance of the battery degradation mechanism and the complex inter-linked factors inside of the machine-learning models may result in irrational prediction outcomes, and the approach requires careful verification and review for even a slightly different batch of products. Many researchers focused on modelling the battery capacity fading at different operating temperatures [9,21], different SOC [22], or their combinations [23]. These modelling methods require a significant amount of battery test data [17,24] to cover different current rates, temperatures, and depth of discharges, but the time and efforts needed to conduct these costly experiments remain a significant challenge.

In developing a hybrid electric propulsion system, determining the optimal powertrain component sizes, particularly of the engine and battery ESS, and developing the optimal power control and energy management strategy (EMS) are significant challenges. The design of other powertrain and power system components, such as electric machines and power electronic converters, are directly related to the engine and the ESS. The growing level of powertrain hybridization with increased battery capacity for achieving better fuel efficiency and emission improvements further demands the optimal sizing of powertrain components and optimal EMSs. The performance degradation of the LIB heavily influences the system performance and LCC of the propulsion system. However, there is not yet a systematic approach to link battery performance degradation to the optimal design and control of hybrid electric propulsion systems at present, due to the lack of an accurate battery performance degradation model [25,26].

This research focuses on the introduction of an accurate LIB performance degradation and life prediction model that can be used to support optimal component sizing and energy management of hybrid electric propulsion systems. Specifically, the special cycling experiments and data analysis of a typical LFP battery is presented; a battery performance degradation and life prediction model using the obtained experimental data and other supplementary data is introduced, and the use of the newly introduced battery performance degradation and life prediction model in optimizing the ESS design and EMS development of a hybrid electric propulsion system of a ferry ship is demonstrated. The LCC improvements in the globally optimized hybrid propulsion system compared to the pure energy efficiency driving hybrid system (without considering battery performance degradation), and compared to the conventional mechanical propulsion system are presented to demonstrate the benefits of the new method.

2. Battery Performance Degradation Experiments

2.1. Design of the Experiments

In this work, sample commercial lithium-ion phosphate/graphite (LFP/C) prismatic cells with 18 Ah nominal capacity, produced by Liyuan New Energy, were used in the extensive battery cycling tests at the State-assigned Electric Vehicle Power Battery Testing Center in Beijing, China. These cells are designed and built for hybrid and pure electric vehicular and marine propulsion applications. The detailed specifications of tested LFP batteries are given in Table 1.

Table 1. Tested lithium-ion phosphate/graphite (LFP/C) battery specifications.

Parameter	Value
Nominal Voltage	3.2 V
Nominal Capacity	18 Ah
Energy Density	120 Wh/kg
Charge/Discharge Cut-off Voltage	3.6 V/2.5 V
Max. Normal/Fast Charge Current Rate	1 C/2 C
Max. Continuous/Max. 30-sec Discharge Current Rate	3 C/15 C

The main purposes of the tests include (a) measuring battery capacity deterioration through repeated cycling profiles; (b) supporting battery modelling for the quantitative prediction of battery performance degradation rates under different use patterns; and, (c) accurately predicting the (remaining) operation life of the battery in a hybrid electric propulsion system under a given load profile. Two types of charge and discharge profiles were used in these tests, the cycling test profile and the capacity test profile. Both types of tests were conducted repeatedly in the environmental chambers with a controlled temperature at 25 °C. These battery test cycles consisted of four steps (a) charging at the designed constant current (CC) followed by a constant voltage (CV) until the current reaches 0; (b) resting for 20 minutes; (c) discharging at the designed CC until the cut-off voltage is reached; and (d) resting for 20 minutes. The cut-off voltage, as shown in Table 1, is 3.6 V for charging and 2.5 V for discharging.

The cycling tests were performed at 1 C charge and 2 C discharge current rates and the capacity tests were performed at 1/3 C charge and 1/3 C discharge rates, respectively. As usual, 1 C means fully discharging the battery in 1 hour. The higher current rate (or C_{rate}) can reduce experimental time, while the lower C_{rate} can better measure the available battery capacity. The tested battery went through 25 cycling tests, followed by a capacity test until the battery reaches its end-of-life (EOL). Following the general rule used by the automotive industry, the battery's EOL is defined as the state at which the measured charge/discharge capacity of the battery falls to 80 percent of its nominal capacity, and the battery needs to be replaced. The cycling test only needs about 25 percent of the time for conducting a capacity test, leading to much-reduced experiment time. Moreover, the cut-off voltage during 2 C discharge in the cycling tests was defined at 3.0 V to prevent potential over-discharge under the high current rate.

2.2. Experiment Results and Data Analysis

The open-circuit voltages, V_{oc}, of fresh LFP batteries under charges and discharges were measured under low current rate (1/50 C) before the start of the cycling tests. Based on the measured voltage variations at such a low current, the differential voltage for capacity analysis could be carried out. The voltage variation vs. capacity variation, dV/dQ, at different battery SOCs showed distinct peaks as illustrated in Figure 1. These peaks revealed the anode (C) and cathode (LFP) material phase transitions during ion's intercalation and de-intercalation, and also visually qualified the total capacity that this battery can store at its beginning-of-life (BOL). As the battery is cycled from its BOL to the EOL, these peaks would shift due to the structural deterioration of the active material and capacity reduction.

The tested battery terminal voltage was measured in each cycle under designed testing profiles. Testing data obtained during the first 2000 test cycles were used in the performance degradation modelling and validation. A few typical capacity cycling test data under the 1/3 C current rate were plotted in Figure 2. The experiment results showed that with the increase in cycling number, experimental time became shorter, indicating that the battery's maximum capacity, Q_{max}, became lower and therefore less energy could be stored. The reduced cell voltage with higher cycling numbers indicated increased inner resistance.

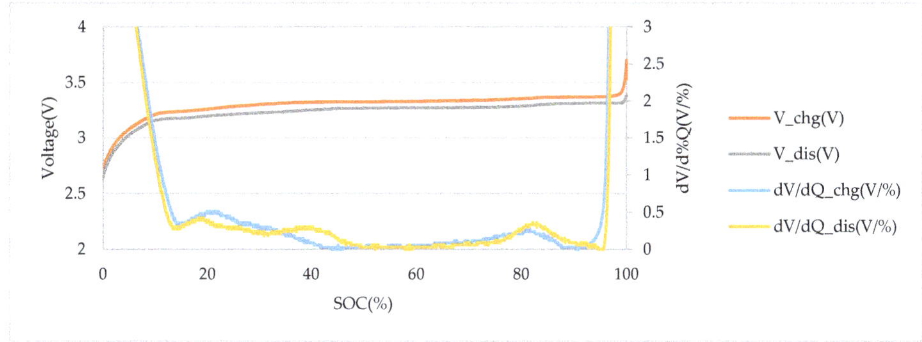

Figure 1. Battery beginning-of-life (BOL) voltage curves.

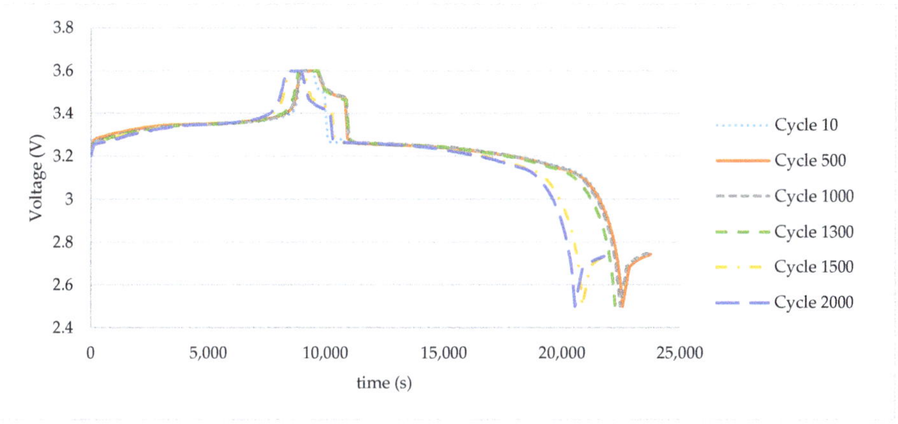

Figure 2. Measured voltage variances under 1/3 C charge and discharge in cycling experiments.

The capacity loss during battery cycle life tests can also be revealed by the differential voltage analysis, i.e., the *dV/dQ* ratio of the discharge profiles recorded at a different number of cycles. Figure 3 shows the variations of the differential voltage of the tested C-LFP battery, cycled from its BOL to EOL at 1/3 C rate. The small peaks of the *dV/dQ* curves at the bottom shifted to the center as the number of the charge/discharge cycle increased, leading to a narrower span of the curve with a reduced distance between the two boundaries, indicating capacity reduction [14].

The measured discharge capacity of tested LIB exhibited a clear decreasing trend as the cycling numbers grew, as shown in Figure 4.

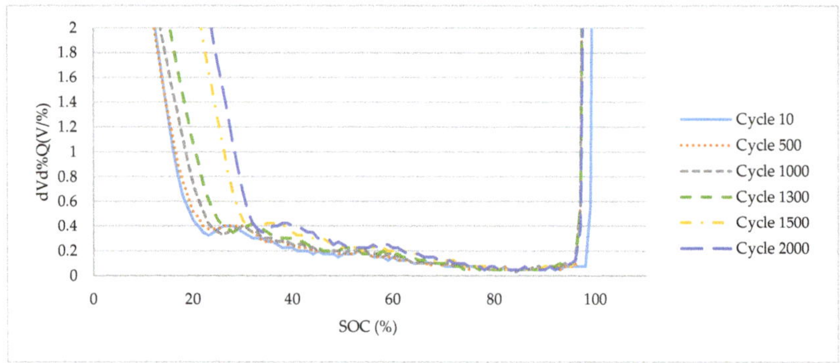

Figure 3. Differential voltages at different number of cycles.

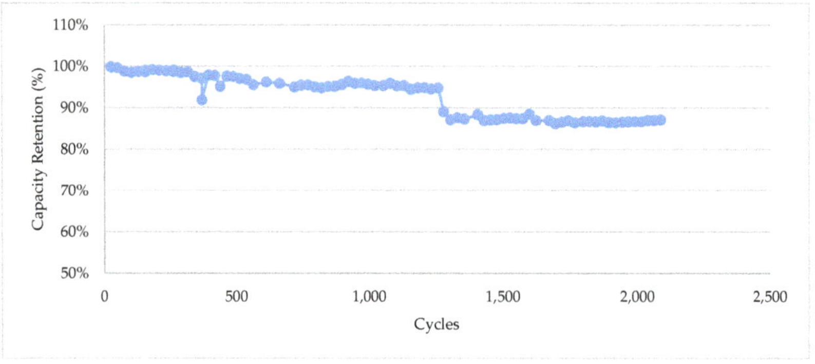

Figure 4. Battery capacity measured in different cycle numbers.

The level of battery performance degradation is indicated using a measure of battery state of health (SOH) in this work. The SOH value measures the ratio of the actual maximum capacity Q_{max} of a battery at time t over the rated capacity Q_{rated} of the fresh cell. When the SOH decreases to 80%, the battery is considered as a dead battery for the transportation applications and is to be replaced to meet the required power and/or energy demands of the applications.

$$SOH = \frac{Q_{max}}{Q_{rated}} \times 100\%$$ (1)

where Q_{max} is the maximum available discharge capacity a battery can provide when fully charged at 100% SOC. Q_{rated} is the rated capacity specified by the battery manufacturer, which is 18 Ah in this study.

The SOC of a battery indicates the remaining available capacity and can be affected by the current maximum capacity. The actual maximum capacity Q_{max} will gradually decrease during the usage due to its ageing phenomena, as shown in Figure 5. For a fresh battery, Q_{max} is equal to the rated capacity Q_{rated}.

$$SOC = SOC_0 - \int_{t_0}^{t_f} \frac{I(t)}{3600 Q_{max}} dt \times 100\%$$ (2)

where t_0 to t_f are the start and end time of each cycle (s); dt is the time step (s); SOC_0 is the initial SOC at the beginning; $I(t)$ is the current (A) which is a function of time, assuming discharge current is positive and charge current is negative; *and* Q_{max} is the maximum battery capacity (Ah).

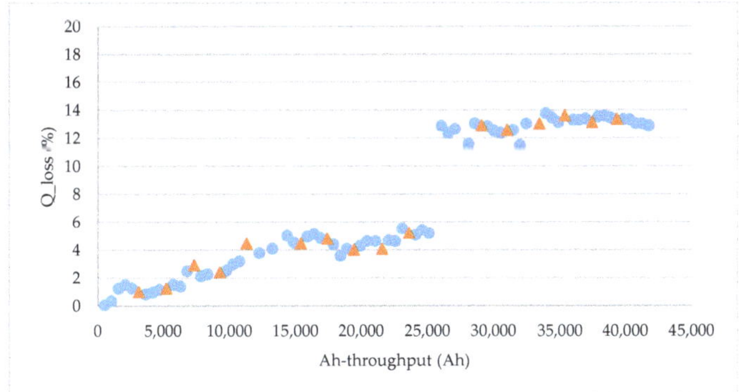

Figure 5. Cycle-induced capacity loss plotted as a function of ampere-hour throughput (Ah-throughput). The blue dots indicate data used for building the performance degradation model, while the orange triangles indicate data used for model validation.

The cycle-induced capacity loss Q_{loss}, as the ratio of the battery's reduced capacity to its nominal capacity, is directly linked to the battery's ampere-hour throughput (Ah-throughput, or Ah_{th}). Ah-throughput represents the total amount of electric charge of the battery during cycling [24], and acts as a critical factor in measuring the cyclic capacity loss. The value of Ah_{th} is calculated by:

$$Ah_{th} = Q_{max} \cdot DOD \cdot N \tag{3}$$

where DOD is the depth of discharge $DOD = 100 - SOC$, and N is the cycle numbers.

The relation of measured capacity loss and Ah-throughput from test data is plotted in Figure 5, showing a clear trend of battery performance degradation.

In the experiments, 2095 cycles of battery testing data have been obtained, including about 2019 sets of cycling tests under a 1 C/2 C charge/discharge current and 76 sets of capacity test data under a 1/3 C charge and discharge current. The 76 sets of capacity test data were used to build the battery performance degradation model which will be discussed in detail in the following section. Among them, about 80% of the data (60 cycles) have been used for building the performance degradation model, shown as the blue dots in Figure 5, while the rest 20% of the test data (16 cycles) were used for modelling accuracy validation, shown as the orange triangles in Figure 5. The sudden change of the cycle-induced capacity loss and sharp capacity loss at around 25,000 Ah might be caused by the cycle-induced material deterioration of the specific LFP battery that was tested.

3. Battery Performance Degradation and Life Prediction Model

The purpose of introducing a battery performance degradation model is to accurately predict the operation life and resulting lifecycle cost of the battery under different given operating temperatures and use patterns. In recent years, considerable research efforts have been devoted to understanding the influence of temperature on battery performance and operating life, and to develop effective thermal management techniques to allow the battery ESS to operate within the desired range of temperature. This work thus focuses on modelling the influence of use patterns on the operating life and resulting lifecycle cost of batteries. Due to the many influencing factors and the not yet fully understood performance decay mechanism of batteries' charge and discharge operations, generation of the model is largely based on experimental data, either by fitting a semi-empirical, multiphysics model [12,19], or by training an artificial neural network (ANN) using machine learning techniques [13]. The semi-empirical, multiphysics model combines generic formula related to the degradation mechanism and detailed model parameters determined by the battery test data, providing a relatively accurate and straightforward modelling method as used in this work. The RUL of a battery under different use patterns is modelled by combining the battery testing data obtained in this research and the results from the literature [24].

The development of performance degradation models for Li-ion batteries has been reported in many pieces of literature, forming three main categories:

1. The empirical modelling method that is primarily used in the early study stage of LIB development due to its simplicity [12]. This modelling method requires a large amount of experimental data and does not have broad applications.

2. The equivalent circuit model that is capable of capturing the dynamic behaviors of the battery using resistance and capacitance to represent battery charge/discharge characteristics [27]. These models are of the semi-empirical, multiphysics type.

3. The electrochemical models that simulate the electrochemical reactions using time and space coupled partial differential equations to describe the ion diffusion process, overall potential variation, and current distribution during the charge and discharge of the battery. This type of model includes the Doyle–Fuller–Newman model [28], the pseudo-two-dimensional model [29], and the single particle model [30]. However, the use of these more detailed and more accurate models require extensive numerical computations, thus they are not suitable to serve as an element in the algorithms for hybrid powertrain control and battery energy management.

The two-order, equivalent circuit battery performance model is thus amended to form the new model for predicting battery performance degradation in this work. This amended equivalent circuit performance model is used later for the design and control optimization of hybrid electric powertrain systems. As shown in Figure 6, the variation of battery capacity and resistance can affect battery voltage and SOC calculation in each cycle carried out using the equivalent circuit performance model. The calculation results are fed into the semi-empirical life prediction model to obtain the estimated remaining useful cycle life.

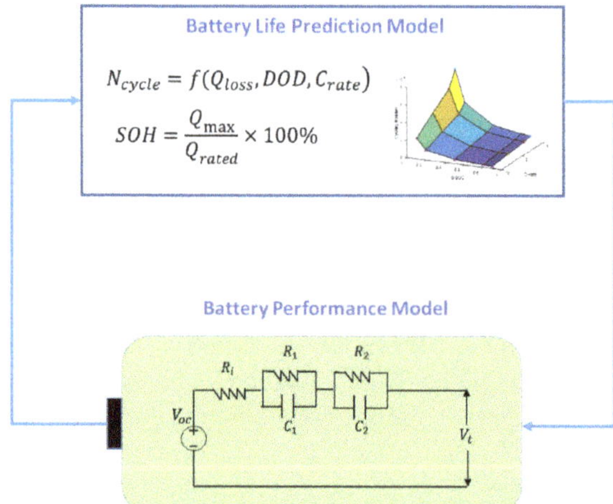

Figure 6. Illustration of battery performance degradation and life prediction model.

3.1. Amended Equivalent Circuit Performance Model

In the equivalent circuit model, the two resistor-capacitor (RC) electrical circuits represent the activation and concentration depolarizations during battery charge and discharge operations. The

Ohmic resistance is decided by the internal resistance, R_i. At any required current, I, the voltage drops for each element can be calculated by:

$$\dot{V_1} = -\frac{V_1}{R_1 C_1} + \frac{I}{C_1} \tag{4}$$

$$\dot{V_2} = -\frac{V_2}{R_2 C_2} + \frac{I}{C_2} \tag{5}$$

$$V_i = IR_i \tag{6}$$

where V_1, V_2, and V_i are voltage drops caused by the first RC circuit (R_1 and C_1), the second RC circuit (R_2 and C_2) and the inner resistance (R_i).

The battery output voltage (V_t) is determined by the Kirchhoff's law:

$$V_t = V_{oc} - V_i - V_1 - V_2 \tag{7}$$

where V_{oc} is the open-circuit voltage. The discharge and charge V_{oc} of an LFP battery is plotted in Figure 1.

The six parameters of the equivalent circuit model, Q_{max}, R_i, R_1, R_2, C_1, and C_2, are determined by fitting the battery testing data with minimum root-mean-squared error (RMSE) between the measured voltage and model output voltage.

$$\min_{x} \sqrt{\frac{1}{n} \sum_{i=1}^{n} (V_{meas,i}(x) - V_{sim,i}(x))^2} \text{subject to}: \ g(x) \leq 0 \tag{8}$$

where V_{meas} is the measured output voltage, V_{sim} is the model simulated output voltage, $x = [Q_{max}, R_i, R_1, R_2, C_1, C_2]'$ are the unknown parameters, i is the time step from 1 to n, and $g(x)$ is the constraint on the design variables.

The genetic algorithm (GA) heuristic global optimization algorithm was used to solve the RMSE minimization problems and to deal with the noise of the test data, as is widely reported [28,30,31]. The algorithm searched all possible solutions and the best results were identified for all unknown parameters. The variation of battery maximum capacity, Q_{max}, and inner resistance, R_i, at different numbers of cycles is plotted in Figure 7.

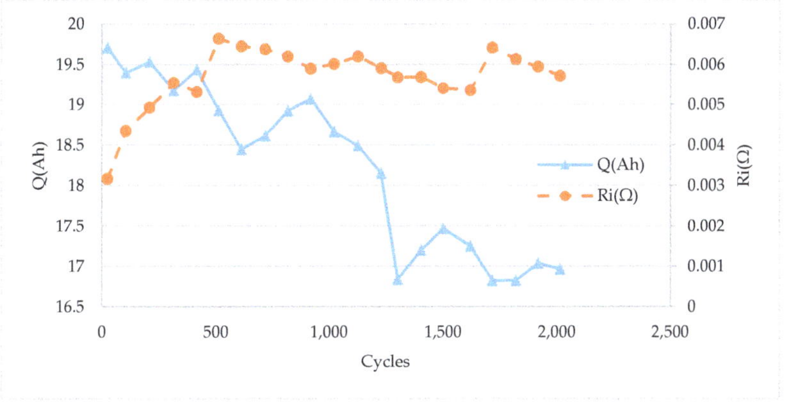

Figure 7. Variations of lithium-ion battery (LIB) maximum capacity and inner resistance.

3.2. Battery Remaining Lifetime Prediction

The combined study of battery SOC (in each mission cycle with varying Q_{max}) and SOH (over the battery lifespan) is important to estimate battery RUL under different use patterns. The cycle life experiment in this study is for battery cycled at 1C charge and 2C discharge for 100 percent DOD, therefore, the RUL prediction may not be valid for other charges/discharge patterns. However, to conduct experiments at a continuously varying current (e.g., from 0.5C to 3C), DOD (e.g., from 10 to 100 percent) and operating temperature would cost years of testing, causing such experiments to become infeasible. The lack of complete experimental data would be an obstacle to building accurate battery performance and life prediction models.

This research utilized the 2000 cycling data acquired above and other accessible experimental results of LFC/C battery from well-cited literatures to deal with the scarcity of battery performance degradation data, including cycle-life experiments from Wang, Liu, Hicks-Garner, Sherman, Soukiazian, Verbrugge, Tataria, Musser and Finamore [24], Deshpande, Verbrugge, Cheng, Wang and Liu [14], and Han, *et al.* [32] to build the LFP battery cycle life prediction model. These researchers have identified and successfully illustrated the key features of cycle-induced battery performance degradation under different cycling patterns. According to previous studies, the calendar life of the battery has a minor influence on performance degradation [33,34], therefore, it has not been considered in the new model. It was assumed that the operating temperature of the batteries could be adequately controlled by the advanced thermal management system of the hybrid electric vehicles and vessels.

The proposed model has considered both voltage and capacity decay in estimating the remaining cycling numbers of a battery. Based on Arrhenius kinetics, the capacity fading rate is affected by the previously discussed factors. The earlier study [33] modelled the battery capacity loss by:

$$Q_{loss} = A \cdot e^{\left(\frac{-E_a + B \cdot C_{rate}}{RT}\right)} (Ah_{th})^z \tag{9}$$

where, A and E_a are pre-determined coefficients; B is the coefficient of C_{rate}; R is ideal gas constant; and, T is the temperature in K.

After combining the capacity loss model with a previously defined Ah-throughput equation, the remaining cycling number of battery (N) can be derived based on the previous function:

$$N = \left(\frac{Q_{loss}}{A \cdot e^{\left(\frac{-E_a + B \cdot C_{rate}}{RT}\right)}}\right)^{\frac{1}{z}} \frac{1}{Q_{max} \cdot DOD} \tag{10}$$

The newly introduced battery life prediction model was implemented in MATLAB/Simulink. The result of total cycling number for the LFP battery, as a function of C_{rate} and DOD, is shown in Figure 8.

The performance degradation rate in the LIB cycling lifespan can be predicted using the battery performance model, and the RUL can be calculated using the resulting life prediction model. The former model calculates the battery performance (include current, voltage, SOC, etc.) under a given charge/discharge profile with updated maximum available capacity Q_{max}. The predicted results are fed into the life prediction model to estimate the remaining cycling numbers under the accumulated deterioration.

The performance degradation of the battery ESS under different use patterns was compared and shown in Figure 9. Harsh use of the battery with a discharge current rate of 2C at 100% DOD would result in fast capacity decay, as shown by the blue dash curve. The life of the battery would be extended if the battery was used gently by reducing the discharge current rate or operating time, as shown by the other two curves. This quantitative model shows that a more conscious use of the battery ESS may be an effective way to strike the best balance between reducing engine fuel consumption and lowering the cost induced by the shortening of battery life due to aggressive battery charges/discharges in a hybrid electric propulsion system. The quantitative model also supports the more appropriate engine and battery ESS sizing in a hybrid powertrain to form the globally optimal design solution considering the investment, operation, and replacement costs over the entire lifecycle.

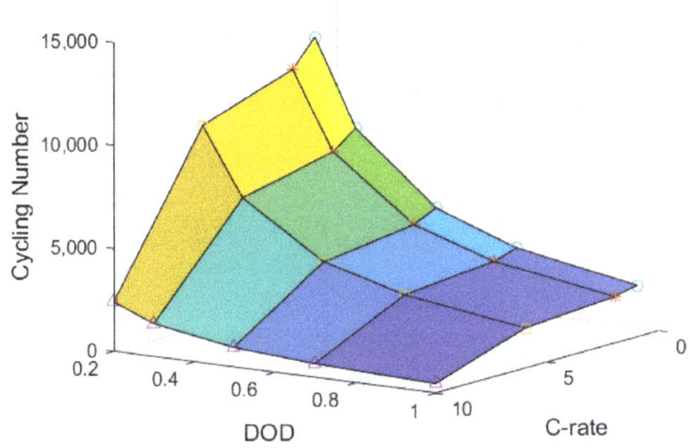

Figure 8. Prediction of remaining cycling numbers under different C_{rate} and depth of discharge (DOD).

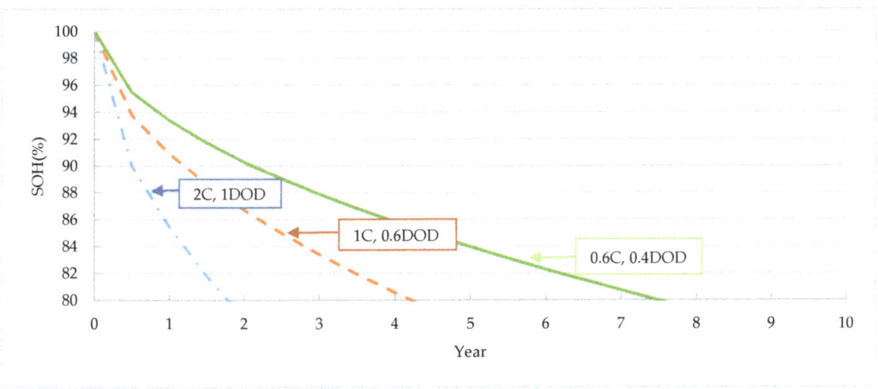

Figure 9. Battery state of health (SOH) variation under different usage patterns.

3.3. Model Validation

In this work, about 2000 sets of experimental cycle data of a commercialized LFP battery have been acquired and used. A large portion (80%) of the capacity test data have been used to build the battery performance degradation model, and the remaining 20% of data have been used to validate the accuracy of the introduced models. The life prediction model under different discharging C_{rate} and DOD has been built for calculating the remaining useful cycle life of the LFP battery. The predicted capacity loss (Q_{loss}) from the RUL model and the battery testing data are shown in Figure 10. In this figure, the data points labelled as blue dots were used to build the model, and the data points represented by the yellow triangles were the original test data for model validation.

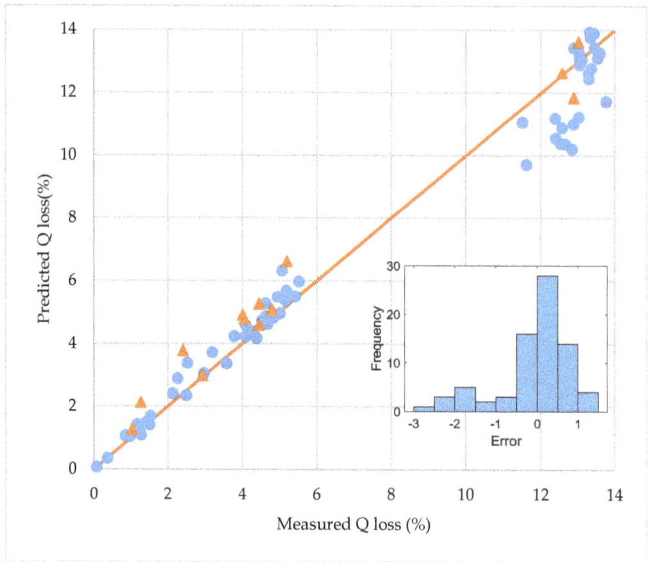

Figure 10. Accuracy of battery performance degradation and lifetime prediction model: blue dots indicate modelling results, orange triangles indicate validation results.

The predicted capacity losses at the very beginning of battery life were slightly higher than the measured results. When the capacity deteriorates over 10%, the predicted results showed slightly lower than the measured values. Ideally, these results would be equal to the measured data, as indicated by the ideal diagonal line. Overall, the absolute mean percentage error of developed LIB performance degradation and the life prediction model is about 13%.

4. An Application Example–Lifecycle Cost-based Design Optimization of a Hybrid Electric Ferry Ship

Li-ion battery ESSs are one of the critical components in a hybrid electric marine propulsion system. With the battery performance degradation model, the actual life of a given size ESS under the specific operation profile of the vehicle or vessel can be predicted. Thus, the initial investment cost, replacement cost, operation cost, and recycling cost of the battery ESS can be calculated. It is, therefore, possible to perform the optimal design and energy management of the hybrid powertrain system to achieve minimum LCC of the hybrid electric vehicle/vessel, in addition to higher system efficiency and lower emissions. This section will compare the total LCC of a hybrid electric propulsion system with or without using the newly introduced battery performance degradation model.

4.1. Design of a Hybrid Electric Marine Propulsion System

The design optimization of the hybrid electric propulsion system for a medium-size vehicle and passenger ferry, Skeena Queen, operated by BC Ferries in B.C. Canada, is used as a test platform. The general information of the ship is given in Table 2.

Daily operation data from the ferry have been collected as shown in Figure 11. The average sailing speed is about 15 knots as shown in Figure 11a. The total propulsion power requested during a roundtrip sailing from the four diesel engines has been measured and plotted in Figure 11b, where P1 to P4 represents the power outputs from engine number 1 to 4.

Table 2. General information about ferry ship Skeena Queen.

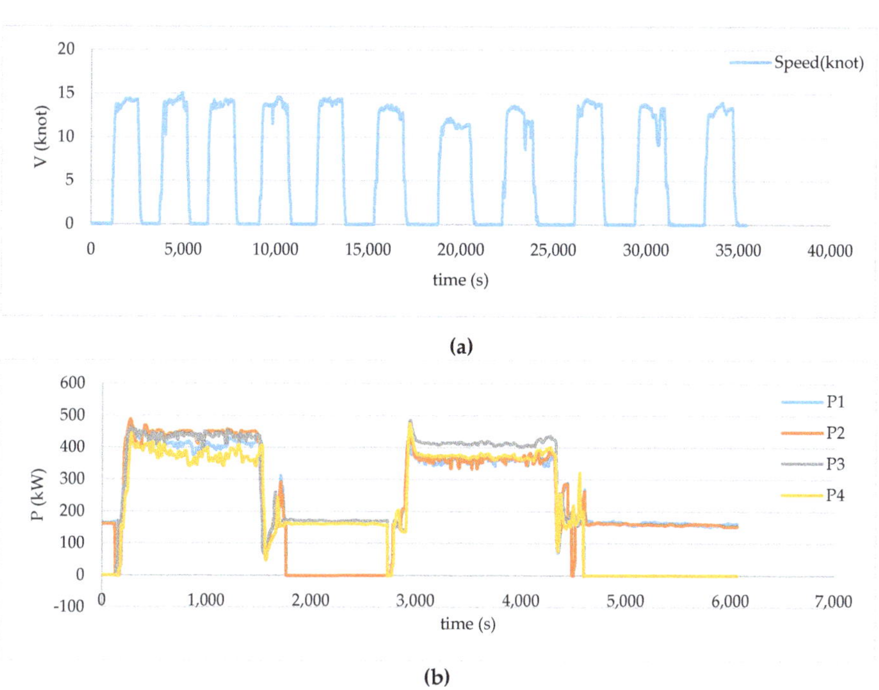

	Skeena Queen
Overall Length	110 m
Car Capacity	92
Passenger and Crew Capacity	450
Maximum Power	4474 kW
Cross Distance	5 nautical miles

(a)

(b)

Figure 11. Ferry ship operation profile. (**a**) Ferry operation speed profile in one day, (**b**) Power measured from the four engine shafts in one roundtrip.

The new hybrid electric propulsion system design for this ferry is aimed at improving its fuel efficiency, reducing emissions, and achieving the minimum LCC. Reducing battery ESS replacements and extending battery ESS lifetime would contribute to a lower LCC of hybrid conversion. The proposed hybrid electric propulsion system for the ferry is shown in Figure 12. The conversion is aimed at producing a series of hybrid powertrains powered by the diesel gen-sets and propelled by electric motors, using the battery ESS as an energy buffer to store and supply electric power. The new hybrid propulsion system would improve engine fuel economy and system efficiency, provide more flexible operation and redundant power, lower fuel consumption and emissions, eliminate the original auxiliary gen-sets, and reduce engine operation time during docking.

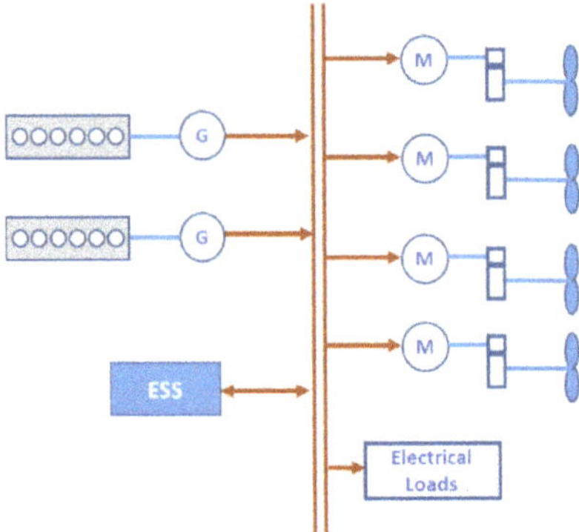

Figure 12. Integrated hybrid electric propulsion system designed for the ferry ship.

4.2. LCC Model Developed for the Hybrid Electric Marine Propulsion System

The new LIB performance degradation and life prediction model can help evaluate the battery investment, replacement and residual costs during the entire life cycle of the hybrid propulsion systems. The LCC of the proposed hybrid electric marine propulsion, considering all costs from engines, ESS, and other electrical machines, can be developed as part of the total ownership costs (TOC) of the vessel. The main elements of the LCC model include the capital cost (C_{cap}), operation cost (C_{ope}), and residual cost (C_{resd}):

$$LCC = C_{cap} + C_{ope} + C_{resd} \tag{11}$$

The capital cost (C_{cap}) includes all the purchase costs for the main propulsion components. The reinvestment cost of Li-ion battery ESS must be considered due to its short lifespan compared to engines and other powertrain components.

$$C_{cap} = C_{eng} + C_{hyb} + C_{ess} + C_{rin} \tag{12}$$

where, C_{eng} is the engine cost; C_{ess} is the battery ESS cost; C_{hyb} is the cost for hybridization and electrification, including purchasing the electric motors/generators and power converters; C_{rin} is the reinvestment cost due to the replacing of battery ESS; C_{rin} is the reinvestment cost, counting for the replacement cost of battery ESS due to the reduced lifetime. The operation life of battery ESS (L_{bat}) is calculated based on the developed battery life prediction model in the previous section.

$$C_{rin} = \sum_{t=0}^{N_t} \frac{C_{ess}}{(1+r)^t} k_t \tag{13}$$

where k_t is the replacement frequency, which is a function of the battery lifetime (L_{bat}). r is the annual inflation rate. L_{bat} is the key parameter that determines the reinvestment capital costs. The optimal result of L_{bat} must be determined at the system level considering both engine and ESS operation conditions.

$$k_t = f(L_{bat}) = \begin{cases} 1, & m = n{\cdot}L_{bat}, \ m \le N_t \\ 0, & otherwise \end{cases} \tag{14}$$

where m is the year when replacement occurs in the whole lifespan N_t, i.e., when the battery life is ended. n is integer numbers, $n = 1, 2, 3 \ldots$ When the battery needs to be replaced in year m, then $k_t = 1$, otherwise, k_t is 0.

The system operation cost consists of fuel consumption and engine maintenance costs. Other costs related to ship insurance, registration, etc., are excluded.

$$C_{ope} = \sum_{i=0}^{N_t} \frac{C_{fuel} + C_{maint}}{(1+r)^i} \tag{15}$$

where C_{fuel}, C_{maint} are the cost of fuel consumption and engine maintenance. r is the annual inflation rate. i is the year from 0 to N_t.

The fuel cost is mainly determined by the operating efficiency of the engine that can be affected by the system design, component sizing, and power control. In this work, it is assumed that engine maintenance cost is closely related to its working time and the engine size [35].

The residual value (or salvage cost) of replaced Li-ion batteries is nontrivial for this expensive component. Retired batteries from hybrid vehicles with 80% remaining capability could be reused for residential energy storage and load levelling in a smart power grid [36]. In this study, the residual cost is the remaining value in the replaced battery ESS, which is also determined by the replacement time and residual price.

$$C_{resd} = \sum_{i=0}^{N_t} \frac{p_r Q_r}{(1+r)^i} k_i \tag{16}$$

where p_r is the price for the remaining value (\$/kWh), Q_r is the residual capacity (kWh), and r is the annual inflation rate.

More detailed information of LCC model, such as the price of marine fuels, the evaluation of fuel consumption cost and engine maintenance cost, etc. can be found in [35].

4.3. A Comparison of Different ESS Sizing Methods

Two different propulsion system and ESS sizing methods are used and compared, including the traditional power-demand-based and fuel consumption minimization-driven approach, and the new minimum LCC-based design using the battery performance degradation model.

The power-demand-based and fuel-economy-driven hybrid electric propulsion design is the predominant method for designing and sizing the hybrid electric propulsion system [37,38] due to the lack of an accurate battery performance degradation model. It determines the battery ESS size based upon the electrical energy required to achieve the best engine fuel economy. Specifically, the hybrid electric propulsion is design to allow the engine to operate at a higher power level (above 80% of maximum continuous rated power) to improve fuel economy and reduce air pollutants. For this purpose, the requested power from battery ESS is needed to substitute engine operation whenever the engine is operated below an 80% load. Due to the lack of optimal sizing of the ESS, the investment cost and replacement cost of battery could rise since no considerations have been made to ease the battery degradation in usage. The approach aims entirely at meeting the electrical energy requirement, the size of the ESS used in the hybrid ferry is then to be at least 500 kWh with about 1C discharge rate and 80% DOD usage in one roundtrip voyage. Based on acquired experimental results, the battery can last for 3.8 years. Therefore, the battery ESS would need to be replaced about every four years during the ship's operational lifetime.

The use of the LIB performance degradation model in the hybrid electric propulsion system design enables the optimization-based ESS sizing to achieve the minimum system LCC. The LCC of the hybrid marine propulsion system includes the initial investment cost of the main powertrain components; the replacement cost of battery ESS that can be determined by the battery performance degradation model; the residual cost of battery when it is recycled; the total operational costs over the 20 years of ship operation with engine fuel consumption and maintenance costs. The optimal design of the ship's hybrid powertrain involves different size combinations of major powertrain components. In this study, the size of the battery ESS is the primary variable that needs to be optimized, considering both system performance and the total LCC, as battery replacements may occur during the vessel's lifespan. A multi-objective optimization problem is formulated as:

$$\min_{x} f(x) = w_1 \cdot LCC\left(C_{fuel}, L_{bat}, N_t\right) + w_2 \cdot m_{emission}.$$
$$\text{subject to}: \quad E_{min} \leq x_1 \leq E_{max}$$
$$DOD_{min} \leq x_2 \leq DOD_{max} \tag{17}$$
$$SOC_{end} = SOC_{target}$$

where $x = [x_1, x_2]'$, x_1 is the ESS size and x_2 is the DOD in one trip; w_1 and w_2 are user-defined weighted factors for total LCC and emissions; LCC is the propulsion system lifecycle cost over 20 years' operation, which is affected by fuel consumptions (C_{fuel}), battery lifetime (L_{bat}) and operational time (N_t); $m_{emission}$ is the total mass of emissions (including equivalent CO_2, PM and SO_2); E_{min} and E_{max} are the minimum and maximum battery ESS capacity; DOD_{min} and DOD_{max} are the minimum and maximum battery DOD variation, respectively.

The optimal size of battery ESS, for this case, is 670 kWh. With this design, the vessel will consume 284 kg fuel during a round trip sailing, and battery ESS will sustain 7.5 years. This optimized hybrid propulsion system led to reduced investment cost due to minimum LLC battery ESS size optimization, supported by the performance degradation model of the LIB. The larger ESS led to a longer lifetime and improved LCC. Both hybrid propulsion system designs require a higher total capital cost compared to the traditional mechanical propulsion system. However, the optimized hybrid system requires less additional investment costs. Under the constrained SOC variation, C_{rate} and DOD, the design leads to extended battery life and lower placement cost.

A comparison of the LCC for the two different hybrid propulsion design approaches is presented in Table 3. As a reference, the capital and operational costs of a traditional mechanical propulsion system were also evaluated and listed. The initial investment cost, battery ESS replacement cost and battery residual cost (shown as a negative value) compose the total investment cost. The design with optimal battery ESS size, requires less additional investment cost and lower operating costs over 20 years. Overall, the total LCC of the battery performance degradation considering optimal hybrid electric propulsion system is 26 percent lower than the original mechanical propulsion system and represents a 12 percent additional cost-saving over the traditional power-demand-based and fuel consumption minimization-driven hybrid electric system design.

Table 3. LCC breakdown and comparison for the optimal hybrid propulsion system.

Cost/Increment	Mechanical Propulsion	Hybrid Propulsion (Minimizing Fuel Consumption)		Optimal Hybrid Propulsion (Minimizing Total LCC)	
	($M)	($M)	(%)	($M)	(%)
Initial Investment($M)	2	1.88		1.97	
Battery Replacement($M)	0	1.38		0.68	
Battery Residual($M)	0	−0.53		−0.37	
Total Investment	2	2.73	+37%	2.28	+14%
Total Operation (20 yrs)	17.94	14.38	−20%	12.44	−31%
Total LCC	19.94	17.11	−14%	14.72	−26%

5. Conclusions

The optimizations of the size and EMS of the battery ESS in a hybrid electric propulsion system have been significant interest and focal point of research for years. These optimizations cannot be achieved without an accurate model for predicting the performance degradation and operating life of the battery under different use patterns. Traditional and present power-demand based and fuel-economy driven ESS sizing and EMS optimization methods often led to shortened battery operation life and higher overall lifecycle cost of the propulsion system. The Li-ion battery performance degradation model and its supported battery ESS size optimization, introduced in this work, can effectively address this issue.

The new semi-empirical, amended equivalent circuit model is introduced based on a large set of 18 Ah LiFePO$_4$ battery cycling experiment data, and learning from previous research, in order to predict battery capacity decay and resistance increase during its lifespan and the remaining useful cycle life under various use patterns. The method for calculating the investment, operation, replacement and recycling costs of the battery ESS using the new model over its lifetime operation under given use patterns is presented. Validation of the new model using battery cycling experimental data showed good accuracy with about 13 percent error.

To demonstrate the use and benefits of the newly introduced LFP battery performance degradation model, the LCC of a hybrid electric passenger and vehicle ferry design using the traditional power-demand based and fuel-economy driven optimal ESS sizing method and the new overall LCC minimization method are compared. With the LIB model supported, minimum LCC battery sizing, the optimized hybrid propulsion system has 12 percent less LCC. The research forms a foundation for the optimal sizing and EMS development of hybrid electric vehicles and marine vessels to achieve minimum lifecycle costs.

Author Contributions: Conceptualization, L.C. and Z.D.; Data curation, Y.T.; Formal analysis, L.C.; Funding acquisition, Z.D.; Methodology, L.C. and Z.D.; Validation, L.C.; Writing – original draft, L.C.; Writing – review and editing, Z.D. All authors have read and agreed to the published version of the manuscript.

Funding: This research was funded by the Natural Science and Engineering Research Council of Canada, Transport Canada, and Washington Foundation.

Acknowledgments: The authors gratefully acknowledge the assistance from the British Columbia Ferry Service Inc. and Haijia Zhu from our UVic Clean Transportation Research team for acquiring the ferry ship operating data.

Conflicts of Interest: The authors declare no conflict of interest.

References

1. Nelson, P.A.; Gallagher, K.G.; Bloom, I.D.; Dees, D.W. *Modeling the Performance and Cost of Lithium-Ion Batteries for Electric-Drive Vehicles*; Argonne National Laboratory (ANL): Lemont, IL, USA, 2012.
2. Nitta, N.; Wu, F.; Lee, J.T.; Yushin, G. Li-ion battery materials: Present and future. *Mater. Today* **2015**, *18*, 252–264. [CrossRef]
3. Scrosati, B.; Garche, J. Lithium batteries: Status, prospects and future. *J. Power Sources* **2010**, *195*, 2419–2430. [CrossRef]
4. Dai, X.; Zhou, A.; Xu, J.; Lu, Y.; Wang, L.; Fan, C.; Li, J. Extending the High-Voltage Capacity of LiCoO2 Cathode by Direct Coating of the Composite Electrode with Li2CO3 via Magnetron Sputtering. *J. Phys. Chem. C* **2016**, *120*, 422–430. [CrossRef]
5. Arora, P.; White, R.E.; Doyle, M. Capacity fade mechanisms and side reactions in lithium-ion batteries. *J. Electrochem. Soc.* **1998**, *145*, 3647–3667. [CrossRef]
6. Ramadass, P.; Haran, B.; White, R.; Popov, B.N. Mathematical modeling of the capacity fade of Li-ion cells. *J. Power Sources* **2003**, *123*, 230–240. [CrossRef]
7. Ashwin, T.; Chung, Y.M.; Wang, J. Capacity fade modelling of lithium-ion battery under cyclic loading conditions. *J. Power Sources* **2016**, *328*, 586–598. [CrossRef]
8. Ramadass, P.; Haran, B.; White, R.; Popov, B.N. Capacity fade of Sony 18650 cells cycled at elevated temperatures: Part I. Cycling performance. *J. Power Sources* **2002**, *112*, 606–613. [CrossRef]
9. Broussely, M.; Herreyre, S.; Biensan, P.; Kasztejna, P.; Nechev, K.; Staniewicz, R. Aging mechanism in Li ion cells and calendar life predictions. *J. Power Sources* **2001**, *97*, 13–21. [CrossRef]
10. Linde, D.; Reddy, T. Lithium-Ion Batteries. In *Handbook of batteries*, 3rd ed.; McGraw-Hill: New York, NY, USA, 1995; p. 265.
11. Vetter, J.; Novák, P.; Wagner, M.; Veit, C.; Möller, K.-C.; Besenhard, J.; Winter, M.; Wohlfahrt-Mehrens, M.; Vogler, C.; Hammouche, A. Ageing mechanisms in lithium-ion batteries. *J. Power Sources* **2005**, *147*, 269–281. [CrossRef]

12. Bloom, I.; Cole, B.; Sohn, J.; Jones, S.; Polzin, E.; Battaglia, V.; Henriksen, G.; Motloch, C.; Richardson, R.; Unkelhaeuser, T. An accelerated calendar and cycle life study of Li-ion cells. *J. Power Sources* **2001**, *101*, 238–247. [CrossRef]

13. Severson, K.A.; Attia, P.M.; Jin, N.; Perkins, N.; Jiang, B.; Yang, Z.; Chen, M.H.; Aykol, M.; Herring, P.K.; Fraggedakis, D. Data-driven prediction of battery cycle life before capacity degradation. *Nat. Energy* **2019**, *4*, 383–391. [CrossRef]

14. Deshpande, R.; Verbrugge, M.; Cheng, Y.-T.; Wang, J.; Liu, P. Battery cycle life prediction with coupled chemical degradation and fatigue mechanics. *J. Electrochem. Soc.* **2012**, *159*, A1730–A1738. [CrossRef]

15. Chen, S.; Wan, C.; Wang, Y. Thermal analysis of lithium-ion batteries. *J. Power Sources* **2005**, *140*, 111–124. [CrossRef]

16. Pesaran, A.A.; National Renewable Energy Laboratory (U.S.). Electrothermal analysis of lithium ion batteries. In *Nrel/Pr 540-39503*; National Renewable Energy Laboratory: Golden, CO, USA, 2006; p. 26.

17. Wang, J.; Purewal, J.; Liu, P.; Hicks-Garner, J.; Soukazian, S.; Sherman, E.; Sorenson, A.; Vu, L.; Tataria, H.; Verbrugge, M.W. Degradation of lithium ion batteries employing graphite negatives and nickel–cobalt–manganese oxide+ spinel manganese oxide positives: Part 1, aging mechanisms and life estimation. *J. Power Sources* **2014**, *269*, 937–948. [CrossRef]

18. Doyle, M.; Fuller, T.F.; Newman, J. Modeling of galvanostatic charge and discharge of the lithium/polymer/insertion cell. *J. Electrochem. Soc.* **1993**, *140*, 1526–1533. [CrossRef]

19. Ramadass, P.; Haran, B.; Gomadam, P.M.; White, R.; Popov, B.N. Development of first principles capacity fade model for Li-ion cells. *J. Electrochem. Soc.* **2004**, *151*, A196–A203. [CrossRef]

20. Ren, L.; Zhao, L.; Hong, S.; Zhao, S.; Wang, H.; Zhang, L. Remaining useful life prediction for lithium-ion battery: A deep learning approach. *IEEE Access* **2018**, *6*, 50587–50598. [CrossRef]

21. Ning, G.; Popov, B.N. Cycle life modeling of lithium-ion batteries. *J. Electrochem. Soc.* **2004**, *151*, A1584–A1591. [CrossRef]

22. Ecker, M.; Gerschler, J.B.; Vogel, J.; Käbitz, S.; Hust, F.; Dechent, P.; Sauer, D.U. Development of a lifetime prediction model for lithium-ion batteries based on extended accelerated aging test data. *J. Power Sources* **2012**, *215*, 248–257. [CrossRef]

23. Smith, K.; Saxon, A.; Keyser, M.; Lundstrom, B.; Cao, Z.; Roc, A. Life prediction model for grid-connected Li-ion battery energy storage system. In Proceedings of the American Control Conference (ACC), Seattle, WA, USA, 24–26 May 2017; pp. 4062–4068.

24. Wang, J.; Liu, P.; Hicks-Garner, J.; Sherman, E.; Soukiazian, S.; Verbrugge, M.; Tataria, H.; Musser, J.; Finamore, P. Cycle-life model for graphite-LiFePO4 cells. *J. Power Sources* **2011**, *196*, 3942–3948. [CrossRef]

25. Hou, C.; Ouyang, M.; Xu, L.; Wang, H. Approximate Pontryagin's minimum principle applied to the energy management of plug-in hybrid electric vehicles. *Appl. Energy* **2014**, *115*, 174–189. [CrossRef]

26. Ebbesen, S.; Elbert, P.; Guzzella, L. Engine downsizing and electric hybridization under consideration of cost and drivability. *Oil Gas Sci. Technol. Rev. d'IFP Energies Nouv.* **2013**, *68*, 109–116. [CrossRef]

27. Liaw, B.Y.; Nagasubramanian, G.; Jungst, R.G.; Doughty, D.H. Modeling of lithium ion cells—A simple equivalent-circuit model approach. *Solid State Ion.* **2004**, *175*, 835–839.

28. Forman, J.C.; Moura, S.J.; Stein, J.L.; Fathy, H.K. Genetic identification and fisher identifiability analysis of the Doyle–Fuller–Newman model from experimental cycling of a LiFePO 4 cell. *J. Power Sources* **2012**, *210*, 263–275. [CrossRef]

29. Jokar, A.; Rajabloo, B.; Désilets, M.; Lacroix, M. Review of simplified Pseudo-two-Dimensional models of lithium-ion batteries. *J. Power Sources* **2016**, *327*, 44–55. [CrossRef]

30. Ahmed, R.; El Sayed, M.; Arasaratnam, I.; Tjong, J.; Habibi, S. Reduced-order electrochemical model parameters identification and SOC estimation for healthy and aged Li-ion batteries Part I: Parameterization model development for healthy batteries. *IEEE J. Emerg. Sel. Top. Power Electron.* **2014**, *2*, 659–677. [CrossRef]

31. Hassan, R.; Cohanim, B.; De Weck, O.; Venter, G. A comparison of particle swarm optimization and the genetic algorithm. In Proceedings of the 46th AIAA/ASME/ASCE/AHS/ASC Structures, Structural Dynamics and Materials Conference, Austin, TX, USA, 18–21 April 2005; p. 1897.

32. Han, X.; Ouyang, M.; Lu, L.; Li, J.; Zheng, Y.; Li, Z. A comparative study of commercial lithium ion battery cycle life in electrical vehicle: Aging mechanism identification. *J. Power Sources* **2014**, *251*, 38–54. [CrossRef]

33. Liu, P.; Wang, J.; Hicks-Garner, J.; Sherman, E.; Soukiazian, S.; Verbrugge, M.; Tataria, H.; Musser, J.; Finamore, P. Aging mechanisms of LiFePO4 batteries deduced by electrochemical and structural analyses. *J. Electrochem. Soc.* **2010**, *157*, A499–A507. [CrossRef]

34. Bourlot, S.; Blanchard, P.; Robert, S. Investigation of aging mechanisms of high power Li-ion cells used for hybrid electric vehicles. *J. Power Sources* **2011**, *196*, 6841–6846. [CrossRef]

35. Chen, L. Integrated Design and Control Optimization of Hybrid Electric Marine Propulsion Systems Based on Battery Performance Degradation Model. Ph.D. Thesis, University of Victoria, Victoria, BC, Canada, 2019.

36. Heymans, C.; Walker, S.B.; Young, S.B.; Fowler, M. Economic analysis of second use electric vehicle batteries for residential energy storage and load-levelling. *Energy Policy* **2014**, *71*, 22–30. [CrossRef]

37. Raustad, R. *Electric Vehicle Life Cycle Cost Assessment*; Report No. FSEC-CR-1984-14; University of Central Florida: Orlando, FL, USA, 2014.

38. Solem, S.; Fagerholt, K.; Erikstad, S.O.; Patricksson, Ø. Optimization of diesel electric machinery system configuration in conceptual ship design. *J. Mar. Sci. Technol.* **2015**, *20*, 406–416. [CrossRef]

Article

Real-time Energy Management Strategy for Oil-Electric-Liquid Hybrid System based on Lowest Instantaneous Energy Consumption Cost

Yang Yang [1,2,*]**, Zhen Zhong** [1,2]**, Fei Wang** [1,2]**, Chunyun Fu** [1,2] **and Junzhang Liao** [1,2]

[1]   State Key Laboratory of Mechanical Transmission, Chongqing University, Chongqing 400044, China;
20183213006t@cqu.edu.cn (Z.Z.); 17353227101@163.com (F.W.); fuchunyun@cqu.edu.cn (C.F.);
20142371@cqu.edu.cn (J.L.)

[2]   School of Automotive Engineering, Chongqing University, Chongqing 400044, China

*   Correspondence: yangyang@cqu.edu.cn; Tel.: +86-136-0831-1819

Received: 10 January 2020; Accepted: 8 February 2020; Published: 11 February 2020

Abstract: For the oil–electric–hydraulic hybrid power system, a logic threshold energy management strategy based on the optimal working curve is proposed, and the optimal working curve in each mode is determined. A genetic algorithm is used to determine the optimal parameters. For driving conditions, a real-time energy management strategy based on the lowest instantaneous energy cost is proposed. For braking conditions and subject to the European Commission for Europe (ECE) regulations, a braking force distribution strategy based on hydraulic pumps/motors and supplemented by motors is proposed. A global optimization energy management strategy is used to evaluate the strategy. Simulation results show that the strategy can achieve the expected control target and save about 32.14% compared with the fuel consumption cost of the original model 100 km 8 L. Under the New European Driving Cycle (NEDC) working conditions, the energy-saving effect of this strategy is close to that of the global optimization energy management strategy and has obvious cost advantages. The system design and control strategy are validated.

Keywords: oil–electric–hydraulic hybrid system; lowest instantaneous energy costs; energy management; global optimization

---

1. Introduction

With the rise and boom of the automobile industry, the number of automobiles has been increasing, but the related problem of environmental pollution has also been growing. At present, pure electric vehicles are considered to be the cleanest automobiles, but their core technologies, such as motors and power batteries, are difficult to make great breakthrough in a short period of time, which has severely restricted their development. On the other hand, hybrid electric vehicles do not have such problems and are thus gradually being favored by more people. The main problem that needs to be solved in hybrid electric vehicles is determining how to make a reasonable allocation among the power sources under the premise that the demand torque is known. At present, the energy management strategy for hybrid electric vehicles can be roughly divided into a rule-based energy management strategy, instantaneous optimization of energy management strategies, and global optimization of energy management strategies.

Zhou et al. proposed a rule-based energy management strategy that uses dynamic programming (DP) to select control parameters. The fuel consumption per 100 km of the strategy is 12.7 L, which is very close to the global optimal value of 12.4 L [1]. Li et al. on the other hand, proposed a logic threshold strategy optimized via the pseudospectral method, which achieves the goal of reducing battery energy loss by making supercapacitors perform better with a high specific power performance [2]. Whereas

Qin et al. proposed an energy management control strategy based on working condition identification, which reduced fuel consumption by 12.77% compared with that of the strategy without working condition identification [3]. Meanwhile, Yin et al. proposed a dual-planetary hybrid electric vehicle as an object of engine torque control. This strategy can optimize the engine operating point while keeping the final battery state of charge (SOC) value within a reasonable range [4]. Although this type of energy management strategy has a simple structure and strong practicability, its advantages and disadvantages are easily affected by the experience of engineering personnel and the working conditions are poor.

For the instantaneous optimization of energy management strategy, Jiao et al. proposed an adaptive equivalent fuel consumption minimum strategy (A-ECMS), which obtains the equivalent factor under current driving conditions based on the equivalent factor map in energy distribution. The fuel consumption is minimized throughout the driving route, and the battery state of charge (SOC) is kept within a reasonable range [5]. On the other hand, Zhang et al. proposed an energy management strategy based on the minimum equivalent fuel consumption. Compared with the rule-based energy management strategy, it has a significant improvement in terms of fuel economy [6]. Meanwhile, Wang et al. proposed an energy control strategy that allows both the engine and the motor to operate in an efficient region to improve fuel economy [7]. Compared with the globally optimized energy management strategy, this strategy has a small amount of calculation and fast speed, but can only achieve instantaneous optimization.

For the global optimization energy management strategy, Xiang Zhu proposed a DP-based energy management strategy. Through online simulation, the solution of the multi-neural network model is determined to be close to the optimal solution obtained by the global optimization algorithm, and the real-time application of dynamic programming is greatly improved [8]. Meanwhile, Wang et al. considered the discrete solutions of related variables and the boundary problems of feasible domains when solving the optimal control problem of hybrid electric vehicle, and systematically studied the relationship between the optimization accuracy and the computational complexity of the dynamic programming algorithm. Compared with that of the traditional control strategy, the fuel economy based on the dynamic planning control strategy increased by about 20% [9]. Although this strategy can achieve global optimization, it needs to obtain the entire driving conditions in advance, and the amount of calculation is large, which is difficult to apply to real vehicles.

In this article, Firstly, the oil–electric–hydraulic system requires one to install a hydraulic energy storage system on the rear axle of the existing oil-electric hybrid vehicle structure, which is proposed in this article and uses a timely four-wheel-drive structure with independent driving of the front and rear axles. Secondly, based on this structure, this study focuses on a steady-state energy management strategy in the driving and braking process, proposes a logic threshold energy management strategy based on the optimal working curve, and selects the relevant threshold according to the steady-state efficiency characteristic curve of the key components. The genetic algorithm is used to jointly optimize the powertrain parameters and logic threshold energy management strategy parameters. Thirdly, for the driving mode, considering that this article mainly focuses on the fuel economy of the entire vehicle, and in the logic threshold energy management strategy, the setting of the threshold value is susceptible to expert experience, the working conditions are poor, the global optimization energy management strategy has a large amount of calculation, the driving conditions need to be known, and practical problems, a real-time energy management strategy based on the lowest energy consumption cost is proposed, whereas for the braking mode, based on the traditional four-wheel vehicle braking force distribution strategy, a braking-force allocation strategy based on the highest energy recovery is proposed. Furthermore, a global optimization energy management strategy based on dynamic programming is used as the basis for evaluating the advantages and disadvantages of other strategies. Finally, the stateflow-based control strategy model is implemented into the forward simulation model to verify the effectiveness of the strategy, and the two strategies are simulated and compared.

2. Oil–Electric–Liquid Hybrid Power System Structure

Unlike a pure electric vehicle, an oil–electric hybrid electric vehicle retains the engine and reduces the power of the battery. Although the vehicle's range is increased, the disadvantage of a reduced energy recovery rate is ignored. Under the same conditions, although the accumulator has a low energy density, it also has a high power density, which not only can quickly recover and release energy, but also has higher energy efficiency and can provide greater auxiliary power for the vehicle. If the characteristics of the high energy density of the storage battery and high power density of the accumulator are combined, not only can the vehicle's cruising range be extended, but the energy recovery rate can also be improved. Therefore, the traditional configuration is equipped with a motor and an external battery pack on the front axle, and a hydraulic energy storage system on the rear axle. In addition, continuously variable transmission (CVT) can not only adjust the operating point of the engine and motor, save fuel consumption, but also improve the ride and stability of the vehicle. Therefore, this article decided to use CVT transmission. As shown in Figure 1, the new setup is composed mainly of an integrated starter generator (ISG) motor, high-pressure accumulator, low-pressure accumulator, hydraulic pump/motor, battery, and continuously variable transmission (CVT). There are clutches at the connection between the engine and the motor, the hydraulic pump/motor, and the rear axle main reducer. The clutch status of the front and rear axles can be controlled to make the vehicle work in different modes. The vehicle working mode is outlined in Table 1.

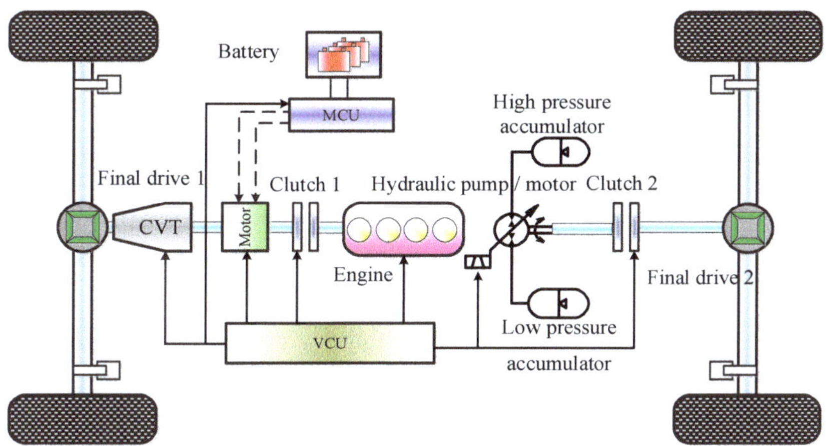

Figure 1. Structure of oil–electric–hydraulic hybrid power system.

Similar to for a traditional automobile, the maximum demand power of an oil–electric hybrid electric vehicle is also determined according to the vehicle dynamics index [10]. This study uses the vehicle's basic parameters and dynamic indicators of the original model. Based on the vehicle parameters and different driving conditions, the maximum required power of the vehicle can be calculated. From these calculations, the total power of the initial power source is determined to be 120 kW.

In addition, the theoretical calculation method and the comprehensive analysis method based on the cycle condition are used to match the parameters of each key component. The matching results are listed in Table 2.

Table 1. Working modes of the hybrid system.

Working Mode	Clutch Status		Description
	C1	C2	
Motor drive alone	○	○	Start and low-speed working conditions, Accumulator pressure reaches the lower limit
Hydraulic drive alone	○	●	Start and low-speed working conditions, Improve vehicle efficiency
Engine drive alone	●	○	Medium and high-speed working conditions. Increase driving distance
Electro-hydraulic hybrid drive	○	●	Improve vehicle traffic
Oil-hydraulic hybrid drive	●	●	High load conditions such as rapid acceleration and climbing
Oil-electro hybrid drive	●	○	High load with high battery power and low and medium load with low battery power
Oil-electro-hydraulic hybrid drive	●	●	Large power demand and more energy in batteries and accumulators
Regenerative braking mode	●	●	Motor or hydraulic pump meets ECE regulations
Friction brake	○	○	Emergency braking. Provide braking torque for as much energy recovery as possible

Note: ○ means the clutch is disengaged, ● means the clutch is engaged.

Table 2. Basic parameters of each key component.

Component	Project	Parameter	Component	Project	Parameter
Engine	Peak power/kW	60	Pump/motor	Peak power/kW	40
	Peak torque/Nm	140		Peak torque/Nm	105
Motor	Peak power/kW	30	Accumulator	Maximum working pressure/MPa	25
	Rated power/kW	15		Minimum working pressure/MPa	15
	Peak speed r/min	7000		Volume/L	35
	Rated speed r/min	2000			
Battery	Voltage level/V	251	Transmission system	CVT speed ratio range	[0.83,2.5]
	Power/KW	33		Front axle final drive speed ratio	6
	Capacity/Ah	48		Rear-axle final drive speed ratio	3

3. Joint Optimization of Energy Management Strategy and Power System Component Parameters

3.1. Logic Threshold Energy Management Strategy based on Optimal Working Curve

This study is based on the gasoline engine's universal characteristic curve, and the research object is the rechargeable oil–electric–hydraulic hybrid vehicle. P_{emin_eco} and P_{emax_eco} are used as the logic threshold parameters for charge-sustaining (CS) stage engine operation to optimize the working area of the engine. $(P_{emin\_eco}, P_{emax\_eco}) = (9 \text{ kW}, 57 \text{ kW})$ is initially selected, and based on the battery SOC model, the battery SOC = 0.3 is initially taken as its lower working limit.

In the parallel hybrid system, which includes a variety of working modes, to ensure that the hydraulic pump can provide sufficient regeneration capacity and improve energy recovery efficiency during braking, this study chooses the control strategy of preferential use of hydraulic energy and electric energy. The specific mode selection logic is shown in Figure 2.

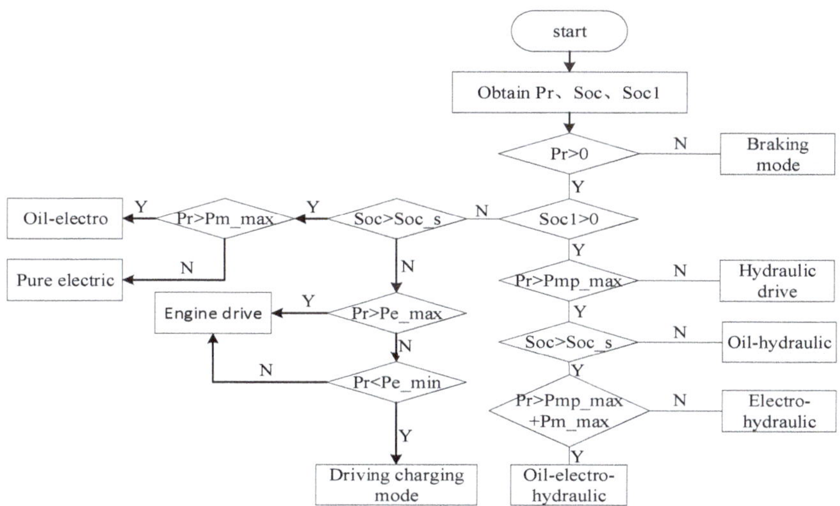

Figure 2. Mode selection logic diagram.

Under the premise of satisfying the power requirements, obtaining the best fuel economy for the whole vehicle is one of the goals of the hybrid electric vehicle energy management strategy. Firstly, the optimal working point corresponding to different power requirements in different modes is obtained via offline optimization, and a MAP table is made. Based on the result of mode selection, the optimal working point that meets the current vehicle power demand is then determined and applied, thereby achieving the optimal power system efficiency.

When the hybrid electric vehicle is operating in the hydraulic pump/motor single drive mode, its operating point is directly determined according to the required power, because the CVT transmission efficiency model was established by interpolation, and the CVT speed ratio range was obtained, whereas when the hybrid electric vehicle is operating in the engine or motor alone drive mode, the engine or motor operating point can be adjusted via continuously variable transmission (CVT). While the demand power is satisfied, the efficiency of itself is optimized, and the working point corresponding to the optimal efficiency is the optimal working point of the engine or the motor. Under the condition that the demand power is satisfied and each key component of the power source is constrained by itself, the optimization problem, wherein the transmission system efficiency is the objective function, is solved, and the optimal working curves of the motor and the engine, when either is working alone, can be obtained. These curves are shown in Figures 3 and 4, respectively.

For hybrid systems, if the CVT efficiency loss is neglected when the engine and motor work together, Equation (1) is used.

$$\begin{aligned}
P_r &= P_e + P_m = T_e\omega_e + T_m\omega_m \\
\omega_e &= \omega_m = \omega_r \\
\omega_{e_min} &\leq \omega_e \leq \omega_{e_max} \\
\omega_{m_min} &\leq \omega_m \leq \omega_{m_max} \\
T_{s_min} &\leq T_e \leq T_{e_max} \\
T_{m_min} &\leq T_m \leq T_{m_max}
\end{aligned} \tag{1}$$

P_r, P_e, and P_m represent the vehicle demand power, engine power, and motor power, respectively, ω_e and ω_m denote the engine and motor speeds, respectively, i_{cvt} indicates the transmission speed ratio, and T_e and T_m refer to the engine and motor torques, respectively.

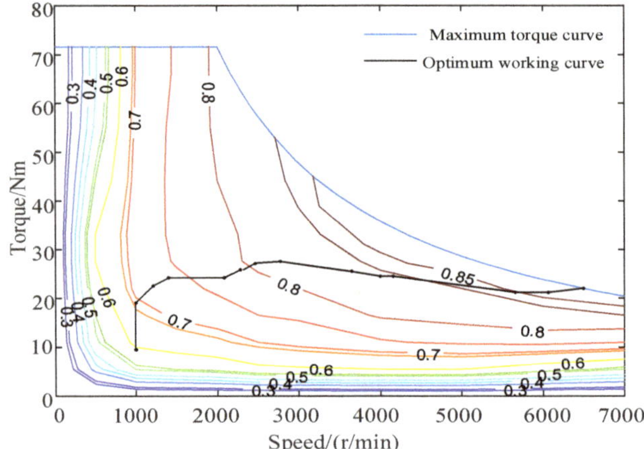

Figure 3. Optimal operating curve of motor.

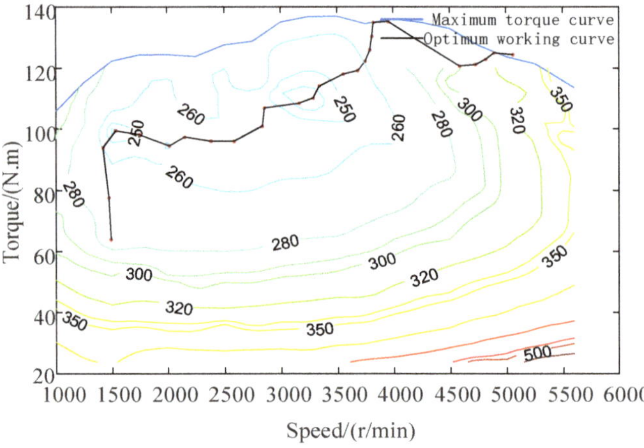

Figure 4. The optimal operating curve of the engine.

When the motor is operating in the drive mode, the efficiency of the hybrid system can be expressed as

$$\eta_1 = \frac{P_m + P_e}{\frac{P_m \eta_d}{\eta_m} + \frac{P_e}{\eta_e}} = \frac{(T_m + T_s)}{\left(\frac{T_m \eta_d}{\eta_m} + \frac{T_e}{\eta_e}\right)} \tag{2}$$

In the formula, η_e, η_m, and η_d are engine efficiency, motor efficiency, and battery discharge efficiency, respectively.

When the motor is operating in the power generation mode, the efficiency of the hybrid system can be expressed as

$$\eta_2 = \frac{[P_e - P_m(1 - \eta_m \eta_c)]}{(P_e/\eta_e)} = \frac{[T_e - T_m(1 - \eta_m \eta_c)]}{(T_e/\eta_e)} \tag{3}$$

η_c is the battery charging efficiency

According to Equation (1), there are multiple combinations of (ω_r, T_r) on the premise that the engine and motor torques and speeds meet their constraints. The torques and speeds of the engine and motor in each combination, and the charge and discharge efficiencies of the engine, motor, and battery at this operating point can be substituted into Equation (2) or Equation (3) to calculate the total efficiency corresponding to each group of operating points. The most efficient combination (ω_e, T_e), (ω_m, T_m) represents the best operating points of the engine and motor. In this way, the best working curves of the engine and the motor can be obtained. For the charge-depleting (CD) mode, the best working curves of the engine and motor combined drive are shown in Figures 5 and 6, respectively.

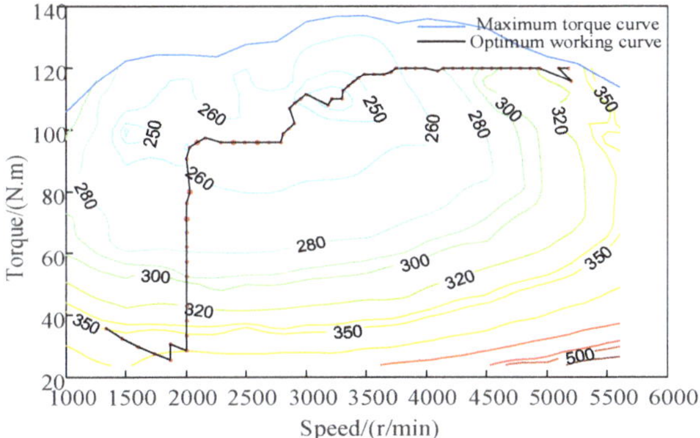

Figure 5. Optimal operating curve of engine.

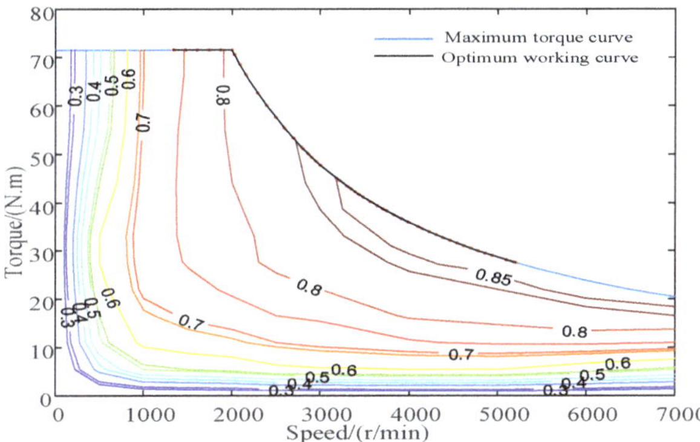

Figure 6. Optimal operating curve of motor.

3.2. Multi-objective optimization problems and their conversion

Under the premise of satisfying the vehicle power performance, achieving the Pareto optimal solution of the two objective functions is equivalent to achieving the two optimal goals for hybrid electric vehicle energy consumption and vehicle manufacturing cost. This study uses a linear weighting method to convert a multi-objective function based on energy consumption and vehicle manufacturing cost into a single objective function:

$$F(x) = \omega_1 \frac{fuel(x)}{fuel_{unopt}} + \omega_2 \frac{cost(x)}{cost_{unopt}} \tag{4}$$

where $fuel_{unopt}$ and $cost_{unopt}$ represent the initial energy consumption and vehicle manufacturing cost, respectively, before optimization, and ω_1, ω_2 are weighted values, wherein $\omega_1 = 0.8$ and $\omega_2 = 0.2$.

In terms of optimizing variable selection and constraint setting, to make the battery capacity meet the electric vehicle mileage index, the hydraulic pump/motor has to have sufficient regenerative braking force, but its corresponding cost should be reduced as much as possible. The optimization problem then becomes more convenient to solve. This study selects engine peak power P_{emax}, motor peak power P_{mmax}, ratio max_factor of maximum operating power of the engine to its peak power, ratio min_factor of engine minimum operating power to peak power, and CD–CS mode switching value Soc_s. That is, $X = [P_{emax}, P_{mmax}, max\_factor, min\_factor, Soc\_s]$ are optimized variables for the joint optimization of energy management strategy parameters and dynamic system parameters. The vehicle's dynamic index (maximum speed is 180 km/h, 0~100 km/h acceleration time is 12.46 s, and the maximum climbable gradient is 40%) is used as an optimization constraint to ensure that the optimization results meet the vehicle power requirements.

3.3. Energy Management Strategy and Optimization of Power System Components Parameters

After the multi-objective problem is transformed into a single-objective problem, this article defines the fitness function, which simplifies the manufacturing cost of the whole vehicle power system to the cost of the engine and the motor. The following Equation (5) is obtained [11–13],

$$\cos t(X) = 849 + 12.236P_{emax} + 10.888P_{mmax} \tag{5}$$

In the formula, P_{emax} and P_{mmax} are the peak powers of the engine and the motor, respectively.

Genetic algorithms are then used to optimize energy management strategies and power system parameters:

1. Because the optimization variables X are continuous variables, the real coding method is selected to encode the variables, and the upper and lower limits of each variable are set, as listed in Table 3.
2. The algorithm parameters are set, such that the maximum evolution algebra is 40, the population size is 100, the number of elites is 8, the crossover probability is 0.3, and the original population is randomly generated.
3. The number of iterations is checked for whether it reaches the maximum. If not, the vehicle simulation model is run to output the optimal fitness value and average fitness value of the contemporary population, and the process is continued. Otherwise, the optimal solution to the previous generation is outputted, and the process is ended.
4. The optimal fitness value is checked for whether it is less than or equal to the set target fitness value. If yes, the optimal fitness value and its corresponding optimal individual are outputted, and the process is ended. Otherwise, the current population is selected, crossed, and mutated.

Table 3. Optimization variable interval.

Variable	Unit	Description	Optimization Interval
P_{emax}	kw	Engine peak power	[50,70]
P_{mmax}	kw	Motor peak power	[20,40]
max_factor	kw	Engine maximum operating power factor	[0.7,0.95]
min_factor	kw	Engine minimum operating power factor	[0.15,0.3]
SOC_s	-	CD-CS Mode switching value (battery)	[0.3,0.4]

The optimization results of the genetic algorithm are shown that the fitness value decreases with the evolution of the population, and finally converges to 0.93. The corresponding optimal individuals are $\left(P_{emax}, P_{mmax}, max\_factor, min\_factor, Soc\_s\right) = (57.32, 32.68, 0.89, 0.16, 0.32)$

According to the comparison of the simulation results of the unoptimized and GA-optimized in Table 4, the manufacturing cost of the whole vehicle power system is reduced by 1.7%, and the energy cost per 100 km is reduced by 8.3%.

Finally, to verify whether the parameter matching result of the power system is reasonable, the vehicle facing-forward simulation model is established based on the MATLAB/Simulink platform, and dynamic simulation results showed that the acceleration time to 100 km is 11.8 s, and the maximum speed is 177 km/h. The maximum grade is 40.24%, and the speed is 30km/h. At the time, the maximum gradeability can reach 39.78%; in summary, the optimized hybrid system parameters can meet the vehicle dynamic performance requirements.

Table 4. Comparison of simulation results.

	Engine Peak Power	Motor Peak Power	Engine Minimum Factor	Engine Maximum Factor	Mode Switch SOC	Power System Cost	Energy Cost
Before optimization	60	30	0.15	0.95	0.3	13,227	32.3
After optimization	57.3	32.7	0.16	0.89	0.32	13,003	29.6

4. Energy Management Strategy based on the Lowest Instantaneous Energy Cost

4.1. Energy Management under Driving Conditions

In this article, the minimum instantaneous energy consumption cost is the objective function; the vehicle travel demand torque T_r, vehicle speed v, hydraulic accumulator Soc_1, and battery Soc are the state variables; the hydraulic pump/motor torque T_{pm}, motor torque T_m, engine torque T_e, and CVT speed ratio i_{cvt} are the control variables. Because this study deals with not only hydraulic regenerative braking, but also motor regenerative braking [14], the hydraulic energy is equivalent to electric energy when the cost of hydraulic energy consumed is calculated, and the instantaneous cost is

$$\text{Cost} = \frac{1}{3600}\left(j_f \frac{P_e b_e}{1000\rho} + j_e\left(\frac{P_m}{\eta_m \eta_b} + \frac{P_{pm}}{\eta_{pm}}\right)\right) \tag{6}$$

where Cost is the sum of the costs of fuel, electricity, and hydraulic energy consumed per unit time (yuan/s), j_f is the price of gasoline (yuan/L), and j_e is the price of electrical energy (yuan/kw·h). P_e, P_m, P_{pm} represents the output power of the engine, motor, hydraulic pump/motor (kw). b_e is the fuel consumption rate $(g/(kw\cdot h))$; ρ is the density of gasoline (g/cm^3); η_m and η_b represent the motor, battery efficiency; η_{pm} represents the mechanical efficiency of the hydraulic pump/motor in motor mode.

The objective function and constraints can be expressed as

$$\min\left(\text{Cost}\left(T_{pm}, T_m, T_e, i_{cvt}\right)\right) \tag{7}$$

$$\left\{\begin{array}{c} (T_e + T_m)\cdot i_{cvt} i_0 \eta_{cvt} + T_{pm}\cdot i_1 = T_{req} \\ 0 \leq n_e \leq n_{emax} \\ 0 \leq n_{pm} \leq n_{pmmax} \\ T_e \leq T_{emax}(n) \\ |T_m| \leq T_{mmax}(n) \\ T_e + T_m \leq T_{cvt\_in\_max} \\ T_{pm} \leq T_{pmmax}(n) \\ \frac{P_m}{(\eta_m \eta_b)} \leq P_{bmax} \\ 0.83 \leq i_{cvt} \leq 2.50 \end{array}\right. \tag{8}$$

To obtain the optimal values of each power source and transmission under different vehicle conditions, the grid traversal algorithm is used to solve any set of state variables, and the MAP table is made to facilitate real-time control. The algorithm flow is shown in Figure 7.

For different drive modes, the optimization results are different. In the single power source mode, because the hybrid system scheme adopted in this study does not have a transmission on the rear axle of the vehicle, the operating point cannot be optimized in the hydraulic pump/motor drive mode. Therefore, this study examines only the optimization of energy cost in the purely electric mode and engine driving mode. In the purely electric mode, the parts related to the engine and the hydraulic pump/motor in the optimization Algorithm 6 and Constraint Condition 8 are omitted, and the results shown in Figure 8a,b can be obtained via offline optimization. Similarly, for the engine mode, the optimization results are shown in Figure 9a,b.

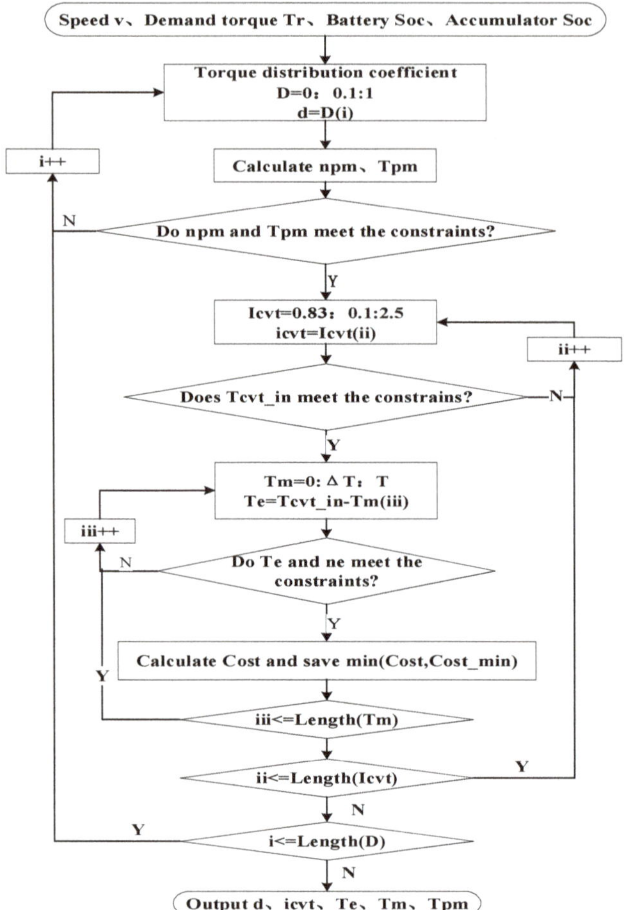

Figure 7. Grid traversal algorithm flow chart.

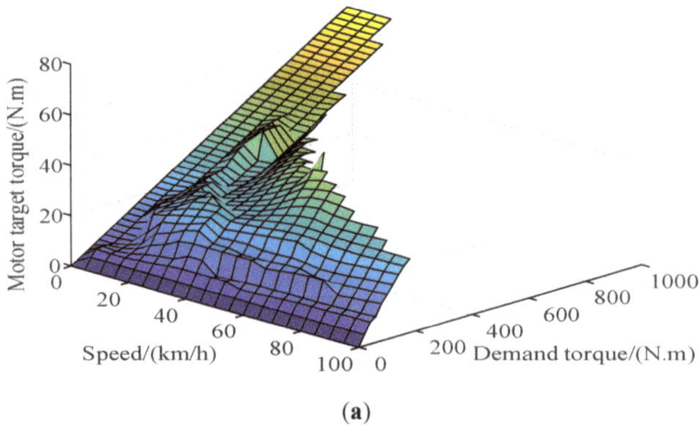

Figure 8. *Cont.*

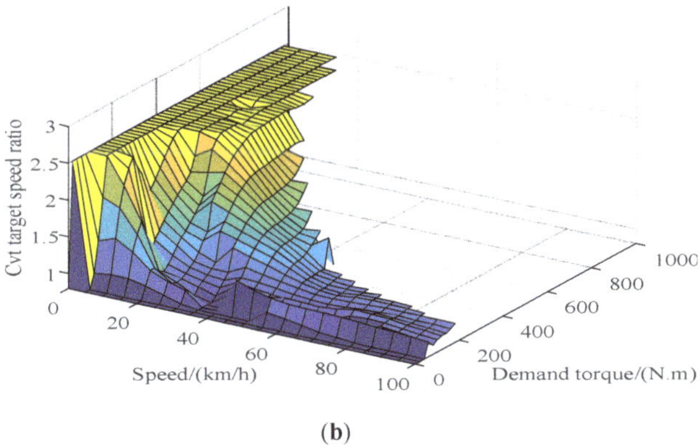

(**b**)

Figure 8. Optimization results in the purely electric mode: (**a**) target torque of motor; (**b**) target speed ratio of continuously variable transmission (CVT).

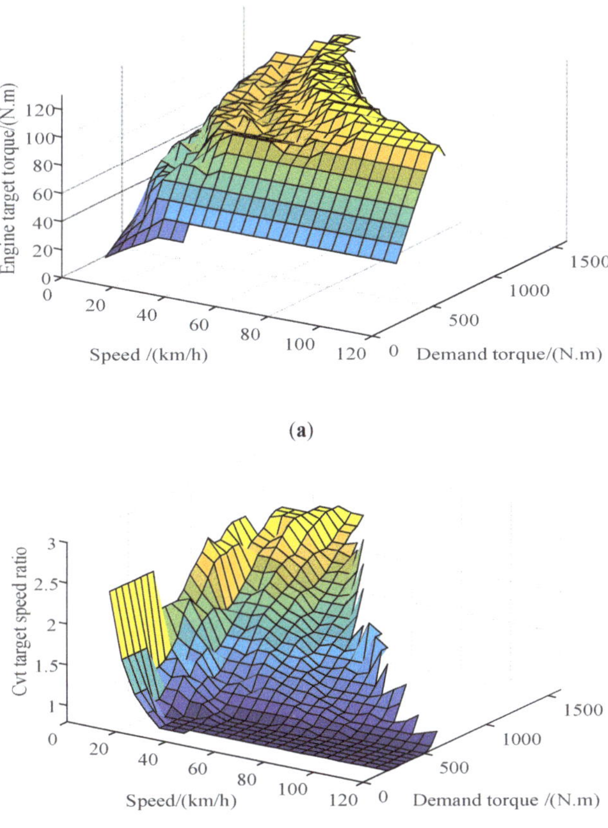

(**b**)

Figure 9. Optimization results in the engine mode: (**a**) target torque of engine; (**b**) target speed ratio of CVT.

The hybrid drive mode includes the electro–hydraulic hybrid drive, oil–electric hybrid drive, and oil–electro–hydraulic hybrid drive. For each mode, corresponding changes are similarly made to the optimization algorithm and constraints, and offline optimization is performed to produce the results shown in the Figures 10–12.

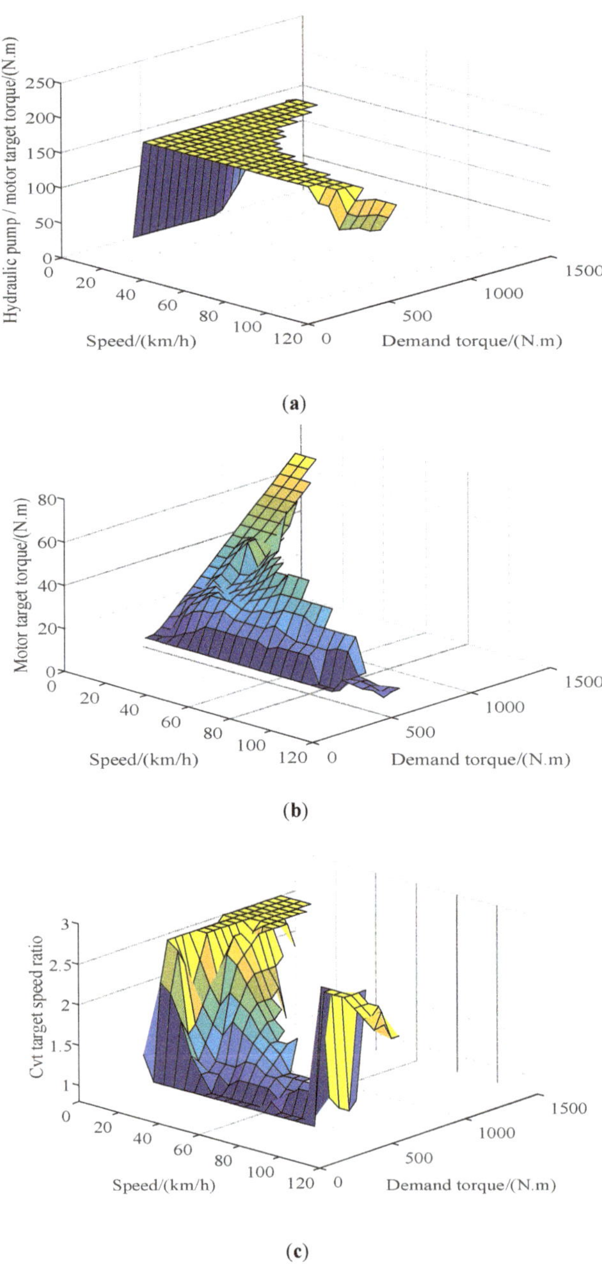

Figure 10. Optimization results in the electro-hydraulic hybrid drive mode: (a) hydraulic pump/motor target torque; (b) motor target torque; (c) target speed ratio of CVT.

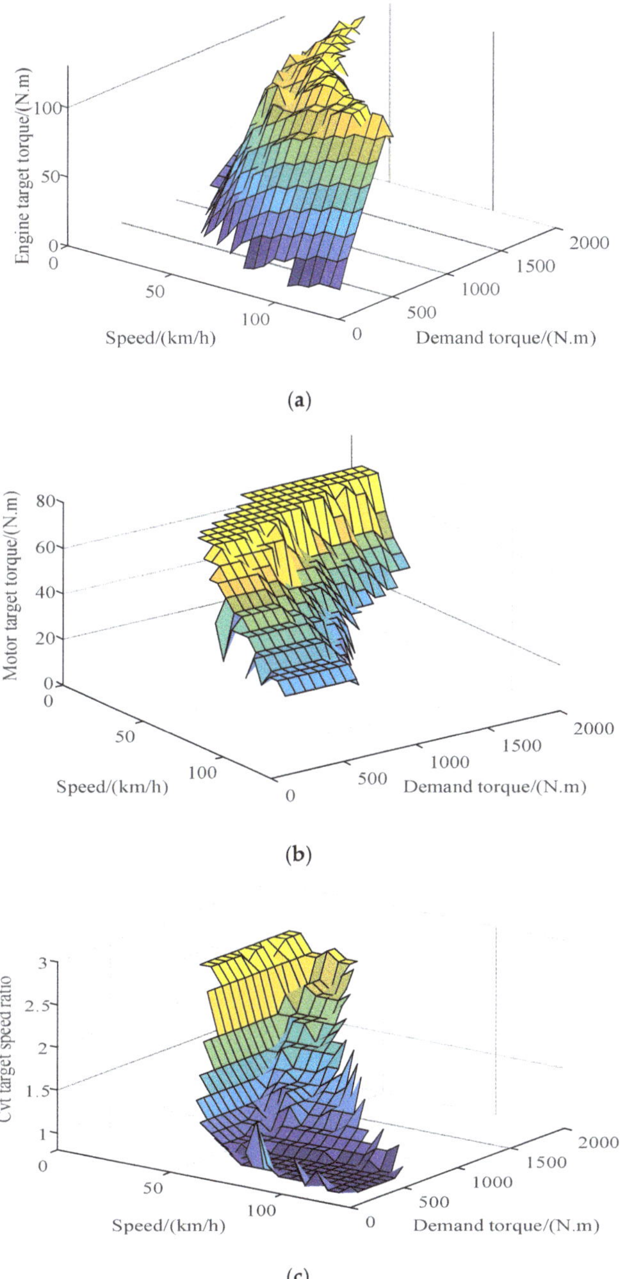

Figure 11. Optimization results in oil–electro hybrid drive mode: (**a**) engine target torque; (**b**) motor target torque; (**c**) CVT of target speed ratio.

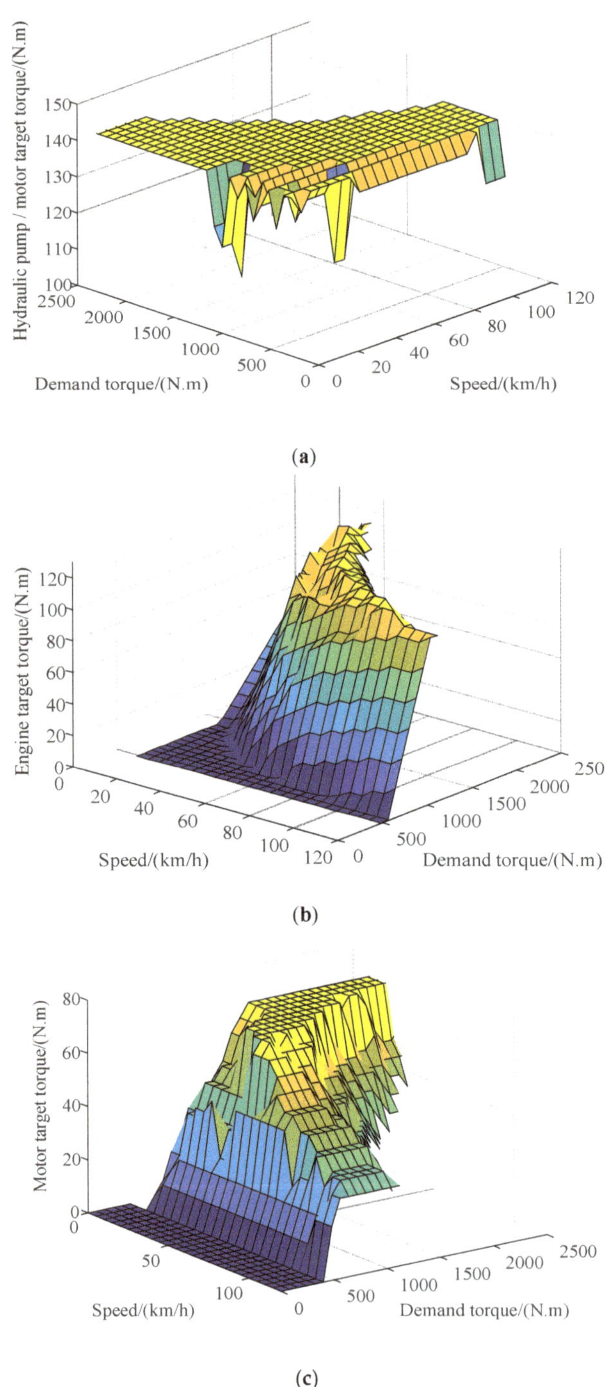

(**a**)

(**b**)

(**c**)

Figure 12. *Cont.*

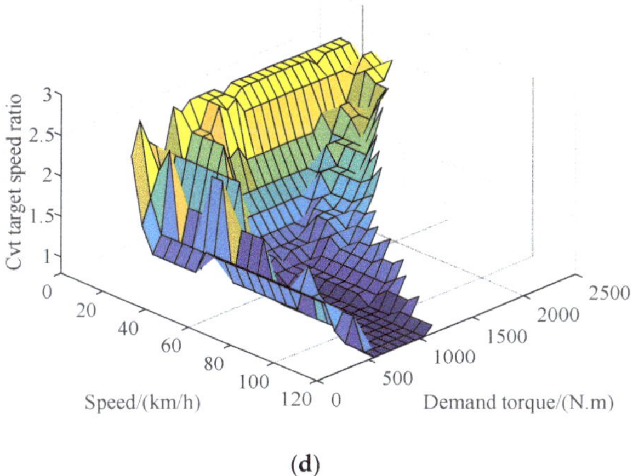

(**d**)

Figure 12. Optimization results in oil-electro-hydraulic hybrid drive mode: (**a**) hydraulic pump/motor target torque; (**b**) engine target torque; (**c**) motor target torque; (**d**) CVT target speed ratio.

4.2. Energy Management under Braking Conditions

For a hybrid electric vehicle with a regenerative braking system, regarding the distribution of the braking force, it is necessary to solve the problems of not only the distribution of the braking force of the front and rear axles but also the distribution of the regenerative braking force and the frictional braking force. Thus, considering the high efficiency of the hydraulic regenerative braking system for recovering energy, this study uses the hydraulic pump/motor as the main way and the motor as the auxiliary way of providing the regenerative braking force. At the same time, the friction braking force is used to coordinate and to meet the driver's demand braking force to achieve maximum energy recovery.

Figure 13 shows the braking force distribution curve designed for the oil–electric–hydraulic hybrid electric vehicle. The OABCD curve is a braking force distribution curve for when the hydraulic accumulator is $SOC_1 < 1$ and the battery SOC > 0.9, the $OA'BB'CD$ curve for when the hydraulic accumulator is $SOC_1 < 1$ and the battery SOC ≤ 0.9, and the $OA'B'CD$ curve for when the hydraulic accumulator is $SOC_1 = 1$ and the battery SOC < 0.9. Point A indicates the maximum braking force that can be transmitted to the rear wheel when the hydraulic pump/motor is working alone; point A' indicates the maximum braking force that can be transmitted to the front wheel when the ISG motor is working alone. Meanwhile, point B indicates the sum of the maximum braking forces that can be transmitted to the wheels when the hydraulic pump/motor and motor are simultaneously operating. B', C, D are the intersections of the I curve and the braking strengths.

When the structural scheme of the hybrid system presented in this article is analyzed, the working point of the motor can be adjusted via the CVT transmission, but the working point of the hydraulic pump/motor cannot be adjusted, and thus the operating point of the motor can only be optimized. This inference considers that in the regenerative braking mode in which all motors participate, only the braking force distribution strategy is different and that there is no influence on the optimization process. Therefore, this section needs only to optimize the motor operating point in the motor regenerative braking mode. In this section, the kinetic energy recovered is used as the objective function, and the target torque of the generator and the CVT target speed ratio are optimized.

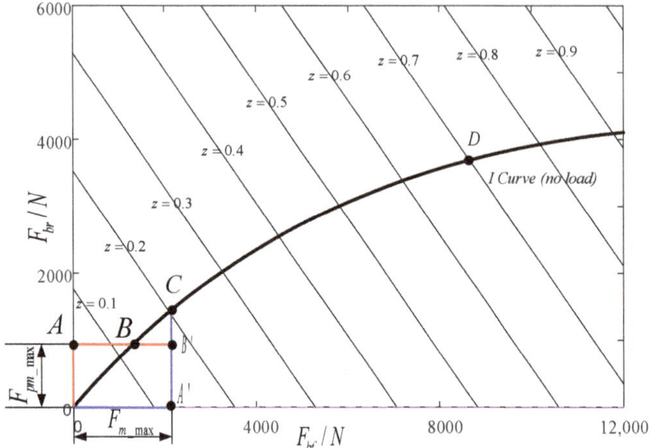

Figure 13. Braking force distribution curve of hybrid vehicles.

The objective function is

$$\min\left(P_b \eta_{cvt} \eta_m \eta_b\right) \tag{9}$$

In the formula, P_b represents the braking power required to be transmitted from the motor to the wheel.

The constraints are

$$\begin{cases} T_m = T_{mb} / \left(i_{cvt} i_1 \eta\right) \\ 0 < n \leq 6000 \\ |T_m| \leq T_{max}(n) \\ \left|P_b \eta_{cvt} \eta_m \eta_b\right| \leq P_{bmax} \\ 0.83 \leq i_{cvt} \leq 2.50 \end{cases} \tag{10}$$

The optimization results are shown in Figure 14.

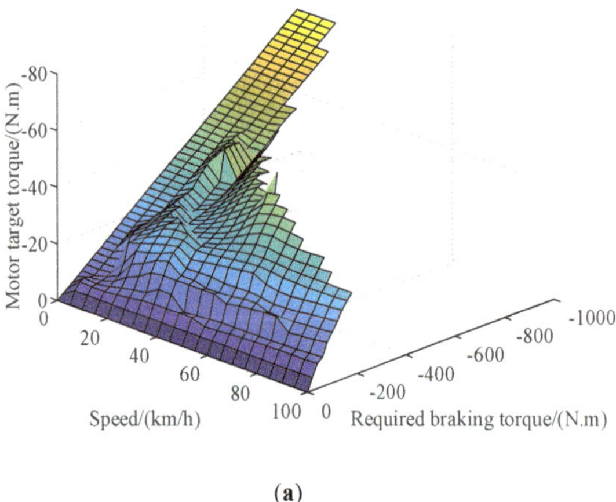

(**a**)

Figure 14. *Cont.*

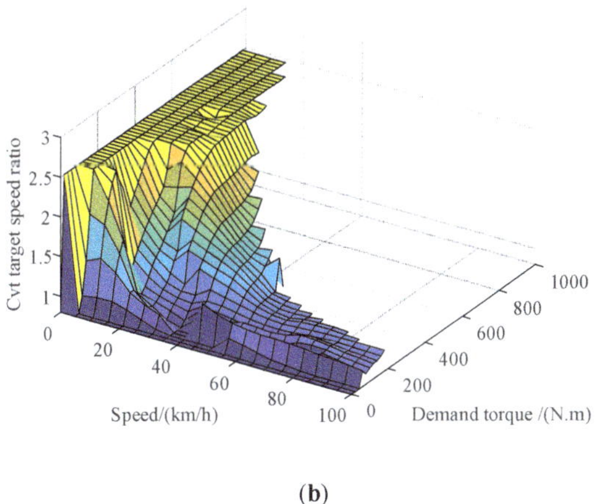

(b)

Figure 14. Optimization results in the regenerative braking mode: (**a**) target torque of motor; (**b**) target speed ratio of CVT.

4.3. Analysis of Simulation Results

To verify that the energy management strategy proposed in this article is effective in each mode, this study simulates under a driving cycle composed of multiple New European Driving Cycle (NEDC) working conditions. The results are shown in Figure 15.

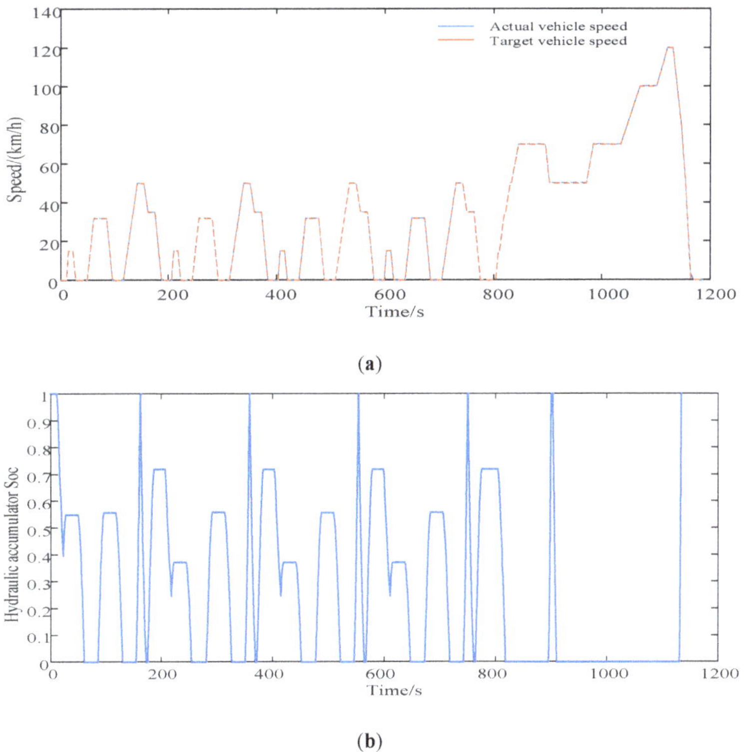

(a)

(b)

Figure 15. *Cont.*

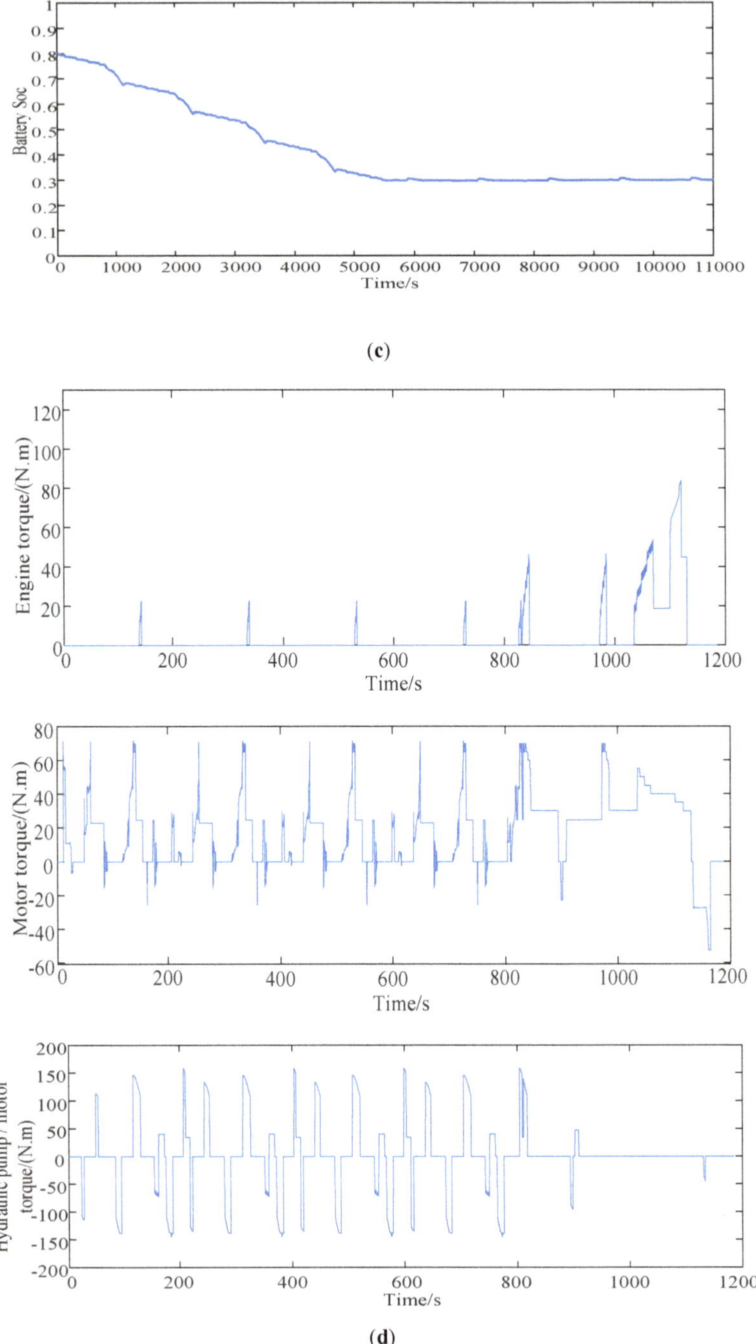

(c)

(d)

Figure 15. Simulation result: (**a**) vehicle speed following curve; (**b**) hydraulic accumulator state of charge (SOC) trajectory curve; (**c**) battery SOC trajectory curve; (**d**) torque distribution curve.

As can be seen from Figure 15a, the driver model based on the PI controller has higher control accuracy. From a comparison of (a) and (b), it can be found that during the driving process, when the hydraulic accumulator $SOC_1 > 0$, the hydraulic accumulator releases energy. Therefore, the hydraulic mode-based drive mode selection strategy proposed in this article achieves the expected control effect. From a comparison of (a), (b), and (c), it can be seen that during the braking process, the hydraulic accumulator SOC_1 or the battery SOC has a significant rise, which indicates that the braking force distribution strategy not only can meet the braking demand but also can fully recover energy. It can also be seen from (c) that the battery SOC can still be maintained within a reasonable range after falling to a certain value. It can also be seen from (a), (d) that the torque distribution of the engine, the motor, and the hydraulic pump/motor can satisfy the torque demand of the entire vehicle. In the whole simulation process over the driving distance, because the hydraulic accumulator SOC_1 returns to the initial state, the energy consumption includes only electric energy and fuel, wherein the electric energy is 5.88 degrees, the fuel is 4.99 L, and the total energy consumption cost is 38 yuan. Compared with the fuel consumption cost of the original model 100 km 8 L, the strategy proposed in this article saves costs by about 32.14%.

4.4. Simulation Comparison under Two Different Strategies

To better compare the proposed strategy with the minimum energy consumption cost strategy, the initial Soc of the battery is selected to be 0.8, and the initial value of the hydraulic accumulator Soc_1 is set to 1. Furthermore, the DP-based global optimized energy management strategy is simulated under the NEDC working conditions. The results are shown in Figure 16.

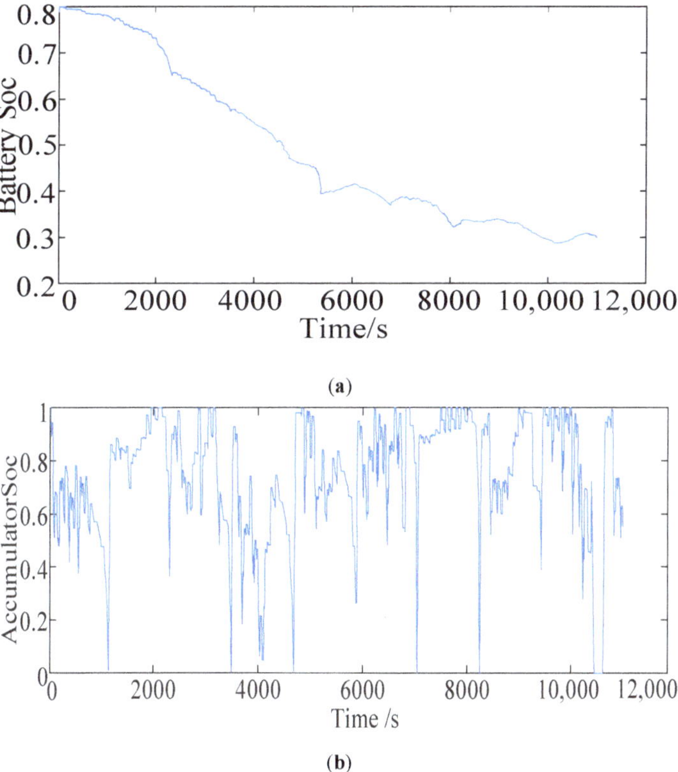

(a)

(b)

Figure 16. *Cont.*

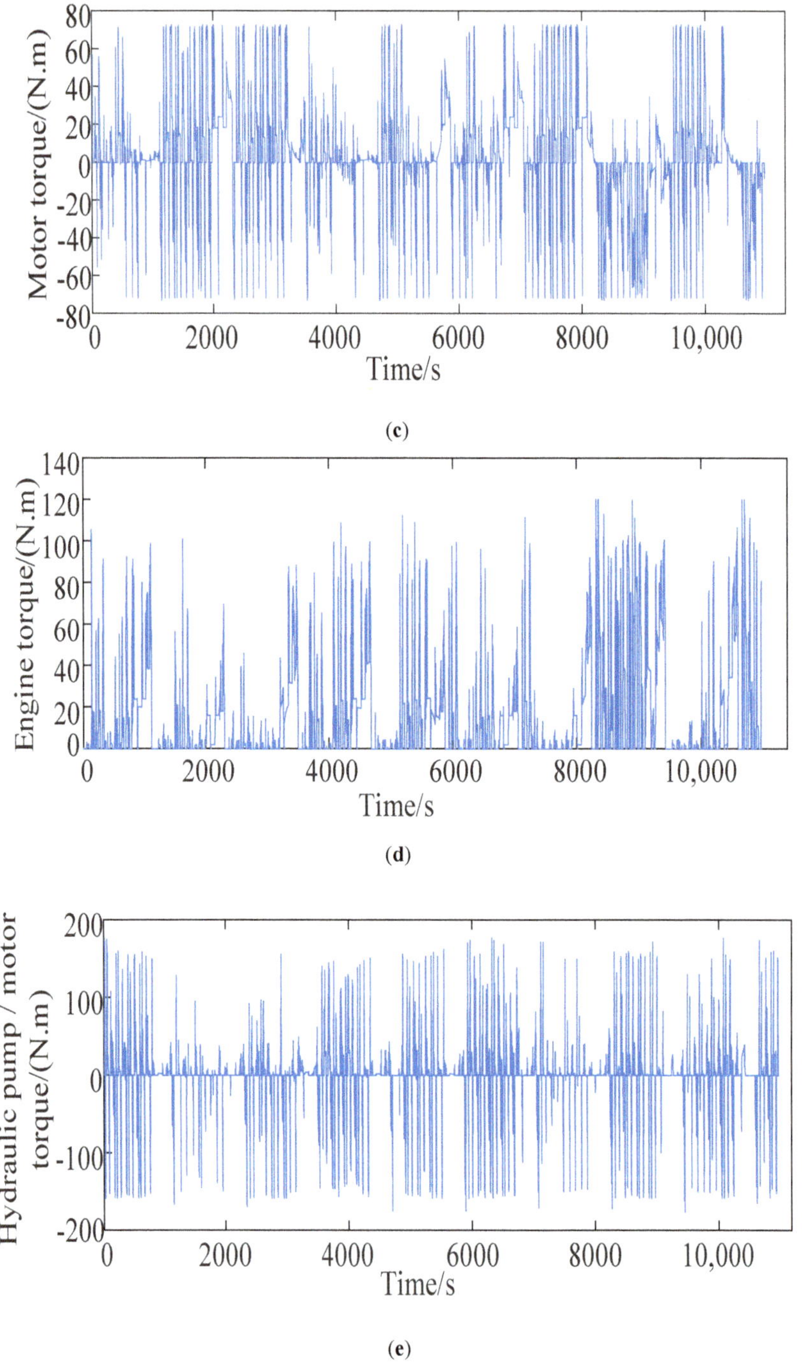

Figure 16. *Cont.*

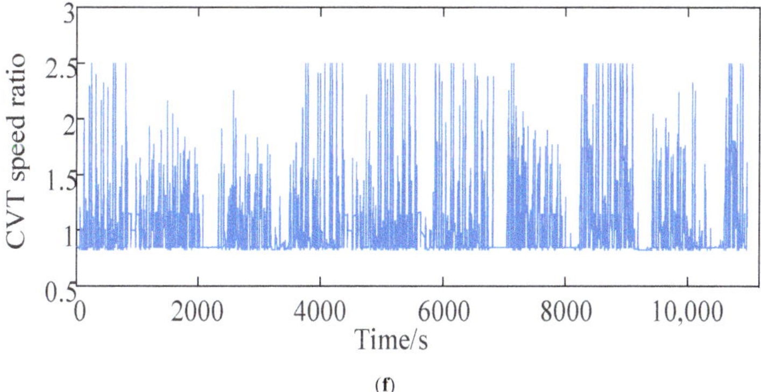

(**f**)

Figure 16. Simulation results of the energy management strategy based on dynamic programming (DP): (**a**) battery Soc; (**b**) accumulator Soc; (**c**) motor torque; (**d**) engine torque; (**e**) hydraulic pump/motor torque; (**f**) CVT speed ratio.

The simulation result Figure 16 shows that the DP-based energy management strategy can extend the cruising range by rationally utilizing the electric energy and can also maintain the balance when the battery Soc is low. The control effect of the strategy is also good, which can provide a certain evaluation point for the advantages and disadvantages of other strategies. Therefore, this strategy is compared with the instantaneous energy consumption cost minimum energy management strategy. The simulation results are shown in Figure 17.

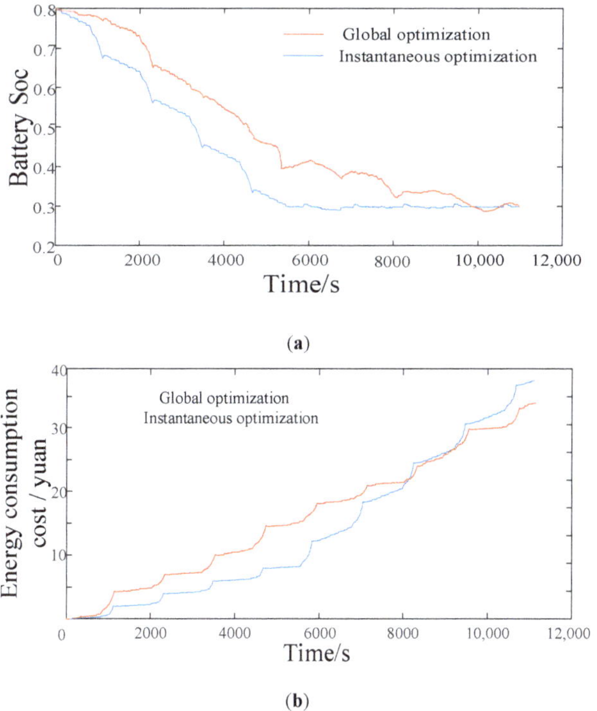

(**a**)

(**b**)

Figure 17. Comparison of simulation results under two strategies: (**a**) battery SOC versus time curve; (**b**) energy consumption cost versus time curve.

It can be seen from Figure 17 that the instantaneous optimized energy management strategy has a faster rate of lowering the Soc in the pre-simulation battery, uses more power, and lowers the energy consumption cost, which has obvious cost advantages compared with the global optimized energy management strategy. When the battery Soc drops to around 0.3, its value is balanced, and the energy consumption cost increases significantly and gradually exceeds the energy consumption cost under the global optimized energy management strategy.

5. Conclusions

In this study, a new type of oil–electric–hydraulic hybrid power system is examined as the research object, and a driving mode based on hydraulic energy and electric energy is selected. A logic threshold energy management strategy based on the optimal working curve is proposed, and then the linear weight method is adopted. The multi-objective function, which aims at the energy consumption cost and the manufacturing cost of the whole vehicle power system, is converted into a single objective function, the optimization variables are selected, and the constraints are set. The genetic algorithm is used to optimize the energy management strategy parameters and power system components. The optimized power system parameters can meet the power performance requirements of the vehicle.

Aiming at managing energy when the vehicle is under driving condition, a real-time energy management strategy based on the lowest instantaneous energy consumption cost is proposed. The strategy uses the instantaneous energy consumption cost in the single power source driving mode or the hybrid driving mode as the objective function and utilizes the grid. The ergodic method solves the target values of different vehicle demand torques and vehicle speeds to form a MAP table for real-time control. For braking conditions, based on the braking force distribution strategy and ECE regulations for traditional four-wheel-drive vehicles, a braking force distribution strategy based on the highest energy recovery is proposed. The simulation results show that the energy management strategy proposed in this article can achieve reasonable distribution of torque and achieve the expected control effect, and saves about 32.14% compared with the fuel consumption cost of the original model 100 km 8 L.

Simulation analysis of global optimization energy management strategy based on dynamic programming is performed, and the results prove that this strategy can be used as the basis for evaluating other strategies. The simulation comparisons under the NEDC working conditions show that the energy-saving effect of the real-time energy management strategy based on the minimum instantaneous energy consumption cost is similar to that of the global optimized energy management strategy.

Author Contributions: Y.Y. designed the oil–electric–hydraulic system and proposed the energy management control strategies; Z.Z. conducted model building, calculation, and analysis based on proposed control strategies; F.W. and C.F. matched the parameters of the oil–electric–hydraulic system and analyzed system performance. J.L. helped Z.Z. verify the energy management control strategies, and organized the manuscript format. All authors have read and agreed to the published version of the manuscript.

Funding: This research was funded by the National Key R&D Program of China (Grant No. 2018YFB0106100) and the National Natural Science Foundation of China (Grant No. 51575063).

Conflicts of Interest: The authors declare no conflict of interest.

References

1. Zhou, H.C.; Xu, Z.P.; Liu, L.; Liu, D.; Zhang, L.L. A Rule-Based Energy Management Strategy Based on Dynamic Programming for Hydraulic Hybrid Vehicles. *Rev. Math. Probl. Eng.* 2018. [CrossRef]
2. Li, J.Q.; Fu, Z.J.; Jin, X. Rule based energy management strategy for a battery/ultra-capacitor hybrid energy storage system optimized by pseudospectral method. In *Review of 8th International Conference on Applied Energy (Icae2016), Beijing Inst Technol, Beijing, PEOPLES R CHINA, OCT 08–11, 2016*; Yan, J., Sun, F., Chou, S.K., Desideri, U., Li, H., Campana, P., Xiong, R., Eds.; ELSEVIER SCIENCE BV: Amsterdam, The Netherlands, 2017; Volume 105, pp. 2705–2711.

3.	Liu, Y.G.; Xie, Q.B.; Qin, D.T. Energy Management Strategy Optimization of Hybrid Electric Vehicle Based on Working Condition Identification. *Mech. Transm.* **2016**, *40*, 64–69. [CrossRef]

4.	Yin, C.F.; Wang, S.H.; Yu, C.Q.; Li, J.X.; Zhang, S. Fuzzy optimization of energy management for power split hybrid electric vehicle based on particle swarm optimization algorithm. *Rev. Adv. Mech. Eng.* **2019**, *11*, 1–12. [CrossRef]

5.	Jiao, X.H.; Li, Y.; Xu, F.G.; Jing, Y. Real-time energy management based on ECMS with stochastic optimized adaptive equivalence factor for HEVs. *Rev. Cogent Eng.* **2018**, 5. [CrossRef]

6.	Zhang, F.Q.; Xi, J.Q.; Langari, R. An Adaptive Equivalent Consumption Minimization Strategy for Parallel Hybrid Electric Vehicle Based on Fuzzy PI. In *Review of 2016 IEEE Intelligent Vehicles Symposium (Iv), Gothenburg, SWEDEN, JUN 19–22 2016*; IEEE: New York, NY, USA, 2016; pp. 460–465.

7.	Wang, Y.Q.; Wu, Z.; Chen, Y.Y.; Xia, A.Y.; Guo, C.; Tang, Z.Y. Research on energy optimization control strategy of the hybrid electric vehicle based on Pontryagin's minimum principle. *Rev. Comput. Electr. Eng.* **2018**, *72*, 203–213. [CrossRef]

8.	Zhu, X. Research on Energy Management Strategy of Parallel Hybrid Electric Vehicle Based on Dynamic Programming. Master's Thesis, Hefei University of Technology, Hefei, China, 2017.

9.	Wang, X.M.; He, H.W.; Sun, F.C.; Zhang, J.L. Application Study on the Dynamic Programming Algorithm for Energy Management of Plug-in Hybrid Electric Vehicles. *Rev. Energ.* **2015**, *8*, 3225–3244. [CrossRef]

10.	Xu, F. Development of plug-in hybrid electric vehicle. Master's Thesis, Jilin University, Jilin, China, 2013.

11.	Zeng, Y.P. Parameter optimization of plug-in hybrid electric vehicle based on quantum genetic algorithm. *Rev. Clust. Comput. J. Netw. Softw. Tools Appl.* **2019**, *22* (Suppl. S6), 14835–14843. [CrossRef]

12.	Chen, Z.; Zhou, L.Y.; Sun, Y.; Ma, Z.L.; Han, Z.Q. Multi-objective parameter optimization for a single-shaft series-parallel plug-in hybrid electric bus using genetic algorithm. *Rev. Sci. China Technol. Sci.* **2016**, *59*, 1176–1185. [CrossRef]

13.	Li, P.X. Research on Vehicle Electro-hydraulic Hybrid Transmission System. Master's Thesis, Chongqing University, Chongqing, China, 2016.

14.	Wang, C. Coordinated control strategy for plug-in hybrid vehicle braking mode switching. Master's Thesis, Chongqing University, Chongqing, China, 2018.

Article

Efficiency Optimization and Control Strategy of Regenerative Braking System with Dual Motor

Yang Yang [1,2,*], Qiang He [2], Yongzheng Chen [2] and Chunyun Fu [1,2]

1 State Key Laboratory of Mechanical Transmission, Chongqing University, Chongqing 400044, China; fuchunyun@cqu.edu.cn
2 School of Automotive Engineering, Chongqing University, Chongqing 400044, China; heqiangcqdx@163.com (Q.H.); cyz19930106@163.com (Y.C.)
* Correspondence: yangyang@cqu.edu.cn; Tel.: +86-136-0831-1819

Received: 10 January 2020; Accepted: 5 February 2020; Published: 6 February 2020

Abstract: The regenerative braking system of electric vehicles can not only achieve the task of braking but also recover the braking energy. However, due to the lack of in-depth analysis of the energy loss mechanism in electric braking, the energy cannot be fully recovered. In this study, the energy recovery problem of regenerative braking using the independent front axle and rear axle motor drive system is investigated. The accurate motor model is established, and various losses are analyzed. Based on the principle of minimum losses, the motor control strategy is designed. Furthermore, the power flow characteristics in electric braking are analyzed, and the optimal continuously variable transmission (CVT) speed ratio under different working conditions is obtained through optimization. To understand the potential of dual-motor energy recovery, a regenerative braking control strategy is proposed by optimizing the dynamic distribution coefficient of the dual-electric mechanism and considering the restrictions of regulations and the I curve. The simulation results under typical operating conditions and the New York City Cycle (NYCC) proposed conditions indicate that the improved strategy has higher joint efficiency. The energy recovery rate of the proposed strategy is increased by 1.18% in comparison with the typical braking strategy.

Keywords: electric vehicle; dual-motor energy recovery; regenerative braking system; CVT speed ratio control; motor minimum loss; energy consumption and efficiency characteristics; braking force distribution

1. Introduction

Given the limitations of oil resources and the importance of environmental protection, governments around the world have enacted stringent regulations on fuel consumption and emissions. Electric vehicles, as environmentally friendly vehicles, have attracted a considerable amount of attention from researchers and corporations, and regenerative braking technology as one of the key technologies of energy conservation and emission reduction has been widely studied and applied [1–3]. The regenerative braking system can use the motor to convert the braking kinetic energy into electric energy and store it in the battery. This electric energy can be released during the driving process, which can not only improve the energy utilization rate and extend the driving range but also reduce the driver's range anxiety. Therefore, maximization of the braking energy recovery under safe braking conditions has been the focus and challenge of energy management of electric vehicles.

Zhang et al. proposed an improved regenerative braking control strategy for rear-drive electric vehicles. In the deceleration braking test, the improved regenerative braking efficiency could reach 47% [4]. Cheng et al. verified a new series control strategy, and the experimental results confirmed that the steady and dynamic contribution of the strategy to the improvement of energy efficiency reached 58.56% and 69.74%, respectively [5]. Itani et al. compared flywheels with supercapacitors as the second energy source of front axle driven electric vehicles, and the results demonstrated that ultra-capacitors performed better in weight, specific energy and specific power. It was more convenient to reuse the braking energy and provided a solution to reduce the damage of the large current to batteries during regenerative braking [6]. For the control of specific components, Yuan et al. proposed a new scheme of the line control dynamic system considering the functional requirements of regenerative braking in the structural development stage and adopted the current amplitude modulation control to improve the accuracy of hydraulic regulation and eliminate vibration noise. The maximum regeneration efficiency of the bench test was 46.32% of the total recoverable energy [7]. Chen proposed a feedback hierarchical controller that tracked the desired speed and distributed the braking torque to four wheels to improve the energy recovery [8]. In terms of overall optimization, Deng et al. analyzed the relationship between the battery, motor, CVT and comprehensive efficiency, and proposed a regenerative braking control strategy for the CVT hybrid electric vehicle. In comparison with the typical strategy, the average power generation efficiency of the motor increased by 2.91% [9]. Shu et al. developed a maximum energy recovery energy management strategy and used the sequential quadratic programming (SQP) algorithm to optimize the CVT ratio control strategy, which achieved a good control effect [10]. To expand the scope of braking energy recovery, Bera et al. used the motor and hydraulic system to jointly adjust the braking process of an anti-lock braking system (ABS) and obtained a good effect [11]. The above literatures have all conducted relevant studies on the improvement of energy recovery in the regenerative braking process, which has improved the regenerative braking performance of vehicles. However, there are a greater number of studies on a single model than on a joint model and more studies on regenerative braking of a single motor than on regenerative braking of vehicles with a dual-motor drive system.

As a key device of the regenerative braking system, the efficiency of the motor directly affects energy recovery. Hence, improving the efficiency of the motor is conducive to the increase of energy recovery. Many scholars have conducted relevant studies on improving motor efficiency. Tripathi et al. conducted a detailed study on the model-based loss minimization algorithm (MLMA), and the results confirmed that this method could not only effectively improve motor efficiency but also exhibit good dynamic performance [12]. Uddin et al. used a model-based loss minimization algorithm (LMA) to compare the efficiency of permanent magnet synchronous motors based on direct torque flux linkage control (DTFC) and vector control (VC). The simulation results showed that the former had higher efficiency [13]. Inoue et al. studied the control performance of the permanent magnet synchronous motor (PMSM) drive system based on current control and direct torque control. Their results showed that the latter, combined with the control law of the M-T framework, had the advantages of control stability [14]. Wang et al. introduced the integral balance of the sine value of the torque angle such that the speed and the electromagnetic torque could be controlled to converge at the same time by adjusting the speed only once and to obtain the optimal dynamic response of the speed [15]. Vido and Le Ballois [16] and Lee et al. [17] also conducted relevant studies and improved the efficiency of the motor to a certain extent. The above literatures have conducted research on the efficiency of the motor and obtained various results. However, in the process of regenerative braking, it is necessary to analyze the influencing factors of motor loss to maximize system efficiency.

To minimize the power loss in the process of electric braking, this study analyses the automobile with an independent motor drive system of the front and rear axles. First, the accurate motor model is established, and various losses are analyzed. Based on the principle of minimum loss, the motor control strategy is designed. The characteristics of power flow in the electric braking process are analyzed, and the combined efficiency model of the front and rear axles is established. The optimal transmission ratio of CVT under different working conditions is obtained through optimization, and the input and output characteristics of the front and rear axles are analyzed. Finally, by optimizing the brake force distribution coefficient of the front and rear motors and considering the ECE regulations and I curve as the limit, a new control strategy of dual motor regenerative braking is proposed to maximize the energy recovery.

2. Hybrid Electric Vehicle System Structure and Parameters

In comparison with pure electric vehicles and fuel cell vehicles, hybrid electric vehicles are widely used in production and by consumers without the disadvantages of short driving range, long charging time, high fuel cell price and difficult hydrogen re-filling [18]. The structural schematic diagram of the hybrid vehicle system studied here is shown in Figure 1, and it should be noted that the schematic diagram is not the layout of the real vehicle.

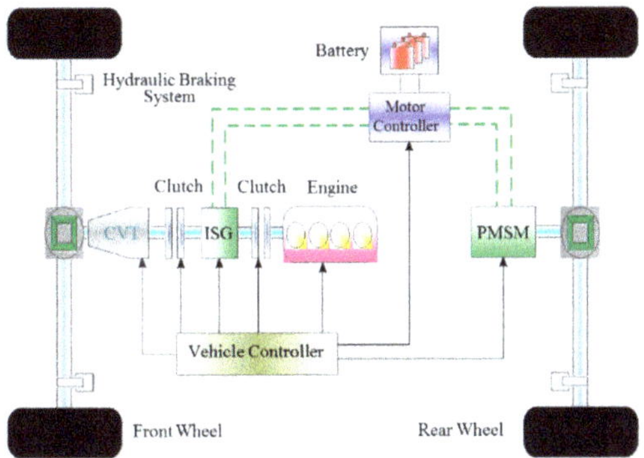

Figure 1. Schematic diagram of dual motor hybrid electric vehicle. ISG—Integrated Starter Generator.

This configuration can be driven by the engine alone or by the motor. During high power demand, the motor and the engine can work simultaneously to meet the needs of the vehicle. The front axle and rear axle of this configuration have motors, which can make the vehicle exhibit better dynamic performance in pure electric mode and can recover more energy when braking. The vehicle controller is responsible for collecting the speed, brake pedal, brake master cylinder pressure and other signals and corresponding responses. When the brake pedal signal is detected, the driving state of the car is quickly determined, and the control signal is sent to the lower controller through the controller area network (CAN) bus. The lower controller makes correlation identification according to the control signal and sends signals to the hydraulic control unit and motor control unit according to the established algorithm to complete the driver's instructions. The vehicle parameters and component parameters are shown in Table 1.

Table 1. Vehicle data and main components parameters.

Name	Description	Value
Vehicle	Curb weight	1800 kg
	Windward area	2.5 m^2
	Wheel radius	0.335 m
	Wheelbase	2.7 m
The ISG motor	Peak power	28 kW
	Rated power	14 kW
	Maximum torque	89.13 N·m
	Number of pole pairs	6
	Armature resistance	0.017 ohm
	d/q axis inductance	0.00021 H
	Magnet flux linkage	0.037 Wb
	iron losses resistance	0.008 w + 1.8 ohm

Table 1. *Cont.*

Name	Description	Value
Rear axle motor	Peak power	27 kW
	Rated power	13.5 kW
	Maximum torque	171.9 N·m
	Pole of pairs	8
	Armature resistance	0.012 ohm
	d/q axis inductance	0.00012 H
	Magnet flux linkage	0.042 Wb
	Iron losses resistance	0.011 w + 1.9 ohm
Lithium-ion battery pack	Rated capacity	38.43 Ah
CVT	Speed ratio range	[0.4, 2.5]

3. Motor Loss Model and Control Strategy

3.1. Motor Loss Model

Permanent magnet synchronous motors with the high-power density and high-efficiency advantages of small volume and light quality have been widely used in new energy vehicles [19]. To obtain a more accurate model, it must be considered that the iron loss in the model is important. Hence, the equivalent iron loss resistance is introduced parallel to the magnetizing branch in the circuit [20], as depicted in Figure 2. Certain idealized conditions are assumed; for example, saturation is ignored, and the electromotive force is sinusoidal [18]. Motor losses mainly include mechanical losses, copper losses, iron losses and stray losses. Since stray losses are difficult to measure and control and account for a small percentage of the total loss [21], they are not considered in this study.

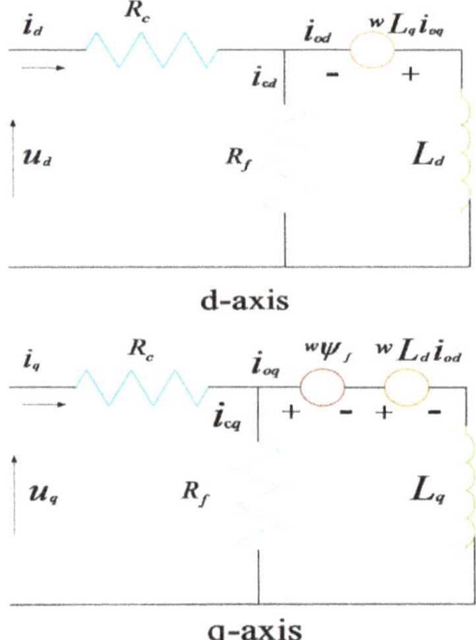

Figure 2. d-q axes equivalent circuits for the PMSM model with iron losses.

In steady-state, the voltage balance equation of the d-q axis is as follows:

$$u_d = R_c i_d - \omega \psi_q \tag{1}$$

$$u_q = R_c i_q + \omega \psi_d \tag{2}$$

Here, R_c is the stator winding resistance, u_q and u_d are the d-q axis components of the stator voltage, i_d and i_q are the d-axis and q-axis current components, respectively, ω is the angular velocity of the stator, and ψ_d and ψ_q are the d-q axis components of the stator flux, respectively. The permanent magnet flux ψ_a has the following relationship:

$$\psi_d = L_d i_{od} + \psi_a \tag{3}$$

$$\psi_q = L_q i_{oq} \tag{4}$$

The electromagnetic torque can be calculated using Equation (5).

$$T_e = \frac{3}{2} p \left(\psi_d i_{oq} - \psi_q i_{od} \right) = \frac{3}{2} p \left[\psi_a i_{oq} + \left(L_d - L_q \right) i_{oq} i_{od} \right] \tag{5}$$

where p is the number of pole pairs, i_{od} and i_{oq} are the d-q axis magnetization current components, respectively, and L_d and L_q are the d-q axis inductance components, respectively. For surface-mounted permanent magnet synchronous motors, $L_d = L_q$.

Then, copper loss and iron loss can be calculated by the Equations (6) and (7), respectively.

$$P_{cu} = \frac{3}{2} R_c \left[\left(i_{od} - \frac{\omega L_q i_{oq}}{R_f} \right)^2 + \left[i_{oq} + \frac{\omega(\psi_a + L_d i_{od})}{R_f} \right]^2 \right] \tag{6}$$

$$P_{Fe} = \frac{3}{2} R_f \left(i_{cd}^2 + i_{cq}^2 \right) = \frac{3}{2} R_f \left[\left(-\frac{\omega L_q i_{oq}}{R_f} \right)^2 + \left(\frac{\omega(\psi_a + i_{od} L_d)}{R_f} \right)^2 \right] \tag{7}$$

When the motor is working, the load and power factor are the key factors influencing the size of the copper loss. Therefore, when the current speed and torque are given, the current optimal i_{od} (minimum loss) can be obtained:

$$i_{od} = -\frac{\psi_a L_d \omega^2 \left(R_c + R_f \right)}{R_c R_f^2 + \omega^2 L_d^2 \left(R_c + R_f \right)} \tag{8}$$

Mechanical loss has an approximately linear relationship with motor speed [22]. By setting the value of K as constant, the mechanical loss model of permanent magnet synchronous motor can be obtained:

$$P_M = Kn \tag{9}$$

3.2. Motor Control Strategy

The motor control method has an important influence on motor performance. Hence, it is necessary to improve it to get higher motor efficiency. In comparison with most conventional proportional-integral-derivative (PID) control method, to obtain better performance and reduce energy losses here, the motor speed loop adopts the sliding mode control and the current loop uses the minimum loss control method (LMA). The total efficiency of the former motor for η_{isg} is set as

$$\eta_{isg}\left(i_{oq}\right) = \frac{T_e \omega}{T_e \omega + P_{cu} + P_{Fe} + P_M} = \frac{\frac{3}{2} p \psi_a i_{oq} \omega}{\frac{3}{2} p \psi_a i_{oq} \omega + P_{cu}\left(i_{oq}\right) + P_{Fe}\left(i_{oq}\right) + P_M} \tag{10}$$

It can be observed that the total efficiency is a quadratic function of the stator q-axis excitation current. By using the mathematical method, it is observed that there is always a value of i_{oq}, which can minimize the total loss under different torque and electric angular speed. When the motor loss is set to be the lowest, it is as follows:

$$\gamma = \frac{\partial P_{isg\_loss}}{\partial i_{oq}} \frac{\partial T}{\partial i_{od}} - \frac{\partial P_{isg_{loss}}}{\partial i_{od}} \frac{\partial T}{\partial i_{oq}} = 0 \tag{11}$$

The constraints are

$$\begin{cases} w_1 = T = \frac{3}{2}p\left[\psi_a i_{oq} + \left(L_d - L_q\right)i_{oq}i_{od}\right] \\ w_2 = \gamma = \frac{\partial P_{isg\_loss}}{\partial i_{oq}} \frac{\partial T}{\partial i_{od}} - \frac{\partial P_{isg_{loss}}}{\partial i_{od}} \frac{\partial T}{\partial i_{oq}} \end{cases} \tag{12}$$

The voltage state equation of the d-q axis can be obtained by calculating the time derivative of each side of the loss constraint as follows:

$$\begin{pmatrix} \dot{w}_1 \\ \dot{w}_2 \end{pmatrix} = \begin{pmatrix} X_{11} & X_{12} \\ X_{21} & X_{22} \end{pmatrix}\begin{pmatrix} \overline{U_d} \\ \overline{U_q} \end{pmatrix} + \begin{pmatrix} Y_1 \\ Y_2 \end{pmatrix} \tag{13}$$

The elements X_{11}, X_{12}, X_{21}, X_{22}, Y_1 and Y_2 are, respectively:

$$X_{11} = \frac{3PR_c\left(L_d - L_q\right)i_{oq}}{2L_d(R_s - R_c)} \tag{14}$$

$$X_{12} = \frac{3pR_c\left[\psi_a i_{oq} + \left(L_d - L_q\right)i_{oq}i_{od}\right]}{2L_d(R_s + R_c)} \tag{15}$$

$$X_{21} = \frac{9P}{2L_dR_c}\left[2\left(\frac{R_fR_c^2}{R_f + R_c} + L_q^2\omega^2\right)\left(L_d - L_q\right)i_{od} + \frac{R_fR_c^2}{R_f + R_c}\psi_a + L_d\left(2L_d - L_q\right)\psi_a\omega^2\right] \tag{16}$$

$$X_{22} = \frac{9P\left(L_d - L_q\right)}{2L_dR_c}\left(\frac{R_fR_c^2}{R_f + R_c} + L_q^2\omega^2\right)i_{oq} \tag{17}$$

$$Y_1 = \frac{3P}{2L_dR_c}\left[\begin{array}{c} -\frac{R_fR_c\left(L_d^2-L_q^2\right)}{L_dL_q\left(R_fR_c\right)}i_{oq}i_{od} - \frac{R_fR_c}{R_f+R_c}\psi_a i_{oq} \\ +\frac{\left(L_dL_q\right)\left(L_d^2i_{oq}^2-L_q^2i_{od}^2\right)\omega}{i_d i_q} - \frac{\psi_a\omega\left(\psi_a+2L_d i_{oq}-L_d i_{od}\right)}{L_q} \end{array}\right] \tag{18}$$

$$\begin{aligned} Y_2 = X_{21}\left[-R_f i_{od} + \frac{L_q\omega i_{oq}\left(R_f+R_c\right)}{R_c}\right] \\ + X_{11}\left[-R_f i_{od} - \frac{L_d\omega i_{oq}\left(R_f+R_c\right)}{R_c} - \frac{\omega\psi_a\left(R_f+R_c\right)}{R_c}\right] \end{aligned} \tag{19}$$

If all the above influential elements depend on the motor parameters and state, assuming that the X and Y elements meet the braking requirements, the output equation of the controller can be expressed as follows:

$$\begin{pmatrix} \overline{U_d} \\ \overline{U_q} \end{pmatrix} = \frac{\begin{pmatrix} X_{22} & -X_{12} \\ -X_{21} & X_{11} \end{pmatrix}\begin{pmatrix} \dot{w}_1-Y_1 \\ \dot{w}_2-Y_2 \end{pmatrix}}{\begin{vmatrix} X_{11} & X_{12} \\ X_{21} & X_{22} \end{vmatrix}} \tag{20}$$

To obtain the stable torque closed-loop output, the PI (Proportional-Integral) algorithm is used as follows:

$$\begin{pmatrix} \dot{w}_1 \\ \dot{w}_2 \end{pmatrix} = \begin{pmatrix} K_{Pt}\Delta T + K_{It}\int \Delta T dt \\ -K_{Py}\Delta\gamma + K_{Iy}\int \Delta\gamma dt \end{pmatrix} \tag{21}$$

where $\Delta T = T^* - T$ and $\Delta\gamma = \gamma^* - \gamma$ can be brought into the above equation to get

$$U_d = \frac{\left(K_{Pt}\Delta T + K_{It}\int \Delta T dt - Y_1\right)X_{22} + \left(K_{P\gamma}\Delta\gamma + K_{I\gamma}\int \Delta\gamma dt - Y_2\right)X_{12}}{X_{11}X_{22} - X_{12}X_{21}} \tag{22}$$

$$U_q = \frac{-\left(-K_{P\gamma}\Delta\gamma + K_{I\gamma}\int \Delta\gamma dt - Y_1\right)X_{21} + \left(-K_{Pt}\Delta T - K_{It}\int \Delta T dt - Y_2\right)X_{11}}{X_{11}X_{22} - X_{12}X_{21}} \tag{23}$$

The optimal PI parameters can be obtained after multiple debugging. The overall control model of the motor is shown in Figure 3.

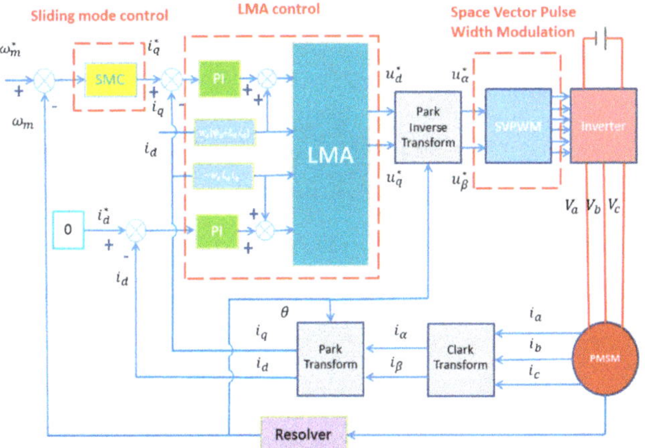

Figure 3. Minimum loss control model of motor.

The analysis shows that when the motor runs without load, the copper loss accounts for a small proportion, and the iron loss increases linearly with the increase of the speed. When the motor is loaded, the copper loss of the motor increases in square shape relative to the load torque, while the iron loss increases slowly. The efficiency of the former PMSM can be simulated in Simulink, as depicted in Figure 4.

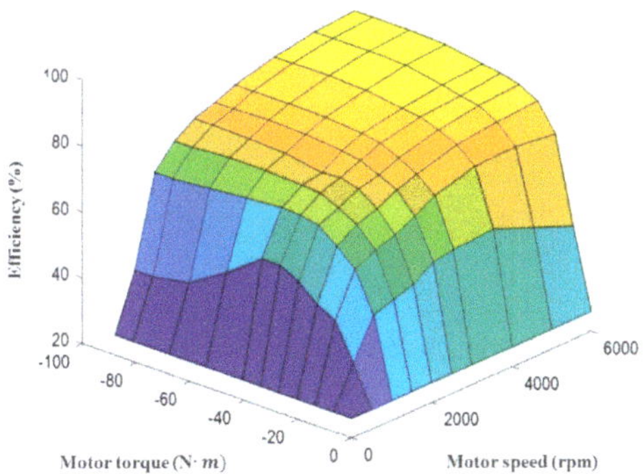

Figure 4. Efficiency map of the ISG.

4. Optimization of the Electric Braking Power Flow Efficiency and the Braking Force Distribution

4.1. Optimization of the Electric Braking Power Flow Efficiency

4.1.1. Power Flow Analysis of Electric Braking

To improve the recovery of braking energy, it is necessary to analyze the loss of power flow in the process of electric braking. Here, the electric braking system is mainly composed of a front and rear motor, battery pack, CVT transmission, clutch and other vehicle parameters. The parameters of each component are shown in Table 1. Furthermore, as both front and rear motors can participate in the process, it means that more energy can be recovered, and the driving range can be effectively increased. The power flow of the vehicle's electric braking loss is shown in Figure 5.

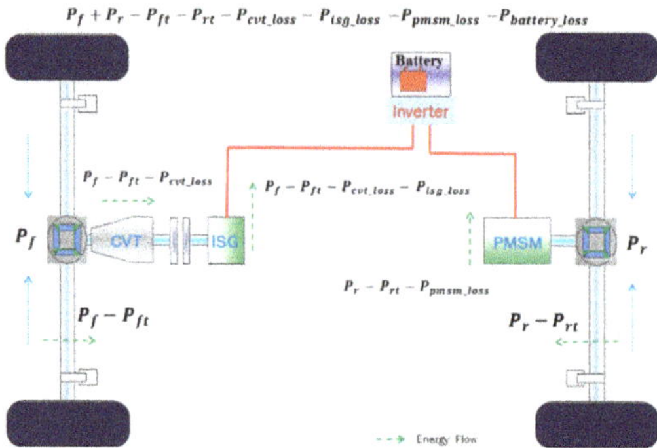

Figure 5. Schematic diagram of power loss during regenerative braking.

The overall efficiency of the vehicle's electric brake energy recovery can be calculated using Equation (24):

$$\eta = \frac{\left(P_f + P_r\right) - P_{ft} - P_{rt} - P_{isg\_loss} - P_{pmsm\_loss} - P_{cvt\_loss} - P_{battery\_loss}}{P_f + P_r} \tag{24}$$

Among them, η is the total efficiency of the electric brake of the whole vehicle, P_f and P_r are the front and rear axle braking input powers, respectively; P_{ft} and P_{rt} are the front and rear axle transmission power losses, respectively; P_{isq_loss} and P_{pmsm_loss} are the power loss of the front and rear motors, respectively; P_{cvt_loss} is the CVT transmission loss; $P_{battery\_loss}$ is the battery charging power loss.

As the power loss of the driving system is primarily related to the speed, whereas the loss of the motor and the inverter is a function of the electric angular speed and torque and the CVT loss is related to the input torque and the speed ratio, the equation can be rewritten as Equation (25).

$$\eta = \frac{\left[\beta T\omega_f + (1-\beta)T\omega_r\right] - Q_{f\_loss}(\beta, \omega_f, i) - Q_{r\_loss}(\beta, \omega_r) - Q_{t\_loss}(v) - Q_{b\_loss}(SOC)}{\beta T\omega_f + (1-\beta)T\omega_r} \tag{25}$$

where T is the braking torque of the vehicle; β is the distribution coefficient of forward and backward torque; ω_f and ω_r are the front and rear motor angular velocity, respectively; i is the CVT transmission speed ratio; v is the speed; $Q_{f\_loss}(\beta, \omega_f, i)$ is the CVT-ISG combined loss; $Q_{r\_loss}(\beta, \omega_r)$ is the loss of the rear motor; $Q_t(v)$ is the loss of the transmission system; and $Q_b(SOC)$ is the loss of the battery.

4.1.2. Establishment of the Joint Efficiency Model and Optimization of the CVT Speed Ratio

The combined front axle model is mainly composed of CVT and front motor and hence, its loss is calculated as Equation (26):

$$P_{f\_loss} = P_{cvt\_loss}(T_{cvt\_in}, \omega_{cvt\_in}) + P_{isg\_loss}(\omega_{cvt\_out}, T_{cvt\_out}) \tag{26}$$

P_{f_loss} is the total power loss of the front axle. T_{cvt_out} is the input torque of the front motor. ω_{cvt_in} and ω_{cvt_out} are CVT input and output speed, respectively. The torque loss of the CVT mainly includes the slip loss of the steel belt, the loss caused by the deformation of the belt wheel and the slip loss of the metal sheet [23]. When the speed is fixed at 2000 rpm, its efficiency changes, as shown in Figure 6. It can be observed that the efficiency of CVT is mainly related to the speed ratio. When the speed ratio is approximately 1, the efficiency reaches a maximum, but when it is less than 1, there is a significant decline in the efficiency.

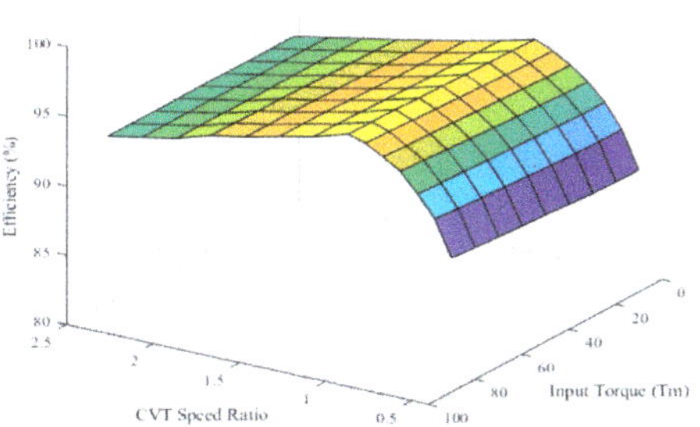

Figure 6. Transmission efficiency of the CVT.

Through the established CVT-ISG joint efficiency model, the CVT speed ratio with the highest joint efficiency can be determined. When the number ratio is 1.5, the joint efficiency changes are shown in Figure 7.

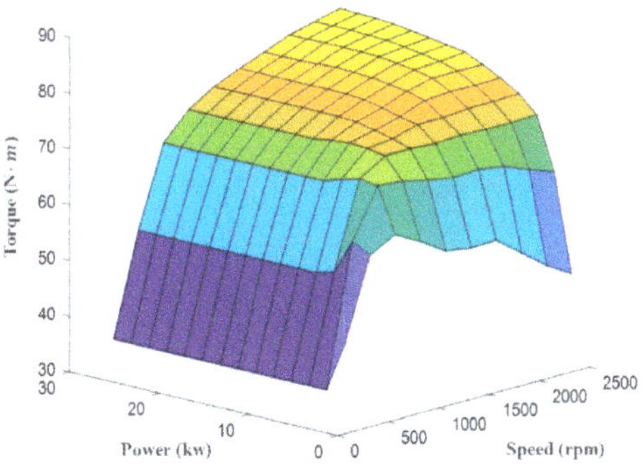

Figure 7. CVT–ISG motor combined efficiency at a speed ratio of 1.5.

It can be found that the efficiency of the combined model in the region with high speed and low torque is lower than that of the single motor model. Since the CVT has lower efficiency in the region with low torque, it results in lower overall efficiency. Further, under different torques and rotating speeds, the combined efficiency changes with the CVT speed ratio. By seeking the CVT speed ratio that makes the combined efficiency reach maximum, the system efficiency can be maximized. The results are shown in Figure 8.

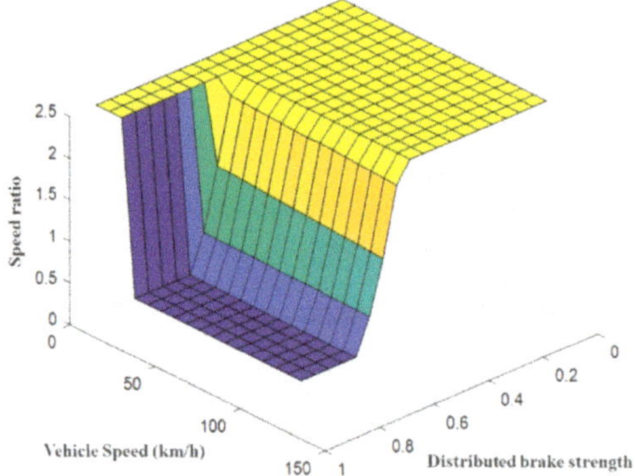

Figure 8. Optimal CVT ratio under different working conditions.

Therefore, by calculating the braking power of the motor through the pedal's opening degree, the optimal CVT speed ratio under this braking torque can be obtained under the combined efficiency model. However, it should be noted that when the vehicle starts, the motor speed should be set greater than 500 rpm, and the speed ratio should be adjusted to the maximum to protect the motor from irreversible damage. When the vehicle is in an emergency braking state ($z > 0.7$), the CVT speed ratio should be adjusted to the minimum to ensure the safety and stability of the vehicle.

The rear axle joint model is mainly composed of a motor, which is relatively simple and similar to the front axle motor model. Therefore, it is not to be introduced separately.

4.1.3. Input and Output Characteristics of the Front and Rear Axis Joint Models

According to the joint model established above, the input and output characteristics of the front and rear axles are analyzed to provide a basis for formulating the braking force distribution strategy.

The braking strength allocated by the front axle during simulation is set to 0.3, and the energy recovery and energy consumption rate of the front axle braking system at different initial velocities are depicted in Figure 9.

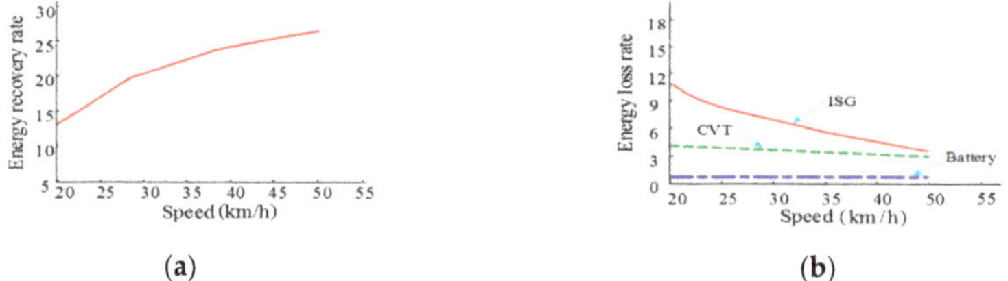

(a)   (b)

Figure 9. (**a**) Energy recovery rate and (**b**) energy loss rate of front axle braking system at different vehicle speeds.

It can be found that the higher the initial braking speed, the higher the energy recovery rate. This is because when the vehicle is at a higher speed, the motor is in an efficient working area, and the energy recovered is more than when it is at a lower speed. With the increase of the initial braking speed, the energy loss rate of the motor, CVT and battery decreases slowly, and the biggest loss is the motor loss. This indicates that the loss of the front axle is relatively small at higher speeds.

Under the same conditions, the characteristics of the rear axle joint model are depicted in Figure 10.

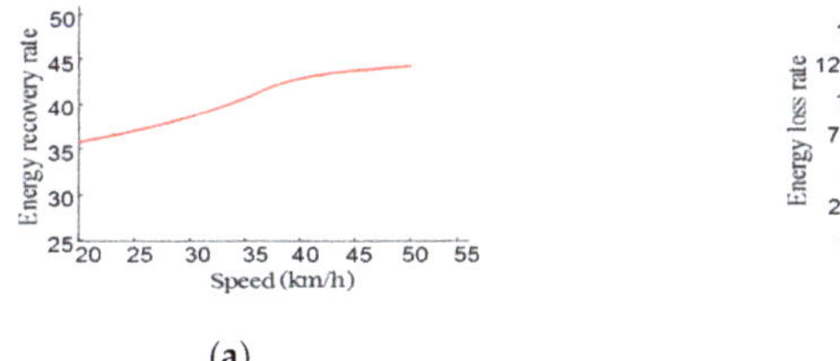
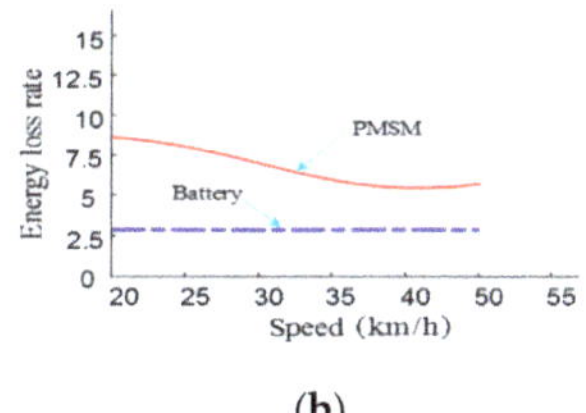

(a)   (b)

Figure 10. (**a**) Energy recovery rate and (**b**) energy loss rate of rear axle braking system at different vehicle speeds.

In comparison with the combined loss model of the front shaft, the recovery rate of the rear shaft is relatively higher, because there is no CVT to affect the efficiency of the motor, so the recovery rate is higher. Further, the loss rate of the rear shaft is lower than that of the front shaft, but the larger torque will cause the larger charging current of the battery, larger battery loss and lower charging efficiency.

4.2. The Braking Force Distribution Strategy with the Maximum Joint Efficiency

4.2.1. Front and Rear Motors Braking Force Distribution

Since the front and rear motors are different, it implies that the optimal operating range of the motor is different during the braking process. It is necessary to adjust the braking force of the front and rear motors to achieve a higher recovery rate. The utilization efficiency of regenerative braking of front and rear shafts is defined as follows:

$$\eta_{sys} = \frac{P_{inf}\eta_{inf} + P_{inr}\eta_{inr}}{P_{inf} + P_{inr}} \tag{27}$$

where P_{inf} and P_{inr} are the braking power of the front and rear shafts, respectively. η_{inf} and η_{inr} are the combined braking efficiency of front and rear axles, respectively. A biaxial regenerative braking model was established.

$$\text{Max } \eta_{sys} = \frac{P_{inf}\eta_{inf} + P_{inr}\eta_{inr}}{P_{inf} + P_{inr}} = \frac{T_{inf}\eta_{inf} + T_{inr}\eta_{inr}}{T_{inf} + T_{inr}} \tag{28}$$

$$s.t\begin{cases} 0 \leq P_{inf} \leq P_{inf\_max} \\ 0 \leq P_{inr} \leq P_{inr\_max} \\ 0 \leq T_{inf} \leq T_{inf\_max} \\ 0 \leq T_{inr} \leq T_{inr\_max} \\ q = \frac{T_{inr}}{T_{reg}} \\ P_{inr} = (1-q) \times P_{reg} \\ n_{inf} = f(T_{inf}, n_f) \\ n_{inr} = f(T_{inr}, n_r) \end{cases} \tag{29}$$

where q is the braking force distribution coefficient of the rear axle; P_{inf_max} and P_{inr_max} are, respectively, the maximum braking power that the front and rear axles can provide. P_{reg} is the total regenerative braking power. The distribution coefficient of the optimal posterior axis is calculated as shown in Figure 11.

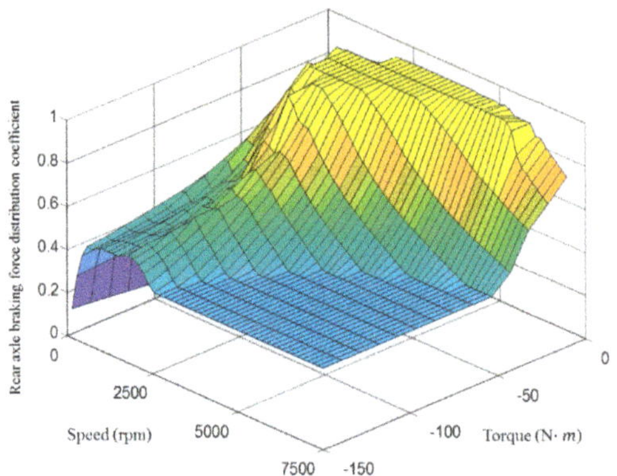

Figure 11. Optimal rear axle distribution coefficient.

It can be observed that the surrounding dark blue part is the separate working area of the front motor during regenerative braking, the middle bright yellow part is the separate working area of the rear motor during regenerative braking and the remaining part is the joint working area of the front and rear axle joint model.

4.2.2. Vehicle Braking Force Distribution Strategy

Based on the above analysis, the mode switching point of regenerative braking of the front and rear axles can be obtained by fitting the boundary of the separate working area of the rear axles. Thus, the relationship between braking torque and speed is

$$T_q(v) = 19.06 \cdot \cos\left(v \times 1.052 \times 10^{-3}\right) - 13.15 \sin\left(v \times 1.052 \times 10^{-3}\right) - 41.29 \tag{30}$$

As illustrated in Figure 12, when the regenerative braking torque of the vehicle is located in the envelope region of the curve and the coordinate axis, i.e., when $\left|T_q(v)\right| > |T_b|$, the rear axis is used for braking alone. When the braking torque is outside the curve, that is, $\left|T_q(v)\right| < |T_b|$, the braking force is allocated according to the p-value, and the peak power peak torque of the front and rear motors should be limited by the threshold value to prevent overload of the front and rear motors. Considering the braking stability and regulatory restrictions, the braking force distribution strategy is as follows:

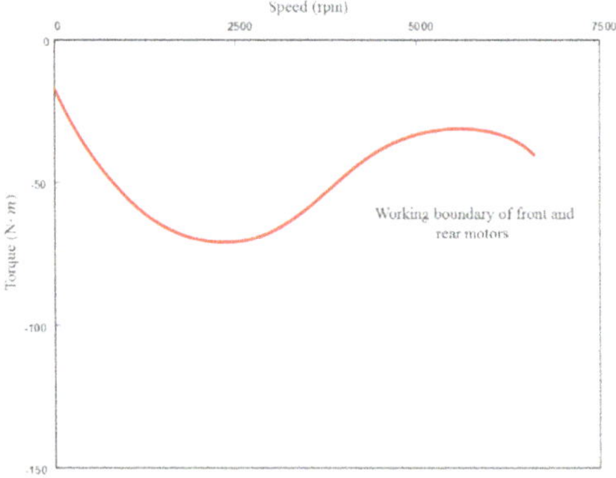

Figure 12. Front and rear axle braking force distribution coefficient boundary curve.

(1) When $z < 0.2$, the braking force is distributed by the distribution coefficient of the rear shaft.

① Braking torque $|T_b| \le |T_q(v)|$

$$\begin{cases} F_r = \frac{T_b}{r} \\ F_f = 0 \\ F_{\mu f} = 0 \\ F_{\mu r} = 0 \end{cases} \tag{31}$$

F_r is the rear axle braking force; F_f is the braking force of the front axle; $F_{\mu f}$ is the hydraulic braking force of the rear shaft; $F_{\mu r}$ is the hydraulic braking force of the front shaft; r is the vehicle radius; v is the speed of the vehicle. At this point, the braking force will be provided by the rear motor alone. The front motor and the hydraulic braking system of the front and rear shafts do not participate in the braking.

② Braking torque $|T_b| > |T_q(v)|$

$$\begin{cases} F_{regr} = q \times \frac{T_b}{r} \\ F_{regf} = (1-q) \times \frac{T_b}{r} \\ F_{\mu f} = 0 \\ F_{\mu r} = 0 \end{cases} \tag{32}$$

The braking force is distributed through the distribution coefficient q of the rear shaft. At this point, the front motor starts to participate in the regenerative braking, while the hydraulic braking system still does not participate in the braking.

When the braking strength is between 0.15 and 0.8, the Economic Commission of Europe (ECE) regulations stipulate that the curve of the rear axle using the adhesion coefficient should not be above the front axle. Hence, if the set distribution is reasonable, it should be considered here. According to the braking force distribution strategy in this study,

$$\beta = 1 - q \tag{33}$$

The relationship between the braking value and ECE braking regulations can be obtained [22], and the relation curve between the braking force distribution coefficient and the braking intensity z can be illustrated as shown in Figure 13. When the speed is 30 km/h and 100 km/h, it can be seen that the curve changes within the range permitted by regulations.

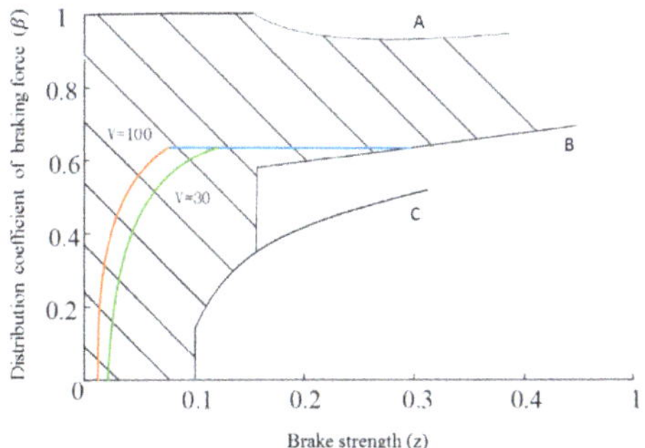

Figure 13. The relationship of the β and z when no-load.

The upper limit curve A is to ensure that the adhesion coefficient of the front axis meets the requirements. Curve B is to limit the locking order of the front and rear wheels of the car. When the β value appears above curve B, the front wheels can always be locked to the rear wheels in braking. However, when the β value is lower than the curve C, the adhesion coefficient of the rear axis will be insufficient and hence, the contact value should be kept above the curve C at all times.

(2) $0.2 < z \leq 0.5$

At this point, the braking force will be distributed according to the I curve. If the braking torque provided by the front and rear motors is insufficient to meet the braking task, the remaining braking power required will be supplemented by the hydraulic braking system.

$$
\begin{cases}
F_r = \frac{T_b \cdot i}{r} \\
F_f = mgz - F_r \\
F_{\mu r} = F_r - F_{regr} \\
F_{\mu f} = F_f - F_{regf}
\end{cases}
\tag{34}
$$

where i represents the braking force distribution coefficient under the I curve.

(3) $z > 0.5$

When the braking strength is greater than 0.5, the braking stability is most important. Therefore, reducing the braking force of the motor at a constant speed gradually withdraws the motor from the braking work. Simultaneously, the missing braking force is supplemented by the hydraulic pressing force to ensure that when $z = 0.7$, the motor completely exits the braking, without affecting the hydraulic pressure to provide the full braking force in case of emergency braking. The specific allocation strategy is shown in Figure 14.

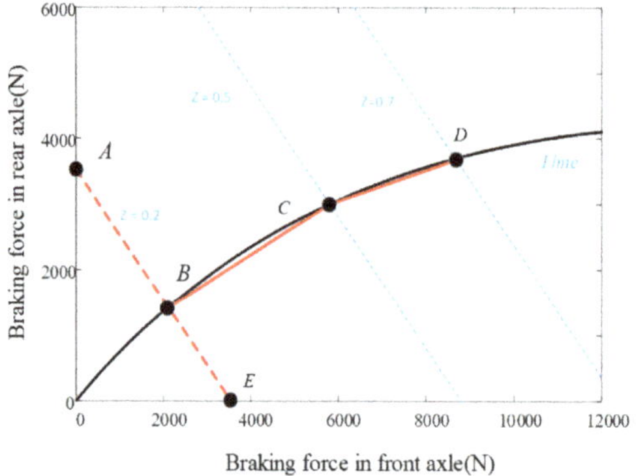

Figure 14. Braking force distribution diagram.

5. Vehicle Performance Simulation and Analysis

Based on the joint loss model, the simulation model of the whole system was established in Simulink/MATLAB, as shown in Figure 15. The simulation analysis was conducted under typical working conditions and cyclic working conditions, respectively, to verify the effectiveness of the strategy.

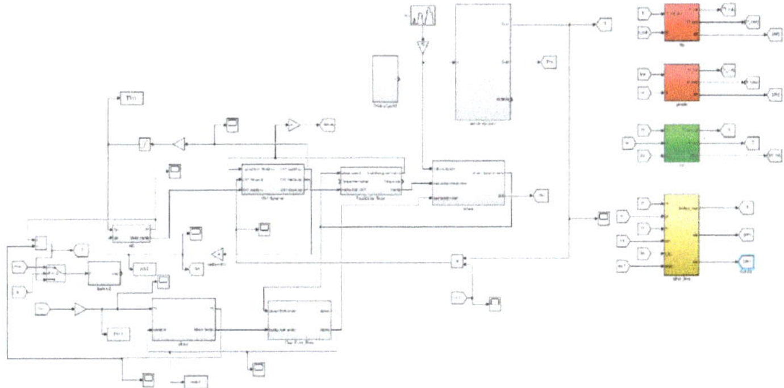

Figure 15. Vehicle simulation model.

5.1. Simulation of Typical Braking Conditions

The initial condition of the vehicle speed is 100 km/h and the SOC (State of charge) value of the power battery is 0.7. In addition, the influence of other resistances other than braking force, such as wind resistance, is not considered temporarily in the braking process. According to the analysis of power flow on the above analysis, the loss of each component in the braking process is made into an energy consumption diagram as shown in Figure 16.

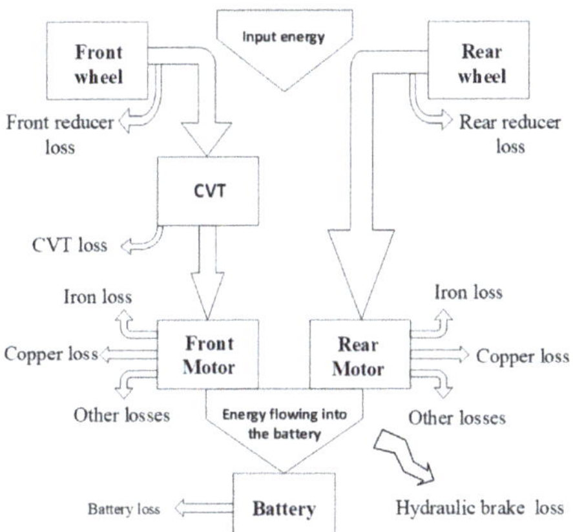

Figure 16. Brake energy flow diagram.

5.1.1. Braking Strength $z = 0.2$

When the braking strength is 0.2, the change in SOC and the overall efficiency during the entire process from the beginning of braking to the end are depicted in Figure 17a and the loss of key components is depicted in Figure 17b.

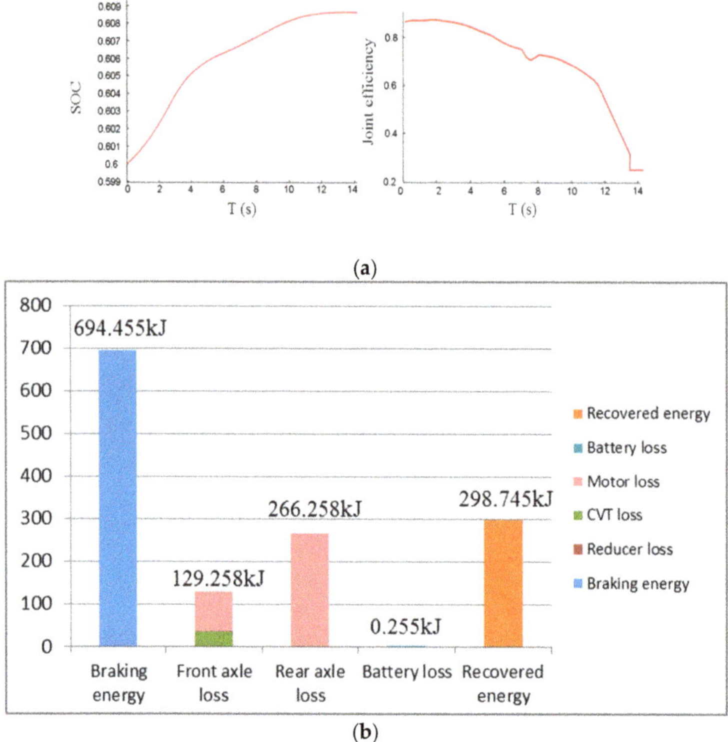

Figure 17. Simulation results when the $z = 0.2$. (**a**) Change in the SOC and joint efficiency; (**b**) Energy loss of the key components.

It can be found that at the initial time, the joint efficiency decreases slowly, the efficiency is higher, and the energy can be fully recovered. According to the data in the figure, at this time, the loss of braking energy mainly comes from the motor. Since the front motor has a short working time, the focus is on the rear motor, which is the same as the CVT loss. It can be seen that 298.745 kJ energy has been recovered from the driver stepping on the brake pedal to the vehicle parking, 395.71 kJ energy has been lost and the recovery rate has reached 43.02%.

5.1.2. Braking Strength $z = 0.4$

When the braking strength is 0.4, the change in SOC and the overall efficiency during the entire process from the beginning of braking to the end are depicted in Figure 18a and the loss of key components is depicted in Figure 18b.

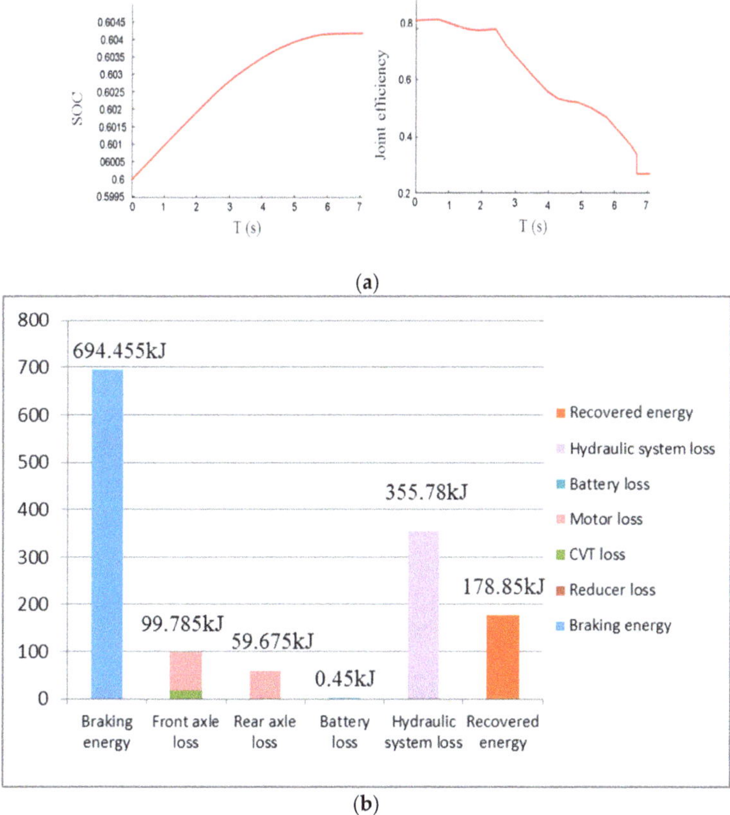

Figure 18. Simulation results when the $z = 0.4$ (**a**) Change in the SOC and joint efficiency (**b**) Energy loss of the key components.

When the braking starts, the regenerative braking efficiency of the dual motors has a short period of platform area, and the efficiency is relatively high. As the speed decreases, the electric braking efficiency decreases, while the mechanical braking proportion increases. According to the data in the figure, due to the addition of hydraulic braking, the energy loss of most regenerative braking is hydraulic braking loss accounting for 69.06%. Both the front and rear motors are in the peak operating state. The loss of the front motor is higher than that of the rear motor due to the CVT, and the loss of the front and rear motors is smaller than that of the rear motor when the braking strength is 0.2 because the motor has a shorter working state.

5.1.3. Braking Strength $z = 0.6$

When the braking strength is 0.6, the change in SOC and the overall efficiency during the entire process from the beginning of braking to the end are depicted in Figure 19a and the loss of key components is depicted in Figure 19b.

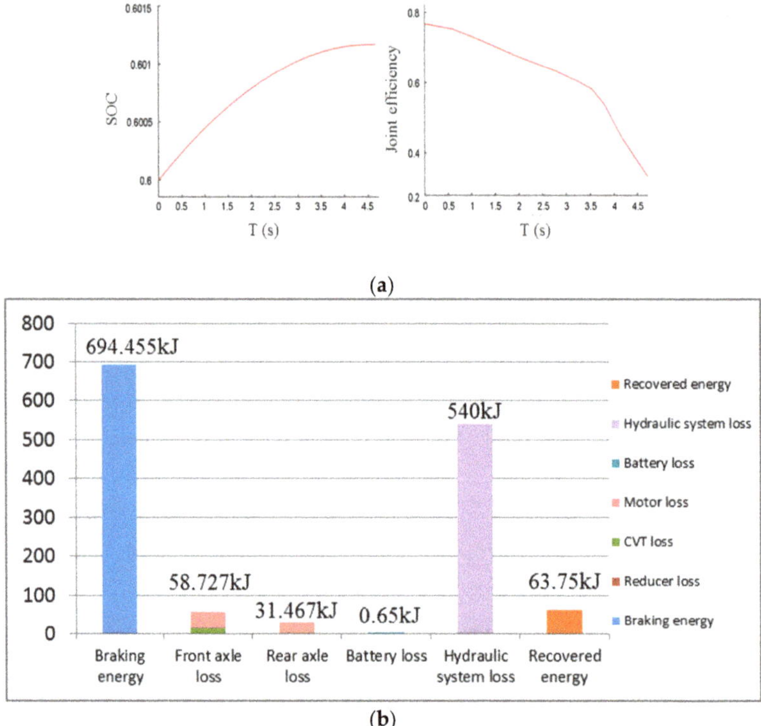

Figure 19. Simulation results when the $z = 0.6$. (**a**) Change in the SOC and joint efficiency; (**b**) energy loss of the key components.

It can be observed that at this point, due to the gradual withdrawal of the motor braking, the increase in SOC is not large. According to the data in the figure, the hydraulic braking loss accounts for a larger proportion, accounting for 85.64%. Furthermore, due to the short braking time, the overall loss of the front and rear motors decreases in comparison with the braking strength, and the CVT loss also decreases.

From the simulation of typical working conditions, it can be observed that despite the braking strength of 0.2, 0.4 or 0.6, the SOC increases to different degrees during the braking process, and the lower the braking strength and the longer the braking time under the same speed, the more energy will be recovered.

5.2. Cycle Simulation

To verify the distribution strategy in this study, NYCC was selected for cycle simulation, and the ideal braking force distribution method of motor first braking was compared. The braking torque, power, total system efficiency and SOC of the front and rear motors are analyzed. NYCC has the characteristics of low speed, high acceleration and frequent braking, and its braking environment and braking strength can be obtained as shown in Figure 20 [24].

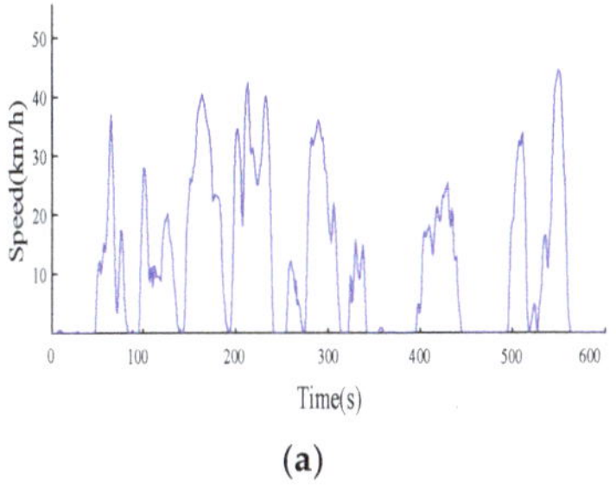 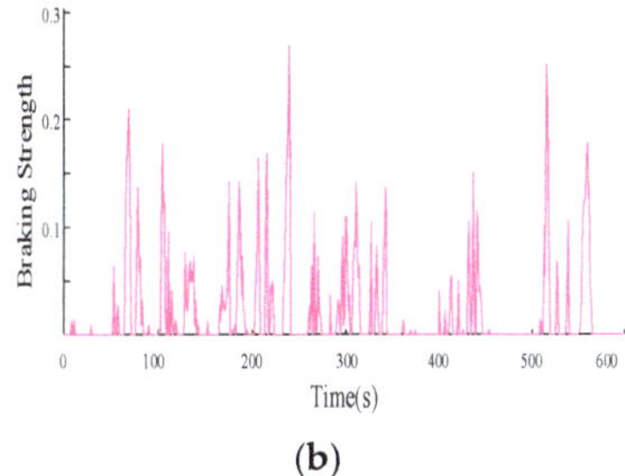

(a) **(b)**

Figure 20. NYCC (**a**) operating speed and (**b**) braking strength.

The braking torque changes of the front and rear motors are depicted in Figure 21. To recover energy more efficiently, the rear motors often work in the state of peak torque, whereas the front motors often work in the state below the rated torque, so as to not be involved in braking as frequently as the rear motor.

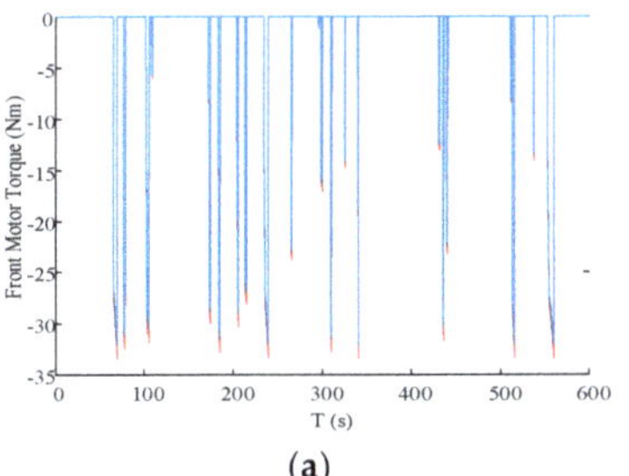 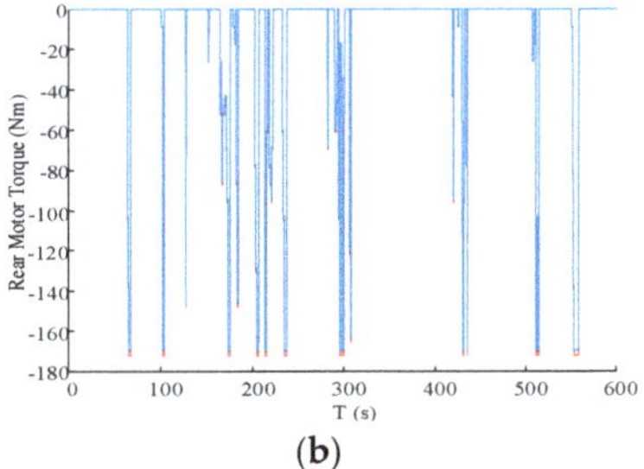

(a) **(b)**

Figure 21. The torque of (**a**) the front and (**b**) rear motors.

The braking power changes of the front and rear motors are shown in Figure 22. It can be seen that the maximum power of the front motor is approximately 6.7 kW and that of the rear motor is approximately 14.8 kW. The braking frequency of the rear motor is relatively large. In comparison with the ideal braking strategy, the front motor did not participate in the braking in the early stage and the rear motor braking power increased.

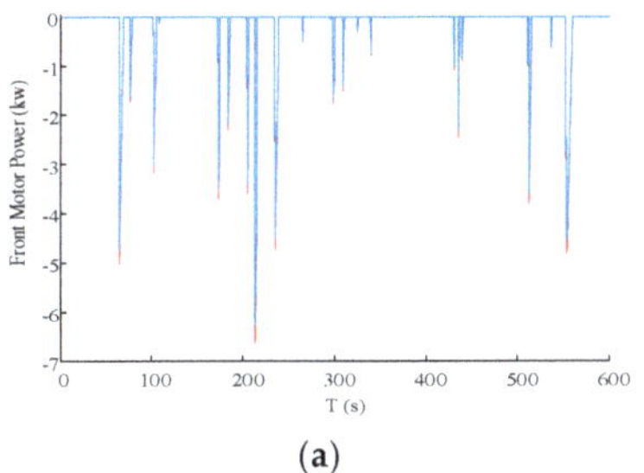 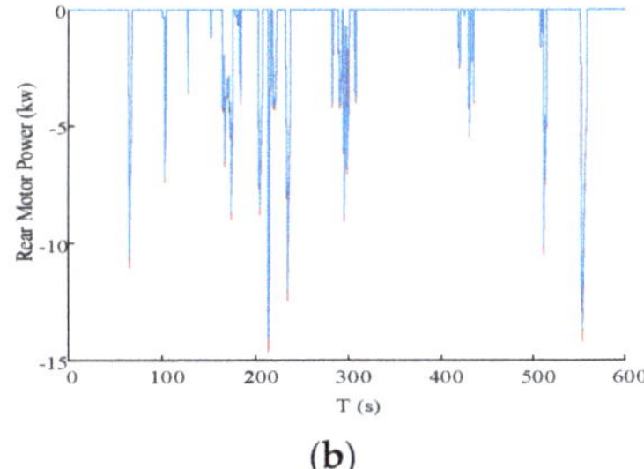

(a) **(b)**

Figure 22. The power of (**a**) the front and (**b**) rear motors.

The efficiency and SOC changes are depicted in Figure 23. The overall efficiency of the rear motor can reach approximately 0.8 when it works alone. As the selection of the CVT speed ratio can adjust the efficiency of the front motor, the overall efficiency of the front and rear motors is higher when they work together to effectively recover energy. After the complete working condition, the SOC rises by approximately 0.003.

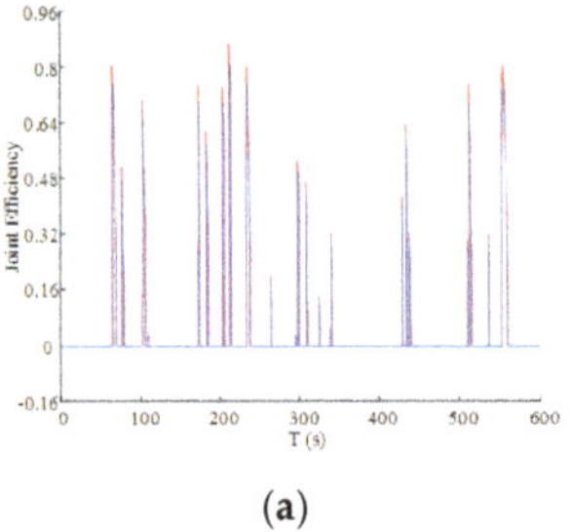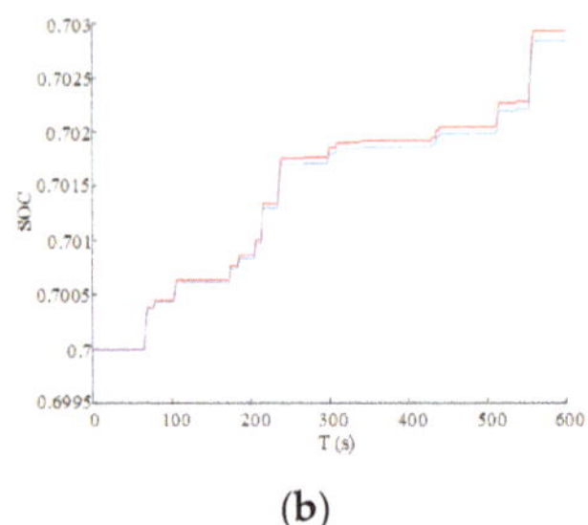

(a)   (b)

Figure 23. The changes in (**a**) joint efficiency and (**b**) SOC.

From the changes in efficiency and SOC, it can be observed that under the condition of low braking strength, the braking force distribution method discussed here has a high recovery efficiency and can recover maximum energy. Additionally, the motor loss minimization algorithm was adopted to maximize the use of the front and rear motors, system efficiency and SOC were improved, and the energy recovery of NYCC increased by 1.18% in comparison with the typical braking strategy.

6. Conclusions

(1) In this study, the front axle and rear axle independent motor drive system vehicle was considered as the research objective, the accurate motor model was established, various losses were analyzed and a new motor control method was proposed based on the principle of minimum loss.

(2) The characteristics of power flow in the process of electric braking were analyzed in detail, the combined efficiency model of the front axle (CVT–ISG) and rear axle (PMSM) was established, the braking force distribution of the front and rear motors was optimized based on the input and output characteristics of the front and rear axles. It was found that the optimal braking force distribution coefficient of the front and rear axles will change with the change of the working conditions. According to this change rule, a dual-motor regenerative braking force distribution strategy based on the optimal braking energy recovery was designed.

(3) In the MATLAB/Simulink simulation platform, the double motor regenerative braking system model was developed, and the simulation analyze was carried out under three typical braking conditions and NYCC conditions, respectively. It was observed that when the braking strength was 0.2, the braking energy recovery rate could reach 43.02%, and the energy recovery rate of the improved strategy was 1.18% higher than that of the typical braking strategy under NYCC conditions, which verify the effectiveness of the strategy proposed in this study.

Author Contributions: Conceived the research ideas and put forward the research methods, Y.Y.; Build the model and simulated it, Q.H. and Y.C.; Analyzed the simulation data, C.F. All authors have read and agreed to the published version of the manuscript.

Funding: The research was supported by (1) the National Key R&D Program of China (grant No. 2018YFB0106100) (2) the National Natural Science Foundation of China (grant No. 51575063). The authors would also like to acknowledge the support from the State Key Laboratory of Mechanical Transmission of Chongqing University, China.

Conflicts of Interest: The authors declare no conflict of interest.

References

1. Hannan, M.A.; Azidin, F.A.; Mohamed, A. Hybrid Electric Vehicles and Their Challenges: A Review. *Renew. Sustain. Energy Rev.* **2014**, *29*, 135–150. [CrossRef]

2. Khodaparastan, M.; Mohamed, A.A.; Brandauer, W. Recuperation of Regenerative Braking Energy in Electric Rail Transit Systems. *IEEE Trans. Intell. Transp.* **2019**, *20*, 2831–2847. [CrossRef]

3. Tang, X.L.; Zhang, D.J.; Liu, T.; Khajepour, A.; Yu, H.S.; Wang, H. Research on the Energy Control of a Dual-Motor Hybrid Vehicle during Engine Start-Stop Process. *Energy* **2019**, *166*, 1181–1193. [CrossRef]

4. Zhang, J.; Li, Y.; Lv, C.; Yuan, Y. New Regenerative Braking Control Strategy for Rear-Driven Electrified Minivans. *Energy Convers. Manag.* **2014**, *82*, 135–145.

5. Qiu, C.; Wang, G.; Meng, M.; Shen, Y.J.E. A Novel Control Strategy of Regenerative Braking System for Electric Vehicles under Safety Critical Driving Situations. *Energy* **2018**, *149*, 329–340. [CrossRef]

6. Itani, K.; De Bernardinis, A.; Khatir, Z.; Jammal, A. Comparative Analysis of Two Hybrid Energy Storage Systems Used in a Two Front Wheel Driven Electric Vehicle during Extreme Start-Up and Regenerative Braking Operations. *Energy Convers. Manag.* **2017**, *144*, 69–87. [CrossRef]

7. Yuan, Y.; Zhang, J.Z.; Li, Y.T.; Li, C. A Novel Regenerative Electrohydraulic Brake System: Development and Hardware-in-Loop Tests. *IEEE Trans. Veh. Technol.* **2018**, *67*, 11440–11452. [CrossRef]

8. Chen, J.; Yu, J.Z.; Zhang, K.X.; Ma, Y. Control of Regenerative Braking Systems for Four-Wheel-Independently-Actuated Electric Vehicles. *Mechatronics* **2018**, *50*, 394–401. [CrossRef]

9. Deng, T.; Lin, C.S.; Chen, B.; Ma, C.F. Regenerative Braking Control Strategy Based on Joint High-Efficiency Optimization. *Adv. Mech. Eng.* **2016**, *8*. [CrossRef]

10. Shu, H. Regenerative Braking Energy Management Strategy for Mild Hybrid Electric Vehicles. *J. Mech. Eng.* **2009**, *45*, 167–173. [CrossRef]

11. Bera, T.K.; Bhattacharya, K.; Samantaray, A.K. Bond Graph Model-Based Evaluation of a Sliding Mode Controller for a Combined Regenerative and Antilock Braking System. *Proc. Inst. Mech. Eng. I-J. Sys.* **2011**, *225*, 918–934. [CrossRef]

12. Tripathi, S.M.; Dutta, C. Enhanced Efficiency in Vector Control of a Surface-Mounted PMSM Drive. *J. Franklin Inst.* **2018**, *355*, 2392–2423. [CrossRef]

13. Uddin, M.N.; Zou, H.B. Comparison of DTFC and Vector Control Techniques for PMSM Drive with Loss Minimization Approach. In Proceedings of the 2014 IEEE 27th Canadian Conference on Electrical and Computer Engineering (CCECE), Toronto, ON, Canada, 4–7 May 2014.

14. Inoue, Y.; Morimoto, S.; Sanada, M. Comparative Study of PMSM Drive Systems Based on Current Control and Direct Torque Control in Flux-Weakening Control Region. *IEEE Trans. Ind. Appl.* **2012**, *48*, 2382–2389. [CrossRef]

15. Wang, Y.; Geng, L.; Hao, W.J.; Xiao, W.Y. Control Method for Optimal Dynamic Performance of DTC-Based PMSM Drives. *IEEE Trans. Energy Convers.* **2018**, *33*, 1285–1296. [CrossRef]

16. Vido, L.; Le Ballois, S. A Simple Method for Optimal Control of PMSM with Loss Minimization Including Copper Loss and Iron Loss. In Proceedings of the 2017 Twelfth International Conference on Ecological Vehicles and Renewable Energies (EVER), Monte Carlo, Monaco, 11–13 April 2017.

17. Lee, J.; Nam, K.; Choi, S.; Kwon, S. Loss-Minimizing Control of PMSM With the Use of Polynomial Approximations. *IEEE Trans. Power Electron.* **2009**, *24*, 1071–1082.

18. Estima, J.O.; Cardoso, A.J.M. Performance Analysis of a PMSM Drive for Hybrid Electric Vehicles. In Proceedings of the International Conference on Electrical Machines, Rome, Italy, 6–8 September 2010.

19. Estima, J.O.; Cardoso, A.J.M. Performance Evaluation of DTC-SVM Permanent Magnet Synchronous Motor Drives under Inverter Fault Conditions. In Proceedings of the 2009 35th Annual Conference of IEEE Industrial Electronics, Porto, Portugal, 3–5 November 2009.

20. Urasaki, N.; Senjyu, T.; Uezato, K. An Accurate Modeling for Permanent Magnet Synchronous Motor Drives. In Proceedings of the APEC 2000 Fifteenth Annual IEEE Applied Power Electronics Conference and Exposition (Cat. No. 00CH37058), New Orleans, LA, USA, 6–10 Feburary 2000.

21. Deng, W.W.; Zhao, Y.; Wu, J. Energy Efficiency Improvement via Bus Voltage Control of Inverter for Electric Vehicles. *IEEE Trans. Veh. Technol.* **2017**, *66*, 1063–1073. [CrossRef]

22. Xiong, H.Y.; Zhu, X.L.; Zhang, R.H. Energy Recovery Strategy Numerical Simulation for Dual Axle Drive Pure Electric Vehicle Based on Motor Loss Model and Big Data Calculation. *Complexity* **2018**, *2018*. [CrossRef]

23. Yang, Y.; He, X.L.; Zhang, Y.; Qin, D.T. Regenerative Braking Compensatory Control Strategy Considering CVT Power Loss for Hybrid Electric Vehicles. *Energies* **2018**, *11*, 497. [CrossRef]

24. Geller, B.M.; Bradley, T.H. Analyzing Drive Cycles for Hybrid Electric Vehicle Simulation and Optimization. *J. Mech. Des.* **2015**, *6328*, 137. [CrossRef]

Article

Energy Control Strategy of Fuel Cell Hybrid Electric Vehicle Based on Working Conditions Identification by Least Square Support Vector Machine

Yongliang Zheng [1], Feng He [1,*], Xinze Shen [1] and Xuesheng Jiang [2]

[1] Department of Automotive Engineering, School of Mechanical Engineering, Guizhou University, Guiyang 550025, China; gs.ylzheng17@gzu.edu.cn (Y.Z.); xz_shen_gzuedu@outlook.com (X.S.)
[2] Guizhou Changjiang Automobile Co., Ltd., Guiyang 550025, China; xs_jiang_CJQC@outlook.com
*   Correspondence: hef@gzu.edu.cn

Received: 23 November 2019; Accepted: 12 January 2020; Published: 15 January 2020

Abstract: Aimed at the limitation of traditional fuzzy control strategy in distributing power and improving the economy of a fuel cell hybrid electric vehicle (FCHEV), an energy management strategy combined with working conditions identification is proposed. Feature parameters extraction and sample divisions were carried out for typical working conditions, and working conditions were identified by the least square support vector machine (LSSVM) optimized by grid search and cross validation (CV). The corresponding fuzzy control strategies were formulated under different typical working conditions, in addition, the fuzzy control strategy was optimized with total equivalent energy consumption as the goal by particle swarm optimization (PSO). The adaptive switching of fuzzy control strategies under different working conditions were realized through the identification of driving conditions. Results showed that the fuzzy control strategy with the function of driving conditions identification had a more efficient power distribution and better economy.

Keywords: fuel cell hybrid electric vehicle; least squares support vector machines (LSSVM); driving conditions identification; power distribution

---

1. Introduction

The introduction of a power battery can make up for the shortcomings of fuel cell hybrid electric vehicles (FCHEV), such as the inability to recover braking energy, slow start speed and soft output characteristics. The dual power source (fuel cell and battery pack) can make the fuel cell hybrid electric vehicles (FCHEVs) produce a better power performance, but how to make the power source power distribution more reasonable and better improve the economy is a research difficulty. Based on previous experience, researchers developed rule-based energy management algorithms, such as thermostatic control strategy (TCS) [1] and a power following control strategy (PFCS) [2,3]. Fuzzy control strategy (FCS) [4–6] and fuzzy control strategy optimized by other algorithms [7] can adapt to the requirements of vehicle nonlinear control and effectively distribute the power between the power sources of fuel cell hybrid vehicles. However, due to the lack of road condition information, they are difficult to further improve the working efficiency and the economy of power sources in complex working conditions. Another control strategy based on optimization, such as dynamic programming (DP) [8–10], are widely used in hybrid electric vehicle energy management strategy because they can achieve global optimization. However, those methods will increase the computational burden and make it difficult to realize the online application. In order to simplify the calculation, some strategies, such as equivalent consumption minimization strategy (ECMS) [11–13], Pontryagin minimum principle strategy (PMPS) [14,15] and stochastic dynamic programming (SDP) [16], further improve the energy management performance on the basis of effectively reducing the calculation amount. For some intelligent algorithms, such as particle swarm optimization (PSO) [17] and genetic algorithm (GA) [18], the fuel economy can also be improved by optimizing some relevant parameters based on the rule-based control strategy.

Working conditions have a profound impact on the economy and power source performance of FCHEVs. Ahmadi et al. [19] investigated the influence of driving patterns, and they found that various driving patterns under different conditions could affect the degradation of a fuel cell, and then affect the economy of the fuel cell vehicles. Raykin et al. [20] investigated the influence of driving patterns under different working conditions and an electric power supply on the well-to-wheel energy use and greenhouse gases of a plug-in hybrid electric vehicle (PHEV). When formulating the FCHEVs' energy control strategy, some references mentioned that they took single working condition into account, and there were certain limitations in improving the economy under different working conditions. Moreover, they did not consider the efficient working area of a fuel cell (FC) and battery pack to give full play to their respective advantages. Under the condition that working conditions can be identified, the energy management strategy of FCHEV should be adjusted according to the actual situation to achieve efficient and reliable power distribution among power sources, improve economy and extend the service life of power sources.

A lot of scholars have studied working condition identification. References [21–24] based on a fuzzy control recognizer, realized the identification of driving conditions. However, membership functions and rules of the fuzzy controller were selected and formulated based on personal experience, and the ideal effect could be achieved after multiple debugging. Clustering methods also play a role in the field of driving conditions recognition [25,26]. In [25], working conditions were divided into five typical working conditions by way of a clustering analysis method, then working conditions were identified by a Euclid approach degree. Yu et al. [26] identified high impact factors affecting pattern characteristics from static and quasi-static environment and traffic information, then proposed a trip/route division algorithm based on data clustering method. However, the selection of initial clustering center affected the clustering analysis results. Recently, machine learning has been further applied. Neural networks, such as back-propagation (BP) neural network [27] and learning vector quantization (LVQ) neural network [28,29], involve first, characteristic parameters that have an important influence on driving conditions being selected as the input, then, the identification period of the working condition samples are classified. After training the samples, the prediction of future working conditions can be realized. However, the accuracy of neural network depends on its structure. Chen [30] et al. proposed an improved hierarchical clustering algorithm to divide the driving cycle data into four groups, and then applied a support vector machine (SVM) to predict driving conditions based on the clustering results.

The least square support vector machines (LSSVM) based on support vector machines (SVM), compared with SVM, can complete a prediction in a shorter time and has a great generalization ability. Moreover, LSSVM is not subject to the set of algorithm structures and has good robustness in handling regression and classification problems.

In order to improve the performance of FCHEV, this paper proposes a driving condition recognizer. By extracting feature parameters and segmenting recognition segments from driving conditions information, LSSVM optimized by CV is used to realize working condition recognition. Energy management controllers based on a fuzzy control under different working conditions are established and optimized. Combined with the driving conditions identification, the energy management controller adopts corresponding fuzzy control strategy according to driving conditions to improve the performance of FCHEV.

2. Vehicle Structure and Parameters

The FCHEV was a front-drive vehicle with the structure shown in Figure 1. The fuel cell system was connected to the Controller Area Network (CAN) bus through a one-way DC/DC converter, while the battery pack was directly connected to the CAN bus. The motor drives the vehicle through the final drive and differential. The complete vehicle parameters of a fuel cell hybrid electric vehicle are shown in Table 1.

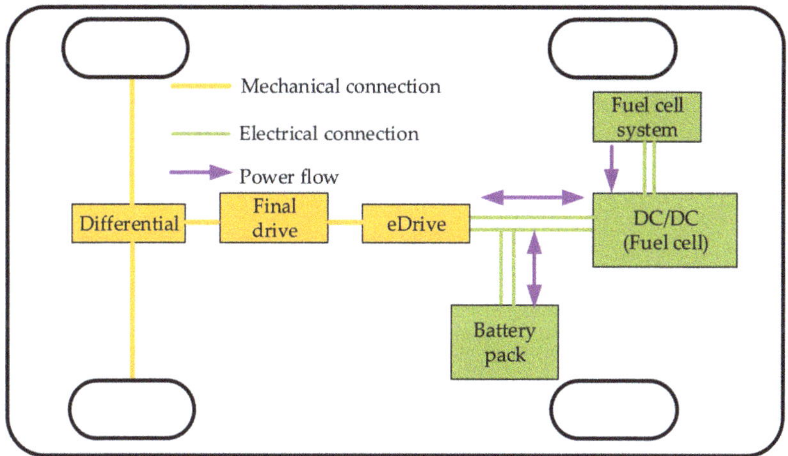

Figure 1. Fuel cell hybrid electric vehicle transmission structure diagram.

Table 1. Vehicle parameters.

Parameters	Value
Vehicle mass (kg)	1315
Vehicle size (mm)	$4760 \times 1815 \times 1530$
Wind resistance coefficient, C_D	0.264
Frontal area, A (m^2)	1.97
Rolling resistance coefficient, f	0.018
Battery pack:	
Rated capacity (Ah)	24
Rated Voltage (V)	450
Fuel cell stack:	
Peak output power (kW)	60
Rated voltage (V)	150
Rated current (A)	200

In this paper, the vehicle model of FCHEV was established in AVL Cruise, as shown in Figure 2, and the control strategy model was established in Matlab/Simulink, shown in Figure 3. In Figure 2, the overall simulation model includes driver module, fuel cell system, power battery pack, motor and controller, one-way DC/DC converter, final drive, and energy management module. The blue line and red line represent mechanical connection electrical connection, respectively.

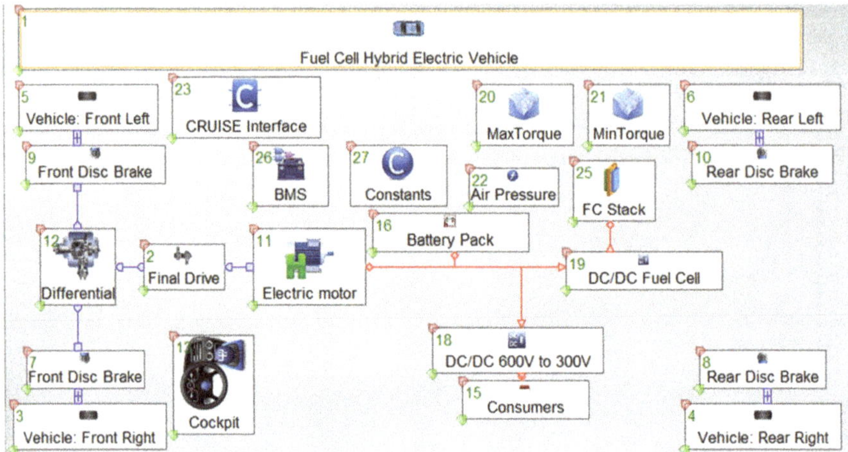

Figure 2. Vehicle structure diagram in AVL Cruise.

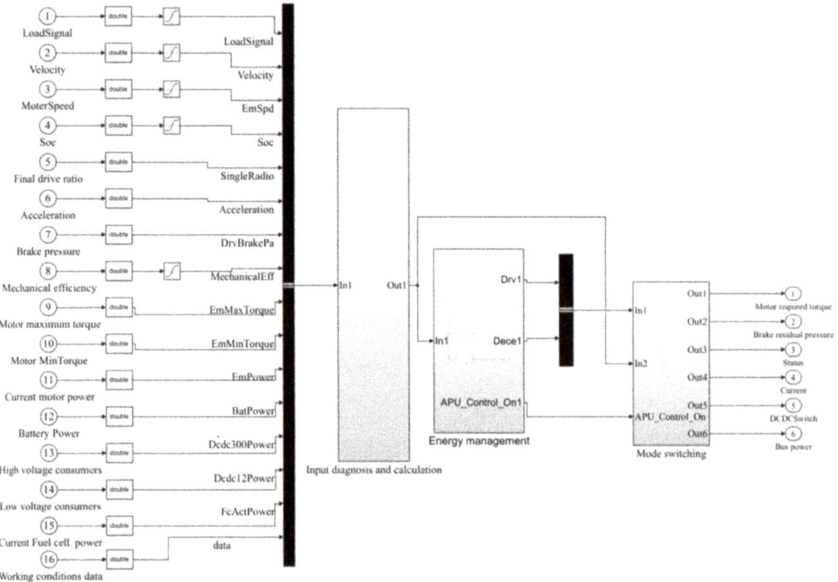

Figure 3. Control module in Simulink.

2.1. Fuel Cell Module

The fuel cells in this paper were proton exchange membrane fuel cells (PEMFC), and they were built out of membrane electrode assemblies (MEA), which included the electrodes, electrolyte, anode catalyst layer, cathode catalyst layer (CCL), and gas diffusion layer (GDL). The detailed modeling process is found in references [31,32]. In the fuel cell component, in addition to the fuel cell, there was a simple compressor model, and its properties are shown in Table 2. The compressor delivered hydrogen continuously to the fuel cell stack, which generated electricity to drive the motor.

Table 2. Compressor properties.

Parameters	Value
Compressor pressure ratio	1.1
Compressor response time (s)	2.0
Compressor idle mass flow (kg/s)	0.002
Compressor efficiency (%)	91.5

The voltage of the fuel cell electrochemical model is calculated as follows:

$$U_{fc} = U_{oc} - \eta_0 - j_0 R \tag{1}$$

$$\eta_0 = V_{act} + V_{CCL} + V_{GDL} \tag{2}$$

$$j_0 R = \frac{I_{st}}{A_{area}} R \tag{3}$$

where U_{fc} is the output voltage, U_{oc} is the ideal open circuit voltage, η_0 is the cathode voltage loss, V_{act} is the activation over potential, V_{CCL} is the voltage loss caused by the oxygen transmission loss in the cathode catalyst layer (CCL), V_{GDL} is the voltage loss caused by the oxygen transmission loss in the anode catalyst layer, j_0 and I_{st} are the electric flow density and current of the stack, while A_{area} is the effective area of the fuel cell, R is the ohmic internal resistance of the fuel cell. The activation loss can be defined as follows.

$$V_{act} = b_{Tf} \cdot \text{arcinh}\left(\frac{\left(\frac{j_0}{j_a}\right)^2}{2\frac{c_{cc}}{c_{ci}}\left(1 - \exp\left(\frac{-j_0}{2j_*}\right)\right)}\right) \tag{4}$$

where b_{Tf} is the Tafel slope which describes the speed of the chemical reaction, and c_{cc} is the oxygen concentration in the cathode channel, while c_{ci} is the oxygen concentration at the channel inlet. Moreover, j_a and j_* can be defined as

$$j_a = \sqrt{2i_* S_{pc} b_{Tf}} \tag{5}$$

$$j_* = S_{pc} b_{Tf} / l_{CCL} \tag{6}$$

where i_* is the volumetric exchange current density, and S_{pc} is the CCL proton conductivity, in addition, l_{CCL} is the thickness of the CCL.

The voltage loss V_{CCL} can be defined as

$$V_{CCL} = \frac{\frac{S_{pc} b_{Tf}^2}{4FD_{CCL} c_{cc}}\left(\frac{j_0}{j_*} - \ln\left(1 + \frac{j_0^2}{j_*^2 B^2}\right)\right)}{1 - \frac{j_0}{j_l^* \frac{c_{cc}}{c_{ci}}}} \tag{7}$$

where F is the Faraday constant, D_{CCL} is the oxygen diffusion coefficient in the CCL. $j^* l$ and B can be defined as

$$j_l^* = \frac{4FD_{GDL} c_{ci}}{l_{GDL}} \tag{8}$$

$$B = 2\arctan\left(\frac{\hat{j}_0}{2\arctan\left(\frac{\hat{j}_0}{2\arctan\left(\frac{\hat{j}_0}{2\arctan\left(\frac{\hat{j}_0}{\frac{\sqrt{2\hat{j}_0}}{2}}\right)}\right)}\right)}\right) \tag{9}$$

where D_{GDL} is the oxygen diffusion coefficient in the GDL, while l_{GDL} is the thickness of GDL.

The voltage loss V_{GDL} can be defined as

$$V_{\text{GDL}} = -b_{\text{Tf}} \ln\left(1 - \frac{j_0}{j_l^* \frac{c_{cc}}{c_{ci}}}\right) \tag{10}$$

Assuming that the fuel cell stack consists of n fuel cell cells, the output power of the fuel cell stack is

$$P_{\text{fc}} = n \times (U_{\text{fc}} \times I_{\text{st}}) \tag{11}$$

The efficiency of fuel cell stack can be expressed as follows:

$$\eta_{\text{fc}} = (U_{\text{oc}} - U_{\text{fc}})/U_{\text{oc}} \tag{12}$$

The single fuel cell properties are shown in Table 3.

Table 3. The properties of a single fuel cell.

Properties	Value	Properties	Value
Nominal voltage (V)	0.6	Ohmic resistance (Ohm)	1.08×10^{-4}
Cell area (m^2)	0.01	Oxygen diffusion coefficient in the GDL (m^2/s)	3.4×10^{-6}
Ideal open circuit voltage (V)	1.23	Oxygen diffusion coefficient in the CCL (m^2/s)	3×10^{-7}
Tafel slope (V)	0.03	Crossover current (A/m^2)	1.05×10^{-4}
CCL proton conductivity (S/m)	3.0	Volumetric exchange current density (A/m^3)	736.974
Catalyst layer thickness (m)	1.0×10^{-5}	GDL thickness (m)	2.5×10^{-4}

2.2. Power Battery Pack

The lithium battery selected in this paper had a capacity of 24 Ah and a rated voltage of 3.3 V, and its specific parameters are shown in Table 4. Its equivalent circuit model adopted the Rint model, as shown in Figure 4a. The voltage of the battery output to the CAN bus is:

$$U_{\text{out}} = U_{\text{ocv}} - I_{\text{b}}R_0 \tag{13}$$

where U_{OCV} is the open circuit voltage of lithium battery, U_{Out} is the output voltage, I_{b} and R_0 are the current and ohmic internal resistance of lithium battery respectively.

Table 4. Battery parameters.

Type	Lithium Iron Phosphate Battery
Normal Voltage	3.3 V
Normal Capacity	24 Ah
Upper/lower cut-off voltage	3.65 V/2 V
Operating temperature	-5–$50\,°C$

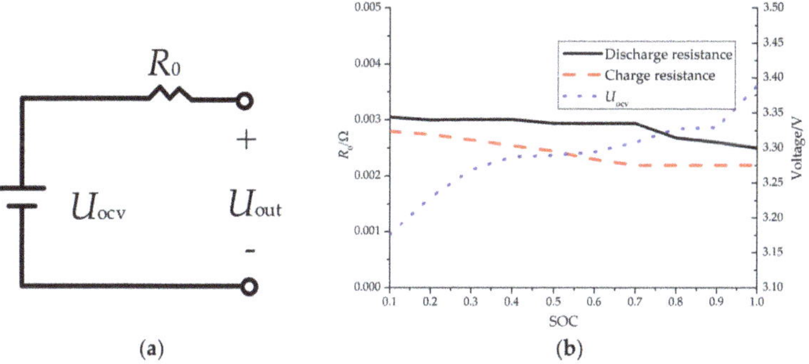

(a)   (b)

Figure 4. Battery model and parameters relationship: (**a**) Rint equivalent circuit model; (**b**) relationship of the relevant parameters of the lithium battery.

SOC, an important parameter of a lithium battery, is expressed by the following equation:

$$\text{SOC}(t) = \text{SOC}_0 - \frac{\eta I \Delta t}{C_\text{P}} \tag{14}$$

where η is the coulomb efficiency, in this paper, $\eta = 1$, SOC_0 was the initial value, sampling time $\Delta t = 1$ s, and C_P was the actual capacity of the battery. Through experiments, the parameters relationships of the battery are shown in Figure 4b.

3. Typical Driving Conditions

Working conditions of a vehicle have an important impact on economy and power distribution. Therefore, a more efficient energy management strategy can be developed by predicting the future working conditions.

In this paper, three typical driving conditions were selected, namely UDDS (Urban Dynamometer Driving Schedule), EUDC (Extra Urban Driving Cycle) and US06 (Highway Driving Schedule), as shown in Figure 5, which corresponded to an urban condition, suburban condition and highway condition, respectively. In an urban working condition, the vehicle speed is low and frequent parking occurs. The average vehicle speed is less than 35 km·h^{-1}, moreover, the vehicle is in a state of low power output. The speed is fast in highway conditions, and the average speed is about 70 km·h^{-1}, in addition, the output power of the car is relatively large. The suburban working condition is in the middle of the two, with an average speed of about 60 km·h^{-1}.

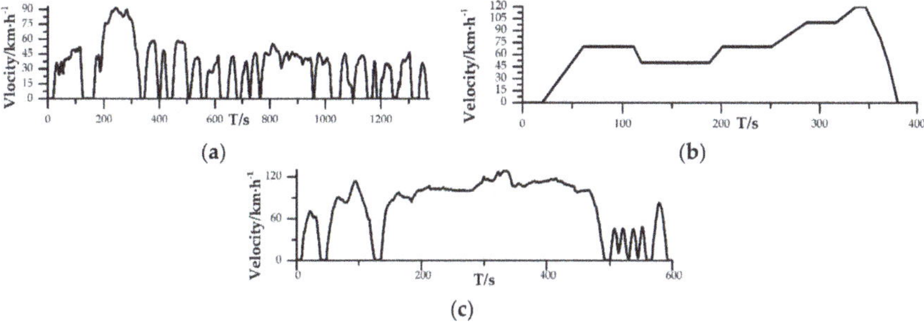

(a)   (b)

(c)

Figure 5. Typical driving cycles: (**a**) city driving cycle—UDDS; (**b**) rural driving cycle—EUDC; (**c**) highway driving cycle—US06.

3.1. Selection of Working Condition Characteristic Parameters

The selection of characteristic parameters of working conditions is the key to accurately identifying future working conditions. In principle, more characteristic parameters is more helpful for prediction, but that requires high computational power. In contrast, too few characteristic parameters cannot cover the information of working conditions, which may lead to a large prediction deviation. Many scholars have studied the selection of characteristic parameters of driving conditions [25,30,33–35]. Based on some research and the importance of each parameter in driving conditions identification, six common characteristic parameters were selected, i.e., acceleration time/total time (r_c), deceleration time/total time (r_{dc}), time of uniform speed/total time (r_u), average speed (v_a), average acceleration (a_c) and average deceleration (a_{dc}). Six characteristic parameters of three working conditions are shown in Table 5.

Table 5. Characteristic parameters of typical working conditions.

Types of Driving Conditions	r_c/%	r_{dc}/%	r_u/%	v_a/(km·h^{-1})	a_c/(m·s^{-2})	a_{dc}/(m·s^{-2})
UDDS	34.30	28.55	18.21	31.51	0.58	−0.73
EUDC	28.12	11.25	45.25	62.63	0.35	−0.89
US06	33.83	32.67	27.67	77.32	0.86	−0.91

3.2. Dividing of Working Condition Samples

The time length of the working conditions samples, namely, the identification cycle and update of identification cycle, will also have an impact on the working condition recognition. The specific segmentation of the working condition recognition samples is shown in Figure 6. ΔT is the identification period, therefore, six characteristic parameters in this period of time can be calculated to identify the working conditions of this sample. While Δs is the update of the period, that is, the time difference between the beginning of the previous cycle segmentation and the beginning of the current cycle segmentation. If ΔT is too long, although it contains more information, it will increase useless information and calculation burden, which will reduce the effect of recognition. If ΔT is too short, it will not accurately reflect the real situation of working conditions. Similarly, a too small Δs leads to frequent cycle switching, which will cause a burden on the processor, while a too large Δs is not conducive to the timely switching of working conditions. References [35,36] studied in detail the effect of ΔT and Δs on the accuracy of working conditions identification. Based on considering the accuracy and calculation cost, $\Delta T = 100$ s, $\Delta s = 3$ s.

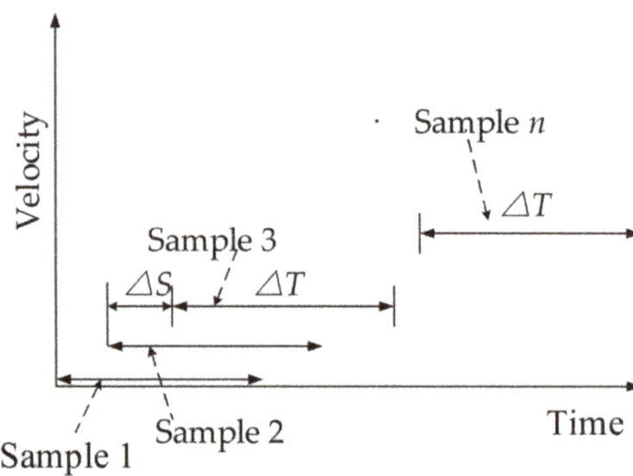

Figure 6. Selection of working condition samples.

4. Working Condition Identification Model Based on LSSVM

4.1. Least Squares Support Vector Machine

LSSVM is able to classify samples by mapping them into high-latitude feature Spaces. LSSVM replaces the inequality constraints of problems in SVM with a set of linear equality constraints, thus simplifying the solution of Lagrange multipliers. A training set is considered with n data samples to be (X_i, y_i), where input data $X_i \in R^n$, output data $y_i \in R$. A linear function in the high-level feature space will be used to fit the samples.

$$y(X_i) = \omega^T \varphi(X_i) + b \tag{15}$$

where $\varphi(X)$ is a nonlinear mapping function, ω is the weight vector in the feature space, and b is the bias term.

According to the principle of structural risk minimization and taking into account the complexity of function and fitting error, the optimization problem of LSSVM can be expressed as:

$$\min_{\omega,b,\xi} J(\omega,\xi) = \tfrac{1}{2}\omega^T\omega + \tfrac{1}{2}C\sum_{i=1}^{n}\xi_i^2$$
$$s.t \quad y(X) = \omega^T\varphi(X) + b + \xi_i \ i = 1,2,\ldots,n \tag{16}$$

where ξ_i is the error variable and C is the penalty factor.

Converting Equation (16) to unconstrained functions by building Lagrange functions and solving this Lagrange function, the classification prediction model of LSSVM can be obtained, as shown in Equation (8), and its structure is shown in Figure 7. Combined with Section 3, six characteristic parameters are taken as the input of the LSSVM, and the output is the working condition categories:

$$y = \text{sign}\left(\sum_{i=1}^{n}\alpha_i y_i K(X, Xi) + b\right) \tag{17}$$

where the radial basis function (RBF) is selected as the kernel function, namely $K(X, Xi) = exe(-\|X - Xi\|^2/(2\sigma^2))$, α is the Lagrange multiplier.

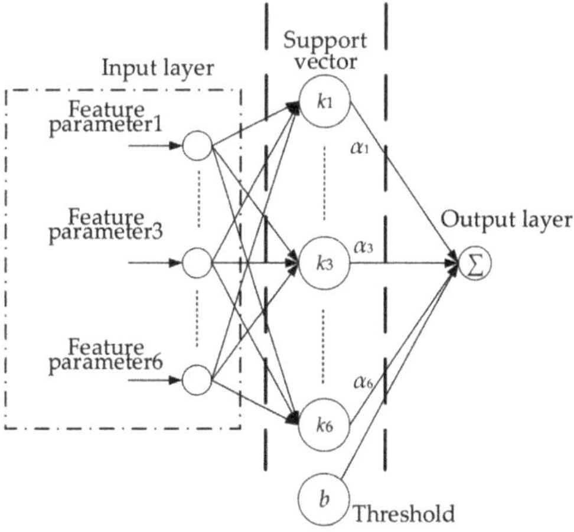

Figure 7. The LSSVM structure diagram.

4.2. The Influence of Key Parameters on the Accuracy of LSSVM

If $\sigma\to0$, then $K(X, X_i)\to0$, which means that all the mapped points have the same distance from each other, that is, there is no clustering phenomenon. However, If $\sigma\to\infty$, then $K(X, X_i)\to1$, which means that all sample points will be divided into the same class and cannot be distinguished. As for the penalty factor C, if C is too large, $\xi_i\to0$, the tolerance of samples between boundaries is very low, and there are less misclassifications, which means the fitting of samples is good, however, the prediction effect is not always good; on the other hand, if the value of C is too small, there are more samples between two boundaries, resulting in a greater possibility of misclassification, and the fitting of samples decreases.

The accuracy of LSSVM's model depends on the kernel parameter σ and the penalty factor C. A too large σ will reduce the model's accuracy, but a too small σ will lead to overfitting. The penalty factor C will affect the error and complexity of the model. Therefore, in this study, the cross-validation method was used to obtain the optimal parameters.

4.3. The K-Fold Cross-Validation for Optimizing LSSVM

Cross-validation has been widely used to estimate prediction errors. In this work, K-fold cross-validation combined with grid search was applied to optimize LSSVM, which could overcome the limitations of the holdout validation [37]. The steps to optimize LSSVM were as follows:

(1)　Establish grid coordinates. Let $a = [-10, 10]$, $b = [-10, 10]$, and the step size is 0.5, then the mesh points of the model parameters are $\sigma = 2^a$ and $C = 2^b$ respectively. In this work, the exponential function was selected to divide the grid, which would ensure that the parameter value was not negative.

(2)　Divide the sample data and calculate the test error. The training data are divided into K subsets ($K = 10$, which means that the CV is 10-fold cross-validation method). For each group (σ, C) in the grid, a 10-fold cross validation method was applied to iterate the training data 10 times, and the mean value of the mean square error (MSE) of the test results under this group of parameters could be obtained.

(3)　Get the optimal combination of parameters. Repeat (2) to replace the parameter σ and C, and calculate the mean square deviation of the training model under all the parameter combinations in the grid in turn. After comparing one by one, the parameter combination corresponding to the minimum mean square deviation is the optimal parameter combination in the grid interval.

In order to present equidistant grid search results more clearly, grid coordinates (σ, C) are converted to logarithmic coordinates $(\log_2\sigma, \log_2C)$.

5. Fuzzy Energy Management Strategy Based on Working Condition Identification

Fuzzy control based on the theory of fuzzy mathematics, fuzzes the actual input and output, and formulates rules through experience. These kinds of simulation of a human's approximate reasoning and comprehensive decision-making process has good robustness and adaptability. Fuzzy energy management strategies [4–6] developed by some researchers were aimed at a single working condition. In addition, fuzzy control rules based on personal experience are difficult to deal with complex multi-working conditions. Therefore, on the basis of condition identification, three fuzzy energy management strategies were formulated to deal with urban, suburban and expressway conditions, respectively. Besides, PSO is used to optimize the fuzzy control under various working conditions with total equivalent energy consumption as the objective function, and the adaptive switching effect is achieved through the identification of working conditions. It should be noted that the following fuzzy controller and optimization take the urban working condition as an example.

5.1. Fuzzy Controller Design

(1)   Selection of input and output variables of fuzzy controller.

The SOC of the battery pack and the total power demand P_r of FCHEV were selected as the input of the fuzzy controller, while the output is the output power P_{fc} of the fuel cell. The power demand relationship is as follows:

$$P_b = P_r - P_{fc} \tag{18}$$

where P_b is the output power of the battery, and P_r includes the power of the drive motor and the power consumed by accessories.

(2)   Fuzzy distribution of input and output variables.

The range of FCHEV's total power demand P_r is [0, 60] (kw), and its fuzzy subsets are very small, small, medium, large and very large, i.e., {VS, S, M, L, VL}; the SOC range of power battery is [0, 1], and the fuzzy subsets are {VL, L, M, H, VH}, representing very low, low, medium, high and very high; the range of fuel cell's output power P_{fc} is [0, 50] (kw), hence its fuzzy subsets {VL, L, M, H, VH} represent very low, low, medium, high and very high.

(3)   Fuzzy control rules.

The fuzzy control rules of FCHEV are formulated according to the following principles:

①   When the SOC of the power battery is too low, the output power of the fuel cell should not only meet the requirements of driving the vehicle, but also charge the battery to make the SOC of the power battery rise to a reasonable range (SOC = 40–80%).

②   When the SOC of the power battery and the demand power are both medium level, the fuel cell acts as the active power source and changes with the demand power. SOC of battery fluctuates in a reasonable range, which is beneficial to prolonging battery life.

③   When SOC is too high, the power battery acts as the main power source, and the output power of the fuel cell is as small as possible to reduce the SOC to a reasonable range.

④   When the demand power is too large, the power battery and fuel cell provide output power together.

5.2. Fuzzy Controller Optimization Based on PSO

As an optimization algorithm, PSO is a solution to reducing the influence of making fuzzy control strategy based on personal experience. In this paper, the membership functions of the input and output of the fuzzy controller were selected as the parameters to be optimized, and the objective function was total equivalent energy consumption (TEEC) of the power sources, i.e.,

$$\begin{cases} \min E(x) = E_{fc}(x) + E_b(x) \\ \text{s.t. } G_i(x) \geq 0, i = 1,\ldots,m \end{cases} \tag{19}$$

where $E_{fc}(x)$ and $E_b(x)$ are the equivalent electric energy consumption of the fuel cell and electric energy consumption of battery, respectively, while $G_i(x)$ is the constraint condition of the vehicle, such as the time of acceleration and SOC fluctuation range of the battery pack.

The distributions of control rules under urban working condition before and after optimization are shown in Figure 8.

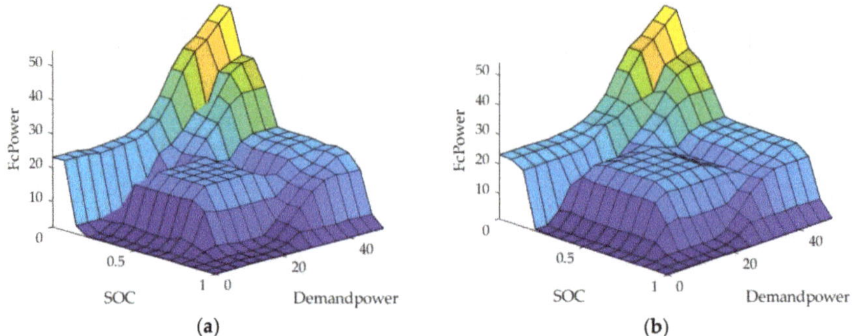

Figure 8. Control rules under a city driving condition. (**a**) before optimization; (**b**) after optimization.

5.3. Fuzzy Energy Management Based on Condition Identification

After identifying working conditions by LSSVM, the corresponding fuzzy control rules are selected by the fuzzy controller according to the working conditions. The flow chart of the energy management strategy based on working conditions identification is shown in Figure 9. Firstly, the characteristic parameters were extracted from the working condition information and sample segmentations were determined, and then working condition identifications were carried out by LSSVM. Fuzzy control strategies were optimized under three working conditions, and corresponding fuzzy control rule was selected under a specific working condition to realize the adaptive switching of the control strategy under complex working conditions.

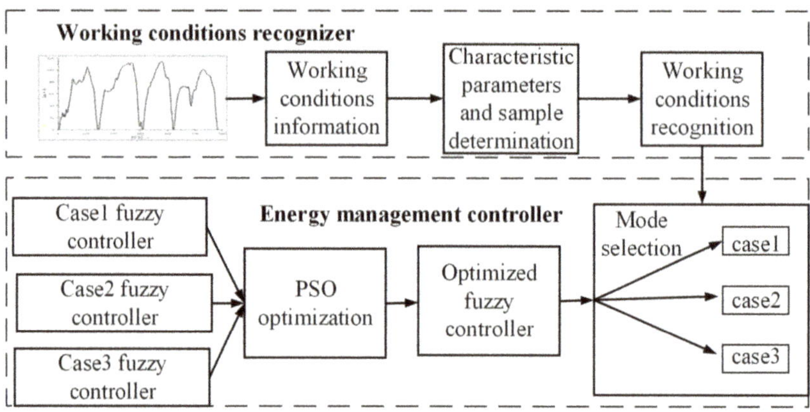

Figure 9. Flow chart of energy management strategy.

6. Results and Discussion

6.1. Results of Working Conditions Identification

The samples of three typical working conditions were divided into 730 samples, of which 547 were training samples and the other 183 were validation samples.

Figure 10a describes the iterative optimization process of LSSVM's parameters under grid search and cross-validation. Among the 183 validation samples shown in Figure 10b, the recognition accuracy reached 98.36%. The key parameters of LSSVM optimized by CV were $\sigma = 2.64$, $C = 25.61$. Figure 10c shows the randomly generated driving conditions, where the LSSVM could identify the random driving conditions with an accuracy of 100%.

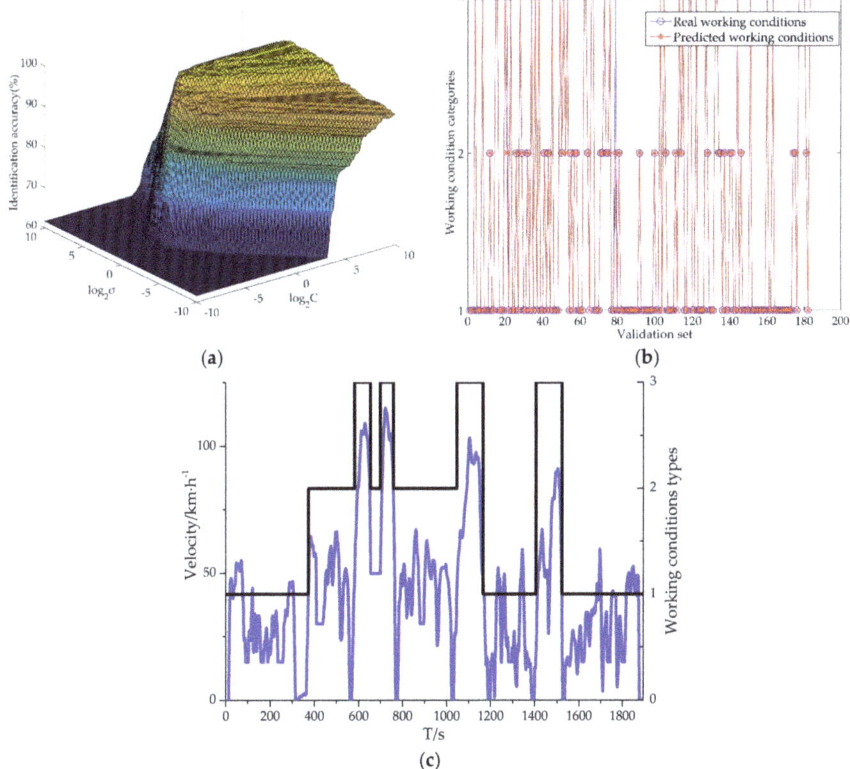

Figure 10. Results of working conditions recognition by LSSVM: (**a**) iterative process of training samples; (**b**) validation samples identification result; (**c**) identification result of mixed working conditions.

6.2. Fuzzy Control Energy Management Strategy Based on Driving Conditions Identification

In order to verify the effectiveness of the proposed energy management strategy, it was compared with the power follow control strategy in the efficiency of fuel cell stack, the SOC fluctuation of battery pack and the economy, at medium SOC level (SOC = 60%) and high SOC level (SOC = 85%).

6.2.1. The Initial SOC of Battery Pack Was 60%

As shown in Figure 11a, the vehicle speed of the proposed fuzzy control strategy (FC1) can well follow the real vehicle speed. In Figure 11b, for total equivalent energy consumption, the power following control strategy (PFCS) was 3.99 (kW·h), which was 5.26% higher than that of the traditional fuzzy control strategy (FC2) (3.78/kW·h). In Figure 11c,d and Table 6, the average efficiency of the fuel cell stack of the FC2 was 67.62%, which was 2.05% higher than that of PFCS. The fluctuation range of SOC of FC2 was 58.56–61.55%, which was gentler than that of PFCS, for the ΔSOC of FC2 improved by 6.67% compared with PFCS.

In Figure 11b, for total equivalent energy consumption, FC1 was 3.65 (kW·h), which was 3.44% lower than that of FC2. In Figure 11c,d and Table 6, the average efficiency of the fuel cell stack for FC1 was 68.71%, which was 1.09% higher than of FC2. The fluctuation range of SOC of the FC1 was 58.15–61.00%. Compared with FC2, the ΔSOC of FC1 improved by 0.47%, which was conducive to the durability of the battery pack. Therefore, when the SOC initial value was 60%, the FC1 was better than the FC2 and PFCS in improving the efficiency and durability of power sources and the economy of FCHEV.

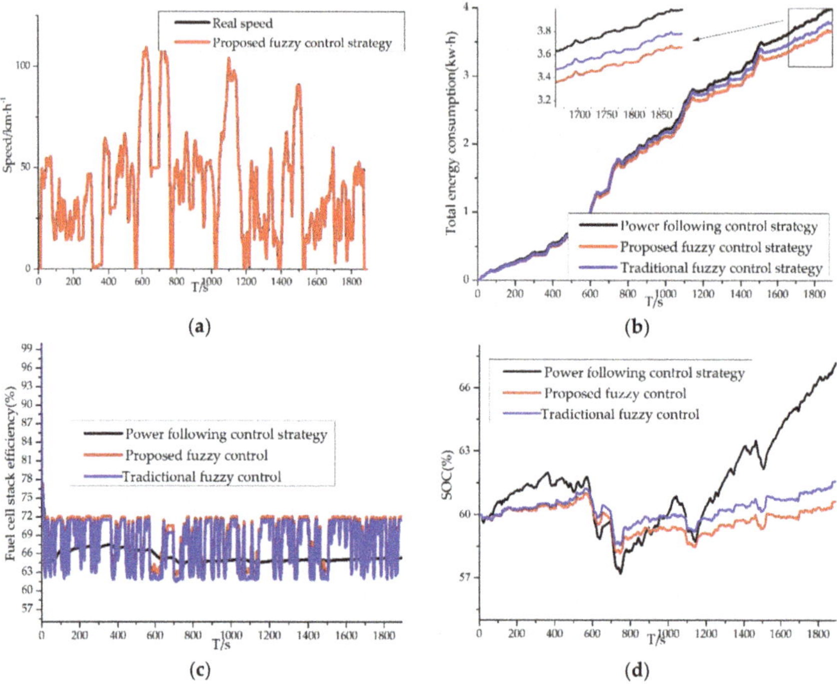

Figure 11. Comparison of results when SOC = 60%: (**a**) comparison of vehicle velocity; (**b**) comparison of overall equivalent energy consumption; (**c**) comparison of fuel cell stack efficiency; (**d**) comparison of SOC.

Table 6. Comparison of results of mixed random driving conditions when SOC = 60%.

Strategies	TEEC (kW·h)	SOC Range (%)	ΔSOC (%)	AEFCS (%)
PFCS	3.99	Min: 57.19 Max: 67.18	9.99	65.57
FC1	3.65	Min: 58.15 Max: 61.00	2.85	68.71
FC2	3.78	Min: 58.56 Max: 61.88	3.32	67.62

Note: Total equivalent energy consumption (TEEC); average efficiency of fuel cell stack (AEFCS); ΔSOC = SOC_{max} − SOC_{min}; power following control strategy (PFCS); proposed fuzzy control (FC1); traditional fuzzy control (FC2).

6.2.2. The Initial SOC of the Battery Pack Is 85%

As shown in Figure 12a, the speed of FC1 can also follow the actual speed well, which is similar to the case of SOC = 60%. In Figure 12b, for total equivalent energy consumption, the PFCS was 4.45 (kW·h), but the figure for FC2 was 3.95 (kW·h), which was 11.24% lower than that of PFCS. In Figure 12c,d and Table 7, the average efficiency of the fuel cell stack of the FC2 was 68.22%, which was 2.66% higher than that of PFCS (65.56%). The fluctuation range of SOC of PFCS was 80.55–86.12%, while the range of SOC for FC2 was 74.80–85.00%, so the ΔSOC of PFCS improved by 4.63% compared with FC2.

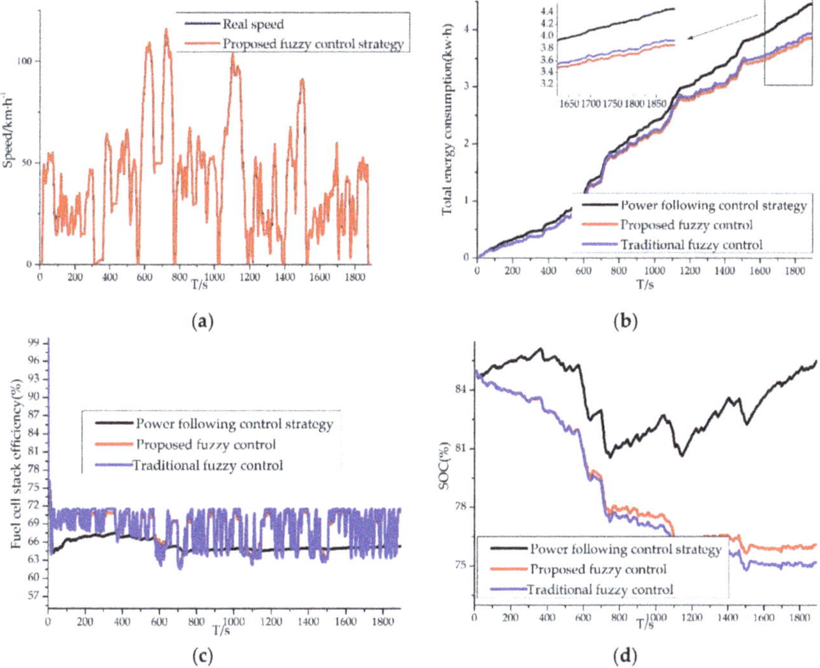

Figure 12. Comparison of results when SOC = 85%: (**a**) comparison of vehicle velocity; (**b**) comparison of overall equivalent energy consumption; (**c**) comparison of fuel cell stack efficiency; (**d**) comparison of SOC.

Table 7. Comparison of results of mixed random driving conditions when SOC = 85%.

Strategies	TEEC (kW·h)	SOC Range (%)	ΔSOC (%)	AEFCS (%)
PFCS	4.45	Min: 80.55 Max: 86.12	5.57	65.56
FC1	3.85	Min: 75.64 Max: 85.00	9.36	68.62
FC2	3.95	Min: 74.80 Max: 85.00	10.20	68.22

In Figure 12b, for total equivalent energy consumption, FC1 was 3.85 (kW·h), which was 2.53% lower than that of FC2. In Figure 12c,d and Table 7, the average efficiency of the fuel cell stack for FC1 was 68.62%, which was 0.40% higher than of FC2. The fluctuation range of SOC of the FC1 was 75.64–85.00%, which meant the ΔSOC of FC1 improved by 0.84% compared with FC2, so FC1 was more conducive to the durability of the battery pack. Therefore, when SOC = 85%, PFCS was better than FC1 and FC2 in controlling the fluctuation of SOC, on the other hand, FC1 showed that it had better performances on the average efficiency of the fuel cell stack and the economy than the other two control strategies.

To summarise, in order to verify the effectiveness of the proposed fuzzy control strategy, it was compared with the traditional fuzzy control and power following control strategy in the case of SOC = 60% and SOC = 85%. It can be seen form Tables 6 and 7, when SOC = 85%, the total equivalent energy consumption of PFCS, FC1 and FC2 were much more than these of SOC = 60%, particularly for PFCS, where the largest difference of the equivalent energy consumption occurred between SOC = 60% and SOC = 85%. At a high SOC level (SOC is above 80%), the battery pack has sufficient power, and the braking energy recovery rate of the FCHEV is low, so as to avoid overcharging of the battery pack. In terms of the operating efficiency of the fuel cell stack, when SOC = 85%, though the average efficiency of the fuel cell stack of FC1 is slightly lower than that of SOC = 60%, the figure for FC1 is still the highest, which indicates the stability of the proposed fuzzy control to maintain the high efficiency of the fuel cell stack. As for the fluctuation of SOC, the expected SOC range of the battery pack of FCHEV was 40–80%, which can prevent the overcharging and over-discharging of the battery pack, thus extending the life of the power battery. It can be seen from Figure 12d that more hydrogen was consumed to reduce the fluctuation of SOC, so the SOC of PFCS was kept above 80%. Although the fluctuation of SOC became smaller, it did not fall rapidly to the expected range, which showed that the power distribution of PFCS was insufficient at a high SOC level. On the contrary, the SOC of FC1 and FC2 decreased from 85% to 75.64% and 74.80% respectively, moreover the fluctuation of SOC for FC1 was smaller, which was more conducive to extending the life of the power battery. It was noted that at a high SOC level, the performance gap between FC1 and FC2 had narrowed. However, at a high SOC level or medium SOC level, the proposed fuzzy control strategy showed the better performances on the working efficiency of fuel cell stack, controlling SOC fluctuation and the economy of FCHEV.

7. Conclusions

In order to deal with the influence of complex working conditions on economy and power distribution between power sources on FCHEV, an energy management strategy based on driving condition identification was developed.

(1) After selecting the characteristic parameters and dividing the working condition samples of three typical driving conditions, working conditions identification were realized by LSSVM.

(2) Fuzzy control rules under different working conditions were formulated, and the total equivalent energy consumption of power sources were taken as the objective function to optimize fuzzy control rules by PSO, and the adaptive switching of the fuzzy control could be realized on the basis of working condition identification.

(3) Simulation results showed that at high SOC level or medium SOC level, the proposed fuzzy control strategy had the ability to recognize the future driving condition, showed a better performance than the traditional fuzzy control strategy and power following the control strategy on improving the working efficiency of the fuel cell stack, controlling the fluctuation of SOC of battery pack and enhancing the economy of FCHEV.

(4) The future work is to establish more complete vehicle driving conditions, and choose different working condition predictors to compare their performance of working conditions prediction, so as to choose a more reliable and efficient working condition predictor.

Author Contributions: Conceptualization, F.H. and Y.Z.; Methodology, Y.Z.; Software, Y.Z.; Validation, X.S., Y.Z.; Formal Analysis, Y.Z.; Resources, X.S.; Data Curation, X.S. and X.J.; Writing—Original Draft Preparation, Y.Z.; Writing—Review & Editing, F.H.; Supervision, F.H. All authors have read and agreed to the published version of the manuscript.

Funding: This work was funded by the Guizhou Province Science and Technology Support Program, grant number (2018)2177.

Conflicts of Interest: The authors declare no conflict of interest.

References

1. Sid, M.N.; Nounou, K.; Becherif, M.; Marouani, K.; Alloui, H. Energy Management and Optimal Control Strategies of Fuel cell/Supercapacitors Hybrid Vehicle. In Proceedings of the International Conference on Electrical Machines (ICEM), Berlin, Germany, 2–5 September 2014. [CrossRef]

2. Corbo, P.; Corcione, F.E.; Migliardini, F.; Veneri, O. Experimental assessment of energy-management strategies in fuel-cell propulsion systems. *J. Power Sources* **2006**, *157*, 799–808. [CrossRef]

3. Geng, C.; Jin, X.F.; Zhang, X. Simulation research on a novel control strategy for fuel cell extended-range vehicles. *Int. J. Hydrogen Energy* **2019**, *44*, 408–420. [CrossRef]

4. Erdinc, O.; Vural, B.; Uzunoglu, M. A wavelet-fuzzy logic based energy management strategy for a fuel cell/battery/ultra-capacitor hybrid vehicular power system. *J. Power Sources* **2009**, *194*, 369–380. [CrossRef]
5. Zhou, D.; Al-Durra, A.; Gao, F.; Ravey, A.; Matraji, I.; Simõesc, M.G. Online energy management strategy of fuel cell hybrid electric vehicles based on data fusion approach. *J. Power Sources* **2017**, *366*, 278–291. [CrossRef]
6. Hemi, H.; Ghouili, J.; Cheriti, A. A real time fuzzy logic power management strategy for a fuel cell vehicle. *Energy Convers. Manag.* **2014**, *80*, 63–70. [CrossRef]
7. Ahmadi, S.; Bathaee, S.M.T. Multi-objective genetic optimization of the fuel cell hybrid vehicle supervisory system: Fuzzy logic and working mode control strategies. *Int. J. Hydrogen Energy* **2015**, *40*, 12512–12521. [CrossRef]
8. Xu, L.F.; Ouyang, M.G.; Li, J.P.; Yang, F.Y. Dynamic programming algorithm for minimizing working cost of a PEM fuel cell vehicle. In Proceedings of the IEEE International Symposium on Industrial Electronics, Hangzhou, China, 28–31 May 2012. [CrossRef]
9. Larsson, V.; Johannesson, L.; Egardt, B. Analytic solutions to the dynamic programming subproblem in hybrid vehicle energy management. *IEEE Trans. Veh. Technol.* **2015**, *64*, 1458–1467. [CrossRef]
10. Fares, D.; Chedid, R.; Panik, F.; Karaki, S.; Jabr, R. Dynamic programming technique for optimizing fuel cell hybrid vehicles. *Int. J. Hydrogen Energy* **2015**, *40*, 7777–7790. [CrossRef]
11. Xu, L.F.; Li, J.Q.; Hua, J.F.; Li, X.J.; Ouyang, M.G. Optimal vehicle control strategy of a fuel cell/battery hybrid city bus. *Int. J. Hydrogen Energy* **2009**, *34*, 7323–7333. [CrossRef]
12. Zhang, W.B.; Li, J.Q.; Xu, L.F.; Ouyang, M.G. Optimization for a fuel cell/battery/capacity tram with equivalent consumption minimization strategy. *Energy Convers. Manag.* **2017**, *134*, 59–69. [CrossRef]
13. Torreglosa, J.P.; Jurado, F.; Garcia, P.; Fernandez, L.M. Hybrid fuel cell and battery tramway control based on an equivalent consumption minimization strategy. *Control Eng. Pract.* **2011**, *19*, 1182–1194. [CrossRef]
14. Odeim, F.; Roes, J.; Wülbeck, L.; Heinzel, P. Power management optimization of fuel cell/battery hybrid vehicles with experimental validation. *J. Power Sources* **2014**, *252*, 333–343. [CrossRef]
15. Zou, Y.; Liu, T.; Sun, F.C.; Huie, P. Comparative study of dynamic programming and Pontryagin's minimum principle on energy management for a parallel hybrid electric vehicle. *Energies* **2013**, *6*, 2305–2318.
16. Elbert, P.; Widmer, M.; Gisler, H.J.; Onder, C. Stochastic dynamic programming for the energy management of a serial hybrid electric bus. *Int. J. Veh. Des.* **2015**, *69*, 8–112. [CrossRef]
17. Trovao, J.P.; Pereirinha, P.G.; Jorge, H.M.; Antunes, C.H. A multi-level energy management system for a multi-source electric vehicles—An integrated rule-based metaheuristic approach. *Appl. Energy* **2013**, *105*, 304–318. [CrossRef]
18. Ryu, J.; Park, Y.; Sunwoo, M. Electric powertrain modeling of a fuel cell hybrid electric vehicle and development of a power distribution algorithm based on driving model recognition. *J. Power Sources* **2010**, *195*, 35–48. [CrossRef]
19. Ahmadi, P.; Torabi, S.H.; Afsaneh, H.; Sadegheih, Y.; Ganjehsarabi, H.; Ashjaee, M. The effects of driving patterns and PEM fuel cell degradation on the lifecycle assessment of hydrogen fuel cell vehicles. *Int. J. Hydrogen Energy* 2019. [CrossRef]
20. Raykin, L.; MacLean, H.L.; Roorda, M.J. Impacts of Driving Patterns on Well-to-Wheel Performance of Plug-in Hybrid Electric Vehicles. *Environ. Sci. Technol.* **2012**, *46*, 6363–6370. [CrossRef]
21. Zhang, S.; Xiong, R. Adaptive energy management of a plug-in hybrid electric vehicle based on driving pattern recognition and dynamic programming. *Appl. Energy* **2015**, *155*, 68–78. [CrossRef]
22. Li, Q.; Wang, T.H.; Dai, C.H.; Chen, W.R.; Ma, L. Power Management Strategy based on Adaptive Droop Control for a Fuel Cell-Battery-Supercapacitor Hybrid Tramway. *IEEE Trans. Veh. Technol.* **2017**, *67*, 5658–5670. [CrossRef]
23. Guo, Q.Y.; Zhao, Z.G.; Shen, P.H.; Zhan, X.W.; Li, J.W. Adaptive optimal control based on driving style recognition for plug-in hybrid electric vehicle. *Energy* **2019**, *186*, 115824. [CrossRef]
24. Chen, Z.Y.; Xiong, R.; Cao, J.Y. Particle swarm optimization-based optimal power management of plug-in hybrid electric vehicles considering uncertain driving conditions. *Energy* **2016**, *96*, 197–208. [CrossRef]
25. Lei, Z.Z.; Cheng, D.; Liu, Y.G.; Qin, D.T.; Zhang, Y.; Xie, Q.B. A dynamic control strategy for hybrid electric vehicles based on parameter optimization for multiple driving cycles and driving pattern recognition. *Energies* **2017**, *10*, 54. [CrossRef]

26. Yu, H.; Tseng, F.; Mcgee, R. Driving pattern identification for EV range estimation. In Proceedings of the Electric Vehicle Conference, Greenville, SC, USA, 4–8 March 2012; pp. 2811–2821. [CrossRef]
27. Li, Y.C.; He, H.W.; Peng, J.K.; Zhang, H.L. Power management for a plug-in hybrid electric vehicle based on reinforcement learning with continuous state and action spaces. *Energy Proc.* **2017**, *142*, 2270–2275. [CrossRef]
28. Song, K.; Li, F.Q.; Hu, X.; He, L.; Niu, W.X.; Lu, S.H.; Zhang, T. Multi-mode energy management strategy for fuel cell electric vehicles based on driving pattern identification using learning vector quantization neural network algorithm. *J. Power Sources* **2018**, *389*, 230–239. [CrossRef]
29. Pham, D.T.; Otri, S.; Ghanbarzadeh, A.; Koc, E. Application of the Bees Algorithm to the Training of Learning Vector Quantisation Networks for Control Chart Pattern Recognition. In Proceedings of the Information & Communication Technologies, Damascus, Syria, 24–28 April 2006.
30. Chen, Z.; Li, L.; Yan, B.J.; Yang, C. Multimode Energy Management for Plug-In Hybrid Electric Buses Based on Driving Cycles Prediction. *IEEE Trans. Intell. Transp. Syst.* **2016**, *17*, 2811–2821. [CrossRef]
31. Kulikovsky, A.A. A physically-Based Analytical Polarization Curve of a PEM Fuel Cell. *J. Electrochem. Soc.* **2014**, *161*, 263–270. [CrossRef]
32. Goessling, S.; Klages, M.; Haussmann, J.; Beckhaus, P.; Messerschmidt, M.; Arlt, T.; Kardjilov, N.; Manke, I.; Scholta, J.; Heinzel, A. Analysis of liquid water formation in polymer electrolyte membrane (PEM) fuel cell flow fields with a dry cathode supply. *J. Power Sources* **2016**, *306*, 658–665. [CrossRef]
33. Dayeni, M.K.; Soleymani, M. Intelligent energy management of a fuel cell vehicle based on traffic condition recognition. *Clean Technol. Environ. Policy* **2016**, *18*, 1945–1960. [CrossRef]
34. Wang, J.; Wang, Q.N.; Zeng, X.H.; Wang, P.Y.; Wang, J.N. Driving cycle recognition neural network algorithm based on the sliding time window for hybrid electric vehicles. *Int. J. Automot. Technol.* **2015**, *16*, 685–695. [CrossRef]
35. Huang, X.; Tan, Y.; He, X. An Intelligent Multifeature Statistical Approach for the Discrimination of Driving Conditions of a Hybrid Electric Vehicle. *IEEE Trans. Intell. Transp. Syst.* **2010**, *12*, 453–465. [CrossRef]
36. Murphey, Y.L.; Chen, Z.H.; Kiliaris, L.; Park, J.M. Neural Learning of Driving Environment Prediction for Vehicle Power Management. In Proceedings of the International Joint Conference on Neural Networks (IEEE World Congress on Computational Intelligence, WCCI 2008), Hong Kong, China, 1–6 June 2008. [CrossRef]
37. Hastie, T.; Tibshirani, R.; Friedman, J. *The Elements of Statistical Learning: Data Mining, Inference, and Prediction*, 2nd ed.; Springer: Stanford, CA, USA, 2008; pp. 241–247. ISBN 978-0-387-84858-7.

Article

Recurrent Neural Network-Based Adaptive Energy Management Control Strategy of Plug-In Hybrid Electric Vehicles Considering Battery Aging

Lu Han, Xiaohong Jiao * and Zhao Zhang

Institute of Electrical Engineering, Yanshan University, Qinhuangdao 066004, China;
hanlu@stumail.ysu.edu.cn (L.H.); 201831030018@stumail.ysu.edu.cn (Z.Z.)
* Correspondence: jiaoxh@ysu.edu.cn

Received: 21 November 2019; Accepted: 27 December 2019; Published: 1 January 2020

Abstract: A hybrid electric vehicle (HEV) is a product that can greatly alleviate problems related to the energy crisis and environmental pollution. However, replacing such a battery will increase the cost of usage before the end of the life of a HEV. Thus, research on the multi-objective energy management control problem, which aims to not only minimize the gasoline consumption and consumed electricity but also prolong battery life, is necessary and challenging for HEV. This paper presents an adaptive equivalent consumption minimization strategy based on a recurrent neural network (RNN-A-ECMS) to solve the multi-objective optimal control problem for a plug-in HEV (PHEV). The two objectives of energy consumption and battery loss are balanced in the cost function by a weighting factor that changes in real time with the operating mode and current state of the vehicle. The near-global optimality of the energy management control is guaranteed by the equivalent factor (EF) in the designed A-ECMS. As the determined EF is dependent on the optimal co-state of the Pontryagin's minimum principle (PMP), which results in the online ECMS being regarded as a realization of PMP-based global optimization during the whole driving cycle. The time-varying weight factor and the co-state of the PMP are map tables on the state of charge (SOC) of the battery and power demand, which are established offline by the particle swarm optimization (PSO) algorithm and real historical traffic data. In addition to the mappings of the weight factor and the major component of the EF linked to the optimal co-state of the PMP, the real-time performance of the energy management control is also guaranteed by the tuning component of the EF of A-ECMS resulting from the Proportional plus Integral (PI) control on the deviation between the battery SOC and the optimal trajectory of the SOC obtained by the Recurrent Neural Network (RNN). The RNN is trained offline by the SOC trajectory optimized by dynamic programming (DP) utilizing the historical traffic data. Finally, the effectiveness and the adaptability of the proposed RNN-A-ECMS are demonstrated on the test platform of plug-in hybrid electric vehicles based on GT-SUITE (a professional integrated simulation platform for engine/vehicle systems developed by Gamma Technologies of US company) compared with the existing strategy.

Keywords: hybrid electric vehicles (HEVs); battery life; multi-objective energy management; adaptive equivalent consumption minimization strategy (A-ECMS); pontryagin's minimum principle (PMP); particle swarm optimization (PSO); recurrent-neural-network (RNN)

---

1. Introduction

Nowadays, the growing energy dilemma and environmental problem are initiating a revolution and innovation within the automobile industry. Hybrid electric vehicles (HEVs) have more of a degree of freedom for vehicle power distribution thanks to invertible energy storage devices and electric machines [1]. To ensure that all hybrid components work cooperatively, a lot of energy management strategies have been proposed. Generally speaking, an energy management control strategy can be categorized as a rule-based (RB) control strategy [2] and an optimization control strategy. The former is realized easily but the fine control performance is not guaranteed. The latter usually is categorized as instantaneous optimization, such as the equivalent consumption minimization strategy (ECMS) [3], model predictive control (MPC) [4], and global optimization including dynamic programming (DP) [5,6] and pontryagin's minimum principle (PMP) [7–9]. DP can obtain a globally optimal solution in theory, but has a serve computational burden and cannot be used in real-time due to the requirement of knowing the global driving cycles in advance. PMP converts energy management to a minimizing Hamilton function, thus greatly reducing the computational burden and making it easier to implement. Despite this, it is still difficult to solve the numerical solution because the dynamic of the co-state is a function of the battery's state of charge (SOC), which is nonlinear. In order to overcome the unknown driving information in advance, MPC is used to solve the energy management optimal control problem [10], which can predict driving information in the fixed prediction horizon. With the development of artificial intelligence, many intelligence algorithms have been applied to predict the driving information in MPC in order to obtain a closer to global optimum solution. For example, the neural network is used in [11] to predict demand power, which makes full preparation for the design of the energy management strategy. The radial basis function neural network (RBF-NN) is trained in [12] using engine working points that is optimized offline utilizing a distributed genetic algorithm. ECMS is applied to search the instantaneous minimum cost function and can be applied in real-time, which is evolved in PMP. Thus, ECMS can obtain more closely to a globally optimal solution by appropriately choosing the equivalent factor. [13] employed the shooting method of PMP to gain the initial co-state and then used the proportional integral (PI) controller to adjust the equivalent factor to guarantee that the SOC has a better trajectory in real time.

Most of the literature only regard minimizing fuel consumption as the optimization objective of energy management problem. In reality, the fading of a battery capcity and the shortening of its life due to the frequenct charge and discharge of a battery when the HEV is running, is inevitable. Furthermore, changing a battery before the HEV is scrapped will significantly increase the usage cost of a HEV. Many studies also have shown that battery life is responsible for the fuel economy of HEVs [14]. Thus it is necessary to consider battery life when designing an energy management control strategy. There are many factors that impact the battery life, such as a battery's thermal management, driving conditions, environment temperature, regional climate, and so on [15]. For this, many ways have been proposed to prolong the battery life for HEVs. In [16], in order to extend the service life of the battery, ultra-capacitors are also equipped to protect the battery by optimizing the distribution of current and using ultra-capacitors to buffer the excessive charge and discharge flow of the battery. [17,18] took fuel consumption and battery capacity loss together into the cost function and solved this multi-objective optimization problem for HEV using PMP and DP, respectively. [19] derived a causal energy management strategy under consideration of battery life for HEV, which effectively reduces battery wear with a reasonable penalty on fuel economy by using ECMS. The distinguishing features of plug-in HEVs (PHEV) over the conventional HEVs are a large variation range of SOC and the repeated charge and discharge of the battery. Moreover, it has been shown that a large depth of discharge (DOD) and frequent use can accelerate the decay of battery life [20]. Consequently, research on the energy management control problem for PHEV considering the battery life is more attractive. [21] established an electrochemical mechanism model for the battery capacity attenuation of PHEV, and formulated a multi-objective optimal energy allocation problem that can be solved by shortest-path stochastic dynamic programming (SP-SDP) while achieving satisfactory vehicle fuel economy and extending battery life. [22] used a genetic algorithm to optimize the energy management strategy aimed at minimum fuel consumption and battery capacity degradation. [23] presented a model predictive control (MPC) strategy and analyzed the Pareto optimal front of the cost function comprised by the equivalent fuel consumption and battery capacity fade during the charge sustaining mode of the battery. [24] further provided the impact of the estimated SOC by the battery management system on the performance of MPC. [25] studied the nonlinear model predictive control for the energy management of a power-split hybrid electric vehicle (HEV) to improve battery aging while maintaining the fuel economy at a reasonable level. [26] employed the shooting method of PMP to obtain the optimal depth of discharge (DOD) and constructed a reference SOC with the optimal DOD, and then, a model predictive controller was used to optimize the conflict between the energy consumption cost and the equivalent battery life loss cost in a moving horizon. [27] used SDP and particle swarm optimization (PSO) to numerically solve the multi-objective optimal control problem under the consideration of the tradeoff between energy consumption and battery loss. The dynamic loop nest optimization of PSO and SDP was used to obtain offline an optimal solution according to the statistical characteristic of the real historical traffic data. The optimal solution was constructed as look-up mappings on different road segments and battery SOC so that in the online implementation of the management control strategy the power demand assignment can be obtained by these mappings without computational burden according to the current driving mode, system states, and road information.

It should be noted that it is necessary to keep a balance between fuel consumption and battery capacity loss in the design of the energy management control strategy for the economy of PHEV, while the designed management strategy should be integrated to global near optimization and the real-time performance. For this, both a globally sub-optimal and implementable energy management strategy, so-called recurrent neural network-based adaptive equivalent consumption minimization strategy (RNN-A-ECMS), is proposed in this paper for a power-split PHEV considering the battery life. Three efforts have been made. Firstly, RNN with long short term memory (LSTM) is trained utilizing the historical global optimal SOC trajectory that can be obtained by DP and real historical traffic data. Secondly, the maps of the weighting factor and main component of the ECMS' equivalent factor (EF) depending on power demand and battery SOC are obtained by PMP and PSO utilizing the real historical traffic data. The PSO is employed to search the weight factor and co-state of PMP, ensuring the vehicle's optimal economic performance, and the map of the PMP's co-state is converted into the main component's map of the ECMS's EF. Thirdly, the maps of weight factor and the main component of the EF and the model of the well-trained offline RNN are inserted into the core structure of A-ECMS to carry out the energy management control strategy responsibilities for online implementation.

The remainder of this paper is organized as follows. The PHEV mathematical model is given in Section 2. The optimal control problem is formulated in Section 3. Then, RNN-A-ECMS is designed in Section 4. The simulation result and the comparison with SDP-PSO are given to demonstrate the effectiveness of this approach in Section 5. Finally, the conclusion is summarized in Section 6.

2. PHEV Model Description

The power-split PHEV with a planetary gear set (PGS), shown in Figure 1, is analyzed in this paper. The powertrain of the PHEV mainly consists of the engine, the battery, motor, and generator.

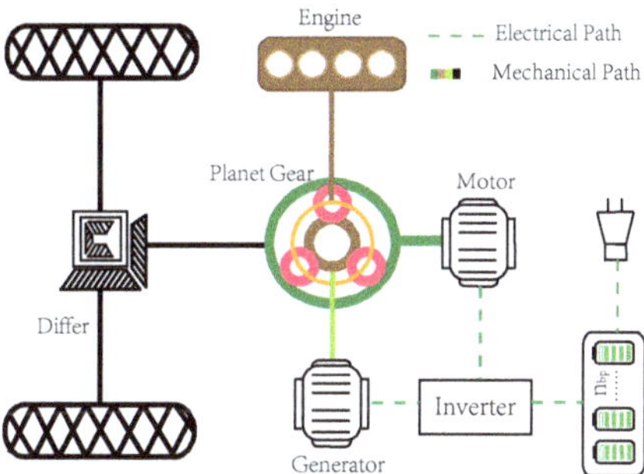

Figure 1. Plug-in hybrid electric vehicle (PHEV) powertrain system architecture.

Considering the aerodynamic drag and rolling friction force, the longitudinal dynamic equation of the vehiclecan be written as [28]:

$$M\dot{v} = \frac{\eta_f T_{trac} - T_{br}}{R_{tire}} - Mg\left(\mu_r \cos\alpha + \sin\alpha\right) - \frac{1}{2}\rho A C_d v^2, \tag{1}$$

where M is the PHEV mass. v denotes the velocity of the PHEV. η_f is the transmission efficiency of the differential gear. T_{trac} and T_{br} are the traction torque and brake torque, respectively. g is the gravity acceleration. μ_r is the coefficient of rolling resistance. α is the grade of the road. ρ is the air density. A is the frontal area of vehicle. C_d is the drag coefficient.

The PGS containing the sun gear, carrier gear, and ring gear respectively connecting to the generator, engine, and motor is a core component of PHEV, which allows the PHEV to run not only the series mode in which the engine provide power to generator to charge the battery or to motor through an inverter to drive the vehicle but also the parallel mode in which the engine directly propels the vehicle together with the motor. Under the assumption of rigid connections in the powertrain and without friction loss, the PGS speed and torque relationships are described as:

$$T_r = \frac{R_r}{R_r + R_s}T_c, \quad T_s = \frac{R_s}{R_r + R_s}T_c, \tag{2}$$

$$(R_r + R_s)\,\omega_c = R_r\omega_r + R_s\omega_s, \tag{3}$$

where T_r, T_c, and T_s are the torques of ring gear, carrier gear, and sun gear, and ω_r, ω_c, and ω_s are the speeds of ring gear, carrier gear, and sun gear, respectively. R_s and R_r are the teeth number of sun gears and ring gears, respectively. With the assumption that the connecting shafts are rigid, the speed relationship between planetary gear set and powertrain is described as follows:

$$\omega_c = \omega_e, \omega_r = \omega_m, \omega_s = \omega_g, \tag{4}$$

where ω_e, ω_m, and ω_g are the speeds of engine, motor, and generator, respectively. In addition, the motor speed can be computed by the following equation:

$$\omega_m = \frac{g_f}{R_{tire}}v, \tag{5}$$

where R_{tire} and g_f are the tire radius of PHEV and the ratio of differential shaft.

Energy consumption chosen as one part of optimization objective contains fuel consumption $\dot{m}_f$ and electricity consumption $\dot{m}_{elec}$ which are defined as follows.

$$\dot{m}_{fuel} = \text{BSFC}(\omega_e, T_e) \cdot T_e \cdot \omega_e \cdot 10^{-5}/36, \tag{6}$$

$$\dot{m}_{elec} = s \cdot P_{elec}/H_l, \tag{7}$$

where T_e is the torque of engine. BSFC is brake specific fuel consumption. H_l is the lower heating value of the fuel. s is EF. P_{elec} is the total battery power, which can be expressed as follows:

$$P_{elec} = P_b + P_l = P_b + I_b^2 \cdot R_b, \tag{8}$$

where P_l is the internal loss power of the battery, I_b and R_b are the current and equivalent internal resistance, respectively. P_b is the electrical load at the terminals, which can be written as follows:

$$P_b = \eta_m^{k_m} T_m\omega_m + \eta_g^{k_g} T_g\omega_g, \tag{9}$$

where η_m and η_g are the efficiency of the motor and generator, respectively. T_m and T_g are the motor torque and generator torque, respectively. $k_m/k_g = 1$ when the motor/generator is in a discharging state, otherwise $k_m/k_g = -1$.

Moreover, it has been shown the battery performance is affected by both storage time and usage, often categorized as Calendar life and Cycle life. Calendar life is the ability of the battery to withstand degradation over time, which may be independent of how much or how hard the battery is used. While, cycle life includes deep cycle life and shallow cycles. Deep cycle life is the number of discharge-recharge cycles the battery can perform in the charge-depleting (CD) mode. For example, one complete deep discharge with starting at 90% SOC, ending at 30% SOC, and recharging back to 90% SOC would complete one full cycle. Shallow cycles refer to SOC variations of only a few percent. These smaller variations occur throughout CD and charge-sustaining (CS) mode because the battery frequently takes in electric energy from the engine via a generator and from regenerative braking, and passes energy to the electric motor to power the vehicle. These frequent shallow cycles cause less degradation than deep cycles, but still affect battery life. Therefore, the management to the battery shallow cycle in operating modes should also be considered in order to minimize the battery life degradation in the discharge/charge cycles together with the minimization of energy consumption. There are many factors affecting battery life, such as temperature, Ah-throughput, and depth of discharge. With usage, battery performance in power and energy capacity can degrade. To get the depletion degree of battery capacity, the effective Ah-throughput (Ah_{eff}) is defined as [14,20]:

$$Ah_{eff}(t) = \int_0^t \sigma(I_b, T_{batt}, \text{SOC}) \cdot |I_b(t)|\, dt, \tag{10}$$

where T_{batt} is the battery temperature. $\sigma(\cdot)$ is called as severity factor, which describes the aging effect of any cycle the battery undergoes with respect to the nominal cycle, which is described as follows.

$$\sigma(I_b, T_{batt}, \text{SOC}) = \frac{\gamma(I_b, T_{batt}, \text{SOC})}{\Gamma} = \frac{\int_0^{EOL} |I_b(t)|\, dt}{\int_0^{EOL} |I_{nom}(t)|\, dt}, \tag{11}$$

where $\gamma(\cdot)$ is the battery duration (Ah-throughput) corresponding to a given sequence of current, temperature, and SOC. Γ is the total Ah-throughput corresponding to the nominal cycle, called as the nominal battery life, which is expressed as:

$$\Gamma = \int_0^{EOL} |I_{nom}(t)|\, dt, \tag{12}$$

where I_{nom} is the current profile under nominal conditions and EOL denotes the battery end of life. The battery life is regarded as the end when $Ah_{eff} = \Gamma$. Then, it may be regarded that the capacity loss is as $Q_{loss}\% = Ah_{eff}/\Gamma$. Thus, prolonging the battery life is equivalent to decreasing the depletion degree of battery capacity, which is also equivalent to minimizing the effective Ah-throughput. It should be worth mentioning that the severity factor $\sigma(\cdot)$ in this paper is obtained by the same approach as in [27], the fitting based on the experimental data.

Moreover, SOC dynamics in the energy management problem can be described as follows:

$$\dot{\text{SOC}} = -\frac{I_b}{Q_b} = -\frac{P_b}{Q_b \cdot U_{OC}}, \tag{13}$$

where U_{OC} is open circuit voltage, and Q_b is the battery capacity.

3. Optimization Problem Formulation

The energy management control for PHEVs is actually an optimization problem, which in this paper is to distribute power among the engine, motor, and generator meeting the power demand of the driver while minimizing energy consumption and prolonging battery life. To this end, the whole objective function can be written as follows:

$$J = \int_0^{t_f} \left\{ (1 - \theta(t)) \frac{\dot{m}_f(t) + \dot{m}_{elec}(t)}{\Omega} + \theta(t) \frac{\sigma(t) \cdot |I_b(t)|}{\Lambda} \right\} dt, \tag{14}$$

where $\theta(t) \in [0, 1]$ is a weight factor balancing two contradict objectives. Ω and Λ are introduced to make normalization, which are the optimal energy consumption with no consideration of battery loss, and the target effective Ah-throughput only the considering battery loss, respectively. The optimization is to calculate the control input u of the motor torque and the generator speed:

$$u^* = \left[T_m^*, \omega_g^* \right] = \arg \min_{u \in U} J \tag{15}$$

subject to the dynamic constraint in Equation (13) and the physical conditions:

$$\begin{cases} SOC_{\min} \leq SOC \leq SOC_{\max} \\ \omega_{e,\min} \leq \omega_e \leq \omega_{e,\max} \\ \omega_{m,\min} \leq \omega_m \leq \omega_{m,\max} \\ \omega_{g,\min} \leq \omega_g \leq \omega_{g,\max} \\ T_{e,\min}(\omega_e) \leq T_e \leq T_{e,\max}(\omega_e) \\ T_{m,\min}(\omega_m) \leq T_m \leq T_{m,\max}(\omega_m) \\ T_{g,\min}(\omega_g) \leq T_g \leq T_{g,\max}(\omega_g) \end{cases} \tag{16}$$

It should be noted from Equations (7) and (14) that the determinations of two factors θ, s play an important role in satisfying the objective and seeking the optimal solution in the actual operation of PHEV.

The weight factor $\theta(t)$ should balance the energy consumption and the battery aging in real time according to the operating mode and current state of the vehicle, which yields a Pareto front. Consequently, the determination of $\theta(t)$ is also an optimization problem. For this, in this paper, $\theta(t)$ will be offline optimized by PSO and PMP utilizing historical traffic data, and then established as a map table about the battery SOC and power demand.

The EF $s(t)$ should be chosen to be linked to the optimal co-state of the PMP for guaranteeing more closely the global optimal solution of the energy management control problem. For this, in this paper, the major component of the $s(t)$ is determined by the optimal co-state of the PMP that is optimized offline by PSO, and the tuned component will then be obtained by the PI controller of the deviation between the actual SOC and the reference trajectory of he SOC from a well-trained RNN by historical traffic data.

4. RNN-Based Adaptive Energy Management Strategy

In order to ensure that the designed management strategy is integrated, for global near optimization and real-time performance, in this paper, ECMS is used as the core algorithm of energy management. Considering the uncertainty of the driving condition, a RNN-based adaptive ECMS (RNN-A-ECMS) energy management strategy is designed. The design process of RNN-A-ECMS can be divided into the offline design part and the online implementation part.

The offline design includes two parts. One is the offline training of the recurrent neural network (RNN), in which the base set of the reference SOC is first generated by the DP algorithm using historical speed profile data, and then the RNN is trained by the reference SOC from the base set, as well as road and vehicle speed information extracted from historical traffic information. The other is offline optimizing the weight factor $\theta(t)$ and the co-state of the PMP corresponding to the major component of EF $s(t)$ using PSO, and then to establish the maps of the weight factor and the major component of the EF of A-ECMS.

In the online part, the implemented energy management control strategy includes three parts: The core ECMS with the weight factor $\theta(t)$, mapping table considering the battery life, the adaptive mechanism of EF $s(t)$ combining the main component mapping table and a PI controller, and a well-trained RNN generating the SOC reference according to current traffic information and vehicle states. The sketch of the design process is shown in Figure 2.

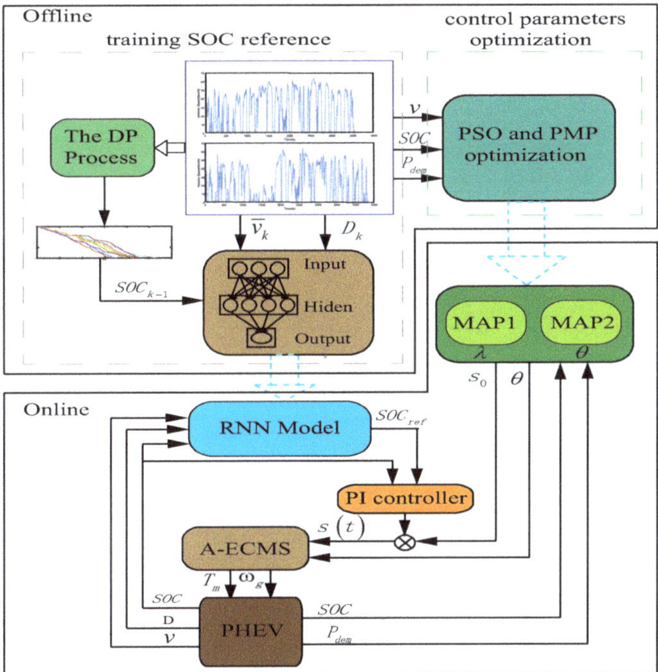

Figure 2. The sketch of the design process of the recurrent neural network- (RNN) based adaptive equivalent consumption minimization strategy (A-ECMS).

It is worth mentioning that in the actual operation of PHEV the energy management control is not complicated and has no computing burden, but the near-global optimality can be guaranteed because of the existence of RNN and the adaptability of the equivalent factor and weighted factor.

The details of the offline design for RNN predicting the SOC reference and maps of the weight factor and the major component of EF are as follows.

4.1. Prediction Model of SOC Reference of RNN

PHEV is different from HEV with the distinguishing feature of a larger battery capacity and being recharged from the power grid. Thus, it is favorable for fuel consumption and battery health to plan ahead a reasonable reference SOC trajectory, which is obtained based on the historical data and current traffic information. More specifically, the average speed $\bar{v}$ of current driving information, distance D, and the SOC of the previous step are chosen as the input of RNN. The average speed ($\bar{v}$) is equal to the distance traveled every ten seconds of PHEV divided by the time. The SOC of the previous step is obtained by the DP optimization and historical traffic data. DP is the global optimization algorithm that can convert the continuous optimization control problem into finding an optimal decision problem for n segments under the known driving cycles. Therefore, by using the historical driving cycles of traffic data, the optimal SOC trajectory obtained from the optimal control sequence of DP optimization can be served as the required SOC of the previous step. At this point, the detail is as follows:

Optimization is regarded as the search for a control decision variable so as to minimize the cost function in Equation (17) while satisfying the constraint condition of Equation (16):

$$J = \sum_{k=0}^{n-1} L\left[x(k), u(k), k\right] = \sum_{k=0}^{n-1} \dot{m}_{fuel}(k) \cdot \Delta t, \tag{17}$$

where n and L represent the duration of the driving cycle and instantaneous fuel consumption, respectively. Δt is the increment of time step and chosen as 1s in order to alleviate the computational burden. According to the Behrman's optimal principle, the optimal control problem described as Equation (17) can be decomposed into a series of single level decision problems. The specific steps are as follows:

$$Step\ n-1: J^*\left[x(n-1)\right] = \min_{u(n-1)}\left\{L\left[x(n-1), u(n-1), n-1\right] + G\left[x(n)\right]\right\}, \tag{18}$$

$$Step\ k(0 \le k \le n-2): J^*\left[x(k)\right] = \min_{u(k)}\left\{L\left[x(k), u(k), k\right] + J^*\left[x(k+1)\right]\right\}, \tag{19}$$

where $J^*\left[x(k)\right]$ is the optimal cost of the step k. $x(k)$ is the state variable SOC. $G\left[x(n)\right]$ is the cost of the step n. The solution of DP algorithm can be divided into the backward and forward process. In the backward process, optimal control u of each step is solved reversely according to the Equations (18) and (19). In the forward process, the optimal control sequence solved by the backward process is substituted into the system state equation to calculate the optimal SOC trajectory, which is regarded as the base sets of training RNN prediction model.

As SOC, distance, and vehicle speed are a sequence about time, RNN is chosen as the predictive model. Although it might be difficult to learn long-term dependence due to the vanishing gradient problem resulting from the gradient propagation of the recurrent network over many layers, Long Short Term Memory (LSTM) can overcome the gradient disappearance in the basic RNN when it introduces a forgetting mechanism. Thus, LSTM can more accurately predict the SOC reference than the basic RNN.

The structure of RNN with LSTM is shown in Figure 3, where x_i represents the input of LSTM including the average speed $\bar{v}_i$, distance D_i, and the SOC of the previous step. SOC_i is the output of LSTM representing the current SOC reference. The bottom of Figure 3 is the relationship among hidden layers which are named long and short term memory units.

The memory cell c_t that retains data of the time step $(t-1)$ plays a important role in the LSTM model. Keeping the value or resetting the value of the cell c_t is managed by several gates. Specially, forgetting, reading, and outputting the new cell value are decided by the forget gate (f_t), input gate (i_t), input modulation gate (g_t), and output gate (o_t), which are defined as follows:

$$f_t = \mathcal{F}\left(\begin{pmatrix} w_{hf} & w_{xf} \end{pmatrix} \cdot \begin{pmatrix} h[t-1,:] \\ x[t,:] \end{pmatrix} + b_f\right), \tag{20}$$

$$i_t = \mathcal{F}\left(\begin{pmatrix} w_{hi} & w_{xi} \end{pmatrix} \cdot \begin{pmatrix} h[t-1,:] \\ x[t,:] \end{pmatrix} + b_i\right), \tag{21}$$

$$g_t = \tanh\left(\begin{pmatrix} w_{hg} & w_{xg} \end{pmatrix} \cdot \begin{pmatrix} h[t-1,:] \\ x[t,:] \end{pmatrix} + b_g\right), \tag{22}$$

$$o_t = \mathcal{F}\left(\begin{pmatrix} w_{ho} & w_{xo} \end{pmatrix} \cdot \begin{pmatrix} h[t-1,:] \\ x[t,:] \end{pmatrix} + b_o\right). \tag{23}$$

Moreover, the calculations on the cell update and output are defined as follows:

$$c_t = f_t \odot c_{t-1} + i_t \odot g_t, \tag{24}$$

$$h_t = o_t \odot \tanh(c_t), \tag{25}$$

where $\odot$ denotes the multiplying each element and the w matrices are the network important parameters. h_t is the hidden state and employed to compute the current output and the next step hidden state. The LSTM can perfect to solve the vanishing gradients. The activation function $\mathcal{F}$ and tanh are the nonlinearity functions of logistic sigmoid and hyperbolic tangent, respectively.

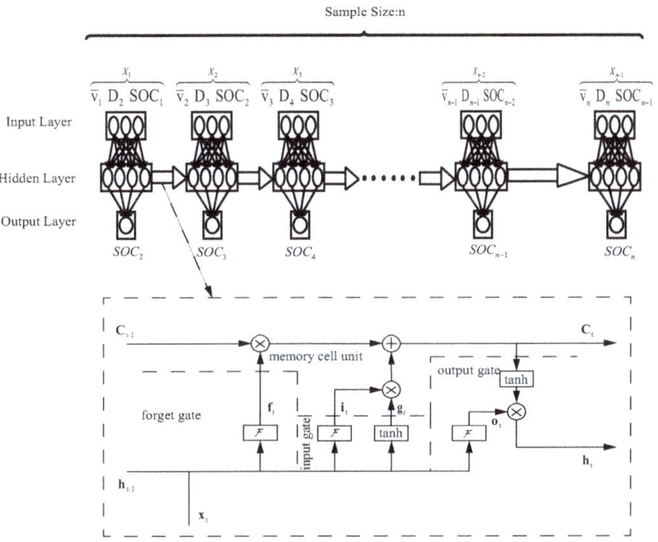

Figure 3. Structure of Long Short Term Memory (LSTM).

4.2. Control Parameters Optimization Based on PSO-PMP

To improve adaptability to the changes in driving conditions, a kind of adaptive EF $s(t)$ is selected, which can be updated by a PI controller in real-time. The specific formulation is as follows:

$$s(t) = s_0 + \delta s = s_0 + \left[K_p(\text{SOC}_{ref} - \text{SOC}) + K_i \int_0^t (\text{SOC}_{ref} - \text{SOC})dt \right], \tag{26}$$

where s_0 is the initial value of equivalent factor $s(t)$ (major component of EF) and δs is the tuning component of EF. SOC, SOC_{ref} represent the actual SOC and the reference SOC trajectory from the RNN, respectively. K_p, K_i are proportional and integral coefficients, respectively, which are determined using the estimation method for the upper and lower bounds of the EF presented in [29] and calibration with trials similar to [13].

In order to more closely obtain the global optimal solution of the energy management control problem, the s_0 is determined by the co-state $\lambda(t)$ of the PMP optimized offline by PSO. In the PMP, minimizing the objective function is converted to minimizing the Hamiltonian function:

$$H(\text{SOC}(t), u(t), \lambda(t), \theta(t)) = (1 - \theta(t))\frac{\dot{m}_f(t)}{\Omega} + \theta(t)\frac{\sigma(t) \cdot |I_b(t)|}{\Lambda} + \lambda(t) \cdot \dot{\text{SOC}}, \tag{27}$$

where $\lambda(t)$ is the co-state, and its dynamics can be described as:

$$\dot{\lambda}(t) = -\frac{\partial H(\text{SOC}(t), u(t), \lambda(t), \theta(t))}{\partial \text{SOC}}. \tag{28}$$

The optimal control trajectory is given by:

$$u^* = \left[T_m^*, \omega_g^* \right] = \arg\min_{u \in U} H(\mathrm{SOC}(t), u(t), \lambda(t), \theta(t)), \tag{29}$$

where the weighting factor $\theta(t)$ and the co-state $\lambda(t)$ are obtained offline by PSO. The flowchart of PSO is shown in Figure 4.

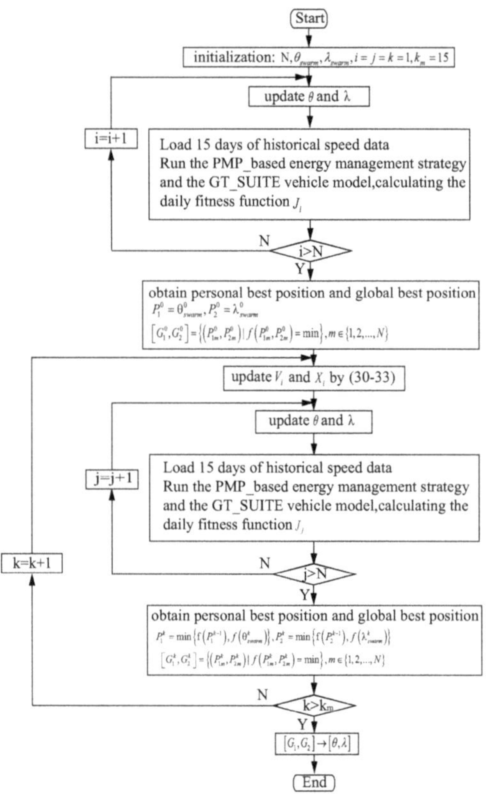

Figure 4. Particle swarm optimization (PSO) flowchart.

In PSO, considering the optimization time and convergence efficiency, the number of swarm N and the maximum iteration k_m is set as 5, 15, respectively. In order to facilitate online looking-up table, SOC is discretized as $[0.3, 0.35, 0.4, 0.45, 0.5, 0.55, 0.6, 0.65, 0.7, 0.75, 0.8, 0.85, 0.9]$. As the range of power demand of the studied PHEV is 0 kW to 50 kW and usually locates in 0 kW to 30 kW, power demand is discretized as $[0, 2, 4, 6, 8, 10, 12, 14, 16, 18, 20, 22, 24, 26, 28, 30, 35, 40, 45, 50]$. Thus, in the PSO algorithm, the swarm is defined as $X_1 = \theta_{swarm}$, $X_2 = \lambda_{swarm}$, and the initialization swarm θ_{swarm}^0 and λ_{swarm}^0 are denoted as two $[20 \times 13 \times N]$-dimensional tensors. The velocities of θ_{swarm}, λ_{swarm} are V_1 and V_2 which are also two $[20 \times 13 \times N]$-dimensional tensors. The update principles of the velocities V_1, V_2 and positions X_1, X_2 during the iterations are described as follows:

$$V_{1i}^{k+1} = wV_{1i}^k + c_1 r_1 \left(P_{1i}^k - \theta_{swarm}^k \right) + c_2 r_2 \left(G_{1i}^k - \theta_{swarm}^k \right), \tag{30}$$

$$V_{2i}^{k+1} = wV_{2i}^k + c_1 r_1 \left(P_{2i}^k - \lambda_{swarm}^k \right) + c_2 r_2 \left(G_{1i}^k - \lambda_{swarm}^k \right), \tag{31}$$

$$X_{1i}^{k+1} = X_{1i}^k + V_{1i}^{k+1}, \tag{32}$$

$$X_{2i}^{k+1} = X_{2i}^k + V_{2i}^{k+1}, \tag{33}$$

where $i = 1, 2, ..., N$ is the current number of swarm and $k = 1, 2, ..., k_m$ is the current iteration step. w, c_1, and c_2 are particle inertia and acceleration constants, respectively. $r_1, r_2 \in [0, 1]$ are uniformly distributed random values. P_1/P_2 is the individual extremum and G_1/G_2 is global extremum.

Figure 5 shows the optimization result of the fitness function, which can converge to a constant. The maps of θ and λ according to the battery SOC and the power demand are shown as Figure 6.

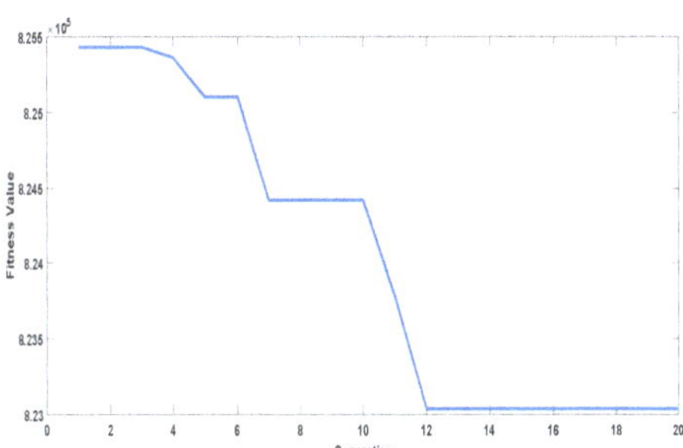

Figure 5. Fitness value.

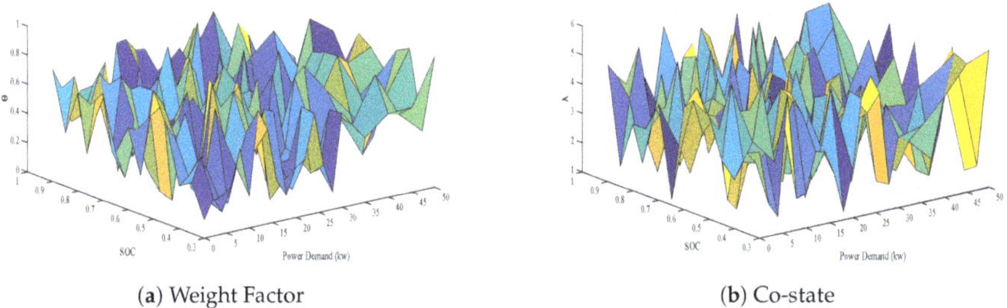

(**a**) Weight Factor   (**b**) Co-state

Figure 6. Maps of weighting factor and co-state.

From Figure 6, it can be seen that both weighting factor θ and co-state λ have a complex functional relationship with SOC and power demand. However, some qualitative conclusions can be drawn, for example, when the power demand is higher than 30 kW and SOC is less than 0.5, the weight factor has larger value. Meanwhile, when the power demand is lower than 10 kW and SOC is larger than 0.7, the co-state has a smaller value. According to the objective function defined in Equations (14) and (7), the evaluation of battery aging is more important than that of energy consumption in optimization than if the θ had a larger value, and then the engine would be used more. Similarly, if the co-state had a smaller value, namely, the corresponding EF was smaller, it means that the evaluation of fuel consumption would be more important than that of electrical consumption in the optimization of energy consumption, which would lead to more use of the battery-powered motor.

5. Simulation Verification on GT-SUITE Test Platform

The effectiveness of the proposed energy management strategy in this paper is demonstrated on the GT-SUITE test platform. The detail of the PHEV system with the proposed energy management control strategy in the simulation is shown in Figure 7. Where the PHEV model is established in GT-SUITE environment so as to more realistically simulate the real PHEV powertrain. The proposed energy management control strategy is constructed in MATLAB/Simulink (MathWorks, Natick, MA, USA) which computes the required torque of motor T_m^* and the required speed of generator ω_g^*. Then, it can send the two control variable to PHEV in real-time. In Figure 7, the module of cost function consists of energy consumption and battery life characterized as effective Ah-throughput, which are balanced by the weight factor θ.

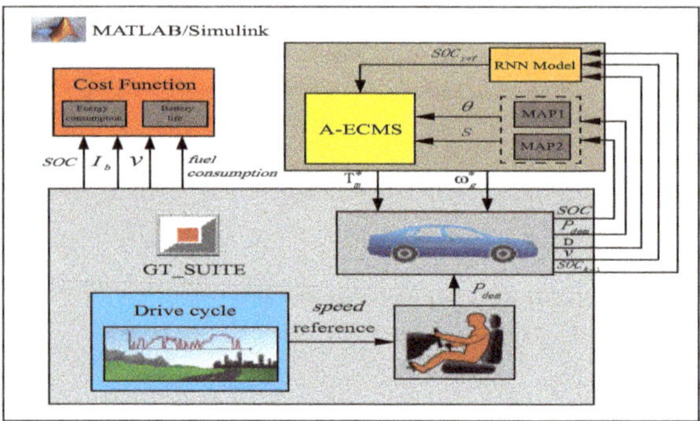

Figure 7. Simulation environment.

In simulation, the parameters and specifications of model components are listed in Table 1.

Table 1. PHEV model parameters.

Notation	Meaning	Parameters
M	Vehicle mass	1460 kg
A	Frontal area	3.8 m^2
C_d	Air drag coefficient	0.33
ρ	Air density	1.293 kg/m^3
$R_t ire$	Radius of the tire	0.2982 m
μ	Coefficient of rolling resistance	0.015
Q_b	Battery maximum charge capacity	6.5 Ah
P_m, max	Motor max power	50 kw
P_g, max	Generator max power	30 kw
J_e	Inertia of the engine crankshaft	0.16 kg $\cdot$ m^2
J_m	Inertia of the motor	0.035 kg $\cdot$ m^2
J_g	Inertia of the generator	0.0265 kg $\cdot$ m^2
g_f	Final differential gear ratio	40,113
R_s	Sun gear teeth number	30
R_r	Ring gear teeth number	78

Moreover, the historical traffic data used in training RNN and offline optimization of DP and PSO is from the real driving cycles provided by [30]. Then, the driving cycles, used in testing RNN and verifying the effectiveness of the online implementation of the designed energy management control strategy, are also from the historical traffic data provided by [30]. Additionally, the Urban Dynamometer Driving Schedule (UDDS) and New European Driving Cycle (NEDC) were chosen as the test driving cycles. However, it is worth pointing out that the actual traffic routes of the HEV [30] is 28 km from home to office and back from office. In research, it is assumed that the PHEV is charged once after a day's commute, namely, SOC is the maximum of 0.9 before going to work in the morning and SOC is about 0.4 after 28 km, close to the minimum 0.3. Accordingly, the distances of both UDDS and NEDC are too short to fully verify the effectiveness and practicability of the proposed control strategy, because the SOC change is too small for such a short distance. For this regard, both UDDS and NEDC are repeated up to 28 km.

Firstly, because the precision of the RNN model has a significant impact on the adaptability of the energy management strategy, the constructed RNN is verified by a comparison between the basic RNN and LSTM. The third week of traffic data not used in training is chosen as the testing set. Figure 8 shows the comparison between the two in three evaluation indexes, mean absolute error (MAE), mean radial error (MRE), and root mean square error (RMSE), which are defined as follows:

$$\text{MAE} = \frac{\sum\limits_{i=1}^{I} |F_i - T_i|}{I},$$ (34)

$$\text{MRE} = \frac{\sum\limits_{i=1}^{I} |F_i - T_i|/T_i}{I},$$ (35)

$$\text{RMSE} = \sqrt{\frac{\sum\limits_{i=1}^{I} |F_i - T_i|^2}{I}},$$ (36)

where I is the total number of prediction point, F_i is the prediction value of the battery SOC, and T_i is the true value of the battery SOC.

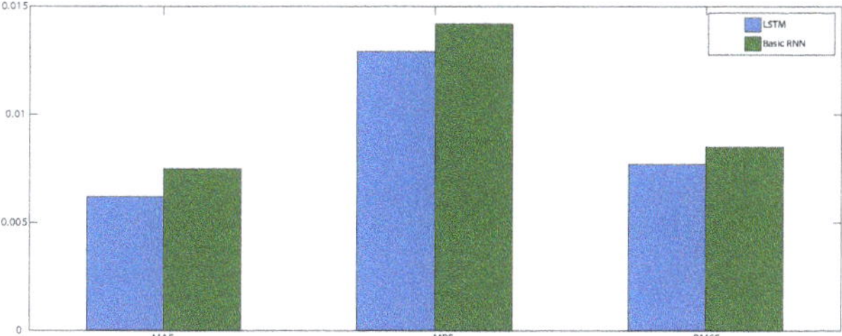

Figure 8. Comparison of basic RNN and LSTM.

It is obvious from Figure 8 that the LSTM has better precision than the basic RNN. That is because LSTM introduces the forgetting mechanism based on the basic RNN, which can capture long-term dependencies.

To verify the applicability of the proposed strategy on a real driving cycle on the third Monday week dating from [30], Urban Dynamometer Driving Schedule (UDDS) and New European Driving Cycle (NEDC) were chosen as the test driving cycles. Meanwhile, to demonstrate the advantage of the proposed energy management control strategy in performance improvement, the comparison results among the proposed strategy (RNN-A-ECMS), the SDP-PSO energy management strategy proposed in [27], and the conventional charge depleting-charge sustaining (CD-CS) strategy are given based on the same simulation environment. In the CD-CS mode strategy, the threshold SOC switching from charge depleting (CD) to charge sustaining (CS) mode was set as 0.35 instead of the lowest value 0.3 so as to avoid excessive discharging of the battery.

Figures 9–11 show the simulation results and comparisons among the three strategies under the three driving cycles, respectively. Figure 9a–d are the operating results of the PHEV with the proposed RNN-A-ECMS strategy, where it can be seen that the actual vehicle speed profile could greatly track the reference speed profile, see Figure 9a, which guarantees the drivability due to low power demand (Figure 9b), the engine not working long (Figure 9c), and the torque and speed of engine, motor, and generator matching with the power demand of the driver and the PHEV working in pure electric mode most of the time (Figure 9d). Figure 9e–h are the comparisons on SOC trajectory, fuel consumption, effective Ah-throughput, and equivalent fuel consumption including electricity consumption. It indicates that the proposed energy management strategy had a better control performance: The actual SOC curve was closest to the reference SOC predicted by RNN (Figure 9e). Fuel consumption, effective Ah-throughput, and equivalent fuel consumption all were much lower than that of CD-CS although the effective Ah-throughput of the proposed RNN-A-ECMS was a little more than that of the

SDP-PSO, the fuel consumption of RNN-A-ECMS was much lower than that of the SDP-PSO, so that the equivalent fuel consumption of the RNN-A-ECMS was the lowest.

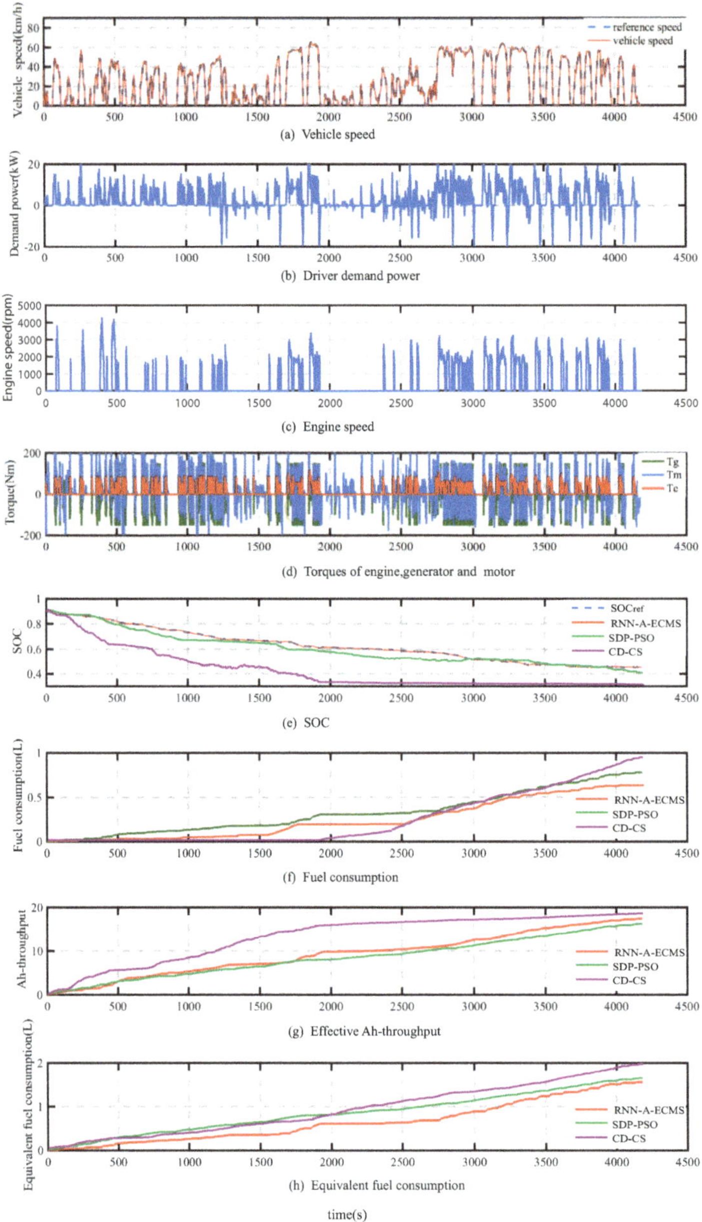

Figure 9. Simulation and comparison results under a real driving cycle.

Similar simulation results can be seen from Figures 10 and 11. Although UDDS and NEDC are not in the database used by RNN, the driveability and control performance are also satisfied in these two conditions.

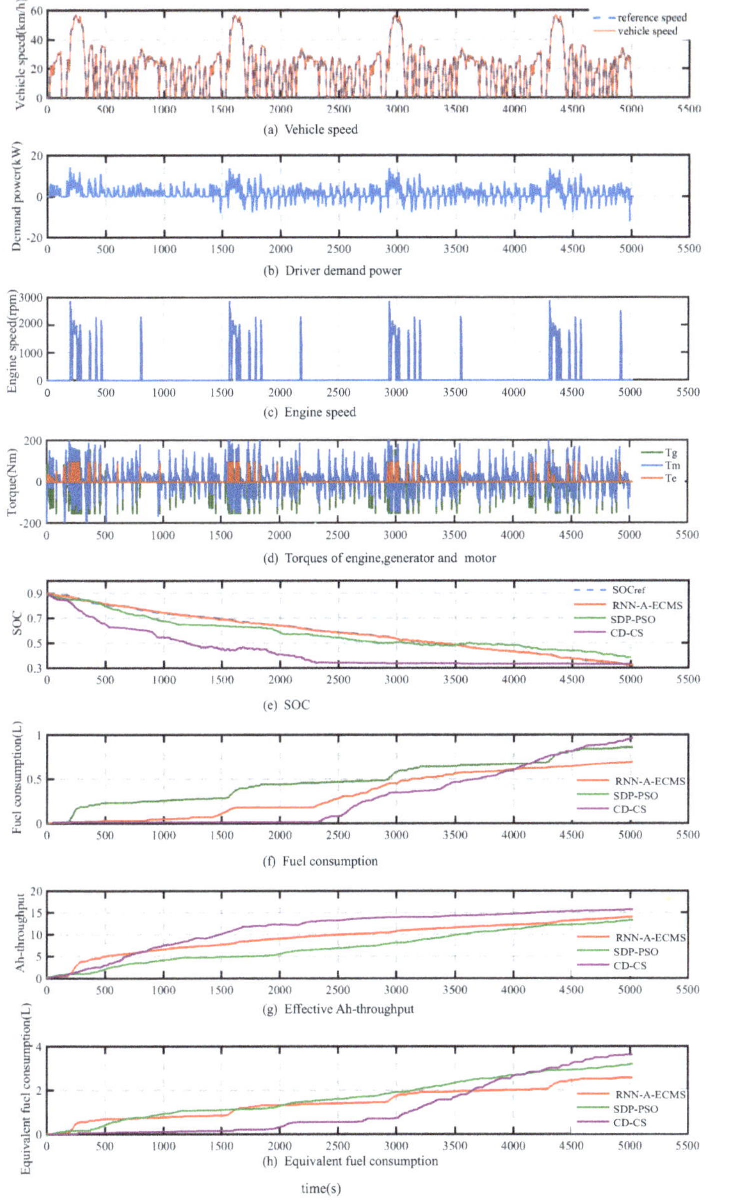

Figure 10. Simulation and comparison results under Urban Dynamometer Driving Schedule (UDDS).

It shows that RNN with LSTM have a stronger generalization ability than the basic RNN. Moreover, the weight factor and the initial value of equivalent factor are converted into a 2-dimension table, which can obtain the different θ and s_0 according to the real-time traffic information to adapt the different driving conditions and get the optimal solution and better control performance.

To further demonstrate the advantage of the proposed energy management control strategy in the performance improvement, the comparison results among the RNN-A-ECMS, the SDP-PSO [27], and the CD-CS strategy are given under multiple driving cycles.

Firstly, the simulation result in another driving cycle on the second Monday week dating from [30] is shown in Figure 12. Table 2 shows the simulation comparison results of the three strategies in the driving cycles of the workdays in the second week, which include the fuel consumption per hundred kilometers FC [L/100 km], battery Q_{loss}, and final SOC SOC_{fin}.

From Figure 12, it can be seen that the final SOC of the RNN-A-ECMS was 0.42 which is lower than the final SOC of the SDP-PSO strategy, which was 0.49, and the final SOC of the CD-CS strategy was always 0.35. The lower average final SOC could reflect that RNN-A-ECMS was more dependent on battery for driving than the SDP-PSO. It may lead to that the battery Q_{loss} of RNN-A-ECMS was slightly higher than that of the SDP-PSO. However, the fuel consumption of the RNN-A-ECMS is greatly lower than that of the SDP-PSO because the battery is frequently involved in driving in RNN-A-ECMS. It indicates that RNN-A-ECMS sacrifices small battery loss to greatly increase fuel consumption. Without the optimization management for either fuel consumption or the battery aging in the CD-CS strategy, no matter what driving conditions the electrical power is first used to propel the vehicle until the CD-CS switching threshold value of the SOC and then engine works while retaining the threshold level of SOC, as a result, both the battery Q_{loss} and the fuel consumption are the highest.

Table 2. Simulation comparison results.

Day	CD-CS			SDP-PSO			RNN-A-ECMS		
	FC	Q_{loss}	SOC_{fin}	FC	Q_{loss}	SOC_{fin}	FC	Q_{loss}	SOC_{fin}
Mon	3.175	6.477×10^{-4}	0.35	2.614	6.185×10^{-4}	0.48	2.496	6.203×10^{-4}	0.32
Tue	3.798	7.615×10^{-4}	0.35	3.311	7.343×10^{-4}	0.52	3.203	7.412×10^{-4}	0.36
Wed	3.054	7.025×10^{-4}	0.35	2.745	6.831×10^{-4}	0.43	2.599	6.865×10^{-4}	0.39
Thu	3.531	6.902×10^{-4}	0.35	3.029	6.579×10^{-4}	0.41	2.925	6.688×10^{-4}	0.35
Fri	2.967	7.143×10^{-4}	0.35	2.450	6.712×10^{-4}	0.39	2.314	6.836×10^{-4}	0.33
ave	3.305	7.032×10^{-4}	0.35	2.829	6.730×10^{-4}	0.45	2.707	6.800×10^{-4}	0.35

According to the average of the second week in Table 2, it can be calculated that the fuel consumptions of the RNN-A-ECMS and the SDP-PSO were reduced by 18.1% and 14.4% compared with that of the CD-CS strategy, respectively. The battery losses of the two strategies were also reduced by 3.3% and 4.3%, respectively. Meanwhile, the fuel consumption was reduced by 4.3%, however, battery loss only sacrificed 1.03% between the two strategies. It means that the RNN-A-ECMS calculating different weight values for different SOC and power demand could be a better solution to the multi-objective optimization problem than the SDP-PSO.

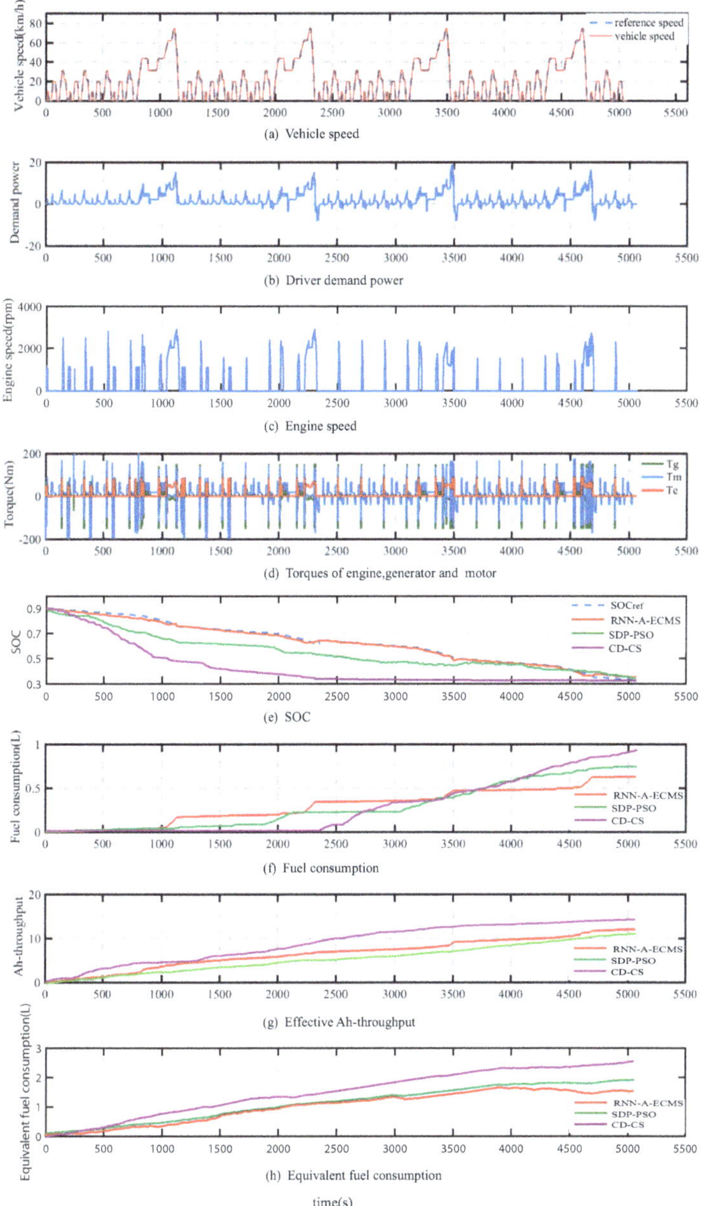

Figure 11. Simulation and comparison results under the New European Driving Cycle (NEDC).

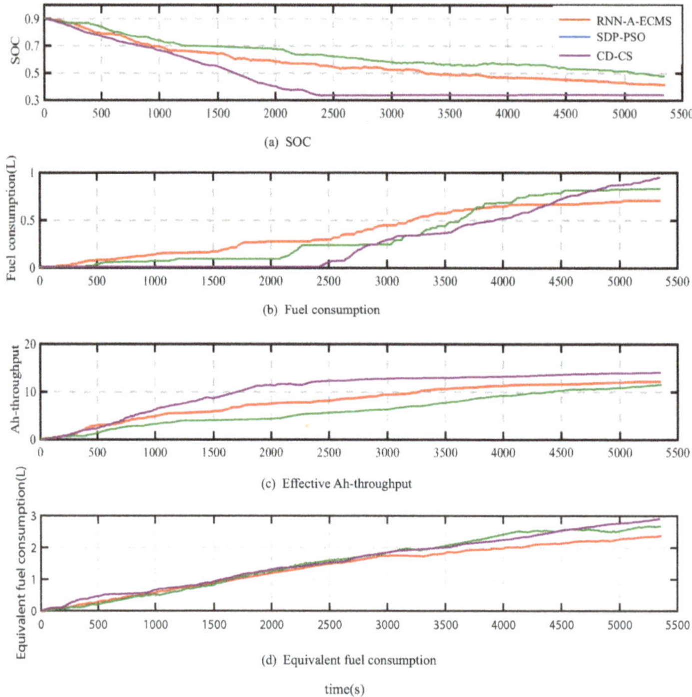

Figure 12. Comparison result between the proposed RNN-A-ECMS and the stochastic dynamic programming (SDP)-PSO strategy.

6. Conclusions

This paper proposed a novel sub-optimal and real-time energy management control strategy RNN-A-ECMS to distribute power demand between the engine and electric machines while considering the fuel consumption and battery aging. Prolonging the battery life and decreasing the fuel consumption were contradictory. Thus, the energy management strategy including battery aging should be regarded as a multi-objective optimization problem. In order to gain the Pareto optimal set, PMP and PSO were used in this paper to solve the multi-objective optimal problem offline, and the time-varying weight factor and the major component the ECMS's EF were obtained as two maps depending on power demand and SOC. In order to enhance adaptation to uncertain driving conditions, RNN with LSTM was trained offline using historically optimal SOC trajectory resulting from DP, and a PI controller was used to form the adaptive mechanism of the adaptive EF. In the implementation of the control strategy, the values of weighting factor and the major component of equivalent factor were generated online by looking up the two maps according to the current SOC of the battery and power without computational burden. Meanwhile, the equivalent factor was adjusted by the PI controller in order to make the actual SOC trajectory close to the optimal SOC trajectory, which could ensure that the real-time energy management strategy was closer to the optimal energy management strategy. The simulation verification and comparison with the existing strategy, which were implemented on GT-SUITE test platform, showed that the proposed energy management strategy in this paper possessed the effectiveness and adaptability to various driving cycles and had the advantage in compromising multi-objective of decreasing the fuel consumption and prolonging battery life.

Author Contributions: L.H. and X.J. conceived of and designed the framework of control methods. L.H. wrote the initial draft, drew the figures, and performed the simulations. X.J. was responsible for supervising this research and involved in exchanging ideas and reviewing the article draft. Z.Z. provided assistance with the architecture of RNN. All authors have read and agreed to the published version of the manuscript.

Funding: This research was funded by the National Natural Science Foundation of China (Grant No. 61573304 and No. 61973265) and the Natural Science Foundation of Hebei Province (Grant No. F2017203210).

Acknowledgments: The authors would like to express their thanks to Yuji Yasui and Masakazu Sasaki for the vehicle speeds and battery degradation data support for this research, respectively.

Conflicts of Interest: The authors declare no conflict of interest.

Nomenclature

M	Vehicle mass [kg].
g	Gravity acceleration [m/s^2].
A	Frontal area [m^2].
C_d	Air drag coefficient [−].
ρ	Air density [kg/m^3].
R_tire	Radius of the tire [m].
μ_r	Coefficient of rolling resistance [−].
α	Grade of the road [−].
v	Velocity of the vehicle [m/s].
η_f	Transmission efficiency of the differential gear [−].
g_f	Gear ratio of differential shaft [−].
T_{trac}	Traction torque [Nm].
T_{br}	Brake torque [Nm].
R_s/R_r	Sun/Ring gear teeth number [−].
$T_s/T_r/T_c$	Torque of sun/ring/carrier gear [Nm].
$T_e/T_m/T_g$	Torque of engine/motor/generator [Nm].
$T_{e,max}/T_{m,max}/T_{g,max}$	Max torque of engine/motor/generator [Nm].
$T_{e,min}/T_{m,min}/T_{g,min}$	Min torque of engine/motor/generator [Nm].
$\omega_s/\omega_r/\omega_c$	Speed of sun/ring/carrier gear [rad/s].
$\omega_e/\omega_m/\omega_g$	Speed of engine/motor/generator [rad/s].
$\omega_{e,max}/\omega_{m,max}/\omega_{g,max}$	Max speed of engine/motor/generator [Nm].
$\omega_{e,min}/\omega_{m,min}/\omega_{g,min}$	Min speed of engine/motor/generator [Nm].
η_m/η_g	Efficiency of the motor/generator [−].
$P_{m,max}/P_{g,max}$	Max power of motor/generator [kW].
P_m/P_g	Power of motor/generator [kW].
$J_e/J_m/J_g$	Inertia of engine/motor/generator [kg · m^2].
BSFC	Brake specific fuel consumption [g/kWh].
$\dot{m}_{fuel}$	Fuel consumption [g/s].
$\dot{m}_{elec}$	Electricity consumption [g/s].
H_l	Lower heating value of the fuel [J/g].
P_{elec}	Total battery power [W].
P_b	Output power of the battery [W].
P_l	Internal loss power of the battery [W].
I_b	Current of the battery [A].
R_b	Equivalent internal resistance of the battery [Ω].
U_{OC}	Open circuit voltage of the battery [V].
Q_b	Battery maximum charge capacity [Ah].
Γ	Nominal battery life [−].
I_{nom}	Battery current profile under nominal conditions [A].
EOL	Battery end of life [−].
Ah_{eff}	Effective Ah-throughput [Ah].

σ	Severity factor of relative aging effects of battery $[-]$.
T_{batt}	Battery temperature $[^\circ C]$.
SOC	State of charge of the battery $[-]$.
SOC_{max}/SOC_{min}	Max and min SOC of the battery $[-]$.
$\bar{v}$	Average speed of the vehicle of current driving information $[m/s]$.
D	Distance the vehicle traveled $[m]$.
c_t	Memory cell of RNN $[-]$.
h_t	Hidden state of RNN $[-]$.
f_t	Forget gate of RNN $[-]$.
i_t	Input gate of RNN $[-]$.
g_t	Input modulation gate of RNN $[-]$.
o_t	Output gate of RNN $[-]$.
$\mathcal{F}$	Nonlinearity functions of logistic sigmoid $[-]$.
$tanh$	Nonlinearity functions of hyperbolic tangent $[-]$.
θ	Weight factor $[-]$.
λ	Co-state factor of PMP $[-]$.
s	Equivalence factor $[-]$.
s_0	Major component of equivalence factor $[-]$.
δs	Tuning component of equivalence factor $[-]$.
SOC_{ref}	Reference of SOC $[-]$.
K_p	Proportional coefficient of PI $[-]$.
K_i	Integral coefficient of PI $[-]$.
X_1, X_2	Position of particle $[-]$.
V_1, V_2	Velocity of particle $[-]$.
P_1, P_2	Individual extremum of particles $[-]$.
G_1, G_2	Global extremum of particles $[-]$.
c_1, c_2	Acceleration constants in update principles of PSO $[-]$.
w	Inertia constant in update principles of PSO $[-]$.
$r1, r2$	Uniformly distributed random values in update principles of PSO $[-]$.
k_m	Maximum iteration step of PSO $[-]$.

References

1. Zhang, S.; Xiong, R. Adaptive energy management of a plug-in hybrid electric vehicle based on driving pattern recognition and dynamic programming. *Appl. Energy* **2015**, *155*, 68–78. [CrossRef]
2. Padmarajan, B.V.; Mcgordon, A.; Jennings, P.A. Blended rule-based energy management for PHEV: System structure and strategy. *IEEE Trans. Veh. Technol.* **2016**, *65*, 8757–8762. [CrossRef]
3. Yang, C.; Du, S.; Li, L.; You, S.; Yang, Y.; Zhao, Y. Adaptive real-time optimal energy management strategy based on equivalent factors optimization for plug-in hybrid electric vehicle. *Appl. Energy* **2017**, *203*, 883–896. [CrossRef]
4. Huang, Y.; Wang, H.; Khajepour, A.; He, H.; Ji, J. Model predictive control power management strategies for HEVs: A review. *J. Power Sources* **2017**, *34*, 91–106. [CrossRef]
5. Xu, L.; Yang, F.; Li, J.; Ouyang, M.; Hua, J. Real time optimal energy management strategy targeting at minimizing daily operation cost for a plug-in fuel cell city bus. *Int. J. Hydrog. Energy* **2012**, *37*, 15380–15392. [CrossRef]
6. Zhang, Y.; Jiao, X.; Li, L.; Yang, C.; Zhang, L.P.; Song, J. A hybrid dynamic programming-rule based algorithm for real-time energy optimization of plug-in hybrid electric bus. *Sci. China Technol. Sci.* **2014**, *57*, 2542–2550. [CrossRef]
7. Serrao, L.; Onori, S.; Rizzoni, G. A comparative analysis of energy management strategies for hybrid electric vehicles. *J. Dyn. Syst. Meas. Control* **2011**, *133*, 031012. [CrossRef]
8. Onori, S.; Tribioli, L. Adaptive Pontryagin's minimum principle supervisory controller design for the plug-in hybrid GM Chevrolet Volt. *Appl. Energy* **2015**, *147*, 224–234. [CrossRef]
9. Hou, C.; Ouyang, M.; Xu, L.; Wang, H. Approximate Pontryagin's minimum principle applied to the energy management of plug-in hybrid electric vehicles. *Appl. Energy* **2014**, *115*, 174–189. [CrossRef]

10. Borhan, H.; Vahidi, A.; Phillips, A.M.; Kuang, M.L.; Kolmanovsky, I.V.; Di Cairano, S. MPC-based energy management of a power-split hybrid electric vehicle. *IEEE Trans. Control Syst. Technol.* **2012**, *20*, 593–603. [CrossRef]

11. Li, T.; Liu, H.; Ding, D. Predictive energy management of fuel cell supercapacitor hybrid construction equipment. *Energy* **2018**, *149*, 718–729. [CrossRef]

12. Liu, H.; Li, X.; Wang, W.; Han, L.; Xiang, C. Markov velocity predictor and radial basis function neural network-based real-time energy management strategy for plug-in hybrid electric vehicles. *Energy* **2018**, *152*, 427–444. [CrossRef]

13. Xie, S.; Li, H.; Xin, Z.; Liu, T.; Wei, L. A Pontryagin minimum principle-based adaptive equivalent consumption minimum strategy for a plug-in hybrid electric bus on a fixed route. *Energies* **2017**, *10*, 1379. [CrossRef]

14. Serrao, L.; Onori, S.; Sciarretta, A.; Guezennec, Y.; Rizzoni, G. Optimal energy management of hybrid electric vehicles including battery aging. In Proceedings of the American Control Conference, San Francisco, CA, USA, 29 June–1 July 2011; pp. 2125–2130.

15. Yuksel, T.; Litster, S.; Viswanathan, V.; Michalek, J.J. Plug-in hybrid electric vehicle LiFePO4 battery life implications of thermal management, driving conditions, and regional climate. *J. Power Sources* **2017**, *338*, 49–64. [CrossRef]

16. Masih-Tehrani, M.; Ha'Iri-Yazdi, M.; Esfahanian, V.; Safaei, A. Optimum sizing and optimum energy management of a hybrid energy storage system for lithium battery life improvement. *J. Power Sources* **2013**, *244*, 2–10. [CrossRef]

17. Tang, L.; Rizzoni, G.; Onori, S. Energy management strategy for HEVs including battery life optimization. *IEEE Trans. Transp. Electrif.* **2015**, *1*, 211–222. [CrossRef]

18. Tang, L.; Giorgio, R. Energy management strategy including battery life optimization for a HEV with a CVT. In Proceedings of the Transportation Electrification Conference, Busan, Korea, 1–4 June 2016; pp. 549–554.

19. Ebbesen, S.; Elbert, P.; Guzzella, L. Battery state-of-health perceptive energy management for hybrid electric vehicles. *IEEE Trans. Veh. Technol.* **2012**, *61*, 2893–2900. [CrossRef]

20. Onori, S.; Spagnol, P.; Marano, V.; Guezennec, Y.; Rizzoni, G. A new life estimation method for lithium-ion batteries in plug-in hybrid electric vehicles applications. *Int. J. Power Electron.* **2012**, *4*, 302–319. [CrossRef]

21. Moura, S.; Stein, J.; Fathy, H. Battery-health conscious power management in plug-in hybrid electric vehicles via electrochemical modeling and stochastic control. *IEEE Trans. Control Syst. Technol.* **2013**, *21*, 679–694. [CrossRef]

22. Hu, J.; Hu, Z.; Niu, X.; Bai, Q. Research on energy management strategy considering battery life for plug-in hybrid electric vehicle. *Adv. Mech. Eng.* **2018**, *10*, 1–12. [CrossRef]

23. Sockeel, N.; Shi, J.; Shahverdi, M. Pareto front analysis of the objective function in model predictive control based power management system of a plug-in hybrid electric vehicle. In Proceedings of the IEEE Transportation Electrification Conference and Expo (ITEC), Chicago, IL, USA, 13–15 June 2018; pp. 971–976.

24. Sockeel, N.; Shahverdi, M.; Mazzola, M. Impact of the state of charge estimation on model predictive control performance in a plug-in hybrid electric vehicle accounting for equivalent fuel consumption and battery aging. In Proceedings of the IEEE Transportation Electrification Conference and Expo (ITEC), Detroit, MI, USA, 19–21 June 2019.

25. Ming, C.; Bo, C. Nonlinear model predictive control of a power-split hybrid electric vehicle with consideration of battery aging. *J. Dyn. Syst. Meas. Control* **2019**, *141*, 081008. [CrossRef]

26. Xie, S.; Hu, X.; Qi, S.; Tang, X.; Lang, K.; Xin, Z.; Brighton, J. Model predictive energy management for plug-in hybrid electric vehicles considering optimal battery depth of discharge. *Energy* **2019**, *173*, 667–678. [CrossRef]

27. Wang, Y.; Jiao, X. Energy management strategy in consideration of battery health for PHEV via stochastic control and particle swarm optimization algorithm. *Energies* **2017**, *10*, 1894. [CrossRef]

28. Jiao, X.; Shen, T. SDP Policy Iteration-Based Energy Management Strategy Using Traffic Information for Commuter Hybrid Electric Vehicles. *Energies* **2014**, *7*, 4648–4675. [CrossRef]

29. Rezaei, A.; Burl, J.B.; Zhou, B. Estimation of the ECMS equivalent factor bounds for hybrid electric vehicles. *IEEE Trans. Control Syst. Technol.* **2018**, *26*, 2198–2205. [CrossRef]

30. Yasui, Y. JSAE-SICE benchmark problem 2: Fuel consumption optimization of commuter vehicle using hybrid powertrain. In Proceedings of the 10th World Congress on Intelligent Control and Automation, Beijing, China, 6–8 July 2012; pp. 606–611.

Article

Analysis of the Electric Bus Autonomy Depending on the Atmospheric Conditions

Călin Iclodean *, Nicolae Cordoș * and Adrian Todoruț

Department of Automotive Engineering and Transports, Technical University of Cluj-Napoca,
Cluj-Napoca 400001, Romania; adrian.todorut@auto.utcluj.ro
* Correspondence: calin.iclodean@auto.utcluj.ro (C.I.); nicolae.cordos@auto.utcluj.ro (N.C.);
Tel.: +40-743-600-321 (C.I.)

Received: 4 November 2019; Accepted: 27 November 2019; Published: 28 November 2019

Abstract: The public-transport sector represents, on a global level, a major ecological and economic concern. Improving air quality and reducing greenhouse gas (GHG) production in the urban environment can be achieved by using electric buses instead of those operating with internal combustion engines (ICE). In this paper, the energy consumption for a fleet of electric buses Solaris Urbino 12e type is analyzed, based on the experimental data taken from a number of 22 buses, which operate on a number of eight urban lines, on a route of approximately 100 km from the city of Cluj-Napoca, Romania; consumption was monitored for 12 consecutive months (July 2018–June 2019). The energy efficiency of the model for the studied electric buses depends largely on the management of the energy stored on the electric bus battery system, in relation to the characteristics of the route traveled, respectively to the atmospheric conditions during the monitored period. Based on the collected experimental data and on the technical characteristics of the electric buses, the influence of the atmospheric conditions on their energy balance was highlighted, considering the interdependence relations between the considered atmospheric conditions.

Keywords: electric bus; battery; energy efficiency; environmental conditions

---

1. Introduction

Urban public transport plays a very important role in society, as it is the current means of transport serving a significant number of people every day. The sustainable tendency of urban mobility is to transport as many people daily as possible, with ecological, nonpolluting means of transport, which will have a direct effect not only on the reduction of the greenhouse gases (GHG), but also on the reduction of the environmental noise, traffic congestion, and the infrastructure vibration due to the vehicles equipped with internal combustion engines (ICE).

Most electric vehicles adopt AC motors due to their higher reliability and longer service life. Various electric motors and batteries used in electric vehicles are still subject to research, and innovative strategies are explored to compete with thermal-engine technology [1].

Due to the tendency of the big cities agglomeration, there is a need to increase the number of buses in the public transport fleets, and if the bus fleet is not renewed with environmentally friendly, nonpolluting means of public transport, it will result an increase of the environmental pollution (chemical and acoustic) that would affect the health of the population. Also, by renewing the bus fleet park of the public transport companies, the aim is to encourage the use of the environmentally friendly, nonpolluting means of transport, to the detriment of using personal cars in urban traffic.

In [2], Grijalva et al. noted that a bus used at the nominal occupancy level could replace up to 40 personal cars from urban traffic.

Regarding the problems mentioned above, the main solution for eliminating them is to replace the classic buses equipped with ICE with silent and nonpolluting electric buses.

However, this solution has two major drawbacks: the high purchase price of an electric bus (which can be double compared to a classic bus with the same capacity [3], but which can be compensated by accessing European non-reimbursable funds [4]), respectively the autonomy of the electric bus, which depends on the capacity of the batteries that equip this bus and on the charging strategy (fast charging between buses route, respectively slow charging overnight) [5]. Because the batteries are the most expensive elements of an electric bus [6,7] and also have a considerable mass (between 1500 and 3000 kg) [8,9], the energy store in them must to be used to the maximum. The energy consumption for the electric bus varies according to a large number of parameters (the technology used in the construction of the electric bus, the experience of the driver, the traffic conditions, atmospheric conditions, the altitude profile of the route, the degree of the boarded passengers, etc.) between 1.0 and 3.5 kWh/km [10–14].

The batteries of the electric buses are recyclable, and their major advantage is the operational costs with electricity that is much cheaper than conventional fuels [3], respectively the maintenance costs that practically do not exist for a period of up to 10 years [15]. The most important feature of a battery pack is to store a maximum amount of energy in a volume and at a minimum mass, in order to ensure the maximum autonomy [16].

In [17], Demircali et al. studied a virtual model for a battery of an electric vehicle, directly dependent on the ambient temperature, showing that, with the increase or decrease of this temperature, the energy stored in the battery is changed.

The energy consumption for the electric vehicles is influenced by the atmospheric conditions, not only from the point of view of the direct influence on the storage capacity of the batteries, but also from the point of view of the increase of the energy consumption due to the supply of the auxiliary systems (heating, ventilation, and air-conditioning in the vehicle), as demonstrated by Iora et al., in [18].

In [5], Vepsäläinen et al. showed that the optimal energy consumption of an electric bus takes place at an ambient temperature of 20 °C. However, the studies of Qian et al. [19] showed that ambient temperature plays an important role in the battery life of an electric vehicle and, therefore, implicitly on the energy storage capacity. Thus, the increase or decrease in ambient temperature above certain thresholds lead to the more frequent use of cooling or heating, resulting in premature aging of the battery and the reduction of its storage capacity.

In different climatic zones, the ambient temperature can directly influence the efficiency of an electric vehicle, having the effect of increasing the energy consumption due to the use of air-conditioning systems for cooling or heating [10,20–25]. In [24], Yuksel et al. showed, by analyzing the energy consumption according to the ambient temperature during more than 7000 trips in six regions of the US, that the energy consumption for extreme values of the ambient temperature can be doubled (–30 °C to +40 °C), but it is kept at optimum values for a thermal regime between +17 and +27 °C.

Zhu et al. [26] showed that, under extreme temperature conditions (–30 °C to +40 °C), the energy efficiency of the electric bus batteries is low and, at the same time, the degradation of the battery characteristics is accelerated. These authors demonstrate the importance of the thermal management of the batteries which, regardless of the atmospheric conditions, must ensure an optimum temperature on the surface of the batteries around +30 °C.

The studies conducted by Jardin et al. in [27] showed that the optimum operating temperature for an electric bus based on energy consumption (kWh/km) is in the range between +20 and +25 °C, the maximum consumption being at low ambient temperatures.

The main objective of this work is to highlight the energy consumption and, respectively, the energy recovered for a fleet of 22 electric buses, Solaris Urbino 12e type, which operate on eight urban lines, on a route about 100 km from the city Cluj-Napoca, Romania. The consumption was monitored for 12 consecutive months (July 2018–June 2019). The temperate climate that characterizes most of the cities located in the continental area of Europe implies the existence of four seasons with extreme differences of environment temperatures (from –30 °C in the winter months and up to +40 °C in the summer months) [28–30], differences that can have a significant impact on the electric buses autonomy.

2. Materials and Methods

2.1. Electric Bus Model

2.1.1. Electric Bus Model Data

Electric buses use electric propulsion based on an electric motor powered by rechargeable batteries via a voltage inverter. The battery-charging strategy involves the use of slow-charging stations (overnight), with a power of 40 kW and a full battery life between 4 and 6 hours, respectively, fast-charging stations (between races) with a power of 230 kW, and a charging time to ensure the autonomy required for a 10 to 20 minute ride [13].

The values for the main technical characteristics of the electric buses Solaris Urbino 12e model (Figure 1) that were used to carry out the study presented in this paper are listed in Table 1 [9,31].

Figure 1. Solaris Urbino 12e electric bus fleet (author photo).

Table 1. Technical characteristics of the electric buses.

Parameters	Unit	Value
Length/Width/Height	m	12.00/2.55/3.25
Distance from hitch to front axle	m	9.30
Wheelbase	m	5.90
Nominal/Loaded Weight	kg	13,800/18,000
Number of passenger's seats/total	-	23/70
Motor (type)	-	Electric asynchronous
Maximum engine power	kW	160
Maximum engine torque	Nm	1450
Inverter input voltage	V_{cc}	687
Inverter output voltage	V_{ca}	3×380
Batteries (type)	-	$Li_4Ti_5O_{12}$ (LTO)
Battery capacity	kWh	210
Battery voltage	V_{cc}	687
Energy consumption summer/winter	kWh/km	1.00/2.00
Autonomy (producer)	km	105
Autonomy (real)	km	75–150

2.1.2. Battery Pack Data

The battery pack acts as a chemical storage unit for the electricity required to operate the motors that provide the bus propulsion. At the same time, the energy from the batteries must supply the auxiliary systems (cooling, heating, ventilation, lights, multimedia, telematics, etc.) under extreme ambient temperature values, ranging from –30 °C to +40 °C.

The basic unit of the battery pack is the cell. A number of "n" cells arranged in a series-parallel grouping form a module [32,33]. In the present case, for the studied Solaris Urbino 12e buses, the standard battery pack consists of eight modules interconnected in parallel that provide the nominal voltage of 690 V_{cc} at the terminals of the battery pack and which is applied as the input voltage to the system of the voltage inverter. The inverter converts the direct current (DC) voltage into a three-phase alternating current (AC) voltage ($3 \times 380\ V_{ac}$) which supplies the electric propulsion motors. To increase the autonomy, the electric buses are powered by a system of 3 to 6 battery packs, connected in parallel, to increase the value of the current and thus of the stored energy [34].

The $Li_4Ti_5O_{12}$ (LTO) batteries that equip the electric buses (Figure 2) are batteries with superior thermal stability, high energy storage capacity in cells, and a high lifespan (expressed through charge–discharge cycles) [9,35]. LTO batteries have the following advantages: operating safety, longevity, performance, and power density, but have a low energy density and are expensive [36].

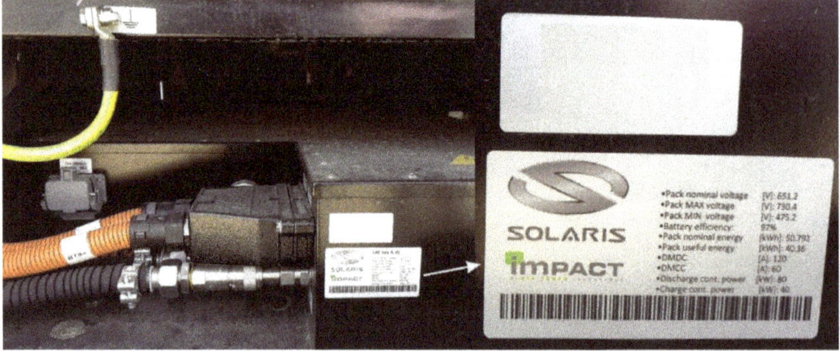

Figure 2. LTO batteries that equip the Solaris Urbino 12e electric buses (author photo).

The virtual model for the basic cell of the battery pack is shown in Figure 3, and, based on this model, the electrical equations that describe the characteristics of cells, modules, and of the battery packs were formulated [34,37–42].

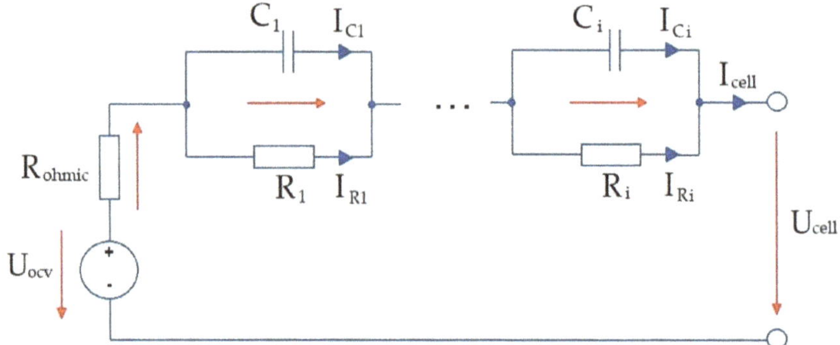

Figure 3. Equivalent circuit models (ECM) of the battery cell.

The voltage at the terminals of a cell (U_{cell}) is calculated with the following equation [38]:

$$U_{cell} = U_{ocv} - I_{cell} \cdot R_{ohmic} - \sum_{i=1}^{n} \frac{Q_i}{C_i},$$ (1)

where U_{ocv} (V) is the voltage of the open circuit of the open-circuit voltage cell (OCV), I_{cell} (A) is the current at the cell terminals, R_{ohmic} (Ω) is the internal resistance of the cell, Q_i (W) is the load capacity of the cell, and C_i (F) is the capacity of the cell capacitor.

The load capacity of the cell is expressed by the following equation [38]:

$$Q_i = \int I_{C_i} \cdot dt = \int \left(I_{cell} - \frac{Q_i}{R_i \cdot C_i} \right) \cdot dt,$$ (2)

where I_{C_i} (A) is the current through the cell capacitor, and R_i (Ω) is the resistance of each resistor–capacitor (RC) element of the cell.

The electrical voltage at the terminals of a battery module (U_{module}) is calculated according to the number of cells connected in series (n_{series}), using the following relation:

$$U_{module} = U_{cell} \cdot n_{series}.$$ (3)

The electric current at the terminals of a battery module (I_{module}) is calculated according to the number of cells connected in parallel ($n_{parallel}$), using the following relation:

$$I_{module} = I_{cell} \cdot n_{parallel}.$$ (4)

The state of charge (SOC) of a battery module Q_{module} (%) is calculated based on the number of cells and on the charge level of each cell Q_{cell} (%), respectively, on the number of cells connected in parallel ($n_{parallel}$), using the following relation [38]:

$$Q_{module} = Q_{cell} \cdot n_{parallel} = Q_{max} \cdot SOC \cdot n_{parallel},$$ (5)

where Q_{max} (%) is the maximum loading degree of each RC element of the cell.

The power of a battery module P_{module} (W) is calculated according to the number of cells and to the power of each cell P_{cell} (W), respectively, according to the number of cells connected in series (n_{series}), using the following relationship:

$$P_{module} = P_{cell} \cdot n_{series} = U_{cell} \cdot I_{cell} \cdot n_{series}.$$ (6)

The lost power of a battery module ($P_{loss, module}$ (W)) is calculated using the following relation:

$$P_{loss,module} = \left(I_{cell}^2 \cdot R_{ohmic} + \sum_{i=1}^{n} I_{R_i}^2 \cdot R_i \right) \cdot n_{series}.$$ (7)

Instant charge of the battery is given by the following relation, based on the *Coulomb-Counting* algorithm [43]:

$$Q(t) = Q_0 - \int_0^t I_{batt}(t)dt,$$ (8)

where Q_0 is the initial battery charging status, and I_{batt} (A) is the current at the battery.

The SOC of the battery is expressed as a percentage of the maximum charge capacity, Q_{max} [43,44]:

$$SOC(t) = 100\% \cdot \frac{Q(t)}{Q_{max}}.$$ (9)

Similarly, depth of discharge (DOD) of the battery is expressed as a percentage of total capacity consumed [43]:

$$DOD(t) = 100\% \cdot \frac{Q_{max} - Q(t)}{Q_{max}}.$$ (10)

The functional characteristic of the battery cell is shown in Figure 4, and the values of the main characteristic parameters of the battery pack that equip the electric buses are shown in Table 2 [39,45].

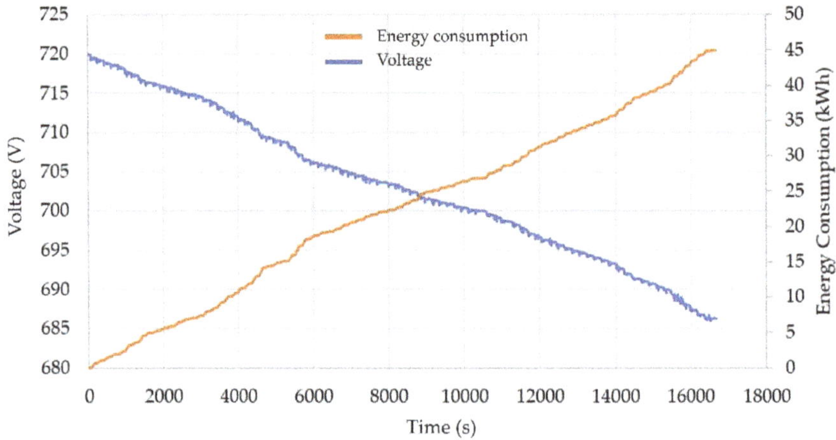

Figure 4. Battery characteristics.

Table 2. Technical characteristics of the electric batteries.

	Parameters	Unit	Value
	Nominal cells capacity	Ah	26
Cells	Nominal cells voltage	V	3.7
	Nominal cells energy	kWh	0.0962
	Nominal modules energy	kWh	6.25
Modules	Number of modules	-	8
	Mass of module	kg	42
	Nominal packs capacity	Ah	58.8
	Nominal packs voltage	V	687
Pack	Nominal packs energy	kWh	41.22
	Number of modules	-	3–6
	Mass of pack	kg	485

2.1.3. Electric Machine

Electric buses are powered by an electric motor, asynchronous motor (ASM) type, which operates in electric-motor mode, consuming battery power, or in generator mode, recovering the energy when descending slopes or in particular braking situations. The operating characteristic of the electric propulsion system is described by the motor torque diagram vs. speed, for all the possible traction regimes (Figure 5) [38,46,47].

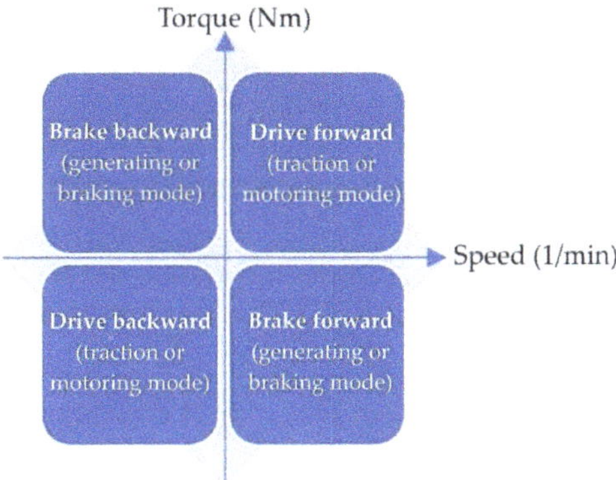

Figure 5. Motor speed-torque diagram (traction modes).

The torque of the electric motor is maximum from the start and remains constant until it reaches a constant speed corresponding to the cruise speed. The power of the electric motor increases linearly until maximum, and then it descends simultaneously with the decrease of the motor torque (Figure 6) [48,49].

Regenerative braking involves the partial recovery of the kinetic energy and the storage of this energy in the battery to increase the range of the electric buses. During the regenerative braking, the electric motor turns into a generator and charges the batteries. The kinetic energy of the electric buses depends on their mass and speed, but only half of this energy, at most, can be recovered, and this depends on the generator's ability to produce electricity, respectively, on the battery's charging capacity [50,51].

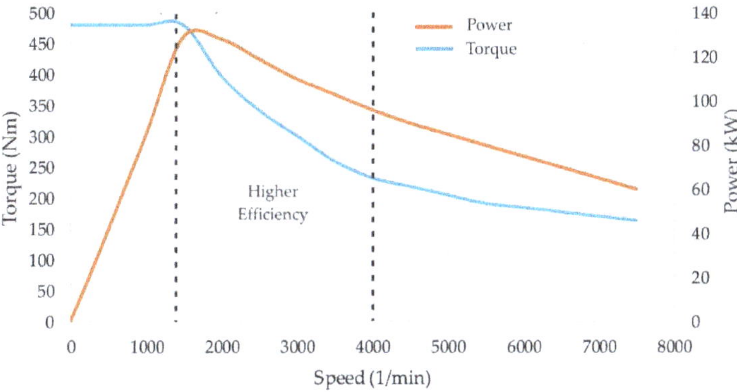

Figure 6. Motor speed-torque/power characteristic.

2.2. Environment Model

2.2.1. Ambient Temperature

The meteorological data regarding the temperature history for Cluj-Napoca, related to the monitored period (July 2018–June 2019), were taken from the archive of the weather station rp5.ru [52]. The average daily values for temperature were calculated as the average values for the electric buses operating at hourly intervals, from the beginning of the program (5:00 a.m.) to the end of the program (11:00 p.m.), based on daily records of variations in temperature values, obtaining the results that are presented in Figure 7.

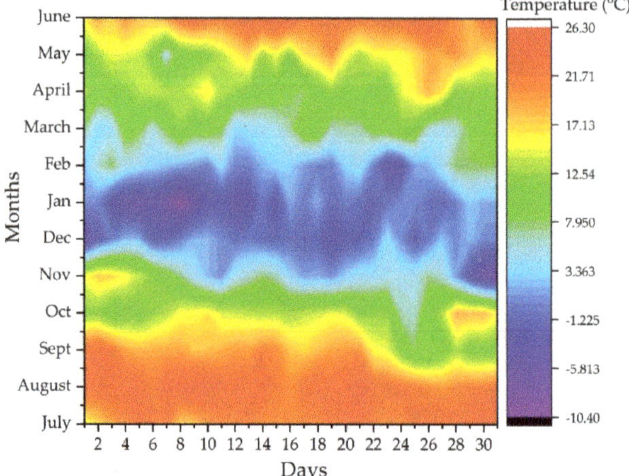

Figure 7. The average daily temperature values recorded for the monitored period.

2.2.2. Ambient Humidity

Similar to the temperature history (See Section 2.2.1), compared to the monitored period (July 2018–June 2019), data were taken regarding the values for the atmospheric humidity [52]. The average values for the atmospheric humidity were calculated as the average values for the electric buses operating at hourly intervals, based on daily records of the variations for the atmospheric humidity values, obtaining the results that are presented in Figure 8.

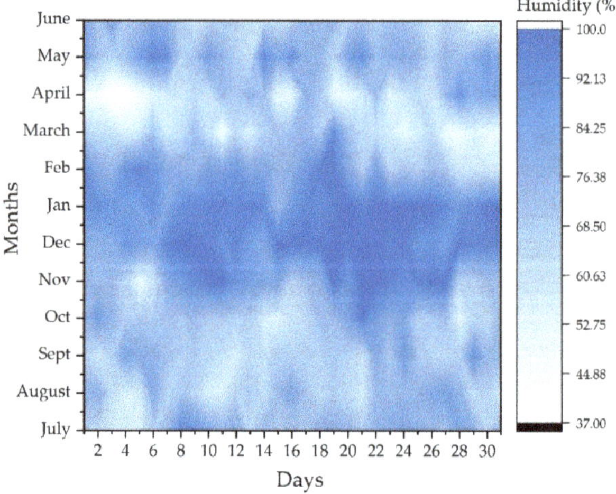

Figure 8. The average daily atmospheric humidity values recorded for the monitored period.

2.2.3. Pressure and Air Density

Similar to the temperature history (see Section 2.2.1), compared to the monitored period (July 2018–June 2019), data were taken regarding the atmospheric pressure values [52]. The average values for the atmospheric pressure were calculated as the average values for the electric buses operating at hourly intervals, based on the daily records of the atmospheric pressure values' variations, obtaining the results that are presented in Figure 9.

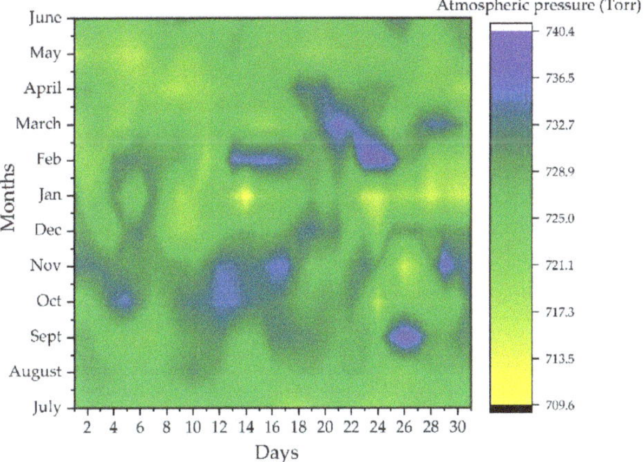

Figure 9. The average daily atmospheric pressure values recorded for the monitored period.

The average daily air-density value (Figure 10) was calculated based on the average daily recordings of the thermal values, using the following relation [53]:

$$\rho_{air} = \frac{p - 0.378 \cdot u \cdot p_s}{287.05 \cdot T},$$ (11)

where ρ_{air} is the average daily air density (kg/m^3), p is the average daily atmospheric pressure (Pa), u is the average daily relative humidity (-), p_s is the saturation pressure (Pa), and T is the absolute temperature (K) relative to the average daily ambient temperature (T (K) = t (°C) + 273.15).

The saturation pressure was calculated using the following relation [54]:

$$p_s = \frac{e^{(77.3450 + 0.0057 \cdot T - \frac{7235}{T})}}{T^{8.2}}.$$ (12)

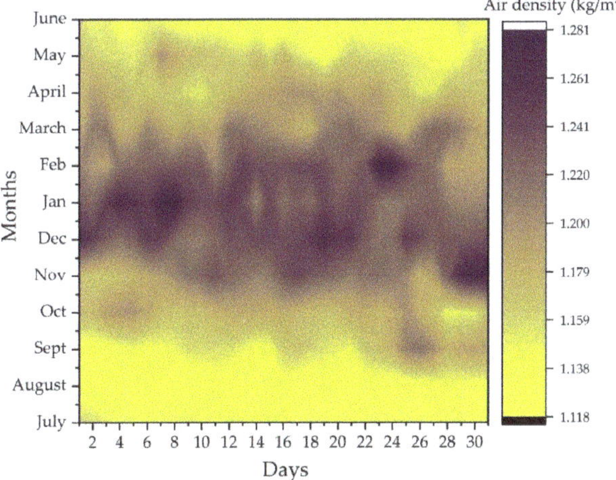

Figure 10. The calculated average daily air density values for the monitored period.

2.3. Energy Balance

2.3.1. Energy Consumption

The energy consumption of the electric buses is influenced by a number of factors, such as: increasing the total mass of buses by loading the passengers, the consumption generated by the auxiliary systems that cause a significant increase in the amount of energy consumed by the batteries (air conditioner equipment, multimedia equipment, lighting equipment, telematics equipment, auxiliary equipment), some of these factors being not dependent on the distance traveled by the electric buses [55–57]. At the same time, the altitude profile of the route can influence the energy consumption. This increases during periods of acceleration or ascent of the ramps and decreases when descending the slopes or when the bus decelerates, reaching negative values (the energy is transferred from the electric motor that works in generator mode, to the battery). The data collected for the monitored period (July 2018–June 2019) showed an average daily distance traveled between 100 km (one driver/electric bus) and 200 km (two drivers/one electric bus).

In addition to these factors, there are climatic parameters (ambient temperature, atmospheric humidity, air pressure, and density) that have a major influence on the energy consumption of the electric buses. The experimental data recorded during the monitored period include the atmospheric conditions characteristic for all seasons with extreme values during a calendar year (ambient temperatures between –15 °C in January 2019 and +32 °C in August 2018, respectively humidity values between 14% in November 2018 and 100% in most months of the year).

Yuan et al. in [58], consider that it is difficult to obtain real traffic data on the energy consumption for electric buses, and standards for defining this consumption and registration procedures are used to evaluate the energy consumption (ISO 8714-2002 Electric road vehicles—Reference energy consumption and range—Test procedures for passenger cars and light commercial vehicles [59], respectively, GB 18386-2017 Electric vehicles—Energy consumption and range—Test procedures [60]).

In this paper, the data on the energy consumption were recorded by real-time monitoring of bus operation in the city of Cluj-Napoca, which is achieved through the system of tracking and monitoring the traffic Thoreb [61], a system that allows the observation of buses in real-time by monitoring on a digital map based on the signals generated by the Global Positioning System (GPS) modules installed on the buses and transmitted to the dispatchers using the General Packet Radio Service (GPRS) technology. At the same time, from the bus controller area network (CAN), data regarding the technical status of the buses, the distance traveled, the energy consumption, the number of passengers transported, etc. are collected and sent to the dispatchers [13].

The evaluation of the energy consumption data on the electric buses for the monitored period (July 2018–June 2019) was performed with a Boxplot graph (Figure 11 and Table 3) which, based on the initial data, generates a statistical model for each monitored month.

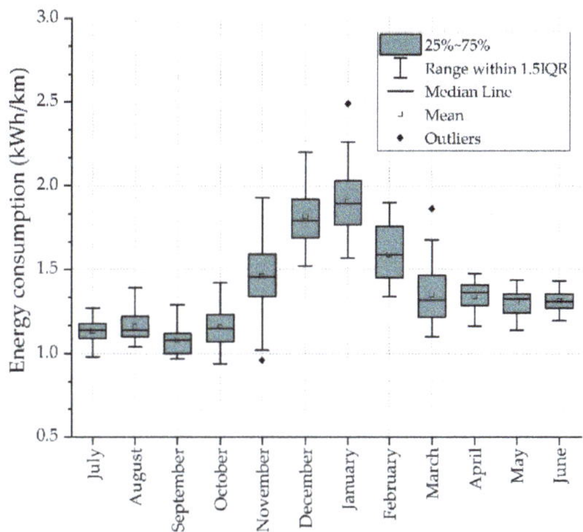

Figure 11. Boxplot analysis of energy consumption (kWh/km).

Table 3. Boxplot analysis of energy consumption (kWh/km).

Month	μ	SD	Min	Q1	Med	Q3	Max
July 2018	1.13	0.07	0.98	1.09	1.13	1.18	1.27
August 2018	1.16	0.08	1.04	1.10	1.16	1.22	1.39
September 2018	1.08	0.08	0.97	1.00	1.08	1.12	1.29
October 2018	1.16	0.11	0.94	1.07	1.16	1.23	1.42
November 2018	1.48	0.25	1.02	1.34	1.48	1.60	1.93
December 2018	1.82	0.17	1.52	1.69	1.82	1.92	2.20
January 2019	1.91	0.19	1.56	1.76	1.89	2.02	2.26
February 2019	1.59	0.17	1.33	1.44	1.59	1.77	1.90
March 2019	1.35	0.18	1.10	1.20	1.33	1.45	1.67
April 2019	1.34	0.10	1.16	1.27	1.34	1.41	1.47
May 2019	1.31	0.07	1.14	1.24	1.31	1.35	1.43
June 2019	1.31	0.06	1.19	1.26	1.31	1.35	1.43

2.3.2. Regenerative Braking Energy

Similar to the energy consumption of the electric buses, the energy recovered through the regenerative braking was recorded for the monitored period (July 2018–June 2019).

The evaluation of the data regarding the energy recovered by the regenerative braking of the electric buses for the monitored period was performed with a Boxplot graph (Figure 12 and Table 4) which, based on the initial data, generates a statistical model for each monitored month.

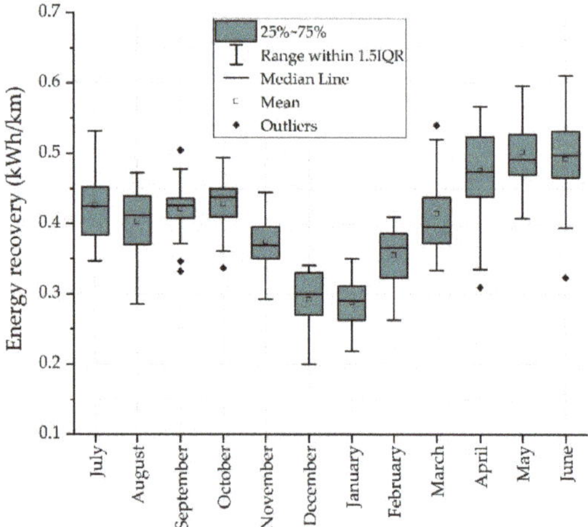

Figure 12. Amount of the energy recovered by regenerative braking (kWh/km).

Table 4. Amount of the energy recovered by regenerative braking (kWh/km).

Month	μ	SD	Min	Q1	Med	Q3	Max
July 2018	0.43	0.05	0.35	0.38	0.42	0.46	0.53
August 2018	0.40	0.05	0.29	0.37	0.41	0.44	0.47
September 2018	0.42	0.02	0.39	0.41	0.43	0.44	0.46
October 2018	0.43	0.03	0.36	0.41	0.44	0.45	0.49
November 2018	0.37	0.04	0.29	0.35	0.37	0.40	0.45
December 2018	0.29	0.04	0.20	0.27	0.30	0.33	0.34
January 2019	0.29	0.03	0.22	0.26	0.29	0.31	0.35
February 2019	0.35	0.04	0.26	0.32	0.37	0.39	0.41
March 2019	0.41	0.05	0.33	0.37	0.39	0.44	0.52
April 2019	0.48	0.05	0.34	0.44	0.48	0.53	0.57
May 2019	0.50	0.05	0.41	0.47	0.49	0.53	0.60
June 2019	0.50	0.05	0.39	0.47	0.50	0.53	0.61

2.3.3. Total Energy Balance

The energy balance recorded during the monitored period (July 2018–June 2019) for the Solaris Urbino 12e electric buses, depending on the atmospheric conditions (ambient temperature, atmospheric humidity, air pressure, and density) is shown in Table 5 and Figure 13.

Table 5. The energy balance of the Solaris Urbino 12e electric bus.

Value		2018						2019					
		July	August	September	October	November	December	January	February	March	April	May	June
Temperature (°C)	Daily minim	9.6	11.6	0.6	1.0	−10.0	−12.5	−15.0	−9	−4.8	−0.7	2.3	11.1
	Monthly minim	15.2	19.8	8.0	5.9	−7.0	−7.2	−10.4	−4.4	3.0	7.6	4.7	16.3
	Average	21.0	23.2	18.6	13.4	5.0	-0.2	−8	2.2	8.2	11.8	15.1	22.8
	Monthly maxim	24.4	25.5	23.7	19.1	17.6	3.6	2.7	8.0	12.4	18.8	20.3	26.3
	Daily maxim	29.7	31.9	31.2	25.2	23.6	6.0	5.5	15.5	20.4	26.3	26.6	32.0
Humidity(%)	Daily minim	35.0	31.0	27.0	25.0	14.0	62.0	59.0	34.0	27.0	21.0	34.0	28.0
	Monthly minim	55.5	55.1	54.7	52.0	47.7	78.4	71.8	58.1	46.3	37.1	61.5	51.8
	Average	75.5	68.1	66.5	69.6	84.7	95.9	94.7	80.4	62.2	59.5	78.3	68.3
	Monthly maxim	97.4	85.3	91.4	97.9	100.0	100.0	100.0	97.8	95.8	93.5	97.3	87.7
	Daily maxim	100.0	100.0	100.0	100.0	100.0	100.0	100.0	100.0	100.0	100.0	100.0	100.0
Atmospheric Pressure (Torr)	Daily minim	718.3	720.6	721.0	713.8	713.5	715.6	707.2	716.5	715.8	716.5	714.8	722.8
	Monthly minim	718.8	721.6	722.5	716.0	714.4	719.1	709.6	717.8	719.0	717.3	716.3	723.7
	Average	723.8	727.3	728.7	729.0	730.3	728.7	721.5	729.8	726.2	725.2	722.2	726.9
	Monthly maxim	726.9	731.0	740.4	735.6	738.1	733.5	731.2	739.9	737.3	733.6	728.7	732.5
	Daily maxim	727.7	731.8	746.5	736.8	738.9	734.4	733.9	741.4	738.4	734.8	729.5	733.7
Consumption (kWh/km)	Daily minim	0.87	0.88	0.83	0.80	0.88	0.91	1.20	1.00	0.90	0.79	0.83	0.95
	Monthly minim	0.87	0.88	0.83	0.80	0.88	0.91	1.20	1.00	0.90	0.79	0.83	0.95
	Average	1.13	1.16	1.07	1.16	1.45	1.78	1.90	1.60	1.30	1.30	1.28	1.33
	Monthly maxim	1.58	1.54	1.67	1.71	2.22	2.49	2.90	2.30	2.30	2.63	2.09	1.78
	Daily maxim	1.58	1.54	1.67	1.71	2.22	2.49	2.90	2.30	2.30	2.63	2.09	1.78
Recover (kWh/km)	Daily minim	0.21	0.16	0.10	0.19	0.13	0.09	0.10	0.20	0.20	0.00	0.00	0.12
	Monthly minim	0.21	0.16	0.10	0.19	0.13	0.09	0.10	0.20	0.20	0.00	0.00	0.12
	Average	0.43	0.40	0.42	0.43	0.38	0.29	0.30	0.30	0.40	0.48	0.50	0.49
	Monthly maxim	0.69	0.72	0.96	0.77	0.70	0.56	0.60	0.60	0.70	1.22	0.99	0.89
	Daily maxim	0.69	0.72	0.96	0.77	0.70	0.56	0.60	0.60	0.70	1.29	0.99	0.89

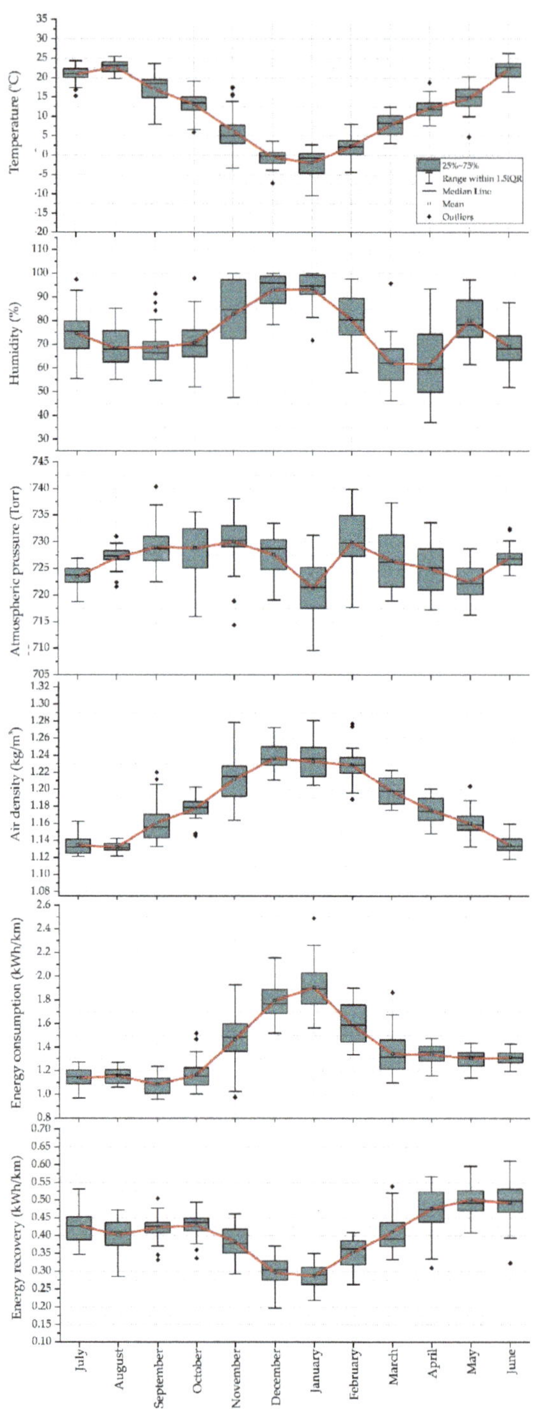

Figure 13. The energy balance of the Solaris Urbino 12e electric bus.

3. Results

As the costs associated with the integration into the urban transport system of the electric buses are high, it is necessary to carry out studies and research in order to provide the best solution in relation to their optimal use, taking into account the particularities imposed by the zoning characteristics (length and the gradient of routes, the flow of the transported passengers, the loading infrastructure, the volume of traffic, the characteristics of the environment, etc.).

The study of the energy performances for the 22 electric buses during the monitored period (July 2018–June 2019) highlighted their behavior at different values of the climatic parameters (temperature, humidity, atmospheric pressure, and air density).

The climatic parameters were monitored during a calendar year and give a clear picture regarding the atmospheric conditions and their influence on the energy consumption, respectively, on the energy recovery during the operation of the electric buses.

The parameters resulting from the behavior of the driver are invariable, difficult to estimate and impossible to generalize because they are psychological factors specific to each individual. However, taking into account the data on variations in atmospheric pressure and air density, there is the possibility to correlate human behavior with these variations, so that in the situation of increasing these values, the behavior of drivers becomes more active, and in the situation of decreasing these values, behavior becomes more passive.

From the results captured in Table 5 and in Figure 13, the average annual values of the climatic parameters can be obtained, as follows: the average annual temperature, 11.6916 °C; the average annual humidity, 75.3083%; the average annual atmospheric pressure, 726.6333 Torr; average annual energy consumption of 1.3716 kWh/km; average annual energy recovery, of 0.4016 kWh/km. Taking these into account, in Figures 14–18 the variations of the respective parameters are presented, as monthly average values obtained, compared with their annual average values.

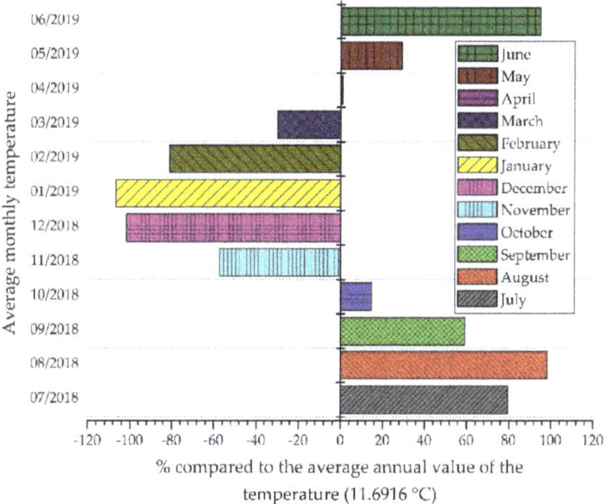

Figure 14. Monthly average temperature variation vs. average annual temperature.

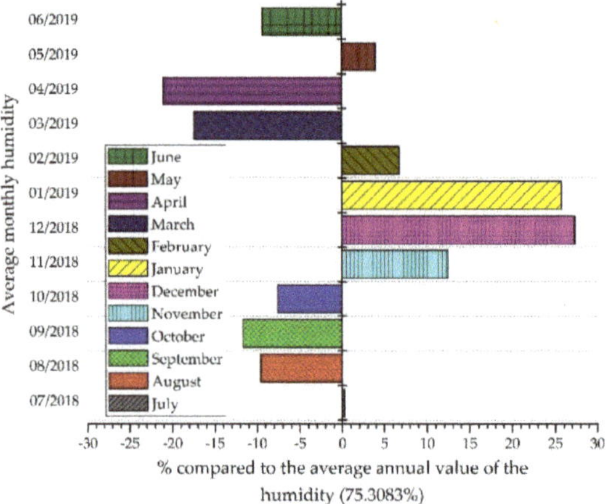

Figure 15. Monthly average humidity variation vs. average annual humidity.

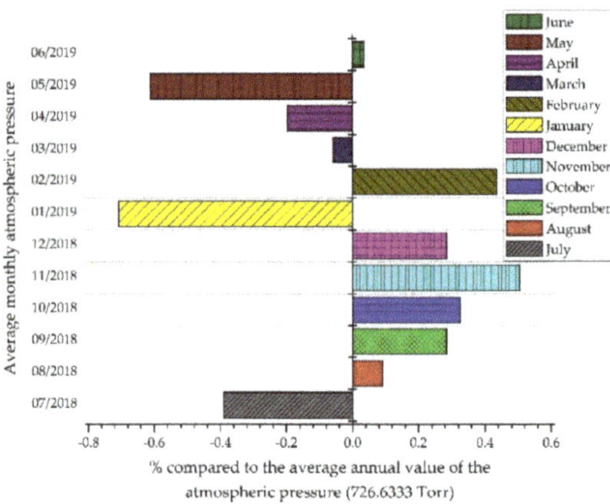

Figure 16. Monthly average atmospheric pressure variation vs. average annual atmospheric pressure.

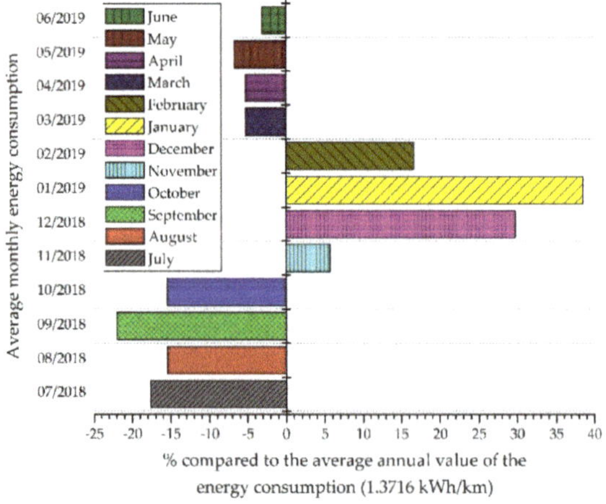

Figure 17. Variation of average monthly energy consumption vs. average annual energy consumption.

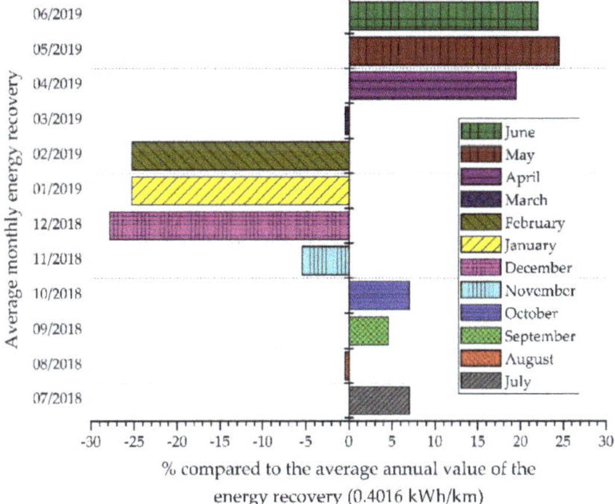

Figure 18. Variation of average monthly energy recovery vs. annual average energy recovery.

A summary of the obtained results, as the interdependence between them and the influence parameters on them, is captured in Figure 19. Thus, for each considered month from the monitored period (July 2018–June 2019), the influence of the atmospheric conditions on the energy efficiency of the studied electric buses was highlighted. Also, for each considered month, the reciprocal link between temperature, humidity, atmospheric pressure, and air density is captured in Figure 19. Thus, it was found that with the increase of the temperature, there is a reduction in air density, a slight decrease in atmospheric pressure, and the recorded humidity showed a reduce tendency. The recorded values of the humidity show that, with its increase, the air density increases, and the atmospheric pressure is reduced. It also notes that the increase in pressure indicates an increase in air density.

Regarding the energy consumption of the electric buses, it can be seen that it increases with decreasing the temperature and the atmospheric pressure, but the same tendency exists in the situation of increasing the air humidity and the air density (Table 6 and Figure 19). The energy recovered by the regenerative braking of the electric buses increases with the increase of the temperature and decreases with the increase of the air humidity, air density, and atmospheric pressure. Also, it can be seen that energy recovery largely compensates for the energy consumption of the electric buses.

Table 6. Data for energy consumption and energy recovery vs. atmospheric conditions.

Energy Consumption (kWh)	Energy Recovery (kWh)	Temperature (°C)	Humidity (%)	Atmospheric Pressure (Torr)	Air Density (kg/m^3)
1.1384	0.4293	20.8806	74.7451	723.4645	1.1346
1.1582	0.4024	22.9064	68.6548	727.0193	1.1321
1.0859	0.4233	17.0466	68.9433	729.0666	1.1620
1.1645	0.4288	13.1774	70.6838	728.8581	1.1775
1.4642	0.3844	5.9866	82.6733	730.0066	1.2136
1.7943	0.2959	−0.4645	93.0290	727.6258	1.2367
1.9076	0.2877	−1.9354	93.4451	721.3612	1.2331
1.5870	0.3532	2.2250	80.7000	729.9035	1.2246
1.3475	0.4098	7.8096	62.1258	726.4741	1.1979
1.3383	0.4758	12.2600	61.7166	724.9633	1.1755
1.3061	0.5018	14.6064	80.0064	722.5193	1.1602
1.2719	0.4924	22.2733	69.4200	727.0833	1.1351

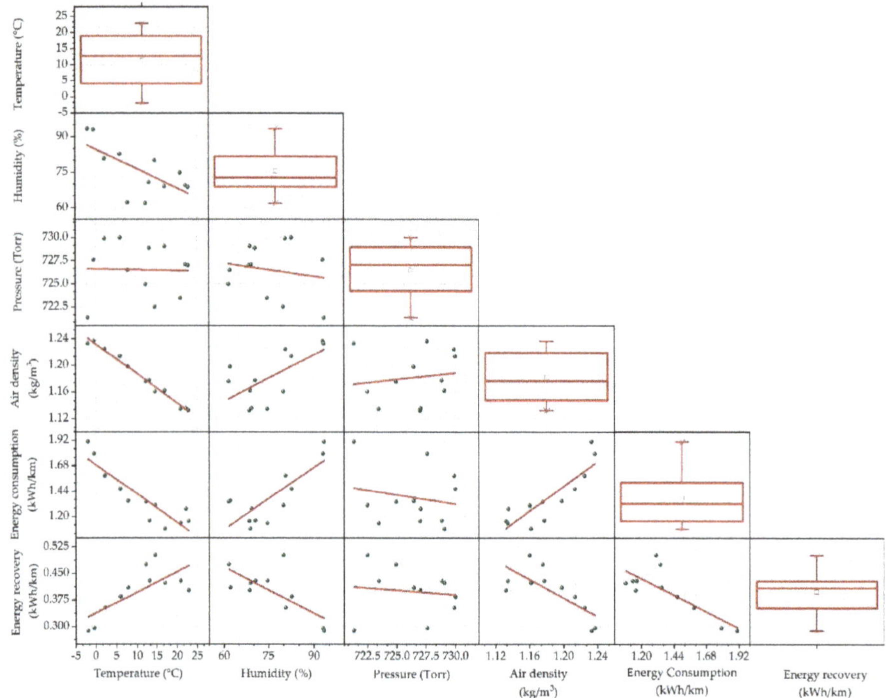

Figure 19. Matrix scatter plot.

4. Discussion

In the summer months (July 2018, August 2018, and June 2019), it resulted in high energy consumption compared to the average monthly values, due to the use of the air-conditioning system, when the ambient temperature exceeded 30 °C. This temperature threshold was imposed from the construction of the electric buses and results in the automatic operation of the air-conditioning until the ambient temperature in the passenger compartment drops below 25 °C. Regarding the recovered energy, an increase of it with the increase of the ambient temperature was noticed, under the conditions of maintaining low values of the atmospheric humidity, which facilitates the braking capacities of the electric buses on a dry road, respectively the increase of the electrical resistance of the braking system, preventing the energy losses through braking rheostats. In the summer months, characterized by the average monthly temperature values (see Table 5) higher than the annual average, with 79.61% in July 2018, 79.61% in August 2018, and 98.43% in June 2019 (see Figure 14), the atmospheric humidity is higher than the annual average by 0.25% in July 2018, lower by 9.57% in August 2018, and with 9.31% in June 2019 (see Figure 15), and the atmospheric pressure is lower than the annual average by 0.39% in July 2018, higher by 0.09% in August 2018, and with 0.04% in June 2019 (see Figure 16), resulting in lower energy consumption compared to the average annual consumption, by 17.62% in July 2018, 15.43% in August 2018, and 3.04% in June 2019 (see Figure 17), and a higher amount of energy recovered compared to its annual average, with 7.05% in July 2018, 21.99% in June 2019, and lower by 0.41% in August 2018 (see Figure 18).

From the recorded results, it can be seen that the autumn months (September 2018 and October 2018) are the months with low energy consumption, mainly due to the thermal conditions, with values of temperature of approx. 20 °C, but also with low values of the atmospheric humidity, being generally dry weather, which facilitates the movement of the electric buses with a minimum of energy consumed and a maximum recovered energy. During these months, the monthly average temperatures (see

Table 5) were higher than the annual average, with 59.09% in September 2018 and with 14.61% in October 2018 (see Figure 14), with lower atmospheric humidity compared to that annual average with 11.69% in September 2018 and with 7.58% in October 2018 (see Figure 15), and the atmospheric pressure higher than the annual average, with 0.28% in September 2018 and with 0.33% in October 2018 (see Figure 16), resulted in a lower energy consumption than the average annual consumption, with 21.99% in September 2018 and with 5.43% in October 2018 (see Figure 17), and a higher amount of energy recovered compared to its annual average, with 4.56% in September 2018 and with 7.05% in October 2018 (see Figure 18).

Since November 2018, the cold season has started, which, due to the decrease of the ambient temperature, especially in the time periods from the beginning of the work program (from 5:00 a.m. to 8:00 a.m.), respectively in the periods after the end of the work program (from 8:00 p.m. to 11:00 p.m.), resulted in an accelerated increase of the average energy consumption. Due to the increase of the atmospheric humidity and the reduction of the grip due to the wet road, the value of the recovered energy decreased. November 2018 was characterized by monthly average values of the temperature lower than the annual average value with 57.23% (see Figure 14), the atmospheric humidity higher than the annual average with 12.47% (see Figure 15), and the atmospheric pressure compared to the annual average by 0.50% higher (see Figure 16), resulted a higher energy consumption than the average annual consumption, by 5.71% (see Figure 17), and a lower amount of recovered energy compared to its annual average, with 5.39% (see Figure 18).

The winter months (December 2018 and January 2019) were the months with the lowest temperatures in the entire monitored period. In addition to the negative temperatures of the day, the start of the working program for the electric buses took place below the freezing threshold. Electric buses, which are connected overnight to the external slow-charging stations, and the batteries are heated by the thermal management system, consume, on the cold winter days (with temperatures below $-10\ ^{\circ}C$) up to 10% of the energy from the batteries, for heating the interior passenger compartment, only in the interval of preparation for the buses' departure on the route. Thus, combined with the increased of the daily consumption and the wet road conditions, it is reached that, during the cold season, the consumption increase by 2 to 2.5 times compared to the periods with the lowest values of consumption, and the energy values recovered will near zero. During these months, the monthly average temperatures (see Table 5) were lower than the annual average by 101.71% in December 2018 and by 106.84% in January 2019 (see Figure 14), the humidity was higher than the annual average with 27.34% in December 2018 and with 25.75% in January 2019 (see Figure 15), and the atmospheric pressure was higher than the annual average, with 0.28% in December 2018 and lower, with 0.71% in January 2019 (see Figure 16). The result was a higher energy consumption than the average annual consumption, with 29.77% in December 2018 and 38.52% in January 2019 (see Figure 17), and a reduced amount of energy recovery compared to its annual average, by 27.80% in December 2018 and by 25.31% in January 2019 (see Figure 18).

The transition from extremely low winter temperatures to thermal relaxation took place in February 2019, which started with low temperatures, especially in the early morning, but also with a slight increase in temperatures in the second half of the day, which have to lead to a reduction in energy consumption. Due to low values of the atmospheric humidity and the lack of precipitation, respectively, on a dry road, the amount of energy recovered during this period also increased. February 2019 was characterized by monthly average values of the temperature lower than the annual average value by 81.18% (see Figure 14), the atmospheric humidity higher than the annual average by 6.76% (see Figure 15), and the atmospheric pressure compared to the annual average higher with 0.44% (see Figure 16). In February it resulted in higher energy consumption than the average annual consumption, with 16.65% (see Figure 17), and a reduced of the energy recovery compared to its annual average, with 25.31% (see Figure 18).

The spring months (March 2019 and April 2019) began with accelerated warming of the weather, with the reduction of the atmospheric humidity, environmental aspects that led to the gradual reduction of the energy consumption to normal values and to the increased of the energy recovery. Even though, during the two months, the mornings were colder, with temperatures around 0 °C, the thermal regime did not influence the electric bus batteries when the buses started on the route. During these months, the monthly average temperatures (see Table 5) were lower than the annual average by 29.86% in March 2019 and higher by 0.93% in April 2019 (see Figure 14), with lower atmospheric humidity compared to the annual average, with 17.41% in March 2019 and with 20.99% in April 2019 (see Figure 15), and the atmospheric pressure was lower than the annual average, with 0.06% in March 2019 and with 0.20% in April 2019 (see Figure 16), and it resulted in lower energy consumption compared to the average annual consumption, with 5.22% in each of the two months (see Figure 17), and a reduced amount of energy compared to its annual average by 0.41% in March 2019 and by 19.50% higher in April 2019 (see Figure 18).

The transition to the hot season at summer temperatures began in May 2019, a month characterized by positive thermal values throughout the operating range of the electric buses, respectively by the average values of the atmospheric humidity. These climatic factors allowed us to achieve average energy consumption within the normal limits specified by the manufacturer, cumulating with maximum energy recovery for the entire monitored interval. May 2019 was characterized by monthly average values of the temperature higher than the annual average value, with 29.15% (see Figure 14), the atmospheric humidity higher than the annual average by 3.97% (see Figure 15), and the atmospheric pressure lower than the annual average by 0.61% (see Figure 16). In May, it resulted in lower energy consumption than the average annual consumption, by 6.68% (see Figure 17), and an amount of recovered energy higher than its annual average, with 24.48% (see Figure 18).

5. Conclusions

The energy efficiency of the electric buses was evaluated for 12 consecutive months (July 2018–June 2019), based on the weather conditions in Cluj-Napoca city, Romania, conditions that are specific to the vast majority of continental Europe, taking into account other area features, such as the loading infrastructure, traffic conditions, altitude profile of the traveled routes, the degree of loading with passengers, and the traffic management.

Based on our findings, the following conclusions can be drawn:

- The climatic parameters influence the consumption and energy recovery for the electric buses;
- The energy consumption of the electric buses increases with the decreasing of the temperature and the atmospheric pressure, but the same tendency exists even if the humidity and density of the air increase;
- The energy recovered by the regenerative braking of the electric buses increases with the increase of the temperature and decreases with the increase of the air humidity, its density, and the atmospheric pressure;
- Energy recovery largely compensates the energy consumption of the electric buses;
- The variations of the monitored parameters, obtained as monthly average values, compared with their annual average values, highlight the interdependence of these quantities;
- Average energy consumption, within the normal limits specified by the manufacturer and maximum energy recoveries, was obtained on the transition to the warm season;
- The energy consumption of the electric buses increases when the buses accelerate or when the buses climb the ramps, and it decreases when the buses decelerate or descend from the ramp.

The authors intend to develop various models to describe other influences on the energy balance of the electric buses, to use them in their modeling, simulation, and exploitation. In addition, based on the collected experimental data and on the technical characteristics of the real model of the electric buses, the authors have already highlighted, with promising results, the influence of the atmospheric conditions on their energy balance, taking into account the interdependence of the climatic parameters (ambient temperature, atmospheric humidity, air pressure, and air density).

Author Contributions: The authors have contributed equally to this article.

Funding: This research received no external funding.

Acknowledgments: The authors wish to thank CTP Cluj-Napoca SA for providing the data from the Opportunity Studies.

Conflicts of Interest: The authors declare no conflict of interest.

References

1. Jayakumar, A.; Chamber, A.; Lie, T.T. Review of prospects for adoption of fuel cell electric vehicles in New Zealand. *IET Electr. Syst. Transp.* **2017**, *7*, 259–266. [CrossRef]
2. Grijalva, E.R.; López Martínez, J.M. Analysis of the Reduction of CO_2 Emissions in Urban Environments by Replacing Conventional City Buses by Electric Bus Fleets: Spain Case Study. *Energies* **2019**, *12*, 525. [CrossRef]
3. Varga, B.O.; Iclodean, C.; Mariașiu, F. *Electric and Hybrid Buses for Urban Transport Energy Efficiency Strategies*, 1st ed.; Springer International Publishing: Cham, Switzerland, 2016; pp. 85–103.
4. Official Journal of the European Union. Directive (EU) 2018/2001 of the European Parliament and of the Council of 11 December 2018 on the Promotion of the Use of Energy from Renewable Sources (Recast). 2018. Available online: https://eur-lex.europa.eu/legal-content/EN/TXT/PDF/?uri=CELEX:32018L2001&from=ro (accessed on 14 October 2019).
5. Vepsäläinen, J.; Ritari, A.; Lajunen, A.; Kivekäs, K.; Tammi, K. Energy Uncertainty Analysis of Electric Buses. *Energies* **2018**, *11*, 3267. [CrossRef]
6. Bak, D.B.; Bak, J.S.; Kim, S.Y. Strategies for Implementing Public Service Electric Bus Lines by Charging Type in Daegu Metropolitan City, South Korea. *Sustainability* **2018**, *10*, 3386. [CrossRef]
7. Topal, O.; Nakir, I. Total Cost of Ownership Based Economic Analysis of Diesel, CNG and Electric Bus Concepts for the Public Transport in Istanbul City. *Energies* **2018**, *11*, 2369. [CrossRef]
8. Bagotsky, V.S.; Skundin, A.M.; Volfkovich, Y.M. *Electrochemical Power Sources: Batteries, Fuel Cells, and Supercapacitors*; John Wiley & Sons, Inc.: Hoboken, NJ, USA, 2015; pp. 91–103.
9. Electric Buses: Everything is Changing, Solaris Bus & Coach. Available online: http://www.oradea.ro/fisiere/module_fisiere/26259/1)%20RO_Solaris_Electric_2017_RO-1.pdf (accessed on 14 October 2019).
10. Li, W.; Stanula, P.; Egede, P.; Kara, S.; Herrmann, C. Determining the main factors influencing the energy consumption of electric vehicles in the usage phase. *Procedia CIRP* **2016**, *48*, 352–357. [CrossRef]
11. Kivekäs, K.; Lajunen, A.; Vepsäläinen, J.; Tammi, K. City Bus Powertrain Comparison: Driving Cycle Variation and Passenger Load Sensitivity Analysis. *Energies* **2018**, *11*, 1755. [CrossRef]
12. Raab, A.F.; Lauth, E.; Strunz, K.; Göhlich, D. Implementation Schemes for Electric Bus Fleets at Depots with Optimized Energy Procurements in Virtual Power Plant Operations. *World Electr. Veh. J.* **2019**, *10*, 5. [CrossRef]
13. Studiu Oportunitate 2018. Primăria și Consiliul Local Cluj-Napoca. Study of Opportunity 2018. Available online: https://storage.primariaclujnapoca.ro/userfiles/files/studiu-oportunitate-autobuze-final.pdf (accessed on 9 April 2019). (In Romanian).
14. Vellucci, F.; Pede, G. Fast-Charge Life Cycle Test on a Lithium-Ion Battery Module. *World Electr. Veh. J.* **2018**, *9*, 13. [CrossRef]
15. Cho, J.; Jeong, S.; Kim, Y. Commercial and research battery technologies for electrical energy storage applications. *Prog. Energy Combust. Sci.* **2015**, *48*, 84–101. [CrossRef]
16. Broussely, M.; Pistoia, G. *Industrial Applications of Batteries: From Cars to Aerospace and Energy Storage*, 1st ed.; Elsevier Science: Amsterdam, The Netherlands, 2007; pp. 203–271.
17. Demircali, A.; Sergeant, P.; Koroglu, S.; Kesler, S.; Oztürk, E.; Tumbek, M. Influence of the temperature on energy management in battery-ultracapacitor electric vehicles. *J. Clean. Prod.* **2018**, *176*, 716–725. [CrossRef]
18. Iora, P.; Tribioli, L. Effect of Ambient Temperature on Electric Vehicles' Energy Consumption and Range: Model Definition and Sensitivity Analysis Based on Nissan Leaf Data. *World Electr. Veh. J.* **2019**, *10*, 2. [CrossRef]

19. Qian, K.; Zhou, C.; Yuan, V.; Allan, M. Temperature Effect on Electric Vehicle Battery Cycle Life in Vehicle-to-Grid Applications. In Proceedings of the CICED 2010 China International Conference on Electricity Distribution, Nanjing, China, 13–16 September 2010.

20. Transit Cooperative Research Program. *Battery Electric Buses—State of the Practice*; The National Academies Press: Washington, DC, USA, 2018; Available online: https://www.nap.edu/catalog/25061/battery-electric-buses-state-of-the-practice (accessed on 1 October 2019).

21. Lindgren, J.; Lund, P.D. Effect of extreme temperatures on battery charging and performance of electric vehicles. *J. Power Sources* **2016**, *328*, 37–45. [CrossRef]

22. Waldmann, T.; Wilka, M.; Kasper, M.; Fleischhammer, M.; Wohlfahrt-Mehrens, M. Temperature dependent ageing mechanisms in Lithium-ion batteries—A Post-Mortem study. *J. Power Sources* **2014**, *262*, 129–135. [CrossRef]

23. Yi, Z.; Smart, J.; Shirk, M. Energy impact evaluation for eco-routing and charging of autonomous electric vehicle fleet: Ambient temperature consideration. *Transp. Res. Part C Emerg. Technol.* **2018**, *89*, 344–363. [CrossRef]

24. Yuksel, T.; Michalek, J.J. Effects of Regional Temperature on Electric Vehicle Efficiency, Range, and Emissions in the United States. *Environ. Sci. Technol.* **2015**, *49*, 3974–3980. [CrossRef] [PubMed]

25. Zhang, Z.; Li, W.; Zhang, C.; Chen, J. Climate control loads prediction of electric vehicles. *Appl. Therm. Eng.* **2017**, *110*, 1183–1188. [CrossRef]

26. Zhu, T.; Min, H.; Yu, Y.; Zhao, Z.; Xu, T.; Chen, Y.; Li, X.; Zhang, C. An Optimized Energy Management Strategy for Preheating Vehicle-Mounted Li-ion Batteries at Subzero Temperatures. *Energies* **2017**, *10*, 243. [CrossRef]

27. Jardin, P.; Esser, A.; Givone, S.; Eichenlaub, T.; Schleiffer, J.E.; Rinderknecht, S. The Sensitivity in Consumption of Different Vehicle Drivetrain Concepts Under Varying Operating Conditions: A Simulative Data Driven Approach. *Vehicles* **2019**, *1*, 69–87. [CrossRef]

28. Hu, X.; Huang, B.; Cherubini, F. Impacts of idealized land cover changes on climate extremes in Europe. *Ecol. Indic.* **2019**, *104*, 626–635. [CrossRef]

29. Reckiena, D.; Salvia, M.; Pietrapertosa, F.; Simoes, S.G.; Olazabal, M.; De Gregorio Hurtado, S.; Geneletti, D.; Krkoška Lorencová, E.; D'Alonzo, V.; Krook-Riekkola, A.; et al. Dedicated versus mainstreaming approaches in local climate plans in Europe. *Renew. Sustain. Energy Rev.* **2019**, *112*, 948–959. [CrossRef]

30. Soares, M.B.; Alexander, M.; Dessai, S. Sectoral use of climate information in Europe: A synoptic overview. *Clim. Serv.* **2018**, *9*, 5–20. [CrossRef]

31. Rymarz, J.; Niewczas, A.; Stoklosa, J. Reliability Evaluation of the City Transport Buses Under Actual Conditions. *Transp. Telecommun.* **2015**, *16*, 259–266. [CrossRef]

32. Briec, E.; Müller, B. (Eds.) *Electric Vehicle Batteries: Moving from Research towards Innovation, Reports of the PPP European Green Vehicles Initiative*; Springer International Publishing: Cham, Switzerland; Berlin, Germany, 2015.

33. Qin, N.; Brooker, R.P.; Raissi, A. Electric Bus Systems. EVTC Report Number: FSEC-CR-2060-17. 2017. Available online: http://www.fsec.ucf.edu/en/publications/pdf/FSEC-CR-2060-17.pdf (accessed on 2 October 2019).

34. Linden, D.; Reddy, T.B. (Eds.) *Handbook of Batteries*, 3rd ed.; McGraw-Hill: New York, NY, USA, 2002.

35. Nitta, N.; Wu, F.; Lee, J.T.; Yushin, G. Li-ion battery materials: Present and future. *Mater. Today* **2015**, *18*, 252–264. [CrossRef]

36. Łebkowski, A. Studies of Energy Consumption by a City Bus Powered by a Hybrid Energy Storage System in Variable Road Conditions. *Energies* **2019**, *12*, 951. [CrossRef]

37. Andreev, A.A.; Vozmilov, A.G.; Kalmakov, V.A. Simulation of lithium battery operation under severe temperature conditions. *J.Proeng* **2015**, *129*, 201–206. [CrossRef]

38. *AVL CRUISE, Users Guide*; VERSION 2011, Document No. 04.0104.2011, Edition 06/2011; AVL LIST GmbH: Graz, Austria, 2011.

39. Solaris Electric Buses experience and Further Development, Electrické Autobusy pro Město II. Available online: http://www.proelektrotechniky.cz/pdf/SeminarEbusyII/Figaszewski_Solaris_Ebus.pdf (accessed on 3 October 2019).

40. Liu, H.; Wang, C.; Zhao, X.; Guo, C. An Adaptive-Equivalent Consumption Minimum Strategy for an Extended-Range Electric Bus Based on Target Driving Cycle Generation. *Energies* **2018**, *11*, 1805. [CrossRef]

41. Soltani, M.; Ronsmans, J.; Kakihara, S.; Jaguemont, J.; Van den Bossche, P.; Van Mierlo, J.; Omar, N. Hybrid Battery/Lithium-Ion Capacitor Energy Storage System for a Pure Electric Bus for an Urban Transportation Application. *Appl. Sci.* **2018**, *8*, 1176. [CrossRef]

42. Yildiz, M.; Karakoc, H.; Dincer, I. Modeling and validation of temperature changes in a pouch lithium-ion battery at various discharge rates. *Int. Commun. Heat Mass Transf.* **2016**, *75*, 311–314. [CrossRef]

43. Mallon, K.R.; Assadian, F.; Fu, B. Analysis of On-Board Photovoltaics for a Battery Electric Bus and Their Impact on Battery Lifespan. *Energies* **2017**, *10*, 943. [CrossRef]

44. Wu, X.; Lv, S.; Chen, J. Determination of the Optimum Heat Transfer Coefficient and Temperature Rise Analysis for a Lithium-Ion Battery under the Conditions of Harbin City Bus Driving Cycles. *Energies* **2017**, *10*, 1723. [CrossRef]

45. Bloomberg New Energy Finance, Electric Buses in Cities—Driving Towards Cleaner Air and Lower CO_2. 29 March 2018. Available online: https://data.bloomberglp.com/professional/sites/24/2018/05/Electric-Buses-in-Cities-Report-BNEF-C40-Citi.pdf (accessed on 10 October 2019).

46. Du, Y.; Zhao, Y.; Wang, Q.; Zhang, Y.; Xi, H. Trip-oriented stochastic optimal energy management strategy for plug-in hybrid electric bus. *Energy* **2016**, *115*, 1259–1271. [CrossRef]

47. Emadi, A. (Ed.) *Advanced Electric Drive Vehicles (Energy, Power Electronics, and Machines)*; CRC Press, Taylor & Francis Group, LLC: Boca Raton, FL, USA, 2015; pp. 491–517.

48. Kumar, L.; Jain, S. Electric propulsion system for electric vehicular technology: A review. *Renew. Sustain. Energy Rev.* **2014**, *29*, 924–940. [CrossRef]

49. Varga, B.O.; Mariașiu, F.; Moldovanu, D.; Iclodean, C. *Electric and Plug-In Hybrid Vehicles, Advanced Simulation Methodologies*; Springer International Publishing: Cham, Switzerland, 2015; pp. 116–137.

50. Luin, B.; Petelin, S.; Al-Mansour, F. Microsimulation of electric vehicle energy consumption. *Energy* **2019**, *174*, 24–32. [CrossRef]

51. Li, L.; Zhang, Y.; Yang, C.; Yan, B.; Martinez, C.M. Model predictive control-based efficient energy recovery control strategy for regenerative braking system of hybrid electric bus. *Energy Convers. Manag.* **2016**, *111*, 299–314. [CrossRef]

52. Arriva Meteo în Cluj-Napoca. Available online: https://rp5.ru/Arhiva_meteo_%C3%AEn_Cluj-Napoca (accessed on 1 October 2019).

53. Air Density Calculator: Altitude, Temperature, Humidity and Barometric Pressure of Weather Systems and Their Impact on Local Atmospheric Air Density. Available online: https://barani.biz/apps/air-density/ (accessed on 10 October 2019).

54. The Engineering ToolBox, Air Psychrometrics. Available online: https://www.engineeringtoolbox.com/air-psychrometrics-properties-t_8.html (accessed on 10 October 2019).

55. Delos Reyes, J.R.M.; Parsons, R.V.; Hoemsen, R. Winter Happens: The Effect of Ambient Temperature on the Travel Range of Electric Vehicles. *IEEE Trans. Veh. Technol.* **2016**, *65*, 4016–4022. [CrossRef]

56. Panchal, S.; Dincer, I.; Agelin-Chaab, M.; Fraser, R.; Fowler, M. Thermal modeling and validation of temperature distributions in a prismatic lithium-ion battery at different discharge rates and varying boundary conditions. *Appl. Eng.* **2016**, *96*, 190–199. [CrossRef]

57. Scarinci, R.; Zanarini, A.; Bierlaire, M. Electrification of urban mobility: The case of catenary-free buses. *Transp. Policy* **2019**, *80*, 39–48. [CrossRef]

58. Yuan, X.; Zhang, C.; Hong, G.; Huang, X.; Li, L. Method for evaluating the real-world driving energy consumptions of electric vehicles. *Energy* **2017**, *141*, 1955–1968. [CrossRef]

59. ISO 8714-2002, Electric Road Vehicles—Reference Energy Consumption and Range—Test Procedures for Passenger Cars and Light Commercial Vehicles. Available online: https://www.iso.org/standard/34087.html (accessed on 1 October 2019).

60. GB/T 18386-2017, Electric Vehicles—Energy Consumption and Range—Test Procedures. Available online: http://www.gbstandards.org/GB_standards/GB-T%2018386-2017.html (accessed on 1 October 2019).

61. Thoreb, CTP Cluj-Napoca. Available online: http://www.thoreb.com/index.php/ro/referinte/31-europe/70-ctp-cluj-napoca (accessed on 1 October 2019).

Article

Lithium-Ion Polymer Battery for 12-Voltage Applications: Experiment, Modelling, and Validation

Yiqun Liu, Y. Gene Liao * and Ming-Chia Lai

Mechanical Engineering, Electric-drive Vehicle Engineering, Wayne State University, Detroit, MI 48202, USA; yiqun.liu@wayne.edu (Y.L.); mclai@wayne.edu (M.-C.L.)
* Correspondence: geneliao@wayne.edu

Received: 12 January 2020; Accepted: 27 January 2020; Published: 3 February 2020

Abstract: Modelling, simulation, and validation of the 12-volt battery pack using a 20 Ah lithium–nickel–manganese–cobalt–oxide cell is presented in this paper. The cell characteristics influenced by thermal effects are also considered in the modelling. The parameters normalized directly from a single cell experiment are foundations of the model. This approach provides a systematic integration of actual cell monitoring with a module model that contains four cells connected in series. The validated battery module model then is utilized to form a high fidelity 80 Ah 12-volt battery pack with 14.4 V nominal voltage. The battery cell thermal effectiveness and battery module management system functions are constructed in the MATLAB/Simulink platform. The experimental tests are carried out in an industry-scale setup with cycler unit, temperature control chamber, and computer-controlled software for battery testing. As the 12-volt lithium-ion battery packs might be ready for mainstream adoption in automotive starting–lighting–ignition (SLI), stop–start engine idling elimination, and stationary energy storage applications, this paper investigates the influence of ambient temperature and charging/discharging currents on the battery performance in terms of discharging voltage and usable capacity. The proposed simulation model provides design guidelines for lithium-ion polymer batteries in electrified vehicles and stationary electric energy storage applications.

Keywords: battery modelling and simulation; battery testing cycler; battery thermal model; lithium-ion polymer battery; SLI battery

---

1. Introduction

Lead–acid-based batteries have a long-term historical usage in the automotive and stationary standby power market, ranging from 12-volt high-power such as automotive starting–lighting–ignition (SLI) applications, low-power applications such as emergency lighting or uninterruptible power supplies (UPSs) for individual computers, to high-power, high-voltage electric energy storage in renewable energy systems or UPSs telecommunications facilities. Typical lead–acid batteries have several problems including high self-discharge rate, relatively heavy and large, and shallow depth of discharge (DOD). For the past decades, lithium-ion batteries have been widely used in portable electronics due to their features of high energy density, high discharge power, and long cycle life. The emerging applications of the lithium-ion batteries to electric-drive vehicles and large-scale energy storage systems for renewable energy make them a promising solution for challenges of environmental preservation and resource conservation [1,2]. The lithium-ion battery is also a suitable replacement for the conventional 12-volt SLI lead–acid battery; for example, Porsche offers an option of a lithium-ion SLI battery [3], and some medium-duty truck manufacturers use a lithium-ion battery for 12/24 V electrical systems [4,5]. The lithium-ion polymer battery uses a high conductivity semisolid (gel) polymer electrolyte instead of a liquid electrolyte. The battery cell voltage depends on the electrode material chemistries. The lithium–metal–oxide-based (such as $LiCoO_2$) cell has 2.5–2.8 V fully discharged voltage and 4.2 V fully charged voltage, while the lithium–iron–phosphate-based (such as $LiFePO_4$) cell has 1.8–2.0 V fully discharged voltage and 3.6–3.8 V fully charged [6,7]. The lithium polymer battery has higher specific energy than do other lithium-based batteries [6]. The polymer electrolyte gives the lithium polymer battery more stable performance under vibration conditions. These two features have led to the promotion of lithium polymer batteries in electric-drive vehicle applications.

The lithium-ion batteries, however, still encounter some roadblocks that complicate their applications. One of the major roadblocks is temperature influence on the operation of lithium-ion batteries. The operating temperature of a battery is the result of ambient temperature augmented by the heat generated by an electrochemical reaction. Operating temperatures from −20 °C to 60 °C is a typically acceptable range for lithium-ion batteries [8]. Pesaran et al. [9] presented that the optimal temperature range for lithium-ion batteries is from 15 °C to 35 °C, which is similarly comfortable for humans. To avoid a severe temperature gradient that might lead to different degradation rates and unbalanced cells, 5 °C should be set as the maximum temperature difference from cell to cell within a module [9,10]. The impacts of temperature can generally be considered as low and high temperature effects [11]. At low operating temperatures, the lithium-ion batteries experience slow chemical reaction and charged transfer-rate, which decrease ionic conductivity and diffusivity [12,13]. Therefore, the battery energy capacity and power are reduced at low temperatures. At high temperatures, the energy capacity and power are degraded, respectively, due to loss of the reduction of active materials and increase of internal resistance [14]. Self-ignition and even explosion caused by thermal runaway may happen if the temperature is too high. The effects caused by low battery temperature mostly occur during low ambient temperatures, while the effects induced by high battery temperature could occur either in low or high ambient temperatures. For an example, the battery temperature could highly increase at a large discharging current even in a low ambient temperature environment. The ambient temperature dominates in low temperature effects, and the battery internal temperature during operations plays a more important (than ambient temperature) role in high temperature effects.

The battery cell characteristics are determined by the electrode materials, electrolyte materials, cell size and shape, as well as the operating conditions including temperature, charging, discharging current, etc. The cell characteristics are essential parameters in battery pack design, thermal management system design, and battery management control. The battery cell characteristics typically are acquired through a series of charging and discharging experimental tests, which are time-consuming and require several pieces of equipment, such as a cycler, temperature chamber, and device for data acquisition. Analytical approach uses certain numbers of cell parameters gathered from less experimental tests to form a mathematical model. The battery model is also helpful for predicting parameters that cannot be directly measured by any sensors, such as state of charge (SOC), state of health (SOH), and state of life (SOL). Model-based estimation algorithms are usually used to compute or estimate theses parameters [15,16]. Nevertheless, a high-fidelity battery model is required to obtain accurate simulation results. Many battery models have been developed ranging from simple models with a few parameters to complex models having a large number of parameters [17–27]. The common battery modelling approaches are electrochemical, mathematical or analytical, and electric circuit-based model [28,29].

This paper describes the development and validation of an electric circuit-based Simulink model of the lithium–nickel–manganese–cobalt–oxide (LiNiMnCoO2)-based cell with 3.6 V nominal voltage and 20 Ah capacity. The thermal effects on cell characteristics are also considered in the model. The experiments apply several charging and discharging currents to the battery cell and module that are enclosed in a chamber with controlled temperatures of −20 °C, −10 °C, 0 °C, 20 °C, and 50 °C as the ambient temperatures. The experimental data are used to calibrate the model parameters. A 12-volt battery pack (14.4 V, 80 Ah) model is built based on validated simulation models of a battery cell and module. This SLI-type pack has four parallel-connected modules where each module (14.4 V, 20 Ah) consists of four cells connected in series. As the 12-volt lithium-ion battery packs might be ready for mainstream adoption in automotive SLI, micro-hybrid (or stop–start engine idling elimination), and

UPS applications, this paper investigates the ambient temperature effect on the battery performance in terms of discharging voltage and usable capacity. The proposed simulation model provides design guidelines for lithium-ion polymer batteries in electric-drive vehicle and stationary UPS applications.

2. Battery Modelling from Cell to Pack

A high-fidelity single cell model is a foundation to form a reliable battery module and pack with statistical confidence. The equivalent circuit technique is commonly used for electrochemical impedance characterizations in a cell model. This study uses parameters normalized directly from single cell experiments, which provide a systematic integration of actual cell monitoring with a module model. The approach begins with single cell model development and validation. A module with four cells connected in series is also validated. A high fidelity SLI battery pack model is then achieved.

2.1. An Enhanced Equivalent Electric Circuit Cell Model

The equivalent electric circuit approach has been adopted by many researchers to model battery cells ranging from lead–acid to lithium-ion batteries. The most commonly used equivalent electric circuit models are the Thevenin-based model [17–19], impedance-based model [20–22], and the runtime-based model [23,24]. The Thevenin-based model can predict battery response to the transient load at a certain SOC due to a series resistor and resistor–capacitor parallel network in the model. An impedance-based model is formed by an AC-equivalent impedance model in the frequency domain and the electrochemical impedance spectroscopy method. The runtime-based model utilizes a capacitor and controllable current source to predict battery capacity, SOC, runtime, and open circuit voltage (OCV). The battery operation time and DC voltage response under a constant discharge C-rate also can be simulated by the runtime-based model. The advantages of the runtime-based model and Thevenin-based model are combined in a model presented by [26], as shown in Figure 1. Based on [27], an equivalent electric circuit model with improved features is presented in this paper. In the developed Simulink model shown in Figure 2, three inputs (discharging current, initial SOC ranging from 0 to 1, and battery capacity) replace the battery runtime model. Since the initial SOC is an input variable, the developed model can simulate batteries that are not fully-charged. The OCV is calculated according to real-time SOC, which is predicted from three inputs to the model. Subtracting both voltages of the resistor-capacitor (RC) parallel networks and series resistor (R_S) from the OCV gives the cell terminal voltage (Vt), which is an output of the developed model. The real-time SOC, OCV, RC value, RC parallel network voltages, and series resistor voltage are calculated by five developed submodels.

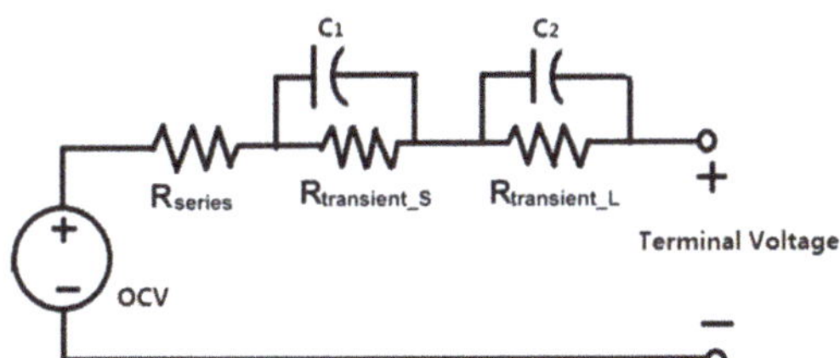

Figure 1. Thevenin with runtime-based model.

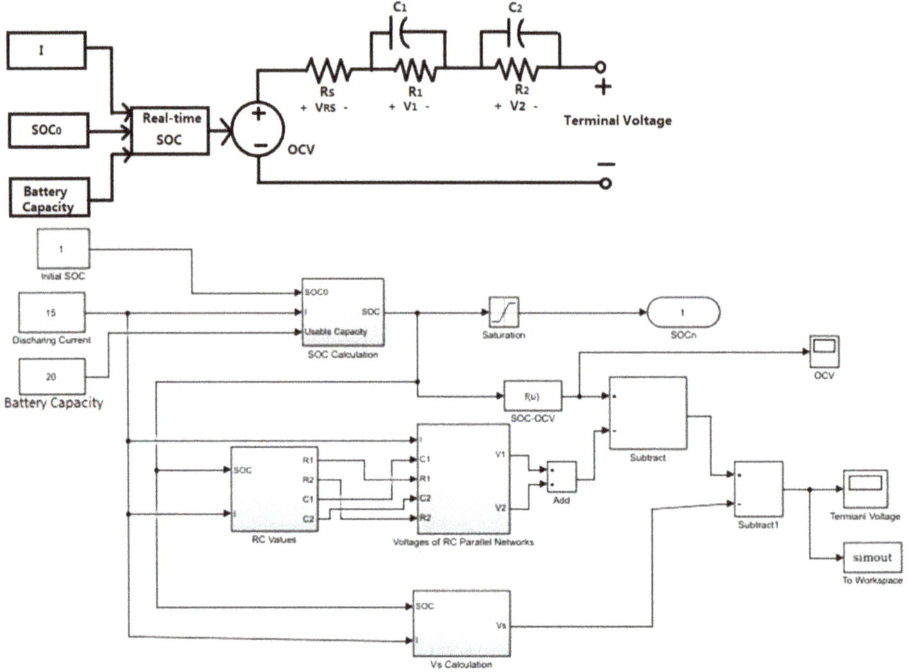

Figure 2. Developed cell model with three inputs.

Equation (1) calculates the real-time SOC, in which SOC_0 denotes the initial SOC, I denotes the discharging current, and UC denotes the usable capacity. A submodel calculating SOC is presented in Figure 3a, in which three inputs are SOC_0, I, and UC and output is real time SOC. Equation (2) is derived from numbers of experimental discharging curves using the method presented in [26] that provides relationship between the SOC and OCV. The interpolation–extrapolation lookup method is applied to calculate and determine the most suitable RC values, as a submodel presented in Figure 3b. The transient response of the battery cell voltage in the developed model is computed by the voltages of RC parallel networks. Equation (3) calculates the voltages of RC parallel networks in the s-domain, as a submodel shown in Figure 3c. For a typical lithium–metal–oxide polymer cell, the series resistor is 0–0.01 ohms in the 20%–100% SOC range, and 0.01–0.06 ohms within the 0%–20% SOC range [26]. Therefore, the developed model uses 0.001 ohms for 20%–100% SOC and 0.03 ohms for 0%–20% SOC in all discharging currents. The voltage on the series resistor (V_S) is calculated by Equation (4) where R_S is the series resistor resistance, and a submodel is presented in the Figure 3d. Equation (5) calculates the terminal voltage (V_t) of the battery cell. A more detail description of the submodels and their paraments is presented in [30].

$$\text{SOC} = \text{SOC}_0 - \int \frac{I}{UC \times 3600} dt \tag{1}$$

$$\text{OCV} = -1.031 e^{-35 \times \text{SOC}} + 3.685 + 0.2156 \times \text{SOC} - 0.1178 \times \text{SOC}^2 + 0.3201 \times \text{SOC}^3 \tag{2}$$

$$V = \left(\frac{1}{s}\right)\left[\frac{I}{C} - \frac{V}{RC}\right] \tag{3}$$

$$V_S = I \times R_S \tag{4}$$

$$V_t = \text{OCV} - V_1 - V_2 - V_S \tag{5}$$

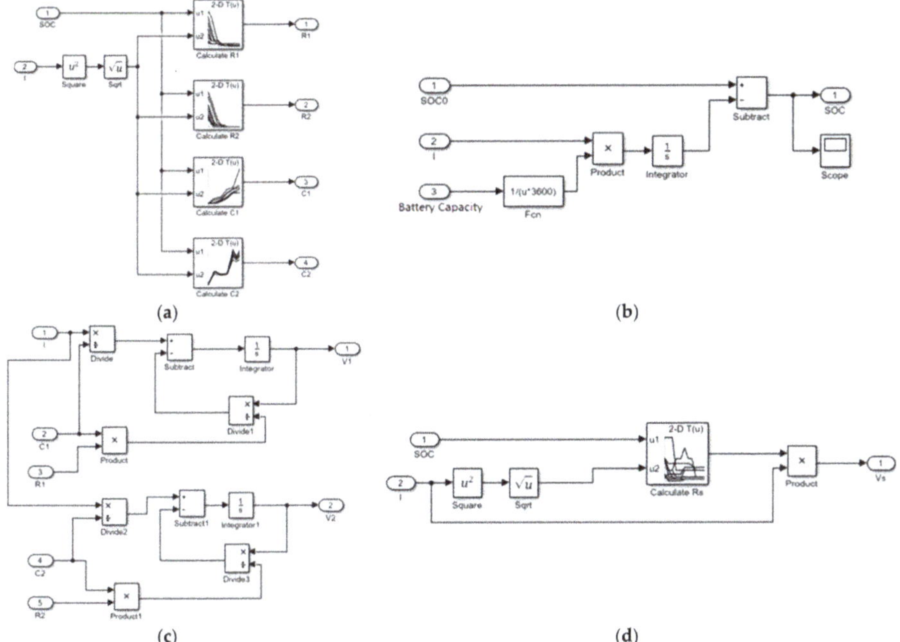

Figure 3. Simulink submodels for developed cell model: (**a**) Calculation of R and C value; (**b**) Calculation of SOG; (**c**) Calculation of RC parallel voltage; (**d**) Calculation of V_S.

2.2. Cell Thermal Model

The lithium-ion polymer cell thermal model was built in the Simulink battery block platform, which implemented similar equations as those discussed in Section 2.1 with thermal effects. In the discharge model (i∗ > 0), Equations (6)–(11) are implemented to represent the temperature effect on the battery model parameters [31]. The temperature tab requires several parameters, which are determined by battery discharging test under 20 °C ambient temperature. The initial cell temperature is set to the ambient temperature because each cell is cooled down or warmed up to the ambient temperature before starting the discharging test. The "nominal ambient temperature T1 (°C)" parameter is the ambient temperature during nominal operations. It is assumed that all parameters in the parameters tab are obtained at 20 °C ambient temperature. The procedures of establishing the cell thermal model in the Simulink platform are presented in [32]. Figure 4 shows the Simulink battery cell discharging model considering ambient temperature effects.

$$f_1(\text{it, i∗, i, T, } T_a) = E_0(T) - K(T)\cdot\frac{Q(T_a)}{Q(T_a) - \text{it}}\cdot(\text{i∗} + \text{it}) + A\cdot\exp(-B\cdot\text{it}) - C\cdot\text{it} \tag{6}$$

$$V_{\text{batt}}(T) = f_1(\text{it, i∗, i, T, } T_a) - R(T)\cdot i \tag{7}$$

$$E_0(T) = E_0\big|_{T_{\text{ref}}} + \frac{\partial E}{\partial T}(T - T_{\text{ref}}) \tag{8}$$

$$K(T) = K\big|_{T_{\text{ref}}}\cdot\exp\left[\alpha\left(\frac{1}{T} - \frac{1}{T_{\text{ref}}}\right)\right] \tag{9}$$

$$Q(T_a) = Q\big|_{T_a} + \frac{\Delta Q}{\Delta T}\cdot(T_a - T_{\text{ref}}) \tag{10}$$

$$R(T) = R\big|_{T_{ref}} \cdot \exp\left[\beta\left(\frac{1}{T} - \frac{1}{T_{ref}}\right)\right] \tag{11}$$

where: T_{ref} (K) nominal ambient temperature, T (K) cell or internal temperature, T_a (K) ambient temperature, E/T (V/K) reversible voltage temperature coefficient, α Arrhenius rate constant for the polarization resistance, β Arrhenius rate constant for the internal resistance, (Ah/K) maximum capacity temperature coefficient, CΔQ/ΔT (V/Ah) nominal discharge curve slope.

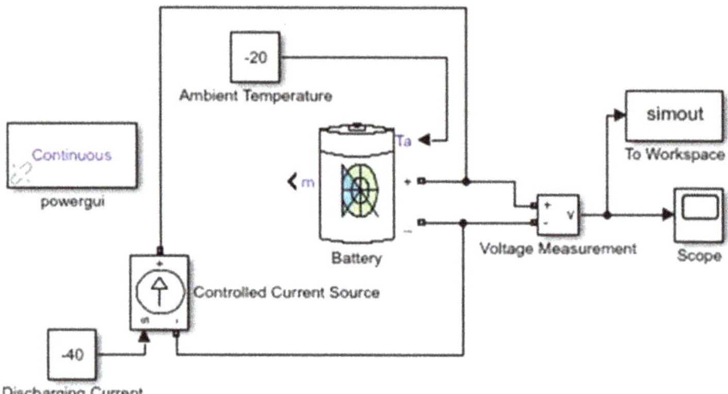

Figure 4. Battery cell thermal model in Simulink.

2.3. Battery Module and Pack Model

A battery module model containing four cells connected in series was created in the Simulink platform, as shown in Figure 5. The controlled current source, four battery cells, breakers with control algorithms to perform the battery management system (BMS) function, and voltage measurement with scopes are the four submodels in the module model. The charging and discharging current to each cell model are generated by the controlled current source sub-model, which has two parameters, namely DC source type and zero initial amplitude (A). The controlled current source block is connected to a constant block for generating a continuously constant charging or discharging current. A value in the constant block is the constant charging/discharging current. The pulse generator block is applied to generate a pulse charging/discharging current. Cell breaker, bypass breaker, cell voltage tag, and cell control tag form a BMS submodel for each cell. When the cell is charged to a voltage higher than 4.3 V or discharged to a voltage lower than 2.3 V, the control tag opens the cell breaker and closes the bypass breaker to prevent the cell from becoming over-charged or over-discharged. Each submodel contains one pair of tags for the breakers. Figure 5 shows only one pair of tags for a better display. Each cell voltage curve is shown in its scope and a terminal voltage is displayed in total voltage scope. All the parameters in this model are determined from the continuous and pulse discharge tests. A more detail description of the submodels and their paraments is presented in [33].

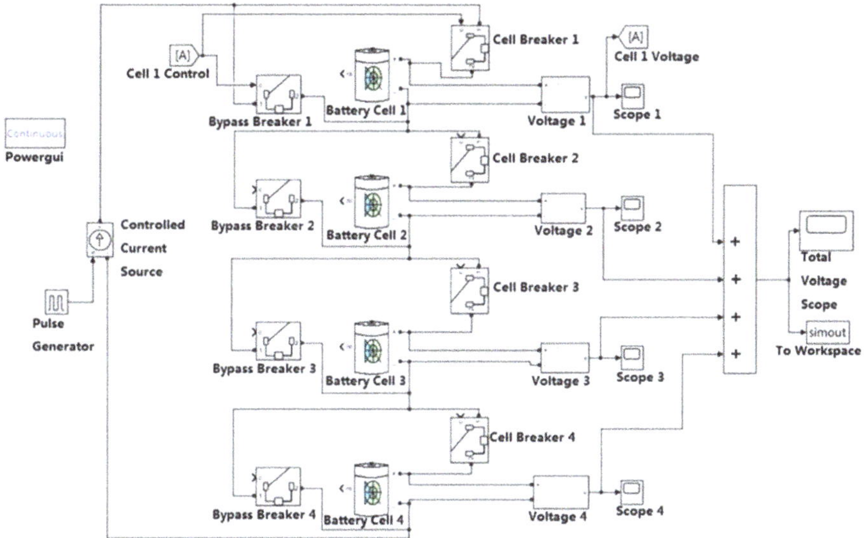

Figure 5. A Simulink battery module model consisting of four series-connected cells.

3. Experiment and Model Validation

The lithium-ion polymer cells used in this study were EiG ePLB C020 lithium–nickel–manganese–cobalt–oxide-based cathode and graphite-based anode with 3.6 V nominal voltage and 20 Ah capacity. Figure 6 shows the test equipment using in this study, namely a Digatron charge/discharge cycler, a computer with Digatron Battery Manager 4 (BM4) software (Battery Manager 4.0, Digatron Power Electronics Inc., Shelton, CT, USA) [34], and an Envirotronics temperature chamber. A fixture was designed to restrain the battery cells and cycler output cables inside the chamber. Three experimental procedures included calibration of battery cell model parameters, validation of the four series-connected battery module, and validation of the battery cell thermal model.

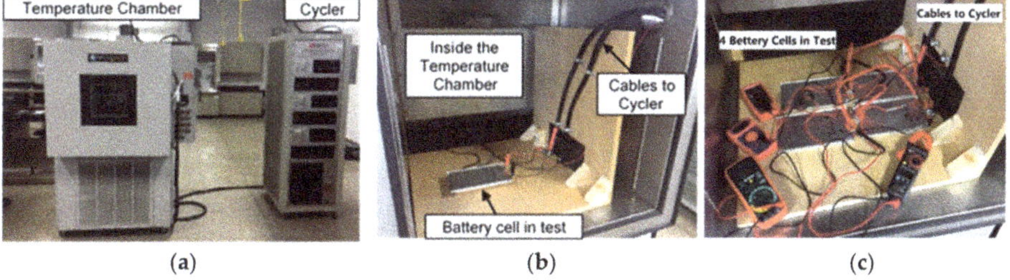

Figure 6. Battery experimental test setup: (**a**) Cycler and chamber; (**b**) cell testing; (**c**) module testing.

3.1. Battery Cell Model Calibration and Validation

An initial battery performance evaluation test was conducted on 27 cells disassembled from a hybrid electric vehicle battery pack. Each cell was charged to 4.17 V and then fully discharged to 2.46 V using 1 C rate in the initial evaluation test. The cell model parameter determination, correlation, and validation utilized six cells with best performance in the initial evaluation test. Figure 7a shows an experimental curve generated by averaging curve data for each testing case (1/3, 1/2, 1, 1.5, and 2 C discharging current). These tests were used to calibrate the cell model parameters. The determination of cell model parameters is presented in [30]. Examples of simulated and experimental discharging curves are shown in Figure 7b. A 7% or less discrepancy between simulations and experiments existed in the 0% to 80% DOD range during constant current discharge.

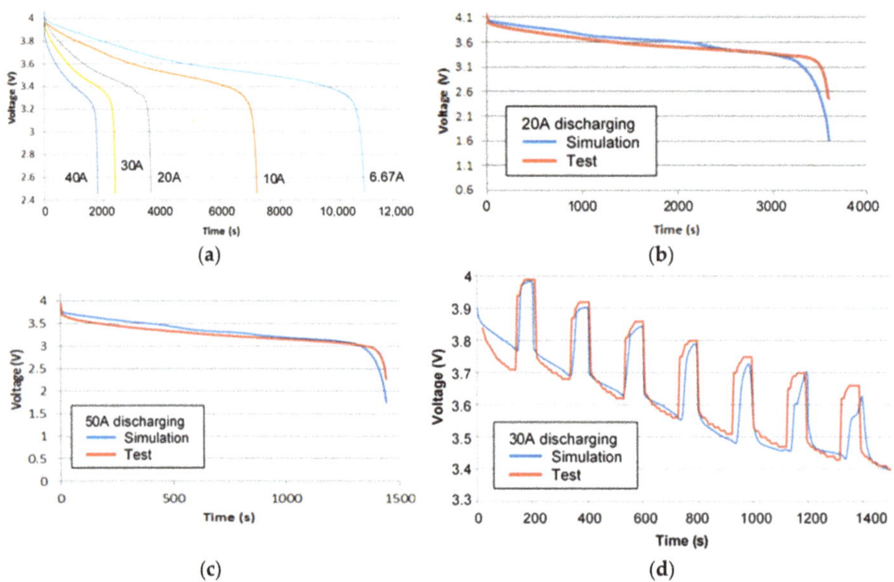

Figure 7. Examples of simulated and experimental discharging curves: (**a**) Averaged constant current discharging curves from tests; (**b**) simulated and test constant current discharging curves; (**c**) simulated and test constant current discharging curves; (**d**) simulated and test pulse discharging.

3.2. Cell Thermal Experiment and Validation

To ensure the battery cell completely cooled down or warmed up to a specific ambient temperature, the battery cell was kept charging by a small current to sustain its voltage in the neighbor of 4.17 V. The temperature chamber was then set to a specific temperature for 15 minutes before applying the discharging current. The validation process consisted of 12 discharging tests, which were 10 A, 20 A, and 40 A constant discharging currents, and each discharging test was conducted under 4 different ambient temperatures (−20, −10, 0, 20, and 50°C). The simulated discharging curves from each discharging current with specified ambient temperature were compared to corresponding experimental discharging curves. Examples of comparisons between simulated and experimental discharging curves using constant currents, 10 A, 20 A, and 40 A, under four different ambient temperatures are shown in Figure 8. A full comparison results is shown in [32]. In each discharging test, the experimental and the simulated discharging curves matched well in the range from 0% to 80% DOD (assuming 100% DOD at 2.5 V). The discrepancy between each comparison was under 7%. From the range of 80% to 100% DOD, the discrepancy became much larger. This large discrepancy might be due to the fact that battery model parameters were acquired in the nominal voltage of 3.6 V. The accuracy of the model was acceptable because most of the batteries only used up to 80% DOD. The experimental and simulated discharging time to reach 2.5 V were correlated in most test cases. Under higher ambient temperature (50 °C), the battery usable capacity increased so the total discharging time was longer than 1 hour for the one C-rate (20 A) discharging test. A similar phenomenon occurred in the one-half C-rate (10 A) test, which showed more than two hours discharging time. However, the Simulink model could not simulate the increase of battery usable capacity under high ambient temperature. Table 1 summarizes two correlated data under different ambient temperatures and discharging currents; discharging voltage at 50% DOD (assuming 100% DOD at 2.5 V) and total discharging time to reach 2.5 V are compared.

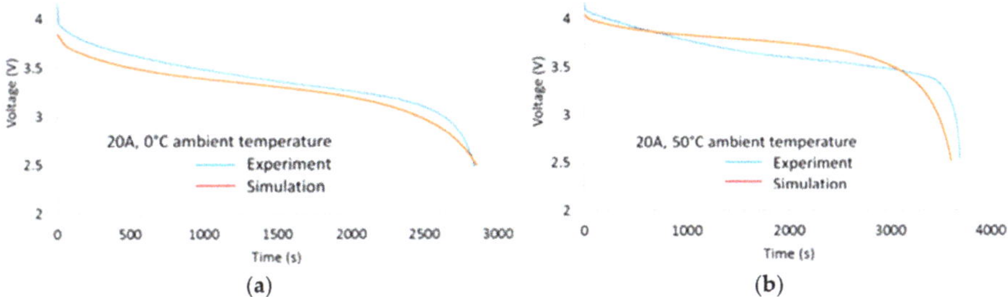

Figure 8. Examples of comparisons between simulated and experimental 20A discharging curves in battery module: (**a**) 0 °C; (**b**) 50 °C.

Table 1. Summary of comparisons between experimental and simulation results.

Ambient	Discharging Current					
	10 A		20 A		40 A	
	Experiment	Simulation	Experiment	Simulation	Experiment	Simulation
Temp (°C)	V (50% DOD)	Dis. time (s)	V (50% DOD)	Dis. time (s)	V (50% DOD)	Dis. time (s)
−20	3.15; 4200	3.15; 4000	3.05; 2072	3.07; 2010	2.95; 1076	3.02; 1020
−10	3.17; 4955	3.22; 4945	3.14; 2410	3.16; 2319	3.07; 1220	3.12; 1235
0	3.32; 5660	3.37; 5893	3.38; 2835	3.32; 2856	3.17; 1427	3.26; 1436
20	3.56; 7143	3.57; 7193	3.54; 3533	3.54; 3595	3.45; 1712	3.46; 1788
50	3.71; 7602	3.80; 7175	3.62; 3701	3.76; 3600	3.52; 1745	3.68; 1799

3.3. Battery Module Experiment and Validation

For the module testing, four series-connected cells formed a battery module, where the four cells were numbered from #1 to #4. This section was abstracted from [33]. The inconsistency of the cells was taken into account in the experiments and analytical models. All tests were conducted at a constant temperature of 25 °C in the chamber. Figure 6c shows the assembled-module experimental setup in the cycler. The four cells selected for the experiments had slightly different aging, internal resistance, and voltage, although they performed very similarly to each other (voltage variation between four cells was smaller than 0.08 V) under loads, as the example shows in Figure 9a. At any specific time, the difference between the highest and lowest voltage among the four cells was always no larger than 0.08 V. Figure 9b indicates that the module terminal voltage measured by the cycler was higher than the summation value of each cell terminal voltage (measured by individual voltmeter) during continuous charging. This phenonium might be due to internal resistances existing in the connecting wires between cells. The difference between cycler measurement and summation voltage became larger as the charging current increased (0.04 V for 10 A, 0.1 V for 20 A, 0.25 V for 30 A, and 0.5 V for 40 A). In continuously constant current discharge tests, each cell had very similar 10 A and 20 A discharging voltage curves. At any timeframe, the difference between the highest and lowest voltage among the four cells was always no larger than 0.15 V, as the example shows in Figure 9c. The internal resistances in the connecting wires between cells resulted in that module voltage measured by the cycler always being lower than the summation value of each cell terminal voltage measured by each individual voltmeter. As the discharging current increased, the voltage difference between the cycler measurement and summation became larger (0.05 V for 10 A, 0.2 V for 20 A, 0.3 V for 30 A, and 0.5 V for 40 A). This discrepancy was clearly observed during the 40 A discharging test shown in Figure 9d.

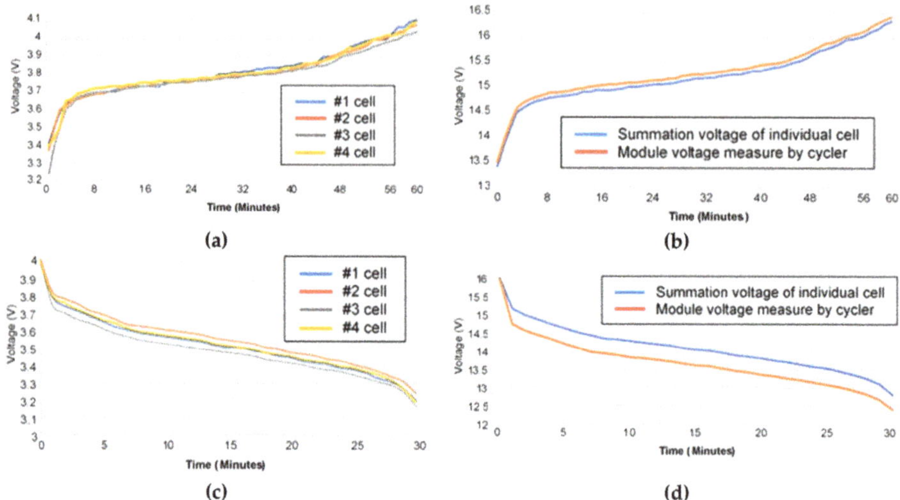

Figure 9. Examples of cell and module charging/discharging curves from tests: (**a**) Individual cell voltage in 20 A charging; (**b**) module voltage in 20 A charging; (**c**) individual cell voltage in 40 A discharging; (**d**) module voltage in 40 A discharging.

Comparing the simulated module voltage curves with experimental voltage curves generated by the summation of four individual cells was conducted for module model validation. The module voltage curve measured by the cycler was not used for comparison because it was affected by the resistance of the connecting wires. Figure 10a–d shows that all the simulated curves matched well with the experimental ones during continuous charging and discharging. An up to 9% discrepancy occurred at the end of each charging and discharging cycle. Figure 10e and f indicate that simulated pulse charging and discharging curves matched with the experimental ones, particularly in the beginning of the cycle. The largest discrepancy was 7.8%, which occurred at the end of the 30 A pulse discharge curve. The validation of the developed battery module model presented an acceptable discrepancy.

The BMS function was simulated in the battery module model with predefined initial conditions. In one example of 25 A continuous charge current to the model, Cell #4 reached 4.3 V earlier than other cells because Cell #4 had a higher initial voltage. Therefore, the BMS opened the cell breaker and closed the bypass breaker to prevent Cell #4 from being overcharged. The voltage of Cell #4 then dropped to 4.08 V, as shown in Figure 11a. In the other example of 35 A continuous discharge current, Cell #4 reached 2.3 V first because it had a lower initial voltage. The BMS opened the cell breaker and closed the bypass breaker to prevent Cell #4 from being overdischarged. The Cell #4 voltage then increased back to 2.8 V.

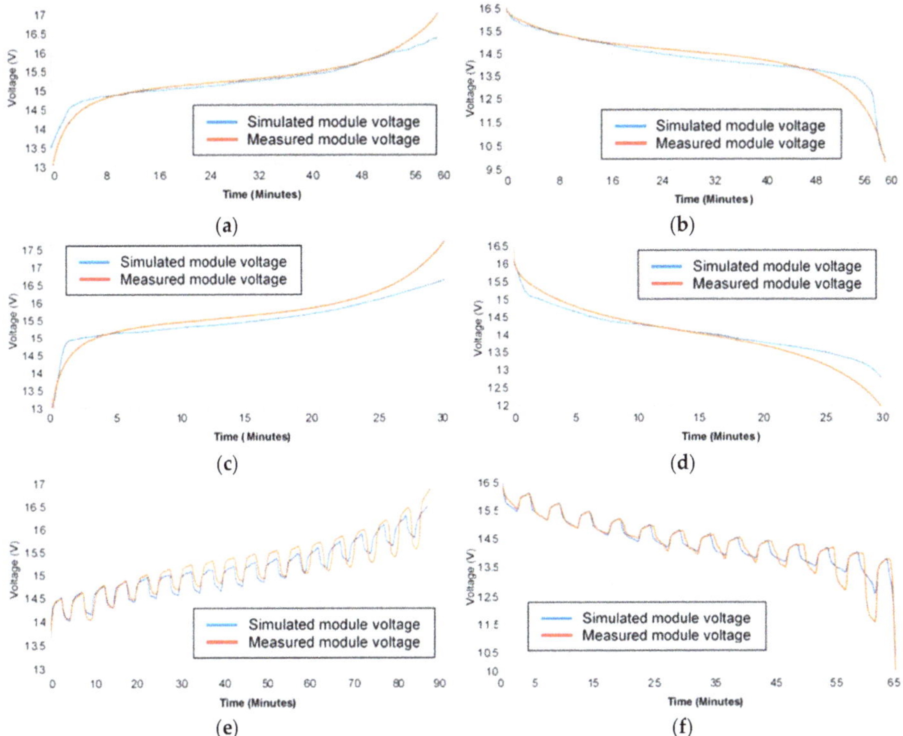

Figure 10. Comparisons of module simulated and experimental charging/discharging curves: (**a**) Simulated and test of 20 A charging in module; (**b**) simulated and test of 20 A discharging in module; (**c**) simulated and test of 40 A charging in module; (**d**) simulated and test of 40 A discharging in module; (**e**) simulated and test of 20A pulse charge in module; (**f**) simulated and test of 30A pulse discharge in module.

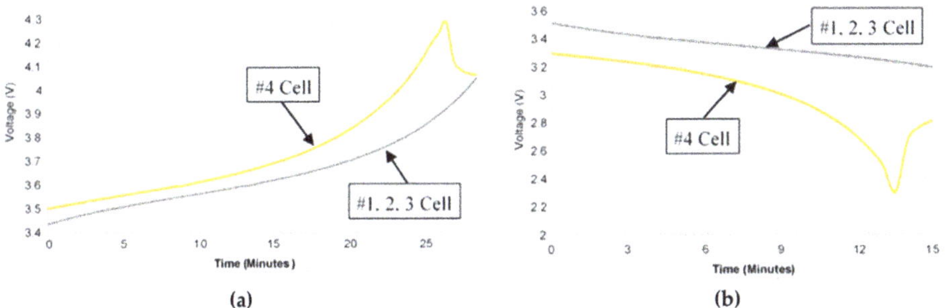

Figure 11. Demonstration of BMS functions: (**a**) 25 A continuous charging; (**b**) 35 A continuous discharging.

4. Twelve-Volt Battery Pack Model

The new designed or future vehicle needs an SLI battery with a higher capacity to support increasing vehicle accessory or auxiliary loads [35,36]. Additionally, the 12-volt battery pack could become a building module to form a high-voltage battery pack (such as 48-volts or higher) used in electrified vehicle and stationary electric energy storage for renewable energy. An 80 Ah SLT-type battery pack with 14.4 V nominal voltage is proposed in this study. This battery pack contains four modules connected in parallel where each module (14.4 V, 20 Ah) has four ePLB-C020 cells connected in series. A Simulink model of the proposed battery pack is shown in Figure 12.

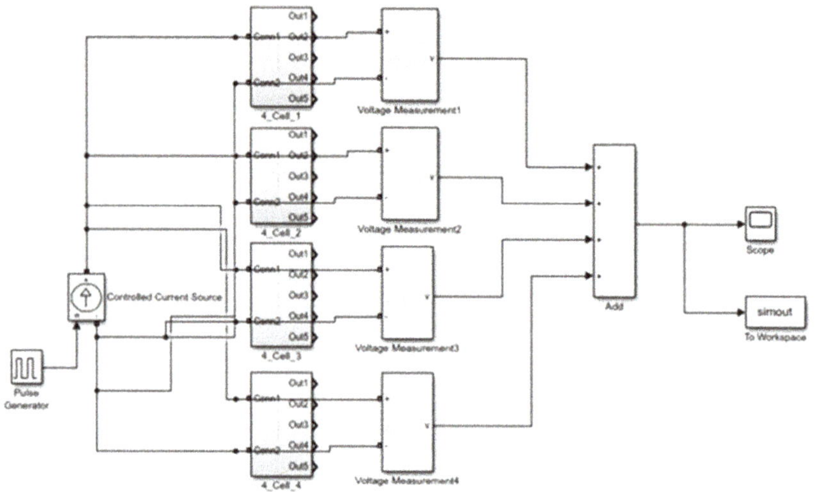

Figure 12. Simulink model of an 80 Ah SLI-type battery pack.

The model scope displays the battery pack voltage, which is a summation of each module voltage. Both simulated battery pack and module have the same shapes of constant current charging/discharging voltage curves. Obviously, the pack has four times charging/discharging durations of the module. The pack voltage curves have large fluctuations in each pulse during 20 A pulse charge simulation (180 seconds charge, 120 seconds pause, and repeat) and 30 A pulse discharge simulation (180 seconds discharge, 120 seconds pause, and repeat), as indicated in Figure 13. Figure 14 shows the simulated discharging curves of a 14.4 V 80 Ah SLI battery with one C-rate (20 A) and two C-rate (40 A) under five ambient temperatures.

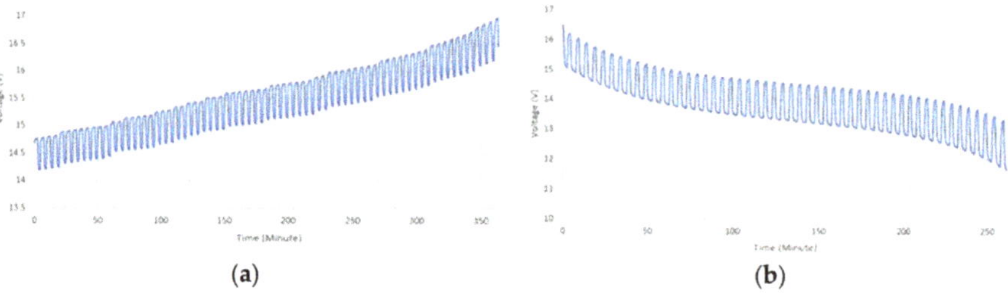

(a)   (b)

Figure 13. Voltage curves of the 80 Ah SLI battery pack during pulse charging and discharging simulations: (**a**) Charging; (**b**) discharging.

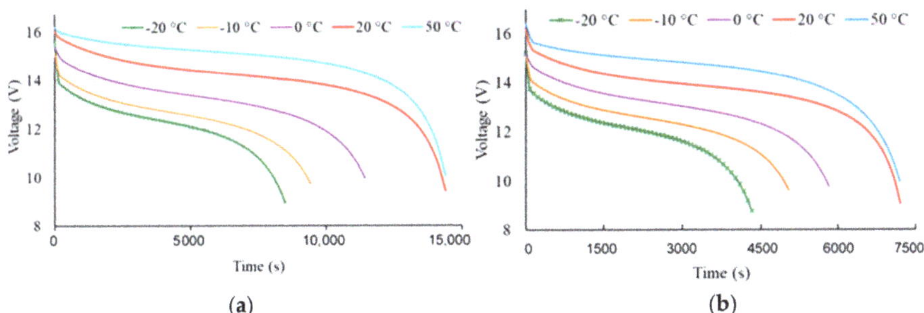

Figure 14. Simulated discharging curves of a 14.4 V 80 Ah SLI battery under five ambient temperatures: (**a**) One C-rate (20 A); (**b**) two C-rate (40 A).

5. Conclusions

Modelling, simulation, and validation of SLT-type 12-volt lithium-ion polymer battery are presented in this paper. The MATLAB/Simulink-based modelling starts from using parameters deduced directly from single cell experiments, which provide convenient integration with actual cell monitoring, to a module containing four cells connected in series. A validated module model is utilized to model a high fidelity 80 Ah SLI-type battery pack with 14.4 V nominal voltage. The battery cell thermal effectiveness and battery management system functions are also considered. The experimental tests are carried out in an industry-scale setup with a charge/discharge cycler, temperature chamber, and computer-controlled software for battery testing.

In the cell-level model validation, either with or without thermal effectiveness, the experimental and the simulated discharging curves match well in the range from 0% to 80% DOD (assuming 100% DOD at 2.5 V). The discrepancy between each comparison is under 7%. From the range of 80% to 100% DOD, the discrepancy becomes much larger. The module model validation indicates a 9% or less discrepancy in all continuous and pulse charge/discharge simulation results. An 80 Ah SLI-type battery pack model with 14.4 V nominal voltage then can be achieved with statistical confidence.

The 12-volt lithium-ion battery packs might be ready for mainstream adoption in automotive SLI, stop–start engine idling elimination, and UPS applications. Additionally, the 12-volt battery pack could become a building module to form a high-voltage battery pack (such as 48-volts or higher) used in electrified vehicle and stationary electric energy storage for renewable energy. The proposed simulation model provides design guidelines for lithium-ion polymer batteries in electric-drive vehicle and stationary energy storage applications.

Author Contributions: Experiment, Software Simulation, Y.L.; Paper Writing, Y.L., Y.G.L.; Paper Review and Editing, Y.G.L., M.-C.L. All authors have read and agreed to the published version of the manuscript.

Funding: This work was supported in part by the National Science Foundation, ATE: Centers, under grant number DUE-1801150.

Conflicts of Interest: The authors declare no conflict of interest.

References

1. Hesse, H.C.; Schimpe, M.; Kucevic, D.; Jossen, A. Lithium-ion battery storage for the grid - a review of stationary battery storage system design tailored for applications in modern power grids. *Energies* **2017**, *10*, 2107. [CrossRef]
2. Larson, S. Energy Storage in Utility Systems. 2019. Available online: https://www.eiseverywhere.com/file_uploads/05612c1b0b638b14d1ede35138c07afc_EnergyStorage_HRS_2019-02-04.pdf (accessed on 20 August 2019).
3. Porsche Offers Li-ion Starter Battery Option. 23 November 2009. Available online: http://www.greencarcongress.com/2009/11/porsche-liion-20091123.html (accessed on 30 May 2019).

4. Ding, Y.; Zanardelli, S.; Skalny, D.; Toomey, L. Technical challenges for vehicle 14V/28V lithium ion battery replacement. In Proceedings of the SAE International 2011 World Congress, Detroit, MI, USA, 12–14 April 2011. Paper No. 2011-01-1375.

5. Nunez, F.; Pugsley, J.; Shields, R.; Heilbrun, R.; Dobley, A. 6T format lithium ion batteries in 12V and 24V configurations. In Proceedings of the Joint Service Power Expo, Cincinnati, OH, USA, 26 August 2015.

6. Reddy, T. (Ed.) *Linden's Handbook of Batteries*, 4th ed.; McGraw Hill: New York, NY, USA, 2010; p. 27.3.

7. Huggins, R. *Energy Storage*; Springer: Cham, Switzerland, 2016; pp. 323–338.

8. Ji, Y.; Zhang, Y.; Wang, C.Y. Li-ion cell operation at low temperatures. *J. Electrochem. Soc.* **2013**, *160*, 636–649. [CrossRef]

9. Pesaran, A.; Santhanagopalan, S.; Kim, G.H. Addressing the impact of temperature extremes on large format Li-ion batteries for vehicle applications. In Proceedings of the 30th International Battery Seminar, Ft. Lauderdale, FL, USA, 10–14 May 2013.

10. Pesaran, A. Battery thermal models for hybrid vehicle simulations. *J. Power Sources* **2002**, *110*, 377–382. [CrossRef]

11. Kim, J.; Oh, J.; Lee, H. Review on battery thermal management system for electric vehicles. *J. Appl. Therm. Eng.* **2019**, *149*, 192–212. [CrossRef]

12. Belt, J.R.; Ho, C.D.; Miller, T.J.; Habib, M.A.; Duong, T.Q. The effect of temperature on capacity and power in cycled lithium ion batteries. *J. Power Sources* **2005**, *142*, 354–360. [CrossRef]

13. Fan, J.; Tan, S. Studies on charging lithium-ion cells at low temperatures. *J. Electrochem. Soc.* **2006**, *153*, 1081–1092. [CrossRef]

14. Ping, P.; Peng, R.; Kong, D.; Chen, G.; Wen, J. Investigation on thermal management performance of PCM-fin structure for Li-ion battery module in high-temperature environment. *J. Energy Convers. Manag.* **2018**, *176*, 131–146. [CrossRef]

15. Li, J.; Mazzola, M.S. Accurate battery pack modelling for automotive applications. *J. Power Sources* **2013**, *237*, 215–228. [CrossRef]

16. Githin, K.; Prasad, G.K.; Rahn, C.D. Reduced order impedance models of lithium ion batteries. *J. Dyn. Sys. Meas. Control* **2014**, *136*, 041012.

17. Putra, W.S.; Dewangga, B.R.; Cahyadi, A. Current estimation using Thevenin battery model. In Proceedings of the Electric Vehicular Technology and Industrial, Mechanical, Electrical and Chemical Engineering Conference, Surakarta, Indonesia, 4–5 November 2015.

18. Barsali, S.; Ceraolo, M. Dynamical models of lead-acid batteries: Implementation issues. *IEEE Trans. Energy Convers* **2002**, *17*, 16–23. [CrossRef]

19. Schweighofer, B.; Raab, K.M.; Brasseur, G. Modelling of high power automotive batteries by the use of an automated test system. *IEEE Trans. Instrum. Meas.* **2003**, *52*, 1087–1091. [CrossRef]

20. Buller, S.; Thele, M.; De Doncker, R.W.; Karden, E. Impedance based simulation models of supercapacitors and Li-ion batteries for power electronic applications. In Proceedings of the Industry Applications Conference, Salt Lake City, UT, USA, 12–16 October 2003.

21. Baudry, P.; Neri, M.; Gueguen, M.; Lonchampt, G. Electro-thermal modelling of polymer lithium batteries for starting period and pulse power. *J. Power Sources* **1995**, *54*, 393–396. [CrossRef]

22. Rao, R.; Vrudhula, S.; Rakhmatov, D.N. Battery modelling for energy-aware system design. *Computer* **2003**, *36*, 77–87.

23. Benini, L.; Castelli, G.; Macii, A.; Macii, E.; Poncino, M.; Scarsi, R. Discrete-time battery models for system-level low-power design. *IEEE Trans. VLSI Syst.* **2001**, *9*, 630–640. [CrossRef]

24. Gold, S. A PSPICE macromodel for lithium-ion batteries. In Proceedings of the Twelfth Annual Battery Conference on Applications and Advances, Long Beach, CA, USA, 14–17 January 1997; pp. 215–222.

25. Camacho-Solorio, L.; Krstic, M.; Klein, R.; Mirtabatabaei, A.; Moura, S.J. State estimation for an electrochemical model of multiple-material lithium-ion batteries. In Proceedings of the ASME 2016 Dynamic Systems and Control Conference, Minneapolis, MN, USA, 12–14 October 2016.

26. Chen, M.; Rincon-Mora, G.A. Accurate electrical battery model capable of predicting runtime and I-V performance. *IEEE Trans. on Energy Convers.* **2006**, *21*, 504–511. [CrossRef]

27. Yao, L.W.; Aziz, J.A.; Kong, P.; Idris, N.R. Modelling of lithium-ion battery using MATLAB/Simulink. In Proceedings of the IECON 2013—39th Annual Conference of the IEEE Industrial Electronics Society, Vienna, Austria, 10–13 November 2013.

28. Garcia-Valle, R.; Peças Lopes, J.A. *Electric Vehicle Integration into Modern Power Networks*; Springer Science+Business Media: New York, NY, USA, 2013; pp. 36–42.
29. Dao, T.; Schmitke, C. Developing mathematical models of batteries in modelica for energy storage applications. In Proceedings of the 11th International Modelica Conference, Versailles, France, 21–23 September 2015.
30. Liu, Y.; Liao, Y.G.; Lai, M.C. Development and validation of a lithium-ion polymer battery cell model for 12V SLI battery applications. In Proceedings of the ASME International Design Engineering Technical Conferences and Computer & Information in Engineering Conference, Quebec City, QC, Canada, 26–29 August 2018.
31. MathWorks. Available online: https://www.mathworks.com/help/physmod/sps/powersys/ref/battery.html (accessed on 1 March 2019).
32. Liu, Y.; Liao, Y.G.; Lai, M.C. Ambient temperature effect on performance of a lithium-ion polymer battery cell for 12-voltage applications. In Proceedings of the ASME International Mechanical Engineering Congress and Exposition, Salt Lake City, UT, USA, 11–14 November 2019.
33. Liu, Y.; Liao, Y.G.; Lai, M.C. Modelling and validation of lithium-ion polymer SLI battery. In Proceedings of the SAE International 2019 World Congress, Detroit, MI, USA, 9–11 April 2019. Paper No. 2019-01-0594. [CrossRef]
34. Battery Manager 4 Manual, Digatron Power Electronics, Inc. 2014. Available online: http://www.digatron.com/nc/en/home/ (accessed on 8 March 2019).
35. Ceraolo, M.; Huria, T.; Pede, G.; Vellucci, F. Lithium-ion starting-lighting-ignition batteries: Examining the feasibility. In Proceedings of the 2011 IEEE Vehicle Power and Propulsion Conference, Chicago, IL, USA, 5–8 September 2011.
36. Fehrenbacher, C. 12V Li-ion batteries—Ready for Mainstream adoption. In Proceedings of the Advanced Automotive Battery Conference, Mainz, Germany, 30 January 30–2 February 2017.

Article

Real Time Design and Implementation of State of Charge Estimators for a Rechargeable Lithium-Ion Cobalt Battery with Applicability in HEVs/EVs—A Comparative Study

Nicolae Tudoroiu [1,*], Mohammed Zaheeruddin [2] and Roxana-Elena Tudoroiu [3]

[1] John Abbott College, Sainte-Anne-De-Bellevue, QC H9X 3L9, Canada
[2] Department of Building, Civil and Environmental Energy Concordia University Montreal, Montreal, QC H3G 1M8, Canada; zaheer@encs.concordia.ca
[3] Department of Mathematics-Informatics, University of Petrosani, 332006 Petrosani, Romania; tudelena@mail.com
* Correspondence: ntudoroiu@gmail.com

Received: 26 March 2020; Accepted: 27 May 2020; Published: 31 May 2020

Abstract: Estimating the state of charge (SOC) of Li-ion batteries is an essential task of battery management systems for hybrid and electric vehicles. Encouraged by some preliminary results from the control systems field, the goal of this work is to design and implement in a friendly real-time MATLAB simulation environment two Li-ion battery SOC estimators, using as a case study a rechargeable battery of 5.4 Ah cobalt lithium-ion type. The choice of cobalt Li-ion battery model is motivated by its promising potential for future developments in the HEV/EVs applications. The model validation is performed using the software package ADVISOR 3.2, widely spread in the automotive industry. Rigorous performance analysis of both SOC estimators is done in terms of speed convergence, estimation accuracy and robustness, based on the MATLAB simulation results. The particularity of this research work is given by the results of its comprehensive and exciting comparative study that successfully achieves all the goals proposed by the research objectives. In this scientific research study, a practical MATLAB/Simscape battery model is adopted and validated based on the results obtained from three different driving cycles tests and is in accordance with the required specifications. In the new modelling version, it is a simple and accurate model, easy to implement in real-time and offers beneficial support for the design and MATLAB implementation of both SOC estimators. Also, the adaptive extended Kalman filter SOC estimation performance is excellent and comparable to those presented in the state-of-the-art SOC estimation methods analysis.

Keywords: lithium-ion cobalt battery; state of charge; state of energy; adaptive EKF SOC estimation; linear observer SOC estimation; MATLAB; Simscape

---

1. Introduction

Currently, hybrid and electric vehicles (EVs) represent a means of transport with low CO_2 emissions. Also, soon, the energy required for these vehicles is expected to be provided by clean, renewable energy sources, such as solar panels. An essential feature of EVs is the recovery of energy they would lose during braking. Of various energy storage systems (ESS), "electrochemical batteries are devices that store chemical energy converted then into electricity to power the electric vehicles; they are preferred over capacitors and flywheels, due to their higher energy density" [1]. Based on a wide range of powers, three main categories are mentioned in [1], namely "EVs light electric vehicles with a power demand of less than several kilowatts, sedan vehicles, including electric sedan hybrid vehicles (HEVs) with a power up to 100 kW and heavy vehicles, used for public transport, with a power exceeding 100 kW". For electric powertrains, the "lithium-ion (Li-ion) battery represents a good choice for EVs/HEVs" [1]. Today, it is already becoming a reality that, "among the batteries with low memory effects", the Li-ion outperforms the most popular nickel-based technologies. They excel by "lighter weight, high density of energy, long life and low self-discharge rate" [1]. However, HEVs/EVs continue to be powered for a long time by both nickel-metal hydride (Ni-MH) batteries and lithium-ion [1–12]. The strengths and weaknesses in "terms of cost, specific energy and power, safety, life span performance for the main different chemistries" are analyzed in [1]. Hard research work is being carried out in the field of lithium-ion batteries to increase their energy density, and to take advantage of the advancement of anode and cathode material technologies. The "common materials used for the positive electrode are cobalt oxide, manganese oxide, iron phosphate, nickel manganese cobalt oxide and nickel cobalt aluminum oxide" [1]. Among them, lithium nickel-manganese-cobalt oxide battery is a suitable choice for EVs since it offers "an excellent trade-off between safety, capacity and performance" [1]. The most popular cobalt Li-ion (Li-ion Co) battery "used in consumer products was believed to be not robust enough"; nevertheless, due to its "high energy density", this "computer battery" power nowadays "the Tesla Roadster and Smart Fortwo ED small cars". The behaviour of the battery changes during deep cycles when "its capacity decreases rapidly" and, it is also sensitive to "high mechanical, thermal or electrical stresses" [6,10]. Specifically, the power of the battery "decreases drastically in cold weather", and "when operating at high temperatures, its performance and life cycle visibly deteriorate" [10]. To avoid these situations and to extend the battery life, a "cooling and heating" system is usually installed. [6,10]. Additionally, the lithium-ion battery is "vulnerable to short-circuiting and overcharging" which could lead to a "combustion reaction, explosion and fire" as is mentioned in [6,10]. Thus, to "prevent overcharging of batteries in hazardous situations", the battery management system (BMS) monitors the battery cells through a "precise voltage control system" [10]. Particular progress is being made today in "lithium-air" and in "nanotechnologies" batteries, as they "have a higher energy density due to oxygen being a lighter cathode and a freely available resource", as mentioned in [13]. Of course, new technology also means high costs, but battery prices are gradually declining over time, as the manufacturing capacity of batteries becomes expanding as well. The Li-ion Co battery is an essential component of BMS which has as its primary function "improving battery performance, extending its life and operating safely" [3,10–12]. Therefore, it must continuously monitor, through the sensors, the internal parameters of the battery, such as the SOC, temperature, cells' currents balance and voltage [3,10].

SOC as an internal battery state, is a priority task for BMS to monitor, as it severely affects battery health and battery life [2,3,5–7,10–12]. In references [2,3,6–12] is defined as an "available battery capacity", which cannot be measured directly; therefore, an estimation technique is needed to prevent hazardous situations and to improve battery performance [3,10,11]. Mostly, the battery SOC estimation techniques are model-based, as is well documented in [7–24].

In conclusion, motivated by some preliminary results of our research, published in [10–12], this article focuses on the selection of the Li-ion Co battery model, the design and implementation of two real-time SOC estimators on a MATLAB simulation platform. The other chapters of this paper are structured as follows. In Section 2 are described the BMS, the selection criteria, the parameters of the battery and identifies the main disturbances that affect the functionality and the battery life. Also, is made a detailed analysis of state of the art on SOC measurement and estimation methods reported in the literature, and at the end of the same section are presented some modelling aspects and validation of Li-ion Co battery. In Section 3 are designed and implemented in MATLAB an adaptive extended Kalman filter (AEKF) and a linear observer (LOE) SOC estimators. The MATLAB simulations result, and rigorous performance analysis are presented at the end of the section. Section 4 is assigned for discussions, and Section 5 concludes the research paper contributions.

2. Lithium-Ion Battery-Cell Modelling and Validation

In this section, we focus our attention on the following topics regarding the Li-Ion batteries cells and packs:

(1)   BMS—definition, multitask safety functions, hardware and software components.
(2)   Battery selection criteria.
(3)   Battery parameters test.
(4)   Disturbances that affect battery functionality and the life span.
(5)   Measurement and estimation.
(6)   Cell modelling and validation test.

2.1. Battery Management System; Definition, Multitask and Safety Functions, Hardware and Software Components

A most comprehensive and mature Battery Management System (BMS) is an analogue-digital multitasking safety functions device integrated into the battery control system structure. The main task is to perform "a variety of safety functions to prevent the voltage, temperature and current in the battery cells from exceeding the specified limits", as stated in [16]. The hardware components include those regarding "the safety circuitry, sensors, data acquisition, charging and discharging, control, communications and thermal management" [12]. In most automotive applications, the BMS performs tasks regarding "the safe operation and reliability of the battery, protecting battery cells and battery systems against damage, as well as battery efficiency and service life" [16]. Besides, it achieves interfacing, protection, control voltage, fault detection diagnosis and isolation (FDDI) and performance management functions, as is revealed in [16]. In a centralized configuration, it combines "into a single printed circuit board (PCB)" up to three module levels into hierarchical architecture, such as at the first level is located "the battery cell monitoring unit" ("data acquisition"), at the second one is the "module management unit" ("cell supervisor circuit" and at the highest level is the "package management unit" ("central management unit") [16]. Thus, the required tasks can be managed and distributed among different subcomponents through PCB connected to the battery cells. Moreover, its advanced modular topology, known as "the master-slave-topology" is an exciting feature [16].

The advantage of this configuration consists of reducing to a minimum "the functions and the elements of the slaves" such that the "master" to implement only the functions related to the battery system [16]. One of the most critical parameters controlled by the BMS is the temperature inside the battery. As it was mentioned in the Introduction section, temperature significantly affects battery performance and, for most cases, when it exceeds the maximum limits, it leads to "fire and explosion", known as "thermal runaway" process; it is an "irreversible process" with a significant heat "dissipated from the battery cells casing" [16]. In a battery pack, the battery's cells can be connected in series, parallel or as mixt combinations to "adapt the voltage level and the battery capacity" to meet the requirements of an HEV/EV or stationary storage applications. Moreover, the primary constraint of the functionality of any BMS, regardless of its chemistry, is the maximum cell voltage measured on the cell monitoring unit. Of the Li-ion batteries, "lithium-iron-phosphate cells are one of the lowest voltage batteries with a maximum of 3.65 V, while for the widespread nickel-manganese-cobalt cells the maximum voltage is 4.2 V" [16]. At least one "communication interface" uses a "CAN-bus communication line for easy interfacing with other controllers in the car environment" [16]. Moreover, currently, the "wireless" devices "operating via wireless networks have promising potential to significantly reduce wiring, connectors and cable effort during assembly" [16].

However, it is possible to disturb the wireless network by "electromagnetic noise inside the car and outside entities, which can create safety and security issues" [16].

For each safety issue, it is necessary to prevent "the deep discharge and the overcharge" of the battery's cell. Thus, the SOC of the battery pack is one of the most critical parameters estimated that keeps the "track" of the energy flow and reacts anytime when it is not operating within the specified range [16]. If SOC exceeds the limits, an alarm signal shall be sent as soon as possible to the "vehicle systems" concerned to prevent possible damage. Also, the battery cells need protection from possible damage generated during a "deep discharge", such as dangerous "internal short circuits" [16].

Like, in the case of an overload, the information is sent to "propulsion controller" that decides to "stop charging the battery" anytime if the "maximum limit value is reached" [16]. Among the main software components of the BMS are highlighted the following [12]:

1.   Estimation and monitoring the battery SOC/SOH.
2.   The currents balancing algorithm.
3.   FDDI estimation techniques.

These components control the hardware operations, receive signals from sensors and "implement in real-time the estimation of SOC, SOH algorithms and of possible faults using FDDI techniques" [10]. Also, the BMS fulfil the task of estimation and monitoring the battery internal and insulation resistances [10,11].

Soft battery failures are detected using FDDI estimation techniques and identify defective components and "abnormal" functionality. In [10] are mentioned the sensor voltage faults (gain and drift) in measured terminal battery voltage, sensor current faults, sensor temperature faults, and fan motor faults. The estimation of sensor faults is particularly useful for improving the "reliability" of BMS [10]. Well, "several faults" have their roots in defective components, "safety component failures or human errors" [10]. Usually, the fault can be persistent, intermittent, unique or overlap with other faults, for which its root cause may be a faulty cable connection, a sensor bias (voltage, current) or a temperature drift [10]. A faulty fan is detected only when a complete dc motor failure occurs.

2.2. Battery Selection Criteria

The main battery selection criteria in all HEV/EVs applications can be found in [10], including "energy and power density, capacity, weight, size, lifespan, cost and memory effect" features that make the difference for selecting any battery. Of these, power and capacity are necessary to optimize the design of the battery, selecting the most suitable cells and package size, able to be adapted to a custom application [10]. Furthermore, given that most HEVs/EVs operate for different climatic conditions of harsh operation and stress caused by abuse and vibration, the size of the battery needs to be adequate to provide a certain amount of energy [10]. Additionally, some constraints can be imposed on the capacity of the battery in terms of "depth of discharge (DOD), SOC, discharging rate and generative braking charge", as is stated in [10].

2.3. Battery Parameters Test

Mainly, the Li-ion battery life span depends significantly on SOC real-time estimation, aging effects, temperature operating conditions and frequency "of the changes in operating cycles" [10].

Also, internal DC resistance and insulation resistance are among the most critical parameters of the battery that have a significant impact on battery life. Related to first battery parameter, in reference [10] the life cycle is defined as "the number of the cycles performed by the battery before its internal resistance increases 1.3 times or double than its initial value when was new". The main factors that affect the internal resistance are revealed in reference [11], including "the conductor and electrolyte resistances, ionic mobility, temperature effects and changes in SOC".

Related to the second battery parameter, in reference [10] is stated that the "high voltages components, electrical motor, battery charger and its auxiliary device deal with a large current and insulation"; thus, the "insulation issues" are under investigation during the" battery design stage" [10]. The harsh "working conditions" detailed in [11], have a significant impact on "fast aging of the power cable and insulation materials", decreasing drastically "the insulation strength" and "endanger the personnel". Thus, it needs to ensure safe operating conditions for personnel are required to evaluate the insulation conditions for entire HEV's BMS. Many details about the insulation standards can you can find in [11]. In conclusion, to ensure the insulation security of on-board BMS, it is necessary to "detect the insulation resistance and raise the alarm in time" as is mentioned in [10].

2.4. Disturbances that Affect the Battery Operation and the Life Span

In "real life", the primary disturbances affecting battery operation and life are well-identified in [10], and include:

Chemical changes—leading to damage to the battery cells.

Active chemicals depletion—take place under different operating conditions, as was mentioned in the Introduction section.

Temperature—battery operation significantly depends on the temperature, which also affects the performance of the battery.

Pressure—is affected by the temperature that increases the internal pressure inside the battery cell.

DOD—is related to SOC and depends on operating temperature conditions and discharge rate, becoming "proportional to the amount of active chemicals" [10].

Charging level limits—the full charge of the battery must be prevented to keep the battery safe.

Charging rate—to keep the battery safe, discharging the battery at high rates should be avoided.

Voltage—to counteract "undesirable chemical reactions" inside the battery cells the values of the battery terminal voltage must be within a specified range [10].

Cell aging—cell aging is mainly affected by the current flow through the battery cells, as well as by the heating and cooling processes of the cells.

Coulombic efficiency (CE)—is an important performance indicator of the charging efficiency of the battery through which the electrons are transferred inside the battery. The CE rating of Li-ion batteries exceeds 99% and is among the highest values of any rechargeable battery.

Electrolyte loss—it has a significant impact on the capacity of the cells whenever there is a reduction in the active chemicals inside the battery.

Internal and insulation resistances—their impact was described in the previous subsection.

2.5. Li-Ion Battery SOC—State of the Art of Measurement and Estimation Methods Reported in the Literature

Basically, "the battery model, estimation algorithm selection, and cells balancing" have a high impact on SOC accuracy and robustness, as is stated in [17]. Also, in [17] the authors investigate several existing SOC estimation techniques reported in the literature field and analyze their "issues and challenges". In reference [17] are well summarized the main Li-ion battery SOC estimation techniques related to HEVs/EVs field, including:

(1) Conventional direct measurement methods.
(2) Adaptive filter algorithms.
(3) Learning algorithms.
(4) Non-linear observers.
(5) Hybrid methods.

Our investigations are motivated by the lack of a sensor capable of measuring the battery SOC and therefore it is necessary to estimate it. Several measurement methods and estimation techniques are well documented and summarized in [12,16–18].

2.5.1. Conventional Direct Measurement Methods

(1) Laboratory tests and chemistry dependent methods. In the literature are reported the main four cell modelling methods that are briefly presented in [10]. Among these we highlight the following:

- A laboratory method for determining SOC—even if it is not suitable for the field of HEV applications, it is still one of "the most accurate SOC measurement" methods [10].

> This method consists of completely discharging the battery, recording the "discharged ampere-hours" and then determining the "remaining cell capacity available".

- Chemistry-dependent methods for other chemistries—unsuitable for Li-ion batteries.
- Open-circuit voltage (OCV) method. This method uses "the stable electromotive force (EMF) of the open circuit" and the SOC relationship to "estimate the SOC value", as is stated in [17]. In [10] are presented in detail some reasons why this method is inadequate for the dynamic estimation of the Li-ion battery SOC. Moreover, since its OCV = f (SOC) characteristic is almost flat for a considerably broad range of SOC values, it isn't easy in this approach to estimate SOC more accurately [18].
- Terminal voltage measurement method. The terminal voltage of the Li-ion battery is "based on the voltage drops on the internal impedances when the battery is discharging. Thus the EMF of the battery is proportional to the terminal voltage" [18]. Moreover, since the "EMF of the battery is approximately linearly proportional to the battery SOC, the terminal voltage is also approximately linearly proportional to SOC" [18]. The "disadvantage of this approach is a large estimated error in the terminal voltage of the battery at the end of battery discharge; it is due to a sharp drop of the terminal voltage" [18].

(2) Electro-chemical method. In this approach, it is "estimated the average amount of Li concentration in the positive or negative electrodes" using "partial differential equations" [17]. These models may achieve an "accurate terminal voltage prediction", but "it would be difficult to measure all the required physical parameters on a cell-by-cell basis in a high-volume consumer product" [8].

(3) Impedance direct measurement method. In this approach, the "measurements provide knowledge of several parameters, the magnitudes of which may depend on the SOC of the battery" [18]. Since the "battery impedance parameters and their variations with SOC are not unique for all battery systems" it is required many impedance experiments to identify its parameters [18].

(4) SOC Estimation spectroscopy method. This method uses the battery impedance and internal resistance "to describe the intrinsic electric characteristic under any current excitation, if temperature, SOC and SOH are fixed", but "it is not suitable for use in HEVs/EVs" applications [17]. The reason you can find in the same reference [17]. According to [17], "it is tough to measure online electrical impedance spectroscopy over a wide range of AC frequencies at the different charge and discharge currents, especially when the SOC and impedance relationship is not stable, and the cost is expensive". Also, is mentioned in the previous subsection that the internal resistance of the Li-ion battery is measured in direct current (DC) and requires the "value of the voltage and current at a small-time interval" [17]. Then, the battery SOC "may be indirectly inferred by measuring present battery cell impedance" [10] and the battery internal resistance correlated "with known impedances at various SOC levels" [8,18].

(5) Ampere-hour counting method. Based on this method is calculated "the amount of charge that flows in and out of the battery" [10]. In this approach, the SOC is estimated directly in an open loop, so the SOC estimate is not accurate due to the current measurement errors. Instead, in a closed loop, the same method can estimate the SOC more accurate [8,10]. The SOC estimate accuracy degrades the accuracy significantly when the battery is not "fully discharged after a complete charging cycle". Since in the most cases, the battery doesn't perform a full charge followed by a full discharge, "a significant drift is difficult to avoid", and thus, since "the signal drifts, the efficiency of coulomb counting decreases" [10]. Also, the Ampere-hour (Ah) counting method becomes "less effective when the battery self-discharges is subject to temperature changes" [10]. The "unknown initial value of battery SOC, capacity fading, self-discharge rate, and current sensor errors are the main sources of errors for Ah counting method" [17]. The presence of "an accurate measurement sensor and a predefined calibration point can overcome the method's drawbacks" [17]. Additionally, the estimation error "can be maintained at a low value by defining a correction factor and a re-calibration point" [10]. It is worth mentioning that this method is more

accurate compared to other SOC calculation methods [10,17]. The most significant advantage of the Ampere-hour counting method is its low power computation cost, and it is secure and reliable when the sensor current measurements are accurate, and a re-calibration point is accessible [10,17].

(6) Model-based method. Since the "previous mentioned methods are not appropriate for online SOC estimation and to achieve an accurate online SOC estimate value, suitable battery models need to be developed" [17]. Among the most suitable models for online SOC estimation are the electrochemical and equivalent circuit models (ECMs) [10,11,17]. More details about ECMs models you can find in [3,4,7–12,17]. In closing, "an ideal ECM should be able to simulate the actual battery terminal voltage to any charging or discharging battery input current", as is stated in [17].

2.5.2. Adaptive Filter Estimators

The adaptive filter estimators improve the "accuracy and the robustness of the battery SOC estimation significantly and reduce" drastically the impact of measurement and process noises on the battery model [17], such those developed in [7–12,14,18]. Among the Kalman filtering estimation techniques in the field literature the Kalman filter [KF], extended Kalman filter (EKF) [4,7–9,11,17,18], adaptive Kalman filter (AEKF) [17], fading Kalman filter (FKF) [17], unscented Kalman filter (UKF) [12,14,18], sigma-point Kalman filter (SPKF) [17] and particle filter (PF) are reported [17].

The KF was developed by Rudolph Kalman in 1960 and currently has become the most popular estimation algorithm. It is an "optimum state estimator and intelligent tool" for linear systems [17]. Its EKF version is also a KF applied to the linearized dynamics of a non-linear system by using the first-order Taylor's series expansion around the current value of the state estimate in each step of the algorithm, as developed in the next section and in [7–9,17]. A combination of KF state estimator and Ah Coulomb counting method can be used to "compensate for the non-ideal factors that can prolong the operation of the battery" [17]. The KF SOC estimator is the most used since it can estimate the battery SOC more accurately even if when the battery is affected by external disturbances mentioned in previous subsections.

Although, if the dynamics of Li-ion battery model are "highly nonlinear", "linearization error may occur due to the lack of accuracy in the first-order Taylor series expansion under a highly non-linear conditions" [17]. The simplicity of the SOC EKF estimator design and implementation motivates researchers to apply this estimation technique for different Li-ion battery models, as in [4,7–9,11,16–24]. In [16] an exciting research project that performs a detailed and rigorous analysis of state of the art on Li-ion BMSs, including also a detailed presentation of the main SOC estimation techniques, among them the adaptive Kalman filtering techniques, is presented. Similarly, [17] presents an intense study of state of the art on Li-ion battery SOC estimation for electric vehicles that completely reviews of all the existing SOC direct measurement and estimation techniques reported in the field literature. Similarly, in [18] a brief review describing the SOC estimating methods for the same Li-based batteries is provided. In [19], the authors proposed a dual EKF for state and parameter estimation for a first-order EMC RC Li-ion battery model. The SOC simulation results reveal an excellent accuracy of the SOC estimate, but the robustness algorithm robustness is not investigated. Comparing the SOC simulation results obtained in current research work and in [19], one can observe an excellent accuracy, and the robustness of the algorithm developed in our research for several scenarios. In [20], the authors use an improved non-linear second-order RC EMC battery model and based on this model have developed an EKF algorithm to estimate the Li-ion battery SOC. The simulations are conducted on the MATLAB platform using two different driving cycles current profiles, namely Urban Dynamometer Driving Schedule (UDDS) and HWFET. The results are compared to those obtained by a Coulomb counting method and reveal an excellent SOC accuracy, but degradation is visible in the robustness performance to changes of battery model parameters values provided by two datasets, compared to the SOC estimator robustness performance designed in our research, for many scenarios introduced in Section 3. In [21] is developed a new application "model-based fault diagnosis scheme to detect and isolate the faults

(FDDI) of the current and voltage sensors applied in the series battery pack based on an adaptive extended Kalman filter (AEKF)". The AEKF algorithm is designed to estimate the magnitude of the faults. The FDDI scheme is validated in the MATLAB/Simulink platform, and the result of the simulations demonstrates the "effectiveness" of the proposed FDDI for "various fault scenarios using the "real-world driving cycles".

The AEKF is an EKF with an adaptive feature, i.e., in the new design the EKF algorithm updates at each step the process and measurement noise covariance matrices to increase the accuracy of EKF SOC estimation. The same feature is also added to the EKF algorithm developed in our research that is very useful to increase the accuracy and the robustness of the SOC EKF estimator. In fact, by updating the noise covariance matrices, a new retuning procedure of the EKF parameters is not more required unlike the time consuming "trial and error" strategy. In the reference [21] you can see the effectiveness of the AEKF algorithm that estimates accurately four injected faults, the first is a fault assigned to a sensor current, and the other three faults are assigned to three different voltage sensors. The robustness of the FDDI technique is demonstrated for a 20% change in the SOC initial value and a current profile corresponding to a UDDS driving cycle. Unfortunately, the MATLAB simulations results don't show the SOC estimated values, very useful to analyze the impact of each fault on the SOC estimation performance. In the reference [22] "an experimental approach is proposed for directly determining battery parameters as a function of physical quantities". The battery model's parameters are dependent on SOC and of the discharge C-rate. This approach is exciting since the battery model's parameters "can be expressed by regression equations in the model" to derive "a continuous-discrete dual EKF SOC state and parameters estimates" [22]. A "standard correction step" of the EKF algorithm is applied to "increase the accuracy of the estimated battery's parameters" [22]. The EKF simulation results with the experimental results for several operating scenarios reveal a high accuracy and the robustness of the estimator for correct identification of the battery parameters. In the reference [23] an adaptive fading EKF (AFEKF) is proposed for Li-ion battery SOC estimation accuracy and convergence speed. The AFEKF SOC estimator combines both structures AEKF and a fading EKF (FEKF). A FEKF "adopts a variable forgetting factor least square (VVFFLS)" to identify the Li-ion battery parameters [23]. The AFEKF estimator can reduce the SOC estimation error of less than 2%. Also, in our research, we add the same feature to the proposed AEKF SOC estimator, and the MATLAB simulation results reveal a high SOC estimation accuracy and robustness for many scenarios including three driving cycles tests UDDS, FTP and EPA-UDDS. Comparing the MATLAB simulation results obtained in [23] to those obtained in our research work, for same UDDS cycle, you can notify that the SOC estimator designed in [23] performs better in terms of accuracy. Instead, the proposed estimator in the current research performs better in terms of robustness and convergence speed. The speed convergence and robustness performance are revealed for a 20% decrease in SOC initial value in [23] and 30% in our case study.

In [24] an exciting online EKF SOC Li-ion internal resistance parameter estimator to "overcome defects from simplistic battery models" is developed. The battery is a first-order ECM RC model for which the internal resistance is dependent on SOC, temperature and aging effects.

For an accurate real-time internal resistance, the EKF estimated values "can be distinguished well" and also "improve the accuracy of SOC and SOH estimation" [24]. The internal resistance test device consists of a dc power supply source, a dc voltmeter, a pulse control switch and a microcontroller unit that controls the testing procedure, the dc power source, the switching time and voltage measurement. The EKF estimator is conceived as parameter estimator. Hence, its model is like for EKF state estimator. Still, in this case, the internal resistance dynamic is given by a slow varying first-order differential equation that has injected a Gaussian process noise. The EKF estimator can also estimate at the same time the SOC of Li-ion battery; thus, it is designed as a dual state-parameter EKF algorithm. The simulation results indicate an excellent accuracy of SOC estimate, for "a repeated current constant-constant voltage of 3200 mA discharge current and 1600 mA charging current, and the estimation error is smaller than 3%" [24]. Unfortunately, a new estimation result from a performance comparison is not possible since the input current profiles used in current research work (UDDS, FTP and EPA-UDDS) are entirely different than the current profile used in [24].

A viable alternative to EKF SOC estimator can be the unscented Kalman filter (UKF) and sigma point Kalman filter (SPKF) that avoid the linearization of nonlinear dynamics of the battery model; thus, they are more accurate and robust than EKF [10,11,14,17]. Also, a particle filter (FP) method is used to estimate the states, estimating the "probability density function" of a nonlinear dynamics of the Li-ion battery model, using a Monte-Carlo simulation technique, such as developed in [12,17].

2.5.3. Learning Methods

In this category the artificial neural networks (ANN), support vector machine (SVM), extreme machine learning (ELM), genetic algorithm (GA) and fuzzy logic (FL), well documented in [17], can be highlighted.

2.5.4. Linear and Nonlinear Observers

The nonlinear observers (NLO), sliding mode observer (SMO) and proportional-integral observer (PIO) are proposed to estimate the SOC of Li-ion batteries, and a detailed description can be found in [17].

2.5.5. Hybrid Methods

The hybrid method is a combination of two or more algorithms' structures, such as an EKF-Ah algorithm, an adaptive EKF (AEKF) and a support vector machine (SVM), like the one developed in [17].

2.6. Li-Ion Battery Cell—Model Selection, Validation and Case Study

In this section, we are focused on the generic Li-ion Co cell model description in a bidimensional continuous and discrete-time state-space representation. Since "the new technologies heavily depend on battery packs, it is therefore important to develop accurate battery cell models that can conveniently be used with simulators of power systems and on-board power electronic systems", such is mentioned in [25]. The Li-ion Co battery model adopted in this research paper is a generic MATLAB/Simscape nonlinear model suggested in [25] and depicted in Figure 1.

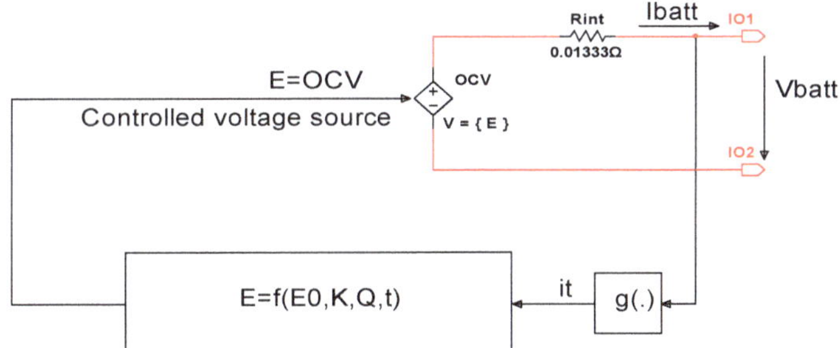

Figure 1. The non-linear Li-ion Co battery generic model (see [25]). (it is common picture met in the literature, it is not copyright issues!).

In this schematic the battery is modeled by a controlled voltage source E, which is a no-load voltage (open circuit voltage (OCV)) [25], given by:

$$E = OCV = f(E_0, K, Q_{max}, t) = E_0 - K\frac{Q_{max}\int idt}{Q_{max} - \int idt} + A_{exp}e^{(-B_{exp}\int idt)} \tag{1}$$

On the internal resistance is dissipated the power losses Ploss, useful to design the thermal model in the next section to simulate the temperature effects on the battery, given by:

$$P_{loss}(t) = R_{int}i^2(t) \tag{2}$$

The battery terminal voltage V_{batt} is related to OCV according to following highly non-linear dynamic relationship:

$$V_{batt}(t) = E - R_{int}i(t) = E_0 - K\frac{Q_{max}\int idt}{Q_{max} - \int idt} + A_{exp}e^{(-B_{exp}\int idt)} - R_{int}i(t), \tag{3}$$

where the meaning of all the variables and coefficients can be found in Table 1. Additionally, we attach the Coulomb counting equation to define the SOC of the battery, which is an important battery internal state supervised by BMS. It delivers a valuable "feedback about the state of health of the battery (SOH) and its safe operation", as is mentioned in [10]. The battery SOC is defined in [10] as:

$$SOC = \frac{\text{Remaining capacity}}{\text{Rated capacity}} \tag{4}$$

Table 1. Description of Li-ion cobalt voltage model variables.

Variable	Description	Unit	Value
E	No-load voltage (OCV)	V	-
E_0	Battery constant voltage	V	8.0259 volts
K	Polarization voltage	V	0.001834 volts
Q_{rated}	Rated battery capacity	Ah	5.4 Ampere-hours
A_{exp}	Exponential zone amplitude	V	0.35904 volts
B_{exp}	Exponential zone time constant inverse	$(Ah)^{-1}$	3/(Ah) = 3/(3600As)
R_{int}	Internal resistance of the battery	Ω	0.0133 ohms
$i(t) = I_{batt}$	Battery direct current (dc) input profile If $i(t) \geq 0$ is discharging current If $i(t) < 0$ is charging current	A	UDDS, EPA UDDS, FTP, constant discharge/charge current
V_{batt}	Battery terminal voltage	V	Vnom = 7.4 V
$\int idt$	Actual battery charge	Ah	-
SOC	Battery state of charge	unitless	0–100%
OCV	Battery open circuit voltage (no-load voltage)	V	It is function of battery SOC
η_{disch}, η_{ch}	Coulombic efficiency coefficients for discharging and charging cycles	unitless	η_{disch} = 0.795 η_{ch} = 0.875

The battery SOC is 100% for a battery fully charged and, 0% for a battery fully empty. Typically, the battery SOC can be defined for a positive current discharging cycle as:

$$SOC(t) = 100(1 - \frac{\eta_{disch}}{Q_{rated}}\int_0^t i(\tau)d\tau) \quad (\%), \quad i(\tau) \geq 0 \tag{5}$$

where η_{disch} is the Coulombic efficiency of the discharging cycle, while Q_{max} represents the maximum capacity of the battery capacity, typically $1.05Q_{rated}$, close to those provided in the battery manufacturer's specs. The relation (5) can be written as a first order differential equation that, together with Equations (1) and (3), will be particularly useful for SOC state estimation in the next section of this research paper, i.e.:

$$\frac{d}{dt}(SOC(t)) = -100\frac{\eta_{disch} \times i(t)}{Q_{rated}}, \quad i(t) \geq 0 \tag{6}$$

It is worth mentioning that for a discharging cycle, the battery current in (6) is positive and for a charging cycle it is negative.

2.6.1. Li-Ion Cobalt MATLAB Simscape Model

A full representation of the generic battery model, dependent on the temperature and aging effects, is developed in MathWorks (Natick, MA, USA; www.mathworks.com) in the MATLAB R2019b/Simulink/Simscape/Power Systems/Extra Sources Library-Documentation. The MATLAB Simscape Li-ion cobalt battery cell specifications are shown in Table 2.

Table 2. Li-ion Cobalt cell specifications.

Lithium-Ion Battery Cell	LiCoO$_2$
Rated capacity (Ah)	5.4
Maximum capacity (Ah)	5.6
Nominal voltage (V)	7.4
Cut-off voltage (V)	5.25
Fully charged voltage (V)	8.307
Nominal discharge current (A)	1.1
Exponential zone [Voltage (V) Capacity (Ah)]	[7.91]
Internal resistance (ohms)	0.0133
Capacity (Ah) at nominal voltage	5.2

The MATLAB Simscape model of a generic battery is beneficial to set up a particular choice of battery chemistry and operation conditions that take into consideration the thermal model of the battery (internal and environmental temperatures) and also its aging effects. The battery terminal voltage, current and SOC can be visualized to monitor and control the battery SOH condition. The nominal current discharge characteristic according to a choice of the Li-ion Co battery having a nominal capacity of 5.4 Ah and a nominal voltage of 7.4 V is shown in Figure 2.

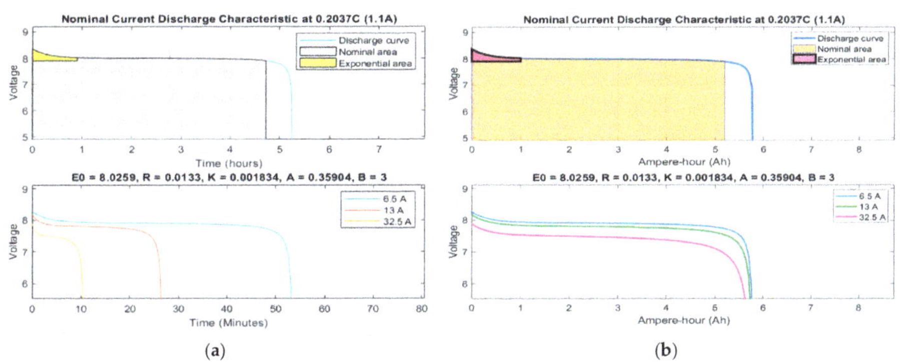

Figure 2. Nominal discharge characteristic of Li-ion Co at 0.2037C-rate (1.1 A)-MATLAB generic model. (**a**) In hours (minutes) (**b**) In Ampere-hours (Ah).

This characteristic corresponds to a constant discharge current of 0, 2037C-rate (0.2037 × 5.4 Ah = 1.1 A). The first battery characteristic from the top side of Figure 2 provides useful information about the estimated coefficients of the open-circuit voltage (OCV) included in Table 1, i.e., E0 = 8.0259 V, Rint = 0.01333 Ω, K = 0.001834 V, Aexp = 0.35904 V, and Bexp = 3 (Ah)-1. To show the evolution of the battery terminal voltage for different input current profiles, at the bottom of same Figure 2 other three nominal current discharge characteristics for three constant discharging currents (6.5, 13 and 32.5 A) are shown. These characteristics reveal that for the highest constant discharging current of 32.5 A, the discharging time of the battery decreases drastically to 10 min compared to 54 min corresponding to the smallest discharging current of 6.5 A. The same trend can be seen in Figure 3, where for a nominal discharging constant current of 1.08 A the Li-ion Co battery needs almost six hours to be fully discharged. The Simscape model of a generic battery set up for a Li-ion Co battery is shown in Figure 4.

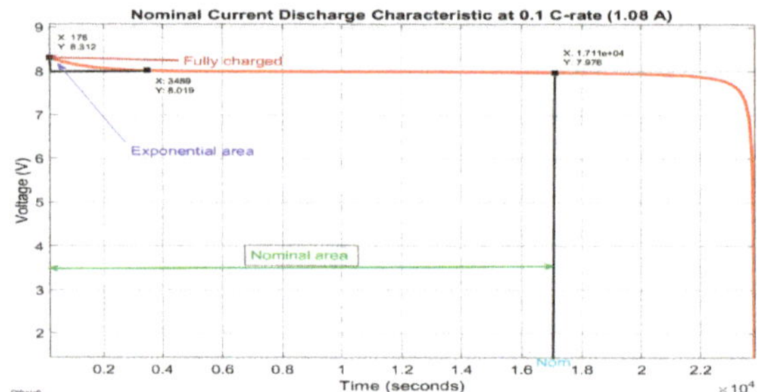

Figure 3. Nominal characteristic for a constant current discharge of 1.08 A—details.

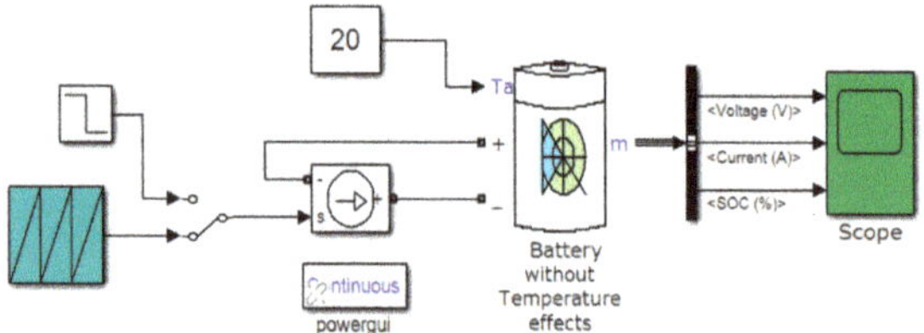

Figure 4. The Simscape model of the generic 5.4 Ah and 7.4 V Li-ion Co battery (without temperature and aging effects).

2.6.2. Li-Ion Cobalt Model in Continuous Time State Space Representation

The purpose of this section is to select and design the most suitable Li-ion Co battery model, which excels in simplicity, accuracy and is easy to implement in the MATLAB real-time simulation environment. Specifically, an accurate battery model is useful to develop in the following section the proposed real-time SOC estimators, which must also be of high precision and robustness. Related to SOC is the DOD, defined in [10] as:

$$DOD(t) = 100(1 - SOC(t)) \quad (\%) \tag{7}$$

The SOH is another internal battery derived parameter defined in [10] "as the ratio of the maximum charge capacity of an aged battery to the maximum charge capacity when this battery was new", as is also mentioned in [2,26]. The "actual operating life of the battery is affected by the charging and discharging rates, DOD, and by the temperature effects" [10]. Also, in [10] is stated that "the higher the DOD is, the shorter is the life cycle", and to attain "a higher life cycle, a larger battery is required to be used for a lower DOD during normal operating conditions", as is stated in [2,11,12]. Another important parameter for BMS in HEVs/EVs is the state of energy (SOE). From "engineering perspective, the SOE is more useful since it takes battery terminal voltage into account, which can predict the available energy for HEVs/EVs" [26].

While SOC indicates "the remaining capacity of the battery, the SOE indicates the remaining energy stored in the battery", as is defined in [26]:

$$SOE(t) = 100(1 - \frac{\eta_{sdisch}}{E_a} \int_0^t V_{batt}(\tau)i_L(\tau)d\tau) \quad (\%), \quad i_L(\tau) \geq 0 \tag{8}$$

or equivalent to:

$$\frac{d}{dt}(SOE(t)) = -100\frac{\eta_{sdisch}V_{batt}(t) \times i_L(t)}{E_a}, \quad i_L(t) \geq 0 \tag{9}$$

where E_a, $i_L(t)$, η_{sdisch} represent the available battery energy, the load current and the "battery energy efficiency" respectively [26]. The input-output battery model Equation (1) is a simplified version of the original Shepherd's combined model that follows the development from [25] and [27–29] replacing:

$$E(t) = E_0 - K\frac{Q_{max}.\int idt}{Q_{max} \quad \int idt} + Ac^{(-B\frac{1}{Q_{max}}\int idt)} \tag{10}$$

by:

$$E(t) = E_0 - \frac{K\int idt}{SOC(t)} + Ae^{(-B(1-SOC(t)))} \tag{11}$$

where A and B are two empirical parameters that are determined by a curve fitting procedure. The advantage of new version is to get a simplified OCV nonlinear model dependent only on SOC, as is developed in [26].

In the case study, we follow the development from [25] corrected by making small changes to increase the model accuracy, as is suggested in [26]. The development from [25] has the advantage to determine the battery model parameters by extracting the values based on simple algebraic manipulations, directly from the battery type OCV curve specifications provided by manufactures [7,10–12,25]. According to (11), the input-output battery generic model Equation (3) in continuous time becomes:

$$V_{batt}(t) = E - R_{int}i(t) = E_0 - \frac{K\int i(t)dt}{SOC(t)} + A_{exp}e^{(-B_{exp}Q_{max}(1-SOC(t)))} - R_{int}i(t), \tag{12}$$

Let's now assign two state variables to the description (12):

$$\begin{aligned} &x_1(t) = SOC, x_2(t) = A_{exp}e^{((-B_{exp}Q_{rated}/\eta_{SOC})\times(1-x_1(t)))}\\ &u(t) = I_{bat}(t) \text{ is the input current profile}\\ &y(t) = V_{bat}(t), \text{ is the battery terminal voltage} \end{aligned} \tag{13}$$

Therefore, a new modelling version is developed for designing and implementing the Li-ion Co battery model. In the new version, the model is described in continuous time in a two-dimensional representation of the state space as:

$$\begin{aligned} \frac{dx_1(t)}{dt} &= -(\frac{\eta_{SOC}}{Q_{rated}}) \times u(t)\\ \frac{dx_2(t)}{dt} &= -B_{exp}x_2(t) \times u(t)\\ \frac{dx_3(t)}{dt} &= -(\frac{\eta_{SOE}}{E_a}) \times V_{batt}(t) \times u(t)\\ y(t) &= E_0 - \frac{K\int i(t)dt}{x_1(t)} + x_2(t) - R_{int}u(t) \end{aligned} \tag{14}$$

and it is implemented in Simulink in the next subsection. The advantage of this representation is the model simplicity, its accuracy and easy to implement in real time.

2.6.3. Li-Ion Model in Discrete Time State Space Representation

To design both SOC estimators based on the adopted generic Li-ion Co battery model, the state space Equation (13) will be converted in discrete time representation. For SOC estimation purpose, a full Li-ion Co model in discrete time space representation is given in (15) and (16):

$$\begin{aligned} x_1(k+1) &= x_1(k) - T_s(\frac{\eta_{SOC}}{Q_{rated}}) \times u(k)\\ x_2(k+1) &= x_2(k) - T_sB_{exp}x_2(k) \times u(k) \end{aligned} \tag{15}$$

$$y(k) = E_0 - \frac{Ku(k)\Delta t}{x_1(k)} + x_2(k) - R_{int}u(k) \tag{16}$$

$$x_1(k) \triangleq x_1(kT_s), x_2(k) \triangleq x_2(kT_s), u(k) \triangleq u(kT_s),$$
$$y(k) \triangleq y(kT_s), k \in Z^+$$

where $k \in Z^+$, is a positive integer number, $\Delta t = T_s$ is the sampling time, set to 1 s in all MATLAB simulations.

2.6.4. Model Validation on ADVISOR MATLAB Integrated Platform

The validation of the Li-ion Co battery model is tested by using one or more driving cycles under different realistic driving conditions required for battery simulation tests. A collection of such of driving cycles profiles is stored in a large database of the US National Renewable Energy Laboratory (NREL) Advanced Simulator (ADVISOR) integrated into a MATLAB simulation environment [10]. The ADVISOR simulator is recommended by the excellent results obtained in [10] and by the fact that so far it has been one of the most used software design tools in the HEV/EV automotive industry, as mentioned in [11,29–32]. More details about this integrated ADVISOR MATLAB platform can be found in [10]. Among the three options of ADVISOR input battery models we choose a NREL Rint internal resistance installed on a hypothetical car model selected from the ADVISOR database, necessary for the validation of the Li-ion Co battery proposed in the case study, such in [10]. The proposed Li-ion Co battery model given by the Equation (14) and integrated into an HEV BMS structure is validated by using three of the most common driving cycles tests provided by Simulink and ADVISOR database, such as an Urban Dynamometer Driving Schedule (UDDS), Environmental Protection Agency (EPA) UDDS, and FTP/FTP-75 [10]. The Li-ion Co battery SOC tests result compared to those obtained by an NREL's internal resistance Rint lithium-ion battery model SOC installed on a midsize hypothetical car, for the same driving cycles tests and in the same initial conditions, like in [10] for a UDDS driving cycle test. Like [10], the hypothetical midsize car has almost the same characteristics. The "midsize town car is selected as an input vehicle on the integrated platform under same standard initial conditions SOCini = 70%, modelled in Simulink" in Figure A1 (Appendix A), and shown as an "ADVISOR page setup" in Figure 5 [10]. An Urban Dynamometer Driving Schedule (UDDS) test is used to validate the battery model in this section and the other two driving cycles tests, FTP (FTP-75), and UDDS-EPA are used in Section 3 for validation of the MATLAB SOC simulations results for both proposed estimators. The UDDS driving cycle profile in (mph) and the discharging battery current (A) are represented on the top and the bottom graphs from the same Figure 6 [10].

For performance comparison purposes, Figure 7 shows the corresponding SOC curves for the proposed Li-ion Co battery model design (red colour) and the ADVISOR SOC estimator (blue colour) on the same graph. The SOC simulations are performed for the same initial conditions (SOCini = 70%) and reveal an excellent SOC accuracy and an estimation error less than 2% between the battery model selection and NREL ADVISOR Rint battery model. The result confirmed by the second source from the first line of Table 3 (battery model vs. ADVISOR Rint model), for which the mean absolute error (MAE) is 0.0658. Other two sources can confirm the model validation by performing same comparisons for UDDS-EPA driving cycle test that will be developed in Section 3.3.2 with the statistical results shown in Table A1 from Appendix A. The third FTP driving cycle test will be developed in Section 3.3.3 and statistical results are shown in Table A2 from the same Appendix A.

The results reveal an estimate value less than 2%, MAE = 0.0235 (Table A1) and MAE = 0.0285 (Table A2) respectively. The MATLAB simulation results of all three tests for UDDS, UDDS-EPA and FTP driving cycles, for same initial conditions show an excellent accuracy for adopted battery model versus ADVISOR Rint and an estimation error less than 2%, confirmed by the results from Table 3, Tables A1 and A2 from Appendix A. Since from three different sources, the simulation results converge to an average error of less than 2% and show an accurate estimate value, we can conclude that these results validate the Li-ion Co battery. This outstanding result is encouraging to use the validated proposed battery model as a support for building "robust, accurate and reliable real-time battery estimators", both developed in Section 3. Further, in Figure A2a,b shown in Appendix A, you can see the statistics obtained for the SOC generated by the proposed Li-ion Co battery model and for SOC estimated by the generic ADVISOR Rint Li-battery model. Figure 8 shows the Simulink model of the adopted generic model that implements the set of Equation (13).

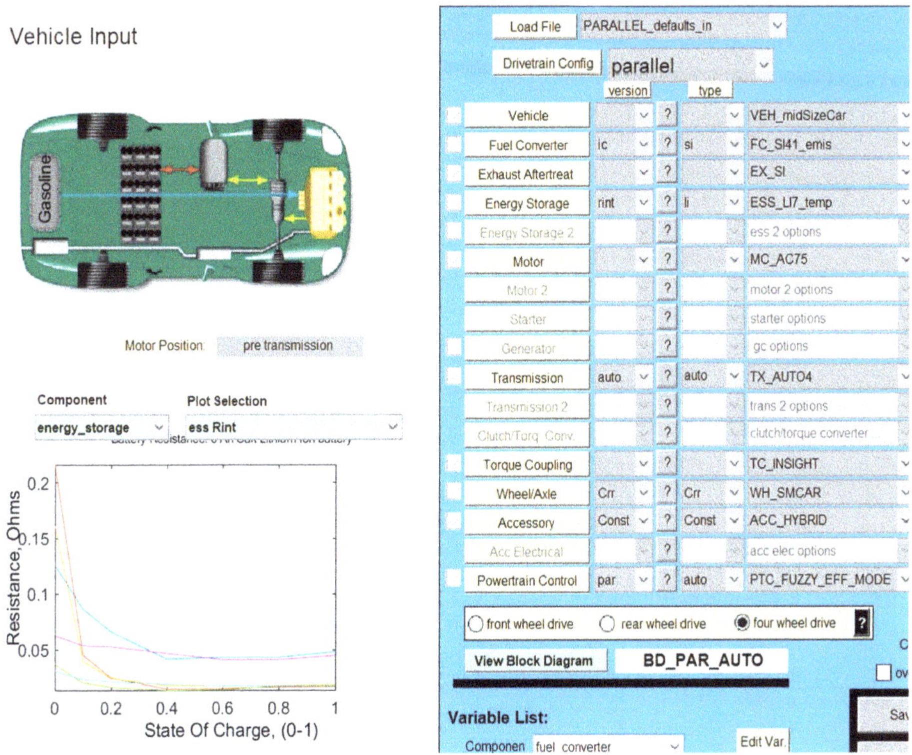

Figure 5. The setup ADVISOR page of the input HEV midsize car.

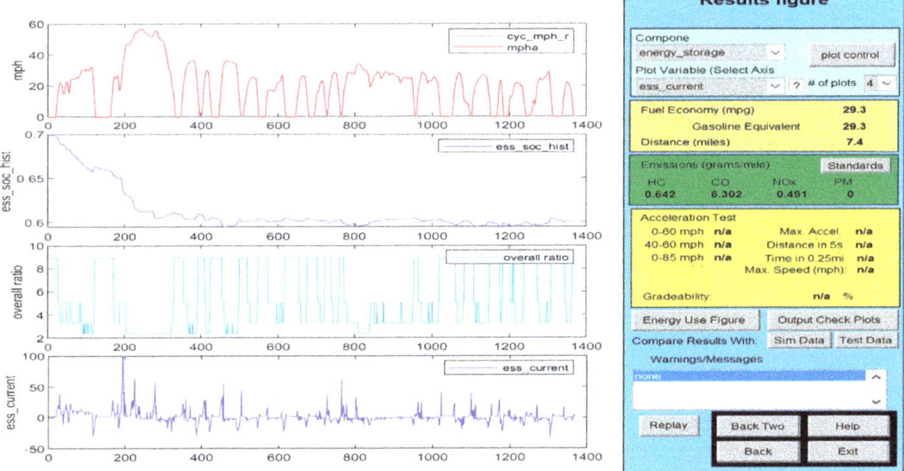

Figure 6. UDDS driving cycle input profile.

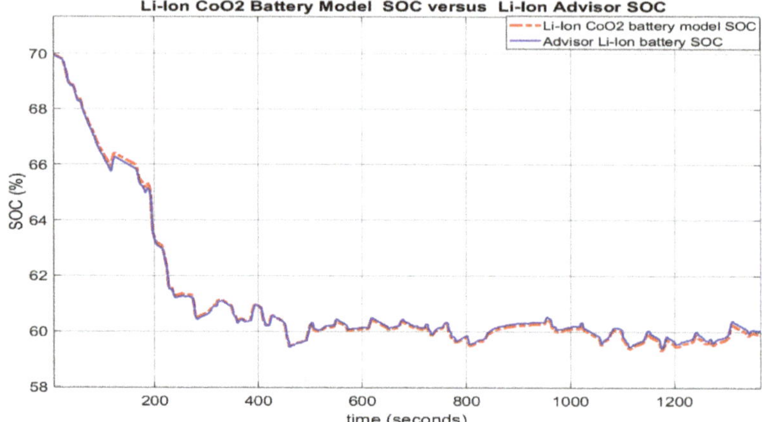

Figure 7. The UDDS test on the ADVISOR 3.2 integrated MATLAB platform. SOC battery model versus Li-ion ADVISOR battery SOC.

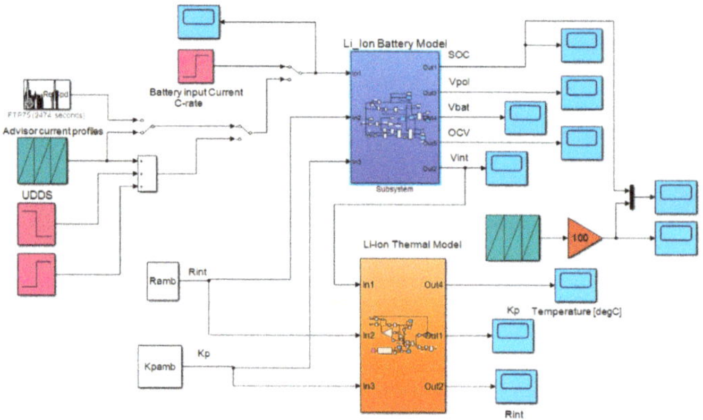

Figure 8. Simulink generic selected battery model.

The block from the top side of Simulink diagram calculates the SOC, OCV and battery terminal voltage Vbatt, shown in detail in Figure 9a,b. For a constant discharging current of 1C-rate (5.4 A), the battery terminal voltage, the OCV-SOC battery characteristic, and SOC are represented in Figure 10a–d. Furthermore, the adopted battery model generates the SOC that is shown in Figure 11a–c for three different driving conditions, namely for a UDDS, an FTP-75 and a constant 1C-rate (5.4 A) discharging current.

It is worth mentioning that a 100 Ah rated pack capacity Li-ion battery model is integrated in a MATLAB-Simulink SimPower Systems library, very helpful to be used for designing and implementation of different HEVs and EVs powertrains configurations, such is suggested in the EV application shown in Figure A3 from Appendix A.

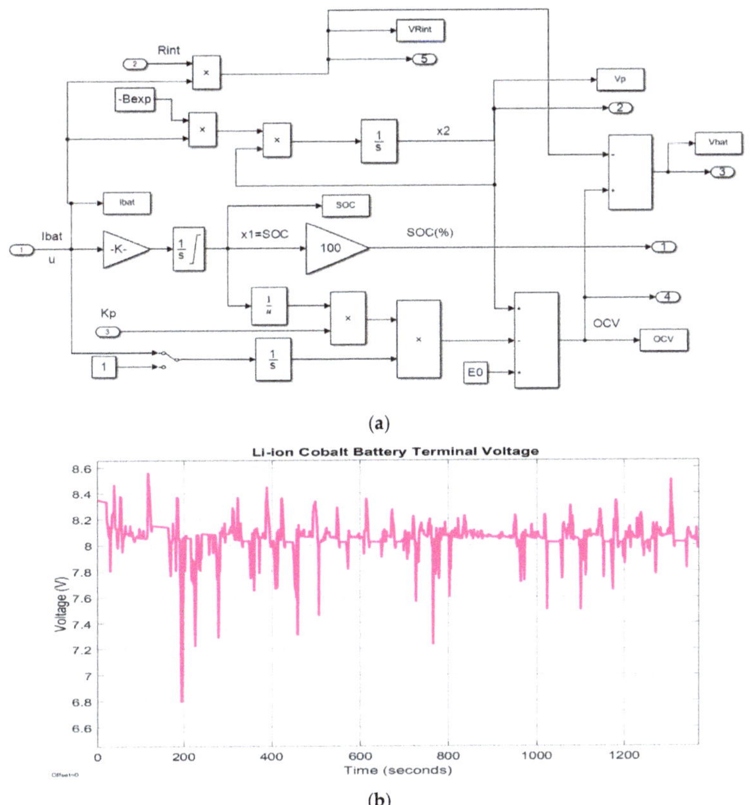

(**a**)

Li-ion Cobalt Battery Terminal Voltage

(**b**)

Figure 9. (**a**) The detailed generic battery Simulink model; (**b**) The generic battery terminal voltage.

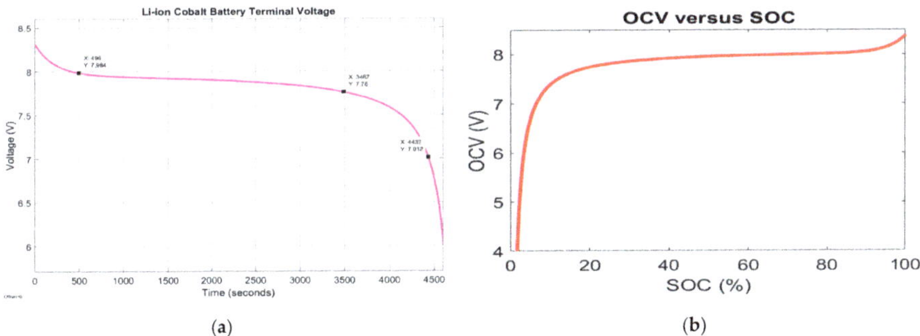

(**a**)

(**b**)

Figure 10. *Cont.*

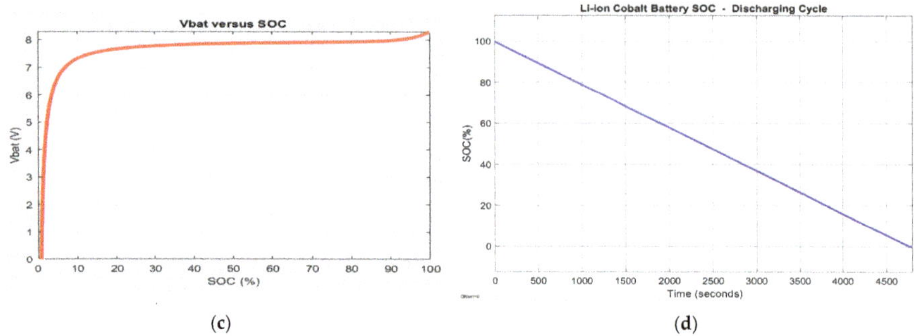

Figure 10. The battery model full discharging cycle at 1C-rate (5.4 A); (**a**) battery terminal voltage; (**b**) OCV vs. SOC characteristic; (**c**) terminal voltage vs. SOC; (**d**) battery SOC.

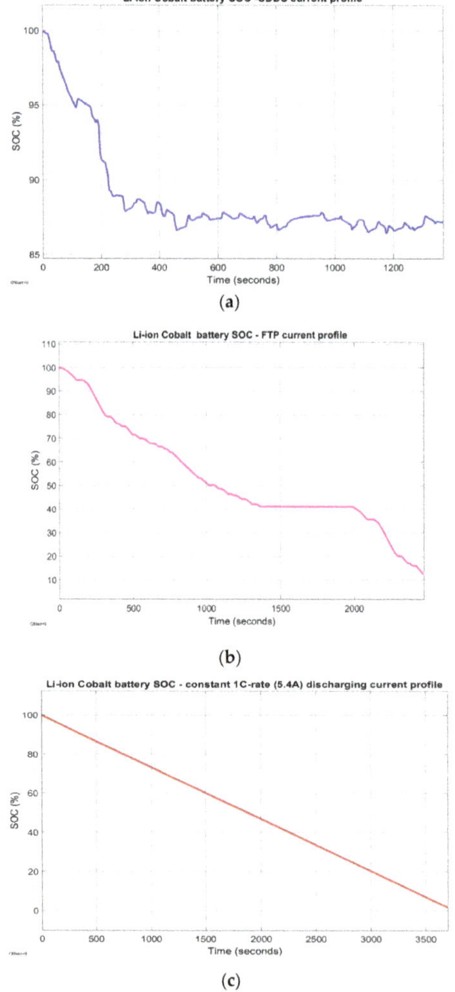

Figure 11. Battery SOC for three different current profiles; (**a**) For UDDS driving cycle current profile; (**b**) For a FTP-75 current profile; (**c**) for a 1C-rate (5.4 A) discharging current profile.

2.7. Li-Ion Cobalt Battery Thermal Model

The dynamics of thermal model block is described by the following equation:

$$mc_p \frac{dT_{cell}(t)}{dt} = hA(T_{amb} - T_{cell}(t)) + R_{int}I^2(t) \tag{17}$$

where m—the mass of the battery cell [kg]; c_p—the specific heat capacity [J/molK]; S—the surface area for heat exchange [m²]; $T_{cell}(t)$—the variable temperature of the battery cell [K]; T_{amb}—the ambient temperature [K]; R_{int}—the value of internal resistance of the battery cell [Ω]; I(t)—the input charging and discharging profile current [A].

For simulation purposes, the battery temperature profile and the robustness of the proposed SOC battery estimators, are tested for the following approximative values, closed to a commercial battery type ICP 18,650 series:

$$S = 4.4E - 3 \,[m^2], m = 0.043 \,[kg], c_p = 925 \,[J/kgK]$$
$$h = 5 \,[w], R_{int} = 0.01333 \,[\Omega], T_{ini} = 293.15 \,[K]$$

An accurate simplified thermal model is given in MATLAB R2019b library, at MATLAB/Simulink/ Simscape/Specialized Power Systems/Electric Drives/Extra Sources/Battery, for a lithium-ion generic battery model, implemented in Simulink as is shown in Figure 12:

$$T_{cell}(s) = \frac{R_{th}P_{loss} + T_{amb}}{T_c s + 1} \tag{18}$$

where $T_{cell}(s)$—the internal temperature of the cell [°K] in complex s-domain (the Laplace transform). R_{th}—thermal resistance, cell to ambient (°C/W). T_c—thermal time constant, cell to ambient (s). $P_{loss} \cong R_{int}I^2$-the overall heat generated (W) during the charge or discharge process [w]. T_{amb}—the ambient temperature set up by the user [K].

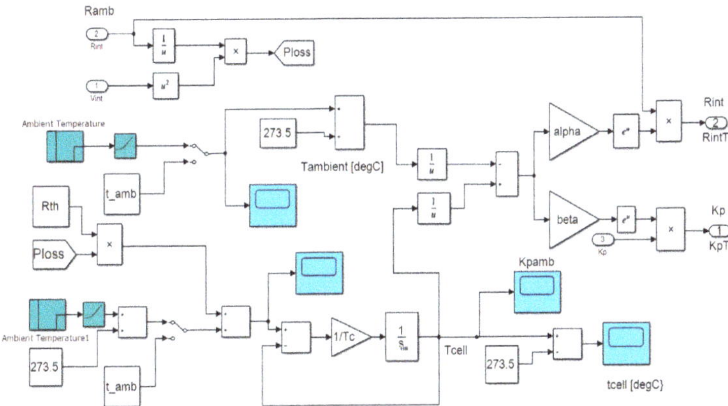

Figure 12. The detailed Simulink diagram of the thermal model block.

The internal resistance and the polarization constant $R_{int}(T)$ and $K(T)$ respectively vary with respect to temperature according to Arrhenius relationships:

$$K(T) = K|_{T_{ref}} \exp\left(\alpha\left(\frac{1}{T_{cell}} - \frac{1}{T_{ref}}\right)\right), \alpha = \frac{E}{RT}$$
$$R_{int}(T) = R_{int}|_{T_{ref}} \exp\left(\beta\left(\frac{1}{T_{cell}} - \frac{1}{T_{ref}}\right)\right), \beta = \frac{E}{RT} \tag{19}$$

where T_{ref}—the nominal ambient temperature, in K. α—the Arrhenius rate constant for the polarization resistance. β—the Arrhenius rate constant for the internal resistance.

For simulation purpose for implemented thermal block in Simulink, we use the following approximative values for the Li-ion battery thermal model parameters:

$$R_{th} = 6\ [°C], T_c = 2000\ [s], \alpha = \beta = \tfrac{E}{R}, E = 20\ [kJ/mol] - \text{activation energy}$$
$$R = 8.314\ [J/molK] - \text{Boltzman constant}$$

The ambient temperature profile and the output temperature of the Simulink thermal model described by the Equations (18) and (19) are shown in Figure 13a,b, respectively. The evolution of the internal resistance of the battery cell R_{int} (T) and of polarization constant K(T), at room temperature Tref = 293.15 [K], is shown in Figure 14a,b.

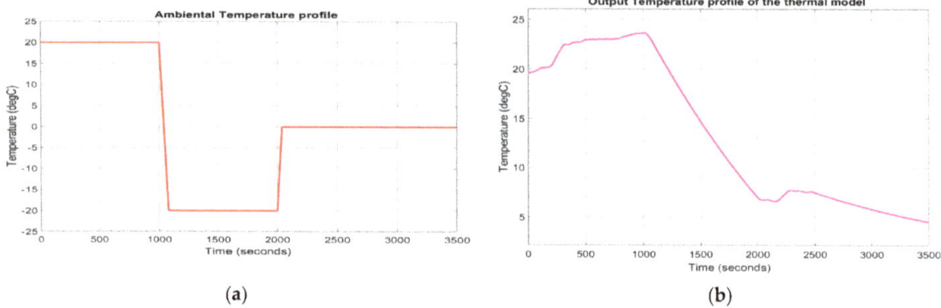

(a)

(b)

Figure 13. (**a**) The ambient temperature profile; (**b**) The output temperature of the thermal model block.

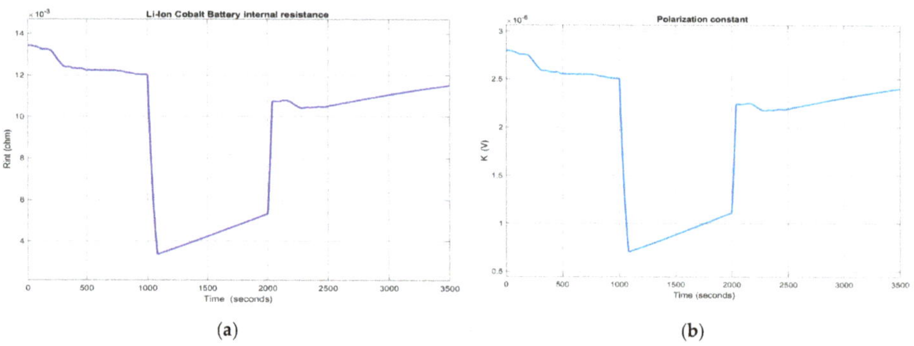

(a)

(b)

Figure 14. (**a**) The internal battery Rint at ambient temperature (20 degC); (**b**) The polarization constant at ambient temperature (20 degC).

The output temperature of the thermal model for changes in ambient temperature is shown in Figure A4a, and the effects on internal battery resistance Rint and polarization constant K are presented in Figure A4b,c, shown in Appendix A.

3. Li-Ion Co Battery State of Charge Estimation Algorithms

Almost all BMS HEV/EV systems in the automotive industry have integrated emergency systems that indicate the available battery capacity. As the SOC is not directly measured, its estimation is required. For estimating SOC, several methods for estimating adaptive filtering are developed in the field literature, among which the Kalman filters are the most used. More details about battery modelling, linear and nonlinear Kalman filter estimators, especially for state and parameter estimation, can be found in [4,7–12,14,15,26–37]. For performance comparison purposes, in this actual study, we develop two well-suited real-time SOC estimators, namely an Adaptive Extended Kalman Filter (AEKF) with the process and measurement noises correction, and a linear observer estimator (LOE) with a constant Luenberger gain.

3.1. Li-Ion Cobalt Battery-Adaptive Extended Kalman Filter SOC Estimator

As we mentioned in the previous section, the most suitable method for estimating SOC in real-time is the Coulomb counting method. The main disadvantage of this estimation technique is the difficulty of "predicting" the most appropriate initial SOC value of the battery, which could lead to an increase in time of the SOC estimation error and to a new "SOC calibration" based on "OCV measurement" [5]. However, "it is tough to measure the battery OCV in real-time and, consequently, a small OCV error may lead to a significant battery SOC difference", as is stated in [5].

Thus, one is thinking of improving the Coulomb metering method, a viable alternative is using an EKF SOC real-time estimator, suitable for a wide range of HEV/EVs applications. Besides, the adopted version of an adaptive EKF (AEKF) real-time estimator combines the advantages of both the Coulomb counting method and battery OCV calibration [5]. More precisely, the AEKF SOC estimator is an EKF, as is developed in detail in [7,8,10] with the performance improved in [5,30].

Additionally, the AEKF algorithm makes a recursive correction of the Gaussian process and measurement noises that simplifies the tuning procedure significantly. In [17], the correction is beneficial to calculate the Kalman gain of the AEKF SOC estimator, which leads to optimal results for the SOC estimation, as is shown in [5]. Furthermore, AEKF algorithm can improve its estimation performance by using "a fading memory factor to increase the adaptiveness for the modelling errors and the uncertainty of Li-ion battery SOC estimation, as well as to give more credibility to the measurements", as is stated in [5,7].

As we mentioned in the previous section, the AEKF requires a dynamic state-space representation model of Li-ion Co battery, in order "to develop a simulation model for the emulation of a nonlinear battery" behaviour [17]. The AEKF algorithm is based on the linearized model of the battery, as is developed in [5,7–10,17,26]. In our research paper, for the case study, we adopt the AEKF algorithm developed in [17] and is presented briefly in Table 3. For more details, the reader can refer to the papers [7–9]. The discrete-time state-space representation of the generic Li-ion Co battery model, required to design and implement in real-time the AEKF SOC estimator, is given by the Equations (20) and (21), further simplified to a unidimensional SOC state-space discrete-time representation:

$$x_1(k+1) = x_1(k) - T_s\left(\frac{\eta_{SOC}}{Q_{rated}}\right) \times u(k) \tag{20}$$

$$
\begin{aligned}
&y(k) = E_0 - \frac{Ku(k)\Delta t}{x_1(k)} + A_{exp}e^{((-B_{exp}Q_{rated}/\eta_{SOC})\times(1-x_1(k)))} - R_{int}u(k) \\
&x_1(k) \triangleq x_1(kT_s)\,,\ SOC\,(k) \triangleq SOC\,(kTS) \rightarrow x_1(k) = SOC\,(k) \\
&u(k) \triangleq u(kT_s),\, I_{bat}(k) \triangleq I_{bat}(kT_s)\ \rightarrow u(k) = I_{bat}(k) \\
&y(k) \triangleq y(kT_s),\, V_{bat}(k) \triangleq V_{bat}(kT_s) \rightarrow y(k) = V_{bat}(k) \\
&k \in Z^+
\end{aligned}
\tag{21}
$$

where $I_{bat}(k)$, $V_{bat}(k)$ are the battery input current profile and terminal voltage at the discrete time k, $\Delta t = T_s$ is the sampling time, set to 1 in MATLAB simulations. In this representation the state space Equation (17) and input-output Equation (18) depends only on SOC, the first equation is linear and the second one is highly nonlinear. The proposed algorithm AEKF follows the same steps such in [5,7–9] combined with the approach developed in [17], as is shown below:

AEKF SOC estimation algorithm steps:

[AEKF 1.1] Write Li-ion Co battery discrete-time nonlinear generic model equations:

$$SOC(k+1) = SOC(k) - T_s\left(\frac{\eta_{SOC}}{Q_{rated}}\right) \times u(k) \tag{22}$$

$$y(k) = E_0 - \frac{Ku(k)\Delta t}{SOC(k)} + A_{exp}e^{((-B_{exp}Q_{rated}/\eta_{SOC})\times(1-SOC(k)))} - R_{int}u(k)$$
$$u(k) = I_{bat}(k)\,, y(k) = V_{bat}(k) \tag{23}$$

[AEKF 1.2] Write the unidimensional Li-ion Co battery model in discrete-time state space representation:

$$x(k+1) = x(k) - T_s\left(\tfrac{\eta_{SOC}}{Q}\right)\times u(k) + w(k) = f(x(k), u(k)) + w(k)$$
$$y(k) = E_0 - \tfrac{K\Delta t}{x(k)}u(k) + A_{exp}e^{((-B_{exp}Q/\eta_{SOC})\times(1-x(k)))} - R_{int}u(k) + r(k) = g(x(k), u(k)) + v(k) \tag{24}$$
$$u(k) = I_{bat}(k), y(k) = V_{bat}(k)$$

where the process $w(k)$ and measurement output $v(k)$ are white uncorrelated noises of zero mean and covariance matrices $Q(k)$ and $R(k)$ respectively, i.e.,

$$w(k) \sim (0, Q(k)), v(k) \sim (0, R(k))$$
$$E(w(k)w(j)^T) = Q(k)\delta_{kj}, E(v(k)v(j)^T) = R(k)\delta_{kj}$$
$$\delta_{kj} = \left\{ \begin{array}{ll} 0, & k \neq j \\ 1, & k = j \end{array} \right\} \tag{25}$$

[AEKF 2] Initialization:
The initial value of SOC is estimated as a Gaussian random vector of the mean and covariance values given in (26).
For $k \geq 0$ set

$$\hat{x}_0 = E[x_0] - \text{the initial mean value}$$
$$\hat{P}_{x_0} = E[(x_0 - \hat{x}_0)(x_0 - \hat{x}_0)^T] - \text{the initial state covariance matrix} \tag{26}$$

[AEKF 3] Linearize the Li-ion Co nonlinear dynamics and calculate the Jacobian matrices:
The nonlinear dynamics of Li-ion Co battery is linearized around the most recent estimation state value $\hat{x}(k|k)$ and $\hat{x}(k|k-1)$ respectively, considered as an operating point. The Jacobian matrices of the linearization are given by:

$$A(k) = \tfrac{\partial f(k,x(k),u(k))}{\partial x(k)}|_{\hat{x}(k|k)} = 1$$
$$B(k) = -\tfrac{\eta_{SOC}}{Q}$$
$$C(k) = \tfrac{\partial g(k,x(k),u(k))}{\partial x(k)}|_{\hat{x}(k|k-1)} = \tfrac{K}{x^2(k)}|_{\hat{x}(k|k-1)} + \tfrac{A_{exp}B_{exp}Q}{\eta_{SOC}}\exp(-\tfrac{B_{exp}Q}{\eta_{SOC}}(1-x(k)))|_{\hat{x}(k|k-1)} \tag{27}$$

For $k \in [1, +\infty)$ do
[AEKF 4] Prediction phase (forecast or time update from $(k|k)$ to $(k+1)|k$):

$$\hat{x}(k+1|k) = A(k)\hat{x}(k|k) + B(k)u(k)$$
$$\hat{P}(k+1|k) = A(k)\hat{P}(k|k)A(k)^T + \alpha^{-2k}Q(k) \tag{28}$$

Remark: In this phase, the predicted value of the state vector $\hat{x}(k+1|k)$ is calculated based on the previous state estimate $\hat{x}(k|k)$ and the state covariance positive definite matrices $\hat{P}(k|k)$ and $\hat{P}(k+1|k)$ (unidimensional in the case study) are affected by a fading memory coefficient α.
[AEKF 5] Compute an updated value of Kalman filter gain:

$$K(k) = \alpha^{2k}\hat{P}(k+1|k)H(k)^T(H(k)\alpha^{2k}\hat{P}(k+1|k)H(k)^T + R(k))^{-1} \tag{29}$$

[AEKF 6] Correction phase (analysis or measurement update):
The Li-ion Co battery SOC estimated state is updated when an output measurement is available in two steps:

[AEKF 6.1] Update the SOC estimated state covariance matrix:

$$\hat{P}(k+1|k+1) = (I - K(k)H(k))\hat{P}(k+1|k)\,(I - K(k)H(k))^T + \alpha^{-2k}K(k)R(k)K(k)^T \tag{30}$$

[AEKF 6.2] Update the SOC estimated state variable:

$$\hat{x}(k+1|k+1) = \hat{x}(k+1|k) + K(k)(y(k) - g(\hat{x}(k+1|k), u(k), k) \tag{31}$$

[AEKF 7] Adaptive process and measurement noise covariance matrices correction in two steps:
For k >= L, the length of the window's samples, compute:
[AEKF 7.1] Output variable error and the correction factor:

$$E_{rr}(k) = y_{mes}(k) - g(\hat{x}(k|k), u_k)$$
$$c(k) = \frac{\sum\limits_{i=k-L+1}^{k} E_{rr}(k)E_{rr}^T(k)}{L} \tag{32}$$

[AEKF 7.2] Measurement noise correction:

$$R(k) = c(k) + H(k)P(k|k)H(k)^T \tag{33}$$

[AEKF 7.3] Process noise correction:

$$Q(k) = K(k)c(k)K(k)^T \tag{34}$$

The AEKF estimator is easy to implement since its "recursive predictor-corrector structure that allows the time and measurement updates at each iteration" [5]. The tuning parameters of AEKF SOC estimator are the following: $Q(0)$ and $R(0)$, $\hat{P}_{x_0}$, the fading factor α and the window length L, obtained by a "trial and error" procedure based "on designer's empirical experience" [5]. It is worth noting that step 7 of the estimation algorithm simplifies substantially the procedure of tuning parameters without to affect the AEKF algorithm convergence. Moreover, the covariance matrices $Q(0)$ and $R(0)$ are chosen as positive definite diagonal matrices, and then during MATLAB simulations, both matrices are adaptively updated using the correction Equations (33) and (34). For simulation purposes, to test the effectiveness of the AEKF SOC estimator we set up the Kalman filter estimator parameters for all three driving cycle profile tests to the same values, i.e., $Q(0) = 5E-4, R(0) = 0.2E-3, \alpha = 1.001, \hat{P}_{x_0} = 1E-10, L = 10$ samples.

3.2. Li-Ion Cobalt Battery-Linear Observer SOC Estimator with Constant Gain

Linear and non-linear observers can estimate the states of the control systems. A linear observer estimator can be used to estimate SOC, as it is easy to adapt to the Li-ion Co battery model. Compared to AEKF SOC estimator performance, the proposed linear observer (LOE) SOC estimator seems to have a fast convergence rate and high estimation accuracy, as mentioned in [18]. It is easy for design a MATLAB/Simulink implementation. Besides, it is robust to changes in the initial value of SOC, to changes in the battery internal resistance and polarization constant due to temperature effects. Furthermore, it has a high capability of compensating the effects of nonlinearity and uncertainty exhibited by Li-ion Co battery model. The main drawback of LOE SOC is its inability to filter the measurement noise, so it is not robust to the measurement noise level compared to AEKF that has this great feature. The proposed linear observer relies on the determination of the appropriate feedback that achieves better SOC estimation accuracy. The following equations describe the dynamics of the linear observer estimator (LOE):

$$\frac{dSOC(t)}{dt} = -\left(\frac{\eta_{SOC}}{Q_{rated}}\right) \times u(t)$$

$$V_{batt}(t) = E - R_{int}i(t) = E_0 - \frac{K\int idt}{SOC(t)} + A_{exp}e^{(-B_{exp}Q_{max}(1-SOC(t)))} - R_{int}u(t) \qquad (35)$$

$$u(t) = i(t)$$

Thus, Equation (35) describes the dynamics of Li-ion Co battery generic model that is unidimensional and dependent only on battery SOC. In this development, the input-output battery terminal voltage equation is linearized around a SOCop battery operating point, retaining only the first order term of Taylor series:

$$V_{bat}(SOC) = V_{bat}(SOC_{op}) + \left(\frac{dV_{bat}}{dSOC}\right)|_{SOCop}(SOC - SOC_{op}) \qquad (36)$$

In discrete time, the LOE battery SOC is described by the following equations:

$$SOC(k+1) = SOC(k) - \frac{Ts \times \eta_{SOC}}{Q_{nom}}u(k) = A \times SOC(k) + Bu(k)$$

$$V_{bat}(SOC(k)) = \alpha_{SOC} \times SOC(k) + k_{SOC,u}SOC(k)u(k) + k_u u(k) + E_{0,op} = C \times SOC(k) + g(SOC(k), u(k)) \qquad (37)$$

$$A = 1, C = \alpha_{SOC}, B = -\frac{Ts \times \eta_{SOC}}{Q_{nom}}, g(SOC(k), u(k)) = k_{SOC,u} \times SOC(k)u(k) + k_u \times u(k) + E_{0,op}$$

where the values of the coefficients α_{SOC}, $k_{SOC,u}$, k_u and $E_{0,op}$ depend on the linearization operating point SOC_{op}. For example, if the operating point is $SOC_{op} = 60\%$, these coefficients get the following values: $\alpha_{SOC} = 0.0019$, $k_{SOC,u} = 0.000077$, $k_u = -0.0133$ and $E_{0,op} = 8.0146$. The Equation (37) are showing that the current output battery terminal voltage $V_{bat}(SOC(k))$ and its future evolution are both determined solely by its current state $SOC(k)$ and the battery current input $u(k)$. If the battery generic model system Equation (37) is observable, then the output battery terminal voltage can be used to steer the $SOC(k)$ state of the observer. After linearization, it is easy to see that the pair $(A, C) = (1, \alpha_{SOC})$ is observable, since $\alpha_{SOC} \neq 0$, regardless of the battery operating point. The observer model of the physical system of the Li-ion Co battery is then typically derived from the Equation (37). Additional terms may be included in order to ensure that, on receiving successive measured values of the Li-ion Co battery $u(k) = i(k)$ input and $V_{bat}(SOC(k))$ output, the model's state $\widehat{SOC}(k)$ converges to $SOC(k)$ of the battery. In particular, the output of the observer $\widehat{V_{bat}}(SOC(k))$ may be subtracted from the battery output $V_{bat}(SOC(k))$ and then is multiplied by a constant gain L to produce a so-called Luenberger observer, defined by the following equations:

$$\widehat{SOC}(k+1) = A \times \widehat{SOC}(k) + L(V_{bat}(k) - \widehat{V_{bat}}(k)) + Du(k)$$
$$\widehat{V_{bat}}(k) = C \times \widehat{SOC}(k) + g(\widehat{SOC}(k), u(k)) \qquad (38)$$

The linear observer SOC estimator is asymptotically stable if the SOC state error:

$$e_{SOC}(k) = \widehat{SOC}(k) - SOC(k) \to 0 \quad \text{when} \quad k \to \infty \qquad (39)$$

For a Luenberger observer, the SOC state estimation error satisfies the following relationship:

$$e_{SOC}(k+1) = (A - LC)e_{SOC}(k) \qquad (40)$$

The asymptotically condition (41) is satisfied only if (A-LC) is a Hurwitz matrix, so all the eigenvalues of this matrix are located in z-plane inside of the unit circle $|z| = 1$. For an unidimensional system, such in our case, A-LC must satisfies the relationship:

$$-1 < A - LC < 1 \to -1 < 1 - L * \alpha_{SOC} < 1$$
$$L \in \left(0, \frac{2}{\alpha_{SOC}}\right) \qquad (41)$$

For MATLAB simulations L is set to 1, 10, and 100 to analyze the performance of the LOE estimator in terms of convergence speed, robustness and SOC estimation accuracy for same driving conditions tests, UDDS, UDDS-EPA and FTP-75, like the AEKF SOC estimator developed in the previous subsection. The Simulink models of the LOE SOC, Li-ion Co battery model, thermal model block and the input driving cycles current profiles are shown in Figure 15. The battery model and the LOE SOC estimator block is detailed in Figure 16, and the Simulink model of thermal block is shown in Figure 12.

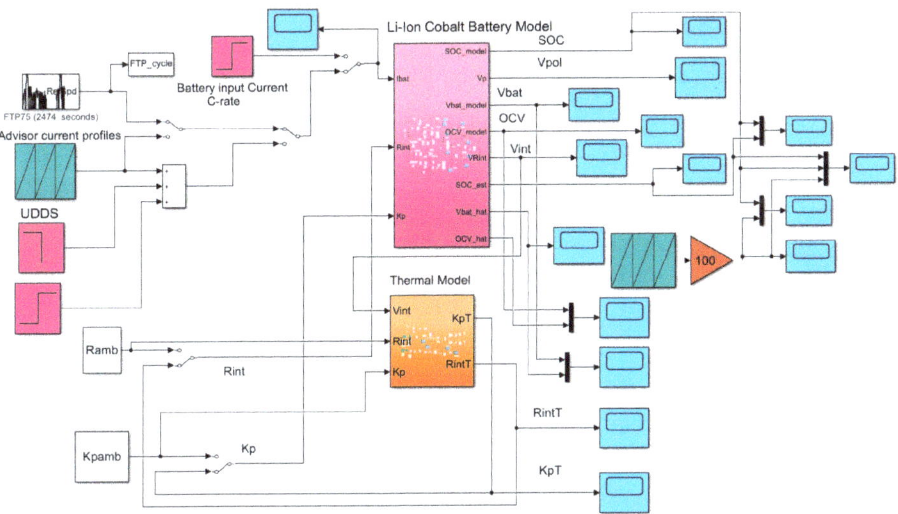

Figure 15. The Simulink diagram of the combined Li-ion Cobalt battery model and LOE block located to the top side and, the Simulink diagram of thermal model block located in the bottom side.

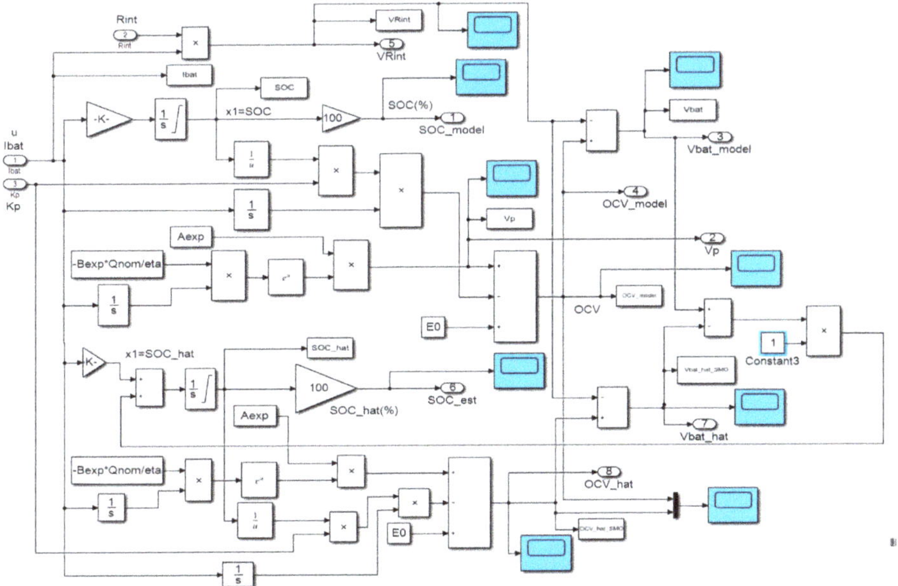

Figure 16. The detailed Simulink diagram of the combined Li-ion Co battery and LOE models.

3.3. Real-Time MATLAB Simulation Results

In this section, an extensive number of simulations, conducted on MATLAB software platform, is performed to validate the battery model and to analyze the performance of both proposed AEKF and LOE SOC estimators. The performance of both, AEKF and LOE SOC estimators is analyzed in terms of accuracy, robustness, convergence speed and real-time implementation simplicity. Robustness is tested for changing driving conditions by performing tests based on each of the three most commonly used driving cycle profiles provided in the ADVISOR-MATLAB platform, namely UDDS, UDDS-EPA and FTP described in Section 2. Furthermore, for each driving cycle profile, the robustness of both SOC estimators are testing the following four scenarios:

o   R1-scenario—changes in SOCini from 70% for UDDS and UDDS-EPA, 80%-FTP, driving cycles tests to:

  o   R11—for SOCini = 100%
  o   R12—for SOCini = 40%

o   R2-scenario—changes in SOC initial value and in measurement sensor noise level

  o   R21—for SOCini = 100%, noise measurement level σ = 0.01(increased from 0.001)
  o   R22—for SOCini = 40%, noise measrement level, σ = 0.01(increased from 0.001)

o   R3-scenario: changes in SOC initial value and in the value of the battery nominal capacity due to aging and/or temperature effects.

  o   R31—for SOCini = 100%, Qnom = 2.7 Ah (decresed from 5.4 Ah)
  o   R32—for SOCini = 40%, Qnom = 2.7 Ah (decresed from 5.4 Ah)

o   R4-scenario: changes in SOC initial values and temperature effects on internal resistance Rint and on polarization constant K

  o   R41—for SOCini = 100%, Rint = Rint(T) changes from Rint = 0.01333 [Ω], K = K(T) changes from K = 0.0099892 [V]
  o   R42—for SOCini = 40%, Rint = Rint(T) changes from Rint = 0.01333 [Ω], K = K(T) changes from K = 0.0099892 [V]

Also, the statistical errors in terms of standard deviation (MATLAB command std, σ), root mean squared error (RMSE), mean squared error (MSE) and mean absolute error (MAE) defined for each driving cycle test by the Equations (42)–(44), are summarized in one table.

$$\mathrm{RMSE} = \sqrt{\frac{\sum_{i=1}^{N}\left(\widehat{\mathrm{SOC}}\,(i) - \mathrm{SOC}_{\mathrm{Battery\_model}}(i)\right)^2}{N}} \tag{42}$$

$$\mathrm{MSE} = \frac{\sum_{i=1}^{N}\left(\widehat{\mathrm{SOC}}\,(i) - \mathrm{SOC}_{\mathrm{Battery\_model}}(i)\right)^2}{N} \tag{43}$$

$$\mathrm{MAE} = \frac{\sum_{i=1}^{N}\left|\widehat{\mathrm{SOC}}\,(i) - \mathrm{SOC}_{\mathrm{Battery\_model}}(i)\right|}{N} \tag{44}$$

N – number of samples.

3.3.1. Test 1-UDDS Driving Cycle Profile

A. Li-ion Co generic model accuracy performance

In Figure 17 the following MATLAB simulation results are shown:

(a)   UDDS driving cycle current profile for SOCini = 70%
(b)   LOE battery SOC estimate versus the battery model SOC and ADVISOR SOC estimate
(c)   AEKF battery terminal voltage estimate versus battery cell model terminal voltage
(d)   AEKF battery SOC estimate versus battery model SOC and AVISOR SOC estimate

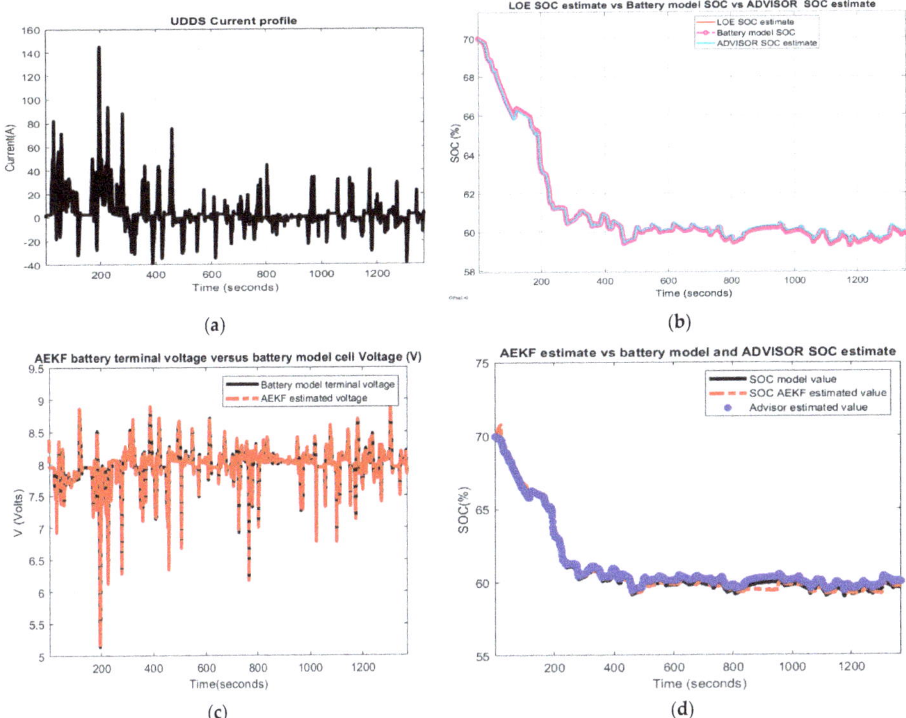

Figure 17. Li-ion Co battery model SOC accuracy performance (**a**) The UDDS current profile; (**b**) LOE SOC estimate vs. battery model SOC and ADVISOR SOC estimate; (**c**) AEKF estimate terminal voltage versus battery model terminal voltage; (**d**) AEKF SOC estimate versus battery model SOC and ADVISOR SOC estimate.

The simulation results reveal that the battery SOC is very accurate with respect to ADVISOR SOC estimate for same SOC initial value, like in Section 2.6.4. Also, the AEKF and LOE SOC estimators are very accurate compared to battery model SOC. Additionally, the Figure 17c reveals a strong ability of the AEKF SOC estimator to predict the battery terminal voltage.

B. Robustness of AEKF and LOE SOC Estimators

- R1-scenario

A great robustness of AEKF and LOE SOC estimators for this scenario is shown in Figure 18a,c, for R11, and in Figure 18b,d for R12.

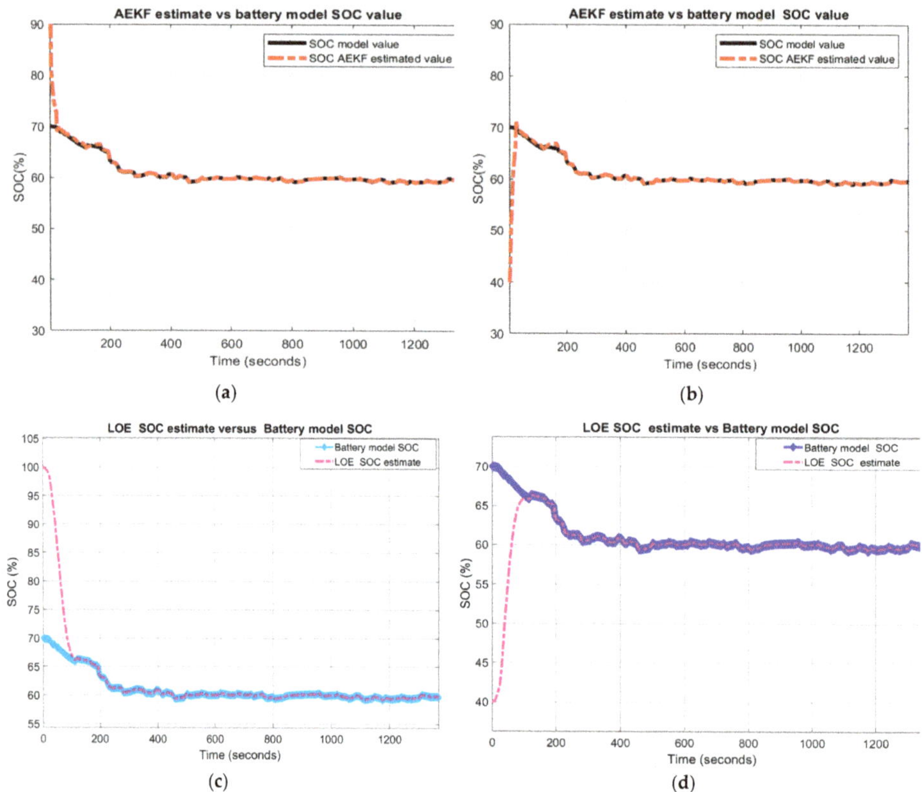

(a)   (b)

(c)   (d)

Figure 18. SOC Estimators robustnsess-Scenario R1; (**a**) AEKF for R11; (**b**) AEKF for R12 (**c**) LOE for R11; (**d**) LOE for R12.

- R2-scenario

The MATLAB simulation results shown in Figure 19a–d indicate a great robustness of AEKF SOC estimator compared to LOE SOC.

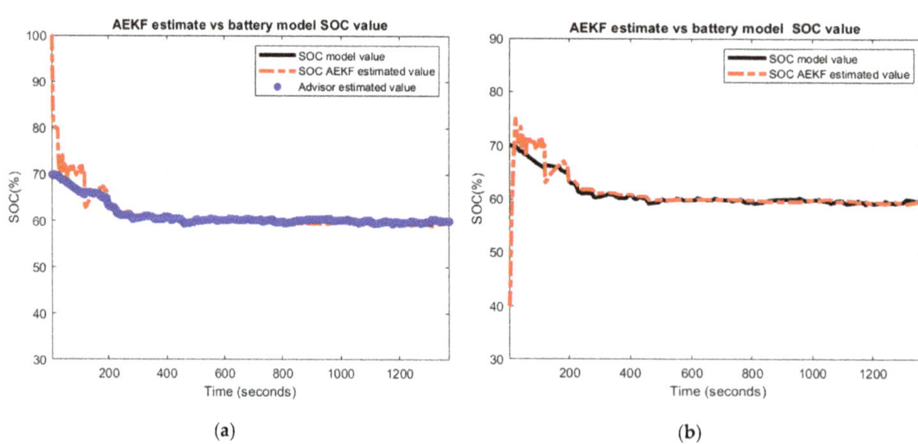

(a)   (b)

Figure 19. *Cont.*

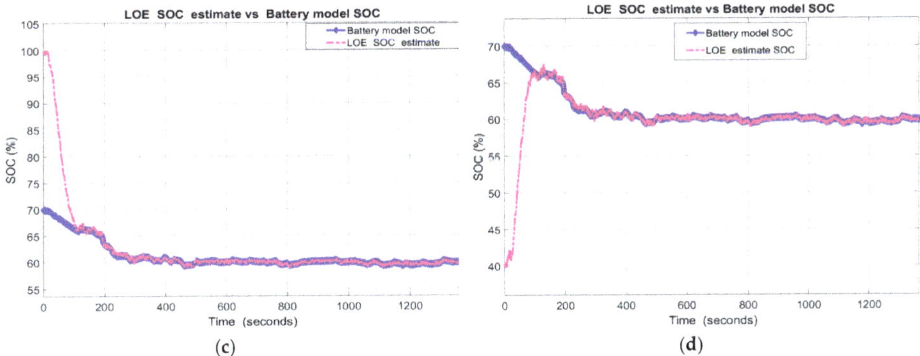

(c)   (d)

Figure 19. The robustness of AEKF and LOE SOC estimators-Scenarion R2 (**a**) AEKF for R21; (**b**) AEKF for R22; (**c**) LOE for R21 (**d**) LOE for R22.

The AEKF SOC estimator has a great ability to filter the measurement noise, thus AEKF SOC estimator outperforms the LOE SOC regarding the robustness performance to changes in noise level.

- R3-scenario

About the MATLAB simulations shown in Figure 20a–d it is worth highlighting the great robustness performance for both SOC estimators for this scenario.

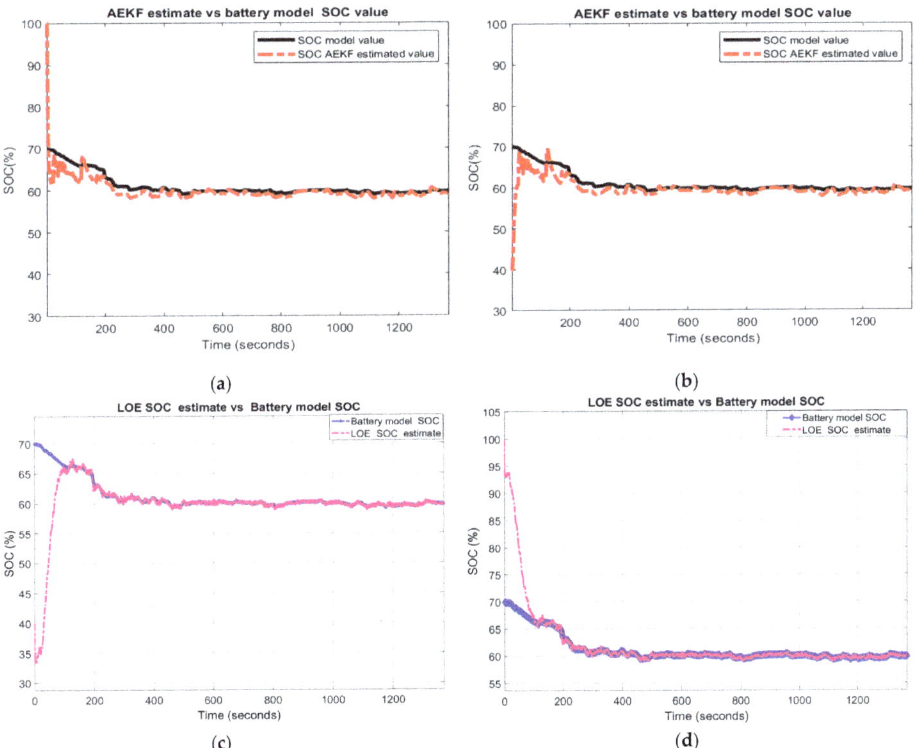

(a)   (b)

(c)   (d)

Figure 20. The SOC estimators' behaviour for R3-scenario (**a**) AEKF for R31; (**b**) AEKF for R32; (**c**) LOE SOC for R31; (**d**) LOE SOC for R32.

- R4-scenario

The output temperature profile of thermal model and the effects of temperature changes on the internal resistance Rint and polarization constant are shown in Figure A4. For this scenario, the MATLAB simulation results depictured in Figure 21a–d show an excellent robustness performance for both SOC estimators.

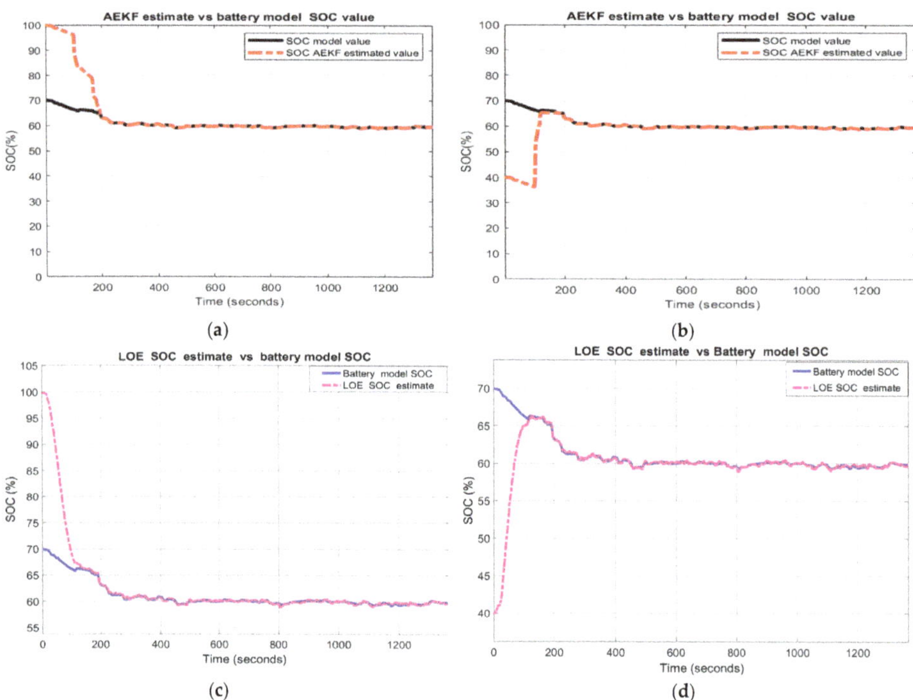

Figure 21. AEKF and LOE SOC estimators robustness performance for R4-scenario; (**a**) AEKF for R41; (**b**) AEKF for R42; (**c**) LOE SOC for R41; (**d**) LOE for R42.

The statistical errors corresponding to all four scenarios developed for UDDS driving cycle test are summarized in Table 3.

Table 3. AEKF and LOE SOC Estimators-statistical errors.

Cycle Test	Accuracy	SOCest Initial Value (%)	Statistic Errors							
			RMSE AEKF	RMSE LOE	MSE AEKF	MSE LOE	MAE AEKF	MAE LOE	Std AEKF (%)	Std LOE (%)
UDDS	Battery model vs. ADVISOR	70	0.4693	0.5500	0.2538	0.3026	0.0029	0.0658	2.66	1.37
	AEKF/LOE	70	0.0325	0	0.0000036	0	0.0018	0	2.6	1.37
	R1-scenario	R11	0.04225	0.1746	0.000384	0.0305	0.0131	0.1289	3.28	13.1
		R12	0.0492	0.1551	0.0033	0.0240	0.0270	0.1057	4.47	10.1
	R2-scenario	R21	0.0463	0.2638	0.00044	0.0696	0.0145	0.2595	3.69	5.3
		R22	0.0576	0.1573	0.0029	0.0247	0.0038	0.1088	4.03	10.14
	R3-scenario	R31	0.0471	0.1832	0.000504	0.0335	0.0168	0.1723	3.6	6.9
		R32	0.0367	0.2811	0.000717	0.0790	0.0124	0.2655	2.97	8.5
	R4-scenario	R41	0.0459	0.2528	0.00041	0.0639	0.01329	0.2473	3.73	5.96
		R42	0.0498	0.1636	0.0054	0.0267	0.0252	0.1195	4.89	9.82

The results of statistical errors performance analysis for all scenarios from Table 3, for UDDS driving cycle test, indicate that the AEKF SOC estimator surpasses the LOE SOC estimator in the competition for robustness performance.

3.3.2. Test 2: UDDS-EPA Charging Current Profile

A. Cobalt Li-ion generic model accuracy and validation

The MATLAB simulation results are shown in Figure 22:

(a)   UDDS-EPA driving cycle current profile
(b)   LOE SOC estimate versus battery model SOC and ADVISOR SOC estimate
(c)   AEKF battery terminal voltage
(d)   AEKF SOC estimate versus battery model SOC and ADVISOR SOC estimate
(e)   A great SOC accuracy of battery model versus ADVISOR SOC estimate is revealed in Figure 22b.

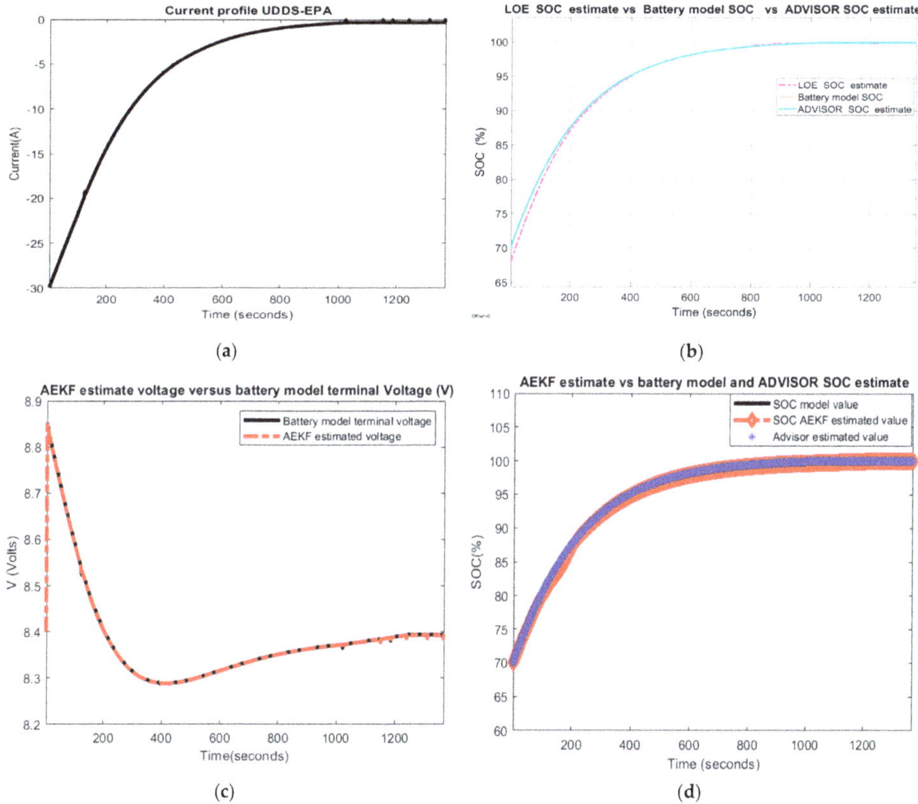

Figure 22. Li-ion Co battery model SOC accuracy performance and validation; (**a**) UDDS-EPA driving cycle; (**b**) LOE SOC estimate vs. battery model SOC and ADVISOR SOC estimate; (**c**) AEKF battery terminal voltage estimate vs. battery model terminal voltage; (**d**)AEKF SOC estimate vs. battery model SOC vs. ADVISOR SOC.

Like UDDS driving cycle, the MATLAB simulation results presented in Figure 22b,d reveal that the Li-ion Co battery model fits very well, within a 2% SOC error, the experimental setup

ADVISOR-MATLAB platform SOC estimate. So, once again these results certainly confirm the validity of the generic lithium-ion cobalt battery model.

B. Robustness of AEKF SOC Estimator

To keep the manuscript lenght reasonable, for the second driving cycle test, we show only the results for SOCini = 40%, i.e., for R12, R22, R32, and R42-scenarios.

- R1-scenario

The MATLAB simulation results shown in Figure 23a,b indicate an excellent robustness performance for both SOC estimators.

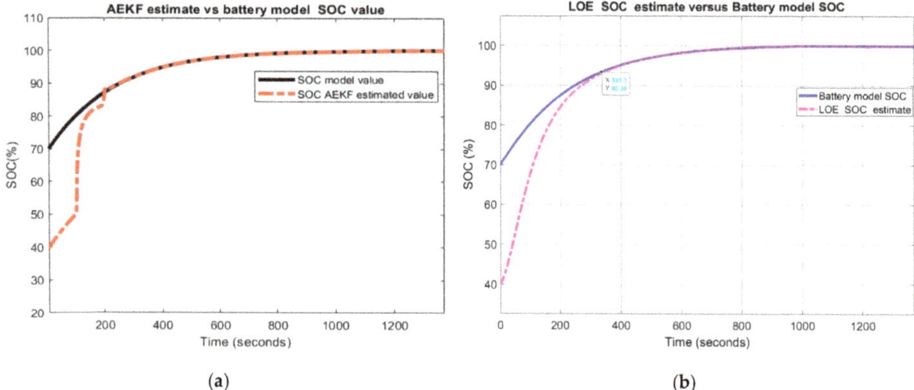

(a)

(b)

Figure 23. The robustness of AEKF and LOE SOC estimators for R1-scenario; (**a**) AEKF for R12 (**b**) LOE for R12.

- R2-scenario

A great robustness for both SOC estimators for this scenario and R22 case is also shown in Figure 24a,b.

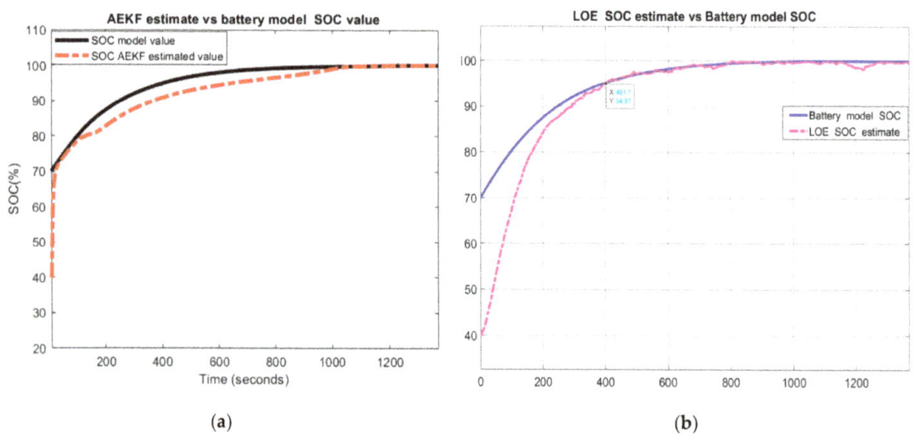

(a)

(b)

Figure 24. The robustness of AEKF and LOE SOC estimators for R2-scenario; (**a**) AEKF SOC for R22 (**b**) LOE SOC for R22.

- R3-scenario

The MATLAB simulation results depicted in Figure 25a,b reveal a great robustness performance for the AEKF SOC estimator compared to the LOE SOC estimator that has small changes in SOC estimate accuracy.

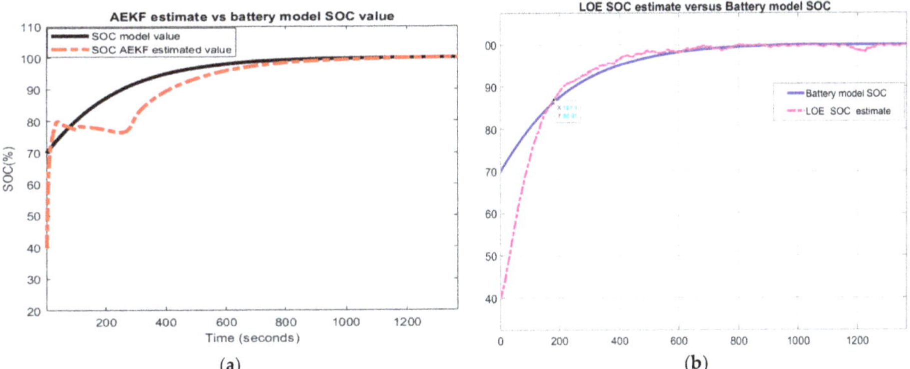

Figure 25. Robustness performance of AEKF and LOE SOC estimators for R3-scenario; (**a**) AEKF SOC for R32; (**b**) LOE SOC for R32.

- R4-scenario:

For this scenario is considered the output temperature profile of thermal model and the effects of temperature changes on the internal resistance Rint and polarization constant K shown in Figure A4. The MATLAB simulation results of AEKF and LOE SOC estimation robustness performance are presented in Figure 26a,b.

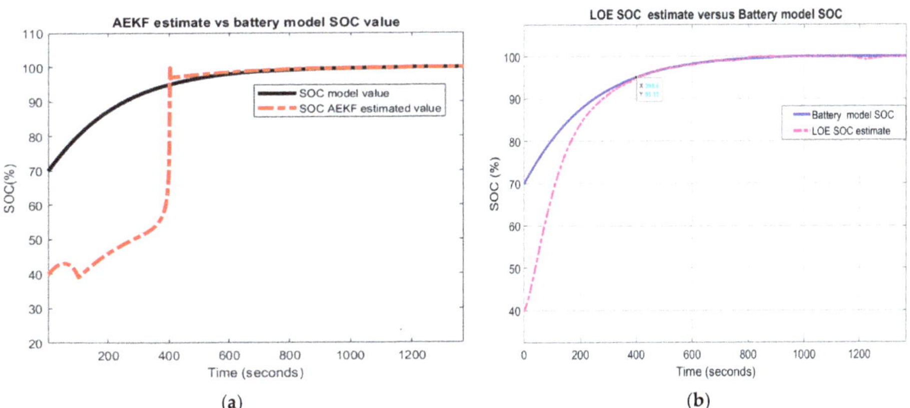

Figure 26. Robustness of AEKF and LOE SOC estimators for R4-scenario (**a**) AEKF SOC for R42; (**b**) LOE SOC for R42.

For this scenario, the simulation results shown in Figure 26 indicate a great robustness performance for both SOC estimators. The statistical errors RMSE, MSE, MAE and standard deviation are summarized in Table A1 in Appendix A.

Like UDDS, the result of the performance analysis, for all the scenarios included in Table A1, indicates once again that the AEKF SOC estimator remains the most suitable SOC estimator as compared to LOE SOC estimator.

3.3.3. Test 3: FTP-ADVISOR Driving Cycle Current Profile

A. Battery model SOC accuracy and model validation

The FTP driving cycle current profile for testing the battery is shown in Figure 27a. For generic battery model validation, the AEKF SOC estimate, the Li-ion Co battery model SOC and the ADVISOR-MATLAB Rint Li-battery SOC estimate are shown on the same graph in Figure 27b.

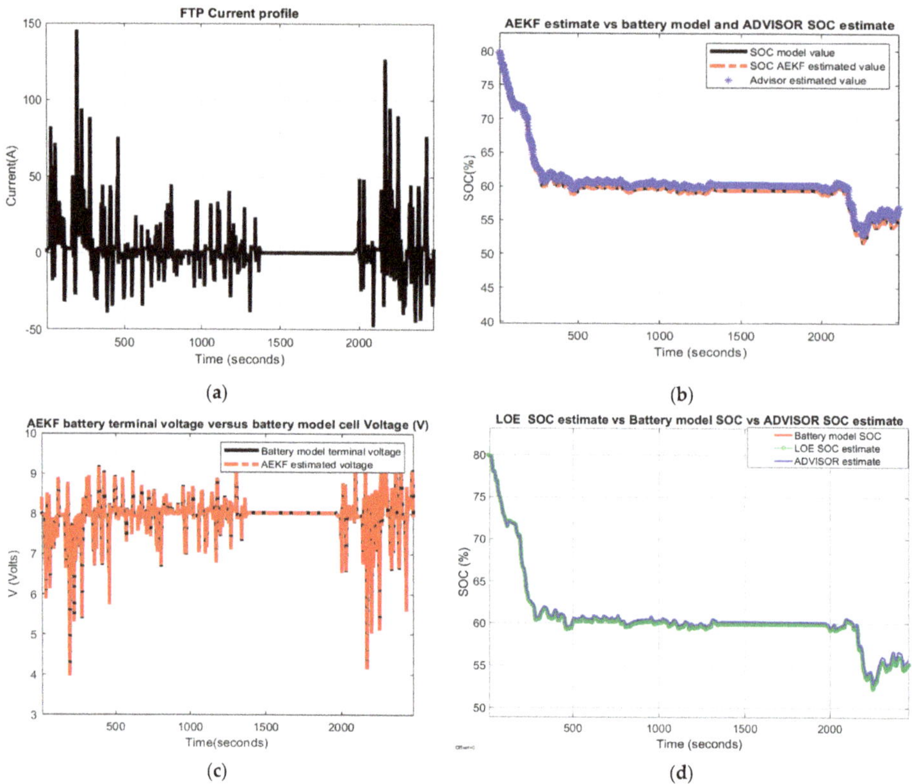

Figure 27. Li-ion Co battery model SOC accuracy performance and validation. (**a**) The FTP driving cycle current profile; (**b**) AEKF SOC estimate vs. battery model SOC; (**c**) AEKF terminal voltage estimate vs. battery terminal voltage; (**d**) LOE estimate vs. battery model SOC vs. ADVISOR SOC estimate.

Similarly, the same graphs related to LOE SOC estimator performance are shown in Figure 27d.

Furthermore, the SOC accuracy of the battery Li-ion Co model revealed by MATLAB simulation results are supported by the experimental results shown in Figure 28 for same FTP driving cycle test performed on the ADVISOR-MATLAB platform.

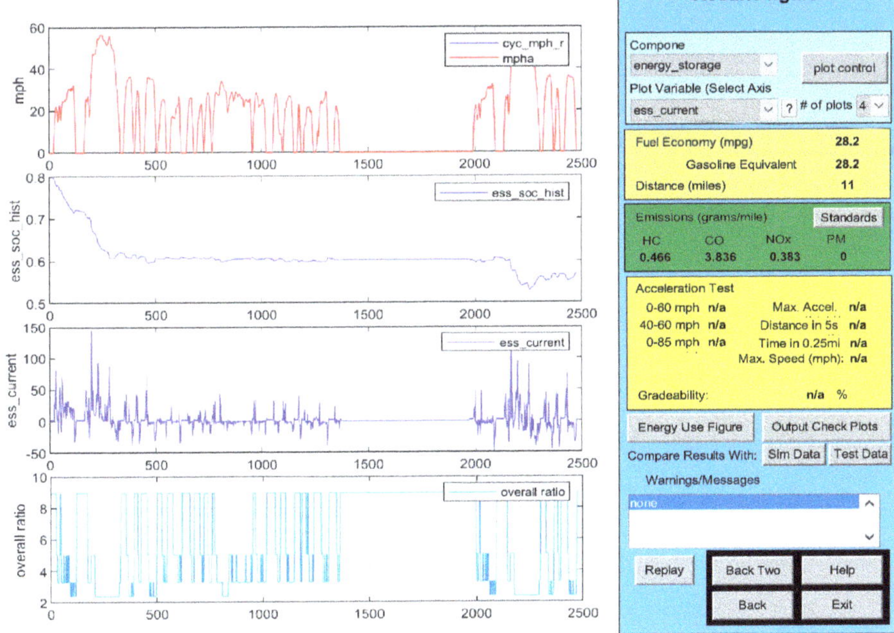

Figure 28. FTP driving speed cycle of the input HEV midsize car; HEV car speed cycle; estimated Rint Li-ion battery SOC on NREL ADVISOR- MATLAB platform, and current profile (from the top to the bottom).

B. AEKF and LOE SOC estimators robustness

- R1-scenario

In Figure 29a,b are depicted the simulation results for both SOC estimators that reveal a great robustness performance for LOE SOC compared to AEKF SOC estimator.

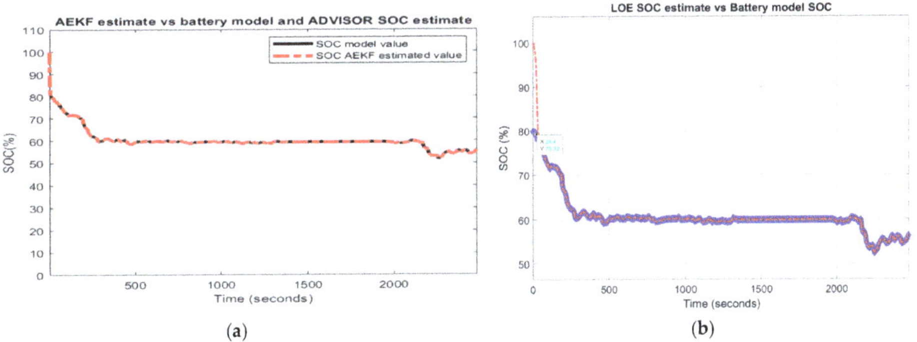

Figure 29. Robustness performance of AEKF and LOE SOC estimators for R1-scenario; (a)AEKF SOC for R12; (b) LOE SOC for R12.

- R2-scenario

For this scenario, the simulation results shown in Figure 30a,b indicate a great robustness for AEKF SOC estimator compared to LOE SOC.

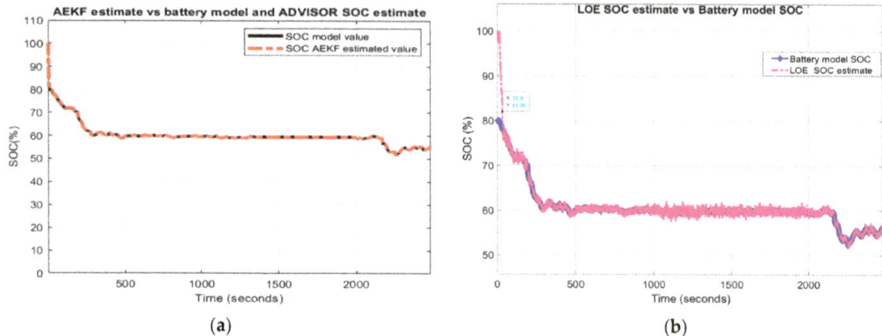

Figure 30. AEKF and LOE SOC estimators - robustness performance for R2-scenario (**a**) AEKF SOC for R22; (**b**) LOE SOC for R22.

- R3-scenario

For the third scenario, the results presented in Figure A5a,b in Appendix A show a slight superiority of the LOE SOC estimator compared to the AEKF SOC estimator.

- R4-scenario

The output temperature profile of thermal model and the effects of temperature changes on the internal resistance Rint and polarization constant are shown in Figure A4a–c, and the results of MATLAB simulations are presented in Figure A6a,b (both figures in Appendix A). From Figure A6, it seems that the LOE SOC estimator performs better than AEKF SOC estimator. Also, the statistical errors for FTP ADVISOR driving cycle are summarized in Table A2 form Appendix A. As in the first two driving cycles, for the FTP driving cycle test the result of the robustness performance analysis based on the statistical errors included in Table A2 confirms again that the AEKF SOC estimator performs better than its competitor LOE SOC estimator. Thus, based on the statistical results of the three tables, it can now decide that the most appropriate SOC estimator for this type of HEV application is the AEKF SOC estimator which shows an absolute superiority compared to the LOE SOC estimator, due to its ability to filtrate the measurement noise, as well as more robust to the aging effects on the Li-ion Co battery.

4. Discussions

During this research, we have substantially enriched our experience in designing, modelling, implementing and validating Li-ion batteries, developing and implementing real-time SOC estimation algorithms in a friendly and attractive MATLAB-Simulink environment. Now we try to summarize some of the most relevant aspects that have captured our attention during this research.

4.1. SOC Estimators' Convergence Speed

The analysis of the convergence speed performance of both SOC estimators can be done visually by examining the graphs strictly related to SOC. In almost all the graphs, the AEKF SOC estimate reaches the true value of Cobalt Li-ion battery model SOC after 40–190 s, when decreasing the SOC initial value from 80% to 40% or 16–150 s for an increase from 80% to 100%, as shown in Figure 31a,b by zooming at the beginning of the transient, which obviously is a rapid convergence speed.

Compared to AEKF SOC estimator, the convergence speed of LOE SOC estimator can be controlled by choosing the most appropriate value for the observer gain. For high gain values, the LOE SOC estimator becomes much faster as can be seen for the FTP driving cycle test, where the observer gain is 100. For the UDDS driving cycle, the observer gain is 10 and, for performance analysis purpose, the observer gain for UDDS-EPA driving cycle is intentionally set to 1. For this case, the LOE SOC estimator reaches the true value of the battery model SOC after 400 s, much higher as compared to AEKF SOC estimator.

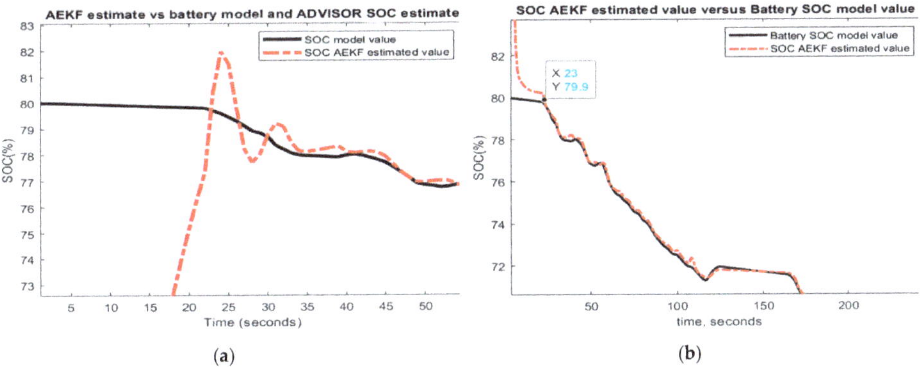

Figure 31. AEKF SOC estimation convergence speed; (**a**) for a decrease SOC initial value; (**b**) for an increase initial value.

4.2. SOC Estimation Accuracy

The MATLAB simulation results shown in the previous section reveal, in most cases, an excellent SOC estimation accuracy of AEKF SOC estimator after the estimate reaches the battery model SOC true value. Still, in some cases, due to unsuitable values for the tuning parameters, the AEKF SOC estimate is biased compared to LOE SOC estimator. On the other hand, the LOE SOC estimator accumulates a significant estimation error during the transient. Regarding the EKF SOC estimator, we observed that the SOC accuracy depends on a "trial and error" empirical adjustment procedure of tuning parameter values. Unfortunately, this procedure takes a lot of time. Moreover, a new readjustment procedure is required when changing the driving conditions and SOC initial value, as well as when aging and temperature effects take place. The adopted version AEKF due to its adaptive features attenuates the tuning procedure of the parameters significantly.

4.3. SOC Estimator-Measurement Noise Filtration

An important aspect that we also observed in this research is the measurement noise filtration by both estimators. Only the AEKF has this ability to filtrate the measurement noise compared to LOE SOC estimator, as you can see, for example, in Figure 30b.

4.4. SOC Estimators-Real Time Implementation

As we mentioned in the previous section due to its "predictor-corrector structure", the AEKF SOC estimator becomes a recursive algorithm, "more simple to implement in real time and computationally efficient" [10]. Also, the LOE SOC structure is simple and easy to design and implement in real time, in particular due to its linearized structure and having a single parameter needed for adjustment. In addition, the proposed generic lithium-ion cobalt battery model is simple, easy to design and quickly to implement in real time, based directly on the manufacturer's battery specifications. MATLAB-Simulink software platform provides a valuable and practical Simscape SimPower Systems library, helpful to be used for designing and implementation of different HEVs and EVs powertrains configurations.

4.5. SOC Estimators-Statistical Errors Analysis

The results from the first line (RMSE, MSE, MAE) provide the accuracy of the battery model SOC and ADVISOR estimate, beneficial for Li-ion Co battery model validation performed in Section 2.6.4 for first UDDS driving cycle test and for third one FTP-75. The validation of the battery model for second UDDS-EPA driving cycle test is proved in Section 3.3 based on the MATLAB simulation results shown in Figure 22. The statistical errors from Table 3, Tables A1 and A2 are valuable to compare the results of both SOC estimators to those obtained in the field literature by similar algorithms SOC estimators, for same driving cycle tests, and same performance error indicators (RMSE, MSE, MAE). In Section 2.5.2 the state of the art analysis focused on adaptive filters SOC estimators reported in the literature is made. For this analysis, Table 3, Tables A1 and A2 provide valuable information to compare the results obtained by AEKF SOC estimator, in terms of accuracy and robustness performance, developed in actual research work to those obtained in [18–24] for similar conditions, especially for same input current cycle profiles. Unfortunately, it was possible to make only a partial analysis since many researchers use different input current profiles and different error indicators that do not match with those used in our research. But, for the cases that match with our current profile, the information collected in all three Table 3, Tables A1 and A2 corresponding to each input current cycle profile can be useful to analyse all similar situations. Thus, the present research work can be a valuable source of inspiration for readers and researchers.

5. Conclusions

In this research paper, among the most relevant contributions the following may be highlighted:

- Model selection—MATLAB Simscape Li-ion cobalt nonlinear model, simple, practical, accurate, easy to implement in real-time (Section 2.6.1)
- Model development in continuous time state-space representation (Section 2.6.2)
- Model development in discrete time state-space representation (Section 2.6.3)
- Model validation based on three different driving cycles tests, using ADVISOR 3.2 software tool (Section 2.6.4)
- Model implementation in MATLAB R 2019b Simulink environment (Simulink diagram from Figures 8 and 9, Section 2.6.4)
- Thermal model design and Simulink implementation (Simulink diagram from Figure 12 Section 2.7)
- Adaptive Extended Kalman Filter SOC estimator with fading feature—Design and MATLAB implementation, Section 3.1)
- Linear observer SOC estimator-Design and Simulink implementation (Simulink model diagram from Figure 16, Section 3.2)
- MATLAB SOC simulations (Section 3.3)
- Performance analysis (SOC accuracy and robustness)—Table 3, Tables A1 and A2 for statistical errors

The case study is a 5.4 Ah Li-ion Cobalt battery, of high simplicity and accuracy, easy to be implemented in real-time and to provide beneficial support to build two real-time AEKF and LOE SOC estimators. For a good insight on the realistic battery life environment, the case of the battery internal resistance and polarization coefficient as parameters temperature-dependent is also investigated. Both parameters are updated dynamically through a simplified thermal model designed in Section 2.7. The robustness and accuracy of both SOC estimators is investigated in detail, for three most used driving cycles tests in the automotive industry (UDDS, UDDS-EPA and FTP) and changes in:

○ SOC initial value ("guess" value)
○ SOC initial value and driving conditions
○ SOC initial value, temperature effects on internal resistance and polarization constant, and driving conditions.

○ SOC initial value, nominal value of battery capacity due to aging effects/temperature effects and driving conditions.

Based on the statistical errors calculated for each driving cycle test in terms of RMSE, MSE and MAE, it was possible to choose from both competitors the most suitable SOC estimator. The result of overall performance analysis indicates that the AEKF SOC estimator performs better than LOE SOC estimator.

In the future work, we continue our investigations on lithium batteries regarding an improved modelling approach by "integrating the effect of degradation, temperature and SOC effects" [10], and for possible extensions to more accurate adaptive neural fuzzy logic SOC estimation techniques.

Author Contributions: R.-E.T. has contributed for algorithms conceptualization, manuscript preparation and writing it. N.T. has contributed for battery model validation, performed the MATLAB simulations and analyzed the results. M.Z. has contributed for a formal analysis of the results, for the visualization, the project supervision and administration. All authors have read and agreed to the published version of the manuscript.

Funding: This research received no external funding.

Acknowledgments: Research funding (discovery grant) for this project from the Natural Sciences and Engineering Research Council of Canada (NSERC) is gratefully acknowledged.

Conflicts of Interest: The authors declare no conflict of interest.

Abbreviations

Ni-Cad	nickel cadmium
Ni-MH	nickel metal hydride
Li-ion Co	lithium-ion cobalt
EV	electric vehicle
HEV	hybrid electric vehicle
BMS	battery management system
ADVISOR	advanced vehicle simulator
EPA	environmental protection agency
UDDS	urban dynamometer driving schedule
FTP-75	Federal test procedure at 75 F
SMO	sliding mode observer
LOE	linear observer estimator
RMSE	root mean squared error
MSE	mean squared error
MAE	mean absolute error
OCV	open-circuit voltage
SOC	state of charge
SOE	state of energy
SOH	state of health
DOD	depth of discharge
NREL	National Renewable Energy Laboratory
UKF	unscented Kalman filter
AUKF	adaptive unscented Kalman filter

Appendix A

Appendix A.1. Figures

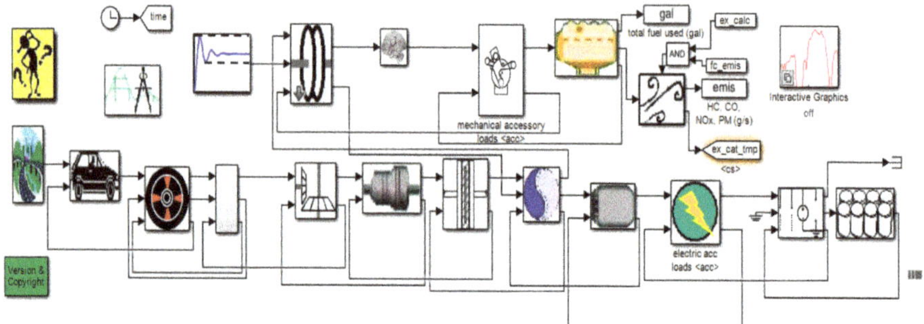

Figure A1. The Simulink block diagram of a hypothetical midsize town car-the diagram includes the following blocks: differential, clutch, gear, battery system, transmission and accessories

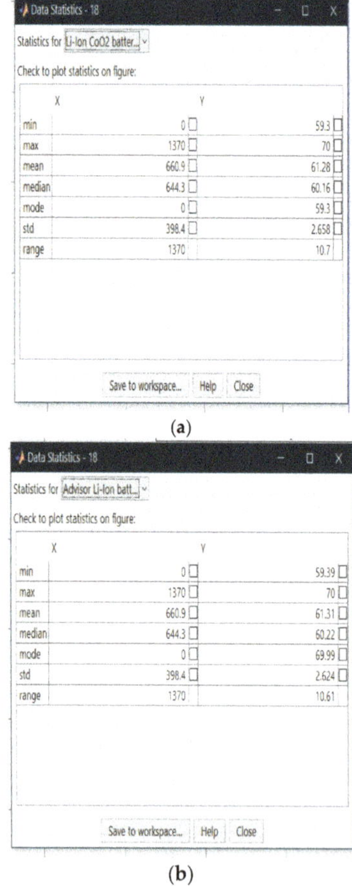

(a)

(b)

Figure A2. The UDDS test on the ADVISOR 3.2 integrated MATLAB platform; (**a**) The plot of statistic errors for Li-Ion CO2 battery model, (**b**)The plot of the statistic errors for Advisor Li battery Rint model.

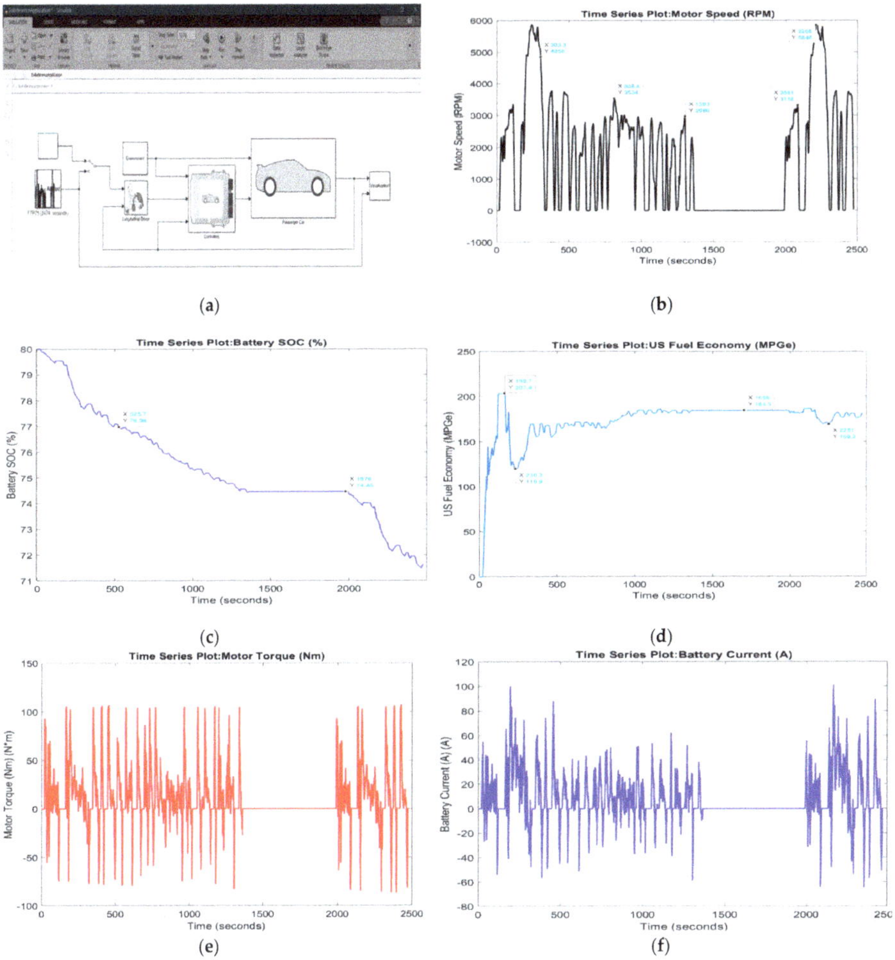

Figure A3. (**a**) Electric Vehicle simulation model application. (**b**) motor speed in RPM; (**c**) battery SOC (%); (**d**), US fuel economy; (**e**) motor torque (Nm); (**f**) discharge battery current profile.

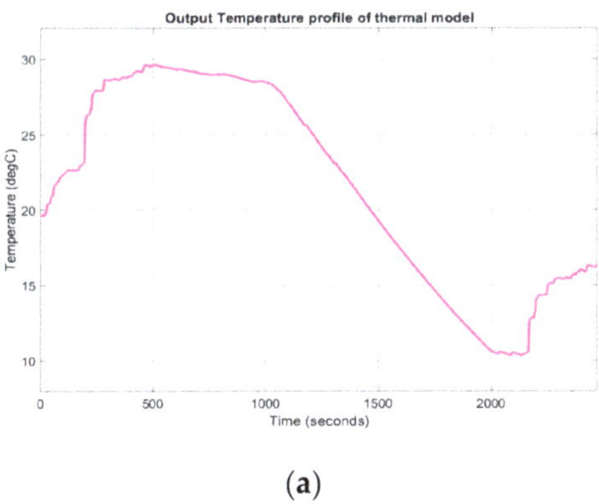

(**a**)

Figure A4. *Cont.*

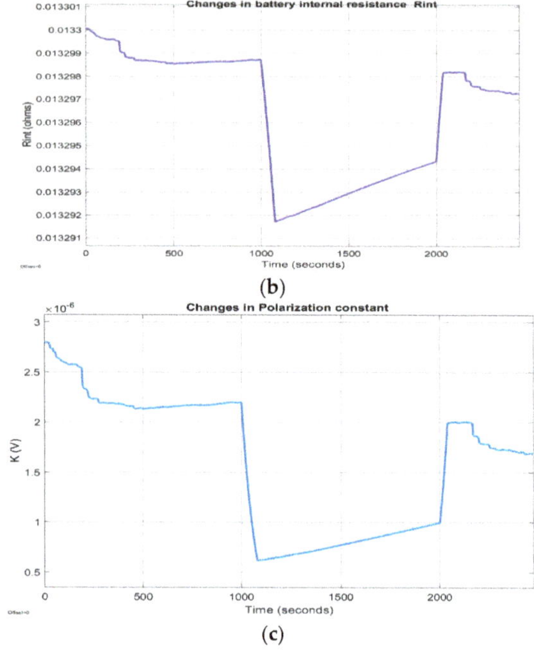

(b)

(c)

Figure A4. Temperature effects on Rint and K. (**a**) output temperature profile; (**b**) internal battery resistance Rint; (**c**) polarization constant K.

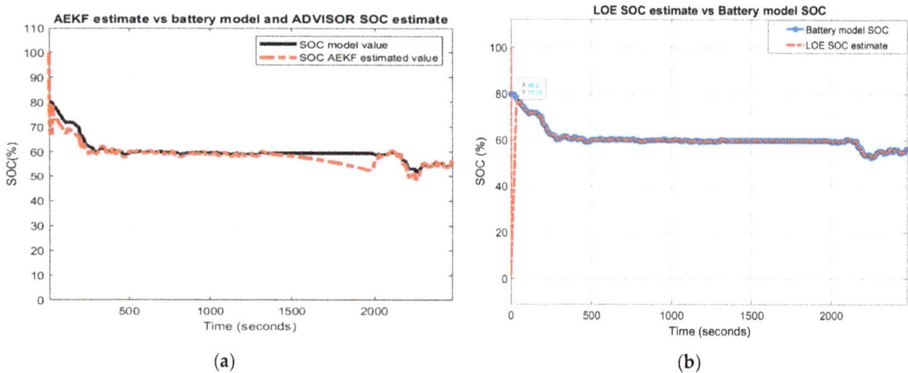

(a)

(b)

Figure A5. Robustness performance of AEKF and LOE SOC estimators for R3-scenario; (**a**) AEKF SOC for R32; (**b**) LOE SOC for R32.

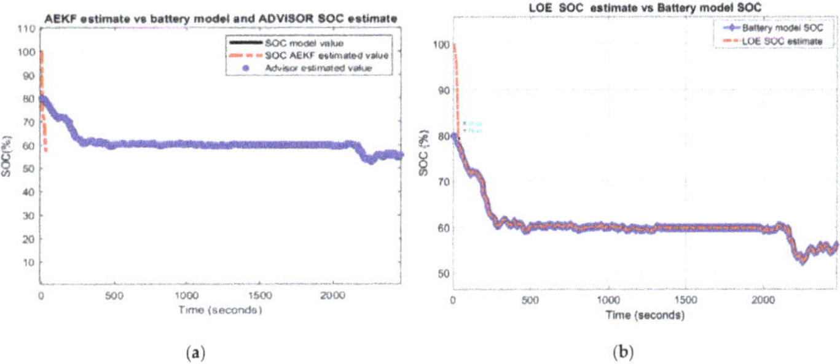

Figure A6. Robustness of AEKF and LOE SOC estimators for R4-scenario; (**a**) AEKF SOC for R42; (**b**) LOE SOC for R42.

Appendix A.2. Tables

Table A1. AEKF SOC Estimator—statistic errors.

Cycle Test	Accuracy	SOCest Initial Value (%)	Statistic Errors							
			RMSE AEKF	RMSE LOE	MSE AEKF	MSE LOE	MAE AEKF	MAE LOE	Std AEKF (%)	Std LOE (%)
UDDS-EPA	Battery model vs. ADVISOR	70	0.2621	0.2630	0.0173	0.0180	0.0224	0.0285	7.22	4.29
	AEKF/LOE	70	0.0325	0.0898	0.0000036	0.0004	0.0018	0.000459	2.6	7.35
	R1-scenario	R12	0.2290	0.2031	0.0212	0.4126	0.0726	0.1896	18.8	11.14
	R2-scenario	R22	0.2289	0.2013	0.2123	0.0405	0.0786	0.1879	18.8	11
	R3-scenario	R32	0.2290	0.2311	0.0212	0.0534	0.0786	0.2251	18.87	7.4
	R4-scenario	R42	0.2341	0.2041	0.0130	0.0416	0.0792	0.1913	19.55	11

Table A2. AEKF and LOE SOC Estimators—statistical errors.

Cycle Test	Accuracy	SOCest Initial Value (%)	Statistic Errors							
			RMSE AEKF	RMSE LOE	MSE AEKF	MSE LOE	MAE AEKF	MAE LOE	Std AEKF (%)	Std LOE (%)
FTP	Battery model vs. ADVISOR	70	0.0819	0.0834	0.0028	0.0043	0.0264	0.0235	4	4.29
	AEKF/LOE	70	0.0011	0.00063	0.00312	0.00006	0.008627	0.00067	4.3	2.82
	R1-scenario	R12	0.2392	0.0.0913	0.4280	0.0083	0.0082	0.0488	4.1	10
	R2-scenario	R22	0.0753	0.1004	0.00037	0.01	0.00169	0.0633	4.34	10.47
	R3-scenario	R32	0.0793	0.2651	0.0006	0.0703	0.0220	0.1297	4.27	21.3
	R4-scenario	R42	0.0683	0.0913	0.000198	0.0083	0.0071	0.0488	5.39	10.17

References

1. Moshirvaziri, A. Lithium-Ion Battery Modelling for Electric Vehicles and Regenerative Cell Testing Platform. Master's Thesis, University of Toronto, Toronto, ON, Canada, 2013.
2. *Electric Vehicle Integration into Modern Power Networks*; Springer Science and Business Media LLC: Berlin, Germany, 2013; pp. 15–26.
3. Xia, B.; Zheng, W.; Zhang, R.; Lao, Z.; Sun, Z. A Novel Observer for Lithium-Ion Battery State of Charge Estimation in Electric Vehicles Based on a Second-Order Equivalent Circuit Model. *Energies* **2017**, *10*, 1150. [CrossRef]
4. Farag, M. Lithium-Ion Batteries, Modeling and State of Charge Estimation. Master's Thesis, McMaster University of Hamilton, Hamilton, ON, Canada, 2013.
5. Cui, X.; Shen, W.; Zhang, Y.; Cungang, H. A Novel Active Online State of Charge Based Balancing Approach for Lithium-Ion Battery Packs during Fast Charging Process in Electric Vehicles. *Energies* **2017**, *10*, 1766. [CrossRef]

6.   Lowe, M.; Tokuoka, S.; Trigg, T.; Gereffi, G. Li-Ion Batteries for Electric Vehicles. The US Chain. Research Report. Center on Globalization Governance and Competitiveness, 5 October 2010; pp. 1–68. Available online: https://unstats.un.org/unsd/trade/s_geneva2011/refdocs/RDs/Lithium-Ion%20Batteries%20(Gereffi%20-%20May%202010).pdf (accessed on 2 March 2018).

7.   Plett, G.L. Extended Kalman filtering for battery management systems of LiPB-based HEV battery packs: Part 1. Background. *J. Power Sources* **2004**, *134*, 252–261. [CrossRef]

8.   Plett, G.L. Extended Kalman filtering for battery management systems of LiPB-based HEV battery packs: Part 2. Modeling and identification. *J. Power Sources* **2004**, *134*, 262–276. [CrossRef]

9.   Plett, G.L. Extended Kalman filtering for battery management systems of LiPB-based HEV battery packs: Part 3. State and parameter estimation. *J. Power Sources* **2004**, *134*, 277–292. [CrossRef]

10.   Tudoroiu, R.-E.; Zaheeruddin, M.; Radu, S.M.; Tudoroiu, R.-E. Real-Time Implementation of an Extended Kalman Filter and a PI Observer for State Estimation of Rechargeable Li-Ion Batteries in Hybrid Electric Vehicle Applications—A Case Study. *Batteries* **2018**, *4*, 19. [CrossRef]

11.   Tudoroiu, R.-E.; Zaheeruddin, M.; Radu, S.M.; Tudoroiu, N. New Trends in Electrical Vehicle Powertrains. *New Trends Electr. Veh. Powertrains* **2019**, *4*. [CrossRef]

12.   Tudoroiu, N.; Radu, S.M.; Tudoroiu, R.-E. *Improving Nonlinear State Estimation Techniques by Hybrid. Structures*, 1st ed.; LAMBERT Academic Publishing: Saarbrucken, Germany, 2017; p. 56. ISBN 978-3-330-04418-0.

13.   Gonzalez, C. The Future of Electric Hybrid Aviation, Machine Design, Technologies-Batteries/Power Supplies, March 2016. Available online: http://www.machinedesign.com/batteriespower-supplies/future-electric-hybrid-aviation (accessed on 24 March 2018).

14.   Simon, J.J.; Uhlmann, J.K. A New Extension of the Kalman Filter to Nonlinear Systems. In Proceedings of the SPIE 3068, Signal Processing, Sensor Fusion, and Target Recognition VI, Orlando, FL, USA, 28 July 1997; p. 3068. Available online: https://people.eecs.berkeley.edu/~{}pabbeel/cs287-fa09/readings/JulierUhlmann-UKF.pdf (accessed on 21 January 2018). [CrossRef]

15.   Xu, J.; Cao, B. Battery Management System for Electric Drive Vehicles – Modeling, State Estimation and Balancing. In *New Applications of Electric Drives*; IntechOpen: London, UK, 2015; pp. 87–113.

16.   Javier Muñoz Alvarez, J.M.; Sachenbacher, M.; Ostermeier, D.; Stadlbauer, H.J.; Hummitzsch, U.; Alexeev, A. *Analysis of the State of the Art on BMS*; Everlasting D6.1 Report; Lion Smart GmbH: Munchen, Germany, 2017.

17.   Zhang, R.; Xia, B.; Li, B.; Cao, L.; Lai, Y.; Zheng, W.; Wang, H.; Wang, W. State of the Art of Lithium-Ion Battery SOC Estimation for Electrical Vehicles. *Energies* **2018**, *11*, 1820. [CrossRef]

18.   Chang, W.-Y. The State of Charge Estimating Methods for Battery: A Review. *ISRN Appl. Math.* **2013**, *2013*, 1–7. [CrossRef]

19.   Lee, S.J.; Kim, J.H.; Lee, J.M.; Cho, B.H. The state and parameter estimation of an Li-Ion battery using a new OCV-SOC concept. In Proceedings of the 2007 Power Electronics Specialists conference, Orlando, FL, USA, 17–21 June 2007; pp. 2799–2803.

20.   Chen, Z.; Fu, Y.; Mi, C.C. State of Charge Estimation of Lithium-Ion Batteries in Electric Drive Vehicles Using Extended Kalman Filtering. *IEEE Trans. Veh. Technol.* **2012**, *62*, 1020–1030. [CrossRef]

21.   He, H.; Liu, Z.; Hua, Y. Adaptive Extended Kalman Filter Based Fault Detection and Isolation for a Lithium-Ion Battery Pack. *Energy Procedia* **2015**, *75*, 1950–1955. [CrossRef]

22.   Diab, Y.; Auger, F.; Schaeffer, E.; Wahbeh, M. Estimating Lithium-Ion Battery State of Charge and Parameters Using a Continuous-Discrete Extended Kalman Filter. *Energies* **2017**, *10*, 1075. [CrossRef]

23.   Zhao, Y.; Xu, J.; Wang, X.; Mei, X. The Adaptive Fading Extended Kalman Filter SOC Estimation Method for Lithium-ion Batteries. *Energy Procedia* **2018**, *145*, 357–362. [CrossRef]

24.   Wang, D.; Bao, Y.; Shi, J. Online Lithium-Ion Battery Internal Resistance Measurement Application in State-of-Charge Estimation Using the Extended Kalman Filter. *Energies* **2017**, *10*, 1284. [CrossRef]

25.   Tremblay, O.; Dessaint, L.-A.; Dekkiche, A.-I. *A Generic Battery Model. for the Dynamic Simulation of Hybrid. Electric Vehicles*; Institute of Electrical and Electronics Engineers (IEEE): New York, NY, USA, 2007; pp. 284–289.

26.   Zhang, Y.; Xiong, R.; He, H.; Shen, W. Lithium-Ion Battery Pack State of Charge and State of Energy Estimation Algorithms Using a Hardware-in-the-Loop Validation. *IEEE Trans. Power Electron.* **2017**, *32*, 4421–4431. [CrossRef]

27.   Syed Mahdi, A. Battery Identification, Prediction and Modelling. Master's Thesis, Colorado State University Fort Collins, Fort Collins, CO, USA, 2018.

28. DiOrio, N.; Dobos, A.; Janzou, S.; Nelson, A.; Lundstrom, B. *Technoeconomic Modeling of Battery Energy Storage in SAM*; US NREL Technical Report; NREL/TP-6A20-64641. Contract No. DE-AC36-08GO28308; NREL: Denver, CO, USA, 2015. Available online: https://www.nrel.gov/docs/fy15osti/64641.pdf (accessed on 3 March 2020).

29. Song, D.; Sun, C.; Wang, Q.; Jang, D. A Generic Battery Model and Its Parameter Identification. *Energy Power Eng.* **2018**, *10*, 10–27. [CrossRef]

30. Johnson, V. Battery performance models in ADVISOR. *J. Power Sources* **2002**, *110*, 321–329. [CrossRef]

31. Wipke, K.B.; Cuddy, R.M. Using an Advanced Vehicle Simulator (ADVISOR) to Guide Hybrid Vehicle Propulsion System Development. Publisher: Golden, CO: National Renewable Energy Laboratory. August 1996. Research gate net. Available online: https://www.nrel.gov/docs/legosti/fy96/21615.pdf (accessed on 1 March 2020).

32. Xing, Y.; Ma, E.W.M.; Tsui, K.-L.; Pecht, M.G. Battery Management Systems in Electric and Hybrid Vehicles. *Energies* **2011**, *4*, 1840–1857. [CrossRef]

33. Jiang, J.; Zhang, C. *Fundamentals and Applications of Lithium-ion Batteries in Electric Drive Vehicles*; Wiley: Singapore, 2015; p. 300.

34. Lakkis, M.E.; Sename, O.; Corno, M.; Bresch, P.D. Combined battery SOC/SOH estimation using a nonlinear adaptive observer. European Control Conference 2015, Linz, Austria. In Proceedings of the European Control Conference, Linz, Austria, 15–17 July 2015. [CrossRef]

35. Battery and Engineering Technologies. Battery Life and Death. Available online: http://www.mpoweruk.com/life.htm (accessed on 18 March 2018).

36. Taesic, K. A Hybrid Battery Model Capable of Capturing Dynamic Circuit Characteristics and Nonlinear Capacity Effects. Master's Thesis, University of Nebraska, Lincoln, NE, USA, July 2012.

37. Siva, R.; Prasanna, M. Modeling, Simulation & Implementation of Li-Ion Battery Powered Electric and Plug-in Hybrid Vehicles. Master's Thesis, University of Akron, Akron, OH, USA, August 2011.

 *energies*

Article

Leveraging Cell Expansion Sensing in State of Charge Estimation: Practical Considerations

Miriam A. Figueroa-Santos *, Jason B. Siegel and Anna G. Stefanopoulou

Department of Mechanical Engineering, University of Michigan, Ann Arbor, MI 48109, USA;
siegeljb@umich.edu (J.B.S.); annastef@umich.edu (A.G.S.)
* Correspondence: miriamaf@umich.edu

Received: 11 April 2020; Accepted: 14 May 2020; Published: 22 May 2020

Abstract: Measurements such as current and terminal voltage that are typically used to determine the battery's state of charge (SOC) are augmented with measured force associated with electrode expansion as the lithium intercalates in its structure. The combination of the sensed behavior is shown to improve SOC estimation even for the lithium ion iron phosphate (LFP) chemistry, where the voltage–SOC relation is flat (low slope) making SOC estimation using measured voltage difficult. For the LFP cells, the measured force has a non-monotonic F–SOC relationship. This presents a challenge for estimation as multiple force values can correspond to the same SOC. The traditional linear quadratic estimator can be driven to an incorrect SOC value. To address these difficulties, a novel switching estimation gain is used based on determining the operating region that corresponds to the actual SOC. Moreover, a drift in the measured force associated with a shift of the cell SOC–expansion behavior over time is addressed with a bias estimator for the force signal. The performance of Voltage-based (V) and Voltage and Force-based (V&F) SOC estimation algorithms are then compared and evaluated against a desired ±5% absolute error bound of the SOC using a dynamic stress test current protocol that tests the proposed estimation scheme across wide range of SOC and current rates.

Keywords: state-of-charge estimation (SOC); linear quadratic estimator; lithium ion battery; iron phosphate; cell expansion; force

---

1. Introduction

The primary function of the battery management system (BMS) is to provide an accurate state of charge (SOC) estimation. The SOC represents the amount of charge in ampere-hours (Ah) remaining in a cell divided by its total capacity [1,2]. The BMS traditionally uses current, voltage, and sometimes temperature measurements to estimate the SOC to plan future actions and to prevent over-charging or discharging of cells. Generally, manufacturers provide conservative estimates of remaining energy, since an overestimation of SOC can leave the vehicle stranded. In the case of unmanned air vehicles (UAV), overestimation of SOC might prevent the vehicle from safe landing, since landing maneuvers require very high power, which typically cannot be achieved at very low SOC levels [3]. Underestimating SOC, on the other hand, wastes valuable resources and adds cost and weight to the vehicles, which is critical for robotic platforms.

From the lithium ion batteries, lithium ion iron phosphate (LFP) has been considered for UAV, hybrid electric vehicles (HEV), and electric vehicles (EV) due to their capacity for fast charging, high power capability, and long cycle life [4]. LFP batteries consists of graphite as the negative electrode and lithium iron phosphate (metal oxide) as the positive electrode [5]. Due to the relative flat half-cell potential the positive electrode exerts (also known as the phase-separating cathode active material) [6], the open circuit voltage (OCV) has a relatively flat slope through most of its operating

SOC range (10–95%), as shown in Figure 1. The phase transitions in the graphite material correspond to the different voltage plateaus with respect to SOC, as shown in Figure 1a. This makes SOC estimation difficult under noisy environments [7,8] and inexpensive sensing, such as robotic and automotive applications. Previous work has suggested strain or stress (pressure or force) measurements to augment terminal voltage for SOC estimation [9]. Specifically, the graphite in the negative electrode expands when the lithium ions are intercalating into it, and the positive electrode contracts as the lithium ions leave it causing a change of thickness in the components of the battery [10]. This change in thickness causes the battery to swell. Therefore, the overall observed cell swelling is the summation of the swelling from the positive and negative electrodes. When the cells are constrained to a fixed displacement, as typical in automotive packs, the battery swelling results in an increased force on the fixture. This swelling force can be measured using a load cell [9].

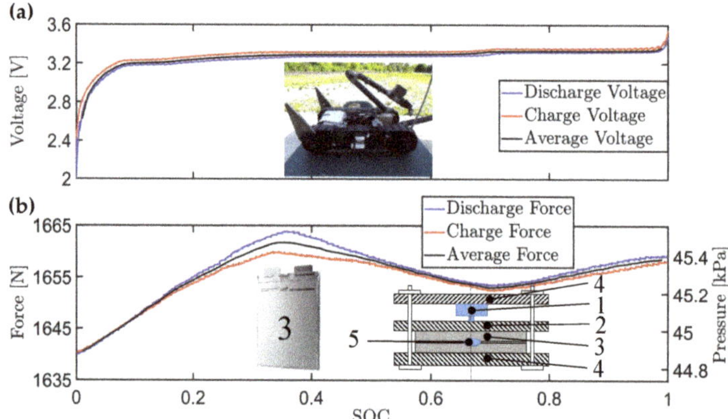

Figure 1. Measured voltage and force for the 20 Ah A123 lithium ion iron phosphate battery cycled under low current rate $C/20$. (**a**) Measured discharge (blue), charge (red), and average of discharge and charge (black) voltage for a $C/20$ cycle. For military robots such as the Packbot shown above, SOC estimation is critical to avoid the robot getting stranded. Lithium ion iron phosphate (LFP) batteries are typically used for operation of this robot due to the high power required. (**b**) Measured discharge (blue), charge (red), and average of discharge and charge (black) force for the under current rate $C/20$. The experimental battery fixture consists of: (1) load cell (force sensor); (2) movable plate; (3) lithium ion battery; (4) two aluminum end-plates; (5) temperature sensor. For generalization of the results to the other cell sizes, the force measurements can be converted to pressure by diving the force by the surface area of our battery (A_b) , which is the width ($w_b = 0.161$ m) times the length ($l_b = 0.227$ m) of the battery, $A_b = w_b \times l_b$ m^2.

The structural changes of this expansion have been studied from the electrode mechanics point of view with respect to strain/OCV coupling [10,11]. The overall volume change of this expansion results in a monotonic function of SOC for a nickel manganese cobalt (NMC) graphite cell [12]. By measuring the force produced by the expansion and including it in the estimation algorithm, improved SOC accuracy can be achieved as compared to voltage based methods [9,13]. The greatest benefits were observed in the 30–50% SOC region, where the voltage slope was relatively flat, but also at the low SOC level, where voltage drops very fast and is challenging to have an accurate model as the battery ages. The change in measured force vs. SOC over many cycles can also inform better estimates of the battery state of health (SOH) [14]. For the LFP-graphite cell studied here, the anode (negative electrode which is graphite) expands during charging but the simultaneous higher contraction rate of the cathode (positive electrode which is LFP) results in a combined cell contraction in the middle SOC range. The overall result is a non-monotonic F–SOC behavior, as shown in Figure 1b. Due to the F–SOC non-monotonicity, SOC estimation based on measured force is challenging since multiple

SOC equilibrium points can correspond to the same measured force. Additionally, all mechanical measurements (force, pressure, or displacement) even after calibration exhibit drift associated with small changes in material and a shift in the SOC–expansion behavior as the battery ages. Loss of cyclable lithium causes a shift in capacity and a change in the SOC–expansion behavior of the cell. This capacity loss shifts the stoichiometric ratio associated with lithium concentration in each electrode and hence changes the electrode expansion [15] as a function of lithium intercalation (Coulombs stored) and the measured force/pressure versus SOC as the unknown drift addressed here. Moreover the LFP pouch cells used here are supported by poron sheets [16] between the cells instead of the spacers used in [13]. Thermal expansion of the battery and fixture and viscoelastic response of the compliant poron pad introduces an additional aging and drift factor. Predicting and modeling this aging behavior requires extensive resources and is currently by-passed by estimating this unknown bias as proposed in this article. As can be appreciated, the drift is a general problem of measuring the mechanical behavior (force, pressure, or displacement) of all batteries and it is not just a battery chemistry (LFP)-related issue.

Battery cell balancing is critical to extend the range of battery powered vehicles, the pack operating lifetime, and charge and discharge power limits [17]. To achieve fast and accurate cell balancing, accurate SOC estimation is needed [18]. Voltage-based SOC estimation for the LFP chemistry is particularly challenging since the voltage is relatively flat with respect to SOC. The proposed method improves the SOC estimation using leveraging information about the cell expansion and contraction during charging and discharging. Cell-to-cell variability due to aging, as well as the resulting changes in the measured force, was not investigated, but should be the focus of future research. The influence of cell balancing or imbalanced cells in a module whose force is measured should also be further investigated following the initial work by [19]. Since there would typically be only one force (or strain) measurement in a pack of series connected cell, cell balancing techniques [20] will be of high importance.

In this paper, we demonstrate the improvement in the SOC estimation of LFP batteries by using mechanical in addition to electrical measurements that can be implemented in packs or modules of both hard encased and pouch cells [21]. A novel solution to the multiple equilibrium SOC points is proposed based on the piecewise linear (PWL) F–SOC characteristic approximation that is further used in a switching gain model-based linear quadratic estimator (LQE) design that consists of a combination of force and voltage measurements [22]. Due to the drift that appears in most force measurements, as shown in Figure 2, the SOC may not be accurately estimated. Therefore a bias state is added to the LQE in order to capture the drift in the force measurement due to un-modeled changes in battery swelling or creep of the plastic materials. Experimental validation is also performed on the model-based (LQE) controller design using the combinations of force and voltage measurements during realistic battery electric vehicle usage profiles including the Dynamic Stress Test (DST). The performance of a controller designed with a "perfect" model is compared to one with model mismatch in the OCV and F–SOC PWL fit. The simulated model mismatch captures the typical modeling uncertainties or changes in the cell expansion and open circuit voltage due to aging [23].

2. Experimental Setup and Force Behavior

The battery considered here for the experimental validation is an A123 20 Ah lithium iron phosphate pouch cell with a voltage range of 3.6–2 V. The fixture, as shown in Figure 1b, consists of an active battery cell (3) (with a temperature sensor on top of it (5)) and a dummy (inactive) cell with a compliant rubber pad in between. The active battery cell and inactive cell consists of a laminated aluminum pouch cell and the rubber pad is a Poron 4701-30 from Rogers, which is 1.14-mm thick. The temperature sensor is a multimeasurand GE sensor that consists of three resistance temperature detectors (RTDs) and one eddy current expansion sensor [21]. The dummy cell and the rubber pad is used as a stress absorber to emulate the conditions a cell might experience in a pack configuration. The purpose of the dummy cell is to simulate the compliance of the whole system. The stiffness of poron is much less than that of the battery, thus the compression of the dummy cell is negligible in terms of the whole system. Therefore, the dummy cell can be used without complicating the force measurement. The dummy cell is held tight by an aluminum bottom end-plate (4), and a movable plate is placed on top of the active cell. This plate has one degree of freedom in the vertical direction with the force sensor (1) placed on top of it. On top of the force sensor, the fixed top aluminum end-plate is bolted to the bottom end-plate to simulate the behavior of a constrained battery pack with fixed distance between the end plates. The force sensor is an Omega (LC305-500) load cell sensor (strain gauge type). The sensor has a 2225 N full scale range with an accuracy of 4.45 N. The load cell and voltage are digitized and recorded by Data Translation DT-9828, which has a voltage accuracy of 2 millivolts. In this paper, the force sensor is used because it is cheaper than the displacement sensors [15], but also less accurate. For example, the accuracy of the displacement sensor used in [15] is 1 μm, which corresponds to 0.35 N of compressive force on the poron, while for the Omega force sensor used the accuracy is listed as a percentage of full scale range (4.45 N). The force only accuracy is the important reason for integrating force and voltage information along with performing the bias estimation.

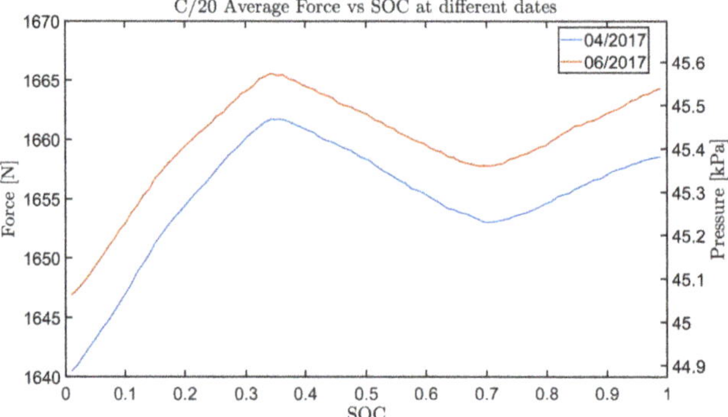

Figure 2. Drift in the force sensor can be observed by comparing the force vs. SOC for two $\frac{C}{20}$ cycles which were conducted two months apart with three cycles done in between them with the same battery. The force measurements have a minimum of 3 N and a maximum of 6 N drift across the entire SOC range. The drift could be caused by thermal expansion of the battery, pad, and fixture [13], capacity change, or a combination of all four throughout the life of the cell and module on which measurements are performed. Adjustments in the fixture and changes in the preload will cause larger changes in measured force but can also be modeled as a sensor drift. For generalization of the results to the other cell sizes, the pressure is shown.

The force plotted against SOC (F–SOC) and the open circuit voltage vs. SOC (OCV–SOC) are measured experimentally using a pulse–relax profile. Specifically, a CCCV charge is applied followed by a pulse–relax discharge profile, with the current rate of $\frac{C}{20}$ for 12 min (which results in a 1% SOC change) followed by a 1 h of rest period to eliminate the influence of the internal resistance and electrolyte polarization of the battery. The data points at the end of the rest are used to obtain the discharge F–SOC and OCV–SOC. The test is then repeated using charge pulses of equal duty cycle. The average between the measurements at the same SOC from the charge and discharge datasets is obtained for the F–SOC, as shown in Figure 1b. The average F–SOC shown in Figure 1b is modeled in Section 3 and is the best fit parameters that match the force inflection points. An average fit is also obtained from data for OCV–SOC and shown in Figure 3a. The values of the fit for the OCV–SOC (average of discharge and charge) and F–SOC can be found in Appendix C.

An important consideration is the drift observed in the measured force between repeated tests as shown in Figure 2. Two average force cycles with an applied current of $\frac{C}{20}$ at different test dates are shown. These cycles have approximately a minimum 3 N and a maximum 6 N of drift in between the second test (performed on 06/2017) and first test (performed on 04/2017). This measurement was taken with the same cell. The drift could be caused by thermal expansion of the battery, pad, and fixture, capacity change or a combination of all four throughout the life of the cell and module on which measurements are performed. That is the reason we have treated the observed bias as an unknown variable. We assume that this unknown/uncertain-origin drift evolves slower than the force measurement from the charge/discharge changes and we therefore can estimate it as an unknown constant bias in real time. This drift in force can affect the SOC estimation if not compensated and, thus, estimation of the sensor bias is used to improve the practical force based SOC estimation.

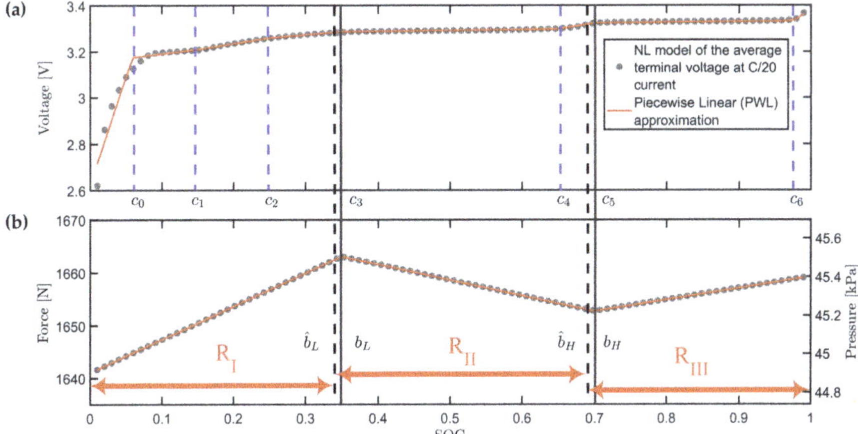

Figure 3. A piecewise linear approximation of the (a) open circuit voltage and (b) cell swelling force was fit to the experimental measured values for the A123 Lithium Iron Phosphate Battery Cell. The gray dots correspond to the average of measured data at each SOC point from the charge and discharge cycles at the $\frac{C}{20}$ rate (see Figure 1). The orange line represents the PWL fit. The horizontal solid lines represent the inflection points of the cell swelling force. The horizontal dashed black lines represent the inflection points of the cell swelling force used for the developed observers. R_I, R_{II}, and R_{III} represent the force slope regions determined by our PWL model. For the PWL force model, we assume $b_L = c_3$ and $b_H = c_5$ since the changes in voltage and force slopes are due to the intrinsic phase transitions in the material of the battery electrodes [15]. Assessment of the SOC estimator robustness is performed by imposing $\hat{b}_L \neq b_L$ and $\hat{b}_H \neq b_H$ to capture model mismatch for the force and voltage behavior. For generalization of the results to the other cell sizes, the pressure is shown.

For the SOC estimation development, tuning, and comparison, two models with two different levels of fidelity are used:

- The simulation model includes the nonlinearities and hysteresis for the electric characteristics detailed in Section 5.
- The observer model ignores hysteresis and uses piecewise linear approximations of the nonlinearities detailed in Section 5.

In the next section, the simulation model is detailed and its efficacy is highlighted in Figure 4 based on a modified DST cycle [24] that is scaled for a 20 Ah Li-ion battery, as shown in Figure 4a. The DST was chosen because it has a current profile that has the combination of the following C-rates and is representative of usage in an electric vehicle: C/4, C/2, 1C, and 2C. If the utilization of the electrode is relatively uniform, the C-rate should not influence the swelling significantly. The electrode expansion depends on the bulk concentration of the electrode solid phase, as opposed to the terminal voltage which depends on the surface concentration [25] and therefore is less sensitive to C-rate. The theory for determining up to what current density the electrode utilization is uniform can be found based on the porous electrode models by Fuller Doyle and Newman [26] and Newman and Tobias [27]. The largest expected contribution of C-rate dependence (or more precisely root mean square (RMS) current) on the result is through thermal expansion of the cell. The experimental profile consists of a Constant-Current/Constant-Voltage (CCCV) charging protocol at a rate of 1C until the battery is fully charged. After a rest period of 30 min, a 1C rate discharge current is applied until it reaches 61% SOC. After the second rest period of 2 h, the modified DST cycle is applied. The resulting voltage, temperature, force, and SOC are shown in Figure 4b–e, respectively. The SOC is calculated by Coulomb counting using a high resolution current sensor and assumed to be the true SOC.

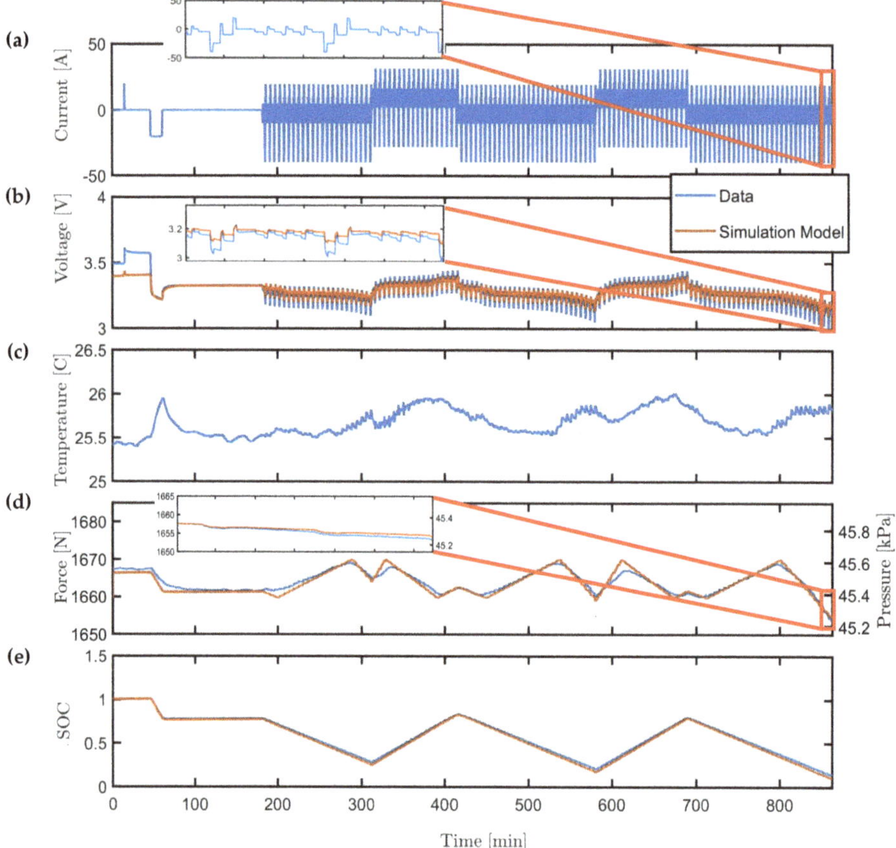

Figure 4. Comparison of open loop simulation model results (denoted by the color orange) and experimental measurement of voltage, temperature, force, and SOC obtained from the 20 Ah battery during the Dynamic Stress Test (DST) inside an environmentally controlled chamber set at 25 °C ambient conditions. (**a**) Current profile scaled for the 20 Ah A123 battery. (**b**) Comparison of the open loop model terminal voltage and measurement. (**c**) Measured battery temperature. (**d**) Comparison of the open loop modeled force and measured force. After 30 min of rest, the un-modeled dynamics in force excited by cycling of the cell to decay to zero. For generalization of the results to the other cell sizes, the pressure is shown. (**e**) Comparison of state of charge (SOC) measurement.

3. Simulation Model

The models described here include the drift present in our force data and higher dynamics such as hysteresis present in our voltage. These models were simulated to validate the robustness of our developed estimators for analysis purposes before the experimental implementation. After this validation, the estimators were used with experimental data and their performance was evaluated. The battery model used for our simulations is presented in the following subsections. The SOC (z) is simply modeled as

$$\frac{dz}{dt} = -\frac{I}{C_b},$$

(1)

where C_b is the cell capacity.

3.1. Cell Swelling Force Model

The force measured at the load cell is modeled using a static non-linearity $F_{sim}(z)$, which is a function of state of charge, with additive bias and noise terms given by

$$F(z) = F_{sim}(z) + v_F + f_d$$

(2)

where $F(z)$ (N) is the measured force that relies on $F_{sim}(z)$, which is the PWL model of the average F–SOC behavior; v_F is the measurement noise; and f_d is a constant drift or bias (assumed to be constant but unknown) value present in our force measurement. A piecewise linear representation of the average F–SOC behavior is given by

$$F_{sim}(z) = C(z)z + C_0(z) = \begin{cases} \alpha_m z + \alpha_{m0}, & \text{if } z \le b_L \\ \beta_m z + \beta_{m0}, & \text{if } b_L < z \le b_H \\ \gamma_m z + \gamma_{m0}, & \text{otherwise} \end{cases}$$

(3)

where α_m, β_m, and γ_m are the slope parameters; α_{m0} is the preload or the force sensed at zero state of charge; and b_L and b_H denote the SOC where a change in the sign of the slope in the PWL model occurs as shown by the solid gray vertical lines in Figure 3. These parameter values can be found in Appendix C and can be adjusted to simulate model uncertainty and mismatch. The parameters β_{m0} and γ_{m0} in the PWL model are uniquely determined from the other parameters via constraints of piecewise continuity

$$\beta_{m0} = (\alpha_m - \beta_m)b_L + \alpha_{m0}$$

(4)

$$\gamma_{m0} = (\beta_m - \gamma_m)b_H + (\alpha_m - \beta_m)b_L + \alpha_{m0}.$$

(5)

The operating regions R represent the force slope regions determined by our PWL model. The first region is defined as $R_I : \hat{z} \in [0, \hat{b}_L)$, the second region is defined as $R_{II} : \hat{z} \in [\hat{b}_L, \hat{b}_H]$, and the third region is defined as $R_{III} : \hat{z} \in (\hat{b}_H, 1]$. The operating regions are shown in Figure 3.

3.2. Terminal Voltage Model

The terminal voltage in volts is modeled as

$$V_T = V_{oc}(z) - IR - V_1 - V_2 - V_h + v_V,$$

(6)

where $R[\Omega]$ is the total equivalent series resistance, I is the discharge current applied to the battery, V_1 and V_2 are the voltages due to the two resistance and capacitance (RC) pairs, and v_V represents the V measurement noise. The OCV characteristic, $V_{oc}(z)$, is SOC dependent and modeled using

$$V_{oc}(z) = V_0 + d(1 - \exp(-fz)) + h\left(1 - \exp\left(-\frac{-k}{1-z}\right)\right) + gz + \sum_{i=1}^{3} a_{v,i} \arctan\left(-\frac{z - b_{v,i}}{c_{v,i}}\right),$$

(7)

where V_0, g, d, f, h, k, $a_{v,i}$, $b_{v,i}$, and $c_{v,i}$ are tuned parameters found in Appendix C. The electric equivalent circuit (EC) battery model [28,29] is used for the simulation in our observer validation. In this study, voltage hysteresis (V_h) [30] is also considered for the simulation model

$$\frac{dV_1}{dt} = \frac{-V_1}{R_1 C_1} + \frac{I}{C_1} \tag{8}$$

$$\frac{dV_2}{dt} = \frac{-V_2}{R_2 C_2} + \frac{I}{C_2} \tag{9}$$

$$\frac{dz}{dt} = -\frac{I}{C_b} \tag{10}$$

$$\frac{dV_h}{dt} = -\left|\frac{\gamma_h I}{C_b}\right| V_h + \left|\frac{\gamma_h I}{C_b}\right| H(z, \mathrm{sgn}(I)) \tag{11}$$

where V_1 and V_2 are the voltages of the RC equivalent circuits, R_1 and R_2 are the resistors, C_1 and C_2 are the capacitors of the RC equivalent circuits, V_h is the hysteresis voltage, γ_h is the hysteresis rate constant, and $H(z, \mathrm{sgn}(I))$ is a function of SOC and the sign of current ($\mathrm{sgn}(I)$) following [30]. The function $H(z, \mathrm{sgn}(I))$ is taken to be half the difference between the charge and discharge OCV measurements, and the parameter values can be found in Appendix A. Although the EC model parameters depend on the battery's SOC and temperature, in this paper, we do not take this dependency into consideration. The constant parameters of the EC model can be found in Table 1. The dynamic equations developed for charge/discharge as well as the measurements in Equations (2) and (6) is used to simulate the battery behavior, and it is numerically discretized with a time step of $T_s = 1$ s.

Table 1. Battery Equivalent Circuit Parameters and its values.

Parameters	Values
R	1.5 mohms
R_1	1.4 mohms
C_1	13,014 farad
R_2	2.7 mohms
C_2	143,000 farad
γ_h	0.00054

4. Voltage Model Parameterization

Before using our model for SOC estimation, the average Force–SOC and OCV–SOC needs to be obtained and model parameterization of the equivalent circuit parameters is required using a pulsed current profile such as the Hybrid Pulse Power Characterization (HPPC) [28]. The pulse current was obtained and the nonlinear programming solver fmincon was used to find the parameters provided in Table 1. To verify that the parameters are correct, different initialization values were used. From the different initialization values used, fmincon converged to the parameters provided in Table 1.

5. SOC Estimation Model

The linear quadratic estimator (LQE) also known as the steady-state Kalman filter is used for state estimation. The goal of the observer is to find a gain K that converges the initial state to the true state of the system using linear filter equation with measurement error feedback

$$\hat{x}_{t+1} = A\hat{x}_t + Bu_t + K(y_t - \hat{y}_t) \tag{12}$$

where the estimated output equation is given by

$$\hat{y}_t = C(\hat{x}_t)\hat{x}_t - C_0(\hat{x}_t) - Du_t \tag{13}$$

with $C(\hat{x}_t)$ the slope of the estimated measurement and $C_0(\hat{x}_t)$ the affine parameter based on our model and shown in Figure 3. The matrix D is the direct transition (or feedthrough) term. The error dynamics are governed by the eigenvalues of $A - KC$, which depends on the chosen gain K so that it is stable and achieves fast convergence of the SOC estimation error $e = z - \hat{z}$. We use the Discrete Algebraic Riccati Equation (DARE) to find our gain K on all our developed observers. The DARE is defined as

$$P = APA^T - APC^T \left[CPC^T + R \right]^{-1} CPA^T + Q \tag{14}$$

and solved for the estimation error covariance matrix P. In this equation, Q is the process noise covariance matrix (size nxn) and $R \in \mathbb{R}^{n_y}$ is the measurement noise covariance matrix (size mxm). The values of the diagonal elements of R are chosen based on actual sensor measurement noise variance and Q is tuned so that the desired transient is achieved. The solution of the estimation error covariance (P) for the Kalman filter converges to the solution of Equation (14) for $t \to \infty$ if (A, C) is detectable. In this case, the asymptotically stable observer gain is then computed [31] as

$$K = \left[CPC^T + R \right]^{-1} CPA. \tag{15}$$

5.1. Voltage Only Observer Design

In the case of voltage only estimation, we neglect the hysteresis dynamic term. This represents a typical model mismatch in voltage measurements. The states of the observer are $\hat{x}_t = [\hat{V}_1, \hat{V}_2, \hat{z}]^T$ given by the discretized version of Equations (8)–(10) and the PWL approximation of the nonlinearities in voltage measurement $\hat{y}_t$ is given by the equations in Appendix B. The values of the parameters in the voltage and force models can be found in Appendix C. The gain K for this observer is given as

$$K_V = \begin{bmatrix} K_1 & K_2 & K_3 \end{bmatrix}^T \tag{16}$$

The values for the gains K_i with $i = 1$–3 are obtained by tuning the Q matrix. The gain K_V is found for the eight regions in which the voltage is divided: R_1: $z \in [0, c_0]$, R_2: $z \in (c_0, c_1]$, R_3: $z \in (c_1, c_2]$, R_4: $z \in (c_2, c_3]$, R_5: $z \in (c_3, c_4]$, R_6: $z \in (c_4, c_5]$, R_7: $z \in (c_5, c_6]$, and R_8: $z \in (c_6, 1]$.

The main challenge in this system is the slow convergence of the estimation error due to the almost zero output gain (in C) especially the C(1,3) that corresponds to the $\frac{dV}{dz}$ in Equation (18). Increasing the K_v gain to compensate for the low state to output gain C governed by $\frac{dV}{dz}$ will amplify voltage sensor noise. Therefore, another observer is developed that uses voltage and force measurement since SOC and bias in the force signal are unobservable by force measurement only.

5.2. Voltage and Force Observer Design

In the case of force and voltage estimation, the states of the observer are $\hat{x}_t = [\hat{V}_1, \hat{V}_2, \hat{z}]^T$ and $\hat{y} = [\hat{V}_T, \hat{F}]^T$ given by the equations in Appendix B. The gains for the V&F observer have the following format

$$K_{VF} = \begin{bmatrix} K_{11} & K_{21} & K_{31} \\ K_{12} & K_{22} & K_{32} \end{bmatrix}^T \tag{17}$$

and the gain K_{VF} is found for the eight regions of the voltage, as explained in the previous subsection. The modeled force used for the state estimator, $\hat{F}$, is given by

$$\hat{F}_{sim}(\hat{z}) = \hat{C}_{dF}(\hat{z})\hat{z} + \hat{C}_0(\hat{z}) = \begin{cases} \alpha_m \hat{z} + \alpha_{m0}, & \text{if } \hat{z} \leq \hat{b}_L \\ \beta_m \hat{z} + \beta_{m0}, & \text{if } \hat{b}_L < \hat{z} \leq \hat{b}_H \\ \gamma_m \hat{z} + \gamma_{m0}, & \text{otherwise} \end{cases} \tag{18}$$

where $\hat{z}$ is the estimated SOC from our observer and $\hat{b}_L$ and $\hat{b}_H$ are the inflection points that are shifted by -10%, as shown by the dashed black lines in Figure 3 to emulate model uncertainty and aging. In the case of a "perfect" model, $\hat{b}_L = b_L$ and $\hat{b}_H = b_H$.

Switching Logic in Observer Design

The gain (K) of the LQE depends on the relationship between the estimated state and the slope of the measurement. In the case of a monotonic function, the estimated state will converge to the true state value when the estimator gain is chosen so that the error dynamics are stable. In the case of a non-monotonic function, the estimated state and the true state can have different slopes in the force output depending on the operating region. Therefore, the estimator gain would have a wrong sign which would lead to the divergence of the estimated state of charge from the true state. This is due to the traditional LQE using the modeled slope at the region of the estimated state. For example, consider the case where the model initialization occurs in the middle section of the SOC range (force decreases and cell contracts as SOC increases), whereas the actual SOC is in the high SOC range (force increases and cell swells as SOC increases); the traditional approach will lead to divergence of the estimated SOC state from the actual SOC due to the difference in slope. To address this difficulty, an algorithm was developed that uses a window of past measurements force data to identify the slope of the non-monotonic F–SOC. The observer gain will need to switch based on a judicious combination of the information at hand, namely

(a) the modeled slope $\hat{C}_{dF}$ from Equation (18) of the F–SOC relation based on the estimated state $\hat{z}$; and
(b) the estimation of the slope by using the measured force with respect to the Coulomb counting based SOC, $\frac{d\tilde{F}}{dz}$ (DFDZ).

Therefore, the output error injection gain K is a function of the state estimate and the estimated force derivative, e.g., $K(\hat{z}, \frac{d\tilde{F}}{dz})$, using two sources of information due to its importance in the convergence of the estimation error. The DFDZ is computed as a line fitting problem based on the moving window of past force measurements and the Coulomb counting based SOC integration ($\tilde{z}$) over the moving window. The fitted DFDZ line $\tilde{F} = \frac{d\tilde{F}}{dz}\tilde{z} + \tilde{F}_0$ parameters are computed using the least-squares estimation as

$$\begin{bmatrix} \frac{d\tilde{F}}{dz} \\ \tilde{F}_0 \end{bmatrix}_k = (\mathbb{L}_k^T \mathbb{L}_k)^{-1} \mathbb{L}_k^T \mathbb{F}_k \tag{19}$$

where $\tilde{F}_0$ is the affine parameter in Equation (18). The moving window of force measurements and the design matrix respectively are defined as follows

$$\mathbb{F}_k = \begin{bmatrix} F_{k-n} \\ F_{k-n+1} \\ \vdots \\ F_k \end{bmatrix}, \quad \mathbb{L}_k = \begin{bmatrix} \tilde{z}_{k-n} & 1 \\ \tilde{z}_{k-n+1} & 1 \\ \vdots & \vdots \\ \tilde{z}_k & 1 \end{bmatrix} \tag{20}$$

where k represents the discrete-time measurement index and n is the number of past samples. The Coulomb counting based SOC integration is computed as

$$\tilde{z}_k = \tilde{z}_{k-1} - \frac{I_{k-1} T_s}{C_b}. \tag{21}$$

The integration is initialized with $\tilde{z}_{k-n} = 0$. The moving window provides the $\frac{d\tilde{F}}{dz}$ computation as the average slope in a window of n prior values of force which causes a delay δ in the switching logic as a function of the window size n, as shown in Figure 5d. We can see that for $n = 150$ the δ is smaller compared to $n = 450$, but, with smaller n, the estimated slope is more susceptible to noise. We chose n here to be 150 samples with a discretized time of 1 s. The reasoning behind choosing this window size is explained in Section 5.3. Due to this delay and the fact that during this time the actual and the estimated state can be in different segments of the SOC, which could cause divergence, the gain is set to zero when the estimated slope $\frac{d\tilde{F}}{dz}$ has a different sign from the modeled slope. Therefore, the system will run open loop for both voltage and force when there is slope mismatch. This is done in order to avoid instability issues.

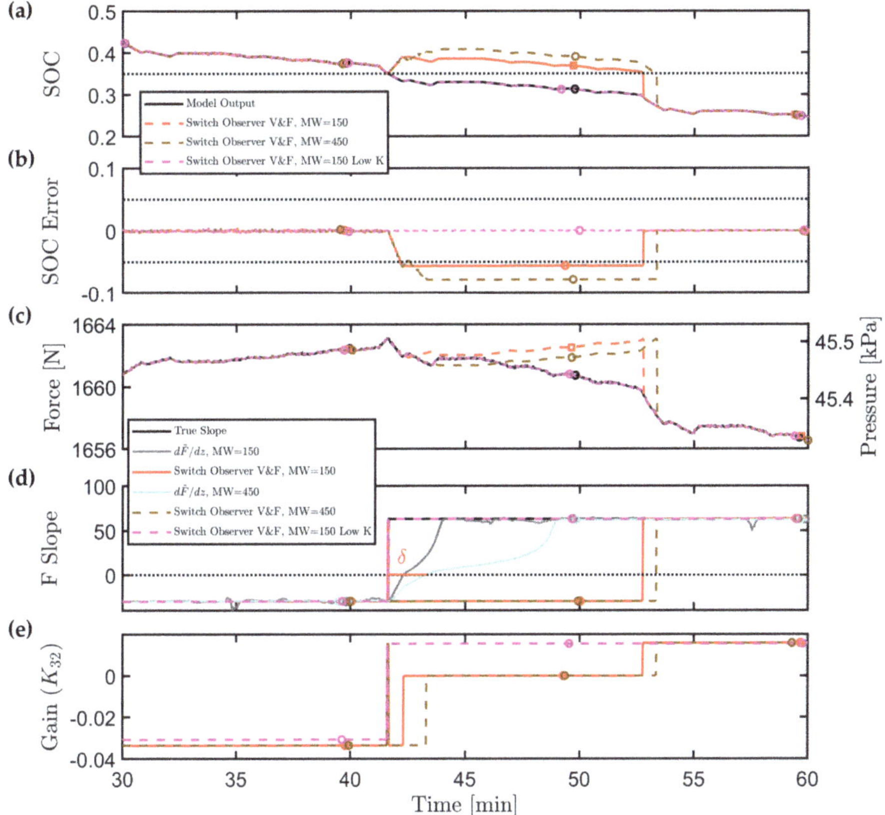

Figure 5. Simulation (without bias in the force measurement) for the Switch Observer V&F developed in our previous work [22]. This figure shows the impact of high gains corresponding to solution of the linear quadratic estimator using $Q_1 = \mathrm{diag}([2, 0.1, 0.1])$ and $R = \mathrm{diag}([1, 1])$ with moving windows of length (MW) = 150 and MW = 450. If the gain is lowered by using $Q_2 = \mathrm{diag}([2, 0.1, 1e-2])$, as shown by the purple dashed line, the error is decreased by 5% with same 150pt moving window length. (**a**) Comparison of the simulated state of charge (SOC). (**b**) Comparison of the state of charge (SOC) error with the dashed lines representing the target ±5% bound. (**c**) Comparison of the simulated force. For generalization of the results to the other cell sizes, the pressure is shown. (**d**) Comparison of the true slope with the estimated slope $\frac{d\tilde{F}}{dz}$ and observer output. Due to the greater delay (δ) before applying the zero gain for the MW = 450 case, a higher error in SOC estimation occurs during the transition from R_{II} to R_I. (**e**) Comparison of the feedback gain K_{32} from force to SOC based on the switching logic. When the observer and estimated slopes mismatch, the gain is set to zero.

The switching logic for our gain $(K(\hat{z}, \frac{d\tilde{F}}{dz}))$ is given by the following rules:

$$K_{VF}(3,2) = \begin{cases} K_{R_I} & \text{if } \frac{d\tilde{F}}{dz} > 0 \quad \text{and} \quad \hat{z} < \hat{b}_L \\ K_{R_{II}} & \text{if } \frac{d\tilde{F}}{dz} < 0 \quad \text{and} \quad \hat{b}_L < \hat{z} < \hat{b}_H \\ K_{R_{III}} & \text{if } \frac{d\tilde{F}}{dz} > 0 \quad \text{and} \quad \hat{z} > \hat{b}_H \\ 0 & \text{otherwise.} \end{cases} \tag{22}$$

Note that the gains K_{R_I}, $K_{R_{II}}$, and $K_{R_{III}}$ are not constant. The set of gains K_{R_I} consists of the four gains that correspond to the four regions in voltage in region R_I, $K_{R_{II}}$ consists of the two gains that correspond to the two regions in voltage in region R_{II}, and $K_{R_{III}}$ consists of the two gains that correspond to the two regions in voltage in region R_{III}, as shown in Figure 3. The gains are therefore a function of $\hat{z}$ and $\frac{d\tilde{F}}{dz}$ and they switch to the corresponding region depending on the value of the estimated slope and estimated SOC.

5.3. SOC Estimation Error during Switching

Going open loop, during the time interval when there is a slope mismatch, is not sufficient to avoid divergence of the estimated state of charge with high feedback gain. To verify this, we analyze the Luenberger Observer for the SOC state using the measured force only. The goal of this analysis is to determine the impact of the gain K on divergence of the state estimate when the true model and state estimate are operating on opposite regions of the output non-linearity. Using Equations (12) and (13), we can write the error dynamic ($e = z - \hat{z}_t$) for the observer assuming $\hat{x}$ is in R_I and x is in R_{II} as:

$$\hat{e}_{t+1} = A\hat{e}_t - K(y_t - \hat{y}_t) = A\hat{e}_t - K(\beta_m x + \beta_{m0} - \alpha_m \hat{x} - \alpha_{m0} + \alpha_m x - \alpha_m x) \tag{23}$$

$$\hat{e}_{t+1} = (A - K\alpha_m)\hat{e}_t - K((\beta_m - \alpha_m)x + \beta_{m0} - \alpha_{m0}). \tag{24}$$

From this error dynamic equation, we notice the error converges to a non-zero quantity so that e tends toward zero only when $x = b_L$, and the steady state error $e_{ss} = ((\beta_m - \alpha_m)x + \beta_{m0} - \alpha_{m0})/\alpha_m$ is independent of the gain K. In the case, where the sign of the model is updated based on the measured slope of the force signal $(\frac{d\tilde{F}}{dz})$, the growth in SOC estimation error is bounded by the number of samples in the filter or the moving window size (MW). If the current is bounded (which it is in our case), then the divergence in the state estimate is bounded by the integral of the current and the switching time δ. When the estimator model is updated to the correct slope, the observer begins to converge again. However, noise in the measurement of force could still result in divergence of the estimate if the gain K is too high. This is shown in Figure 5. We can see that just by changing the $MW = 150$ to $MW = 450$ by using the same gains given by $Q = \text{diag}([2, 0.1, 0.1])$ and $R = \text{diag}([1, 1])$. This is because the gain is large and the delay in $\frac{d\tilde{F}}{dz}$ crossing zero is 1 min greater than $MW = 150$. Therefore, the divergence in SOC, because of the slope mismatch, will grow for a longer period of time with increasing filter length and a higher error is achieved due to the delay in switching the gain to zero.

Now, if we decrease the gain by using $Q = \text{diag}([2, 0.1, 1 \times 10^{-2}])$ and $R = \text{diag}([1, 1])$ without bias estimation and $n = 150$, we notice that at the same region the error decreases by 5%. Therefore, low gains should be used to avoid the divergence and a window size of 150 is chosen since this is the lowest window that provides sufficient noise rejection. The previous error dynamic analysis can be done for the other mismatch slope areas to determine the minimum SOC state estimation error.

5.4. Bias Influence in SOC Estimation

In this section, the proposed estimation is tested under biased force measurements. In Figure 6, the simulated force data have a bias or drift of 3 N present in the force measurement. This drift in force affects the SOC estimation, as shown in Figure 6. Application of the switched model observer based on voltage and the force measurements based on previous work [22] without taking into account the bias estimation results in error larger than 5%. Therefore, we assumed a constant bias state ($\hat{f}_{d,t}$) that is given by

$$f_{d,t+1} = f_{d,t} \tag{25}$$

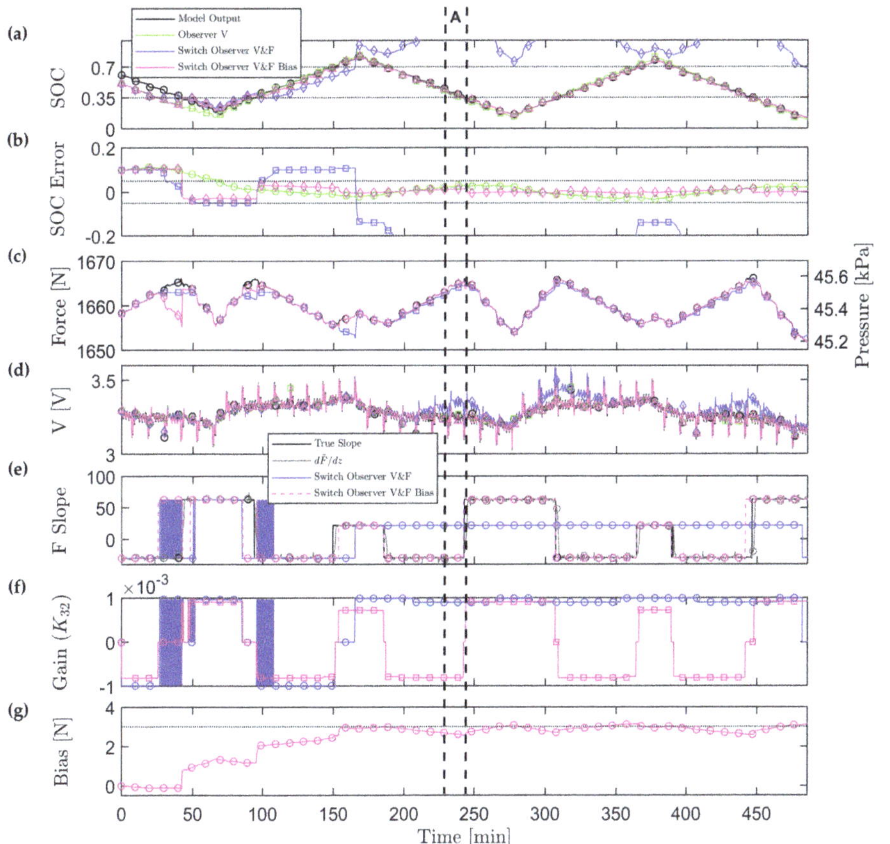

Figure 6. Simulation assuming accurate modeling using Switch Observer V&F Bias observer with an emulated bias of 3 N and 10 % (0.1) initial SOC error. (**a**) Comparison of the simulated state of charge (SOC). Is clear from the results that Switch Observer V&F without bias estimation diverges from the simulated SOC. Therefore, bias estimation is needed in the developed observer. (**b**) Comparison of the state of charge (SOC) error with the dashed lines representing the target ±5% bound. SOC errors greater than 20% are obtained with the Switch Observer V&F without bias estimation. It is shown that with bias estimation the SOC error is within the 5% estimation error bound (EEB). (**c**) Comparison of the simulated force and observer outputs. The bias state permits deviation in the force output, without compromising SOC estimation. For generalization of the results to the other cell sizes, the pressure is shown. (**d**) Comparison of the simulated terminal voltage. (**e**) Comparison of the true slope with the estimated slope $\frac{d\hat{F}}{dz}$ and observer output. (**f**) Comparison of the feedback gain K_{32} from force to SOC based on the switching logic. As shown in region A, denoted by the black dashed horizontal lines, when the observer and estimated slopes mismatch, the gain is set to zero. (**g**) The observer estimates the bias state which converges to the true bias as shown by the black horizontal dashed lines.

where $f_{d,t}$ is the constant bias or drift term that augments the electrical states. The estimator now estimated the electrical states (except hysteresis) and the force drift with $\hat{x}_t = [\hat{V}_1, \hat{V}_2, \hat{z}, \hat{f}_{d,t}]^T$. Therefore, the gains for the V&F Bias observer have the following format

$$K = \begin{bmatrix} K_{11} & K_{21} & K_{31} & K_{14} \\ K_{12} & K_{22} & K_{32} & K_{24} \end{bmatrix}^T \tag{26}$$

The values for K_{1i} and K_{2i} with $i = 1$–4 are obtained by tuning the Q and R matrices.

6. Simulation Results without Model Mismatch in F–SOC and OCV–SOC

The standard Dynamic Stress Test (DST) profile is repeated back to back and modified by adding a constant current to periodically recharge the battery at $1/6$ C rate, exercising a wider range of SOC, as shown in Figure 4a. A measurement noise variance of 5 mV for voltage and 0.05 N for force was chosen based on the variance of the experimental data. As for the drift value, we chose a value of 3 N based on the monthly drift observed between repeated characterization experiments. Our observer works if the initial error in bias is within 3 N. The LQE estimator is initialized with an SOC error of ±10% in order to evaluate convergence. In the case of the DST cycle, the true SOC state may be approaching the estimated value or diverging from the true SOC value depending on the initial SOC estimation error. The objective is to stay within the ±5% estimation error bound (EEB) for SOC denoted by the dashed lines in the Figure 6b. We chose to simulate initial conditions around 60–80% and the SOC swing of the whole cycle around 20–80% SOC. This range was chosen due to the challenge associated with flat OCV–SOC profile present for voltage (around 40–60%) and the negative slope in F–SOC (around 35–70%). Therefore, the simulated "measured" data are initialized at 61% SOC. The weights chosen to obtain the gains K_{1i} with $i = 1$–4 and K_{2i} with $i = 1$–4 are shown in Table 2. The gain K_{32} is shown in Figure 6f to illustrate when the algorithm uses the estimated slope mismatch to zero the observer gain and run open loop.

Table 2. Control weights for the different sensors using the simulated data. V, terminal voltage; V&F, fusion of both sensors; V&F Bias, fusion of both sensors with bias state.

Models	Q	R
V	$\mathrm{diag}(2, 10, 2)$	50
V&F	$\mathrm{diag}(2, 0.01, 0.01)$	diag(50,10,000)
V&F Bias	$\mathrm{diag}(2, 0.1, 0.01, 4)$	diag(50,10,000)

The SOC estimation using the V measurement and V&F measurement with and without the bias state estimation are shown in Figure 6a. In all three cases, a 10% initial estimation error is assumed. The inflection points in the force with respect to SOC are denoted by the dotted horizontal lines in Figure 6a. In this case, the correct values for b_L and b_H are used for the simulated data and the observer. The switched model applies zero gain (as shown in Figure 6f in Region A) when the model slope (based on $\hat{z}$) does not agree with the estimated slope using the least squares on the moving window with Equations (19) and (20). Using the previously developed V&F observer without bias estimation [22], the SOC error does not remain within the ±5% bound with a 3 N bias in the measurement, even though the error in the force signal is small, as shown in Figure 6a,c. For the proposed switching force-and voltage-based LQE with bias (Switch Observer V&F Bias) before $t = 50$ min there is an SOC estimation error of 10%, even though the force estimation is matching our simulated data, due the error in bias state estimation shown in Figure 6g. Between $t = 25$ and $t = 42$ min, there is a slope mismatch in $\frac{d\tilde{F}}{dz}$ due to the delay of our moving window. After $t = 42$ min, the slopes match again, and the correct non-zero feedback gain is applied and the state of charge estimation error convergences within our ±5% EEB, as shown in Figure 6b. Even though both the proposed switching force and voltage based LQE (Switch Observer V&F Bias) and the LQE based on voltage (Observer V) estimate voltage accurately (as shown in Figure 6d, it can be observed that the Switch Observer V&F Bias has faster convergence than the estimate based on V alone, as shown in Figure 6b. The faster convergence is due to the addition of the force signal that produces a lower output error injection gain from V to SOC in the regions where the voltage signal has flat slope. It can be appreciated from Figure 6g that the bias term is able to estimate the drift value slowly in our force measurement. The slow convergence of the bias estimate to the 3 N value denoted by the dotted black line can be seen in Figure 6g. Therefore, as time progresses, the estimated force converges to the modeled data, as shown in Figure 6c. This is due to the chosen inflection points in F–SOC function being the same as the modeled data ($\hat{b}_L = b_L = c_3$ and $\hat{b}_H = b_H = c_5$). We can observe that the estimated bias value oscillates around the true bias value. The model error is being attributed to the bias state by the estimation algorithm. The Switch Observer V&F Bias has lower root mean square error (RMSE), faster time convergence to the denoted SOC EEB (referred to as Time to 5% EEB in Table 3), and reduced maximum absolute SOC error after the force measurement has converged to an SOC error (referred to as Max SOC Error in Table 3) compared to Observer V as shown in Table 3. Therefore, the advantage of Switch Observer V&F Bias is the fast convergence in the region of 40–60% SOC while having more accurate SOC estimation due to the low RMSE values.

Table 3. Comparison of the RMSE index for different initial estimate error and different sensors using the simulated data. V, terminal voltage; V&F Bias, fusion of both sensors with bias state.

Initial SOC Error	Parameters	Observer V	Switch Observer V&F Bias
	Time to 5% EEB [min]	67.81	42.44
+10%	Max SOC Error [%]	3.32	3.22
	RMSE	0.0394	0.0337
	Time to 5% EEB [min]	41.33	7.86
−10%	Max SOC Error [%]	5.15	1.54
	RMSE	0.0321	0.0185

7. Simulation Results with Inflection Point Mismatch in F–SOC and OCV–SOC

In reality, we do not have a "perfect" model that captures the battery data. Moreover, during aging, the capacity loss shifts the inflection points as compared with the fresh cell. Therefore, to simulate a more realistic application model mismatch is included in the F–SOC and OCV–SOC observer by shifting the inflections points b_L and b_H of by −10% ($\hat{b}_L$ and $\hat{b}_H$) to represent capacity loss on the negative electrode [15]. During capacity loss, these inflections points shift for both functions of F–SOC and OCV–SOC because they correspond to the electrochemical and mechanical model having the same phase transitions. The aging effect on the swelling as the battery is cycled is that the inflection points will shift by approximately 10%. The proposed V&F bias SOC method will work if the initial unknown bias is within 3 N. The bias estimator will converge to the true value, when the SOC is outside the middle region (where there is a multiplicity of state of charge). A larger error in the initial bias will be "corrected" by visiting 100% or 0% SOC based on the cell voltage feedback. In the middle region (R_{II}), the feedback of force error is split between the SOC and bias in the middle region without a strong feedback from the terminal voltage and therefore can have a persistent SOC error. Checking the observer performance under inflection points mismatch due to capacity loss is important since it captures the robustness needed as the battery ages. Therefore, we want to check the performance of the developed Switching Observer V&F Bias under this model mismatch and both voltage and force have −10% inflections points shift. We initialize our simulated "measured" data at 61% SOC. The weights chosen to obtain the gains K_{1i} with $i = 1$–4 and K_{2i} with $i = 1$–4 are shown in Table 4. The gain K_{32} for feedback of the force error to the SOC state is shown in Figure 7f. The results from the "perfect" model observer (Switch V&F Bias), as shown in Figure 6, are compared with the observer with inflection mismatch (Switch V&F Bias Mismatch).

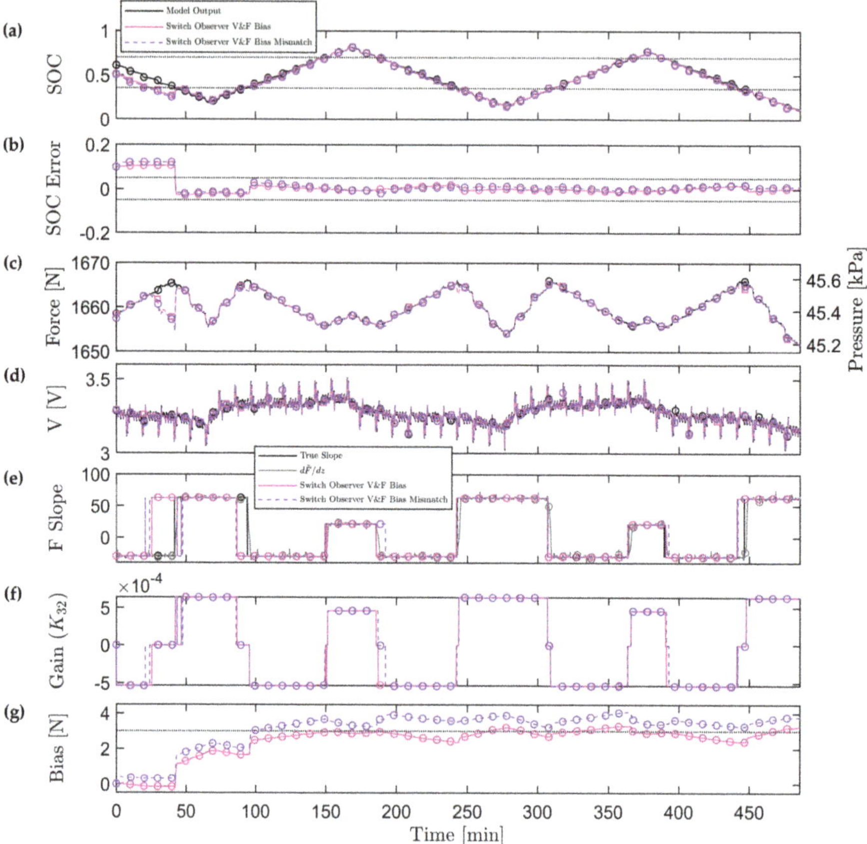

Figure 7. Simulation of the impact of model mismatch in the inflection points of the force vs. SOC curve. A 10% error in $\hat{b}_L$ and $\hat{b}_H$ is tested for the observer (Switch Observer V&F Bias) with an emulated bias of 3 N and 10% (0.1) initial SOC error. (**a**) Comparison of the simulated state of charge (SOC). (**b**) Comparison of the state of charge (SOC) error with the dashed lines representing the target $\pm5\%$ bound. The SOC error converges to within the $\pm5\%$ error bound. The largest SOC errors are observed near the switching points denoted by the horizontal dashed lines in Figure 7a. (**c**) Comparison of the simulated force and observer output. For generalization of the results to the other cell sizes, the pressure is shown. (**d**) Comparison of the simulated terminal voltage and observer output. (**e**) Comparison of the true slope with the estimated slope $\frac{d\hat{F}}{dz}$ and observer output. (**f**) Comparison of the feedback gain K_{32} from force to SOC based on the switching logic. The gain is set to zero when there is mismatch in the estimated slope and that based on the observer SOC. (**g**) The Bias state estimate converges slowly. The impact of model mismatch in the inflection points of the force vs. SOC curve can be seen by comparing the (Switch Observer V&F Bias Mismatch) and (Switch Observer V&F Bias) which uses the correct value. The accurate observer converges to the true value of 3 N, denoted by the dashed horizontal line whereas mismatch leads to a constant over-estimate of about 1 N.

Table 4. Estimation Weights for the different sensors using the simulated data with and without model mismatch. V&F Bias fusion of both sensors without mismatch; V&F Bias Mismatch, fusion of both sensors with model mismatch.

Models	Q	R
V&F Bias	$\mathrm{diag}(9, 0.01, 0.005, 3)$	$\mathrm{diag}(50, 10{,}000)$
V&F Bias Mismatch	$\mathrm{diag}(9, 0.01, 0.005, 3)$	$\mathrm{diag}(50, 10{,}000)$

The SOC estimation using the V measurement and V&F Bias measurement with 10% initial estimation error are shown in Figure 7a. The inflection points in the force with respect to SOC are denoted by the dotted horizontal lines in Figure 7a. As shown in Table 5, we obtain lower RMSE, faster convergence to the desired SOC estimation error bound (EEB), and smaller absolute SOC error after the force measurement has converged to an SOC error value for the V&F Bias observer compared to V&F Bias Mismatch observer when initialized at 10%. For -10% SOC initialization, the V&F Bias Mismatch observer has lower RMSE and lower time convergence to the desired SOC estimation error bound (EEB) compared to V&F Bias, as shown in Figure 7b. The force bias estimate for V&F Bias Mismatch converges faster to the constant bias value of 3N. Due to this faster convergence, the RMSE and the convergence to the desired SOC estimation error bound (EEB) is lower for the model with mismatch due to the initial condition. In both -10% and 10% SOC initialization, the maximum absolute SOC error after the force measurement has converged to an SOC error lower for V&F Bias than V&F Bias Mismatch. The non-zero state estimation error is due to the mismatch present in the F–SOC and OCV–SOC function ($b_L \neq c_3$ and $b_H \neq c_5$). To understand this, we need to analyze the error dynamics equation. We know from Section 3 the form of our model and the observer form is given in Equation (12). Therefore, denoting our error as $e = x - \hat{x}$ and using our model and observer model equations, we obtain the dynamic error equation as

$$\dot{e} = (A - KC)e - K\Delta C\hat{x} - K\Delta C_0 \tag{27}$$

where $\Delta C = C - \hat{C}$ and $\Delta C_0 = C_0 - \hat{C}_0$. From the dynamic error equation, we notice that the bias error and the SOC error will not converge to 0 due to the terms $-K\Delta C_0$. Therefore, the bias will converge to a value that is not the true value of the drift due to this error, as shown in Figure 7g. The SOC error will converge to a value but it will not converge to 0, as shown in Figure 7b, due to the model mismatch. The magnitude of the estimation error varies depending on the force region we are operating in. The SOC estimation error, due to model mismatch, will grow as the force sensor drift increases due to the terms $-K\Delta C_0$. For the given model tuning and 10% shift in the force curve with respect to SOC, the force-based observer only achieves better SOC estimation than the voltage only case if the uncorrected force sensor drift is less than 3 N initially.

Table 5. Comparison of the RMSE index for different initial estimate error and different sensors using the simulated data with and without model mismatch. V, terminal voltage; V&F Bias, fusion of both sensors; V&F Bias Mismatch, fusion of both sensors with model mismatch.

Initial SOC Error	Parameters	Observer V&F Bias	Switch Observer V&F Bias Mismatch
	Time to 5% EEB [min]	42.5	42.55
$+10\%$	Max SOC Error [%]	2.79	3.01
	RMSE	0.0327	0.0377
	Time to 5% EEB [min]	8.38	5
-10%	Max SOC Error [%]	2.38	4.31
	RMSE	0.0200	0.0177

8. Experimental Data Results Using F–SOC and OCV–SOC

With a 10% initial error in our SOC estimate, we obtain the results shown in Figure 8. The observer was initialized to 51% SOC, whereas the true state was 61% SOC, to highlight the performance in the middle SOC region where voltage based techniques are less effective. The current waveform discussed in Section 2 was applied to the battery and the observer. As in the simulated results, the proposed Switch Observer V&F Bias has faster convergence than the SOC estimation based on V alone as shown in Figure 8b. The V&F Bias observer exhibits lower RMSE, faster time convergence to the denoted SOC estimation error bound (EEB), and reduced maximum absolute SOC error, as shown in Table 6 compared to V observer. The bias term oscillates around the estimated bias value of 6.8 N, as shown in Figure 8g. These larger oscillations in the bias could be due to our force data having additional dynamics besides the bias term. According to [13], the force has a dynamic term that is temperature dependent, and the ambient chamber temperature may oscillate within $\pm 1\ ^\circ$C. Therefore, this dynamic term should be added to our force model. There are some large errors in SOC estimation (approximately around 15%) at low SOC. This is due to our piecewise linear (PWL) fit.

Table 6. Comparison of the RMSE index for different initial estimate error and different sensors for data validation. V, terminal voltage; V&F Bias, fusion of both sensors.

Initial SOC Error	Parameters	Observer V	Switch Observer V&F
+10%	Time to 5% EEB [min]	95.5	0.18
	Max SOC Error [%]	11.89	11.79
	RMSE	0.0741	0.0611
−10%	Time to 5% EEB [min]	6.95	6.54
	Max SOC Error [%]	11.89	11.79
	RMSE	0.0594	0.0574

In the area near 20% SOC, the PWL fit is less accurately, as shown in Figure 1, which results in increased SOC estimation error of around 10%. To better capture the non-linearity, the piecewise linear approximation could be further divided into more regions to provide a better fit. The weights chosen to obtain the gains K_{1i} with $i = 1$–4 and K_{2i} with $i = 1$–4 are shown in Table 7. The gain K_{32} that satisfies sign or is zero is shown in Figure 8f. The bias estimation state is initialized at 10 N.

Table 7. Control Weights for the different sensors for data validation: V, terminal voltage; V&F Bias, fusion of both sensors.

Models	Q	R
V	$\mathrm{diag}(5, 0.1, 0.1)$	50
V&F Bias	$\mathrm{diag}(1, 1 \times 10^{-2}, 0.09, 50)$	$R = \mathrm{diag}(50, 10{,}000)$

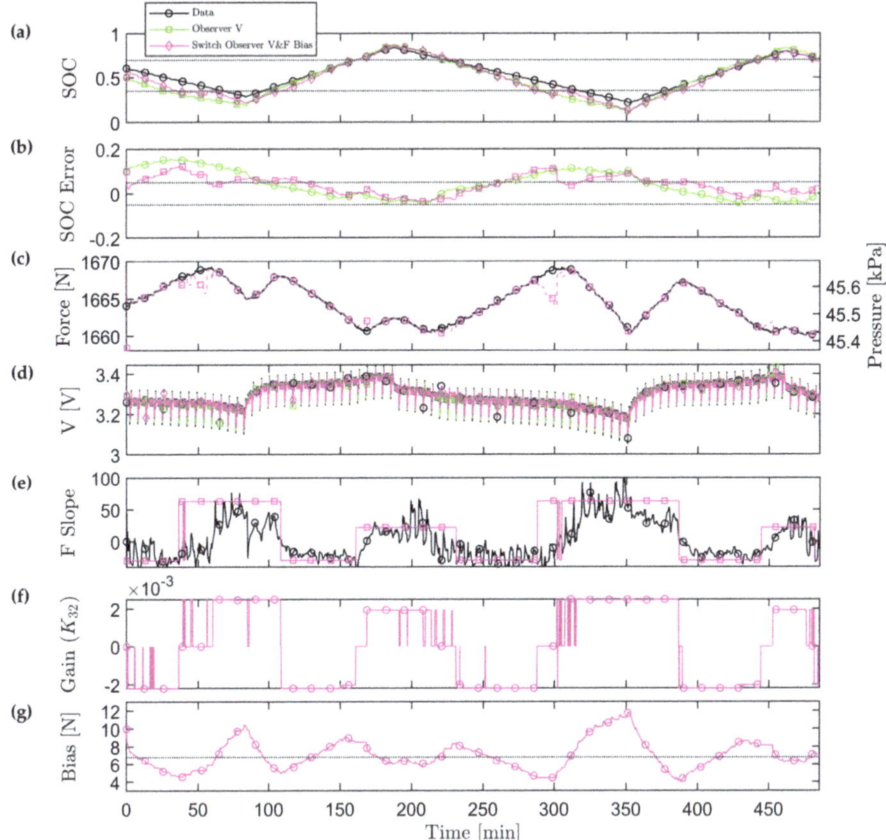

Figure 8. Experimental validation of the developed observer (Switch Observer V&F Bias) with 10% (0.1) initial SOC error. The offline experimentally measured current, voltage, and force data were fed into the model to assess performance. (**a**) Comparison of high accuracy coulomb counting based state of charge (SOC) with observer estimates. (**b**) Comparison of the state of charge (SOC) error with the dashed lines representing the target ±5% bound. The observer with force bias state estimation demonstrates better performance. The large error at low SOC for both observers is due to model mismatch the piecewise linear OCV–SOC fit at low SOC. (**c**) Comparison of measured and estimated force. (**d**) Comparison of the experimental and observer modeled terminal voltage. (**e**) Comparison of the estimated slope $\frac{d\hat{F}}{dz}$ and observer model. (**f**) Comparison of the feedback gain K_{32} from force to SOC based on the switching logic. (**g**) The Bias state estimate fluctuates around the average value (denoted by the horizontal dashed line). This could be due model mismatch, where the force error is exciting the bias state estimate.

9. Conclusions

In this paper, a new switching estimator design for a battery with the lithium ion LFP chemistry that integrates the non-monotonic F–SOC relation is proposed, verified by simulation, and validated using experimental data with respect to the SOC estimation accuracy. The estimator is based on switching PWL models that are scheduled according to the identified slope of the F–SOC operating point with a bias state in order to capture the drift exerted by the fixture and battery in our force measurement. Two different sensor scenarios, namely V and V&F Bias fusion, are compared, where it is concluded that the V&F Bias sensor improves the rate of SOC estimator convergence. This is due to the information added from the steeper, hence more informative, F–SOC characteristic than OCV–SOC relation despite large errors in the voltage and force models. The bounds on SOC estimation accuracy depend on the chosen inflection points of the F–SOC function. If they are correct, then the accuracy is better for the observer with both F and V than V only. Our future work will focus on determining if the drift on our force measurement is due to creep exerted by the poron and thermal expansion of the fixture or due to creep exerted by a degraded battery influenced by compressive stresses [32], or a combination of both. The thermal expansion term and swelling dynamic as a function of temperature will be considered. Data from an aged swelling cell will be used with the developed SOC estimator and the State of Health of the battery (SOH) will be estimated through the capacity fade. The inflection point model mismatch on the F–SOC function will be further studied in an aged cell since the inflection points change as the battery degrades or fades. Due to the tight manufacturing tolerances, the variability in thickness should not be a significant contributor to force measurement uncertainty. In terms of aging, the expected variability in the cell expansion is a subject of future studies. Cell-to-cell variability due to aging in the resulting force should also be investigated in future work based on initial findings from [14,15]. Finally, the measured force is the result of all cell's expansion (summation) in a constrained module, therefore different levels of degradation for individual cells when charged in series would result in smoothing (convolution) of the force sensed on the module-level.

10. Disclaimer

Reference herein to any specific commercial company, product, process, or service by trade name, trademark, manufacturer, or otherwise, does not necessarily constitute or imply its endorsement, recommendation, or favoring by the United States Government or the Department of the Army (DA). The opinions of the authors expressed herein do not necessarily state or reflect those of the United States Government or the DA, and shall not be used for advertising or product endorsement purposes.

Author Contributions: Methodology, Software Simulation, Experimental validation, Visualization, and Writing—original draft preparation, M.A.F.-S.; Conceptualization, Methodology, Supervision, Funding acquisition, and Writing—review and editing, J.B.S.; and Conceptualization, Supervision, Funding acquisition, and Project administration, A.G.S. All authors have read and agreed to the published version of the manuscript.

Funding: This research was funded by the National Science Foundation (NSF) Graduate Research Fellowship Program and the DOE's Advanced Research Projects Agency-Energy (ARPA-E) AMPED program under award number DE-AR0000269.

Acknowledgments: The authors would like to acknowledge funding from the National Science Foundation (NSF), specifically the Graduate Research Fellowship Program, the technical and financial support of the Automotive Research Center (ARC) in accordance with Cooperative Agreement W56HZV-14-2-0001, and the U.S. Army Combat Capabilities Development Command (CCDC) Ground Vehicle Systems Center (GVSC) in Warren, MI.

Conflicts of Interest: The authors declare no conflict of interest.

Abbreviations

The following abbreviations are used in this manuscript:

V	terminal voltage
SOC, z	state of charge
LFP	lithium ion iron phosphate
F	force
LQE	linear quadratic estimator
DST	Dynamic Stress Test
BMS	battery management system
Ah	ampere-hours
UAV	unmanned air vehicles

OCV   open circuit voltage
NMC   nickel manganese cobalt
SOH   state of health
PWL   piecewise linear
CCCV   Constant-Current/Constant-Voltage
DFDZ   force derivative or slope of the measured force with respect to the SOC
N   Newton
EEB   estimation error bound
Max   maximum
RMS   root mean square
RMSE   root mean square error
RTDs   resistance temperature detectors
DARE   Discrete Algebraic Riccati Equation
MW   moving window
HEV   hybrid electric vehicle
EV   electric vehicle
HPPC   Hybrid Pulse Power Characterization

Appendix A. Voltage Hysteresis State H Function

$$H(z, sign(I)) = sign(I)(a_{12}z^{12} + a_{11}z^{11} + a_{10}z^{10} + a_9 z^9 + a_8 z^8 + a_7 z^7 + a_6 z^6 + a_5 z^5$$
$$+ a_4 z^4 + a_3 z^3 + a_2 z^2 + a_1 z + a_0) \tag{A1}$$

where a_i with $i = 0$–12 are the tuned parameters and their values are found in Table A1.

Table A1. Voltage hysteresis state H function parameters and its values.

Parameters	Values	Parameters	Values	Parameters	Values
a_{12}	8662.54	a_8	319,201.70	a_4	9255.46
a_{11}	$-57{,}939.63$	a_7	$-234{,}698.07$	a_3	-1381.64
a_{10}	170,685.13	a_6	117,809.10	a_2	127.23
a_9	$-291{,}398.86$	a_5	$-40{,}316.61$	a_1	-6.53
				a_0	0.16

Appendix B. Discrete Terminal Voltage Model

The discrete terminal voltage measurement equation is defined as

$$V_{T,t} = \tilde{V}_{oc}(z_t) - IR - V_{1,t} - V_{2,t} + v_{V,t} \tag{A2}$$

where $v_{V,t}$ is the V measurement noise. The piecewise linear (PWL) approximation of the OCV characteristic is modeled as

$$\tilde{V}_{oc}(z_t) = \begin{cases} \zeta z_t + \zeta_0 & \text{if } z_t \le c_0 \\ \eta z_t + \eta_0 & \text{if } c_0 < z_t \le c_1 \\ \theta z_t + \theta_0 & \text{if } c_1 < z_t \le c_2 \\ \kappa z_t + \kappa_0 & \text{if } c_2 < z_t \le c_3 \\ \sigma z_t + \sigma_0 & \text{if } c_3 < z_t \le c_4 \\ \mu z_t + \mu_0 & \text{if } c_4 < z_t \le c_5 \\ \varphi z_t + \varphi_0 & \text{if } c_5 < z_t \ge c_6 \\ \lambda z_t + \lambda_0 & \text{if } z_t \ge c_6 \end{cases} \tag{A3}$$

where ζ, η, θ, κ, σ, μ, φ, and λ are the slope parameters; ζ_0 is the minimum voltage sensed at fully discharged state; and c_0, c_1, c_2, c_3, c_4, c_5, and c_6 are the piecewise point parameters. The parameters

η_0, θ_0, κ_0, σ_0, μ_0, and λ_0 are uniquely determined from the other parameters via constraints of piecewise continuity.

$$\eta_0 = (\zeta - \eta)c_0 + \zeta_0 \tag{A4}$$

$$\theta_0 = (\eta - \theta)c_1 + (\zeta - \eta)c_0 + \zeta_0 \tag{A5}$$

$$\kappa_0 = (\theta - \kappa)c_2 + (\eta - \theta)c_1 + (\zeta - \eta)c_0 + \zeta_0 \tag{A6}$$

$$\sigma_0 = (\kappa - \sigma)c_3 + (\theta - \kappa)c_2 + (\eta - \theta)c_1 + (\zeta - \eta)c_0 + \zeta_0 \tag{A7}$$

$$\mu_0 = (\sigma - \mu)c_4 + (\kappa - \sigma)c_3 + (\theta - \kappa)c_2 + (\eta - \theta)c_1 + (\zeta - \eta)c_0 + \zeta_0 \tag{A8}$$

$$\varphi_0 = (\mu - \varphi)c_5 + (\sigma - \mu)c_4 + (\kappa - \sigma)c_3 + (\theta - \kappa)c_2 + (\eta - \theta)c_1 + (\zeta - \eta)c_0 + \zeta_0 \tag{A9}$$

$$\lambda_0 = (\varphi - \lambda)c_6 + (\mu - \varphi)c_5 + (\sigma - \mu)c_4 + (\kappa - \sigma)c_3 + (\theta - \kappa)c_2 + (\eta - \theta)c_1 + (\zeta - \eta)c_0 + \zeta_0 \tag{A10}$$

The OCV characteristic and its PWL approximation are shown in Figure 3. Note that c_3 and c_5 also correspond to the inflection points in the PWL approximation of the force. The reason for this is that both functions of F–SOC and OCV–SOC have the same inflection points due to the electrochemical and mechanical model having the same phase transitions.

Appendix C. Force and OCV Function Values that Represent Average Data

Table A2. Parameter values for the F–SOC function that represent average data. This model is also used for PWL F–SOC without model mismatch (b_L and b_H) and with model mismatch ($\hat{b}_L$ and $\hat{b}_H$).

Parameters	Values	Parameters	Values	Parameters	Values
α_m	63.11	γ_m	21.78	$\hat{b}_L$	0.34
α_{m0}	1641	b_L	0.35	$\hat{b}_H$	0.69
β_m	−29.53	b_H	0.7		

Table A3. Parameter values for the OCV–SOC function in Equation (7).

Parameters	Values	Parameters	Values	Parameters	Values
V_0	−2.4354	$a_{v,1}$	0.0206	$a_{v,3}$	0.0166
d	0.1162	$b_{v,1}$	0.2321	$b_{v,3}$	0.6799
f	5.7469	$c_{v,1}$	0.0626	$c_{v,3}$	0.0306
h	1.2942	$a_{v,2}$	5.6185		
k	3.0014×10^{-4}	$b_{v,2}$	−0.0513		
g	−0.0098	$c_{v,2}$	0.0406		

Table A4. Parameter values for the PWL OCV–SOC function in Equation (A3) without model mismatch. Same values are used with model mismatch except for $c_3 = 0.34$ and $c_5 = 0.69$.

Parameters	Values	Parameters	Values	Parameters	Values
ζ	9.232	μ	0.02836	c_3	0.35
ζ_0	2.622	φ	2.264	c_4	0.6511
η	0.3899	λ	0.03506	c_5	0.7
θ	0.4922	c_0	0.06	c_6	0.9767
κ	0.2724	c_1	0.1516		
σ	0.5515	c_2	0.2528		

References

1. Cheng, K.W.E.; Divakar, B.P.; Wu, H.; Ding, K.; Fai, H. Battery-Management System (BMS) and SOC Development for Electrical Vehicles. *IEEE Trans. Veh. Technol.* **2011**, *60*, 76–88. [CrossRef]
2. Rahimi-Eichi, H.; Ojha, U.; Baronti, F.; Chow, M.Y. Battery Management System: An Overview of Its Application in the Smart Grid and Electric Vehicles. *IEEE Ind. Electron. Mater.* **2013**, *7*, 4–16. [CrossRef]

3. Chen, Z.; Fu, Y.; Mi, C. State of Charge Estimation of Lithium-Ion Batteries in Electric Drive Vehicles Using Extended Kalman Filtering. *IEEE Trans. Veh. Technol.* **2013**, *62*, 1020–1030. [CrossRef]

4. Zubi, G.; Dufo-López, R.; Carvalho, M.; Pasaogluc, G. The lithium-ion battery: State of the art and future perspectives. *Renew. Sustain. Energy Rev.* **2018**, *89*, 292–308. [CrossRef]

5. Satyavani, T.V.S.L.; Kumar, A.S.; Rao, P.S. Methods of synthesis and performance improvement of lithium iron phosphate for high rate Li-ion batteries: A review. *Int. J. Eng. Sci. Technol.* **2016**, *19*, 178–188. [CrossRef]

6. Malik, R.; Abdellahi, A.; Ceder, G. A critical review of the Li insertion mechanisms in LiFePO4 electrodes. *J. Electrochem. Soc.* **2013**, *160*, A3179. [CrossRef]

7. Xiong, R.; He, H.; Sun, F.; Zhao, K. Evaluation on State of Charge Estimation of Batteries With Adaptive Extended Kalman Filter by Experiment Approach. *IEEE Trans. Veh. Technol.* **2013**, *62*, 108–117. [CrossRef]

8. Zhang, C.; Wang L.; Li, X.; Chen, W.; Yin, G.; Jiang, J. Robust and Adaptive Estimation of State of Charge for Lithium-Ion Batteries. *IEEE Trans. Ind. Electron.* **2015**, *62*, 4948–4957. [CrossRef]

9. Mohan, S.; Kim, Y.; Stefanopoulou, A. On Improving Battery State of Charge Estimation Using Bulk Force Measurements. In Proceedings of the ASME 2015 Dynamic Systems and Control Conference, Columbus, OH, USA, 28–30 October 2015.

10. Jones, E.M.C.; Silberstein, M.N.; White, S.R.; Sottos, N.R. In Situ Measurements of Strains in Composite Battery Electrodes during Electrochemical Cycling. *Exp. Mech.* **2014**, *54*, 971–985. [CrossRef]

11. Cannarella, J.; Leng, C.Z.; Arnold, C.B. On the Coupling between Stress and Voltage in Lithium-Ion Pouch Cells. In Proceedings of the SPIE Sensing Technology + Applications Energy Harvesting and Storage: Materials, Devices, and Applications, Baltimore, MD, USA, 5–9 May 2014.

12. Oh, K.; Siegel, J.; Secondo, L.; Kim, S.; Samad, N.; Qin, J.; Anderson, D.; Garikipati, K.; Knobloch, A.; Epureanu, B.; et al. Rate Dependence of Swelling in Lithium-Ion Cells. *J. Power Sources* **2014**, *267*, 197–202. [CrossRef]

13. Mohan, S.; Kim, Y.; Siegel, J.B.; Samad, N.A.; Stefanopoulou, A.G. A phenomenological model of bulk force in a li-ion battery pack and its application to state of charge estimation. *J. Electrochem. Soc.* **2014**, *161*, A2222–A2231. [CrossRef]

14. Samad, N.A.; Kim, Y.; Siegel, J.B.; Stefanopoulou, A.G. Battery capacity fading estimation using a force-based incremental capacity analysis. *J. Electroche. Soc.* **2016**, *163*, A1584–A1594. [CrossRef]

15. Mohtat, P.; Lee, S.; Siegel, J.B.; Stefanopoulou, A.G. Towards Better Estimability of Electrode-Specific State of Health: Decoding the Cell Expansion. *J. Power Sources* **2019**, *427*, 101–111. [CrossRef]

16. PORON® 4701-30 Polyurethane. Available online: https://rogerscorp.com/elastomeric-material-solutions/poron-industrial-polyurethanes/poron-4701-30 (accessed on 26 April 2020).

17. Kim, T.; Qiao, W.; Qu, L. A series-connected self-reconfigurable multicell battery capable of safe and effective charging/discharging and balancing operations. In Proceedings of the 2012 Twenty-Seventh Annual IEEE Applied Power Electronics Conference and Exposition (APEC), Orlando, FL, USA, 5–9 February 2012.

18. Shousha, C.M.; McRae, T.; Prodić, A.; Marten, V.; Milios, J. Design and Implementation of High Power Density Assisting Step-Up Converter With Integrated Battery Balancing Feature. *IEEE Trans. Emerg. Sel. Top. Power Electron.* **2017**, *5*, 1068–1077. [CrossRef]

19. Kim, Y.; Samad, N.A.; Oh, K.Y.; Siegel, J.B.; Epureanu, B.I.; Stefanopoulou, A.G. Estimating State-of-Charge Imbalance of Batteries Using Force Measurements. In Proceedings of the 2016 American Control Conference (ACC), Boston, MA, USA, 6–8 July 2016.

20. Huang, W.; Qahouq, J.A.A. Energy sharing control scheme for state-of-charge balancing of distributed battery energy storage system. *IEEE Trans. Ind. Electron.* **2015**, *62*, 2764–2776. [CrossRef]

21. Knobloch, A.; Kapusta, C.; Karp, J.; Plotnikov, Y.; Siegel, J.B.; Stefanopoulou, A.G. Fabrication of Multimeasurand Sensor for Monitoring of a Li-Ion Battery. *J. Electron. Packag.* **2018**, *140*, 031002. [CrossRef]

22. Polóni, T.; Figueroa-Santos, M; Siegel, J; Stefanopoulou, A. Integration of Non-Monotonic Cell Swelling Characteristic for State-of-Charge Estimation. In Proceedings of the 2018 Annual American Control Conference (ACC), Milwaukee, WI, USA, 27–29 June 2018.

23. Lee, S.; Mohtat, P.; Siegel, J.; Stefanopoulou, A. Beyond Estimating Battery State of Health: Identifiability of Individual Electrode Capacity and Utilization. In Proceedings of the 2018 Annual American Control Conference (ACC), Milwaukee, WI, USA, 27–29 June 2018.

24. Usabc Electric Vehicle Battery Test Procedures Manual, Appendix j-Detailed Procedure. Available online: https://www.uscar.org/guest/article_view.php?articles_id=74 (accessed on 26 April 2020).

25. Domenico, D.D.; Stefanopoulou, A.; Fiengo, G. Lithium-Ion Battery State of Charge and Critical Surface Charge Estimation Using an Electrochemical Model-Based Extended Kalman Filter. *J. Dyn. Syst.-Trans. ASME* **2010**, *132*, 061302. [CrossRef]

26. Forman, J.C.; Moura, S.J.; Stein, J.L.; Fathy, H.K. Genetic parameter identification of the Doyle-Fuller-Newman model from experimental cycling of a LiFePO4 battery. In Proceedings of the 2011 American Control Conference, San Francisco, CA, USA, 29 June–1 July 2011.

27. Pletcher, D. *Electrochemistry: Volume 8*; Royal Society of Chemistry: London, UK, 2007.

28. Hu, X.; Li, S.; Peng, H. A Comparative Study of Equivalent Circuit Models for Li-Ion Batteries. *J. Power Sources* **2012**, *198*, 359–367. [CrossRef]

29. Perez, H.E.; Siegel, J.B.; Lin, X.; Stefanopoulou, A.G.; Ding, Y.; Castanier, M.P. Parameterization and Validation of an Integrated Electro-Thermal Cylindrical LFP Battery Model. In Proceedings of the ASME 2012 5th Annual Dynamic Systems and Control Conference Joint with the JSME 2012 11th Motion and Vibration Conference, Fort Lauderdale, FL, USA, 17–19 October 2012.

30. Plett, G.L. *Battery Management Systems, Volume I: Battery Modeling*; Artech House: Boston, MA, USA, 2015.

31. Lewis, F.L. *Optimal Estimation: With an Introduction to Stochastic Control Theory*; Wiley: New York, NY, USA, 1986.

32. Peabody, C.; Arnold, C.B. The Role of Mechanically Induced Separator Creep in Lithium-Ion Battery Capacity Fade. *J. Power Sources* **2011**, *196*, 8147–8153. [CrossRef]

Article

A Method for the Combined Estimation of Battery State of Charge and State of Health Based on Artificial Neural Networks

Angelo Bonfitto

Department of Mechanical and Aerospace Engineering, Politecnico di Torino, 10129 Torino, Italy;
angelo.bonfitto@polito.it; Tel.: +39-011-090-6239

Received: 25 March 2020; Accepted: 17 May 2020; Published: 18 May 2020

Abstract: This paper proposes a method for the combined estimation of the state of charge (SOC) and state of health (SOH) of batteries in hybrid and full electric vehicles. The technique is based on a set of five artificial neural networks that are used to tackle a regression and a classification task. In the method, the estimation of the SOC relies on the identification of the ageing of the battery and the estimation of the SOH depends on the behavior of the SOC in a recursive closed-loop. The networks are designed by means of training datasets collected during the experimental characterizations conducted in a laboratory environment. The lithium battery pack adopted during the study is designed to supply and store energy in a mild hybrid electric vehicle. The validation of the estimation method is performed by using real driving profiles acquired on-board of a vehicle. The obtained accuracy of the combined SOC and SOH estimator is around 97%, in line with the industrial requirements in the automotive sector. The promising results in terms of accuracy encourage to deepen the experimental validation with a deployment on a vehicle battery management system.

Keywords: battery; state of charge; state of health; artificial intelligence; artificial neural networks; hybrid vehicles; electric vehicles; estimation

---

1. Introduction

The automotive industry is recently dedicating increasing attention to sustainability, with the objective of mitigating the negative effects of vehicular mobility on the environment. Carmakers cope with the always more stringent regulations about CO_2 emissions, focusing their efforts on the development of advanced powertrain architectures [1,2]. Solutions based on the adoption of full electric (battery electric vehicles (BEVs)) powertrains or on the combination of an internal combustion engine (ICE) and electric traction (hybrid/plug-in hybrid electric vehicles (HEVs/PHEVs)) are now established as reliable alternatives to conventional powertrains [3,4]. They exploit batteries as the primary energy source in BEVs or as an auxiliary source in HEVs and PHEVs [5]. In the automotive industry, the most common battery technology exploits lithium because of its remarkable advantages in terms of the energy density, fast charging, low maintenance, and long lifetime allowances. Moreover, lithium-based solutions allow for obtaining powerful, compact, and light configurations together with satisfactory levels of autonomy, which is currently settled in the order of a few hundreds of kilometers [6]. However, the reliability and performance of these type of batteries are strongly influenced by the management of the charging and discharging phases. It is indeed well known that an appropriate handling of these operations is mandatory to avoid the occurrence of overcharging or deep discharging, that would lead to permanent or hardly reversible damages of the pack. A continuous and accurate monitoring of the battery state takes on significant importance to extend the battery lifetime, effectively plan the trip route and charging stops, optimize the energy flow management of HEVs [7,8], and mitigate psychological effects, such as the range anxiety that is commonly experienced by a large number of BEV drivers [6]. The main parameters to be assessed for a correct battery monitoring are the residual available energy in the pack, known as state of charge (SOC), and the degradation suffered by the battery, indicated by the state of health (SOH) [9]. As is well known, these two states cannot be directly measured, since the technology to make a sensor that plays the equivalent role of a fuel gauge is not available. Therefore, the adoption of some estimation techniques becomes mandatory [10,11]. Typically, carmakers exploit look-up tables (LUTs), where the SOC and SOH behavior is mapped during the preliminary experimental characterizations conducted in a laboratory environment. These tests are done following the so-called direct methods, which are based on ampere-hour counting or the measurement of the internal impedance and open circuit voltage of the battery [10,12]. However, the adoption of LUTs may have a high computational cost and imposes the storage of a huge amount of data in the electronic control unit memory, particularly in the case of the SOH estimation. A further class of methods exploits model-based techniques for the real-time assessment of both the SOC and SOH [13]. The most common are the Kalman filter [14] and its derivations, namely the extended (EKF) [15] and unscented Kalman filters (UKF) [16,17], the adaptive particle filter (APF) [18], and the smooth variable structure filter (SVSF) [19]. Although these solutions can be implemented in real time on a vehicle, they may suffer problems of inaccuracies if the reference model is not completely and accurately tuned in all the possible operating conditions. An alternative and promising approach to overcome this limitation is represented by artificial intelligence (AI). In most cases, these solutions adopt artificial neural networks (ANNs) and allow getting rid of the model while obtaining satisfactory levels of accuracy and reliability, provided that the networks are properly trained. An extensive literature is dedicated to the methods for the estimation of the SOC [20–23] or SOH [24–27] with AI. Nevertheless, to the best of the author's knowledge, very few works deal with the combined estimation of the SOC and SOH and most of them describe model-based techniques [28–30].

This paper proposes a technique for the combined estimation of the SOC and SOH with a set of five ANNs: four regression networks dedicated to the SOC estimation and one classification network for the SOH identification. The method is independent by the battery model and is designed with a training phase conducted with datasets obtained from the preliminary laboratory experimental characterizations. The SOC estimation exploits four nonlinear autoregressive neural networks with exogenous input. Each of them is associated with a specific class of ageing (SOH) of the battery. The correct estimation among the four outputs is selected according to the SOH identification, which is obtained separately by a classifier that is done with a pattern recognition neural network. The SOH estimator provides a class of ageing among four possibilities, ranging from 80% to 100% with a step of 5%. A further class is associated to exhausted batteries and covers the range from 0% to 80% of the SOH, where 80% is the degradation threshold in the automotive sector. The output of the SOH classifier is used to select the correct SOC estimation among the four outputs of the regression ANNs, while the SOC estimation is used as an input for the SOH classifier in a closed-loop recursive architecture. The SOH estimator is an algorithm which is triggered only when a specific battery load condition in terms of the mean charging/discharging capacity request in a predefined time window is detected. This procedure allows reducing the training dataset of the SOH neural classifier to only one specific case. This aspect represents a relevant advantage in terms of a size reduction of the training dataset and a consequent time saving during the dataset collection and learning procedures. Additionally, the size of the network is smaller with a consequent reduction of the memory occupation when deployed on the battery management system (BMS).

The paper describes the design of the two estimators and the validation phase is conducted with the adoption of driving cycles acquired on a mild hybrid electric vehicle. The performance of the SOC estimator is evaluated by comparing the temporal evolution of the expected and estimated state of charge, whereas the SOH classifier accuracy is measured by using a confusion matrix, a common evaluation tool of classification algorithms.

The novel contributions of this work are as follows: a) the proposal of a combined estimation of the SOC and SOH with ANNs, allowing to make the method independent from the model and valid for every operating condition, provided that the network training dataset is complete and accurate; and b) the proposal of an SOH estimation method that is triggered only when a specific load condition corresponding to a predefined charging/discharging current profile is detected: this results in a compact algorithm that can be trained with a dataset that is smaller with respect to what would be needed in the case of a reproduction of the whole set of ageing conditions.

2. Method

The proposed method aims to provide a combined estimation of both the SOC and SOH of a battery. The approach is equally valid for a battery pack, module, or for the single cell.

Figure 1 illustrates the overall layout of the method that is composed of two subsystems: the SOC estimator, consisting of four regression ANNs, that is illustrated in the top left dashed box, and the SOH estimator, that exploits a neural classifier, that is reported in the bottom right dotted box. As is well known, the behavior of the two parameters is connected: the SOC of a battery is strongly influenced by the ageing, as well as the SOH estimation needing the information of the SOC variation during the charging/discharging operations. This motivates the adoption of a recursive loop architecture, where the SOC output is provided as an input to the SOH classifier and vice-versa. Both algorithms were trained on the basis of the preliminary experimental characterizations conducted in a laboratory. The two subsystems are described in detail in the following sections.

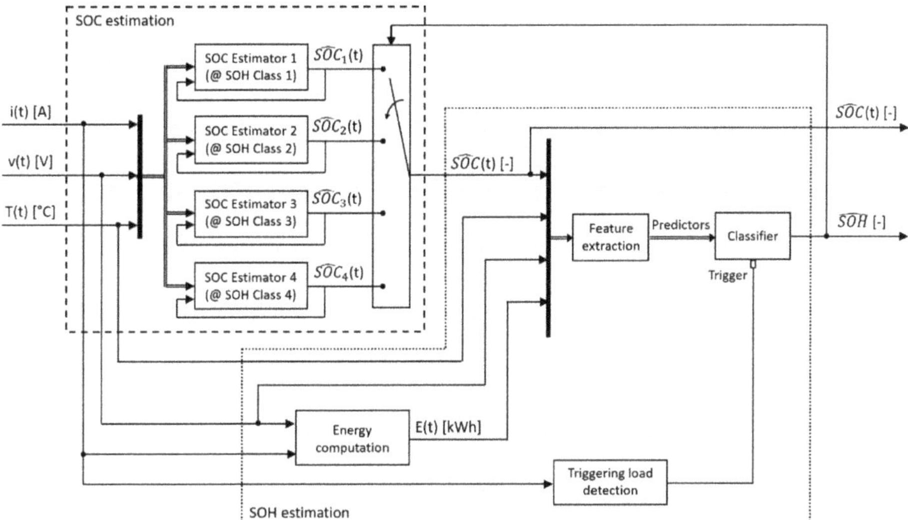

Figure 1. Overall method architecture. Dashed box: state of charge (SOC) estimation. Dotted box: state of health (SOH) estimation. $i(t)$: charging/discharging current. $v(t)$: voltage at battery terminals. $T(t)$: battery temperature. $E(t)$: energy request. SOH classes: 1: $(100 \div 95)\%$; 2: $(95 \div 90)\%$; 3: $(90 \div 85)\%$; 4: $(85 \div 80)\%$.

The battery pack considered for the study is composed of 168 cells (the cell model is Kokam SLPB 11543140H5, its characteristics are reported in Table 1) in the configuration 12p14s (p: parallel, s: series). The pack has a nominal voltage of 48 V, a nominal capacity of 60 Ah, and is designed for a mild hybrid electric vehicle with a peak electric power of around 20 kW, obtained considering a discharge rate of around 7C in nominal conditions.

Table 1. Main characteristics of the battery cell.

Typical Capacity (@0.5C, 4.2 V ÷ 2.7 V, 25 °C)		5 Ah
Nominal Voltage		3.7 V
Cut-off voltage		2.7 V
Continuous current		150 A
Peak current		250 A
Cycle life (Charge/Discharge @ 1C)		>800 cycles
Charge condition	Max. Current	10 A
	Voltage	4.2 V ± 0.03 V
Operating Temperature	Charge	0–40 °C
	Discharge	–20–60 °C
Mass		128.0 ± 4 g
Dimension	Thickness	11.5 ± 0.2 mm
	Width	42.5 ± 0.5 mm
	Length	142.0 ± 0.5 mm

2.1. SOC Estimation

The SOC estimator consists of four parallel regression ANNs (dashed box in Figure 1) working on the same inputs. Each network is associated with a specific ageing condition: SOH class 1 (from 100% to 95%), SOH class 2 (from 95% to 90%), SOH class 3 (from 90% to 85%), and SOH class 4 (from 85% to 80%). The threshold of 80% was decided considering that in the automotive sector, a battery has to be considered exhausted when the capacity or power fading is higher than 20%. The step of 5% is aligned with the typical precision that can be reached when dealing with the SOH estimation problem [31,32].

Each of the four regression ANNs receive, simultaneously, the following signals as inputs: charging/discharging current ($i(t)$ [A]), voltage at battery terminals ($v(t)$ [V]), and temperature ($T(t)$ [C]). They provide four different outputs: $S\hat{O}C_1(t)$, $S\hat{O}C_2(t)$, $S\hat{O}C_3(t)$, and $S\hat{O}C_4(t)$. The final SOC estimation ($S\hat{O}C(t)$) is obtained with a downstream selector that is operated by a signal fed back from the SOH classifier output, that is running separately, as indicated in Figure 1.

The structure of the four SOC estimators is the nonlinear autoregressive neural network with exogenous input (NARX) architecture. Typically, this layout is adopted for prediction tasks and finds an application in industrial engineering fields as well as in other sectors, namely linguistic search engines or weather forecasting. However, its effectiveness has been demonstrated also for estimation tasks and has been presented as an effective solution to estimate the SOC of lithium batteries in [21], where an additional comparison with other ANN architectures in terms of the computational cost and estimation accuracy is provided. The scheme of the NARX is reported in Figure 2, where the two adopted configurations are illustrated: an open-loop configuration (a), often indicated also as the series–parallel (SP) mode, that is adopted during the training procedure, and a closed-loop configuration (b), or equivalently the parallel (P) mode, that is the final architecture adopted for the estimation when the network is deployed on the vehicle for the real-time execution.

The output of the regression is defined as

$$y(n) = \varphi\left[y(n-1), y(n-2), \ldots, y(n-d_y); x(n-1), x(n-2), \ldots, x(n-d_x)\right] \tag{1}$$

where $y(n) \in \mathbb{R}$ and $x(n) \in \mathbb{R}$ denote the output (state of charge) and inputs (current, voltage, and temperature) of the NARX model at the discrete timestep n, respectively, d_x and d_y are the input and output memory delays used in the model, respectively, and φ is the function, generally non-linear, represented by the ANN. During the regression computation, the next value of the dependent output signal $y(n)$ is regressed on the previous d_y values of the output signal and previous d_x values of the independent (exogenous) input signal. In the open-loop configuration, the output regressor is

$$y(n) = \varphi\left[\overline{y}(n-1), \overline{y}(n-2), \ldots, \overline{y}(n-d_y); x(n-1), x(n-2), \ldots, x(n-d_x)\right] \tag{2}$$

A supervised training procedure is conducted using the measured output as the target. This approach allows for enriching the information to be processed by the network and permits using a common static backpropagation algorithm, the Levenberg–Marquardt in this case, for the training process, since the resulting network has a purely feedforward architecture.

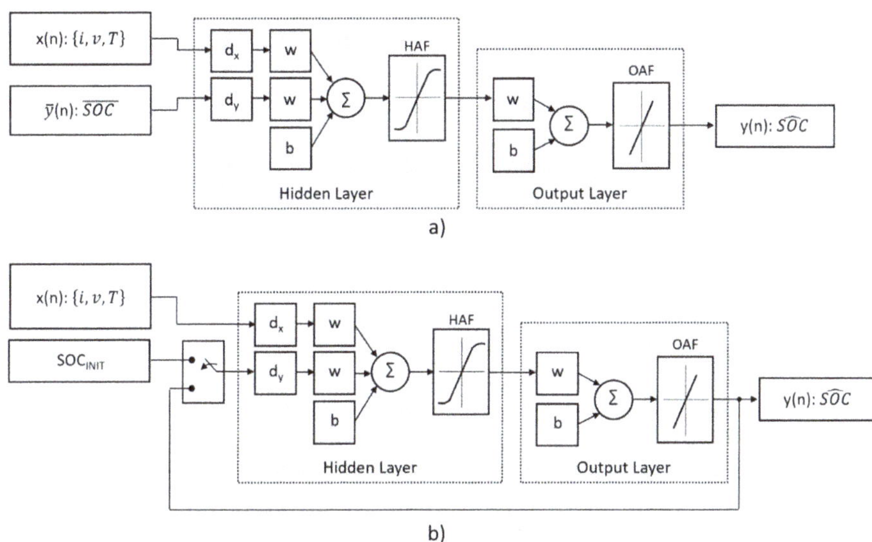

Figure 2. Nonlinear autoregressive neural network with exogenous input (NARX) architecture. (**a**) Series–parallel (SP) mode (open-loop configuration) adopted during the training. (**b**) Parallel (P) mode (closed-loop configuration) adopted for the estimation when the network is deployed. HAF: hidden activation function. OAF: output activation function. *w*: weight. *b*: bias.

In the first second of computation, the value of the algorithm output is not stable and is unpredictable. Therefore, if this value is fed back and provided as input to the ANN, it generates an estimation divergence over time. To avoid the occurrence of this irremediable condition, during the first second of estimation the feedback of the estimated SOC is replaced by the last estimation value (SOC_{INIT} in Figure 2b) recorded on a non-volatile memory at the previous shut down of the vehicle. After 1 second, when the output has become stable, the SOC input of the network switches from the previously recorded value to the real feedback of the estimation so that the regular operation of the algorithm can start.

Referring to Figure 2b and indicating with $n = n_0$ the time instant when the feedback signal switches from SOC_{INIT} to the estimated output, the characteristic equations of the model are written as

$$y(n) = \varphi[SOC_{INIT}; x(n-1, x(n-2), \ldots, x(n-d_x))], \ n < n_0 \tag{3}$$

and

$$y(n) = \varphi\left[y(n-1), y(n-2), \ldots, y(n-d_y); x(n-1), x(n-2), \ldots, x(n-d_x)\right], \ n \geq n_0 \tag{4}$$

The four networks have the same size in terms of layers, neurons, and delays and adopts the same activation functions. All these parameters have been designed with a trial and error approach aimed to maximize the estimation accuracy and avoid the risk of overfitting. Specifically, each network has one layer with eight neurons, the delays d_x and d_y are equal to two, the activation function in the hidden layer (*HAF*) and output layer (*OAF*) are the hyperbolic tangent and linear functions respectively, and the training function is the Levenberg–Marquardt function.

During the design phase, the training precision is evaluated by computing the mean square error (MSE) that reached a value of 1×10^{-13} as indicated in the small box embedded in Figure 3, and the estimation accuracy is measured with the maximum relative error (MRE), that is computed as

$$\text{MRE } [\%] = \max_{1<i<n}\left(\left|\frac{SOC_{exp}(i) - SOC_{est}(i)}{SOC_{exp,max} = 1}\right|\right) \times 100 \tag{5}$$

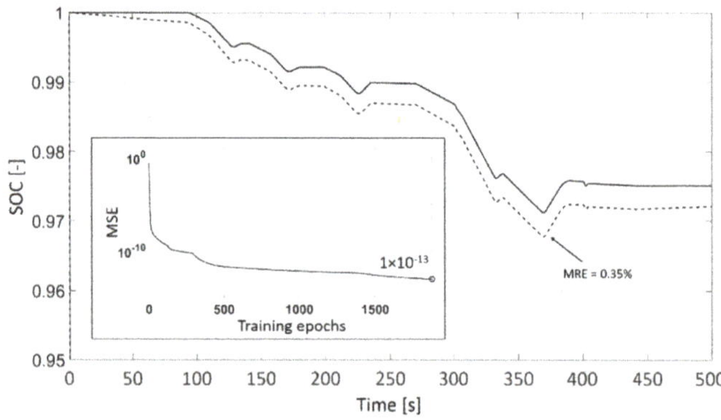

Figure 3. Comparison performance between the estimation (dashed line) and expected values (solid line) of the SOC in the case of an SOH = 100%. The obtained maximum relative error (MRE) is equal to 0.35%. The small box in the bottom left indicates the trend of the mean square error during the training phase.

This parameter reached the value of 0.35% as indicated in Figure 3, where the comparison between the estimation (dashed line) and the expected value (solid line) of the state of charge is reported in the case of a new battery (SOH = 100%). This plot wants to represent an indication of the training evaluation during the design phase.

The time length of the training dataset for the four regression ANNs is 13 h.

A more detailed description of the overall method results is reported in the final section of the paper.

2.2. SOH Estimation

The degradation of the battery is estimated with an algorithm reproducing a pattern recognition classifier with an ANN. Since the algorithm is proposed for the automotive sector, the method considers 20% as the maximum admitted capacity fading. Therefore, the considered life cycle of the battery ranges from an SOH of 100% when the battery is new to an SOH of 80% when the battery has to be considered exhausted. The proposed solution aims at quantifying the degradation suffered by the battery by identifying the five different levels of ageing which correspond to the five classes provided as an output by the classification algorithm. The first class covers the interval of ageing below the level of 80% (assumed as the threshold of the maximum degradation of the battery) of the SOH. The other four classes are equally distributed between 80% and 100% with four intervals of 5%, a percentage that is considered as consistent with the reasonable level of accuracy that can be reached when dealing with the problem of the SOH estimation.

As in the case of the SOC network design, the proposed algorithm for the SOH estimation exploits a preliminary experimental characterization phase conducted on the battery in a laboratory environment. The obtained results are used to build the training dataset to be adopted for the learning phase of the neural classifier. Specifically, the data of interest are recorded in a specific battery load condition corresponding to a mean request of 12 Ah in an interval of time of 120 s. This condition was selected because it can be detected quite frequently during a common driving cycle of an electric or hybrid vehicle. Afterwards, the network is trained with the dataset corresponding to this specific operating condition obtained at different values of temperature. Therefore, when the algorithm is deployed on the vehicle, it is called to estimate the level of ageing whenever the same condition is detected during the real driving cycle. This implies that when driving the vehicle, consecutive buffers of 120 s are analyzed back-to-back by a control logic that is implemented in the "Triggering load detection" block in Figure 1. As soon as the specific load condition of interest (mean capacity request of 12 Ah in 120 s) is detected, the classifier is triggered and provides the SOH classification as an output. Therefore, the estimation rate is not continuous over time, but it is produced in a discrete and not time deterministic way, only in correspondence with the detection of the predefined known load condition. The output of the estimator is kept equal to the last SOH estimation if the triggering condition is not occurring.

Figure 4 reports a part of the ANN training dataset obtained by the preliminary experimental characterization conducted on the battery. Subplot "a)" illustrates the behaviour of the degradation of the battery as a function of the number of discharging cycles at different values of temperature [33]. The discharging is conducted with the predefined load above-mentioned. Subplot "b)" reports the coupling effects between the SOH, capacity, SOC, and battery voltage. In this test, the temperature is set to 25 °C and the variation of the capacity is motivated by the difference in the time needed to discharge the battery at the different levels of ageing.

The time length of the training dataset covering all the considered levels of ageing is equal to 916 h obtained from 27,494 buffers with a duration of 120 s.

The SOH classifier works on discrete inputs, the so-called predictors, that are extracted in the "Feature extraction" block in Figure 1 from the time histories of the following signals: current, voltage, temperature, SOC, and energy. The latter is obtained from the "Energy computation" block in Figure 1 and is defined as

$$E = \int_{t_0}^{t_0+t_b} v(t)i(t)dt \tag{6}$$

where t_0 is the initial time of the buffer and t_b is the time length of the processed buffer that is set equal to 120 s.

The list of the extracted predictors is state of charge variation (-) (ΔSOC), voltage variation (V) (ΔV), requested energy (Wh) (E), and mean temperature (°C) (T).

The architecture of the classifier is illustrated in Figure 5. The training phase of the neural classifier is performed exploiting the scaled conjugate gradient (SCG) backpropagation training function [27]. This algorithm is designed to minimize the cost function including the difference between the estimated and expected outputs. This approach gives a good performance over a large number of pattern recognition problems that may include numerous parameters and guarantees a low performance degradation while reducing the training error. Additionally, this function is characterized by a relatively low computational cost and memory requirements [21], and its ability to provide well-separated classes in data mining and classification problems has been proven in many research works [34].

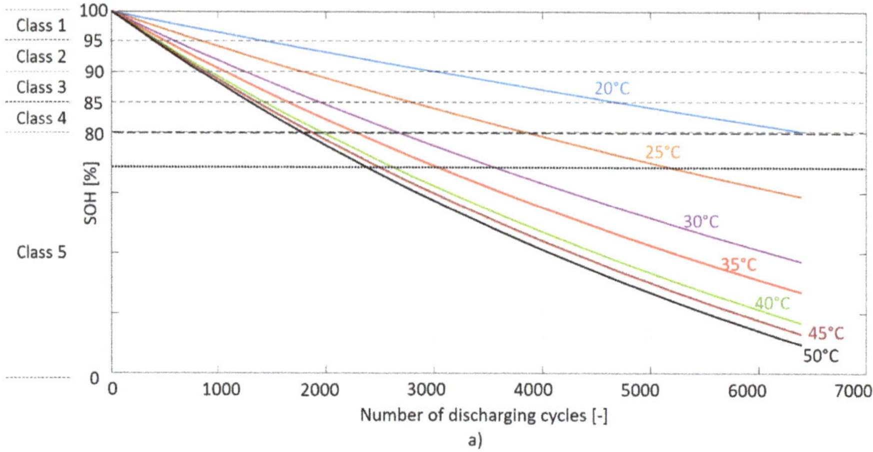

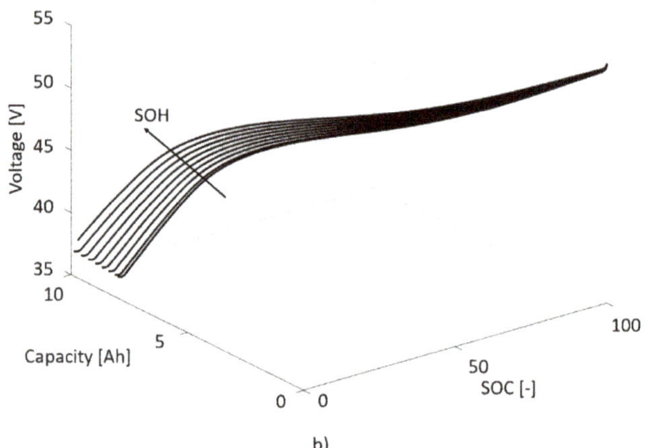

Figure 4. Battery experimental characterization for the SOH estimation. (**a**) SOH as a function of the number of discharging cycles and of the temperature. (**b**) Behavior of the SOH as a function of the voltage, capacity, and SOC. The temperature in this case is set equal to 25 °C.

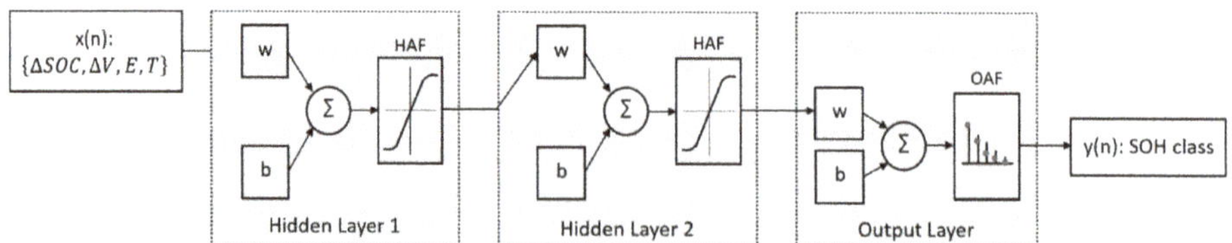

Figure 5. Pattern recognition a feed-forward artificial neural network (ANN) architecture for the SOH classification. HAF: hidden activation function. OAF: output activation function. w: weight. b: bias.

The classifier is composed of one input, two hidden and one output layer. As in the case of the SOC network design, the number and size of the hidden layers is defined heuristically, by means of a trial and error procedure. Specifically, the hidden layers consist of ten neurons each, HAF is a hyperbolic tangent sigmoid, and OAF is a normalized exponential function. The performance of the training process is evaluated by means of the cross-entropy cost function, that at the end of the training process is equal to 1×10^{-3}, after around 3000 training epochs.

3. Results and Discussion

The validation of the method is conducted in two separate phases: (a) an analysis of the performance of the SOH identifier and (b) an evaluation of the accuracy of the overall SOC estimation that includes the ageing classification.

3.1. SOH Classification

As is described above, the classification algorithm is called to identify the class of degradation only when a specific load condition is detected during the driving operations. To evaluate the effectiveness of the method, a profile corresponding to the specific charging/discharging profile was created artificially to have an exhaustive number of occurrences in the different operating conditions to test.

The profile is reported in Figure 6, where it has a duration of 5000 s and includes 42 different consecutive buffers with the time length of 120 s and a mean capacity request of 12 Ah. The profile was cycled until reaching a total duration of 50 h, to sweep the range of the SOC of the battery, corresponding to 1500 buffers of 120 s, for each class of ageing. The resulting timeseries was provided to the LUT representing the battery. This LUT was tuned after the preliminary laboratory experimental characterization and allows for extracting the predictors provided to the classifier in the five ageing conditions.

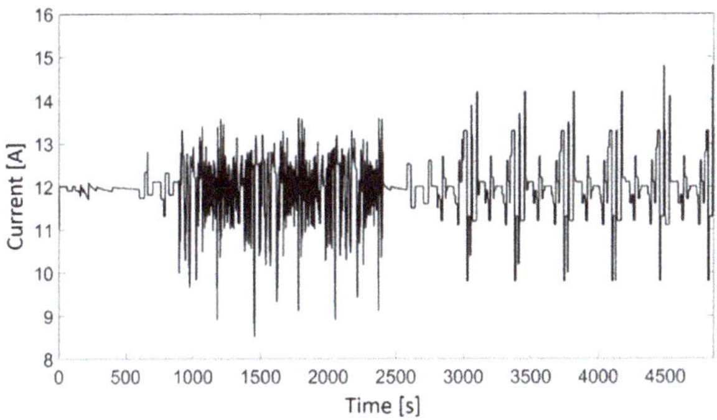

Figure 6. Current profile created to validate the SOH classifier. The profile is replicated until reaching the total duration of 50 h and a number of buffers of 1500 for each class of ageing.

The resulting validation dataset is therefore composed of 7500 different buffers with a time length of 120 s each. The resulting profile represents the different operating conditions at different degradation levels and is given as an input to the classifier.

The tool adopted to evaluate the accuracy of the SOH estimation is the confusion matrix reported in Figure 7. The classified and actual ageing condition instances are reported in the rows and columns, respectively. The values contained in the main diagonal cells indicate the correct classifications, whereas the off-diagonal cells report the number of the misclassifications. The overall obtained estimation accuracy is equal to 2.4%, which is equal to the number of misclassifications (178 buffers) over the total number of tested occurrences (7500 buffers). This result is aligned with the expected accuracy.

	Class 1 100÷95	Class 2 95÷90	Class 3 90÷85	Class 4 85÷80	Class 5 80÷0	
Class 1 100÷95	1469	19	2	0	0	
Class 2 95÷90	21	1459	15	0	0	
Class 3 90÷85	10	21	1467	15	14	
Class 4 85÷80	0	1	16	1472	21	
Class 5 80÷0	0	0	0	23	1465	
Error	2%	2.7%	2.2%	2.5%	2.3%	2.4%

Classification output / Classification target

Figure 7. Evaluation of the SOH classification performance. Confusion matrix obtained for the ANN trained with the scaled conjugate gradient (SCG) algorithm. The cell in the grey background indicates the overall accuracy of the method.

3.2. SOC Estimation

The second part of the validation is dedicated to the evaluation of the accuracy of the SOC estimation. To this end, the profiles illustrated in Figure 8 have been adopted as validation timeseries. The subplot "a)" reports the current profile, and the subplot "b)" illustrates the behavior of the battery terminal voltage at different levels of ageing. The voltage is only an occurrence of the many possibilities that are associated to a class of degradation. The plots in the right part of the figure are zoomed-in areas with a time length of 2000 s. When providing these timeseries to the SOC estimation block (dashed box in Figure 1), the regression ANNs will provide four different outputs. The one corresponding to the correct ageing level of the battery is then selected according to the output of the SOH classifier (dotted box in Figure 1).

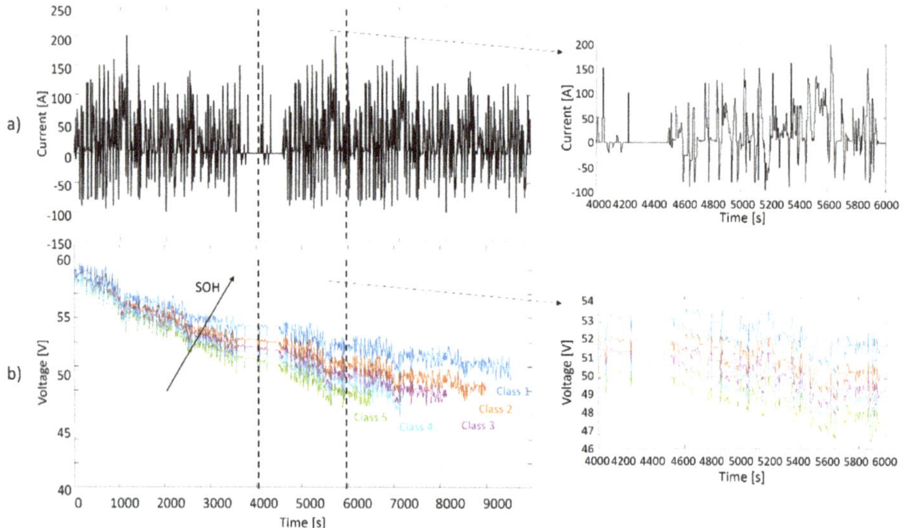

Figure 8. ANN validation datasets recorded from a real mild hybrid vehicle. (**a**) Current $i(t)$. (**b**) Voltage $v(t)$ at four different degradation levels corresponding to the four SOH classes.

The results obtained in the five ageing levels are illustrated in Figure 9, where for each SOH class, the estimated SOC, on the blue line, is compared with the expected value, on the red line. The expected value is the one obtained from the preliminary experimental characterization conducted in the laboratory. The estimation error is reported in the lower subplot for each case. The accuracy of the estimation is demonstrated by the error that is limited to a maximum value of 3%. The results obtained for the class of ageing going from 0% to 80% (subplot "e") demonstrate that the algorithm keeps being valid also under the threshold of 80%. The reported test has been conducted at a temperature of around 25 °C. A more exhaustive validation of the method should be conducted in a climatic test chamber to evaluate the accuracy of the estimation at different environmental conditions.

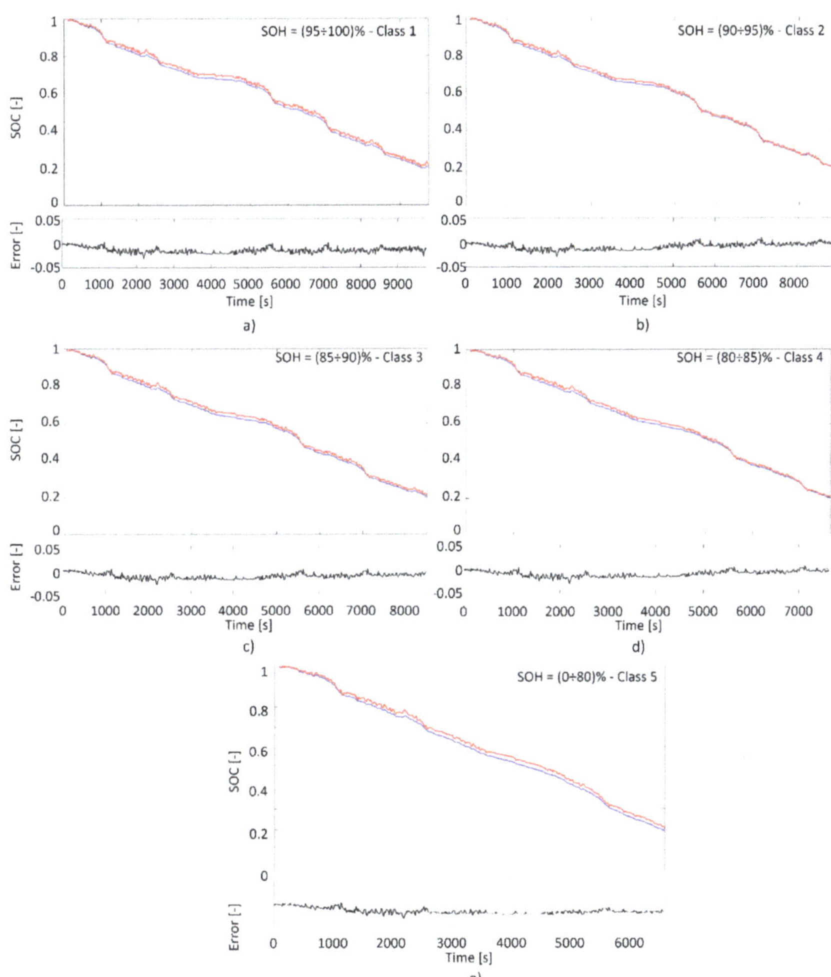

Figure 9. SOC estimation at different degradation levels. Red line: expected value. Blue line: estimation. Error indicates the difference between the estimated and expected values. (**a**): ageing class 1 (SOH: 95 ÷ 100%); (**b**): ageing class 2 (SOH: 90 ÷ 95%); (**c**): ageing class 3 (SOH: 85 ÷ 90%); (**d**): ageing class 4 (SOH: 80 ÷ 85%); (**e**): ageing class 5 (SOH: 0 ÷ 80%).

4. Conclusions

This paper presented a method for the combined estimation of the state of charge and state of health of batteries with artificial intelligence. The technique is valid at the cell, module, and pack levels and is suitable for adoption in the automotive sector in the case of hybrid and full electric vehicles. The design procedure of the algorithm and specifically the training phase of the artificial neural networks were presented. The method was demonstrated to be effective in terms of the estimation accuracy when tested on real driving cycles extracted from the acquisition on-board of an electric vehicle. The estimation error of the combined method is around 3%. The good potential and the promising results encourage the adoption of the proposed method for deployment in a vehicle battery management system for a real-time battery monitoring.

Funding: This research received no external funding.

Acknowledgments: This work was developed in the framework of the activities of the Interdepartmental Center for Automotive Research and Sustainable Mobility (CARS) of Politecnico di Torino (www.cars.polito.it).

Conflicts of Interest: The author declares no conflict of interest.

References

1. Ahman, M. Assessing the future competitiveness of alternative powertrains. *Int. J. Veh.* **2003**, *33*, 309. [CrossRef]
2. Wang, Y.; Diaz, D.F.R.; Chen, K.S.; Wang, Z.; Adroher, X.C. Materials, technological status, and fundamentals of PEM fuel cells—A review. *Mater. Today* **2020**, *32*, 178–203. [CrossRef]
3. Walther, G.; Wansart, J.; Kieckhafer, K.; Schneider, E.; Spengler, T.S. Impact assessment in the automotive industry: Mandatory market introduction of alternative powertrain technologies. *Syst. Dyn. Rev.* **2010**, *26*, 239–261. [CrossRef]
4. Bishop, J.D.K.; Martin, N.P.D.; Boies, A.M. Cost-effectiveness of alternative powertrains for reduced energy use and CO_2 emissions in passenger vehicles. *Appl. Energy* **2014**, *124*, 44–61. [CrossRef]
5. Chan, C.C.; Chau, K.T. *Modern Electric Vehicle Technology*; Oxford University Press: New York, NY, USA, 2002.
6. Onori, S.; Serrao, L.; Rizzoni, G. *Hybrid Electric Vehicles: Energy Management Strategies*; Springer: London, UK, 2016.
7. Anselma, P.G.; Huo, Y.; Roeleveld, J.; Belingardi, G.; Emadi, A. Slope-Weighted Energy-Based Rapid Control Analysis for Hybrid Electric Vehicles. *IEEE Trans. Veh. Technol.* **2019**, *68*, 4458–4466. [CrossRef]
8. Anselma, P.G.; Huo, Y.; Roeleveld, J.; Belingardi, G.; Emadi, A. Integration of On-Line Control in Optimal Design of Multimode Power-Split Hybrid Electric Vehicle Powertrains. *IEEE Trans. Veh. Technol.* **2019**, *68*, 3436–3445. [CrossRef]
9. Vetter, J.; Novák, P.; Wagner, M.; Veit, C.; Möller, K.C.; Besenhard, J.O.; Winter, M.; Wohlfahrt-Mehrens, M.; Vogler, C.; Hammouche, A. Ageing mechanisms in lithium-ion batteries. *J. Power Sources* **2005**, *147*, 269–281. [CrossRef]
10. Chang, W.Y. The State of Charge Estimating Methods for Battery: A Review. *Appl. Math.* **2013**, *2013*, 953792. [CrossRef]
11. Piller, S.; Perrin, M.; Jossen, A. Methods for state-of-charge determination and their applications. *J. Power Sources* **2001**, *96*, 113–120. [CrossRef]
12. Remmlinger, J.; Buchholz, M.; Meiler, M.; Bernreuter, P.; Dietmayer, K. State-of-health monitoring of lithium-ion batteries in electric vehicles by on-board internal resistance estimation. *J. Power Sources* **2011**, *196*, 5357–5363. [CrossRef]
13. Huang, S.; Tseng, K.; Liang, J.; Chang, C.; Pecht, M.G. An Online SOC and SOH Estimation Model for Lithium-Ion Batteries. *Energies* **2017**, *10*, 512. [CrossRef]
14. Wei, Z.; Zhao, J.; Ji, D.; Tseng, K.T. A multi-timescale estimator for battery state of charge and capacity dual estimation based on an online identified model. *Appl. Energy* **2017**, *204*, 1264–1274. [CrossRef]
15. Pérez, G.; Garmendia, M.; Reynaud, J.F.; Crego, J.; Viscarret, U. Enhanced closed loop State of Charge estimator for lithium-ion batteries based on Extended Kalman Filter. *Appl. Energy* **2015**, *155*, 834–845. [CrossRef]
16. Yu, Q.; Xiong, R.; Lin, C. Online estimation of state-of-charge based on H-infinity and unscented Kalman filters for lithium ion batteries. *Energy Procedia* **2017**, *105*, 2791–2796. [CrossRef]

17. Zeng, M.; Zhang, P.; Yang, Y.; Xie, C.; Shi, Y. SOC and SOH Joint Estimation of the Power Batteries Based on Fuzzy Unscented Kalman Filtering Algorithm. *Energies* **2019**, *12*, 3122. [CrossRef]

18. Ye, M.; Guo, H.; Xiong, R.; Yang, R. Model-based state-of-charge estimation approach of the Lithium-ion battery using an improved adaptive particle filter. *Energy Procedia* **2016**, *103*, 394–399. [CrossRef]

19. Kim, T.; Wang, Y.; Sahinoglu, Z.; Wada, T.; Hara, S.; Qiao, W. State of Charge Estimation Based on a Realtime Battery Model and Iterative Smooth Variable Structure Filter. In Proceedings of the IEEE Innovative Smart Grid Technologies—Asia, Kuala Lumpur, Malaysia, 20–23 May 2014.

20. Charkhgard, M.; Farrokhi, M. State-of-Charge Estimation for Lithium-Ion Batteries Using Neural Networks and EKF. *IEEE Trans. Ind. Electron.* **2010**, *57*, 4178–4187. [CrossRef]

21. Bonfitto, A.; Feraco, S.; Tonoli, A.; Amati, N.; Monti, F. Estimation Accuracy and Computational Cost Analysis of Artificial Neural Networks for the State of Charge Estimation in Lithium Batteries. *Batteries* **2019**, *5*, 47. [CrossRef]

22. He, W.; Williard, N.; Chen, C.; Pecht, M. State of charge estimation for Li-ion batteries using neural network modeling and unscented Kalman filter-based error cancellation. *Electr. Power Energy Syst.* **2014**, *62*, 783–791. [CrossRef]

23. Chaoui, H.; Ibe-Ekeocha, C.C. State of Charge and State of Health Estimation for Lithium Batteries Using Recurrent Neural Networks. *IEEE Trans. Veh. Technol.* **2017**, *66*, 8773–8783. [CrossRef]

24. Chang, C.; Liu, Z.; Huang, Y.; Wei, D.; Zhang, L. Estimation of Battery state of Health Using Back Propagation Neural Network. *Comput. Aided Draft. Des. Manuf.* **2014**, *24*, 60.

25. Bonfitto, A.; Feraco, S.; Ezemobi, E.; Tonoli, A.; Amati, N.; Hegde, S. State of Health Estimation of Lithium Batteries for Automotive Applications with Artificial Neural Networks, IEEE. In Proceedings of the 2019 AEIT International Conference of Electrical and Electronic Technologies for Automotive, AEIT AUTOMOTIVE, Turin, Italy, 2–4 July 2019.

26. Lin, H.; Liang, T.; Chen, S. Estimation of Battery State of Health Using Probabilistic Neural Network. *IEEE Trans. Ind. Inform.* **2013**, *9*, 679–685. [CrossRef]

27. Yang, D.; Wang, Y.; Pan, R.; Chen, R.; Chen, Z. A neural network based state-of-health estimation of lithium-ion battery in electric vehicles. In Proceedings of the 8th International Conference on Applied Energy—ICAE 2016, Beijing, China, 8–10 October 2016.

28. Huet, F. A review of impedance measurement for determination of state-of-charge or state-of-health of secondary battery. *J. Power Sources* **1998**, *70*, 59–69. [CrossRef]

29. Wassiliadis, N.; Adermann, J.; Frericks, A.; Pak, M.; Reiter, C. Revisiting the dual extended Kalman filter for battery state-of-charge and state-of-health estimation: A use-case life cycle analysis. *J. Energy Storage* **2018**, *19*, 73–87. [CrossRef]

30. Andre, D.; Appel, C.; Soczka-Guth, T.; Sauer, D.U. Advanced mathematical methods of SOC and SOH estimation for lithium-ion batteries. *J. Power Sources* **2013**, *224*, 20–27. [CrossRef]

31. Zhang, C.W.; Chen, S.R.; Gao, H.B.; Xu, K.J.; Yang, M.Y. State of Charge Estimation of Power Battery Using Improved Back Propagation Neural Network. *Batteries* **2018**, *4*, 69. [CrossRef]

32. Samolyk, M.; Sobczak, J. Development of an Algorithm for Estimating Lead-Acid Battery State of Charge and State of Health. Master's Thesis, Blekinge Institute of Technology, Karlskrona, Sweden, 2013.

33. Leng, F.; Tan, C.; Pecht, M. Effect of Temperature on the Aging rate of Li Ion Battery Operating above Room Temperature. *Sci. Rep.* **2015**, *5*, 12967. [CrossRef]

34. Bonfitto, A.; Feraco, S.; Tonoli, A.; Amati, N. Combined regression and classification artificial neural networks for sideslip angle estimation and road condition identification. *Veh. Syst. Dyn.* **2019**, 1–22. [CrossRef]

Article

Recursive State of Charge and State of Health Estimation Method for Lithium-Ion Batteries Based on Coulomb Counting and Open Circuit Voltage

Alejandro Gismero *, Erik Schaltz and Daniel-Ioan Stroe

Department of Energy Technology, Aalborg University, Pontoppidanstraede 101, 9220 Aalborg, Denmark; esc@et.aau.dk (E.S.); dis@et.aau.dk (D.-I.S.)
* Correspondence: aga@et.aau.dk

Received: 12 March 2020; Accepted: 2 April 2020; Published: 9 April 2020

Abstract: The state of charge (SOC) and state of health (SOH) are two crucial indicators needed for a proper and safe operation of the battery. Coulomb counting is one of the most adopted and straightforward methods to calculate the SOC. Although it can be implemented for all kinds of applications, its accuracy is strongly dependent on the operation conditions. In this work, the behavior of the batteries at different current and temperature conditions is analyzed in order to adjust the charge measurement according to the battery efficiency at the specific operating conditions. The open-circuit voltage (OCV) is used to reset the SOC estimation and prevent the error accumulation. Furthermore, the SOH is estimated by evaluating the accumulated charge between two different SOC using a recursive least squares (RLS) method. The SOC and SOH estimations are verified through an extensive test in which the battery is subjected to a dynamic load profile at different temperatures.

Keywords: Coulomb counting; lithium-ion battery; open circuit voltage; state of charge; state of health; temperature

---

1. Introduction

Lithium-ion (Li-ion) batteries have prevailed over other energy storage types during the last decade due to the longer lifetime, higher efficiency and energy density [1]. This fact has driven its gradual integration in many applications from consumer electronics e.g., smartphones or power banks to renewable storage systems or electric vehicles.

However, Li-ion batteries may experience a fast degradation or even become hazardous if operated out of the limits specified by the manufacturer [2]. Therefore, a battery management system (BMS) to monitor and control the state of the battery is required. Besides keeping the battery within the operative limits, to estimate the state of charge (SOC) and the state of health (SOH) accurately is an essential function of the BMS.

Numerous methods for SOC estimation have been proposed in the literature, with diverse complexity and accuracy [3]. The coulomb counting relies on the current monitoring and calculates the net charge transferred, to estimate the SOC [4]. The relationship between the open-circuit voltage (OCV) and the SOC can also be used as an estimator if the battery is in a long-enough rest period [5]. Equivalent circuit models and extended Kalman filtering can be applied to calculate the OCV and estimate the SOC during the operation of the battery [6,7]. The accuracy of the model parameters affects the performance of the method, hence dual/joint extended Kalman filters method estimates SOC and model parameters at the same time, increasing the method performance but also its complexity [8]. The model parameters are easy to obtain, however, they are affected by factors such as SOC, current rate (C-rate), temperature or degradation level. The combination of these variables may affect the SOC estimation accuracy [9,10]. These methods are most widely used for online battery monitoring, however, more advanced methods are being developed proving their effectiveness under controlled environments. The complexity of the implementation in practical applications and the robustness of these methods is, however, one of the great challenges that still needs to be solved in this area. Meanwhile, Coulomb counting remains as one of the most accepted and widespread methods in the market for all types of applications.

Coulomb counting method stands out for its simplicity, however, its accuracy is compromised by the error accumulation during the incremental calculation of the transferred charge in a long time operation. Moreover, the method depends on the initial value of SOC and the actual capacity. OCV can be used in combination with coulomb counting to minimize the error and provide a reliable starting point for the estimation [11]. The capacity of the battery depends on the C-rate and the temperature, additionally, there are losses during the charging and discharging process and, to a lesser extent, due to the self-discharge. All these factors must be taken into account during the SOC estimation to improve the accuracy.

This paper presents an approach that improves the method described above providing accurate results throughout the range of temperatures and battery load. The OCV-SOC relationship at different temperatures is used for the SOC estimation during the idling periods, minimizing the cumulative error due to the integration of current. The presented method provides better precision results with a very low computational complexity, therefore representing an evolution of the classic Coulomb counting algorithm. The influence of the different operating conditions is analyzed at the beginning of life (BOL) of the battery in order to implement a correction mechanism. The SOH is evaluated with a recursive least squares method [12] and updated for a reliable SOC estimation. The method is verified through an extensive test based on a real driving pattern at different temperatures.

2. Experimental

In this work, Lithium Nickel Manganese Cobalt Oxide (NMC) batteries are used to validate the proposed method. The battery's nominal voltage is 3.6 V and the nominal capacity 3.4 Ah. The upper/lower cutoff voltage is 4.2 V/2.65 V. Figure 1 shows the set-up used to test the batteries. A thermal test chamber is used to control the temperature of the battery. The battery test system charges and discharges the cells with a predefined current profile and the host computer collects the measurements from the battery, with one second resolution.

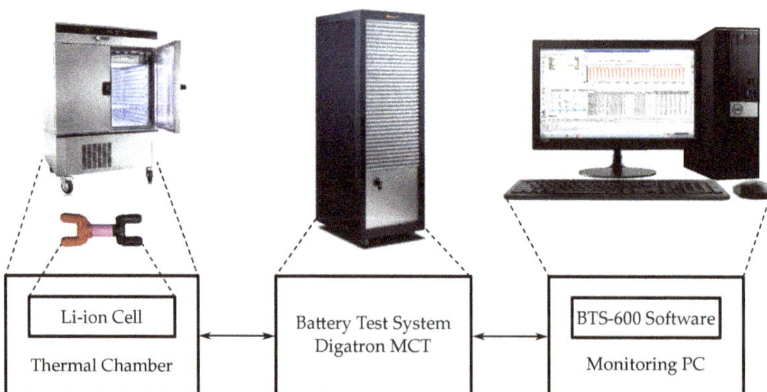

Figure 1. Battery test bench.

The experimental procedure is shown in Figure 2. The tests were designed to age the cells simulating a real scenario of cycling at different temperatures. A series of driving cycles and characterization tests were conducted for this purpose.

To carry out the aging process the cells are distributed in three groups and subjected to successive discharge cycles according to the World Harmonized Light-duty Vehicle Test Procedure (WLTP) for class B vehicles at three different temperatures (15 °C, 25 °C and 35 °C). The batteries are charged to 90% SOC with 0.2C (i.e., 0.68 A) just after each driving cycle followed by a rest period before repeating the aging procedure. As shown in Figure 3 each described discharging/charging cycle takes 2 h to complete. After a week of testing equivalent to 84 driving cycles, the cells are characterized by a reference performance test (RPT). This test consists of a first full cycle at low current to stabilize the battery followed by another cycle to measure the capacity and efficiency of the cell.

The OCV-SOC relationship is measured both at beginning of life state (BOL) and after 350 full cycles, which corresponds to 6.5%, 7.35% and 7.65% capacity fade for the cells aged at 15 °C, 25 °C and 35 °C, respectively.

The battery is charged and discharged at a constant low rate of 0.2C until the upper and lower cutoff voltage is reached. The OCV-SOC curve is obtained by averaging the values of both, charge and discharge curves. The OCV-SOC relationship is measured at 5 different temperatures (5 °C, 15 °C, 25 °C, 35 °C and 40 °C).

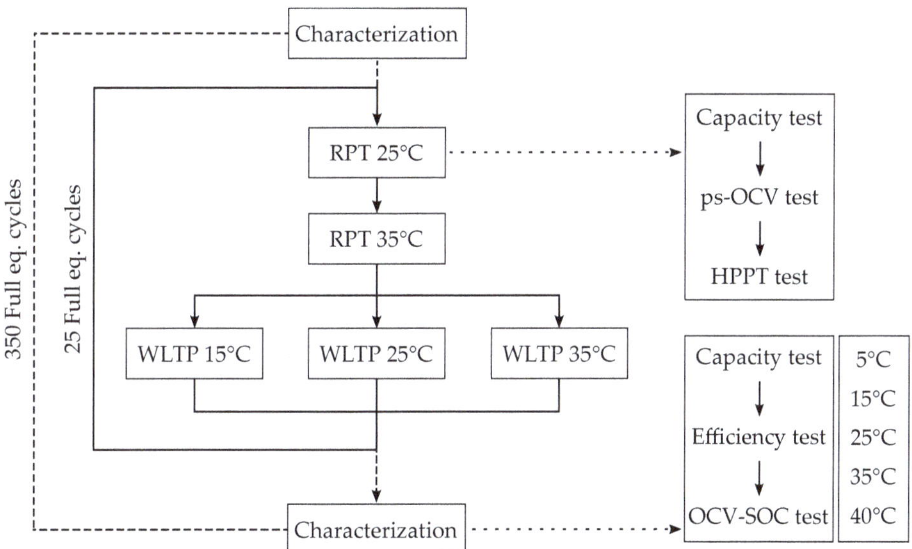

Figure 2. Aging test procedure.

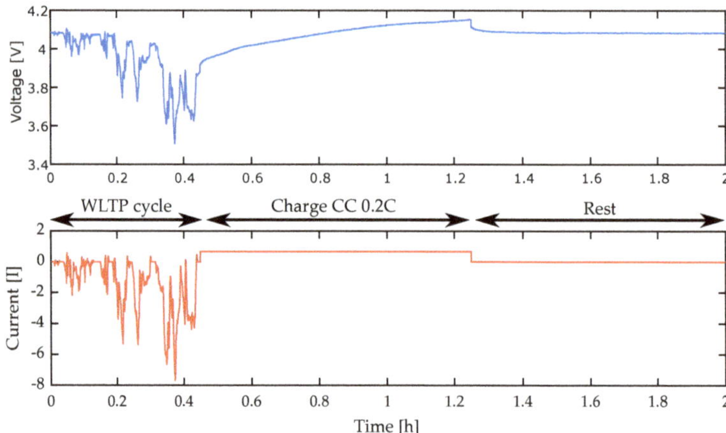

Figure 3. World Harmonized Light-duty Vehicle Test Procedure (WLTP)-based aging cycle. Current profile (red) and voltage response (blue).

3. SOC Determination

Battery capacity can be represented by different terms. The rated capacity (Q_{rated}) represents the capacity that a fresh new battery is capable of delivering from a fully charged state until it is completely discharged, reaching the cut-off voltage under a determined C-rate and ambient temperature.

However, as the battery degrades, its capacity is reduced respect to the BOL condition, so the actual capacity (Q_{act}) is used to describe the capacity at a determined degradation state, C-rate and temperature.

Hence the SOC can be defined as the ratio between the remaining capacity (Q_{rem}) and the actual capacity, usually expressed as a percentage:

$$SOC = \frac{Q_{rem}}{Q_{act}}. \tag{1}$$

The charge variation between two different points in time can be calculated by integrating the current, what is commonly known as Coulomb counting:

$$q = \int_{t_0}^{t} I_b dt \tag{2}$$

Thus the SOC at a determined time can be expressed as the SOC of a previous time plus the charge variation between both points in time.

$$SOC_t = SOC_{t_0} + \frac{q}{Q_{act}} \tag{3}$$

The Coulombic efficiency of the battery during the charging and discharging processes depends on the current applied and the battery temperature during operation. Lower battery temperatures lead to a maximum capacity decrease while C-rate effect would have influence mainly at low temperatures. The actual capacity is established at the reference conditions (0.2C and 25 °C), this causes an inaccuracy when estimating the SOC at any other condition. A compensation factor, denoted as $cf_{i,t}$, is introduced in the current integration process to normalize the SOC estimation to the reference conditions.

$$q = \int_{t_0}^{t} cf_{I,T} I_b dt. \tag{4}$$

Figure 4a,b show the charge and discharge capacity at different temperature and current conditions. While temperatures above 20 °C have a minor influence on the measured capacity, lower temperatures produce a decrease in both charging and discharging capacity. High currents also lead to a capacity reduction, mostly at low temperatures. The compensation factor is defined as the ratio between the capacity at the reference condition and the capacity at the actual operating condition. In this work this relationship is assumed to be constant with the degradation of the battery and it is measured at BOL state. Figure 4c,d show the compensation factor values used by the proposed algorithm, the red dot represents the reference conditions (0.2C and 25 °C) to establish Q_{act}, so that $cf_{0.2C,25\,°C} = 1$.

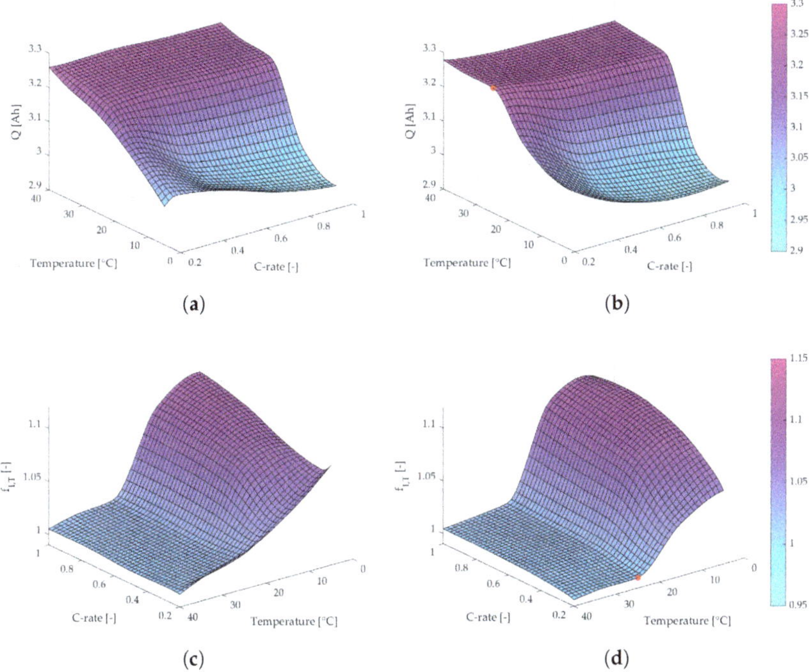

Figure 4. Charge and discharge capacity under different conditions (**a**) charge capacity, and (**b**) discharge capacity. Compensation factor for charging and discharging processes under different conditions (**c**) charge compensation factor, and (**d**) discharge compensation factor. The red dot represents the reference conditions (0.2C and 25 °C).

Once the use of Coulomb counting at different operation conditions is solved and the actual SOC can be estimated from the SOC of a previous time, the challenge is to avoid the cumulative errors produced by the incremental estimation. OCV-SOC relationship is used to set a new estimation starting point when the battery is at a rest period. Generally, a long relaxation time is required to measure the OCV [9], however, measuring it at shorter rest periods can minimize the cumulative error. Figure 5 shows the OCV curves measured at different temperatures. The five curves show the same behavior within the 0.2–0.8 SOC range, however, lower temperatures show higher voltages at the lower end of the curve and lower voltages at the higher end. The error produced by using the OCV-SOC relationship to estimate the SOC at a different temperature could be unacceptable in these SOC ranges, especially for SOC higher than 0.8. Thus is convenient to measure the OCV at different temperatures and interpolate the battery temperature to obtain the SOC.

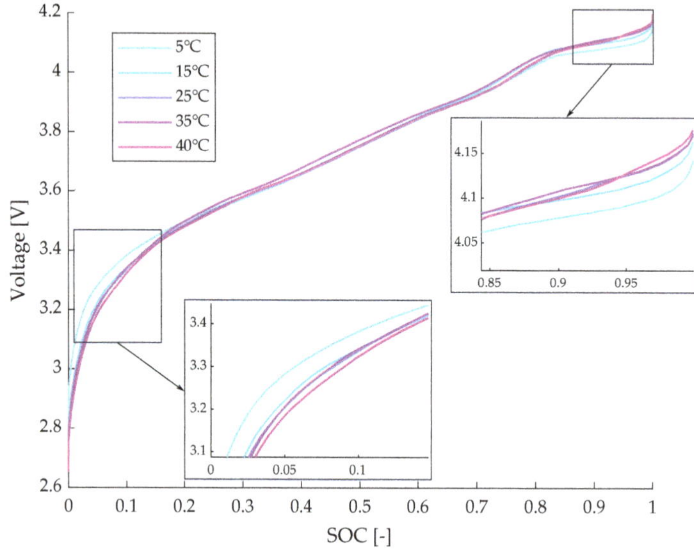

Figure 5. Open-circuit voltage (OCV)-state of charge (SOC) relationship for different temperatures.

4. SOH Determination

In many applications it is of high importance to know the degradation level of the battery, as it determines the maximum capacity the battery is able to deliver. Furthermore, when the SOC estimation method is based on the capacity integration, knowledge of the actual capacity is crucial to ensure an accurate estimation.

Performing a full charge and discharge cycle at low current is the easiest and obvious method to get the actual capacity; however, this procedure is time-consuming and incompatible with the normal use of the battery and thus not suitable for online estimation in most cases.

As an alternative, it is possible to measure the partial capacity between two known SOC levels to infer the total capacity. Therefore, in this work it is proposed the use of Equation (3) taking advantage of the SOC estimated by OCV, resulting in:

$$Q_{est} = \frac{q}{\Delta SOC} \tag{5}$$

where,

$$\Delta SOC = SOC_t^{OCVupdate} - SOC_{t_0}^{OCVupdate}. \tag{6}$$

Hence, it is possible to estimate the actual capacity of the battery by measuring the transferred charge without affecting the normal operation of the battery. The accuracy of the method will depend on the magnitude of ΔSOC as well as on the q measurement error, which is influenced by operating conditions (current and temperature) and the self discharging effect.

In order to minimize the error, a recursive least squares filter is used to determine the actual capacity from previous estimations as addressed in [12]. A forgetting factor is included to give less weight to older samples.

5. Proposed Algorithm and Verification

The flow chart shown in Figure 6 describes the process for estimating the SOC and SOH of the battery. First of all, the actual values of current, voltage and temperature are collected. If the current is different from zero, the algorithm measures the accumulated charge, compensating the temperature and the current, and estimates the SOC from it. If the value of the current is zero for a minimum time of 25 min, it is then considered to be on idle state and a look-up table is used to estimate the SOC from the OCV and temperature. Hence the active periods of the battery begin and end with an OCV-based SOC estimation. If the magnitude of the SOC increment is sufficient for a capacity estimation, the SOH is updated.

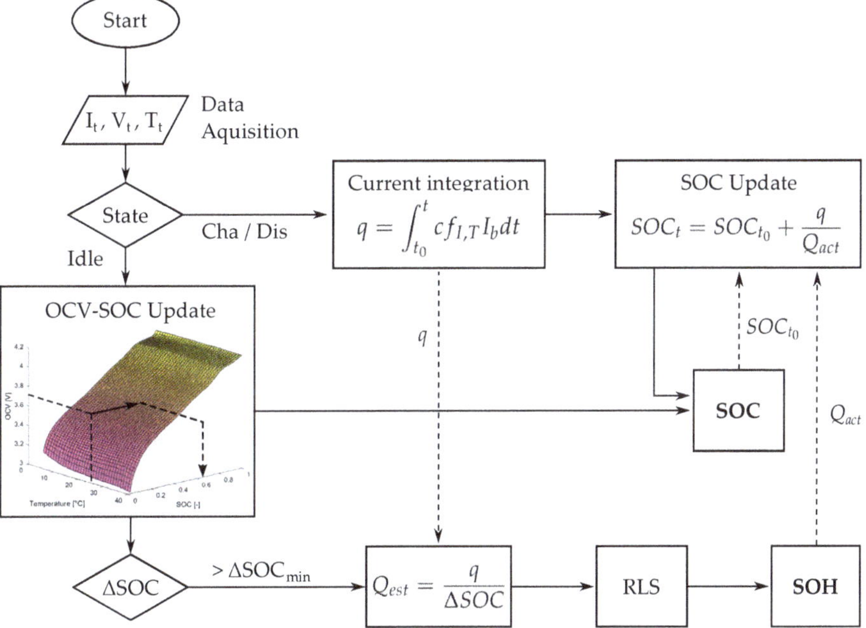

Figure 6. Flowchart of the proposed estimation algorithm.

A verification test using the aged cells was designed to acquire experimental data at dynamic conditions. As shown in Figure 7 a sequence of WLTP cycles at different SOC levels is used to verify the algorithm for SOC and SOH estimation. A total of 58 cycles were performed at different temperatures, with one hour of resting between cycles and charge periods. The voltage, current and temperature of the battery are logged with 1 Hz.

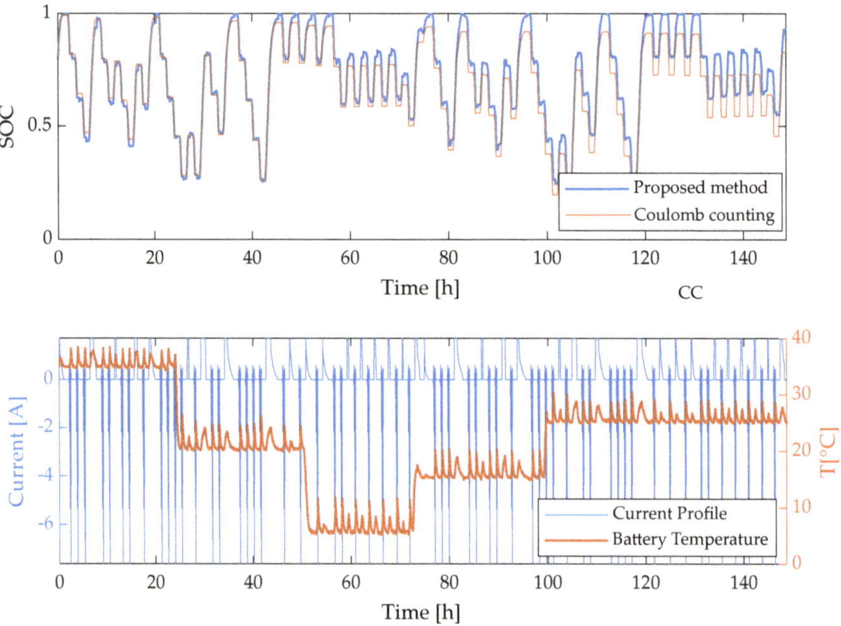

Figure 7. Verification test. SOC estimation comparison (top). Current profile and battery temperature (bottom).

Several times during the test, the battery is fully charged (CC-CV) to the 100% SOC, this state is used to evaluate the error of the method. Figure 8 displays the results for the three cells aged at 35 °C, 25 °C and 15 °C respectively. Figure 8a,c,e compare the absolute SOC error of the traditional Coulomb counting method with the proposed method, before and after updating the estimation using the OCV. The error using only Coulomb counting constantly increases with time as it is accumulated in every iteration. The use of OCV to update the estimate limits the maximum error to 5% while the compensation factor reduces it below 2.5% during the next cycle. The error after the OCV update is negligible although it could be higher in SOC areas where the OCV presents a less steeper slope.

Figure 8b,d,f display the actual capacity estimation during the verification test under different capacity starting values. Throughout the test, the four estimations converge around the same value with a maximum estimation error of 0.05 Ah for the cell aged at 25 °C. An insufficient rest time makes the algorithm over-valuate ΔSOC which results in an underestimation of the capacity. This test was limited to one hour rest time for practical reasons, however, longer relaxation periods are expected to reduce the gap in a real application.

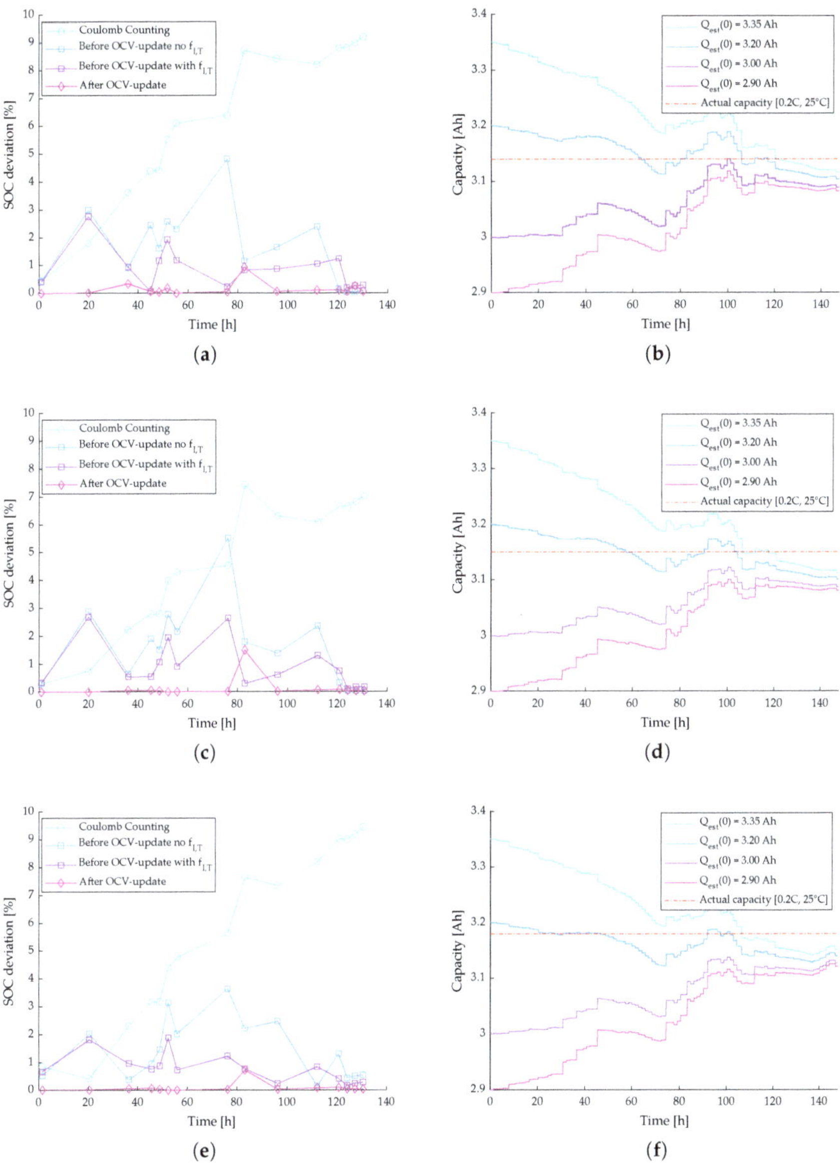

Figure 8. SOC deviation measured at fully charge state (**a**) SOC deviation (35 °C aging), (**c**) SOC deviation (25 °C aging) and (**e**) SOC deviation (15 °C aging). Actual capacity estimation using the proposed method and different starting values (**b**) actual capacity estimation (35 °C aging), (**d**) actual capacity estimation (25 °C aging) and (**f**) actual capacity estimation (15 °C aging).

6. Conclusions

An improved coulomb counting method for estimating the SOC and SOH of Lithium-ion batteries is proposed in this work. The coulombic efficiency of charging and discharging processes of the NMC batteries used was studied for different working conditions, as well as the effect of the temperature on the OCV. The results of this analysis are the key to develop a simple but accurate estimation algorithm.

The use of the OCV-SOC relationship allows us to reset the SOC estimation more frequently and thus to reduce the accumulated error. Moreover, it provides an effective mechanism for those applications which never or rarely reach the low or high ends of the SOC range.

The relaxation period affects the estimation of the SOC and SOH. In this work the effectiveness of the proposed method was demonstrated for resting times as low as one hour, obtaining absolute errors below 3% in the SOC estimation. However, the results also reflect that the relaxation time depends on the operating conditions, requiring longer periods for lower temperatures. In future improvements of this algorithm, the use of dynamic conditions for setting the minimum relaxation time will be considered.

Author Contributions: Conceptualization, A.G., E.S. and D.-I.S.; methodology, A.G.; software, A.G.; validation, A.G.; formal analysis, A.G.; investigation, A.G.; resources, E.S. and D.-I.S.; data curation, A.G.; writing—original draft preparation, A.G.; writing—review and editing, E.S. and D.-I.S.; visualization, A.G.; supervision, E.S. and D.-I.S.; project administration, E.S. and D.-I.S.; funding acquisition, E.S. and D.-I.S. All authors have read and agreed to the published version of the manuscript.

Funding: This research was funded by Electric Mobility Europe Call 2016 (ERA-NET COFUND) and Innovation Fund Denmark grant number 7064-00011B.

Conflicts of Interest: The authors declare no conflict of interest.

Abbreviations

The following abbreviations are used in this manuscript:

BMS	Battery Management System
BOL	Beginning Of Life
CC	Constant Current
CV	Constant Voltage
NMC	Nickel Manganese Cobalt Oxide
RLS	Recursive Least Squares
RPT	Reference Performance Test
OCV	Open Circuit Voltage
SOC	State of Charge
SOH	State of Health
WLTP	World Harmonized Light-duty Vehicle Test Procedure

References

1. Stroe, A.I.; Knap, V.; Stroe, D.I. Comparison of lithium-ion battery performance at beginning-of-life and end-of-life. *Microelectr. Reliab.* **2018**, *88–90*, 1251–1255. [CrossRef]
2. Bubbico, R.; Greco, V.; Menale, C. Hazardous scenarios identification for Li-ion secondary batteries. *Saf. Sci.* **2018**, *108*, 72–88. [CrossRef]
3. Hannan, M.; Lipu, M.; Hussain, A.; Mohamed, A. A review of lithium-ion battery state of charge estimation and management system in electric vehicle applications: Challenges and recommendations. *Renew. Sustain. Energy Rev.* **2017**, *78*, 834–854. [CrossRef]
4. Ng, K.S.; Moo, C.S.; Chen, Y.P.; Hsieh, Y.C. Enhanced coulomb counting method for estimating state-of-charge and state-of-health of lithium-ion batteries. *Appl. Energy* **2009**, *86*, 1506–1511. [CrossRef]

5. Dang, X.; Yan, L.; Jiang, H.; Wu, X.; Sun, H. Open-circuit voltage-based state of charge estimation of lithium-ion power battery by combining controlled auto-regressive and moving average modeling with feedforward-feedback compensation method. *Int. J. Electr. Power Energy Syst.* **2017**, *90*, 27–36. [CrossRef]

6. Lee, J.; Nam, O.; Cho, B. Li-ion battery SOC estimation method based on the reduced order extended Kalman filtering. *J. Power Sources* **2007**, *174*, 9–15. [CrossRef]

7. Zhi, L.; Peng, Z.; Zhifu, W.; Qiang, S.; Yinan, R. State of Charge Estimation for Li-ion Battery Based on Extended Kalman Filter. *Energy Procedia* **2017**, *105*, 3515–3520. [CrossRef]

8. Wassiliadis, N.; Adermann, J.; Frericks, A.; Pak, M.; Reiter, C.; Lohmann, B.; Lienkamp, M. Revisiting the dual extended Kalman filter for battery state-of-charge and state-of-health estimation: A use-case life cycle analysis. *J. Energy Storage* **2018**, *19*, 73–87. [CrossRef]

9. Stroe, D.I.; Swierczynski, M.; Stroe, A.I.; Knudsen Kaer, S. Generalized Characterization Methodology for Performance Modelling of Lithium-Ion Batteries. *Batteries* **2016**, *2*, 37. [CrossRef]

10. Stroe, D.; Swierczynski, M.; Kær, S.K.; Teodorescu, R. Degradation Behavior of Lithium-Ion Batteries During Calendar Ageing—The Case of the Internal Resistance Increase. *IEEE Trans. Ind. Appl.* **2018**, *54*, 517–525. [CrossRef]

11. Einhorn, M.; Conte, F.V.; Kral, C.; Fleig, J. A Method for Online Capacity Estimation of Lithium Ion Battery Cells Using the State of Charge and the Transferred Charge. *Ind. Appl. IEEE Trans.* **2010**, *48*, 736–741. [CrossRef]

12. Plett, G.L. *Battery Management Systems*; Artech House: Norwood, MA, USA, 2016; Volume 2.

Article

Model-Based Adaptive Joint Estimation of the State of Charge and Capacity for Lithium–Ion Batteries in Their Entire Lifespan

Zheng Chen [1,2], Jiapeng Xiao [1], Xing Shu [1], Shiquan Shen [1], Jiangwei Shen [1,*] and Yonggang Liu [3,*]

[1] Faculty of Transportation Engineering, Kunming University of Science and Technology, Kunming 650500, China; chen@kust.edu.cn (Z.C.); xiaolangxjp@163.com (J.X.); shuxing92@kust.edu.cn (X.S.); shiquan219@gmail.com (S.S.)
[2] School of Engineering and Materials Science, Queen Mary University of London, London E1 4NS, UK
[3] State Key Laboratory of Mechanical Transmissions & School of Automotive Engineering, Chongqing University, Chongqing 400044, China
*   Correspondence: shenjiangwei6@163.com (J.S.); andyliuyg@cqu.edu.cn (Y.L.)

Received: 31 January 2020; Accepted: 13 March 2020; Published: 18 March 2020

Abstract: In this paper, a co-estimation scheme of the state of charge (SOC) and available capacity is proposed for lithium–ion batteries based on the adaptive model-based algorithm. A three-dimensional response surface (TDRS) in terms of the open circuit voltage, the SOC and the available capacity in the scope of whole lifespan, is constructed to describe the capacity attenuation, and the battery available capacity is identified by a genetic algorithm (GA), together with the parameters related to SOC. The square root cubature Kalman filter (SRCKF) is employed to estimate the SOC with the consideration of capacity degradation. The experimental results demonstrate the effectiveness and feasibility of the co-estimation scheme.

Keywords: state of charge; available capacity; adaptive model-based algorithm; square root cubature Kalman filter; joint estimation

---

1. Introduction

Nowadays, energy crises and environmental damage have become the main concerns of society, and require being tackled with high attention [1]. Transportation electrification provides a possible manner to reduce emissions and dependence upon fossil fuels. Electric vehicles (EVs) and hybrid EVs (HEVs) are promising solutions, which however, require electrical energy storage systems to completely or partially replace propelling power supplied by traditional internal combustion engines [2]. In this context, applications of lithium–ion batteries have been intensively spurred due to their numerous advantages, such as their wide environmental temperature operation capability, high energy density, long lifespan and their large charge/discharge current [3]. For lithium–ion batteries, the state of charge (SOC) and available capacity, usually provided by battery management systems (BMSs), are crucial parameters for evaluation of the electrical performance of the battery, as well as for the control of the vehicle.

Typically, estimation methods of the SOC can be divided into four categories, including the coulomb counting method, and characterization parameter-based methods such as the open circuit voltage (OCV) method, model-based methods and data-driven methods. Amongst them, the coulomb counting method [4] and OCV method [5] have been widely applied in BMSs of EVs, because of their simplicity and ease of implementation, whereas the former is prone to the production of large accumulated errors, due to interferences or uncertainties of current sensors/transducers and inaccurate initial values, and the latter is not suitable for online estimation, as it usually costs long shelving time to acquire the OCV value. With the development of computation technologies and machine learning, a variety of artificial intelligence-based, data-driven methods, such as neural networks [6] and support vector machines [7], are proposed for SOC estimation by establishing black-box models. Data-driven methods feature a strong nonlinear mapping capability with high accuracy; however, these approaches show high complexity, and require a considerable amount of training data. Alternatively, model-based methods have been widely investigated and applied for SOC estimation, thanks to the capabilities of online application, high precision and the independence of initial values. Conventional modeling manners mainly include electrochemical models and the equivalent circuit models (ECMs). Compared with complicated electrochemical models, ECM is commonly used to describe the electrical behavior of batteries, and subsequently to estimate the SOC due to its simplification and preferable precision. Yanwen Li et al. proposed a multi-model probability fusion algorithm to describe the battery's electrical characteristics, and subsequently estimate the SOC [8]. In model-based methods, the combination of the battery model and the intelligent filtering algorithm is a hotspot in SOC estimation research. The frequently used filtering algorithms include Kalman filtering (KF) [9], the H-infinity filter (HIF) [10], particle filter (PF) [11], and their various extensions. In particular, the extended KF (EKF) is widely employed to execute SOC estimation using a first-order Taylor expansion on the basis of the battery's nonlinear model [12]. Nonetheless, the second and higher order expansion is usually neglected, thus leading to slow a convergence rate, and even divergence. The unscented KF (UKF) is exploited to estimate battery SOC, based on the recursive unscented transformation to approximate the nonlinear observation without Taylor polynomial expansions [13]. The UKF shows better estimation precision and robustness than the EKF in strong, nonlinear systems [14]. On the basis of the radial–spherical cubature criterion, the cubature Kalman filter (CKF) leverages a set of volume points to approximate the mean and covariance of states with additional Gaussian noise [15]. Although CKF outperforms EKF and UKF in terms of filtering divergence and estimation error, it is susceptible to inaccurate, initial difference and disturbances, and is difficult to guarantee a symmetric and nonnegative definition of the covariance matrix all the time. HIF is applied in state estimation and model parameters identification of batteries, due to its good, anti-interference performance in high nonlinear systems [16]. PF exhibits attractive advantages in solving nonlinear, non-Gaussian distribution problems, and highlights more application potential than EKF. Thus, it has been widely developed and applied in multifarious fields, such as batteries, robotics and navigation systems [17]; however, it is limited by strong dependence upon noise and time-varying parameters of the system.

In addition, battery aging is an irreversible process with operation, where the most intuitive appearance is a decline of capacity and the increase of internal impedance [18]. In general, the attenuation process is nonlinear, complex, and even difficult to predict. To attain it, a body of algorithms have been successfully proposed and applied to achieve capacity estimation, mainly including experimental analysis methods and model-based methods. The most direct and easiest manner of evaluating the capacity is to conduct the calibration test [19]. However, it is obviously time-consuming, and only supports offline estimation. Additionally, the battery's impedance variation also highlights the capacity degradation trend [20]. However, the online electrochemical impedance measurement is not suitable for practical applications, due to its exceptional complexity of experiment. Motivated by these difficulties and constraints, incremental capacity analysis (ICA) is introduced to conduct capacity estimation by evaluating the increment of capacity in a certain charging interval [21]. Similar algorithms also include differential voltage analysis (DVA), with the help of analyzing the variation characteristics of voltage curves in predetermined charge/discharge operations [22]. Yes, they can reflect the aging mechanism of batteries, and highlight preferable accuracy; nonetheless, they are intractable to apply in practice, as chances of encountering the interval with predetermined current are seldom. Since it is time-consuming to measure and determine the battery capacity directly; model-based estimation approaches may supply an indirect manner to evaluate it. The model-based methods can leverage adaptive algorithms (such as joint estimation approaches [23] and fuzzy logic algorithms [24]) to identify the battery capacity.

These algorithms are easy to be implemented, and meanwhile demonstrate preferable accuracy; whereas, in the model-based approaches, the battery capacity is regarded as a key parameter or state variable, and is obtained by parameter identification, or estimated together with the SOC, with an established circuit model or electrochemical model. From this point of view, the model accuracy can directly affect the estimation accuracy of our battery capacity.

The above-mentioned state estimation methods are mostly developed for either SOC or capacity estimation individually, rather than for both simultaneously. SOC refers to the residual capacity rate over nominal values, while SOH represents the nominal capacity value with operations. To a certain extent, battery capacity shows the same significance as SOC, and essentially, they are tightly coupled with each other [25]. Apparently, SOC estimation based on the known and unchanged capacity exhibits certain limitations in practice. A common knowledge is that the internal parameters of batteries change with degradation. The internal resistance will increase, and the capacity decreases gradually, thus resulting in the challenges and difficulties of estimating the SOC reliably and robustly. Consequently, it is critical to update the model parameters, particularly the capacity, in a timely manner. Motivated by this, a joint estimation scheme is proposed in this study to improve the estimation accuracy of the SOC and capacity in the entire lifespan of the battery. Firstly, a second-order resistance–capacitance (RC)-based ECM is established, and the co-estimation scheme of SOC and the battery capacity is presented. In it, the square root cubature Kalman filter (SRCKF) algorithm is employed to estimate the SOC; meanwhile, the battery capacity, as one of the key model parameters, is identified by the genetic algorithm (GA), based on the constructed three-dimensional response surface (TDRS). Finally, the estimation results of the SOC and battery capacity are verified by different experimental validations over their entire lifespans. This study dedicates to the following two contributions: 1) A novel capacity estimation method based on a TDRS is proposed, and the model parameters are updated synchronously; and 2) based on the capacity and parameters revision, a co-estimation scheme is established for the SOC and capacity estimation simultaneously against different degradation statuses.

In the remainder of this study, Section 2 details the second-order RC model and the experiment test profiles. In Section 3, the co-estimation scheme of both capacity and SOC is elaborated. The validation results are exhibited and discussed in Section 4. Finally, Section 5 draws the main conclusions and looks to future works.

2. The Lithium–Ion Battery Model and the Experimental Details

2.1. Battery Modeling and Analysis

To better estimate battery states, various mathematical models have been established, including electrochemical models [26] and ECMs [27]. However, they differ greatly in accuracy, computation complexity and reliability. Considering the precision and complexity of models, a second-order RC-based ECM, as shown in Figure 1, is deployed in this work, thanks to its relatively satisfactory precision and acceptable computation intensity [28]. As can be seen, it contains two parallel RC networks connected in series topology to characterize the battery polarization. Based on Figure 1, the circuit equation can be built, as:

$$\dot{U}_s = -\frac{U_s}{R_s C_s} + \frac{I}{C_s} \tag{1}$$

$$\dot{U}_l = -\frac{U_l}{R_l C_l} + \frac{I}{C_l} \tag{2}$$

$$U_t = U_{ocv} - U_s - U_l - R_e I \tag{3}$$

where R_l and C_l indicate the internal resistance and capacitance of electrochemical polarization, while R_s and C_s denote the internal resistance and capacitance of the concentration polarization.

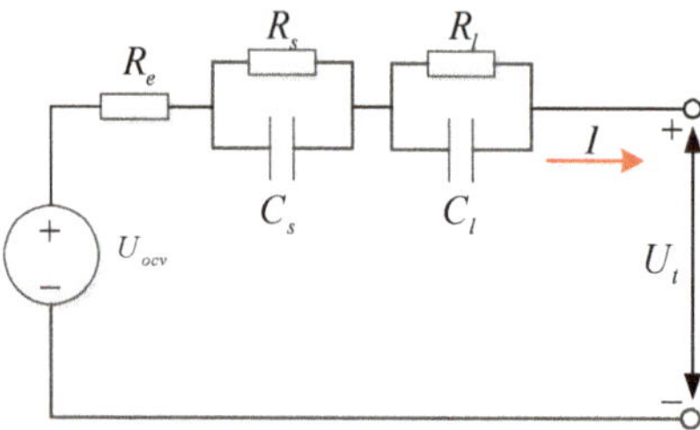

Figure 1. The second-order resistance–capacitance (RC) equivalent circuit model.

U_s and U_l denote the voltage drop across R_sC_s and R_lC_l, respectively; U_t indicates the terminal voltage; I is the loading current; U_{ocv} stands for the open circuit voltage; and R_e represents the internal ohmic resistance. In addition, the SOC denotes the ratio of available remaining capacity over the rated capacity (the maximum available capacity), as:

$$SOC(t) = SOC(t_0) - \frac{\int_{t_0}^{t} \eta_c I(t)dt}{Q_N} \tag{4}$$

where $SOC(t)$ indicates the SOC value at t, respectively; $SOC(t_0)$ stands for the SOC value at t_0; Q_N is the rated capacity of battery; and η_c represents the columbic efficiency.

2.2. Experiments

The basic specifications of the battery are illustrated in Table 1. The battery's nominal voltage is 3.6 V, and the nominal capacity is 2.55 Ampere hour (Ah). To characterize the battery's electrical performance, some prerequisite experiments are conducted, including an accelerated aging test, performance test, and dynamic test. The accelerated cycle life aging test is carried out at 25 °C, which is divided into seven stages. They are separated by the cycles 0, 30, 60, 90, 120, 150 and 180 (defined as cyc0, cyc30, cyc60, cyc90, cyc120, cyc150 and cyc180 hereinafter). The battery cell is charged by means of the constant current-constant voltage (CCCV) scheme with the current rate of 1 C, and discharged by means of constant current (CC), with the current rate of 2 C in each cycle. Here, C denotes the rated capacity value of the battery with the unit of Ah. The performance tests, including the capacity test and hybrid pulse power characterization (HPPC) test, are carried out periodically during the cycle life test. In addition, a typical dynamic test based on the urban dynamometer driving schedule (UDDS) is executed to verify the performance under dynamic operating conditions. Figure 2 shows the decay variation of the discharge capacity. It can be clearly observed that the discharge capacity basically remains unchanged from cyc0 to cyc30, and it tends to decline faster after 30 cycles, and the amount of electric energy decreases faster as the cycle number increases. After 180 cycles, the maximum discharge capacity decreases from the initial 2.614 Ah to 1.129 Ah, which remains only 44.3% of the nominal capacity. The discharge capacity decreases to 1.97 Ah at the fifth stage (cyc120), which is 77.25% of the nominal capacity. When the capacity drops to 80% of the rated value, the battery should be abandoned; and therefore, only the experimental results of the first five aging stages are applied to analyze and verify the performance of proposed algorithm in this work.

Table 1. The specifications of the battery cell.

Material	Ternary Lithium–Ion Battery
Nominal capacity	2.55 Ah
Nominal voltage	3.6 V
End-of-charge voltage	4.2 V
End-of-charge current	51 mA
End-of-discharge voltage	2.5 V

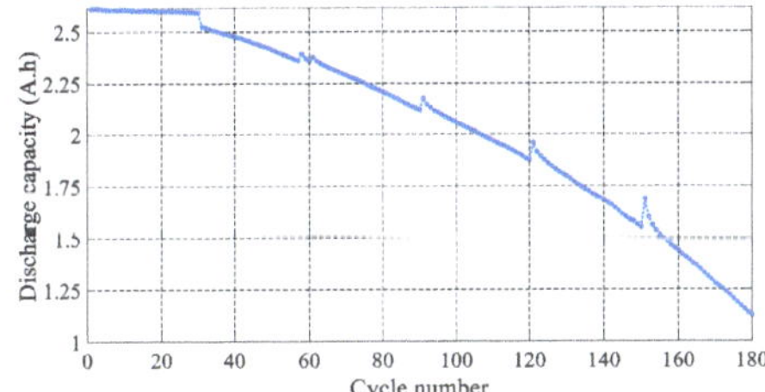

Figure 2. The decay of battery cell discharge capacity under 180 cycles.

Furthermore, battery degradation not only features the capacity reduction, but also embodies the variation of the OCV-SOC relationship. The relationship curve between OCV and SOC at various aging status is presented in Figure 3. It is apparent that the OCV–SOC relationship curve changes gradually as the cycle number increases. When the SOC is more than 20%, the trend of the curves remains basically the same. However, when the SOC is less than 20%, especially under 10%, the OCV changes significantly. Generally, to protect and extend the battery life of EVs, the battery discharge cut-off SOC is usually set to 10% or 20%. When the SOC ranges from 20% to 60%, the OCV at different aging status differs obviously, and the maximum difference value is 20 mV. When the SOC is more than 60%, the OCV values at different aging statuses are relatively close, and the maximum difference is within 10 mV.

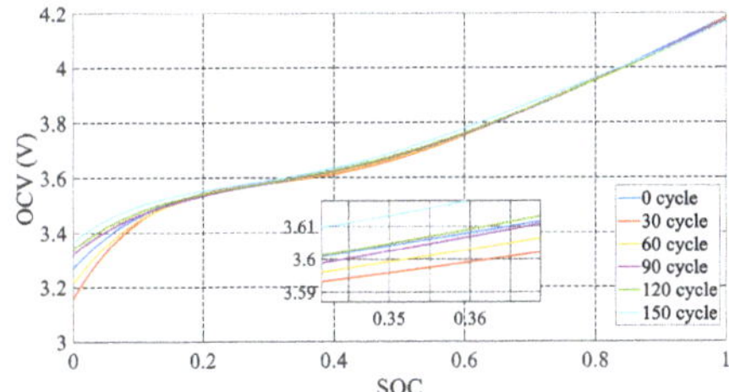

Figure 3. The relationship curve of the open circuit voltage (OCV) and state of charge (SOC) at different aging stages.

3. The Joint Estimation of SOC and Battery Capacity

The framework of the joint capacity and SOC co-estimation is shown in Figure 4. It mainly includes four parts: the strategy module, the modeling module, the capacity and parameters estimation module, as well as the SOC estimation module. First, the strategy module starts to accumulate the experimental data until the length of data is more than the preset threshold. Then, the capacity estimation module employs the GA to conduct the parameter identification and capacity estimation, based on the acquired data and the established model. After finding the model parameters and capacity, the SOC estimation module is triggered to estimate the SOC based on the SRCKF. Note that the modeling and parameters estimation modules are not invoked every time. The detailed co-estimation procedure is elaborated in the following.

3.1. The Capacity Estimation Algorithm

The battery capacity is deemed to be a significant parameter that needs to be identified. Firstly, based on the OCV–SOC curves illustrated in Figure 3, a TDRS with respect to the capacity, SOC and OCV, is constructed, as plotted in Figure 5. Next, the TDRS is imported into the established battery model. Finally, the usable battery capacity is incorporated into the battery model parameters, and is identified by the GA [29].

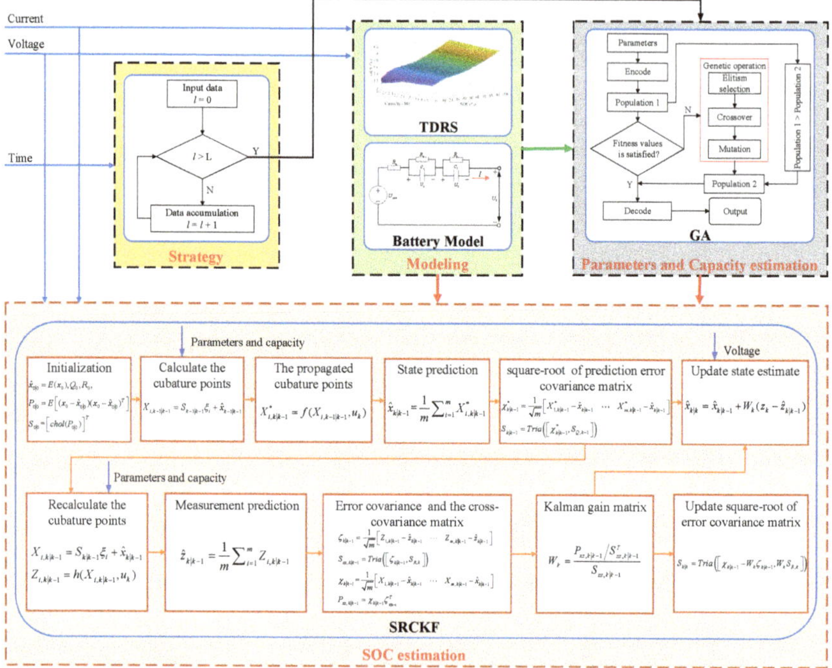

Figure 4. The scheme of the SOC and battery capacity co-estimation algorithm for lithium–ion batteries.

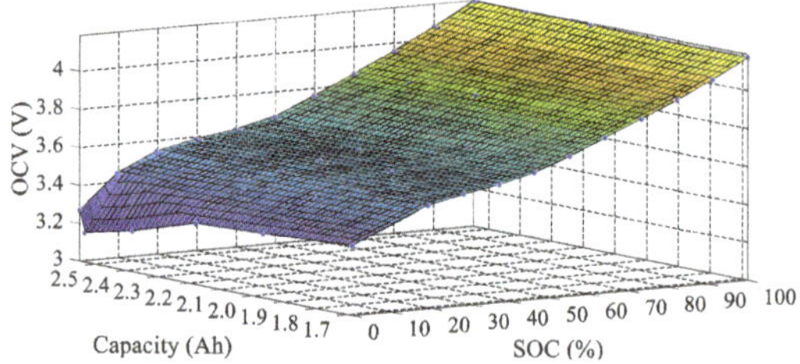

Figure 5. The three-dimensional response surface via capacity–SOC–OCV.

By this manner, the problem of usable battery capacity estimation can be transformed into the problem of searching the optimal OCV–SOC relationship match on the built TDRS by applying the optimization algorithm. It is worth noting that the ambient temperature plays an important influence on battery capacity. However, the temperature influence is not taken into account in this study, as the battery system onboard is generally equipped with a good thermal management system, thereby ensuring the temperature variation is within ±5 °C [30].

In consideration of accuracy and complexity, a fifth-order polynomial function is selected to describe the relationship among OCV, SOC and capacity, as:

$$U_{OCV}(SOC, C_{a,i}) = \alpha_{1,i} \times SOC^5 + \alpha_{2,i} \times SOC^4 + \alpha_{3,i} \times SOC^3 + \alpha_{4,i} \times SOC^2 + \alpha_{5,i} \times SOC + \alpha_{6,i} \quad (5)$$

where $C_{a,i}$ represents the available battery capacity at the ith capacity point.

$\alpha_{j,i}(j = 1, \cdots, 6)$ denotes the fitting coefficient of OCV and SOC at the ith capacity point, which is no longer a constant, and is herein defined as a quadratic function of $C_{a,i}$, as:

$$\alpha_{j,i} = b_{2,i}C_{a,i}^2 + b_{1,i}C_{a,i} + b_{0,i} \tag{6}$$

where $b_{t,i}(t = 0,1,2)$ denotes the capacity coefficient. The above equation can be rewritten into a matrix form, as:

$$[\alpha_{1,i}, \alpha_{2,i}, \alpha_{3,i}, \alpha_{4,i}, \alpha_{5,i}, \alpha_{6,i}]^T = \Gamma \times \begin{bmatrix} C_{a,i}^2 & C_{a,i} & 1 \end{bmatrix}^T \tag{7}$$

where Γ refers to a 6×3 capacity coefficient matrix, which can be obtained by the polynomial fitting, and the results are described in Equation (8).

$$\Gamma = \begin{bmatrix} 14.2473 & -57.3965 & 61.0568 \\ -39.1366 & 155.6827 & -165.5359 \\ 38.9053 & -151.7664 & 161.0539 \\ -16.8787 & 63.8408 & -66.9571 \\ 3.0020 & -10.7412 & 11.42.84 \\ -0.1149 & 0.2951 & 3.2020 \end{bmatrix} \tag{8}$$

Now, according to Equations (5)–(8), a nonlinear relationship can be built among OCV, capacity and SOC. Once the SOC is determined, it will be a deterministic mapping function between OCV and SOC. Together with Equations (1)–(3), the capacity identification can be conducted simultaneously with other parameters, including those of the RC networks. During the parameters identification, the variation of model parameters is eventually reflected by the difference between the estimation result and the terminal voltage. Based on the OCV model, the discrete mathematical expression of the second-order RC-based ECM can be reformulated as:

$$\begin{aligned} U_{t,k} = U_{ocv,k}(SOC_k, C_a) &- \left\{ \exp(-\Delta t/\tau_s)U_{s,k} + R_s[1 - \exp(-\Delta t/\tau_s)]I_k \right\} \\ &- \left\{ \exp(-\Delta t/\tau_l)U_{l,k} + R_l[1 - \exp(-\Delta t/\tau_l)]I_k \right\} - I_k R_e \end{aligned} \tag{9}$$

where Δt indicates the time interval, and both $\tau_s = R_s C_s$ and $\tau_l = R_l C_l$ belong to time constants. As can be seen from Equation (9), the TDRS based on the OCV model in Equation (5) is imported into the ECM. Hence, the difference of relationship between SOC and OCV at different capacity levels is eventually highlighted by the estimation results of the terminal voltage. In this manner, the battery capacity can be added into the model parameter series for identification. To attain it, the GA is employed to find the optimal combination of model parameters and capacity, in which the optimal parameter group to be identified can be expressed as:

$$\theta_{optimal} = [R_e, R_s, R_l, C_s, C_l, C_a] \tag{10}$$

During the identification process, the minimum root mean squared error (RMSE) of the terminal voltage is taken as the fitness function, as:

$$\left\{ \begin{aligned} &\varphi(\theta_{optimal}) = \min|RMSE| \\ &RMSE = \sqrt{\frac{\sum\limits_{k}^{N}[U_{t,k} - \hat{U}_{t,k}(\theta_{optimal})]^2}{N}} \end{aligned} \right. \tag{11}$$

where $U_{t,k}$ and $\hat{U}_{t,k}(\theta_{optimal})$ represent the measured terminal voltage and the estimated terminal voltage at step k, respectively.

In addition, the constraints of optimization algorithm are subject to:

$$\begin{cases} 20\% \leq SOC \leq 100\% \\ 0.005 \ \Omega \leq R_e, R_s, R_l \leq 0.1 \ \Omega \\ 0.5 \ \text{s} \leq \tau_s \leq 1000 \ \text{s} \\ 0.5 \ \text{s} \leq \tau_l \leq 1000 \ \text{s} \end{cases} \tag{12}$$

The setting of these constraints is explained as follows. In practical applications, to avoid the over-charge and over-discharge of the battery, the range of the SOC is generally set to 20% to 100% for guaranteeing proper operation and extending the service life of batteries [31]. The upper limits of internal ohmic resistance R_e, internal resistance R_l of electrochemical polarization and internal resistance R_s of the concentration polarization are determined in terms of the specifications of batteries and the technical parameters supplied by the manufacturers. Their low limits are all determined to be 0.005 Ω, based on the parameter calculation of ECM introduced in [32], as well as the experimental analysis. Meanwhile, the range of C_l and C_s can be deduced to be 100 F to 10^4 F. In addition, τ_l and τ_s are time constants, where $\tau_l = R_l C_l$ and $\tau_s = R_s C_s$. Hence, the range of τ_l and τ_s can be limited with 0.5 s to 1000 s. In summary, when the above-mentioned battery model parameters and capacity are identified, these parameters will be transmitted into the SOC estimation module. However, it is worth noting that the parameter identification based on the GA requires a certain amount of data to obtain an ideal identification result. Therefore, the capacity estimation method proposed in this paper only runs when the data length reaches a pre-set condition, and the determination of data length will be discussed in the next section.

3.2. The SOC Estimation Algorithm

After obtaining the model parameters and battery capacity, the SRCKF is proposed to attain the estimation of battery SOC with the cyclic recurrence based on the established second-order RC ECM. In comparison with the traditional cubature Kalman filter (CKF), the SRCKF can directly perform iterative update in the form of calculating the square-roots of the covariance matrices during the filtering process, which determines the non-negative definite value of the covariance matrix, and avoids the divergence of filter [33]. In general, a discrete nonlinear dynamic system with enhanced noise can be modeled, as:

$$\begin{cases} x_{k+1} = f(x_k, u_k) + w_k \\ z_k = h(x_k, u_k) + v_k \end{cases} \tag{13}$$

where $x_k \in R^n$ and z_k indicate the system state vector and the system output at time k, respectively. $f(\cdot)$ and $h(\cdot)$ denote the nonlinear system state function and nonlinear measurement function, respectively. w_k stands for random process noise indicating uncertain input. v_k denotes the observation noise, which is generally employed to simulate sensor noise affecting the output measurement. Additionally, the corresponding covariance of w_k and v_k are Q_k and R_k, respectively. Based on the established ECM, the time-discrete state equation and measurement equation can be respectively expressed, as:

$$\begin{bmatrix} U_{s,k+1} \\ U_{l,k+1} \\ SOC_{k+1} \end{bmatrix} = \begin{bmatrix} \exp(-\Delta t/\tau_s) & 0 & 0 \\ 0 & \exp(-\Delta t/\tau_l) & 0 \\ 0 & 0 & 1 \end{bmatrix} \begin{bmatrix} U_{s,k} \\ U_{l,k} \\ SOC_k \end{bmatrix} + \begin{bmatrix} R_s(1 - \exp(-\Delta t/\tau_s)) \\ R_l(1 - \exp(-\Delta t/\tau_l)) \\ -\eta_c \Delta t/C_a \end{bmatrix} I_k + \begin{bmatrix} w_{1,k} \\ w_{2,k} \\ w_{3,k} \end{bmatrix} \tag{14}$$

$$U_{t,k} = U_{ocv,k}(SOC_k, C_a) + \begin{bmatrix} -1 & -1 & 0 \end{bmatrix} \begin{bmatrix} U_{s,k} \\ U_{l,k} \\ SOC_k \end{bmatrix} + [-R_e] I_k + v_k \tag{15}$$

where the system state variable $x_k = \begin{bmatrix} U_{s,k} & U_{l,k} & SOC_k \end{bmatrix}^T$, input variable $u_k = I_k$ and system output $z_k = U_{t,k}$. In this study, The SRCKF algorithm is adopted to estimate the SOC, of which the general process is summarized in Table 2, where n is the state dimension, and m denotes the total number of volume points, which number is twice those of the state dimension. The sample $[1]$ indicates a complete set of fully symmetric points, of which the set of points is obtained through the complete permutation of elements of the n-dimensional unit vector $e = [1, 0 \cdots 0]^T$ and the alteration of the element symbol. $[1]_g$ represents that the point is centered at the gth point of $[1]$. $\hat{x}_k$ and $\hat{z}_k$ are the predicted state and measurement, respectively. $S_{Q,k-1}$ and $S_{R,k}$ denote the square-roots of the process noise covariance matrix Q_{k-1} and the measurement noise covariance matrix R_k, respectively.

Table 2. The process of SOC estimation based on the square root cubature Kalman filter (SRCKF) algorithm.

(a) Initialization:

$$\begin{cases} \hat{x}_{0|0} = E[x_0] \\ P_{0|0} = E\left[(x_0 - \hat{x}_{0|0})(x_0 - \hat{x}_{0|0})^T\right] \end{cases} \tag{16}$$

Determine the initial value $S_{0|0}$ of the square roots of the error covariance matrix by the Cholesky decomposition:

$$S_{0|0} = \left[chol(P_{0|0})\right]^T \tag{17}$$

(b) Calculate the basic cubature points and weight:

$$\xi_g = \sqrt{\frac{m}{2}}[1]_g, \, (g = 1, 2, \cdots, m) \tag{18}$$

(c) Iteration:
for $k = 1, 2, \cdots, N$
    Time update:

        Step 1: calculate the cubature points:

$$X_{g,k-1|k-1} = S_{k-1|k-1}\xi_i + \hat{x}_{k-1|k-1} \tag{19}$$

        Step 2: calculate the propagated cubature points:

$$X^*_{g,k|k-1} = f(X_{g,k-1|k-1}, u_k) \tag{20}$$

        Step 3: calculate the predicted state:

$$\hat{x}_{k|k-1} = \frac{1}{m}\sum_{g=1}^{m} X^*_{g,k|k-1} \tag{21}$$

        Step 4: calculate the state-weighted center matrix:

$$\chi^*_{k|k-1} = \frac{1}{\sqrt{m}}\left[X^*_{1,k|k-1} - \hat{x}_{k|k-1} \quad \cdots \quad X^*_{m,k|k-1} - \hat{x}_{k|k-1} \right] \tag{22}$$

        Step 5: calculate the square-root of the prediction error covariance matrix:

$$S_{k|k-1} = Tria\left(\left[\chi^*_{k|k-1}, S_{Q,k-1}\right]\right) \tag{23}$$

    Measurement update:
        Step 1: recalculate the cubature points:

$$X_{g,k|k-1} = S_{k|k-1}\xi_g + \hat{x}_{k|k-1} \tag{24}$$

        Step 2: update the propagated measurement cubature points:

$$Z_{g,k|k-1} = h(X_{g,k|k-1}, u_k) \tag{25}$$

Table 2. *Cont.*

Step 3: estimate the predicted measurement:

$$\hat{z}_{k|k-1} = \frac{1}{m}\sum_{g=1}^{m} Z_{g,k|k-1} \tag{26}$$

Step 4: evaluate the measurement-weighted center matrix:

$$\zeta_{k|k-1} = \frac{1}{\sqrt{m}}\left[\ Z_{1,k|k-1} - \hat{z}_{k|k-1} \quad \cdots \quad Z_{m,k|k-1} - \hat{z}_{k|k-1}\ \right] \tag{27}$$

Step 5: estimate the square root of the innovation covariance matrix:

$$S_{zz,k|k-1} = Tria\left(\left[\zeta_{k|k-1}, S_{R,k}\right]\right) \tag{28}$$

Step 6: update the state-weighted center matrix:

$$X_{k|k-1} = \frac{1}{\sqrt{m}}\left[\ X_{1,k|k-1} - \hat{x}_{k|k-1} \quad \cdots \quad X_{m,k|k-1} - \hat{x}_{k|k-1}\ \right] \tag{29}$$

Step 7: estimate the cross-covariance matrix:

$$P_{xz,k|k-1} = X_{k|k-1}\zeta_{k|k-1}^{T} \tag{30}$$

Moreover, update the Kalman gain, state and square root of the error covariance
Step 1: estimate the Kalman gain matrix:

$$W_k = \frac{P_{xz,k|k-1}/S_{zz,k|k-1}^{T}}{S_{zz,k|k-1}} \tag{31}$$

Step 2: estimate the final updated state:

$$\hat{x}_{k|k} = \hat{x}_{k|k-1} + W_k(z_k - \hat{z}_{k|k-1}) \tag{32}$$

Step 3: update the corresponding square-root of the error covariance matrix:

$$S_{k|k} = Tria\left(\left[X_{k|k-1} - W_k\zeta_{k|k-1}, W_kS_{R,k}\right]\right) \tag{33}$$

End

4. Verification and Discussion

4.1. Verification Study on Different Data Lengths

In this section, different data lengths are selected to investigate the effectiveness of the proposed co-estimation algorithm. Actually, the experimental data at different aging stages can be chosen to verify the SOC estimation of different data lengths. Nonetheless, based on the estimated capacity, the estimation error of battery capacity is maximum at cyc30. Hence, to better verify the performance of the proposed estimation algorithm, the battery after being cycled 30 times is chosen as the test target, and the current schedules acquired based on the UDDS experiment are repetitively operated until the terminal voltage reaches the cut-off voltage designated by the manufacturer. Note that when the data length is less than the pre-set threshold value, the SOC module still uses the previously identified capacity and parameters to conduct the estimation.

Figure 6 shows the current profiles under the UDDS experiment. It can be clearly found that the entire discharging process takes around 554 min. To evaluate the influence incurred by different data lengths when identifying the model parameters, the data with the duration of 65, 84, 130 and 200 min (defined as 65 min, 84 min, 130 min and 200 min, respectively) are randomly selected as the test target, and the remaining data are applied for SOC estimation. Note that when the data length is 554 min (total loading profile process), the SOC is estimated without the update of the capacity value and parameters. Table 3 compares the battery capacity estimation results with respect to different data lengths.

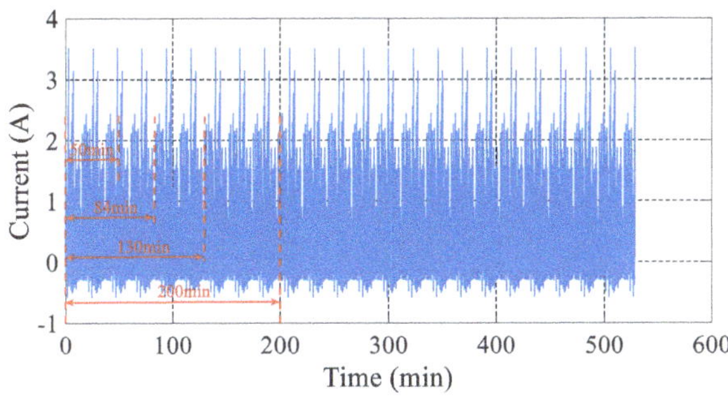

Figure 6. The urban dynamometer driving schedule (UDDS) current.

Table 3. The estimation results and errors of the battery capacity corresponding to various data lengths.

Data Length	Estimated Capacity/Ah	Absolute Error/Ah	Relative Error/%
65 min	2.4366	0.0955	3.7716
84 min	2.4710	0.0611	2.4130
130 min	2.4854	0.0467	1.8443
200 min	2.5065	0.0256	1.0110

The actual capacity 2.5321 Ah is measured through the calibration test, and the estimated capacity ranges from 2.4366 Ah to 2.5065 Ah. It can be observed that data duration shows certain influence on the estimation results, and the estimated error of the battery capacity decreases by 2.7606%, from 3.7716% to 1.011%, after increasing the data length.

Based on the obtained capacity, the detailed results of the SOC and estimation error with different identification data lengths are presented in Figure 7, and the statistic results are provided in Figure 8. As demonstrated in Figure 7, when the date length increases from 54 min to 200 min, the estimated SOC can quickly converge to the reference value according to the updated parameters and capacity, and the maximum absolute error decreases from 3.643% to 0.989%. When the data length reaches 200 min, the estimation errors are restricted within a small range, less than 1%. Besides, Figure 7 also shows the SOC estimation results without considering the capacity's update. The maximum absolute error, the mean absolute error and the RMSE are 2.538%, 1.661% and 1.737%, respectively. It is apparent that the estimated SOC looks more divergent without the capacity update, thereby manifesting the advance of the joint estimation algorithm. From Figure 8, we can find that when the data length increases from 65 min to 200 min, the maximum absolute error, mean absolute error and RMSE decrease from 3.643%, 2.331%, 2.498% to 0.989%, 0.234%, 0.319%, respectively. The results demonstrate that as the calculated data length increases, the estimated SOC becomes closer to the reference value, and this is mainly because the GA shows a global optimization ability, and when more input data samples are referred, the prediction results will be more accurate. Hence, appropriately increasing the duration of data is beneficial for improving the accuracy of capacity identification and SOC estimation. Nonetheless, it is appreciably time-consuming when increasing the amount of data to estimate the battery capacity. To balance the relationship between error and calculation time, the data duration of 200 min is considered as the preferred length. In the following, the estimated results with the data length of 200 min are all adopted for SOC estimation and comparison.

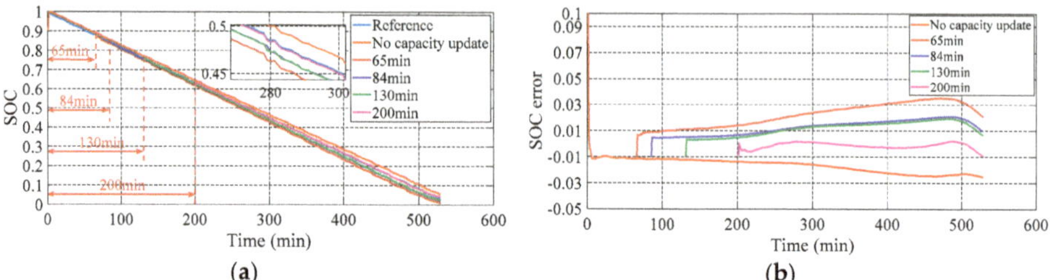

Figure 7. The results of SOC estimation based on various data lengths: (**a**) SOC estimation results; (**b**) SOC error.

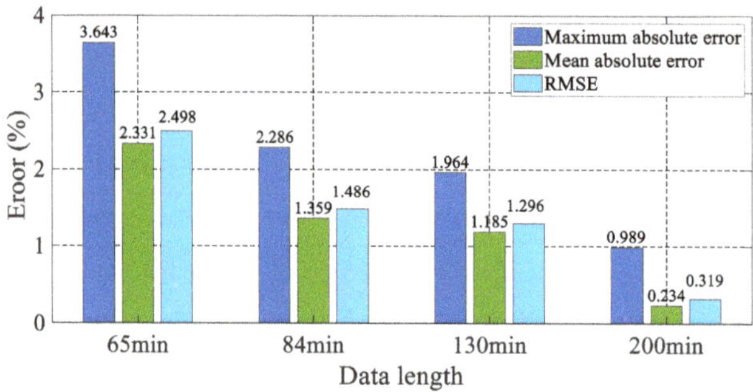

Figure 8. Comparison of the battery SOC estimation results of different data lengths.

4.2. SOC Estimation under Various Degradation Stages

To verify the feasibility of the proposed co-estimation scheme, the battery cells are experimentally and circularly tested with the UDDS current at different aging levels. According to the estimation algorithm of capacity addressed previously, the pre-set data length is 200 min. Table 4 and Figure 9 compare the estimated results of battery capacity at different aging stages (fresh, nearly fresh, slightly cycled, severely cycled and lifespan exceeded), of which the number of cycles ranges from 0 to 120, with 30 as the interval. As illustrated in Table 4, the battery capacity declines with the cycling operation. The proposed algorithm enables that the maximum relative and absolute errors are less than 1.011% and 0.026 Ah when the battery is cycled for 30 times, thereby indicating its preferable capability of estimating the battery capacity at different aging statuses. Furthermore, the estimated results can also commendably reflect the decay trend of battery capacity.

Table 4. The estimated results and errors of the battery during the entire lifespan.

Cycle Number	Actual Capacity/Ah	Estimated Capacity/Ah	Absolute Error/Ah	Relative Error/%
cyc0	2.5478	2.5344	0.0134	0.5259
cyc30	2.5321	2.5065	0.0256	1.0110
cyc60	2.3788	2.3655	0.0133	0.5591
cyc90	2.1779	2.1758	0.0021	0.0964
cyc120	1.9655	1.9551	0.0104	0.5291

After finding the model parameters including the capacity value, the estimated SOC results at different aging status are demonstrated in Figure 10, and the statistic results are summarized in Figure 11. As Figure 10 suggests, it is obvious that when the battery ages, the total discharging time gradually decreases under the same operating conditions. The initial SOC is 20%, with the error of 80%, and the estimated SOC at various aged status can all converge to the reference values. Figure 10 also reveals that the maximum absolute error of SOC is restricted within 1% after the correction of the initial

SOC error, even when the battery is aged, and the convergence time is less than 120 s. As discussed previously, the accuracy of SOC estimation is heavily influenced by the battery degradation. Without considering the update of battery capacity, the SOC estimation error increases towards higher numbers of cycles. In this study, the updated battery capacity is exploited to assist in improving the SOC estimation in the entire lifespan. As can be found in Figure 11, the maximum value of absolute error, mean absolute error and the RMSE are 0.987%, 0.484% and 0.566%, respectively, occurring in cyc30. The reason is that when the cycle number is 30, the estimation error of capacity reaches 1.011%, thus leading to the worst SOC estimation. Even so, the maximum absolute error of SOC is still restricted within 1.1% after the correction of initial SOC. As the number of battery cycles increases, the estimated SOC error does not increase obviously, manifesting that the updated capacity value contributes to the SOC estimation. From this point of view, regular updates of battery capacity in the aging process are imperative to improve the accuracy of SOC estimation.

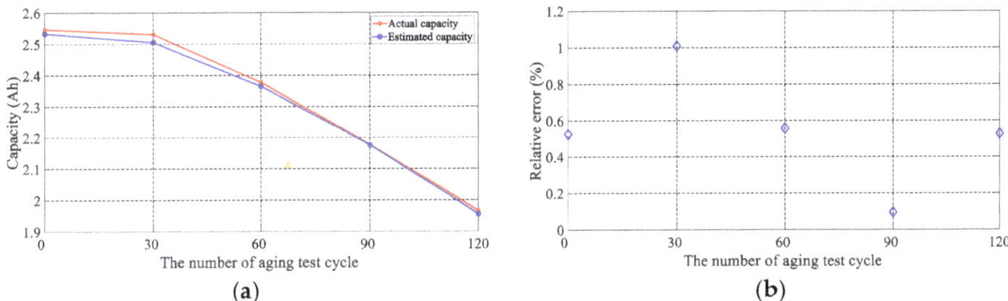

Figure 9. Capacity value of measurement and estimation at various degradation extents. (**a**) Estimation results; (**b**) Relative error.

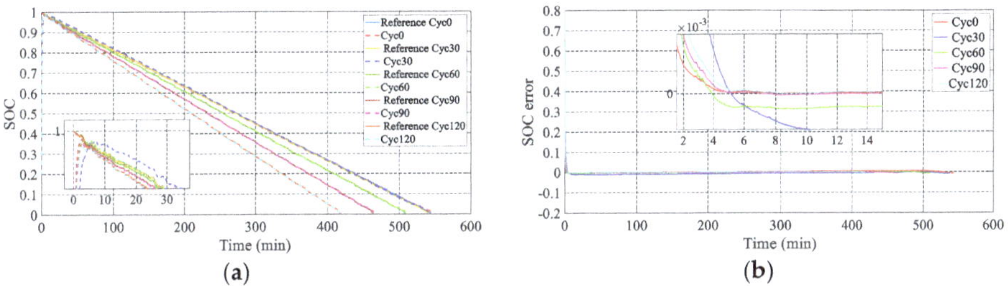

Figure 10. The results of SOC estimation with the aged battery cell: (**a**) SOC estimation results; (**b**) SOC error.

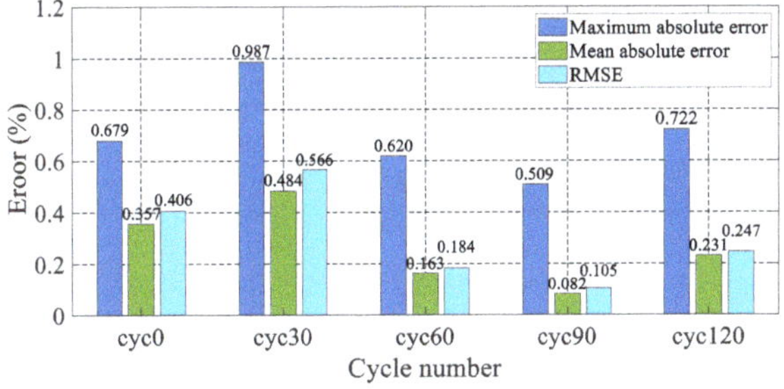

Figure 11. The estimated errors of the SOC corresponding to the aging battery cell.

5. Conclusions

In this study, a model-based adaptive joint estimation algorithm of SOC and capacity is proposed for lithium–ion batteries. The SOC estimation is implemented based on a second-order ECM, with the SRCKF algorithm considering the capacity degradation and parameters variation. The battery capacity is imported into the model parameter group, and it is jointly identified by the GA and the constructed TDRS. After obtaining the parameters and the capacity, the SOC is accurately estimated by the SRCKF. Through the experimental validations in terms of different degradation status, varying duration of recorded data and various dynamic operating conditions, the preferable performance of the proposed method is satisfactorily verified. The experimental results elucidate that the co-estimation approach can improve the SOC estimation accuracy in the entire battery lifespan cycle with the update of capacity, even in the cases of aged batteries and under complicated operating conditions.

In addition, this paper only investigates the SOC and capacity estimation for battery cells. However, the capacity and SOC of battery packs are also particularly critical in practical applications, and they will certainly be our research focus in the future.

Author Contributions: Conceptualization, Z.C. and J.X.; methodology, Z.C.; software, J.X.; validation, Z.C., J.X., X.S., S.S., and Y.L.; formal analysis, J.X.; investigation, Z.C.; resources, X.S.; data curation, J.X., X.S., and J.S.; writing—original draft preparation, Z.C., J.X., X.S. and J.S.; writing—review and editing, Z.C.; visualization, J.X.; supervision, Z.C.; project administration, Z.C.; funding acquisition, Z.C. All authors have read and agreed to the published version of the manuscript.

Funding: This work was supported in part by the National Science Foundation under Grant 61763021 and Grant 51775063, in part by the National Key R&D Program of China under Grant 2018YFB0104000, and in part by the EU-funded Marie Skłodowska-Curie Individual Fellowships Project under Grant 845102-HOEMEV-H2020-MSCA-IF-2018 in part.

Conflicts of Interest: The authors declare no conflict of interest.

References

1. Hu, X.; Feng, F.; Liu, K.; Zhang, L.; Xie, J.; Liu, B. State estimation for advanced battery management: Key challenges and future trends. *Renew. Sustain. Energy Rev.* **2019**, *114*, 109334. [CrossRef]
2. Chen, Z.; Shu, X.; Li, X.; Xiao, R.; Shen, J. LiFePO4 battery charging strategy design considering temperature rise minimization. *J. Renew. Sustain. Energy* **2017**, *9*, 64103. [CrossRef]
3. Chen, Z.; Shu, X.; Xiao, R.; Yan, W.; Liu, Y.; Shen, J. Optimal charging strategy design for lithium-ion batteries considering minimization of temperature rise and energy loss. *Int. J. Energy Res.* **2019**, *43*, 4344–4358. [CrossRef]
4. Zhao, L.; Lin, M.; Chen, Y. Least-squares based coulomb counting method and its application for state-of-charge (SOC) estimation in electric vehicles. *Int. J. Energy Res.* **2016**, *40*, 1389–1399. [CrossRef]
5. Xiong, R.; Yu, Q.; Wang, L.Y.; Lin, C. A novel method to obtain the open circuit voltage for the state of charge of lithium ion batteries in electric vehicles by using H infinity filter. *Appl. Energy* **2017**, *207*, 346–353. [CrossRef]
6. Houlian, W.; Gongbo, Z. State of charge prediction of supercapacitors via combination of Kalman filtering and backpropagation neural network. *IET Electr. Power Appl.* **2018**, *12*, 588–594. [CrossRef]
7. Meng, J.; Luo, G.; Fei, G. Lithium Polymer Battery State-of-Charge Estimation Based on Adaptive Unscented Kalman Filter and Support Vector Machine. *IEEE Trans. Power Electron.* **2016**, *31*, 2226–2238. [CrossRef]
8. Li, Y.; Wang, C.; Gong, J. A multi-model probability SOC fusion estimation approach using an improved adaptive unscented Kalman filter technique. *Energy* **2017**, *141*, 1402–1415. [CrossRef]
9. Campestrini, C.; Heil, T.; Kosch, S.; Jossen, A. A comparative study and review of different Kalman filters by applying an enhanced validation method. *J. Energy Storage* **2016**, *8*, 142–159. [CrossRef]
10. Charkhgard, M.; Zarif, M.H. Design of adaptive H-infinity filter for implementing on state-of-charge estimation based on battery state-of-charge-varying modelling. *IET Power Electron.* **2015**, *8*, 1825–1833. [CrossRef]
11. Pola, D.A.; Navarrete, H.F.; Orchard, M.E.; Rabie, R.S.; Cerda, M.A.; Olivares, B.E.; Silva, J.F.; Espinoza, P.A.; Perez, A. Particle-Filtering-Based Discharge Time Prognosis for Lithium-Ion Batteries with a Statistical Characterization of Use Profiles. *IEEE Trans. Reliab.* **2015**, *64*, 710–720. [CrossRef]

12. Chen, Z.; Fu, Y.; Mi, C.C. State of Charge Estimation of Lithium-Ion Batteries in Electric Drive Vehicles Using Extended Kalman Filtering. *IEEE Trans. Veh. Technol.* **2013**, *62*, 1020–1030. [CrossRef]
13. Wang, Y.; Liu, C.; Pan, R.; Chen, Z. Modeling and state-of-charge prediction of lithium-ion battery and ultracapacitor hybrids with a co-estimator. *Energy* **2017**, *121*, 739–750. [CrossRef]
14. Alkaya, A. Unscented Kalman filter performance for closed-loop nonlinear state estimation: A simulation case study. *Electr. Eng.* **2014**, *96*, 299–308. [CrossRef]
15. Liu, Z.; Dang, X.; Jing, B.; Ji, J. A novel model-based state of charge estimation for lithium-ion battery using adaptive robust iterative cubature Kalman filter. *Electron. Pow. Syst. Res.* **2019**, *177*, 105951. [CrossRef]
16. Yu, Q.; Xiong, R.; Lin, C. Online Estimation of State-of-charge Based on the H infinity and Unscented Kalman Filters for Lithium Ion Batteries. *Energy Procedia* **2017**, *105*, 2791–2796. [CrossRef]
17. Zheng, L.; Zhu, J.; Wang, G.; Lu, D.D.-C.; He, T. Differential voltage analysis based state of charge estimation methods for lithium-ion batteries using extended Kalman filter and particle filter. *Energy* **2018**, *158*, 1028–1037. [CrossRef]
18. Li, S.; Hu, M.; Li, Y.; Gong, C. Fractional-order modeling and SOC estimation of lithium-ion battery considering capacity loss. *Int. J. Energy Res.* **2019**, *43*, 417–429. [CrossRef]
19. Zhang, J.; Lee, J. A review on prognostics and health monitoring of Li-ion battery. *J. Power Sources* **2011**, *196*, 6007–6014. [CrossRef]
20. Waag, W.; Kaebitz, S.; Sauer, D.U. Experimental investigation of the lithium-ion battery impedance characteristic at various conditions and aging states and its influence on the application. *Appl. Energy* **2013**, *102*, 885–897. [CrossRef]
21. Han, X.; Ouyang, M.; Lu, L.; Li, J. A comparative study of commercial lithium ion battery cycle life in electric vehicle: Capacity loss estimation. *J. Power Sources* **2014**, *268*, 658–669. [CrossRef]
22. Feng, X.; Li, J.; Ouyang, M.; Lu, L.; Li, J.; He, X. Using probability density function to evaluate the state of health of lithium-ion batteries. *J. Power Sources* **2013**, *232*, 209–218. [CrossRef]
23. Zou, Y.; Hu, X.; Ma, H.; Li, S.E. Combined State of Charge and State of Health estimation over lithium-ion battery cell cycle lifespan for electric vehicles. *J. Power Sources* **2015**, *273*, 793–803. [CrossRef]
24. He, T.; Xu, W.; Lu, Z.; Yong, H.; Tian, G. Adaptive Fuzzy Logic Energy Management Strategy Based on Reasonable SOC Reference Curve for Online Control of Plug-in Hybrid Electric City Bus. *IEEE Trans. Intell. Transp.* **2018**, *19*, 1607–1617.
25. Hussein, A.A. Capacity Fade Estimation in Electric Vehicle Li-Ion Batteries Using Artificial Neural Networks. *IEEE Trans. Ind. Appl.* **2015**, *51*, 2321–2330. [CrossRef]
26. Jin, N.; Danilov, D.L.; van den Hof, P.M.J.; Donkers, M.C.F. Parameter estimation of an electrochemistry-based lithium-ion battery model using a two-step procedure and a parameter sensitivity analysis. *Int. J. Energy Res.* **2018**, *42*, 2417–2430. [CrossRef]
27. Wang, C.; He, H.; Zhang, Y.; Mu, H. A comparative study on the applicability of ultracapacitor models for electric vehicles under different temperatures. *Appl. Energy* **2017**, *196*, 268–278. [CrossRef]
28. Hu, X.; Li, S.; Peng, H. A comparative study of equivalent circuit models for Li-ion batteries. *J. Power Sources* **2012**, *198*, 359–367. [CrossRef]
29. Liu, L.; Zhang, M.; Buyya, R.; Fan, Q. Deadline-constrained coevolutionary genetic algorithm for scientific workflow scheduling in cloud computing. *Concurr. Comput. Pract. Exp.* **2017**, *29*, e3942. [CrossRef]
30. Jin, L.W.; Lee, P.S.; Kong, X.X.; Fan, Y.; Chou, S.K. Ultra-thin minichannel LCP for EV battery thermal management. *Appl. Energy* **2014**, *113*, 1786–1794. [CrossRef]
31. Yang, R.; Xiong, R.; He, H.; Mu, H.; Wang, C. A novel method on estimating the degradation and state of charge of lithium-ion batteries used for electrical vehicles. *Appl. Energy* **2017**, *207*, 336–345. [CrossRef]
32. Sun, X.; Ji, J.; Ren, B.; Xie, C.; Yan, D. Adaptive Forgetting Factor Recursive Least Square Algorithm for Online Identification of Equivalent Circuit Model Parameters of a Lithium-Ion Battery. *Energies* **2019**, *12*, 2242. [CrossRef]
33. Cheng, S.; Li, L.; Chen, J. Fusion algorithm design based on adaptive SCKF and integral correction for side-slip angle observation. *IEEE Trans. Ind. Electron.* **2017**, *65*, 5754–5763. [CrossRef]

Article

Dual Nonlinear Kalman Filter-Based SoC and Remaining Capacity Estimation for an Electric Scooter Li-NMC Battery Pack

Filip Maletić *, Mario Hrgetić and Joško Deur

Department of Robotics and Automation of Manufacturing Systems, Faculty of Mechanical Engineering and Naval Architecture, University of Zagreb, 10000 Zagreb, Croatia; mario.hrgetic@fsb.hr (M.H.); josko.deur@fsb.hr (J.D.)
* Correspondence: filip.maletic@fsb.hr; Tel.: +385-1-6168-555

Received: 23 December 2019; Accepted: 17 January 2020; Published: 22 January 2020

Abstract: Accurate, real-time estimation of battery state-of-charge (SoC) and state-of-health represents a crucial task of modern battery management systems. Due to nonlinear and battery degradation-dependent behavior of output voltage, the design of these estimation algorithms should be based on nonlinear parameter-varying models. The paper first describes the experimental setup that consists of commercially available electric scooter equipped with telemetry measurement equipment. Next, dual extended Kalman filter-based (DEKF) estimator of battery SoC, internal resistances, and parameters of open-circuit voltage (OCV) vs. SoC characteristic is presented under the assumption of fixed polarization time constant vs. SoC characteristic. The DEKF is upgraded with an adaptation mechanism to capture the battery OCV hysteresis without explicitly modelling it. Parameterization of an explicit hysteresis model and its inclusion in the DEKF is also considered. Finally, a slow time scale, sigma-point Kalman filter-based capacity estimator is designed and inter-coupled with the DEKF. A convergence detection algorithm is proposed to ensure that the two estimators are coupled automatically only after the capacity estimate has converged. The overall estimator performance is experimentally validated for real electric scooter driving cycles.

Keywords: electric vehicle; lithium-ion battery; estimation; Kalman filter; state-of-charge; state-of-health; resistance; open-circuit voltage; battery capacity

---

1. Introduction

Modern battery management systems (BMSs), among other functionalities, include a number of algorithms for estimating key battery state variables such as state-of-charge (SoC) and remaining available charge capacity, and model parameters such as internal resistance [1]. The SoC estimate can be used for predicting the current vehicle range, as well as for identification of current battery operating point which is important from the standpoint of ensuring battery safety. On the other hand, the internal resistance and capacity estimates are the main indicators used for tracking the battery degradation level, i.e., estimation of battery state-of-health (SoH) [2]. Furthermore, almost every battery model parameter is changing with battery degradation, so that for robust SoC and SoH estimation, those changes should be accurately tracked, as well.

Battery state and parameter estimation algorithms are often based on Kalman filters (KF), which in its basic linear version represent an optimal recursive solution for estimating hidden states of a linear, time-varying Gaussian system (i.e., probabilistic inference) [3]. While the Gaussian assumption holds in many cases based on the central limit theorem, the battery model is inherently nonlinear, which calls for application of nonlinear KF forms. Two of the most widely used nonlinear KFs are extended Kalman filter (EKF) and sigma-point Kalman filter (SPKF) [3]. The EKF relies on analytical linearization of the model around a time-varying operating point (i.e., an expected value of the estimated random state), while SPKF statistically linearizes the model around several operating points (depending on the number of states that are estimated).

Topic of state and parameter estimation of Li-ion battery cells has been addressed by many previous studies. One of the first implementations of dual extended Kalman filter-based (DEKF) estimator of SoC and resistance parameters can be found in [4]. Researchers have been upgrading the estimators ever since, e.g., using an adaptation mechanism for process noise variance recalculation [5], or applying more advanced filters such as SPKF [6] or particle filter (PF) [7]. These approaches are based on the assumption of constant model parameters/characteristics such as the SoC-dependent open-circuit voltage (OCV) characteristic $U_{oc}(SoC)$ or battery remaining capacity. Since those parameters are in fact dependent on SoH [8] and temperature [9], they should be estimated as well, for accurate and robust overall estimation.

There are several studies that account for $U_{oc}(SoC)$ variation with SoH by implementing the offline identified response surface model of U_{oc} with respect to SoC and remaining capacity [10–12]. Authors in [13] use the model migration method to adapt an offline trained model. An obvious disadvantage of this approach is related to the need of having a large data set from previously conducted aging experiments on the same cell type, as well as lack of temperature dependency in the response surface model. This disadvantage is tackled in this paper by describing the characteristic $U_{oc}(SoC)$ with a model whose parameters are estimated along the rest of model states and parameters within the DEKF structure. Moreover, this approach includes an adaptation mechanism of $U_{oc}(SoC)$ which allows for identification of $U_{oc}(SoC)$ hysteresis profile.

Remaining capacity estimators based on EKF and PF can be found in [14], while a recursive approximate least-squares approach is proposed in [15]. In both cases the characteristic $U_{oc}(SoC)$ is again considered as a constant-parameter dependence. Dual estimation of SoC and capacity can be found in [16], where authors use multiscale estimation with the online identified model, which can be regarded as a next step towards complete estimator. Certain weaknesses of that approach include: (i) Still an offline identified $U_{oc}(SoC)$ map is used, (ii) capacity estimate shows considerable variations in steady state, and (iii) the capacity estimator needs to be turned on manually after 25 min in order to ensure overall estimator stability. The multiscale estimator presented in this paper improves the capacity estimation accuracy and flexibility by using a more accurate SPKF and automated turning on the capacity estimator by means of applying a convergence detection algorithm.

Finally, a fully-electric scooter-based experimental verification of the proposed battery estimators is conducted, including consideration of different temperature operating points.

2. Experimental Setup

The experimental setup includes the fully-electric scooter Govecs S2.6+, powered by the 3.3 kW BLDC electric motor and the battery pack of 400 Li-Ni$_{0.33}$Mn$_{0.33}$Co$_{0.33}$O$_2$ (Li-NMC) cells, connected in the 20 × 20 matrix [17], with the total nominal voltage of 72 V, and the total energy capacity of 4.1 kWh. Battery pack is equipped with BMS which provides basic battery measurements and estimates accessible through the scooter CAN bus.

Electric scooters became an attractive transportation solution in urban areas with mild climate conditions, thus contributing to the current transport electrification effort aimed at reducing traffic congestion, and air and noise pollution. There are already several strong electric scooter manufacturers in the EU (and worldwide), e.g., Govecs, Ujet, Hrowin, Torrot, etc. NMC-type Li-ion batteries represent a preferred energy storage solution in scooter applications [17], because they offer favorable energy density, while not experiencing high loads (in terms of battery C-rate) and not operating in extreme, particularly low temperature conditions, in those applications.

For the research purposes, the scooter has been equipped with the measurement and telemetry system illustrated in Figure 1. The system is built around the Artronic SkyTrack telemetry module, custom-programmed for the acquisition and storage of measurement data, as well as for communication with the server through GPRS connection in real time. The measurement system consists of voltage and current measurement on battery output nodes, and acquisition of data available from the scooter CAN bus. The battery current is measured by using a precise, low-offset current transducer (LEM CAB 300, [18]), while the battery voltage is measured through a 12-bit analogue input of the telemetry module. Those two measurement values are sampled every 0.1 s and stored in the module. Selected values from the scooter internal CAN bus, such as battery voltage, current and temperature, vehicle's distance travelled, motor on/off flag, as well as the vehicle's current GPS coordinates and longitudinal velocity are stored with the sample rate of 1 s. GPRS connection is used to send data relevant for real-time tracking of scooter, such as its GPS coordinates, battery SoC, and other diagnostic parameters. The whole measurement dataset, including the fast current and voltage measurements, is stored in the telemetry module memory card and can be occasionally downloaded through USB connection to a local PC.

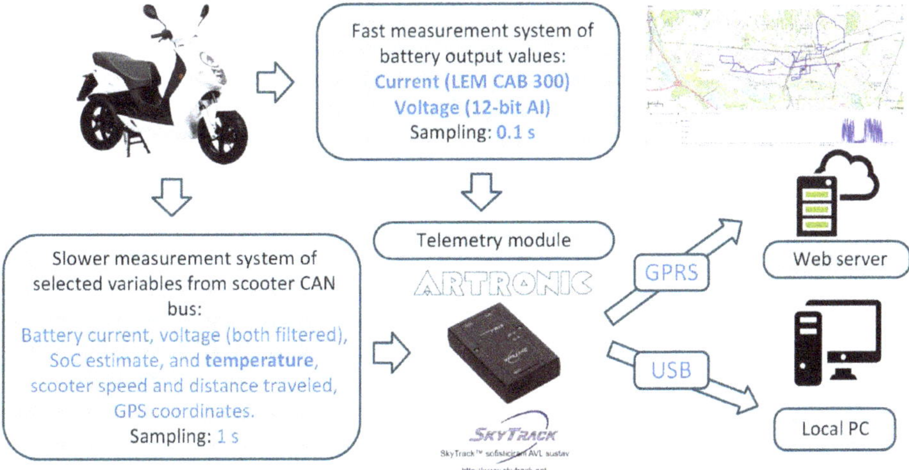

Figure 1. Scooter measurement and telemetry system.

3. Battery Pack Model

This section first presents a battery mathematical model used as a basis for SoC estimator design. Next, models employed for estimation of battery internal parameters used by the SoC estimator are presented. Finally, two offline identification experiments are described, which have been conducted to determine battery model parameters that are considered as constant or used in estimator verification.

3.1. Mathematical Model

The battery model used in this research is based on the equivalent-circuit model (ECM) showed in Figure 2, which consists of (i) a voltage source dependent on the battery SoC, i.e., the OCV characteristic $U_{oc}(SoC)$, (ii) an ohmic resistance R_{ohm} which models voltage drops in the electrolyte and electrical contacts, and (iii) a single polarization RC term (R_p and C_p) which models the slow battery dynamics, i.e., diffusion process. It should be noted that the diffusion process is more accurately modelled with the Warburg element [19] which is here avoided due to the complexity, but it can be approximated by a single or more RC elements connected in series (a single RC element is usually used as a good trade-off between simplicity and accuracy [20]). Moreover, note that the polarization resistance R_p in Figure 2 models all voltage drops that are not related to the ohmic one, including that related to charge transfer.

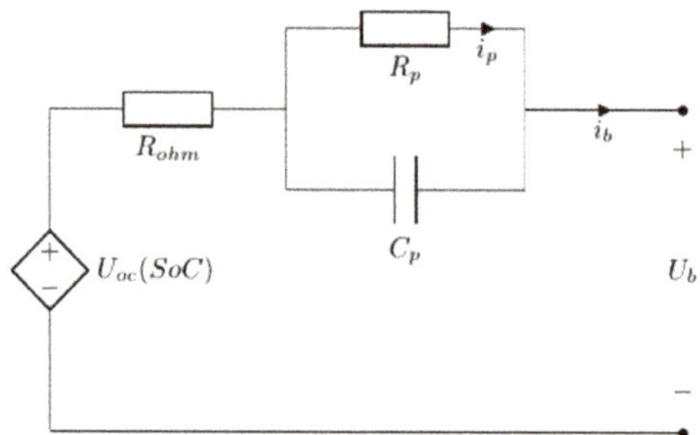

Figure 2. Battery equivalent circuit model used in the DEKF.

The above ECM can be described by the following discrete-time time-varying state-space mathematical model [4]:

$$\begin{bmatrix} SoC(k) \\ i_p(k) \end{bmatrix} = \begin{bmatrix} 1 & 0 \\ 0 & e^{-\frac{T_u}{\tau_p(SoC(k-1))}} \end{bmatrix} \begin{bmatrix} SoC(k-1) \\ i_p(k-1) \end{bmatrix} + \begin{bmatrix} -\frac{T_u}{C_n} \\ 1 - e^{-\frac{T_u}{\tau_p(SoC(k-1))}} \end{bmatrix} i_b(k-1) \tag{1}$$

$$U_b(k) = U_{oc}(SoC(k)) - R_{ohm}(k)i_b(k) - R_p(k)i_p(k) \tag{2}$$

where T_u is the filter sampling time, C_n is the battery capacity, $\tau_p = R_p C_p$ is the polarization term time constant, and k is discrete sample step.

3.2. OCV Model

Since the battery OCV is a nonlinear function of SoC, and to a lower extent temperature [9] and SoH [8], it is desirable to describe it using a parametric model such as the one used in [4]:

$$U_{oc}(SoC) = \begin{bmatrix} K_0 & K_1 & K_2 & K_3 & K_4 \end{bmatrix} \begin{bmatrix} 1 & -\frac{1}{SoC} & -SoC & \ln(SoC) & \ln(1-SoC) \end{bmatrix}^T = k_{oc} x_{oc} \tag{3}$$

where vector k_{oc} contains U_{oc}-model parameters that need to be estimated.

3.3. Model of Internal Resistance Parameters

The presented ECM has two resistance parameters in its model. Both of those resistances are known to depend on SoH and temperature [21]. So, it is important to have them estimated along with the model states. Since there is no resistance model feasible for online estimator implementation, resistances are modelled as random-walk variables:

$$\begin{bmatrix} R_{ohm}(k) \\ R_p(k) \end{bmatrix} = I \cdot \begin{bmatrix} R_{ohm}(k-1) \\ R_p(k-1) \end{bmatrix} + r \tag{4}$$

where I is the identity matrix, and r is the vector containing variances of both resistances. Other variable model parameters, such as those from Equation (3), can be modelled using this approach, as well.

3.4. Identification Experiments

The battery model parameters that are assumed to be constant or used in estimator verification should be determined by means of specific (targeted) offline identification tests.

3.4.1. Battery OCV Curve

The curve $U_{oc}(SoC)$ has been identified during low- and constant-load experiments (during which the vehicle was in rest, while only considerable battery load was scooter headlight), in which case any voltage drop in the battery can be neglected due to the low current (~C/50), so that the measured voltage U_b can be taken as the OCV U_{oc}. The SoC was estimated by Coulomb counting, i.e., by integrating the measured current. The battery capacity was also identified in this experiment by integration of measured current during the process of full battery discharge, which gave $C_n = 49.57$ Ah. Graphical illustration of the identification experiment and related results are shown in Figure 3. The identified curve $U_{oc}(SoC)$ has been used in validation of U_{oc} estimation results (see next sections).

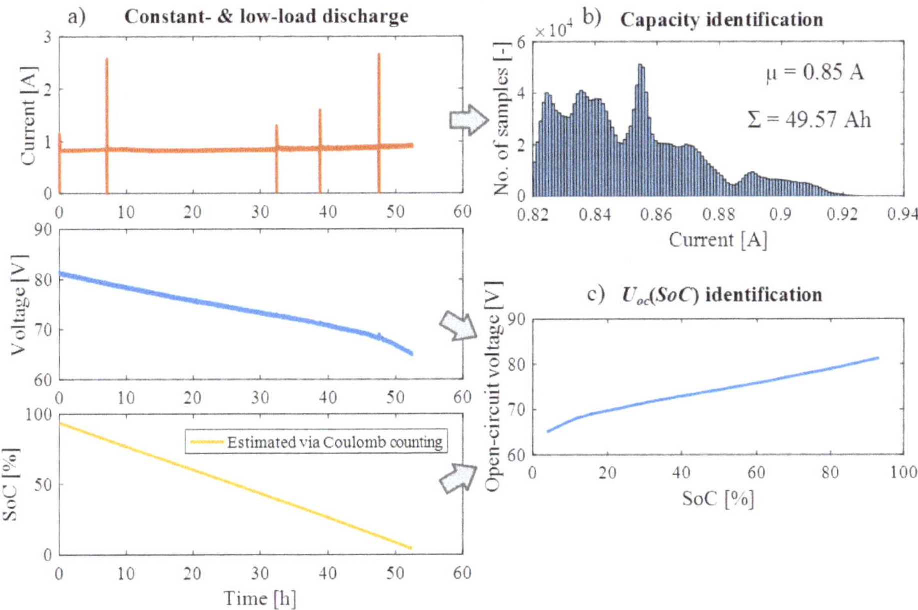

Figure 3. (**a**) Low- and constant-load experiment: Current, voltage, and SoC responses, and illustration of (**b**) capacity and (**c**) $U_{oc}(SoC)$ identification.

Polarization Time Constant

This parameter can be identified during the battery relaxation periods, i.e., parts of driving cycle where current has dropped to zero and remained equal to zero for at least 15 min. The relaxation transients to be identified were extracted from the voltage response (see Figure 4a,b) and approximated with the ECM model shown in Figure 4c.

The identified values of relaxation time constant $\tau_p(SoC)$ are shown in Figure 4d. These values were then approximated with a 3rd-order polynomial in dependence on SoC, and that polynomial was later used for calculation of τ_p at every estimator step based on the current, slowly changing SoC working point.

It is important to note that the polarization time constant can also vary with battery temperature and aging [22,23]. These effects are neglected in the estimator problem formulation in this paper, i.e., parameters of the characteristic $\tau_p(SoC)$ are not estimated online. This is motivated by the following main reasons: (i) τ_p is not directly involved in the ECM voltage equation (see Equation (2)), thus making it weakly observable in the proposed estimator design; ii) error in τ_p will cause an error in voltage modelling during the transient periods (i.e., before voltage has relaxed), so that the polarization dynamics may influence estimator accuracy only in transient conditions. As needed, the slow temperature- and aging-influenced polarization dynamics can be accounted for in the estimator design either by extending the τ_p characteristic with the temperature and SOH inputs or by considering τ_p as an additional parameter to be estimated, which is a subject of future work.

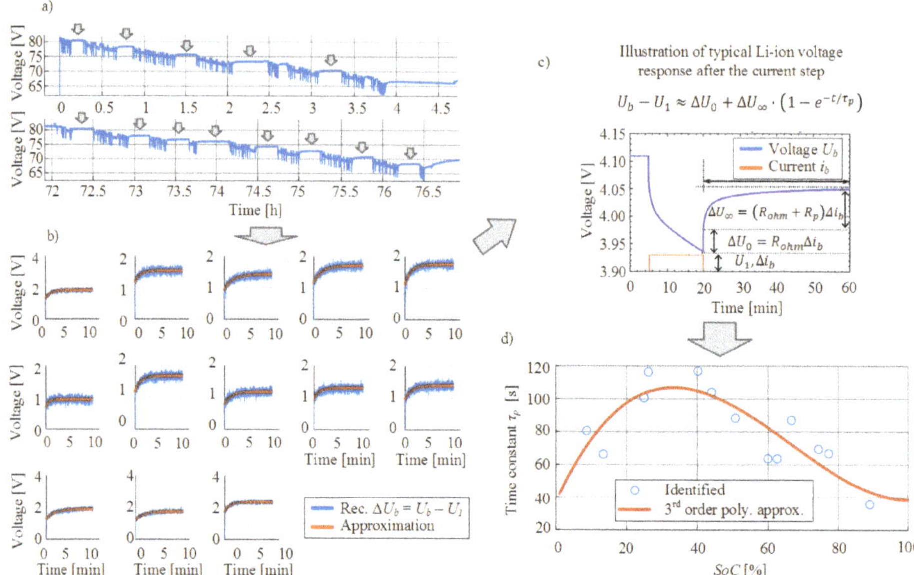

Figure 4. Illustration of τ_p identification procedure: (**a**) Measured battery voltage responses, (**b**) extracted voltage relaxation periods, (**c**) illustration of typical Li-ion cell voltage response after the current steps with the approximation equation, taken from [1], and (**d**) $\tau_p(SoC)$ identification results.

4. State and Parameter Estimator

This section deals with design, parametrization, and verification of the SoC estimator. It is designed as a dual state and parameter estimator, thus allowing for accompanying estimation of selected ECM parameters (i.e., the battery internal resistances and OCV parameters). A special attention is devoted to estimation of battery OCV hysteresis based on two complementary approaches (adapting the OCV parameters to current sign change or using an explicit hysteresis model).

4.1. DEKF-Based State and Parameter Estimator

States and parameters of the ECM are estimated with the DEKF, as a well-known approach in the model-based estimation problems where model states and slowly varying model parameters are to be estimated simultaneously [1]. The DEKF equations are not listed here due to paper size constraints, and they can be found in [24]. DEKF consists of two filters operating in parallel based on the state and parameter models:

State estimator state-space model:

$$x(k) = f(x(k-1), u(k-1), w(k-1))$$
$$y(k) = h(x(k), u(k), \theta(k), v(k))$$

Parameter estimator state-space model:

$$\theta(k) = \theta(k-1) + r(k-1)$$
$$y(k) = h(x(k), u(k), \theta(k), v(k))$$

where x and u are the vectors of model states and inputs, respectively, w is the vector of state variances (with the corresponding covariance matrix Q_x), θ is the vector of model parameters with their variances contained in vector r (with the corresponding covariance matrix Q_θ), h is the model output function (the same output function is used in both state and parameter models), y is the measured model output vector with measurement noise and corresponding covariance matrix denoted by v and R, respectively. The complete, discrete-time state-space model for simultaneous state and parameter estimation then reads (cf. Equations (1)–(4)):

$$\hat{x} = \begin{bmatrix} SoC(k) \\ i_p(k) \end{bmatrix} = \begin{bmatrix} 1 & 0 \\ 0 & e^{-\frac{T_u}{\tau_p(SoC(k-1))}} \end{bmatrix} \begin{bmatrix} SoC(k-1) \\ i_p(k-1) \end{bmatrix} + \begin{bmatrix} -\frac{T_u}{C_n} \\ 1 - e^{-\frac{T_u}{\tau_p(SoC(k-1))}} \end{bmatrix} (i_b(k-1) + w) \tag{5}$$

$$\hat{\theta} = \begin{bmatrix} R_{ohm}(k) \\ R_p(k) \\ k_{oc}^T(k) \end{bmatrix} = I \cdot \begin{bmatrix} R_{ohm}(k-1) \\ R_p(k-1) \\ k_{oc}^T(k-1) \end{bmatrix} + r \tag{6}$$

$$y(k) = U_b(k) = k_{oc}x_{oc} - R_{ohm}(k)i_b(k) - R_p(k)i_p(k) + v \tag{7}$$

The state estimator model, given by Equations (6) and (8), is considered linear in state equation under the assumption that the nonlinearity of function $\tau_p(SoC)$ can be neglected. The only nonlinearity resides in the output equation of the state estimator, related to the x_{oc} term (see Equation (3)), so that an EKF is finally used as a model state estimator. On the other hand, the parameter estimator model, given by Equations (7) and (8), is linear, so that the estimator reduces to KF.

4.2. Estimator Parametrization

The DEKF needs to be properly parametrized. For instance, appropriate statistic parameters such as process and output noise covariances Q and R should be determined offline. Polarization time-constant τ_p was assumed to be degradation-invariant and used as the identified SoC-dependent profile (see previous section), while battery capacity was in this case taken as a constant value that was measured as described in the previous section. This section also describes an estimator adaptation mechanism that indirectly compensates for the influence of unmodelled hysteresis of curve $U_{oc}(SoC)$.

4.2.1. DEKF Covariance Matrices Parametrization

The measurement variable in the DEKF model is the battery output voltage U_b (see Equation (8)). Its measurement noise has been estimated by approximating the voltage measurement error histogram with normal distribution, as shown in Figure 5a. The parameter μ identified in Figure 5a is the voltage noise mean value (expectation), while σ is the standard deviation which, after being squared, yields the measurement covariance $R = \left(53 \cdot 10^{-3}\right)^2$ mV^2. The parameter L_{stat} in Figure 5a is the result of Lilliefors normality test. Further in this paper, we calculate L_{stat} for estimator voltage residuals and compare it to the calculated $L_{stat} = 0.0436$ of voltage sensor noise (see Figure 5a) to check how similar they are, i.e., how accurate is the estimator.

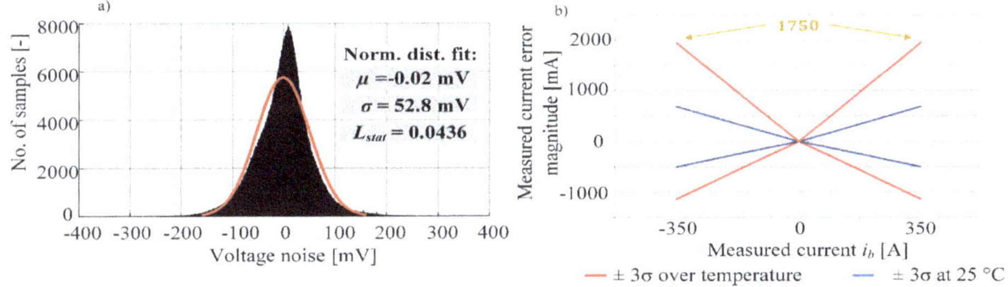

Figure 5. (**a**) Estimated voltage sensor noise, (**b**) amplitude of current sensor noise with respect to measured current, taken from [18].

The process noise relates to the current sensor noise, as can be seen from Equation (6). The current is in this case measured with LEM CAB 300 sensor, whose datasheet specifies a linear relation between measured current and magnitude of its measurement error (see Figure 5b). The standard deviation of current sensor noise can be estimated as a value three times lower than the noise magnitude, and covariance matrix is then the diagonal matrix of current sensor noise variances:

$$\sigma_x = \frac{1.75}{350}\frac{1}{3} \cdot i_b \rightarrow Q_x = diag\left(\sigma_x^2, \sigma_x^2\right) = diag\left(\left(\frac{i_b}{600}\right)^2, \left(\frac{i_b}{600}\right)^2\right) \tag{8}$$

4.2.2. Adaptation Mechanism

The relatively simple battery model given by Equations (1) and (2) does not take into account some secondary, but generally influential effect such as the hysteresis of OCV curve $U_{oc}(SoC)$ [25]. Since the hysteresis cannot be directly measured in this case, an adaptation mechanism is introduced in the form of single-step increase of the elements of parameter covariance submatrix $Q_\theta[3,7;3,7]$ when the start or end of charging is detected. This approach allows faster convergence of the U_{oc} parameters (written in k_{oc}), which abruptly change when the sign of battery current (or SoC derivative) occurs due to the existence of hysteresis of $U_{oc}(SoC)$ curve. Note that the battery current for the given scooter changes its sign only when the scooter is exposed to change from normal driving to charging or vice versa, because it does not incorporate regenerative braking.

4.3. Estimation Results

The presented DEKF was validated based on the recorded scooter real city driving cycle data consisting of seven load cycles (i.e., charge/discharge cycles) lasting for 150 h in total. The obtained estimation results are shown in Figure 6. Since the battery SoC cannot be measured, and there is no fully reliable SoC estimate available, the DEKF accuracy is evaluated by analyzing *a posteriori* voltage residual, i.e., difference between the recorded voltage U_b and the voltage calculated from output Equation (8) using *a posteriori* estimated states and parameters. The perfectly accurate filter would reduce the voltage residual to the voltage sensor noise, i.e., the residual mean value, standard deviation, and L_{stat} would be close to the values from Figure 5a.

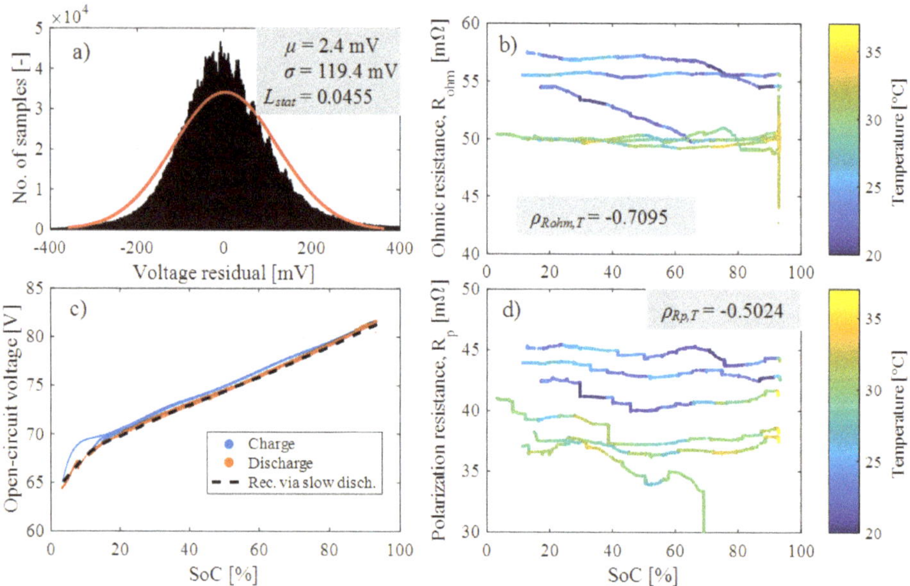

Figure 6. DEKF verification results: (**a**) Voltage residual histogram including normal distribution fit, (**b,d**) estimated resistances R_{ohm} and R_p, (**c**) estimated and recorded $U_{oc}(SoC)$ curves for a long set of real-life discharging and charging cycles.

Figure 6a shows the voltage residuals histogram including the corresponding normal distribution fit and its parameters. Residual mean value is low, while standard deviation and L_{stat} are larger than those of the voltage sensor noise. The estimated values of resistances R_{ohm} and R_p are shown in Figure 6b,d, respectively, vs. SoC and color-mapped with respect to battery temperature. These results point out that both resistances show negative correlation with respect to temperature (note: $\rho_{X,Y}$ stands for correlation coefficient between vectors X and Y, and are obtained by using the MATLAB function `corrcoef`), which is expected for the Li-ion cell resistances [21]. As of the correlation with respect to SoC (based on visual inspection of Figure 6b,d), both R_{ohm} and R_p do not seem to be correlated with SoC, which is an expected result for the particular SoC range, based on the estimator results from the available literature [16,21,26] in which resistances more significantly depend on SoC only at the very low and very large SoC bands. The estimated U_{oc} curves during charging and discharging intervals are shown in Figure 6c, along with the "measured" one adopted from Figure 3c. Evidently, the estimated and "measured" curves are in good agreement, and a relatively small hysteresis is apparent (i.e., the charging and discharging curves do not overlap).

The two sets of estimated $U_{oc}(SoC)$ curves from Figure 6c have been averaged and shown as dotted lines in Figure 7a. Half of the difference between those two curves yields the estimate of battery hysteresis voltage which is shown in Figure 7b. The estimated hysteresis voltage trend is in line with the results from the literature (e.g., [25]), except in the low-SoC region ($SoC < 20\%$), where the estimated hysteresis is larger than what would be expected based on the literature.

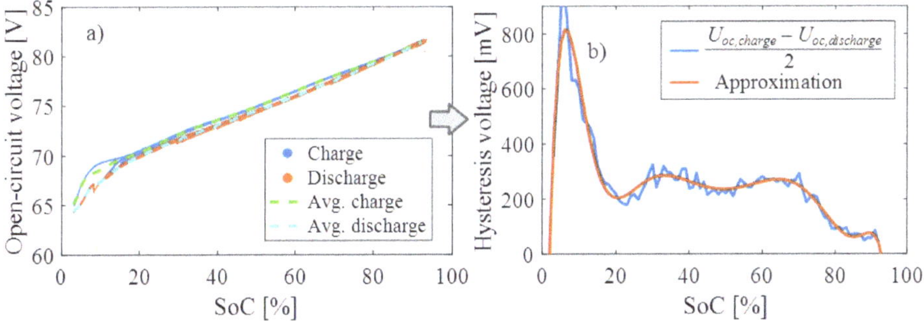

Figure 7. (**a**) Replot of Figure 6c with added average values of estimated $U_{oc}(SoC)$ curves for charge and discharge periods, (**b**) estimated hysteresis voltage.

Now, when the hysteresis voltage is known, the adaptation mechanism may be omitted, and the hysteresis can be accounted for directly through a proper $U_{oc}(SoC)$ model extension. A complex, dynamic hysteresis model [27] is not necessary in this case, because the particular scooter does not support regenerative breaking (i.e., its battery is not exposed to often changes of current sign). A simple, instantaneous hysteresis model can be described by introducing an auxiliary variable s described as [4]:

$$s(k) = \begin{cases} 1, & i_b(k) > 3\sqrt{Q_x} \\ -1, & i_b(k) < -3\sqrt{Q_x} \\ s(k-1), & i_b(k) < \left|3\sqrt{Q_x}\right| \end{cases} \tag{9}$$

(where $3\sqrt{Q_x}$ is the current sensor noise amplitude calculated using the current sensor variance from Equation (9)) and using it to modify the output equation (cf. Equation (8)):

$$y(k) = U_b(k) = U_{oc}(SoC(k)) - R_{ohm}(k)i_b(k) - R_p(k)i_p(k) + s(k)M_0(k) + v \tag{10}$$

where M_0 is the hysteresis voltage value obtained from data shown in Figure 7b by means of 10th-order approximation polynomial. Described hysteresis is used instead of the adaptation mechanism in the rest of the paper.

5. Battery Capacity Estimation

This section presents the battery remaining charge capacity estimator, and its integration into the overall SoC and capacity estimation algorithm. The capacity estimator is supplemented with a convergence detection algorithm to perform automatic coupling of the capacity estimator with the SoC estimator after the capacity estimate has converged. Finally, the complete estimation algorithm is verified for real driving battery load cycles.

5.1. Capacity Estimation Model

Since the battery capacity parameter is not directly involved in the model output equation (i.e., Equation (8)), it is not convenient to estimate it as another random-walk parameter in the DEKF [15]. Instead, the model for capacity estimation could be defined as [15]:

$$C(k) = C(k-L) + r_C \tag{11}$$

$$SoC(k-L+1) - SoC(k) = \frac{T_u}{C(k)} \sum_{j=k-L+1}^{k} i_b(j) + v_{SoC}(k) \tag{12}$$

where $C(k)$ is capacity, r_C is random walk noise for capacity parameter model with the corresponding covariance Q_C, L is the number of basic (DEKF) sampling steps between two capacity estimates, and v_{SoC} is measurement noise of SoC signal difference with the corresponding covariance R_{SoC}.

The model output is the SoC difference between two capacity estimates, while its input is the cumulative sum of battery current between those time instances. The SoC, as an output term, cannot be measured, but can be estimated by using the previously designed DEKF (both, estimates of SoC mean value and its variance are available). By looking at Equation (13) it can be seen that capacity estimate cannot be updated at the same rate as DEKF, because the signal-to-noise ratio of SoC estimate would be too low for the SoC dynamics being much slower than the current dynamics. The capacity estimator is therefore executed every L time steps, where L is in the range of 600–6000, i.e., 1 to 10 min. The overall estimator, i.e., the previously discussed DEKF extended with the capacity estimator, is shown in Figure 8.

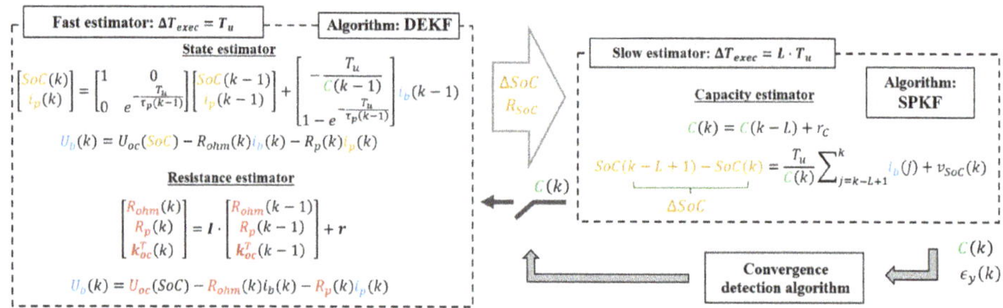

Figure 8. Overall algorithm for dual SoC and remaining charge capacity estimation.

5.2. SPKF-Based Capacity Estimator

Since the battery capacity model output equation is distinctively nonlinear, the EKF-based estimator application has been found to give too noisy estimates with slow convergence rate. This is an expected result since EKF uses analytic linearization through Taylor series expansion around the current operating point, i.e., around the state variable (in this case capacity C) mean value. Another, more coherent approach to this problem is statistical linearization which linearizes the model at multiple points drawn from prior distribution of C. The estimator derived using this approach is called SPKF [3]. There is a couple of SPKF versions which differ in calculation of sigma-points for linearization; in this paper, the method called central-difference Kalman filter (CDKF) is used because it provides simple parametrization without compromising accuracy [3]. Comparison between EKF- and SPKF-based capacity estimation, shown in Figure 9, clearly illustrates the benefits of using SPKF when compared to EKF.

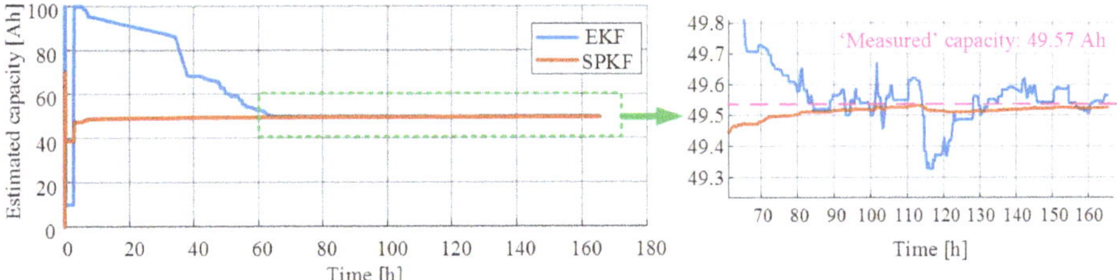

Figure 9. Comparison between EKF- and SPKF-based capacity estimation, where the estimated capacity is not fed back to DEKF-based state and parameter estimator.

It is important to note that in the case shown in Figure 9 the capacity estimates were not fed back into the state model of the DEKF, i.e., into Equation (6). If this were the case, i.e., if the state model of DEKF was updated with capacity estimates every L time stamps, the estimator would not converge to correct estimates, as shown in Figure 10a–c. This is because every model parameter is estimated in a coupled manner, so there are multiple parameter combinations where output voltage residual would be minimized. For instance, Figure 10d shows an estimate of $U_{oc}(SoC)$ which is narrower than the actual curve, because the capacity is estimated higher than the actual one.

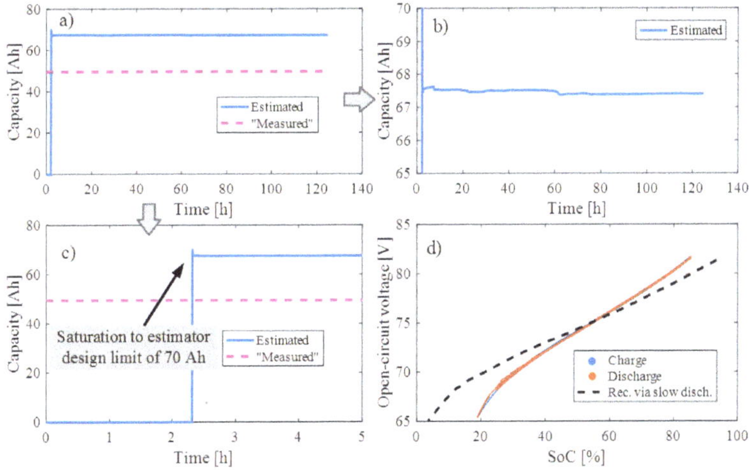

Figure 10. SPKF-based capacity estimation with capacity adaptation of the DEKF from the start, i.e., $t_{start} = 0$: (**a–c**) estimated capacity vs. time with zooms, (**d**) estimated and measured $U_{oc}(SoC)$ curves.

The capacity estimate feedback to the DEKF should be, therefore, turned on with some delay, i.e., until capacity estimate convergence is detected. For that purpose, capacity convergence detection algorithm has been designed, as presented in the next subsection.

5.3. Capacity Convergence Detection Algorithm

The capacity convergence detection algorithm is based on monitoring of the normalized estimation error (NEE) [28]:

$$\epsilon_y(k) = (y(k) - \hat{y}(k))P_y^{-1}(y(k) - \hat{y}(k))^T \tag{13}$$

where $y(k)$ is the SoC estimate generated by the DEKF, $\hat{y}(k)$ is the SoC calculated from the SPKF model output, and P_y is the SPKF innovation matrix (which is regularly calculated as a part of SPKF; note that it is a scalar in the particular case of single estimated parameter—the capacity). The convergence algorithm monitors the NEE, and when it is lower than some predefined value during some predefined number of consecutive time steps, the convergence is claimed.

5.4. Capacity Estimation Results

Results of SPKF-based capacity estimation algorithm with delayed and automatically calculated (through capacity convergence detection algorithm) start of capacity update (i.e., t_{start}) within the DEKF state model (version with hysteresis model included was used) are shown in Figure 11. The capacity estimates plotted versus time are shown in Figure 11a along with the "measured" capacity (see Figure 3b for details about capacity identification). Capacity convergence has automatically been detected after 2.9 h and from that point on, SPKF has been coupled to the DEKF. Figure 11b shows capacity estimates during the discharge periods plotted versus SoC and color-mapped with respect to temperature. Capacity shows expected (based on the [29]) positive correlation with the temperature.

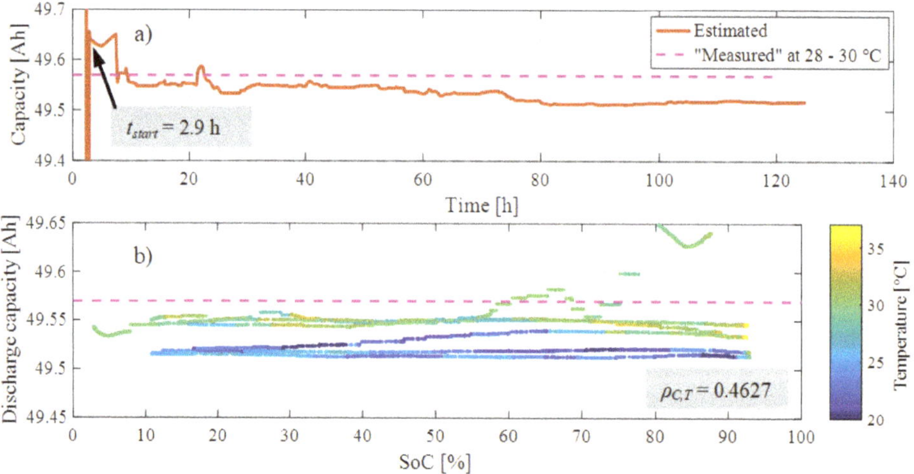

Figure 11. SPKF-based capacity estimation with automatic convergence detection: (**a**) Capacity estimates vs. time, (**b**) capacity estimates vs. SoC and temperature.

Figure 12 shows the same plots as in the case of Figure 6, but instead of using the adaptation mechanism the estimator relies on the explicit hysteresis model and has the capacity estimation included. The voltage residual is shown in Figure 12a together with the usual statistics. This residual has higher L_{stat} value than the one from Figure 6a, which may be explained by the influence of added capacity estimation. The estimates of R_{ohm} and R_p, plotted in Figure 12b,d with respect to SoC and temperature, respectively, are similar to those from Figure 6b,d, but with slightly higher correlation with temperature for both resistances. Finally, it should be noted that there are no distinguishable sets of estimated U_{oc} curves in Figure 12c (unlike in Figure 6c), because estimated $U_{oc}(SoC)$ now describes the central curve while the hysteresis is accounted for in the model (see Figure 7 and Equations (10) and (11)). Estimated $U_{oc}(SoC)$ is slightly larger than the recorded one (see Figure 3c for details about $U_{oc}(SoC)$ identification) because the latter is discharge $U_{oc}(SoC)$ curve while we estimate the average $U_{oc}(SoC)$ since hysteresis is explicitly modelled in this case. The overall estimation algorithm is parametrized as given in Appendix A.

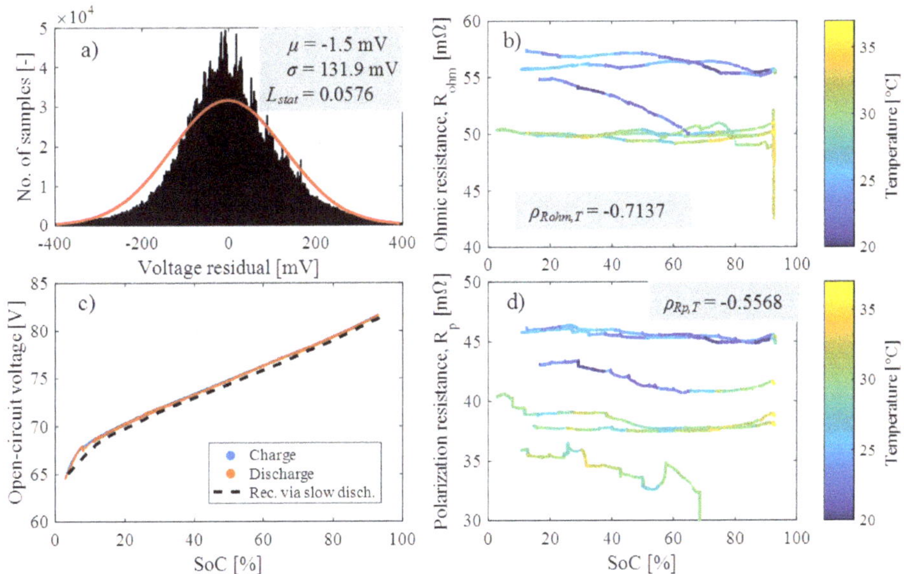

Figure 12. DEKF verification results with added hysteresis model and capacity estimation: (**a**) Voltage residual histogram including normal distribution fit, (**b,d**) estimated resistances R_{ohm} and R_p, (**c**) estimated and recorded $U_{oc}(SoC)$ curves.

6. Conclusions

An algorithm for dual estimation of battery state-of-charge (SoC) and remaining charge capacity has been proposed, which is aimed to be accurate over the whole battery lifetime and real-driving conditions including varying ambient temperatures. This was achieved by simultaneous estimation of relevant battery degradation-dependent parameters such as internal resistances and parameters of open-circuit voltage vs. SoC characteristic, $U_{oc}(SoC)$.

To this end, the dual extended Kalman filter-based SoC estimation algorithm has been extended to estimate parameters of the characteristic $U_{oc}(SoC)$ along with the resistance parameters. This extension allows the DEKF to adapt for $U_{oc}(SoC)$ variations and capture its hysteresis without explicitly modelling it. The latter can be useful in cases when the exact hysteresis profile is not known in advance or when it needs to be updated at the given state-of-health level without a specific identification experiment.

Next, a battery capacity estimator has been designed as a separate estimator, as it is based on a different model than the one that has been used in the DEKF design. Moreover, capacity estimation is meant to be executed on a significantly slower time scale than the DEKF. It has been shown that the EKF-based capacity estimator gives rather inconsistent estimates with a slow convergence rate, which is explained by a distinctively nonlinear capacity model. Capacity estimator has, therefore, been designed by using a sigma-point Kalman filter (SPKF). Furthermore, it has been demonstrated that SoC and capacity estimators (i.e., DEKF and SPKF, respectively) cannot be started in a coupled manner, unless it is ensured that both estimators have converged. A capacity convergence detection algorithm has, therefore, been designed to automatically couple the two estimators.

Finally, the overall estimator has been successfully verified based on real driving cycle data acquired by using a fully electric scooter equipped with a telemetry measurement system. The DEKF output voltage estimation residual distribution was confirmed to be close to the voltage measurement noise, while resistance estimates showed expected correlations with temperature. The estimated capacity was shown to be close to the measured one and expectedly correlated with temperature, as well.

Future work will be directed towards further extensions and verifications of the proposed estimator to account for temperature- and aging-dependent variations of the polarization time constant τ_p and further analyze the sensitivity of estimator for broader operating conditions (e.g., wider temperature range), respectively. The emphasis will be on using the estimator to track battery degradation features in support of modelling the battery degradation process.

Author Contributions: Conceptualization, J.D., F.M., and M.H.; Methodology, F.M., M.H., and J.D.; Software, F.M.; Validation, F.M., J.D., and M.H.; Writing—Original Draft Preparation, F.M.; Writing—Review and Editing, J.D. and M.H.; Supervision: J.D. All authors have read and agreed to the published version of the manuscript.

Funding: It is gratefully acknowledged that this work has been supported through a small-scale grant program of the University of Zagreb, related to the project entitled "Experimental Characterization of Electric Scooter Battery Pack Aging", as well as through project "Advanced Methods and Technologies in Data Science and Cooperative Systems" (DATACROSS) of the Centre of Research Excellence within the University of Zagreb.

Conflicts of Interest: The authors declare no conflict of interest.

Abbreviations

BMS	Battery management system
DEKF	Dual extended Kalman filter
ECM	Equivalent circuit model
EKF	Extended Kalman filter
KF	Kalman filter
NEE	Normalized estimation error
OCV	Open-circuit voltage
PF	Particle filter
SoC	State-of-charge
SoH	State-of-health
SPKF	Sigma-point Kalman filter

Appendix A. Estimator Parameters

The overall estimation algorithm is parametrized as given in Table A1.

Table A1. List of estimator parameters.

Parameter Description and Its Mathematical Notation	Value
Variance of R_{ohm} estimation, $\mathbf{Q}_\theta[1,1]$	$\left(0.85\cdot10^{-8}\right)^2$
Variance of R_p estimation, $\mathbf{Q}_\theta[2,2]$	$\left(0.85\cdot10^{-8}\right)^2$
Variance of $U_{oc}(SoC)$ estimation, $\mathbf{Q}_\theta[3,7;3,7]$	$\left(0.85\cdot10^{-7}\right)^2$
Initial SoC, $SoC(0)$	93
Initial polarization current, $i_p(0)$	0
Initial polarization resistance, $R_{ohm}(0)$	$50\cdot10^{-3}$
Initial polarization resistance, $R_p(0)$	$25\cdot10^{-3}$
Initial U_{oc} parameter, K_0	69
Initial U_{oc} parameter, K_1	$78\cdot10^{-3}$
Initial U_{oc} parameter, K_2	-10
Initial U_{oc} parameter, K_3	0.87
Initial U_{oc} parameter, K_4	-0.88
Scaling factor of submatrix $\mathbf{Q}_\theta[3,7;3,7]$ bump in adaptation mechanism (see Section 4.2.2)	10^{10}
NEE threshold value (see Section 5.3)	100
Consecutive time steps NEE has to be lower than the above threshold (see Section 5.3)	10
Ratio between SPKF and DEKF sampling time, L (see Section 5.1)	3000

References

1. Plett, G.L. *Battery Management Systems, Volume II: Eqivalent-Circuit Methods*; Artech House: London, UK, 2016.
2. Waag, W.; Fleischer, C.; Sauer, D.U. Critical review of the methods for monitoring of lithium-ion batteries in electric and hybrid vehicles. *J. Power Sources* **2014**, *258*, 321–339. [CrossRef]
3. Van Der Merwe, R.; Wan, E. Sigma-Point Kalman Filters for Probabilistic Inference in Dynamic State-Space Models. Ph.D. Thesis, OGI School of Science and Engineering, Portland, OR, USA, 2004.
4. Plett, G.L. Extended Kalman filtering for battery management systems of LiPB-based HEV battery packs—Part 2. Modeling and identification. *J. Power Sources* **2004**, *134*, 262–276. [CrossRef]
5. Xiong, R.; Sun, F.; Gong, X.; He, H. Adaptive state of charge estimator for lithium-ion cells series battery pack in electric vehicles. *J. Power Sources* **2013**, *242*, 699–713. [CrossRef]
6. Plett, G.L. Sigma-point Kalman filtering for battery management systems of LiPB-based HEV battery packs. Part 2: Simultaneous state and parameter estimation. *J. Power Sources* **2006**, *161*, 1369–1384. [CrossRef]
7. Ye, M.; Guo, H.; Cao, B. A model-based adaptive state of charge estimator for a lithium-ion battery using an improved adaptive particle filter. *Appl. Energy* **2017**, *190*, 740–748. [CrossRef]
8. Severson, K.A.; Attia, P.M.; Jin, N.; Perkins, N.; Jiang, B.; Yang, Z.; Chen, M.H.; Aykol, M.; Herring, P.K.; Fraggedakis, D.; et al. Data-driven prediction of battery cycle life before capacity degradation. *Nat. Energy* **2019**, *4*, 383–391. [CrossRef]
9. Baccouche, I.; Jemmali, S.; Manai, B.; Omar, N.; Essoukri Ben Amara, N. Improved OCV model of a Li-ion NMC battery for online SOC estimation using the extended Kalman filter. *Energies* **2017**, *10*, 764. [CrossRef]
10. Chen, X.; Lei, H.; Xiong, R.; Shen, W.; Yang, R. A novel approach to reconstruct open circuit voltage for state of charge estimation of lithium ion batteries in electric vehicles. *Appl. Energy* **2019**, *255*, 113758. [CrossRef]
11. Xiong, R.; Sun, F.; Chen, Z.; He, H. A data-driven multi-scale extended Kalman filtering based parameter and state estimation approach of lithium-ion olymer battery in electric vehicles. *Appl. Energy* **2014**, *113*, 463–476. [CrossRef]
12. Yang, R.; Xiong, R.; He, H.; Mu, H.; Wang, C. A novel method on estimating the degradation and state of charge of lithium-ion batteries used for electrical vehicles. *Appl. Energy* **2017**, *207*, 336–345. [CrossRef]
13. Tang, X.; Wang, Y.; Zou, C.; Yao, K.; Xia, Y.; Gao, F. A novel framework for Lithium-ion battery modeling considering uncertainties of temperature and aging. *Energy Convers. Manag.* **2019**, *180*, 162–170. [CrossRef]
14. Li, S.; Pischinger, S.; He, C.; Liang, L.; Stapelbroek, M. A comparative study of model-based capacity estimation algorithms in dual estimation frameworks for lithium-ion batteries under an accelerated aging test. *Appl. Energy* **2018**, *212*, 1522–1536. [CrossRef]
15. Plett, G.L. Recursive approximate weighted total least squares estimation of battery cell total capacity. *J. Power Sources* **2011**, *196*, 2319–2331. [CrossRef]
16. Wei, Z.; Zhao, J.; Ji, D.; Tseng, K.J. A multi-timescale estimator for battery state of charge and capacity dual estimation based on an online identified model. *Appl. Energy* **2017**, *204*, 1264–1274. [CrossRef]
17. Kasiński, P. GOVECS GROUP Supporting Role of IT for Lithium Ion Batteries Transportation and SOH Monitoring. Available online: http://fplreflib.findlay.co.uk/images/pdf/scms/PrzemekGOVECSBerlinSummit.pdf (accessed on 15 September 2019).
18. LEM CAB 300 Current Transducer Manual. Available online: http://media.ev-tv.me/CAB300-CSP3.pdf (accessed on 30 August 2019).
19. Fleischer, C.; Waag, W.; Heyn, H.M.; Sauer, D.U. On-line adaptive battery impedance parameter and state estimation considering physical principles in reduced order equivalent circuit battery models part 1. Requirements, critical review of methods and modeling. *J. Power Sources* **2014**, *262*, 457–482. [CrossRef]
20. Rodríguez, A.; Plett, G.L. Controls-oriented models of lithium-ion cells having blend electrodes. Part 1: Equivalent circuits. *J. Energy Storage* **2017**, *11*, 219–236. [CrossRef]
21. Waag, W.; Käbitz, S.; Sauer, D.U. Experimental investigation of the lithium-ion battery impedance characteristic at various conditions and aging states and its influence on the application. *Appl. Energy* **2013**, *102*, 885–897. [CrossRef]
22. Tang, X.; Wang, Y.; Yao, K.; He, Z.; Gao, F. Model migration based battery power capability evaluation considering uncertainties of temperature and aging. *J. Power Sources* **2019**, *440*, 227141. [CrossRef]

23. Abdel Monem, M.; Trad, K.; Omar, N.; Hegazy, O.; Mantels, B.; Mulder, G.; Van den Bossche, P.; Van Mierlo, J. Lithium-ion batteries: Evaluation study of different charging methodologies based on aging process. *Appl. Energy* **2015**, *152*, 143–155. [CrossRef]

24. Gibbs, B.P. *Advanced Kalman Filtering, Least-Squares and Modeling*; John Wiley & Sons: Hoboken, NJ, USA, 2011; ISBN 9780470529706.

25. Barai, A.; Widanage, W.D.; Marco, J.; Mcgordon, A.; Jennings, P. A study of the open circuit voltage characterization technique and hysteresis assessment of lithium-ion cells. *J. Power Sources* **2015**, *295*, 99–107. [CrossRef]

26. Wei, Z.; Zhao, J.; Zou, C.; Lim, T.M.; Tseng, K.J. Comparative study of methods for integrated model identification and state of charge estimation of lithium-ion battery. *J. Power Sources* **2018**, *402*, 189–197. [CrossRef]

27. Baronti, F.; Femia, N.; Saletti, R.; Visone, C.; Zamboni, W. Hysteresis modeling in Li-ion batteries. *IEEE Trans. Magn.* **2014**, *50*, 1–4. [CrossRef]

28. Li, X.R.; Zhao, Z.; Jilkov, V.P. *Measures of Performance for Evaluation of Estimators and Filters—Semantic Scholar*; International Society for Optics and Photonics: Bellingham, WA, USA, 2001.

29. Farmann, A.; Waag, W.; Marongiu, A.; Sauer, D.U. Critical review of on-board capacity estimation techniques for lithium-ion batteries in electric and hybrid electric vehicles. *J. Power Sources* **2015**, *281*, 114–130. [CrossRef]

Article

SOC and SOH Joint Estimation of the Power Batteries Based on Fuzzy Unscented Kalman Filtering Algorithm

Miaomiao Zeng, Peng Zhang, Yang Yang, Changjun Xie * and Ying Shi *

School of Automation, Wuhan University of Technology, Wuhan 430070, China
* Correspondence: jackxie@whut.edu.cn (C.X.); yingsh@whut.edu.cn (Y.S.)

Received: 5 July 2019; Accepted: 12 August 2019; Published: 14 August 2019

Abstract: In order to improve the convergence time and stabilization accuracy of the real-time state estimation of the power batteries for electric vehicles, a fuzzy unscented Kalman filtering algorithm (F-UKF) of a new type is proposed in this paper, with an improved second-order resistor-capacitor (RC) equivalent circuit model established and an online parameter identification used by Bayes. Ohmic resistance is treated as a battery state of health (SOH) characteristic parameter, F-UKF algorithms are used for the joint estimation of battery state of charge (SOC) and SOH. The experimental data obtained from the ITS5300-based battery test platform are adopted for the simulation verification under discharge conditions with constant-current pulses and urban dynamometer driving schedule (UDDS) conditions in the MATLAB environment. The experimental results show that the F-UKF algorithm is insensitive to the initial value of the SOC under discharge conditions with constant-current pulses, and the SOC and SOH estimation accuracy under UDDS conditions reaches 1.76% and 1.61%, respectively, with the corresponding convergence time of 120 and 140 s, which proves the superiority of the joint estimation algorithm.

Keywords: power batteries; improved second-order RC equivalent circuit; fuzzy unscented Kalman filtering algorithm; joint estimation

---

1. Introduction

The power batteries serving as the power supply for electric vehicles (EVs) have direct effects on the overall performance of EVs, and the battery overcharge may cause overheating or even an explosion, while the battery over-discharge may result in accelerated aging and permanently reduced capacity [1]. Concerning the issues of safety usage, the state estimation of batteries available for safety precautions can facilitate the elimination of safety hazards, which means the state of charge (SOC) and state of health (SOH) joint estimation is of great significance for the research on power batteries [2].

Lots of scholars have proposed many SOC estimation methods, such as the open circuit voltage method [3,4], the Coulomb counting method [5], the neural network method [6] and the Kalman filtering algorithm [7]. Among them, the open circuit voltage method was to first establish a corresponding function of the open circuit voltage and the SOC and then obtain the SOC by measuring the open circuit voltage after the battery was stationary [8]; the Coulomb integral method, which discretizes the current flowing through the battery and sums it up, and obtains the SOC value by simple division [9]; the neural network method optimizes the relevant parameters of the SOC estimation algorithm and solves complex abstract problems through autonomous learning [7]; a series of Kalman filtering algorithms based on the extended Kalman filtering algorithm optimize autoregressive data processing, which can make the optimal estimation in the minimum variance sense for the state of the dynamic system [10,11].

The estimation methods in the SOH are mainly divided into two categories: One is to start with the characteristic parameters of the battery, and the other is to analyze the aging characteristics and electrochemical reaction characteristics of the battery [12]. The former mainly uses the direct measurement method, obtaining the current SOH by obtaining aging characteristic parameters such as capacity and ohmic internal resistance [8,13]. There are also methods such as neural networks [14] and fuzzy logic [15], which can directly estimate the SOH of the battery through data training without an a priori model. The latter uses an electrochemical model method [12] that models the internal physical and chemical reactions during the charging process and designs an estimator for SOH estimation. There is also a method based on an equivalent circuit model [16] that establishes a circuit that reflects internal variables for SOH estimation.

All of the above algorithms are only a single estimate for the SOC or the SOH, ignoring the close relationship between the SOC and the SOH. The SOC estimate is affected by battery aging—as the battery ages, inaccurate SOC estimates can affect the SOH correction. Therefore, a joint estimate of the SOC and the SOH is necessary. The literature proposes an online SOH estimation method for the lithium battery using the constant-voltage (CV) charge current, as proposed in reference document [17], which can ensure the estimation error of less than 2.5%. However, it is difficult to accurately estimate the true state of the lithium battery by merely estimating the value of the SOH. Another SOC and SOH joint estimation method applicable to the cycle life of lithium-ion batteries for EVs, as proposed in reference document [18], involves an SOC and SOH identification using offline state estimators with different time scales; this requires substantial data to ensure asymptotic convergence without the real-time update.

Reference document [19] analyzed the error sources from the four angles of measurement, model, algorithm and state parameters for the SOC estimation. Finally, the author put forward new concerns in the practical application of SOC estimation. A multi-time-scale observer of the SOC and the SOH for a lithium-ion battery with coupled fast and slow dynamics was proposed in the reference document [16]. The authors used a deterministic transformation of the extended Kalman filter. The paper made an effective estimation of the SOC and the SOH by strictly characterizing the stability of estimation error. Three model-based filtering algorithms [20] were used to estimate the SOC, and the tracking accuracy, calculation time, robustness, etc., were analyzed and compared. Experimental results showed the advantages of three algorithms; the unscented Kalman filtering (UKF) algorithm has a good stability and the Particle filter (PF) algorithm, in the early stage has extreme rapidity. This article gave a combination of the two algorithms to improve the accuracy of the research direction.

In this paper, full consideration was given to the estimation error caused by the change in ohmic resistance during the service of power batteries, and the constant ohmic resistance was replaced by that of gentle variations resistance so as to propose a joint estimation algorithm of the power battery SOC and SOH based on a fuzzy control trace-free Kalman filter. This algorithm uses two complete fuzzy unscented Kalman filtering (F-UKF) algorithms to estimate the SOC and ohmic resistance of the battery at the same time. First of all, the use of a fuzzy controller can effectively reduce the impact of observation noise under complex conditions and to further improve the accuracy of battery SOC estimation. Secondly, the fuzzy controller is used to make a real-time correction of the variance matrix of the observed noise so as to finally realize the estimation of the battery ohmic internal resistance; experiments show that the joint estimation algorithm is not affected by the initial value of SOC, and it still has good convergence speed and tracking accuracy under complex conditions.

The rest of this paper is organized as follows: Section 2 introduces the model of lithium battery, open circuit voltage, SOC calibration experiment, and parameter identification. Section 3 reviews the implementation method of traceless Kalman filtering, fully considers the intrinsic coupling relationship between the SOC and the SOH, puts forward the fuzzy and traceless Kalman filter algorithm on the basis of traceless Kalman filtering, and uses two F-UKF algorithms to estimate the SOC and ohmic internal resistance at the same time. Section 4 discusses the relevant experimental process and conclusions, and Section 5 summarizes the full text.

2. Model for the Lithium Battery

2.1. Setup of Equivalent Circuit Model for the Lithium Battery

An accurate battery model can effectively describe the external features and characteristics of internal electrochemical reactions, which is of great significance for the SOC and SOH evaluation of power batteries [21]. In this paper, the SOC is defined as the ratio between remaining battery capacity and nominal battery capacity under the same environmental conditions and specified discharge rate [22]:

$$SOC = \frac{Q_{res}}{Q_N} \times 100\%$$ (1)

In which Q_{res} is the remaining battery capacity after the discharge of partial electric quantity and Q_N is the nominal battery capacity.

Through the comparative analysis of differences between old and new batteries, the researchers found that the ohmic resistance and actual maximum battery capacity have more significant changes due to the SOH variations, and SOH is defined as follows from the perspective of ohmic resistance [23]:

$$SOH_R = \frac{R_0(end) - R_0(t)}{R_0(end) - R_0(0)} \times 100\%$$ (2)

In which SOH_R is the battery SOH, which defined based on the ohmic resistance R_0; $R_0(end)$ is the ohmic resistance when the actual maximum battery capacity drops to 80% of the nominal battery capacity; $R_0(t)$ is the ohmic resistance of the battery at t; and $R_0(0)$ is the ohmic resistance upon the battery delivery from the factory.

The proposed improved second-order Resistor-capacitor (RC) equivalent circuit model based on the equivalent circuit model [24,25] is shown in Figure 1. The high capacitance C_p and current-controlled current source (CCCS) on the left side characterize the battery capacity, SOC and running time. The second-order RC circuit on the right side simulates the internal polarization characteristics of the battery, R1 and C_1 describe the concentration polarization characteristics of the battery, while R_2 and C_2 describe the electrochemical polarization characteristics of the battery. The voltage-controlled voltage source (VCVS) simulates the nonlinear relationship between the open-circuit voltage and U_{soc}, which links the circuit parts on both sides.

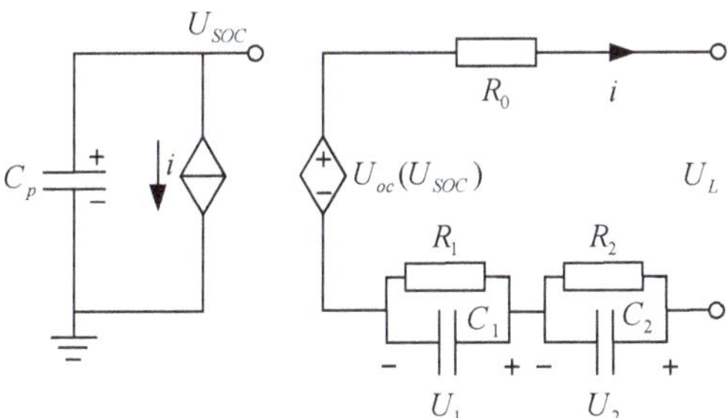

Figure 1. Improved second-order resistor-capacitor (RC) equivalent circuit model.

Based on the improved second-order RC equivalent circuit model for the lithium battery, select $x = [SOC\ U_1\ U_2]^T$ as the state variable to obtain the following continuous state space equation:

$$\begin{bmatrix} \dot{SOC} \\ \dot{U}_1 \\ \dot{U}_2 \end{bmatrix} = \begin{bmatrix} 0 & 0 & 0 \\ 0 & -\frac{1}{R_1 C_1} & 0 \\ 0 & 0 & -\frac{1}{R_2 C_2} \end{bmatrix} \cdot \begin{bmatrix} SOC \\ U_1 \\ U_2 \end{bmatrix} + \begin{bmatrix} -\frac{\eta}{Q_N} \\ -\frac{1}{C_1} \\ -\frac{1}{C_2} \end{bmatrix} \cdot i$$ (3)

In Equation (3), Q_N is the nominal battery capacity and η is the charge–discharge efficiency of the battery. The discretized state equation and observation equation are as follows:

$$\begin{cases} x(k+1) = A \cdot x(k) + B \cdot i(k) \\ U_L(k) = U_{oc}(SOC) - U_1(k) - U_2(k) - R_0 \cdot i(k) \end{cases} \tag{4}$$

In which T is the sampling period of the system, with A and B expressed as follows:

$$A = \begin{bmatrix} 1 & 0 & 0 \\ 0 & \exp(-\frac{T}{R_1 C_1}) & 0 \\ 0 & 0 & \exp(-\frac{T}{R_2 C_2}) \end{bmatrix}, \; B = \begin{bmatrix} -\frac{\eta \cdot T}{Q_N} \\ R_1(1 - \exp(-\frac{T}{R_1 C_1})) \\ R_2(1 - \exp(-\frac{T}{R_2 C_2})) \end{bmatrix} \tag{5}$$

2.2. Open-Circuit Voltage and SOC Setting Experiments

The procedures of open-circuit voltage and SOC setting experiments for the lithium battery at a normal temperature (25 °C) based on the ITS5300 battery test platform (ITECH ELECTRONIC CO., LTD., Nanjing, China) are as follows: Charge the battery until the full-load capacity is reached before the 3-hour standing and record the open-circuit voltage of the battery, discharge the battery for 6 minutes at a discharge rate of 1 C (40 A), and repeat the above steps until the cutoff voltage is reached. The fitting of open-circuit voltage curve corresponding to the SOC variations of lithium battery was completed via the MATLAB software (2017a, The MathWorks, Inc, Natick, MA, USA), which showed that the fitting curve had the minimum root-mean-square error when the polynomial order was 5. The fitting curve is shown in Figure 2, and the function expression is as follows.

$$U_{oc}(SOC) = 3.2821 \cdot SOC^5 - 10.3004 \cdot SOC^4 + 13.0068 \cdot SOC^3 - 7.9724 \cdot SOC^2 + 2.4054 \cdot SOC + 2.9752 \tag{6}$$

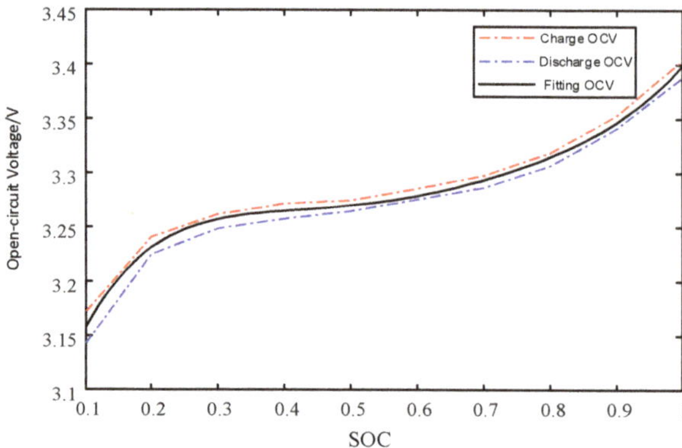

Figure 2. Fitting curves of the open-circuit voltage and state of charge.

2.3. Parameter Identification of the Lithium Battery Model

In accordance with the improved second-order RC equivalent circuit model, the Bayesian identification algorithm based on the least-square equation was adopted for the identification of resistance and capacitance parameters of the equivalent circuit model. Taking the parameters to be estimated as random variables, the Bayesian identification algorithm achieved the optimal estimation indirectly through the observation on other related parameters [26,27].

Kirchhoff's law should be adopted to obtain the following Laplace's equation of the improved second-order RC equivalent circuit model:

$$U_{oc}(s) - U_L(s) = i(s) \cdot \left(\frac{R_1}{R_1 C_1 s + 1} + \frac{R_2}{R_2 C_2 s + 1} + R_0 \right) \tag{7}$$

The equation obtained using the bilinear transformation method is as follows:

$$d(k) = -k_1 d(k-1) - k_2 d(k-2) + k_3 i(k) + k_4 i(k-1) + k_5 i(k-2) \tag{8}$$

In which $i(k)$ is the system input, $d(k) = U_{OC}(k) - U_L(k)$, and the final derivation of the Bayesian identification algorithm is as follows.

$$\begin{cases} \hat{\theta}(k) = \hat{\theta}(k-1) + K(k) \cdot \left[z(k) - H^T(k)\hat{\theta}(k-1) \right] \\ K(k) = P_\theta(k-1)H^T(k) \cdot \left[H^T(k)P_\theta(k-1)H(k) + \frac{1}{\sigma_r^2} \right]^{-1} \\ P_\theta(k) = \left[I - K(k)H^T(k) \right] \cdot P_\theta(k-1) \end{cases} \tag{9}$$

The initial value of $\theta(0)$ is 0, and the initial value of covariance matrix $P_\theta(0)$ is $a \cdot I$, among which a is a small positive number and I is a 5-order unit matrix. Use the recursion Formula (9) of the Bayesian identification algorithm to estimate the model parameters and then calculate the resistance and capacitance values of the model via Equation (8). In practical applications, it is necessary to consider the amount of calculation and the length of time for parameter identification. The joint estimation algorithm designed in this paper has a large amount of computation. Therefore, the mean value of the online identification result was selected as the parameter identification result. The results are shown in Table 1.

Table 1. Online identification result based on the Bayesian identification algorithm.

Model Parameter	Maximum Value	Minimum Value	Average Value
Ohmic internal resistance R_0 (mΩ)	1.704	0.923	1.278
Concentration polarization internal resistance R_1 (mΩ)	0.0603	0.1189	0.0927
Concentration polarization capacitor C_1 (KF)	6.017	3.021	3.821
Electrochemical polarization internal resistance R_2 (mΩ)	0.248	0.176	0.219
Electrochemical polarization capacitance C_2 (KF)	3.281	2.683	2.746

3. SOC and SOH Joint Estimation Based on F-UKF

3.1. Unscented Kalman Filtering Algorithm

The unscented Kalman filtering algorithm adopts the linear Kalman filter framework instead of the traditional linearization for nonlinear functions, with the nonlinear transfer of mean value and covariance completed via unscented transformation in the one-step prediction equation [7,28]. The unscented Kalman filtering algorithm is applicable to the nonlinear dynamic systems described with the following state-space equation:

$$\begin{cases} x(k+1) = f[x(k), u(k)] + e(k) \\ y(k) = g[x(k), u(k)] + v(k) \end{cases} \tag{10}$$

In which f is the function of nonlinear state equation and g is the function of nonlinear observation equation. Assume that $e(k)$ has the covariance matrix Q and $v(k)$ has the covariance matrix R; thus, the essential operation steps of the unscented Kalman filtering algorithm for a random variable X at the different time K are shown in Figure 3.

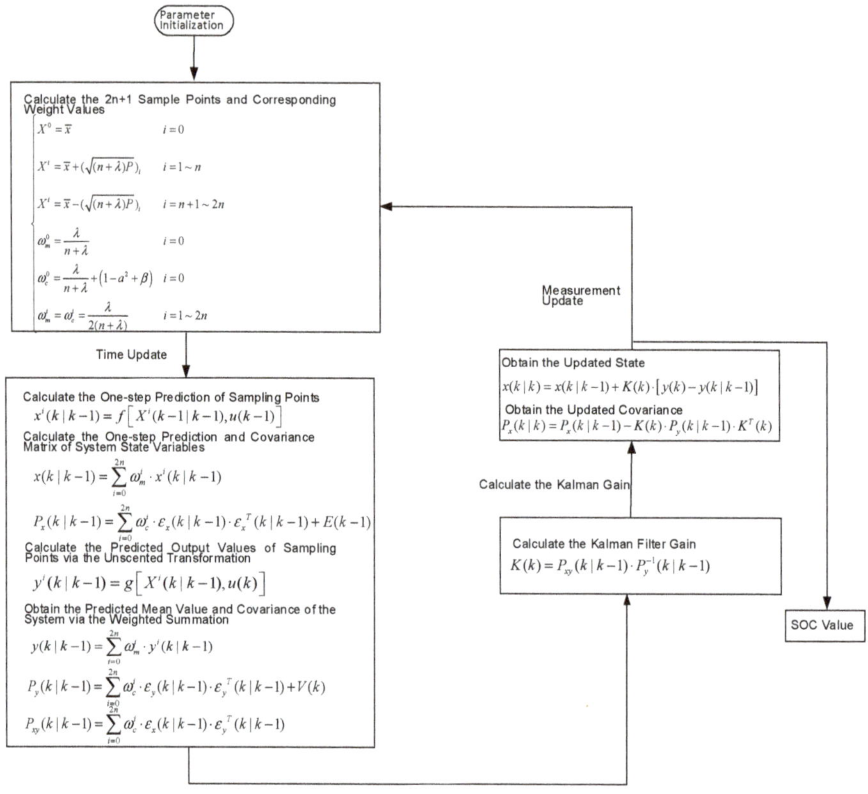

Figure 3. Essential operation steps of the unscented Kalman filtering algorithm.

3.2. Fuzzy Unscented Kalman Filtering Algorithm

The application of the unscented Kalman filtering algorithm should have been based on the already known statistical characteristics of process noises and observation noises; however, the insufficient estimation accuracy caused by the difficulty in noise determination during the use of power batteries required the introduction of adaptive filtering technique for algorithm optimization [29,30]. With reference to the unscented Kalman filtering algorithm, the covariance matching technique based on the fuzzy inference system was adopted in this section to effectively improve the accuracy of real-time observation noise estimation.

Assume that the statistical characteristics of process noises are already known, implement the recursive correction of observation noise variance based on the calculation of real-time ratio between the theoretical and actual covariances of observation errors.

Calculate the theoretical covariance $N(k)$ and actual covariance $M(k)$ of observation errors first, among which $i = k - n + 1$.

$$N(k) = \sum_{i=0}^{2n} \omega_c^i \cdot \varepsilon_y(k|k-1) \cdot \varepsilon_y^T(k|k-1) + V(k) \tag{11}$$

$$M(k) = \frac{1}{n} \sum_i^k [y(i) - y(i|i-1)] \cdot [y(i) - y(i|i-1)]^T \tag{12}$$

The input value $G(k)$ of the fuzzy controller is a ratio between the theoretical and actual covariances of observation errors, while its output value $\alpha(k)$ is the adjustment factor of observation noise variance. Take the adjusted observation noise variance $\hat{V}(k)$ into the unscented Kalman filtering algorithm to calculate the new observation-error covariance matrix $\hat{P}_y(k|k-1)$ and then update the Kalman filter gain and state-error covariance matrix with the updated $\hat{P}_y(k|k-1)$.

$$G(k) = \frac{M(k)}{N(k)} \tag{13}$$

$$\hat{V}(k) = \alpha(k) \cdot V(k) \tag{14}$$

As a kind of uncertainty reasoning method, the fuzzy controller is composed of three parts. First, initiate the fuzzy processing in accordance with the input membership function shown in Figure 4 for the input value $G(k)$ of the fuzzy controller to obtain the corresponding fuzzy index.

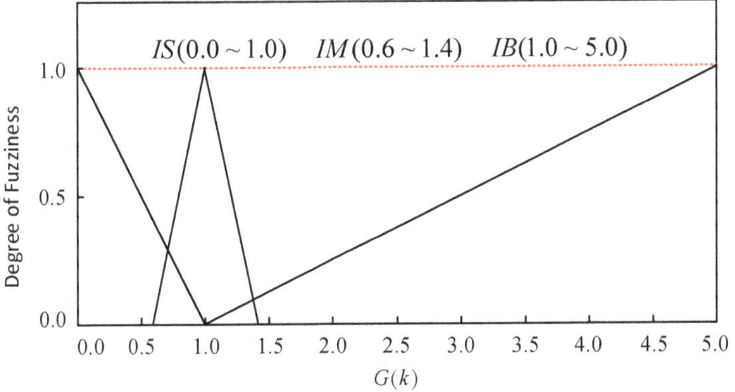

Figure 4. Input membership function.

Second, concerning the fuzzy controller with a single input/output, the correspondent fuzzy rules are relatively simple. The greater observation noise will result in the greater actual covariance $M(k)$ and $G(k)$, while the change in theoretical covariance $N(k)$ is subject to the variation of observation noise variance $V(k)$. In order to keep the variation consistency between $N(k)$ and $M(k)$, adjust $\alpha(k)$ to enlarge $V(k)$ when the observation noise becomes greater, so that the decreased $G(k)$ will approach 1. The decreased observation noise will result in the decreased actual covariance $M(k)$ and $G(k)$. Therefore, $\alpha(k)$ should be adjusted accordingly to decrease $V(k)$ so that the enlarged $G(k)$ will approach 1. The fuzzy rules established in accordance with the above derivation process is shown in Table 2.

Table 2. Fuzzy rules.

Input fuzziness	Input Small (IS)	Input Middle (IM)	Input Big (IB)
Output fuzziness	Output Small (OS)	Output Middle (OM)	Output Big (OB)

Finally, initiate the anti-fuzzy processing in accordance with the output membership function shown in Figure 5 for the output fuzziness to obtain the output value $\alpha(k)$ of the fuzzy controller.

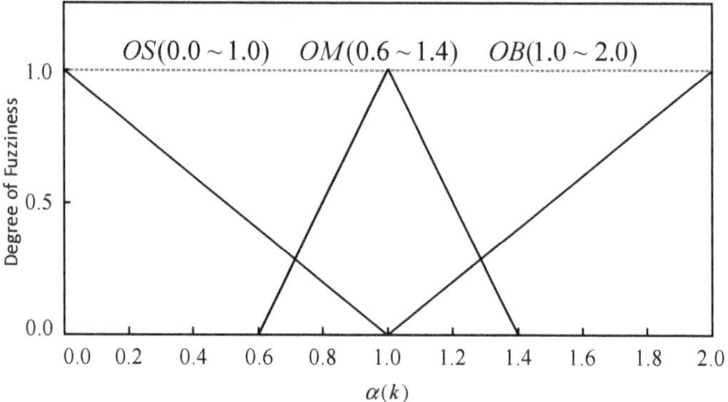

Figure 5. Output membership function.

The diagram of working principle using the fuzzy controller for the adjustment of observation noise variance matrix is shown in Figure 6.

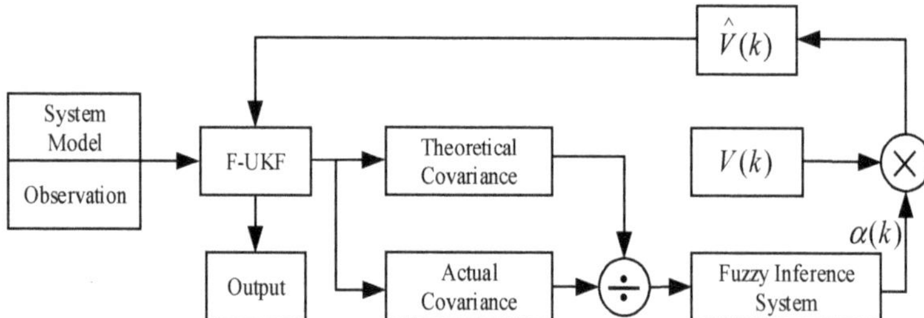

Figure 6. Structure diagram of the fuzzy controller.

3.3. Design and Implementation of the SOC and SOH Joint Estimation Algorithm

The following state-space equation can be used to express the improved second-order RC equivalent circuit model:

$$\begin{cases} x(k+1) = A_x(k) \cdot x(k) + B_x(k) \cdot i(k) + e_x(k) \\ y(k) = U_{oc}(SOC(k)) - U_1(k) - U_2(k) - R_0 \cdot i(k) + v_x(k) \end{cases} \tag{15}$$

In which $x(k)$ is the state variable; $y(k)$ is the predicted terminal voltage of the battery; $e_x(k)$ is the process noise, with the mean value of zero and variance $E_x(k)$; $v_x(k)$ is the observation noise, with the mean value of zero and variance $V_x(k)$; and $E_x(k)$ and $V_x(k)$ are irrelevant.

The gradual increase of ohmic resistance in a non-linear way is unnoticeable within a short period, which means the ohmic resistance of the battery at two adjacent moments can be taken as the constant value. Therefore, the state equation and observation equation for the estimation of ohmic resistance can be expressed as follows:

$$\begin{cases} R_0(k+1) = R_0(k) + e_R(k) \\ y(k) = U_{oc}(SOC(k)) - U_1(k) - U_2(k) - R_0(k) \cdot i(k) + v_R(k) \end{cases} \tag{16}$$

In which $R_0(k)$ is the state variable; $y(k)$ is the predicted terminal voltage of the battery; $e_R(k)$ is the process noise, with the mean value of zero and variance $E_R(k)$; $V_R(k)$ is the observation noise, with the mean value of zero and variance $V_R(k)$; and $E_R(k)$ and $V_R(k)$ are irrelevant.

The optimal estimated value of the battery SOC was adopted in the joint estimation algorithm for the one-step-ahead prediction of ohmic resistance; meanwhile, the optimal estimated value of ohmic resistance was also available for the one-step-ahead prediction of the battery SOC, and the above mutual application facilitated the acquisition of the estimated battery SOC and ohmic resistance closer to the actual values.

The flowchart of the SOC and SOH joint estimation algorithm based on F-UKF is shown in Figure 7.

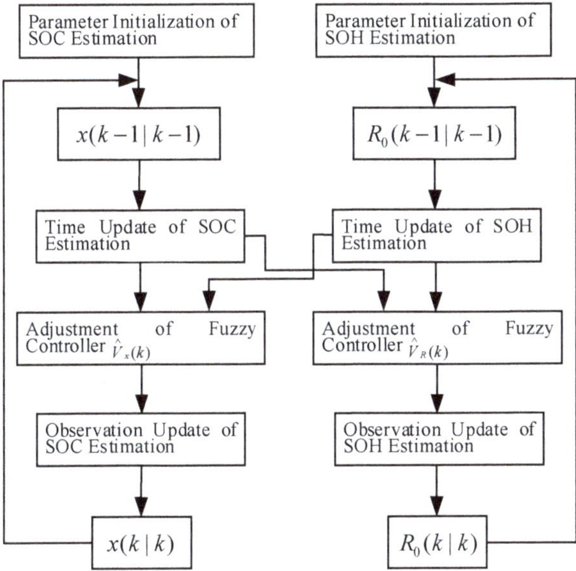

Figure 7. Flowchart of the joint estimation algorithm.

(1) Parameter initialization. First, initialize the corresponding parameters of the F-UKF algorithm for the battery SOC estimation; then, initialize the corresponding parameters of the F-UKF algorithm for the ohmic resistance estimation, and the ohmic resistance should be close to the actual value to ensure the fast convergence of the battery SOC.

(2) Obtain the terminal voltage $U_L(k)$ and working current $i(k)$ of the battery at the time k through the voltage-current acquisition module.

(3) Obtain the estimated value of the battery SOC at the time k through the recursion formula using the F-UKF algorithm based on the above terminal voltage and working current at the time k.

(4) Obtain the estimated value of ohmic resistance at the time k through the recursion formula using the F-UKF algorithm based on the estimated value of battery SOC and working current at the time k.

(5) Take the value of SOC(k) obtained from step (3) into the nonlinear functions of open-circuit voltage and the battery SOC to obtain the open-circuit voltage $U_{OC}(k)$ at the time k; repeat the steps (2), (3), (4) and (5) for the real-time estimation of the battery SOC and ohmic resistance.

4. Experimental Verification and Result Analysis

The ITS5300-based battery test platform available to verify the proposed SOC and SOH joint estimation algorithm is shown in Figure 8. The nominal capacity of a single lithium iron phosphate battery is 40 Ah, and the corresponding performance parameters are shown in Table 3. In order to measure the terminal voltage and working current of the battery, the software of IT9320 battery test system (ITECH ELECTRONIC CO., LTD, Nanjing, China) was used to simulate the discharge conditions with constant-current pulses and the urban dynamometer driving schedule (UDDS) driving cycles, and the MATLAB software was adopted for the simulation verification and analysis of the joint estimation algorithm proposed in this paper.

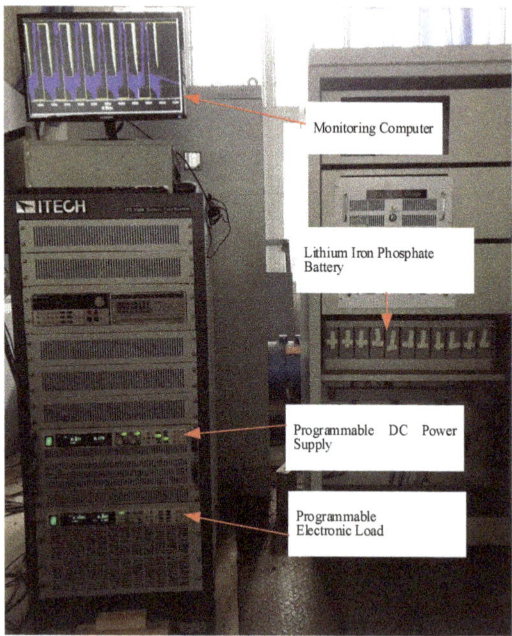

Figure 8. Battery test platform.

Table 3. Parameters of the lithium iron phosphate battery.

Nominal Capacity (Ah)		40
Battery voltage (V)	Charge cutoff voltage	3.6
	Discharge cutoff voltage	2.0
Cycle life (times)	80% DOD	≥2000
	70% DOD	≥3000
Standard charge–discharge current (A)		0.3C
Maximum charge current (A)		3C
Maximum discharge current (A)		4C
Operating temperature (°C)		−25–55

4.1. Sensitivity Verification of the F-UKF Algorithm against Initial Values

Concerning the estimation of the battery SOC and ohmic resistance using the F-UKF algorithm, it was difficult to obtain the accurate initial values of battery SOC, but the values of ohmic resistance were relatively stable without violent fluctuations. The lithium iron phosphate battery was charged until the battery SOC reached 85% of the initial state before the experiment. Under the discharge conditions with constant-current pulses, the different initial values of the battery SOC were set to verify the F-UKF sensitivity against initial values. In the MATLAB software, the respective initial values of the SOC were set to 40%/0% and 85% for the F-UKF algorithm and the ampere-hour integration method, with the sampling period and discharge rate set to 1 s and 0.5 C (20 A) for the 500-second simulation experiment.

The simulations under discharge conditions with constant-current pulses shown in Figures 9 and 10 indicate that the different initial values of battery SOC converged to the vicinity of reference value after a period of filtering iteration. Though the greater deviation of SOC initial values resulted in a longer convergence time, the stabilized values could follow the reference value well, and the estimation error was extremely small. Therefore, the F-UKF algorithm proposed in this paper is insensitive to the initial values.

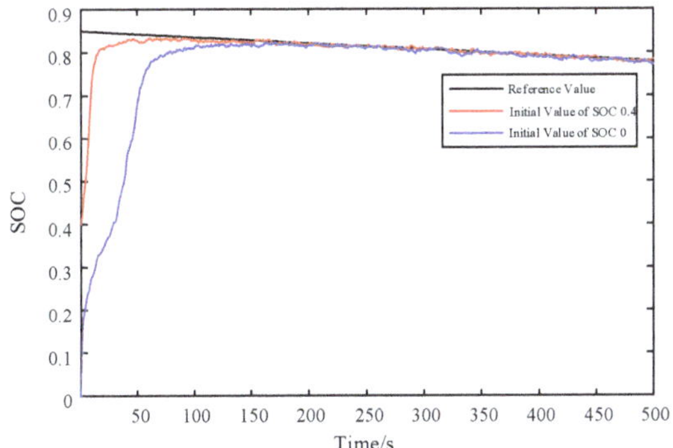

Figure 9. State of charge (SOC) estimation curve based on the fuzzy unscented Kalman filtering algorithm (F-UKF) under discharge conditions with constant-current pulses.

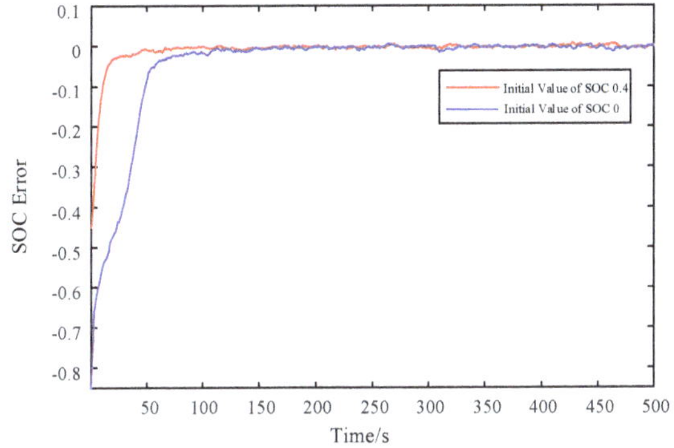

Figure 10. SOC estimation error curve based on F-UKF under discharge conditions with constant-current pulses.

4.2. Joint Simulation Verification of UDDS Driving Cycles

The lithium iron phosphate battery was charged until the battery SOC reached 85% of the initial state before the experiment.

The software of IT9320 battery test system was used to compile the pulse current driving cycles before the acquisition of terminal voltage and working current from UDDS driving cycles. In the MATLAB software, the initial values of SOC were set to 80% for the F-UKF and joint estimation algorithms and 85% for the ampere-hour integration method, respectively. The initial value of ohmic resistance was set to 1.50 mΩ for the joint estimation algorithm, which was greater than the reference value of 1.278 mΩ.

The simulations of UDDS driving cycles shown in Figure 11 indicate that the convergence time of SOC from the initial 80% to the vicinity of reference value based on the F-UKF algorithm was about 170 s, while the UKF algorithm was about 185 s. The SOC estimation error of the F-UKF algorithm after convergence could be controlled within 2.82%, while the estimation error of the UKF algorithm was about 2.93%. Since the noise of UKF was random white noise, the F-UKF algorithm that introduces adaptive technology was to adjust the noise instead of eliminating the noise. It can be seen that the convergence performance of the F-UKF algorithm was not only better than the UKF algorithm, but the estimation accuracy was also relatively improved in complex conditions.

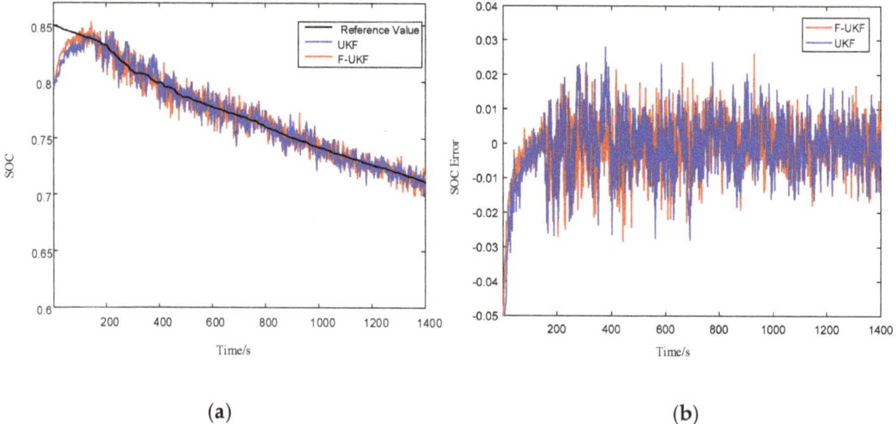

(a)   (b)

Figure 11. (**a**) SOC estimation curve of urban dynamometer driving schedule (UDDS) driving cycles based on the F-UKF and UKF algorithms. (**b**) SOC estimation error curve of UDDS driving cycles based on the F-UKF and UKF algorithms.

The simulations of UDDS driving cycles shown in Figure 12a indicate that the convergence time of SOC from the initial 80% to the vicinity of reference value based on the F-UKF algorithm was about 170 s, while the corresponding convergence time with an increased rising velocity based on the joint estimation algorithm was about 120 s. Therefore, the convergence performance of the joint estimation algorithm was better than that of the F-UKF algorithm under complex conditions. Figure 12b shows that the respective SOC estimation errors of the F-UKF algorithm and the joint estimation algorithm after convergence were less than 2.82% and 1.76%. Therefore, the tracking performance of the joint estimation algorithm was better than that of the F-UKF algorithm in terms of the SOC estimation.

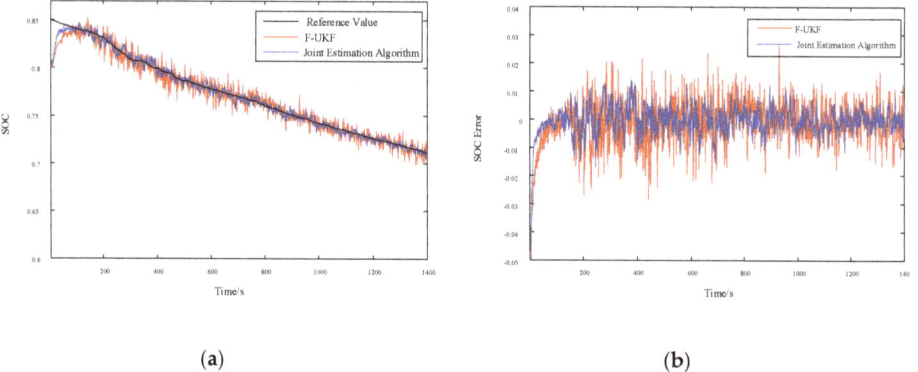

(a)   (b)

Figure 12. (**a**) SOC estimation curve of UDDS driving cycles based on the F-UKF and joint estimation algorithms. (**b**) SOC estimation error curve of UDDS driving cycles based on the F-UKF and joint estimation algorithms.

The simulations of UDDS driving cycles shown in Figures 13 and 14 indicate that the convergence time of ohmic resistance from 1.50 mΩ to the vicinity of reference value (1.278 mΩ) based on the joint estimation algorithm was about 140 s, and the corresponding battery SOH was about 89.87% of the reference value after stabilization. Figure 15 shows that the maximum SOH estimation error based on the joint estimation algorithm was 1.61%, and the SOH estimation error was less than 1.20% over time.

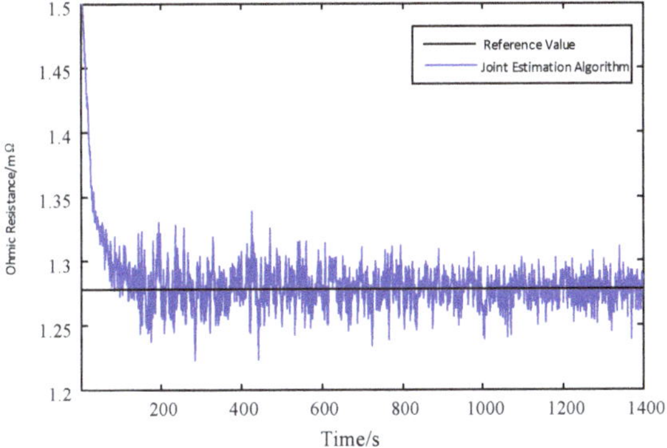

Figure 13. Ohmic resistance estimation curve of UDDS driving cycles based on the joint estimation algorithm.

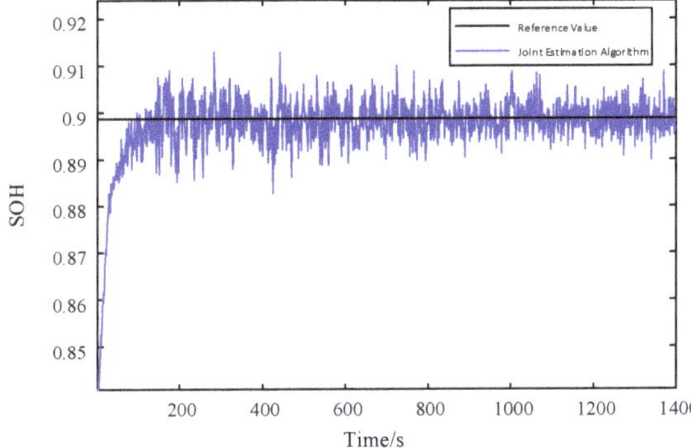

Figure 14. State of health (SOH) estimation curve of UDDS driving cycles based on the joint estimation algorithm.

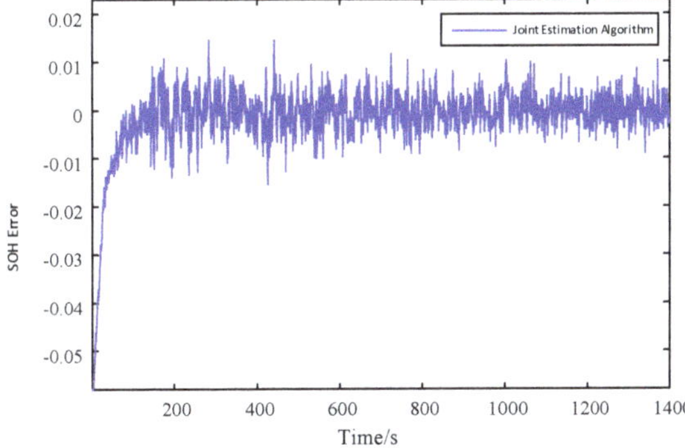

Figure 15. SOH estimation error curve of UDDS driving cycles based on the joint estimation algorithm.

5. Conclusions

In order to implement the real-time state estimation of power batteries for EVs, taking account observation noises and gradually changed ohmic resistance, an improved second-order RC equivalent circuit model was established in this paper for the SOC and SOH joint estimation using the fuzzy unscented Kalman filtering algorithm (F-UKF). The experimental data obtained from a test bench was adopted for the simulation to verify the convergence and stability of the F-UKF algorithm and to achieve the required design effects. The experimental data were obtained by the ITS5300 battery test platform, and the proposed joint estimation algorithm considered the influence of ohmic internal resistance change and noise error.

Author Contributions: Conceptualization, Y.S.; data curation, M.Z.; formal analysis, Y.Y.; funding acquisition, C.X.; investigation, M.Z.; methodology, P.Z.; project administration, Y.Y.; software, P.Z.; supervision, C.X.; validation, Y.S.; writing—original draft, M.Z.; writing—review & editing, C.X. and Y.S.

Funding: This research was funded by the National Natural Science Foundation of China, grant number 51477125", "the Hubei Science Fund for Distinguished Young Scholars, grant number 2017CFA049" and "the Hubei province technological innovation major project, grant number 2018AAA059".

Conflicts of Interest: The authors declare no conflict of interest.

References

1. Zhu, X.; Fan, D.; Xiang, Z.; Quan, L.; Hua, W.; Cheng, M. Systematic multi-level optimization design and dynamic control of less-rare-earth hybrid permanent magnet motor for all-climatic electric vehicles. *Appl. Energy* **2019**, *253*, 113549. [CrossRef]

2. Shen, P.; Ma, O.U.; Lu, L.G.; Li, J.Q.; Feng, X.N. The co-estimation of state of charge, state of health and state of function for lithium-ion batteries in electric vehicles. *IEEE Trans. Veh. Technol.* **2017**, *67*, 92–103. [CrossRef]

3. Chen, Y.J.; Yang, G.; Liu, X.; He, Z.C. A time-efficient and accurate open circuit voltage estimation method for lithium-ion batteries. *Energies* **2019**, *12*, 1803. [CrossRef]

4. Plett, G. Extended kalman filtering for battery management systems of lipb-based hev battery packs: Part 2. modeling and identification. *J. Power Sources* **2004**, *134*, 262–276. [CrossRef]

5. Cui, Y.Z.; Zuo, P.J.; Du, C.Y.; Gao, Y.D.; Yang, J.; Cheng, X.Q.; Ma, Y.L.; Yin, G.P. State of health diagnosis model for lithium ion batteries based on real-time impedance and open circuit voltage parameters identification method. *Energy* **2018**, *144*, 647–656. [CrossRef]

6. Ramadan, H.S.; Becherif, M.; Claude, F. Extended kalman filter for accurate state of charge estimation of lithium-based batteries: a comparative analysis. *Int. J. Hydrogen Energy* **2017**, *42*, 29033–29046. [CrossRef]

7. Meng, J.H.; Luo, G.Z.; Gao, F. Lithium polymer battery state-of-charge estimation based on adaptive unscented kalman filter and support vector machine. *IEEE Trans. Power Electron.* **2015**, *31*, 2226–2238. [CrossRef]

8. Berecibar, M.; Gandiaga, I.; Villarreal, I.; Omar, N.; Mierlo, V.; Bossche, V.D. Critical review of state of health estimation methods of li-ion batteries for real applications. *Renew. Sustain. Energy Rev.* **2015**, *56*, 572–587. [CrossRef]

9. Zhao, L.; Lin, M.Y.; Chen, Y. Least-squares based coulomb counting method and its application for state-of-charge (soc) estimation in electric vehicles. *Int. J. Energy Res.* **2016**, *40*, 1389–1399. [CrossRef]

10. Sun, Q.; Zhang, H.; Zhang, J.R.; Ma, W.T. Adaptive unscented kalman filter with correntropy loss for robust state of charge estimation of lithium-ion battery. *Energies* **2018**, *11*, 3123. [CrossRef]

11. Chen, Z.; Li, X.Y.; Shen, J.W.; Yan, W.S.; Xiao, R.X. A novel state of charge estimation algorithm for lithium-ion battery packs of electric vehicles. *Energies* **2016**, *9*, 710. [CrossRef]

12. Li, J.; Adewuyi, K.; Lotfi, N.; Landers, R.G.; Park, J. A single particle model with chemical/mechanical degradation physics for lithium ion battery state of health (soh) estimation. *Appl. Energy* **2018**, *212*, 1178–1190. [CrossRef]

13. Chen, L.; Lu, Z.Q.; Lin, W.L.; Li, J.Z.; Pan, H.H. A new state-of-health estimation method for lithium-ion batteries through the intrinsic relationship between ohmic internal resistance and capacity. *Measurement* **2018**, *116*, 586–595. [CrossRef]

14. Zou, C.F.; Manzie, C.; Nešić́b, D.; Kallapura, A.J. Multi-time-scale observer design for state-of-charge and state-of-health of a lithium-ion battery. *J. Power Sources* **2016**, *335*, 121–130. [CrossRef]

15. Li, H.; Ravey, A.; N'Diaye, A.; Djerdir, A. Online adaptive equivalent consumption minimization strategy for fuel cell hybrid electric vehicle considering power sources degradation. *Energy Convers. Manag.* **2019**, *192*, 133–149. [CrossRef]

16. Wei, Z.; Lim, T.M.; Skyllas-Kazacos, M.; Wai, N.; Tseng, K.J. Online state of charge and model parameter co-estimation based on a novel multi-timescale estimator for vanadium redox flow battery. *Appl. Energy* **2016**, *172*, 169–179. [CrossRef]

17. Yang, J.F.; Xia, B.; Huang, W.X.; Fu, Y.H.; Mi, C. Online state-of-health estimation for lithium-ion batteries using constant-voltage charging current analysis. *Appl. Energy* **2018**, *212*, 1589–1600. [CrossRef]

18. Zou, Y.; Hu, X.S.; Ma, H.M.; Li, S.E. Combined state of charge and state of health estimation over lithium-ion battery cell cycle lifespan for electric vehicles. *J. Power Sources* **2015**, *273*, 793–803. [CrossRef]

19. Zheng, Y.; Ouyang, M.G.; Han, X.B.; Lu, L.G.; Li, J.Q. Investigating the error sources of the online state of charge estimation methods for lithium-ion batteries in electric vehicles. *J. Power Sources* **2018**, *377*, 161–188. [CrossRef]

20. Yang, F.F.; Xing, Y.J.; Wang, D.; Tsui, K. A comparative study of three model-based algorithms for estimating state-of-charge of lithium-ion batteries under a new combined dynamic loading profile. *Appl. Energy* **2016**, *164*, 387–399. [CrossRef]

21. Hua, Y.; Cordoba-Arenas, A.; Warner, N.; Rizzoni, G. A multi time-scale state-of-charge and state-of-health estimation framework using nonlinear predictive filter for lithium-ion battery pack with passive balance control. *J. Power Sources* **2015**, *280*, 293–312. [CrossRef]

22. Zhang, Y.H.; Song, W.J.; Lin, S.L.; Feng, Z.P. A novel model of the initial state of charge estimation for lifepo4 batteries. *J. Power Sources* **2014**, *248*, 1028–1033. [CrossRef]

23. Deng, Y.W.; Ying, H.J.; Jiaqiang, E.; Zhu, H.; Wei, K.X.; Chen, J.W.; Zhang, F.; Liao, G.L. Feature parameter extraction and intelligent estimation of the state-of-health of lithium-ion batteries. *Energy* **2019**, *176*, 91–102. [CrossRef]

24. Petricca, P.; Shin, D.; Bocca, A.; Macii, A.; Macii, A.; Poncino, M. Automated generation of battery aging models from datasheets. In Proceedings of the 2014 IEEE 32nd International Conference on Computer Design (ICCD), Seoul, South Korea, 19–22 October 2014; pp. 483–488.

25. Bocca, A.; Sassone, A.; Shin, D.; Macii, A.; Macii, E.; Poncino, E. An equation-based battery cycle life model for various battery chemistries. In Proceedings of the 2015 IFIP/IEEE International Conference on Very Large Scale Integration (VLSI-SoC), Daejeon, South Korea, 5–7 October 2015; pp. 57–62.

26. Yan, W.Z.; Zhang, B.; Zhao, G.Q.; Tang, S.J.; Niu, G.X.; Wang, X.F. A battery management system with lebesgue sampling-based extended kalman filter. *IEEE Trans. Ind. Electron.* **2019**, *66*, 3227–3236. [CrossRef]

27. Wei, Z.B.; Zhao, J.Y.; Ji, D.X.; Tseng, K.J. A multi-timescale estimator for battery state of charge and capacity dual estimation based on an online identified model. *Appl. Energy* **2017**, *204*, 1264–1274. [CrossRef]

28. Zhang, W.G.; Shi, W.; Ma, Z.Y. Adaptive unscented kalman filter based state of energy and power capability estimation approach for lithium-ion battery. *J. Power Sources* **2015**, *289*, 50–62. [CrossRef]

29. Dong, G.Z.; Wei, J.W.; Chen, Z.H.; Sun, H.; Yu, X.W. Remaining dischargeable time prediction for lithium-ion batteries using unscented kalman filter. *J. Power Sources* **2017**, *364*, 316–327. [CrossRef]

30. Li, Y.W.; Wang, C.; Gong, J.F. A wavelet transform-adaptive unscented kalman filter approach for state of charge estimation of lifepo4 battery. *Int. J. Energy Res.* **2018**, *42*, 586–600. [CrossRef]

Article

High Reynold's Number Turbulent Model for Micro-Channel Cold Plate Using Reverse Engineering Approach for Water-Cooled Battery in Electric Vehicles

Satyam Panchal [1,2,*], **Krishna Gudlanarva** [2], **Manh-Kien Tran** [3], **Roydon Fraser** [1] and **Michael Fowler** [3]

[1] Mechanical and Mechatronic Engineering Department, University of Waterloo, 200 University Avenue West, Waterloo, ON N2L 3G1, Canada; rafraser@uwaterloo.ca

[2] Detroit Engineering Products, 850 East Long Lake Road, Troy, MI 48085, USA; krishna_g@depusa.com

[3] Chemical Engineering Department, University of Waterloo, 200 University Avenue West, Waterloo, ON N2L 3G1, Canada; kmtran@uwaterloo.ca (M.-K.T.); mfowler@uwaterloo.ca (M.F.)

* Correspondence: satyam.panchal@uwaterloo.ca or satyam_p@depusa.com; Tel.: +1-519-722-4420

Received: 4 March 2020; Accepted: 30 March 2020; Published: 2 April 2020

Abstract: The investigation and improvement of the cooling process of lithium-ion batteries (LIBs) used in battery electric vehicles (BEVs) and hybrid electric vehicles (HEVs) are required in order to achieve better performance and longer lifespan. In this manuscript, the temperature and velocity profiles of cooling plates used to cool down the large prismatic Graphite/LiFePO$_4$ battery are presented using both laboratory testing and modeling techniques. Computed tomography (CT) scanning was utilized for the cooling plate, Detroit Engineering Products (DEP) MeshWorks 8.0 was used for meshing of the cooling plate, and STAR CCM+ was used for simulation. The numerical investigation was conducted for higher C-rates of 3C and 4C with different ambient temperatures. For the experimental work, three heat flux sensors were attached to the battery surface. Water was used as a coolant inside the cooling plate to cool down the battery. The mass flow rate at each channel was 0.000277677 kg/s. The k-ε model was then utilized to simulate the turbulent behaviour of the fluid in the cooling plate, and the thermal behaviour under constant current (CC) discharge was studied and validated with the experimental data. This study provides insight into thermal and flow characteristics of the coolant inside a cooing plate, which can be used for designing more efficient cooling plates.

Keywords: heat and mass transfer; thermal analysis; Lithium-ion battery; micro-channel cooling plate; battery thermal management; MeshWorks; CFD

---

1. Introduction

The collective effects of global warming, environmental degradation, and energy crisis have prompted attention towards clean and sustainable energy [1]. However, there is inconsistency of renewable energy harvesting, since it depends on the effects of climate, which could result in complications in providing sufficient electricity in contrast with traditional nonrenewable energy sources. This has led to an interest in developing large-scale energy storage systems (ESS), for which batteries show promise [2]. Among the available secondary batteries, lithium-ion and lead-acid are broadly considered as effective candidates for energy storage systems. In the automotive industry, plug-in hybrid electric vehicles (PHEVs), hybrid electric vehicles (HEVs), and battery electric vehicles (BEVs) commonly utilize lithium-ion batteries (LIBs) [3]. The widespread use of LIBs is the result of 1) high specific power and energy densities [4]; 2) high nominal voltage and low self-discharge rate [5];

and 3) long cycle-life and no memory influence [6,7]. All these characteristics are required for electric vehicles (EVs) to achieve desirable driving range and vehicle speed [8]. In addition to the driving range and vehicle speed, the life cycle of the LIBs is also a critical factor in EVs. Some important factors in determining the allowable discharging and charging currents (also known as C-rates) and the batteries' life cycle are battery materials, working temperature, and assembling process. Current research on enhancing the life cycle of a battery has mainly been focused on the improvement in the materials and assembling technology, with a specific goal to obtain desired energy density. In addition, little attention has been directed toward the advancement and change of battery cooling systems (BCS) [9].

Pouch-type LIBs are being commonly utilized in new EVs. However, many problems in their safety and life span remain. First, when using the pouch-type lithium-ion battery, particularly in cases of high discharge rate, high heat generation may occur [10]. This type of battery expands due to overheating when the heat is not removed promptly. In some cases, the LIB may even burst and explode. In addition, these pouch-type batteries are connected either in parallel or in series within the LIB packs or modules, which also generate high amount of heat during both discharging and charging. Therefore, a good battery thermal management system (BTMS) is necessary [11]. The heat of LIBs increases when EVs accelerate and experience fast charging. If this generated heat is not adjusted or if it is overtaken by the rate of heat production, the battery pack temperature drastically increases. A high operating temperature of lithium-ion batteries can lead to capacity fade of the battery [12]. The impact of high working temperatures on the execution of a LIB, particularly for cylindrical battery cells (Sony 18650), was researched by Ramadass et al. [13] and the prismatic LIB cell (A123 20 Ah) was explored by Panchal et al. [14]. The authors found that the capacity fading is not the main negative impact related to high working temperatures of the battery; it also may lead to the explosion of the electrolyte. In addition, thermal runaway of a LIB cell can cause the whole LIB pack to fail [15]. In addition, there is a major effect on the electrochemical behaviors in terms of the degradation of electrolyte, electrodes, separator, and the life cycle cost [14,16]. Hence, a robust BTMS is required, to achieve better LIB pack performance in low-temperature conditions and a better lifespan in high-temperature conditions [12,17,18]. A typical operating range is between 20 °C and 40 °C [19], and an extended range is between −10 °C and +50 °C for certain applications [20].

Based on the coolant utilized in the BTMS, the BTMS can be divided into (1) air-, (2) fluid- or liquid-, and (3) phase change material (PCM)-based. For the air-based BTMS, cooling of the batteries is done by airflow passing between the batteries in a module or pack. The stream of air can come from the motion of the vehicle, and these types of systems are called natural air-cooled BTMS. If the airflow is generated by power-operated equipment, then these systems are called forced-air BTMS. The air streaming direction for air-based BCSs is usually 90 degrees to the axes for cylindrical batteries. Fluid-based BTMS utilizes coolants in the liquid phase, and such systems typically use power-operated equipment to move the coolant flow. In the PCM cooling method, the cooling is provided by the latent heat of the PCM, while in air- and liquid-cooling methods, the cooling is provided by sensible cooling. The advantages and disadvantages of air, fluid and PCM cooling are: (1) the air-cooling method is simple and lightweight [21]; (2) the water-cooling method is more efficient since it absorbs more heat, and takes less volume, yet more complexities are involved, including high weight and cost [22]; (3) in comparison with the air-cooling strategy, the liquid-cooling strategy provides better cooling performance owing to the high thermal conductivity of water compared to air [23]; and (4) because of the low thermal conductivity of air [24], a high speed of air is required in order to provide adequate cooling for LIBs [25,26]. The advantage of PCM over fluid and air-based strategies is better temperature consistency throughout the battery pack. Most recent examples of EVs and HEVs that utilize air-based cooling systems are the Toyota Prius, the Nissan Leaf, and the Honda Insight [27]. A very noted use of fluid-cooling systems for thermal management of LIBs is in Tesla vehicles, including the Roadster. These cooling techniques utilize a 50/50 mixture of water and glycol to keep the battery pack temperature within the appropriate limits [28]. Both the Chevrolet Bolt and BMW i3 utilize a bottom-cooling plate in their battery packs, with the cooling medium being a water-glycol solution for the Chevrolet and a refrigerant for the BMW [29]. The objective of this study is to conduct a reverse engineering study to experimentally investigate the design and heat generation of the 20 Ah lithium-ion battery with the cold plate. Subsequently, heat flux data and developed CAD modeling will be used to numerically characterize the thermal and flow behaviors of coolant in the cold plate at different inlet coolant temperatures and higher C-rates. The study provides insight into thermal and flow characteristics of the coolant inside a cooing plate which can be used for future improvements of the cooling plate.

2. State of the Art

According to the research by several scholars from all over the world, there are numerous papers available on BTMS with air, water, and PCM cooling, and battery modeling in the open literature [28–33].

Al-Hallaj et al. [34] were the first to utilize a PCM experiencing a solid to fluid change, and the authors used the PCM to cool down the 18,650 cylindrical-shaped lithium-ion batteries. In the BCS, they utilized a paraffin blend of pentacosane and hexacosane as the PCM. Moreover, for a passive cooling strategy, PCM-based BTMS is less expensive, requires smaller volume, and accomplishes preferred temperature consistency over air and fluid-based BTMS. The authors additionally utilized a commercial finite-element (FE) software, PDEase2D™, to simulate the thermal behavior of EV battery modules with a PCM BTMS. The authors claimed that the research was important for EV performance under cold conditions, or in space applications where the battery working temperature drops greatly when an orbiting satellite moves from the light to the dark side of the earth.

Zhang et al. [35] worked on the simulation of pouch-type LIB with thermal management using a cooling plate approach. The authors used a 22 Ah battery and developed a computational fluid dynamics (CFD) model using ANSYS Fluent. The authors studied the effect of inlet mass flow rate, difference in temperature, and pressure drop at 4C discharge rate, with the end goal of improving the efficiency and economy of the cooling plate. Their results demonstrated that the increase in mass flow rate of coolant reduced the maximum temperature and temperature difference of cells, but when mass flow rate exceeded 0.003 kg/s, then the economy of the cooling plate worsened.

Omkar et al. [36] developed a PCM/cooling-plate-coupled BTMS using CFD. The authors used LiFePO$_4$/C as a battery and determined the heat generation in simulation. They varied several factors in simulating battery and cold plate, such as the inlet mass flow rate, PCM, thermal conductivity, direction of flow, water cooling, to know the impact on the cooling execution of module. The authors concluded that as the space between adjoining batteries increased, the most extreme temperature had little change, yet the temperature field was uniform.

Chen et al. [37] worked on the comparison of four diverse cooling strategies, including air cooling, indirect fluid cooling, direct fluid cooling, and fin cooling for LIB cells. The authors evaluated the effectiveness on the basis of coolant parasitic power utilization, temperature difference in a cell, the most extreme temperature rise, and extra weight utilized for the cooling strategy. The authors found that (1) to keep the same average temperature, an air cooling strategy needed two to three times more energy than other techniques; (2) an indirect liquid cooling strategy had the lowest maximum temperature rise; (3) a fin cooling system includes an approximately 40% higher weight of cell, and (4) indirect fluid cooling is more efficient than direct fluid cooling.

Lu et al. [38] worked on BTMS of thickly pressed EV batteries with forced-air cooling systems to investigate the air cooling capacity on the temperature consistency and hotspots alleviation of a smaller battery pack subject to different airflow rates as well as airflow paths. The authors used 252 cylindrical lithium-ion batteries (32650). Their numerical outcomes demonstrated that the effective improvement in heat transfer exchange zones between air-coolant and battery surfaces could bring down the highest temperature and improve the most extreme temperature variation in the thickly pressed battery box.

In another study, Qian et al. [39] worked on thermal performance of a thermal management system (TMS) of LIBs by using a minichannel cooling method. The authors utilized a fluid cooling technique based on mini-channel cold plate and later a 3D numerical model was developed. They investigated the effects of the direction of flow, the mass flow rate at the inlet, the impact of number of channels, and the channel width on the thermal performance of the pack. Their outcomes demonstrated that at 5C discharge, the TMS based on minichannel cooling plates was effective in cooling productivity and controlling the battery temperature. They also found that a five-channel cold-plate was sufficient to control the temperature by increasing the mass flow rate at the inlet.

Jarrett et al. [40] developed a CFD model of a battery cooling plate while considering the impact of working conditions on the ideal design of electric vehicle battery cooling plates. The authors considered three important performance measurements: (1) average temperature, (2) temperature consistency, and (3) drop in pressure. They identified that out of these three, temperature consistency was the most sensitive to the working conditions, particularly the circulation of the heat flux input and the flow rate of the coolant.

Zou et al. [41] worked on an experimental study on multiwalled carbon nanotube (MWCNT)-based, graphene-based and MWCNT/graphene-based PCM to enhance the thermal performance of lithium-ion BTMS. Their results demonstrated that a composite PCM mass ratio of 3:7 of the MWCNT/graphene could display the best synergistic improvement for the heat transfer effect, for which the thermal conductivity was increased by 31.8%, 55.4% and 124% compared to graphene-based composite PCM, MWCNT-based composite PCM and pure PCM respectively.

Greco et al. [42] developed a heat-pipe-based BTMS arranged in a sandwiched pattern to improve the cooling for EVs. The authors also built a 1D model utilizing the thermal circuit technique. The proposed model was contrasted to an analytical solution in view of variable partition and CFD simulations in 3D. They found that the higher surface contact of the heat pipes allowed a better cooling management compared to forced convection cooling.

Liang et al. [43] researched the thermal execution of a BTMS under various ambient temperatures using heat pipes. The authors examined impacts of environment temperature, coolant flow rate, coolant temperature, and start-up time on the thermal execution of BTMS. They also claimed that the power utilization can be minimized by diminishing run time of HP-BTMS.

Lastly, Wang et al. [44] experimentally examined a high capacity $LiFePO_4$ battery pack at high temperatures and quick discharge using a novel fluid cooling method. They designed and developed thermal silica plate-based BTMS. Their test results demonstrated that adding the thermal silica plates significantly improved the cooling limit. This can enable the most extreme temperature distinction to be controlled at 6.1 °C and decrease the highest temperature by 11.3 °C in the battery module.

In the above paragraphs, various methods of lithium-ion BTMS have been studied, and the cooling plates cooling method demonstrates various practical application prospects. The main research contents of the existing literature on the cooling plate cooling methods include the impact of flow direction and mass flow rate. However, as a key part of the cooling system, the effect of external temperature (or boundary conditions) within mini channels was seldom studied. Therefore, in this paper, a cooling plate design and development was done in such a way that it gives maximum cooling close to the anode and the cathode, as the greatest heat production is close to electrodes of LIBs during high acceleration of EVs. A comprehensive examination and modeling was conducted on the cooling plate in the LIB system. From this, the investigation under the rates of 3C and 4C (constant current discharge), and operating conditions of 5 °C, 15 °C, 25 °C, and 35 °C was assessed in detail. In our previous CFD studies (Panchal et al. [45,46]), we designed and developed a cooling plate with only one channel composed of a single inlet and outlet, and put one on both the top and bottom of the battery to cool it down during the discharging rates of 1C, 2C, 3C, and 4C, and diverse cooling temperatures. In this paper, STAR CCM+ was used for CFD simulation and then the simulated results were validated with experimental data of various temperature and velocity profiles. The results of this research can assist in the design, development and optimization of a cooling-plate cooling system. The data generated is also helpful in battery thermal modeling and EVs development.

The rest of the manuscript is organized as follows. Section 3 introduces the experimental studies including laser scanning, heat flux locations on the surface of the battery, experimental plan and procedure. Section 4 explains the cooling plate physical model in detail, provides geometry and boundary conditions as well as meshing. Section 5 analyzes the results of the numerical calculation, specifically, the effect of inlet mass flow rate on the temperature and velocity profiles of the cooling plate mini channels. Section 6 presents a summary of the conclusions.

3. Experimental Studies

Here, the lab testing details are given through the laser scanning, test set-up, heat flux distributions, and testing plan and procedure.

3.1. Reverse Engineering

For reverse engineering, we used a 25S2P battery pack in an EV. The battery pack and reverse engineering are shown in Figure 1.

Figure 1. Reverse engineering approach: taking out battery cell from pack.

3.2. Laser Scanning

The 3D laser-scanning machine used for this work is shown in Figure 2. It consists of probe, Light Emitting Diode (LED) indicators, wide stable joints, dampener, stable base, etc. The laser probe has an accuracy of ±25 μm (±0.001 in); field depth: 115 mm (4.5 in); width of effective scan: near field 3.1 in (80 mm), far field 5.9 in (150 mm); minimum point spacing: 40 μm (±0.0015 in); scan rate: 280 fps (frames/second), 280 fps × 2000, point/line = 560,000 points/sec; and laser class: 2 M. Arm specifications were measuring range: 1.8 m (6 ft); volumetric accuracy: ±0.034 mm (±0.0013 in); single point repeatability: 0.024 mm (0.0009 in); and seven-axis movement. After scanning the microchannel cooling plate, the image was transferred into CAD using DEP MeshWorks 8.0 software.

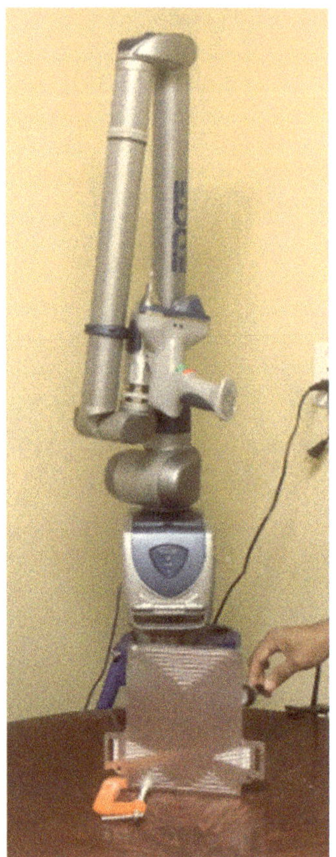

Figure 2. Laser scanning set-up.

3.3. Battery Description, Experimental Set-up and Heat Flux Locations

The test set-up utilized for this work is explained in detail in our previously published paper [47]. In this work, a different cooling plate configuration was used. Three heat flux sensors were attached on the principle surface of the battery (one near the anode, one near the cathode, and one near the mid body) and the sensor measurements were used for simulation. The heat flux sensor location is shown Figure 3. A pouch-type 20-Ah-capacity LIB cell was utilized for the testing and model validation. Table 1 organizes the LIB cell technical details. The battery cell was placed between two cooling plates to form a sandwich structure.

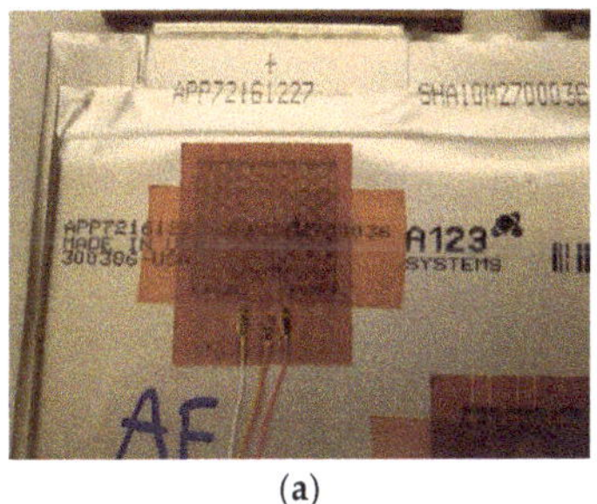

(a)

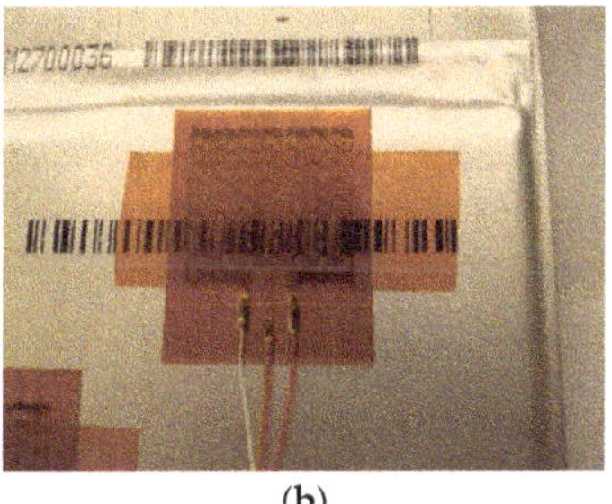

(b)

Figure 3. Heat flux sensors locations near (**a**) cathode and (**b**) anode.

Table 1. Technical details of 20 Ah lithium-ion battery (LIB) cell.

Specification	Value	Unit
Material for electrolyte	Carbonate based	-
Material for anode	Graphite	-
Material for cathode	$LiFePO_4$	-
Voltage (nominal)	3.3	V
Dimensions	7.25 (t) × 160 (w) × 227 (h)	mm
Capacity of the cell (nominal)	20	Ah
Discharge power	1200	W
Energy (nominal)	65	Wh
Specific energy	131	Wh/kg
Energy density	247	Wh/L
Operating temperature	−30 to 55	°C
Mass of the cell	496	g
Specific power	2400	W/kg
Maximum discharge	300	A
Internal resistance	0.5	mΩ
Volume	0.263	L
Storage temperature	−40 to 60	°C
Number of cycles	Min. 300, approx. 2000	Cycles
Maximum charge	300	A

3.4. Test Plan

In the experiment, four different fluid inlet temperatures were chosen for the water-cooling strategy: 5 °C, 15 °C, 25 °C, and 35 °C. Two distinctive discharge currents were selected: 60 A and 80 A (3C and 4C). The charge current is 20 A (1C). The testing sequence is presented in Table 2. The test layout and experimental uncertainty are available in our previously published paper [47].

Table 2. Testing sequence.

Working Fluid	Operating Temperature (°C)	Charge Current	Discharge Current
Water	5	20 A	60 A, 80 A
	15	20 A	60 A, 80 A
	25	20 A	60 A, 80 A
	35	20 A	60 A, 80 A

4. Cold Plate Cooling System Modeling

4.1. Governing Equations

The fluid stream in this test was considered turbulent because the Reynold's number was 8700. As such, the stream was modeled by Reynolds-averaged Navier–Stokes Equations (RANS). STAR CCM+

software was used for the CFD simulation. In this investigation, the realizable k-ε turbulence model was utilized because of the strengths of the model, which include reasonable precision for an extensive variety of flows and its demonstrated capacity in heat transfer and stream examination. The equations used in STAR CCM+ for turbulent kinetic energy and eddy viscosity were:

$$\frac{\partial}{\partial t}(\rho k) + \frac{\partial}{\partial x_i}(\rho k u_i) = \frac{\partial}{\partial x_j}\left[\left(\mu + \frac{\mu_t}{\sigma_k}\right)\frac{\partial k}{\partial x_j}\right] + G_k + G_b - \rho\varepsilon - Y_M + S_k \tag{1}$$

$$\frac{\partial}{\partial t}(\rho\varepsilon) + \frac{\partial}{\partial x_i}(\rho\varepsilon u_i) = \frac{\partial}{\partial x_j}\left[\left(\mu + \frac{\mu_t}{\sigma_\varepsilon}\right)\frac{\partial\varepsilon}{\partial x_j}\right] + C_{1\varepsilon}\frac{\varepsilon}{k}(G_k + C_{3\varepsilon}G_b) - C_{2\varepsilon}\rho\frac{\varepsilon^2}{k} + S_\varepsilon \tag{2}$$

In the above equations, S_k and S_ε are user-defined source terms. Y_M is the contribution of the fluctuating dilatation in compressible turbulence to the overall dissipation rate. G_k is the production of turbulence kinetic energy due to the average speed gradients. G_b is the generation of turbulence kinetic energy due to buoyancy. $C_{1\varepsilon}$, $C_{2\varepsilon}$, and $C_{3\varepsilon}$ represents the model constants, σ_k and σ_ε are the turbulent Prandtl numbers for k and ε, respectively. The turbulent (or eddy) viscosity was computed by combining k and ε as follows:

$$\mu_t = \rho C_\mu \frac{k^2}{\varepsilon} \tag{3}$$

where C_μ is a constant. The model constants $C_{1\varepsilon}$, $C_{2\varepsilon}$, C_μ, σ_k and σ_ε have default values of: $C_{1\varepsilon} = 1.44$, $C_{2\varepsilon} = 1.92$, $C_\mu = 0.09$, $\sigma_k = 1.0$ and $\sigma_\varepsilon = 1.3$.

4.2. CFD Modeling Details Using STAR CCM+

In a CFD simulation, the boundary condition "wall" is considered at locations where the stream cannot penetrate and includes walls, the ceiling, and the floor. The accompanying parameters were chosen for the model development: (1) the stream is incompressible, turbulent, and steady state; (2) water is selected as the working medium with 997.56 kg/m^3 density; (3) The mass flow rate at each channel is 0.000277677 kg/s, while the aggregate mass flow rate at all nine channels is 0.002499003 kg/s; (4) the area at each channel is 5.272×10^{-7} m^2; (5) the dynamic viscosity of 0.00088871 Pa s; (6) the specific heat is 4181.72 J/kg K; (7) the thermal conductivity is 0.62 W/m K; and (8) the turbulent Prandtl number is 0.9. In addition to this, the thermal conductivity of the outlet aluminum cover is 237 W/m K, the density of cover is 2702 kg/m^3, and specific heat is 903 J/kg K. The selected parameters for model set-up were (1) flow: turbulent; (2) fluid: incompressible; (3) time: steady state; (4) realizable K-epsilon (RANS); (5) two-layer wall: y+ wall treatment (y+ $\approx$ 5); (6) solver: segregated; (7) convection: second order; (8) turbulence intensity: 0.01 (default); and (9) turbulent viscosity ratio: 10.0 (default).

4.3. Meshing in DEP MeshWorks 8.0

The meshing of the area was a crucial step because different lattice parameters, such as quality criteria, mesh size, the shape of the elements, and the number of nodes have a significant impact on the result accuracy and the numerical solution. Here, the meshing was done using DEP MeshWorks 8.0 software. The screenshot of DEP MeshWorks during meshing of cooling plate is shown in Figure 4. Meshing in all nine-inlet channels of the cooling plate and meshing in the top portion of the cooling plate, which is specifically designed for this prismatic battery cooling, is shown in Figure 5. This design provides maximum cooling in this region because the heat production is the highest near the electrodes. In order to accurately represent the heat and flow transfer characteristics, the mesh was refined at regions of high geometrical deviations. Furthermore, P1, P2, and P3 areas for all cases for the heat flux sensor (HFS) is shown in Figure 6.

Figure 4. MeshWorks 8.0 screenshot during meshing of cooling plate.

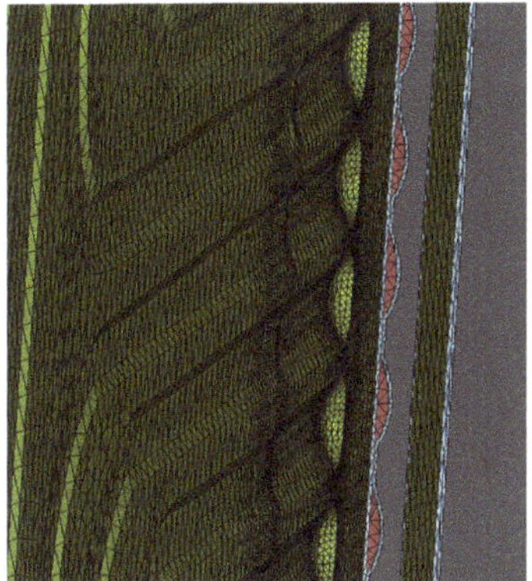

Figure 5. Meshing in inlet channels in DEP MeshWorks 8.0.

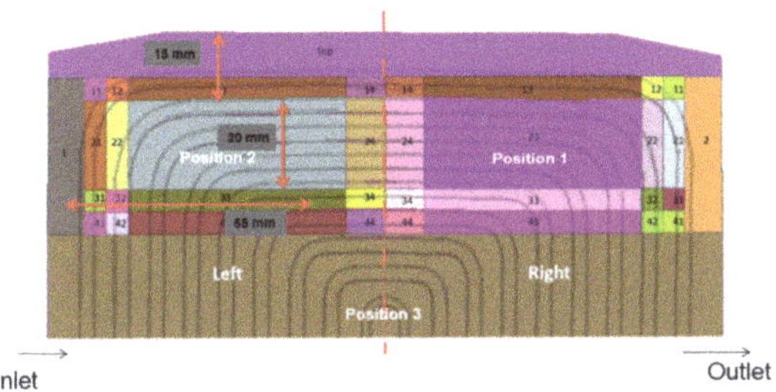

Figure 6. Heat flux sensor (HFS) positions for model development.

5. Results and Discussion

This section presents on the results obtained from investigations for a specific prismatic LIB at various higher discharge currents of 60 A (3C) and 80 A (4C) for water cooling at working conditions of 5 °C, 15 °C, 25 °C, and 35 °C.

5.1. Temperature Contours at 3C (60 A) Discharge

The temperature contours determined from STAR CCM+ CFD at 60 A discharge current and 5 °C, 15 °C, 25 °C, and 35 °C working temperatures are shown in Figures 7–10. As previously mentioned, three heat flux sensors were put on the main surface of the battery: one was situated close to the positive terminal or cathode, the second was situated close to the negative terminal or anode, and the third was situated at the center of the cell. Figure 7 demonstrates the simulation results at 60 A discharge current and 5 °C coolant inlet temperature with heat flux near cathode = 2347.7 W/m^2, anode = 2259.5 W/m^2, and mid surface = 539.3 W/m^2. Similarly, Figure 8 shows temperature contours at 60 A discharge current and 15 °C coolant inlet temperature with heat flux values near cathode = 1711.8 W/m^2, anode= 2351.6 W/m^2, and mid surface = 548.4 W/m^2. It is observed that the temperature contours and trends are similar with the inlet being cold and outlet being hot. During the battery operation at high C-rates, the generated heat from the battery is conducted to the cooling plate. As the coolant flows inside from the inlet, the heat is absorbed continuously with increments in coolant temperature along the flow path. The maximum temperature of coolant is observed at the outlet surface, as expected. Figure 9 shows temperature contours at 60 A discharge current and 25 °C coolant inlet temperature with heat flux values near cathode = 1597.3 W/m^2, anode = 1851.6 W/m^2, and mid surface = 413.0 W/m^2. Figure 10 demonstrates temperature contours at 3C discharge rate and 35 °C coolant inlet temperature with heat flux values near cathode = 1468.4 W/m^2, anode = 1579.9 W/m^2, and mid surface = 340.6 W/m^2.

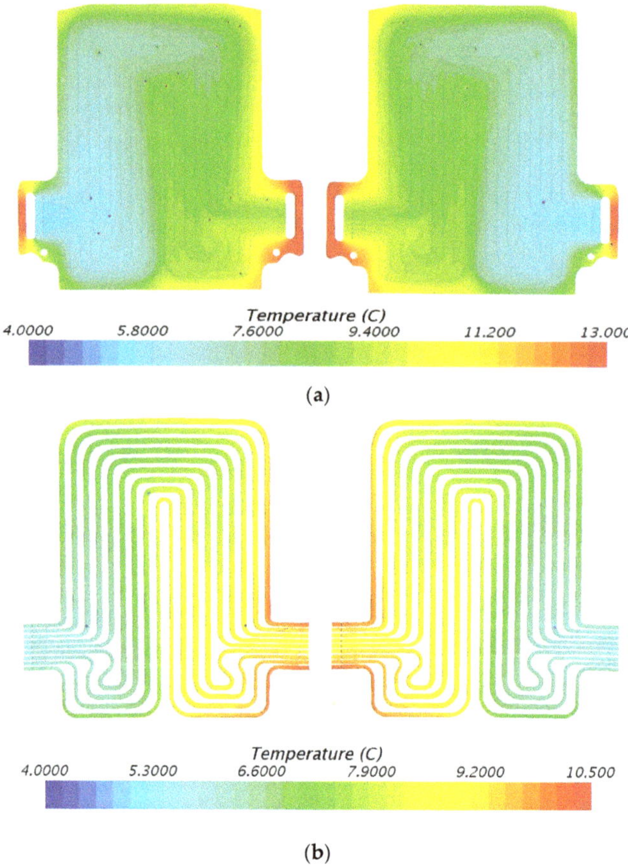

Figure 7. Temperature profile at 60 A and 5 °C with heat flux values near cathode = 2347.7 W/m^2, anode = 2259.5 W/m^2, and mid surface = 539.3 W/m^2. (**a**) Top view and (**b**) bottom view.

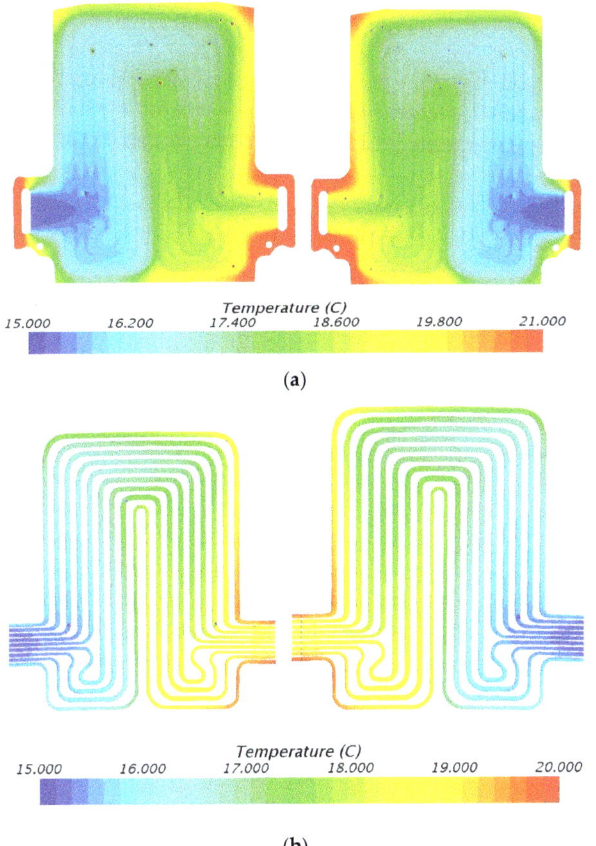

Figure 8. Temperature profile at 60 A and 15 °C with heat flux values near cathode = 1711.8 W/m^2, anode = 2351.6 W/m^2, and mid surface = 548.4 W/m^2. (**a**) Top view and (**b**) bottom view.

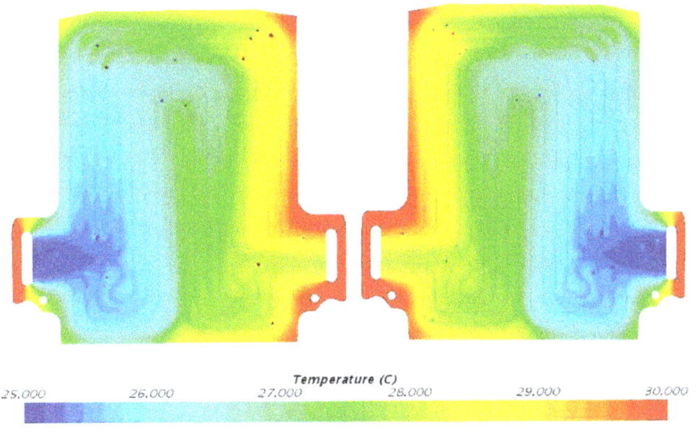

(**a**)

Figure 9. *Cont.*

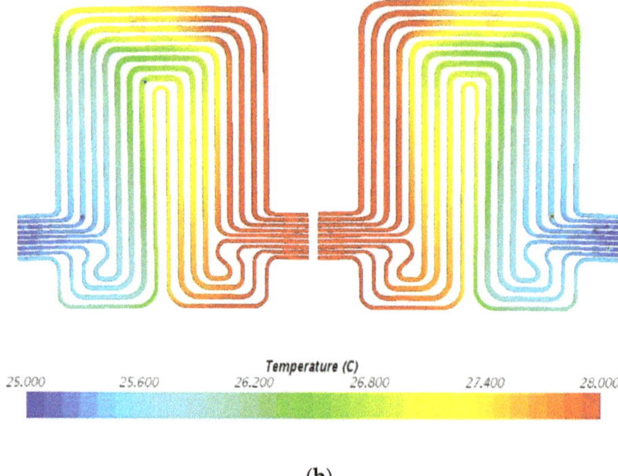

(b)

Figure 9. Temperature profile at 60 A and 25 °C with heat flux values near cathode = 1597.3 W/m^2, anode = 1851.6 W/m^2, and mid surface = 413.0 W/m^2. (**a**) Top view and (**b**) bottom view.

The result for temperature from simulation for 4C discharge rate and ambient temperature of 35 °C showed 38.12 °C, which is 2.85% higher or lower than experimental values for similar boundary conditions. The result for temperature from simulation for 3C discharge rate and ambient temperature of 5 °C showed 7.93 °C, which is 2.19% higher or lower than experimental values for similar boundary conditions. It is also noted that working temperature had a great impact on battery discharge capacity. As the working temperature rose from 5 °C to 35 °C, the temperature contour values also increased for a particular C-rate. The general cooling patterns are identical, demonstrating more noteworthy contrasts at the inlet of the cooling plate where the water is coldest. The temperatures differ with the inlet working condition temperature, yet the overall pattern remains generally consistent. Table 3 gives the summary of inlet and outlet water temperatures at 60 A discharge current and working temperature conditions of 5 °C, 15 °C, 25 °C, and 35 °C. Table 3 also provides the difference in experimental and simulated values obtained from STAR CCM+ software. It is additionally seen that the simulated values are higher than experimental values because of the assumption of ideal wall conditions of adiabatic wall at non-heat-transferring boundaries. In the actual experiment, a small amount of heat transfer could be observed at some places where insulation was provided.

Table 3. Water inlet and outlet temperature at 60 A and 80 A with different working temperature.

Working Fluid	Working Temperature (°C)	Difference between Experimental and Simulated Values	Water Inlet and Outlet Temperature (°C)			
			60 A		80 A	
			Inlet	Outlet	Inlet	Outlet
Water	5	Experimental (°C)	5.7391	7.9307	5.1435	7.7029
		Simulated (°C)	7.93	9.35	7.70	10.25
		Difference (%)	38.17	17.90	49.70	33.07
	15	Experimental (°C)	15.1377	16.7696	15.0906	16.9376
		Simulated (°C)	16.76	18.84	16.93	19.75
		Difference (%)	10.72	12.35	12.55	16.60
	25	Experimental (°C)	25.0992	25.9614	25.0984	26.3445
		Simulated (°C)	25.96	28.47	26.34	29.3
		Difference (%)	3.43	9.66	4.95	11.22
	35	Experimental (°C)	34.4912	34.9092	34.2555	35.2637
		Simulated (°C)	34.90	37.29	35.26	38.12
		Difference (%)	1.19	6.82	2.93	8.10

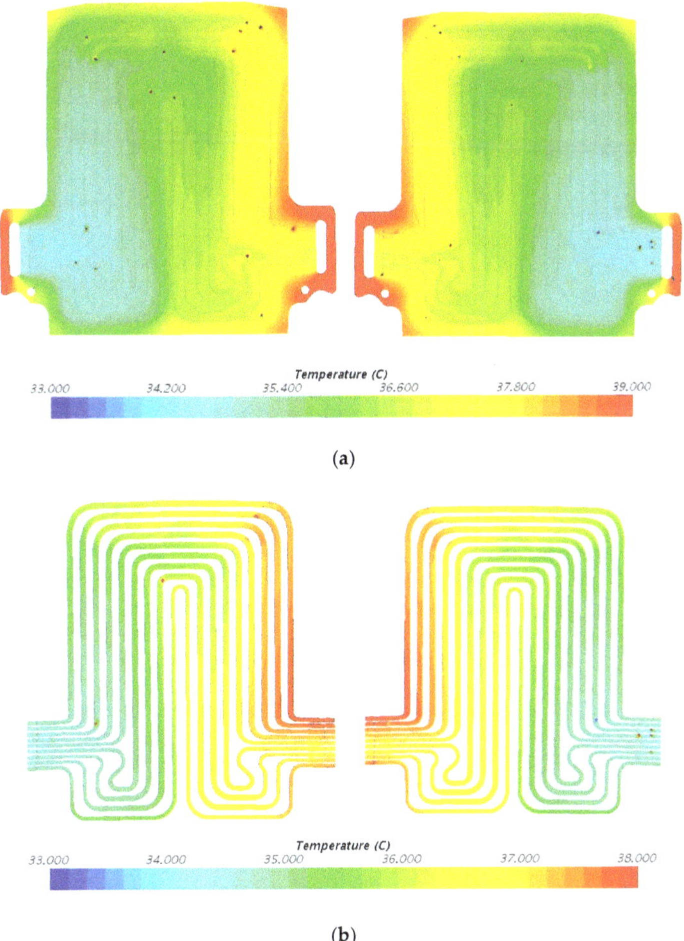

(a)

(b)

Figure 10. Temperature profile at 60 A and 35 °C with heat flux values near cathode = 1468.4 W/m^2, anode = 1579.9 W/m^2, and mid surface = 340.6 W/m^2. (**a**) Top view and (**b**) bottom view.

5.2. Temperature Contours at 4C (80 A) Discharge

Figure 11 shows temperature contours at 80 A discharge current and 5 °C water inlet temperature with heat flux values near cathode = 3112.2 W/m^2, anode = 3072.8 W/m^2, and mid surface = 764.1 W/m^2. Figure 12 shows temperature contours at 80 A discharge current and 15 °C water inlet temperature with heat flux values near cathode = 2419.0 W/m^2, anode = 2887.1 W/m^2, and mid surface = 697.3 W/m^2. Figure 13 demonstrates temperature contours at 80 A discharge current and 25 °C water inlet temperature with heat flux values near cathode = 2309.3 W/m^2, anode = 2648.2 W/m^2, and mid surface = 611.1 W/m^2. Figure 14 shows temperature contours at 80 A discharge current and 35 °C water inlet temperature with heat flux values near cathode = 2160.2 W/m^2, anode = 2101.5 W/m^2, and mid surface = 471.8 W/m^2. It is noted that, as the battery discharged, the flowing water inside the cooling plate got heated because the heat was first conducted to the cooling plate and subsequently transferred to the coolant by convection. The joule heating is the dominant factor for heat generation. As the discharge current changed from 60 A to 80 A, there was an increase in temperature values as well. The pattern observed was that increased discharge currents and increased ambient temperatures resulted in higher temperatures in the cooling plate. Table 3 gives the outline of inlet and outlet water temperatures at 80 A discharge current and different working temperature conditions of 5 °C, 15 °C, 25 °C, and 35 °C. Table 3 also provides the difference in experimental and simulated values obtained from STAR CCM+ software. It was also found that the simulated values were higher than experimental values. In addition, the general cooling patterns were identical, similar to the results discussed in Section 5.1. There were noteworthy temperature differences at the inlet of the cooling plate where the water was coldest.

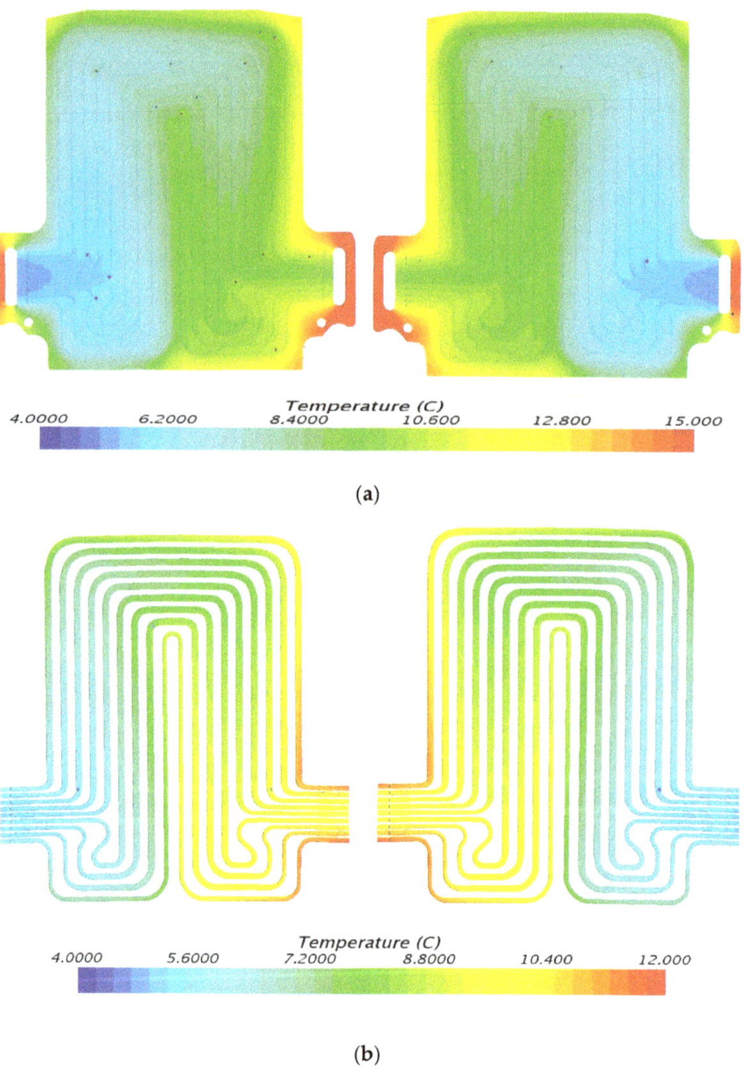

(a)

(b)

Figure 11. Temperature profile at 80 A and 5 °C with heat flux values near cathode = 3112.2 W/m^2, anode = 3072.8 W/m^2, and mid surface = 764.1 W/m^2. (**a**) Top view and (**b**) bottom view.

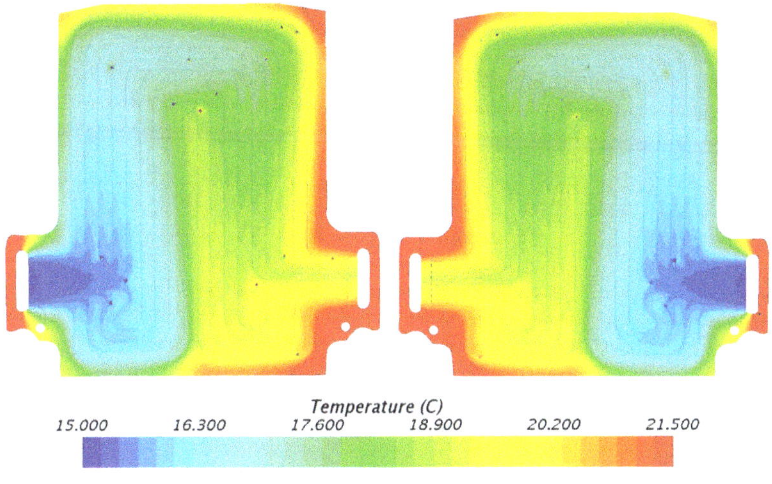

(a)

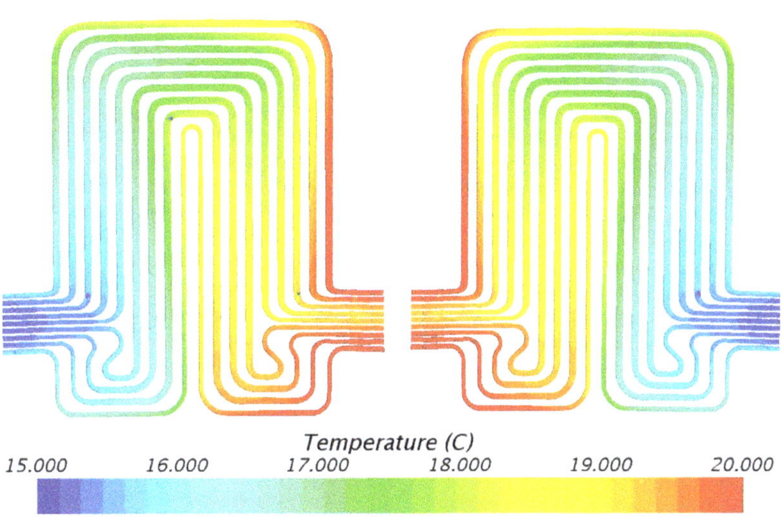

(b)

Figure 12. Temperature profile at 80 A and 15 °C with heat flux values near cathode = 2419.0 W/m^2, anode = 2887.1 W/m^2, and mid surface = 697.3 W/m^2. (**a**) Top view and (**b**) bottom view.

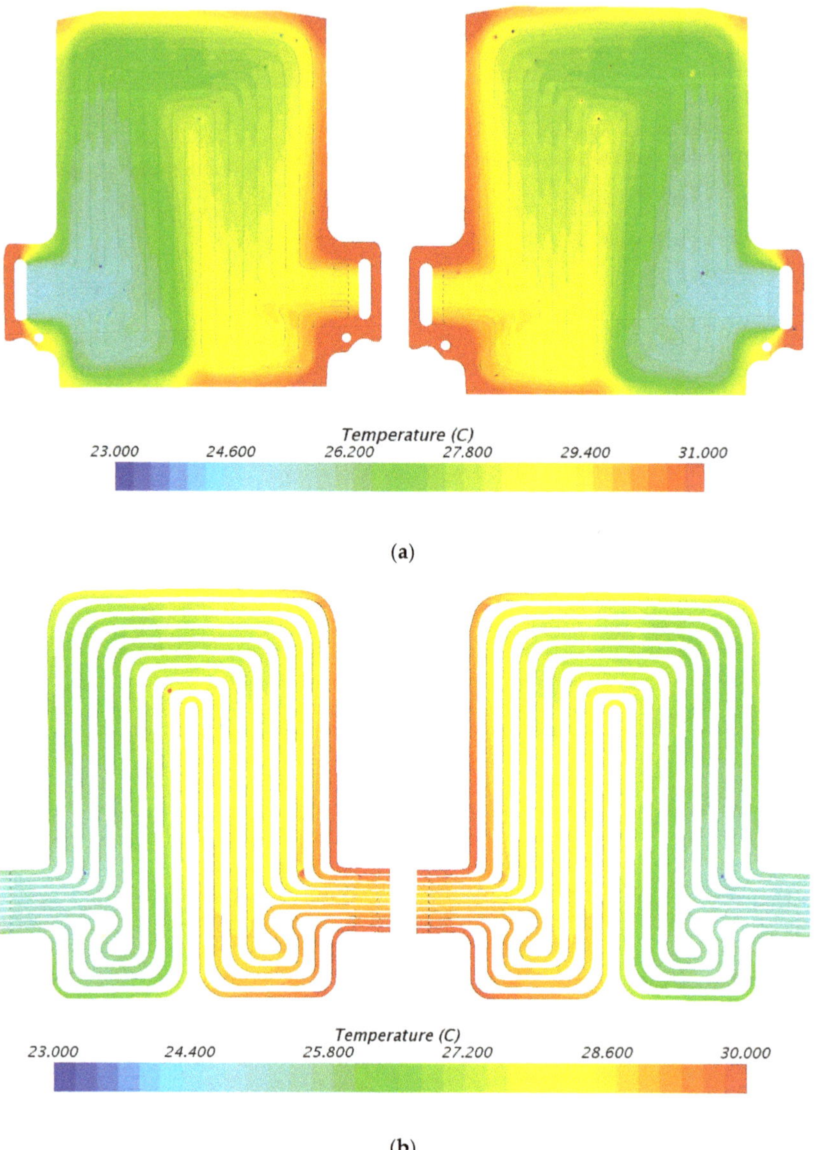

Figure 13. Temperature profile at 80 A and 25 °C with heat flux values near cathode = 2309.3 W/m², anode = 2648.2 W/m², and mid surface = 611.1 W/m². (**a**) Top view and (**b**) bottom view.

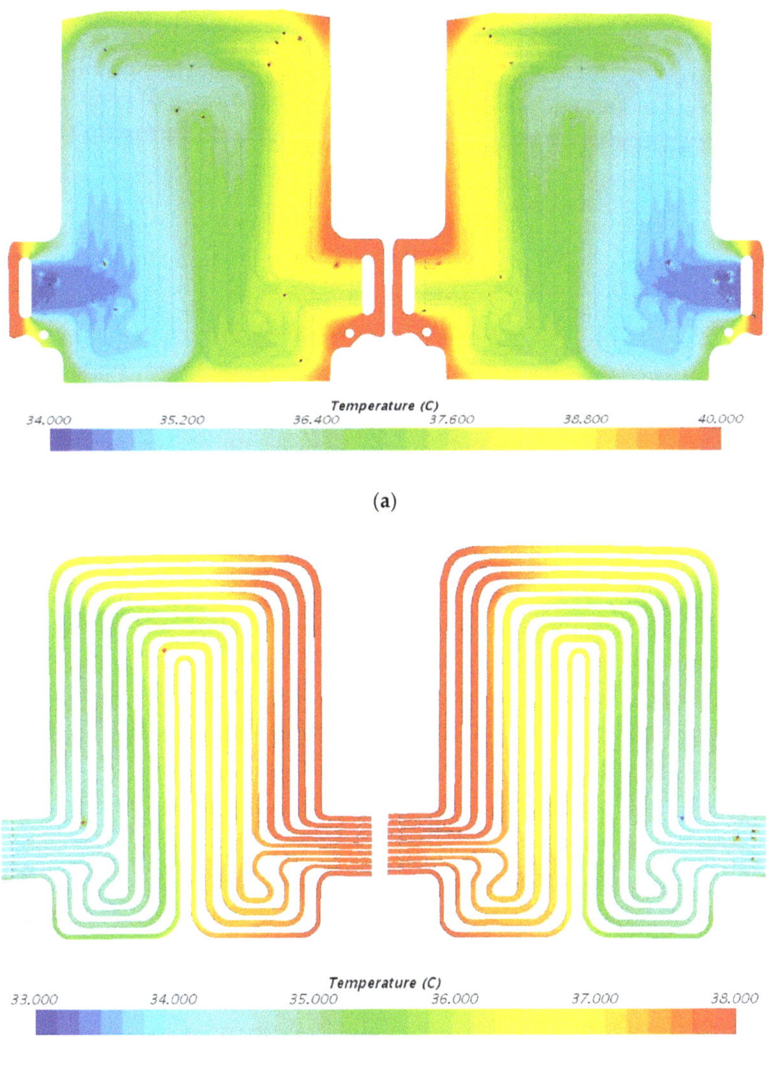

Figure 14. Temperature profile at 80 A and 35 °C with heat flux values near cathode = 2160.2 W/m^2, anode = 2101.5 W/m^2, and mid surface = 471.8 W/m^2. (**a**) Top view and (**b**) bottom view.

5.3. Velocity Contours at 3C (60 A) and 4C (80 A) Discharge

The investigations of velocity contours can provide insights into future design considerations by comparing contour results of velocity and comparing with respective contour results of temperatures. The velocity contours at 60 A and 80 A discharge currents and 5 °C, 15 °C, 25 °C, and 35 °C working temperatures appear in Figures 15 and 16. The velocity contours were identical in all the cases, in accordance with general trends, given the low temperatures associated in the modeling that would have had a minimal impact on the water density. These results might be influenced by the lower y+ value, wall functions and turbulence model utilized. It was also observed that the velocity distribution at the inlet to the cooling plate and the outlet from the cooling plate was curved with relatively higher velocity gradients.

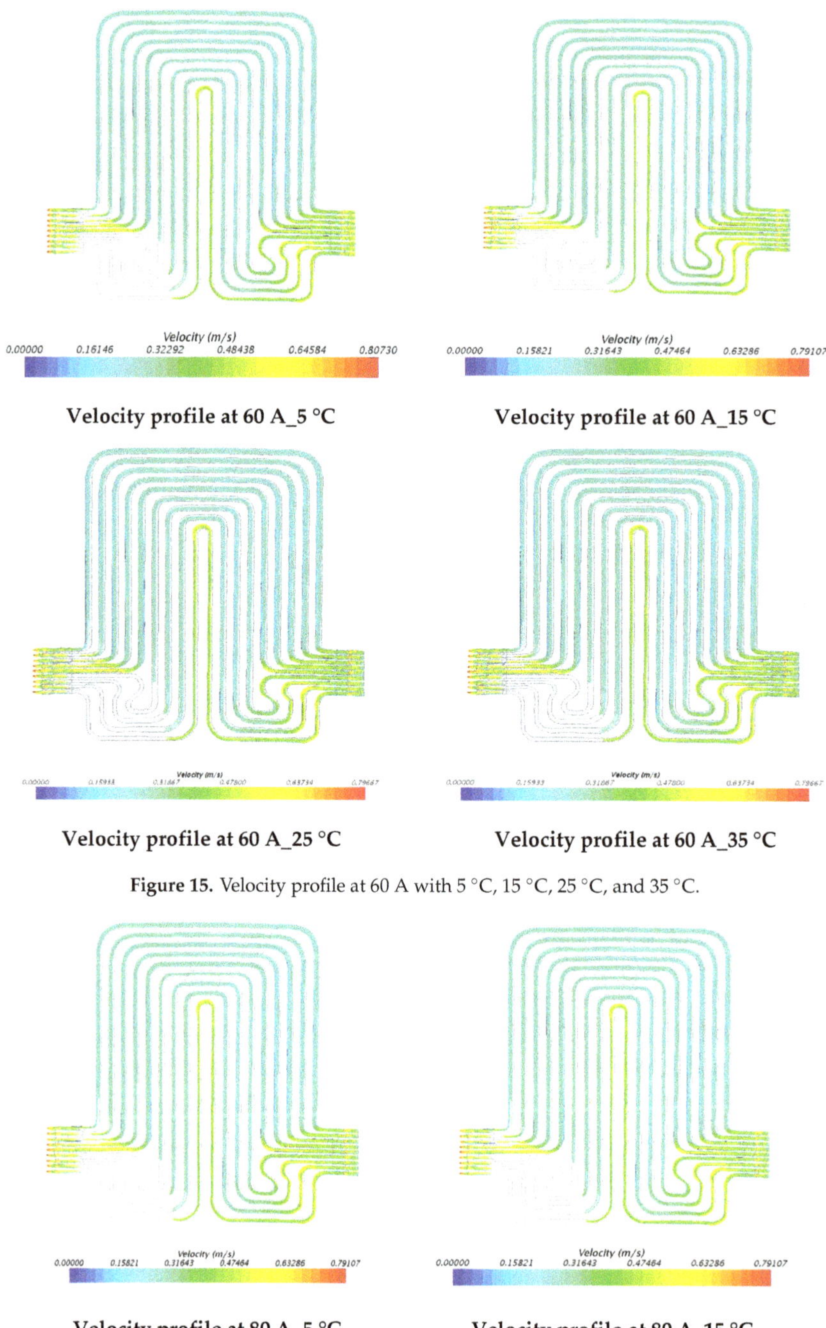

Figure 15. Velocity profile at 60 A with 5 °C, 15 °C, 25 °C, and 35 °C.

Figure 16. *Cont.*

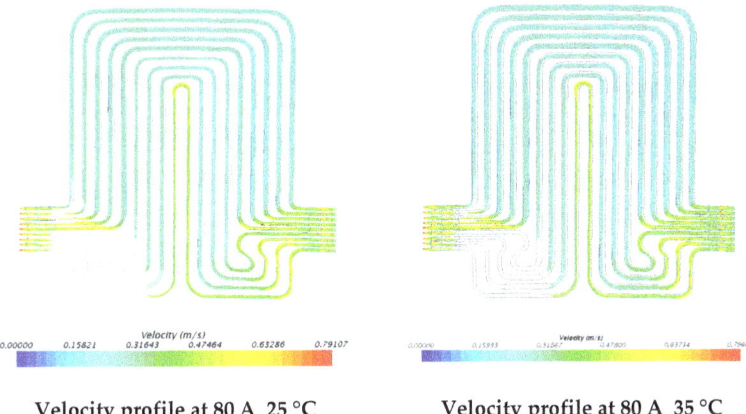

Velocity profile at 80 A_25 °C   **Velocity profile at 80 A_35 °C**

Figure 16. Velocity profile at 80 A with 5 °C, 15 °C, 25 °C, and 35 °C.

5.4. Transient Temperature Profiles of Water Flow and Voltage Distributions

Figures 17 and 18 show the transient behavior of water flowing inside the cooling plates at 60 A and 80 A constant current discharges with various working temperatures of 5 °C, 15 °C, 25 °C, and 35 °C. As discussed earlier, the increase in temperature was due to the joule heating I^2R from the LIB during discharge.

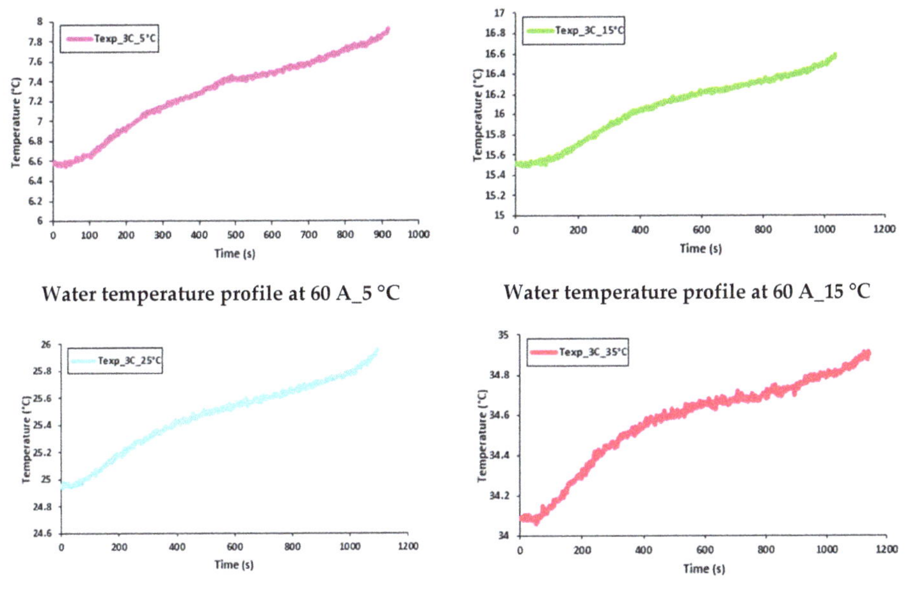

Water temperature profile at 60 A_5 °C   **Water temperature profile at 60 A_15 °C**

Water temperature profile at 60 A_25 °C   **Water temperature profile at 60 A_35 °C**

Figure 17. Transient temperature profile of water flow at 60 A with 5 °C, 15 °C, 25 °C, and 35 °C.

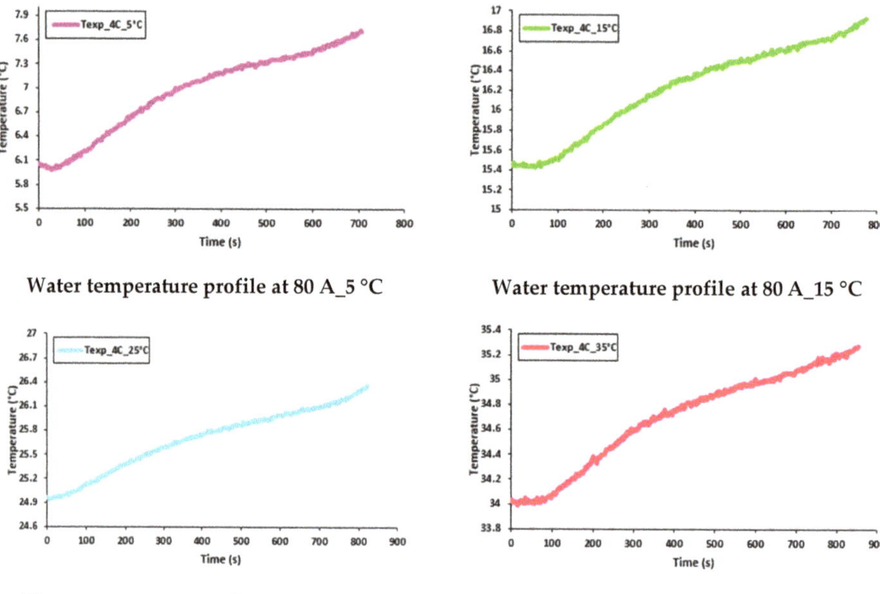

Water temperature profile at 80 A_5 °C   **Water temperature profile at 80 A_15 °C**

Water temperature profile at 80 A_25 °C   **Water temperature profile at 80 A_35 °C**

Figure 18. Transient temperature profile of water flow at 80 A with 5 °C, 15 °C, 25 °C, and 35 °C.

It was discovered that the working temperature had a great impact on the battery performance. At lower discharge currents, the battery capacity was close to the manufacturer's supplied data sheet, but as discharge current increased, there was a decrease in the discharge capacity. Further, when the working temperature changed from 35 °C to 5 °C, there was a more prominent decrease in the discharge capacity. Consequently, it is clear that as the working temperature decreased, the battery discharge capacity also decreased. These effects (reduction in battery discharge capacity) can be seen in Figures 19 and 20, which present the discharge/charge profiles at 60 A and 80 A constant current discharges (and charge current being 20 A) with various working temperatures of 5 °C, 15 °C, 25 °C, and 35 °C.

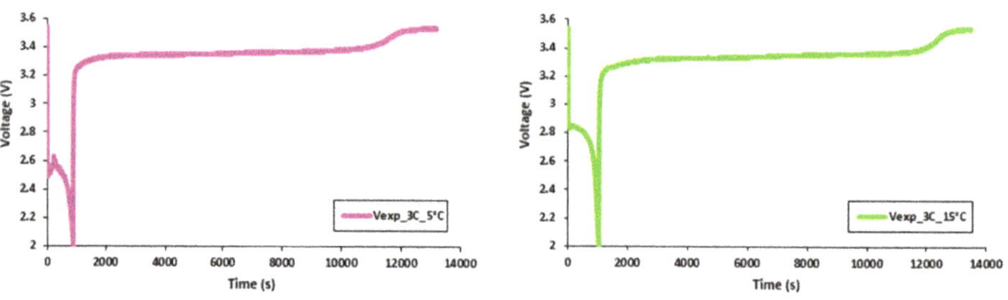

Discharge/charge voltage profile at 60 A_5 °C   **Discharge/charge voltage profile at 60 A_15 °C**

Figure 19. *Cont.*

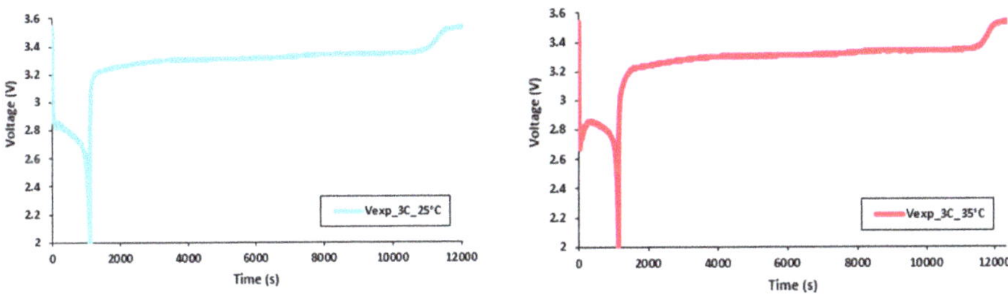

Discharge/charge voltage profile at 60 A_25 °C   **Discharge/charge voltage profile at 60 A_35 °C**

Figure 19. Discharge/charge voltage profile at 60 A with 5 °C, 15 °C, 25 °C, and 35 °C.

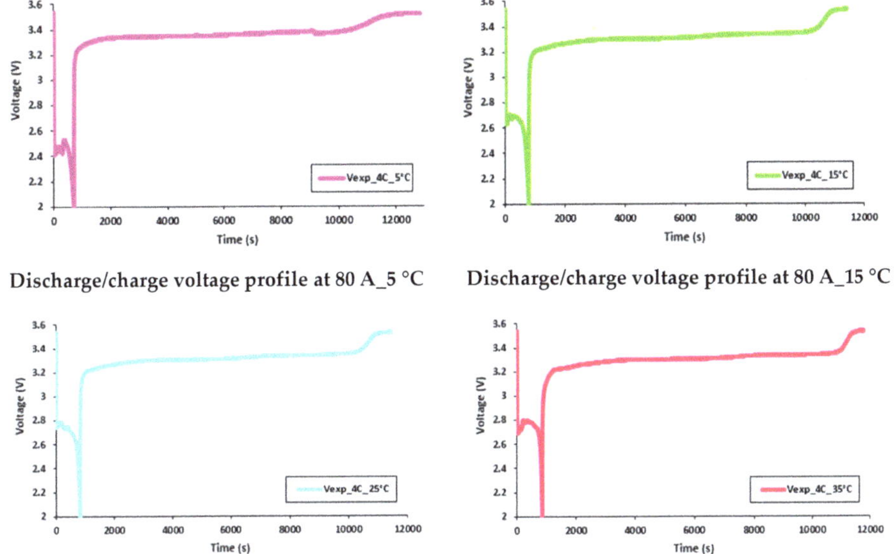

Discharge/charge voltage profile at 80 A_5 °C   **Discharge/charge voltage profile at 80 A_15 °C**

Discharge/charge voltage profile at 80 A_25 °C   **Discharge/charge voltage profile at 80 A_35 °C**

Figure 20. Discharge/charge voltage profile at 60 A with 5 °C, 15 °C, 25 °C, and 35 °C.

6. Conclusions

This paper presented a numerical model using STAR CCM+ for CFD simulations at high C-rates and diverse working temperatures of the liquid (water) with 5 °C, 15 °C, 25 °C, and 35 °C. We discovered that the temperature distributions within cooling plate channels increased with C-rates (3C to 4C). As C-rate increased, the heat flux values measured near the anode, the cathode, and the middle surface also increased. The cooling patterns obtained from simulation were similar to the experimental values with slightly higher values. The velocity plots were identical for all cases. There results provide valuable information on the design considerations that must be made for battery cooling systems in EVs.

Author Contributions: Conceptualization, S.P., Formal analysis, M.-K.T.; Supervision, R.F. and M.F.; Validation, K.G.; Writing—original draft, S.P.; Writing—review and editing, M.-K.T. and S.P. All authors have read and agreed to the published version of the manuscript.

Funding: This research received no external funding.

Acknowledgments: This research work (reverse engineering, CT scanning, MeshWorks and STAR CMM+ software) was done at Detroit Engineering Products (DEP), Michigan, USA, during year 2018–2019. The experimental data was used from University of Waterloo for the model development and validation. This work was supported by equipment and manpower from the Department of Chemical Engineering at the University of Waterloo. Special thanks to Danielle Skeba for the contribution in editing the paper.

Conflicts of Interest: The authors declare no conflict of interest.

Nomenclature

C_μ	constant
C_1, C_2, C_3	constant for model
C	cell potential or cell voltage (V)
I	current (A)
G_k	turbulence kinetic energy generation due to the mean velocity gradients
G_b	turbulence kinetic energy generation due to buoyancy
k	turbulent kinetic energy (J)
L	characteristic dimension (m)
P	pressure (Pa)
Pr	Prandtl number
Pr_t	Turbulent Prandtl number
Re	Reynold's number
S_k and S_ε	user-defined source terms
t	time (s)
T	temperature ($^\circ$C or K)
V	speed (m/s)
$\overline{V}$	average velocity (m/s)
v_s	mean fluid velocity (m/s)
y^+	enhanced wall treatment
ω	turbulent eddy frequency (1/s)
Y_M	the contribution of the fluctuating dilatation in compressible turbulence to the overall dissipation rate
Greek Symbols	
ν	kinematic fluid viscosity (m^2/s)
λ	Reynold's stress
ρ	density (kg/m^3)
μ	dynamic fluid viscosity (Ns/m^2)
∇	gradient operator
σ_k and ε_k	Turbulent Prandtl numbers for k and ε
Subscripts	
sim	simulated
act	actual
Superscripts	
$\circ$	degree
$+$	Related to wall treatment
Acronyms	
BCS	Battery cooling system
BEV	Battery electric vehicle
BTMS	Battery thermal management system
C	Capacity
CC	Constant-current
CV	Constant-voltage
CT	Computed tomography

CAD	Computer aided design
CFD	Computational fluid dynamics
DEP	Detroit Engineered Products, Inc.
EV	Electric vehicle
ESS	Energy storage system
FE	Finite element
HEV	Hybrid electric vehicle
HFS	Heat flux sensor
HPBTMS	Heat pipe battery thermal management system
LIB	Lithium-ion battery
$LiFePO_4$	Lithium iron phosphate
LED	Light emitting diode
MeshWorks	Popular proprietary software package used for CAD and mesh generation
MWCNT	multi-walled carbon nanotubes
PC	Personal computer
PCM	Phase change material
PHEV	Plug-In hybrid electric vehicle
RE	Reverse engineering
RANS	Reynolds-Averaged Navier-Stokes
STAR CCM+	Simulation of Turbulent flow in Arbitrary Regions-Computational Continuum Mechanics + (C++ based)
TMS	Thermal management system
1D	One-dimensional
2D	Two-dimensional
3D	Three-dimensional
18650	IFR 18650 cylindrical valence cells ("I" stands for Li-ion rechargeable, "F" stands for the element "Fe" which is Iron, "R" indicates that the cell shape is round, 18650 means 18 mm diameter and 650 means 65 mm height)
25S2P	25 series 2 parallel

References

1. Walter, M.; Kravchyk, K.V.; Böfer, C.; Widmer, R.; Kovalenko, M.V. Polypyrenes as High-Performance Cathode Materials for Aluminum Batteries. *Adv. Mater.* **2018**, *30*, 1705644. [CrossRef] [PubMed]
2. Angell, M.; Pan, C.-J.; Rong, Y.; Yuan, C.; Lin, M.-C.; Hwang, B.-J.; Dai, H. High Coulombic efficiency aluminum-ion battery using an AlCl3-urea ionic liquid analog electrolyte. *Proc. Natl. Acad. Sci. USA* **2017**, *114*, 834–839. [CrossRef] [PubMed]
3. Crawford, A.J.; Huang, Q.; Kintner-Meyer, M.C.; Zhang, J.-G.; Reed, D.M.; Sprenkle, V.L.; Viswanathan, V.V.; Choi, D. Lifecycle comparison of selected Li-ion battery chemistries under grid and electric vehicle duty cycle combinations. *J. Power Sources* **2018**, *380*, 185–193. [CrossRef]
4. Panchal, S.; Mathewson, S.; Fraser, R.; Culham, R.; Fowler, M. Measurement of Temperature Gradient (dT/dy) and Temperature Response (dT/dt) of a Prismatic Lithium-Ion Pouch Cell with LiFePO4 Cathode Material. *SAE Tech. Pap. Ser.* **2017**, *1*.
5. Panchal, S.; Rashid, M.; Long, F.; Mathew, M.; Fraser, R.; Fowler, M. Degradation Testing and Modeling of 200 Ah LiFePO4 Battery. *SAE Tech. Pap. Ser.* **2018**.
6. Panchal, S. *Impact of Vehicle Charge and Discharge Cycles on the Thermal Characteristics of Lithium-ion Batteries*; University of Waterloo: Waterloo, ON, Canada, 2014.
7. Panchal, S. *Experimental Investigation and Modeling of Lithium-ion Battery Cells and Packs for Electric Vehicles*; University of Ontario Institute of Technology: Oshawa, ON, Canada, 2016.
8. Al-Zareer, M.; Dincer, I.; Rosen, M.A. Heat and mass transfer modeling and assessment of a new battery cooling system. *Int. J. Heat Mass Transf.* **2018**, *126*, 765–778. [CrossRef]
9. Waldmann, T.; Wilka, M.; Kasper, M.; Fleischhammer, M.; Wohlfahrt-Mehrens, M. Temperature dependent ageing mechanisms in Lithium-ion batteries—A Post-Mortem study. *J. Power Sources* **2014**, *262*, 129–135. [CrossRef]

10. Tran, M.-K.; Fowler, M. A Review of Lithium-Ion Battery Fault Diagnostic Algorithms: Current Progress and Future Challenges. *Algorithms* **2020**, *13*, 62. [CrossRef]

11. Panchal, S.; Mathewson, S.; Fraser, R.; Culham, R.; Fowler, M. Experimental Measurements of Thermal Characteristics of LiFePO4 Battery. *SAE Tech. Pap. Ser.* **2015**.

12. Panchal, S.; Mathewson, S.; Fraser, R.; Culham, R.; Fowler, M. Thermal Management of Lithium-Ion Pouch Cell with Indirect Liquid Cooling using Dual Cold Plates Approach. *SAE Int. J. Altern. Powertrains* **2015**, *4*, 293–307. [CrossRef]

13. Ramadass, P.; Haran, B.; White, R.; Popov, B. Capacity fade of Sony 18650 cells cycled at elevated temperatures: Part II. Capacity fade analysis. *J. Power Sources* **2002**, *112*, 614–620. [CrossRef]

14. Panchal, S.; Dincer, I.; Agelin-Chaab, M.; Fowler, M.; Fraser, R. Uneven temperature and voltage distributions due to rapid discharge rates and different boundary conditions for series-connected LiFePO 4 batteries. *Int. Commun. Heat Mass Transf.* **2017**, *81*, 210–217. [CrossRef]

15. Uddin, A.I.; Ku, J. Design and Simulation of Lithium-Ion Battery Thermal Management System for Mild Hybrid Vehicle Application. *SAE Tech. Pap. Ser.* **2015**.

16. Wang, C.H.; Lin, T.; Huang, J.T.; Rao, Z.H. Temperature response of a high power lithium-ion battery subjected to high current discharge. *Mater. Res. Innov.* **2015**, *19*, S2–S156. [CrossRef]

17. Panchal, S.; Dincer, I.; Agelin-Chaab, M.; Fraser, R.; Fowler, M. Experimental and theoretical investigation of temperature distributions in a prismatic lithium-ion battery. *Int. J. Therm. Sci.* **2016**, *99*, 204–212. [CrossRef]

18. Panchal, S.; Dincer, I.; Agelin-Chaab, M.; Fraser, R.; Fowler, M. Thermal modeling and validation of temperature distributions in a prismatic lithium-ion battery at different discharge rates and varying boundary conditions. *Appl. Therm. Eng.* **2016**, *96*, 190–199. [CrossRef]

19. Patil, M.S.; Panchal, S.; Kim, N.; Lee, M.-Y. Cooling Performance Characteristics of 20 Ah Lithium-Ion Pouch Cell with Cold Plates along Both Surfaces. *Energies* **2018**, *11*, 2550. [CrossRef]

20. He, F.; Ma, L. Thermal Management in Hybrid Power Systems Using Cylindrical and Prismatic Battery Cells. *Heat Transf. Eng.* **2015**, *37*, 581–590. [CrossRef]

21. Giuliano, M.R.; Prasad, A.K.; Advani, S. Experimental study of an air-cooled thermal management system for high capacity lithium–titanate batteries. *J. Power Sources* **2012**, *216*, 345–352. [CrossRef]

22. Jin, L.W.; Lee, P.; Kong, X.; Fan, Y.; Chou, S. Ultra-thin minichannel LCP for EV battery thermal management. *Appl. Energy* **2014**, *113*, 1786–1794. [CrossRef]

23. Rao, Z.; Wang, S. A review of power battery thermal energy management. *Renew. Sustain. Energy Rev.* **2011**, *15*, 4554–4571. [CrossRef]

24. Lemmon, E.W.; Jacobsen, R.T. Viscosity and Thermal Conductivity Equations for Nitrogen, Oxygen, Argon, and Air. *Int. J. Thermophys.* **2004**, *25*, 21–69. [CrossRef]

25. Park, H. A design of air flow configuration for cooling lithium ion battery in hybrid electric vehicles. *J. Power Sources* **2013**, *239*, 30–36. [CrossRef]

26. Fan, L.-W.; Khodadadi, J.; Pesaran, A. A parametric study on thermal management of an air-cooled lithium-ion battery module for plug-in hybrid electric vehicles. *J. Power Sources* **2013**, *238*, 301–312. [CrossRef]

27. Kelly, K.J.; Mihalic, M.; Zolot, M. Battery usage and thermal performance of the Toyota Prius and Honda Insight during chassis dynamometer testing. In Proceedings of the Seventeenth Annual Battery Conference on Applications and Advances (Cat. No.02TH8576), Long Beach, CA, USA, 18 January 2002; Institute of Electrical and Electronics Engineers (IEEE): Piscataway, NJ, USA, 2002; pp. 247–252.

28. Berdichevsky, G.; Kelty, K.; Straubel, J.B.; Toomre, E. The Tesla Roadster Battery System. 2007. Available online: http://www.batterypoweronline.com (accessed on 21 July 2019).

29. What Is the Best Electric Vehicle Battery Cooling System? 2017. Available online: https://avidtp.com/what-is-the-best-cooling-system-for-electric-vehicle-battery-packs/ (accessed on 29 June 2019).

30. Samba, A.; Omar, N.; Gualous, H.; Firouz, Y.; Bossche, P.V.D.; Van Mierlo, J.; Boubekeur, T.I. Development of an Advanced Two-Dimensional Thermal Model for Large size Lithium-ion Pouch Cells. *Electrochim. Acta* **2014**, *117*, 246–254. [CrossRef]

31. Li, G.; Li, S.-P. Physics-Based CFD Simulation of Lithium-Ion Battery under the FUDS Driving Cycle. *ECS Trans.* **2015**, *64*, 1–14. [CrossRef]

32. Vyroubal, P.; Kazda, T.; Maxa, J.; Vondrák, J. Analysis of Temperature Field in Lithium Ion Battery by Discharging. *ECS Trans.* **2015**, *70*, 269–273. [CrossRef]

33. Yeow, K.; Teng, H.; Thelliez, M.; Tan, E. Thermal Analysis of a Li-ion Battery System with Indirect Liquid Cooling Using Finite Element Analysis Approach. *SAE Int. J. Altern. Powertrains* **2012**, *1*, 65–78. [CrossRef]

34. Al Hallaj, S.; Selman, J.; King, S.W.; Kern, R.S.; Benjamin, M.C.; Barnak, J.P.; Nemanich, R.J.; Davis, R.F. A Novel Thermal Management System for Electric Vehicle Batteries Using Phase-Change Material. *J. Electrochem. Soc.* **2000**, *147*, 3231. [CrossRef]

35. Zhang, X.; Wang, T.; Jiang, S.B.; Xu, H.G.; Zhang, Y.N. Modelling and Simulation of Pouch Lithium-Ion Battery Thermal Management Using Cold Plate. *Int. J. Simul. Model.* **2018**, *17*, 498–511. [CrossRef]

36. Omkar, D.; Vijaykumar, P. Development of Phase Change Material/Cooling Plate Coupled Battery Thermal Management System Using CFD. *Int. J. Res. Eng. Appl. Manag.* **2018**, *4*, 1–6.

37. Chen, D.; Jiang, J.; Kim, G.-H.; Yang, C.; Pesaran, A. Comparison of different cooling methods for lithium ion battery cells. *Appl. Therm. Eng.* **2016**, *94*, 846–854. [CrossRef]

38. Lu, Z.; Meng, X.; Wei, L.; Hu, W.; Zhang, L.; Jin, L.W. Thermal Management of Densely-packed EV Battery with Forced Air Cooling Strategies. *Energy Procedia* **2016**, *88*, 682–688. [CrossRef]

39. Qian, Z.; Li, Y.; Rao, Z. Thermal performance of lithium-ion battery thermal management system by using mini-channel cooling. *Energy Convers. Manag.* **2016**, *126*, 622–631. [CrossRef]

40. Jarrett, A.; Kim, I.Y. Influence of operating conditions on the optimum design of electric vehicle battery cooling plates. *J. Power Sources* **2014**, *245*, 644–655. [CrossRef]

41. Zou, D.; Ma, X.; Liu, X.; Zheng, P.; Hu, Y. Thermal performance enhancement of composite phase change materials (PCM) using graphene and carbon nanotubes as additives for the potential application in lithium-ion power battery. *Int. J. Heat Mass Transf.* **2018**, *120*, 33–41. [CrossRef]

42. Greco, A.; Cao, D.; Jiang, X.; Yang, H. A theoretical and computational study of lithium-ion battery thermal management for electric vehicles using heat pipes. *J. Power Sources* **2014**, *257*, 344–355. [CrossRef]

43. Liang, J.; Gan, Y.; Li, Y. Investigation on the thermal performance of a battery thermal management system using heat pipe under different ambient temperatures. *Energy Convers. Manag.* **2018**, *155*, 1–9. [CrossRef]

44. Wang, C.; Zhang, G.; Li, X.; Huang, J.; Wang, Z.; Lv, Y.; Meng, L.; Situ, W.; Rao, M. Experimental examination of large capacity liFePO4 battery pack at high temperature and rapid discharge using novel liquid cooling strategy. *Int. J. Energy Res.* **2017**, *42*, 1172–1182. [CrossRef]

45. Panchal, S.; Khasow, R.; Dincer, I.; Agelin-Chaab, M.; Fraser, R.; Fowler, M. Numerical modeling and experimental investigation of a prismatic battery subjected to water cooling. *Numer. Heat Transf. Part A Appl.* **2017**, *93*, 1–12. [CrossRef]

46. Panchal, S.; Khasow, R.; Dincer, I.; Agelin-Chaab, M.; Fraser, R.; Fowler, M. Thermal design and simulation of mini-channel cold plate for water cooled large sized prismatic lithium-ion battery. *Appl. Therm. Eng.* **2017**, *122*, 80–90. [CrossRef]

47. Panchal, S.; Dincer, I.; Agelin-Chaab, M.; Fraser, R.; Fowler, M. Design and simulation of a lithium-ion battery at large C-rates and varying boundary conditions through heat flux distributions. *Measurement* **2018**, *116*, 382–390. [CrossRef]

Review

Thermal Storage Using Metallic Phase Change Materials for Bus Heating—State of the Art of Electric Buses and Requirements for the Storage System

Werner Kraft *, Veronika Stahl and Peter Vetter

German Aero Space Center, Institute of Vehicle Concepts, 70569 Stuttgart, Germany; veronika.stahl@dlr.de (V.S.); peter.vetter@dlr.de (P.V.)
* Correspondence: werner.kraft@dlr.de; Tel.: +49-711-6862-273

Received: 12 March 2020; Accepted: 27 April 2020; Published: 11 June 2020

Abstract: Battery-powered electric buses currently face the challenges of high cost and limited range, especially in winter conditions, where interior heating is required. To face both challenges, the use of thermal energy storage based on metallic phase change materials for interior heating, also called thermal high-performance storage, is considered. By replacing the battery capacity through such an energy storage system, which is potentially lighter, smaller, and cheaper than the batteries used in buses, an overall reduction in cost and an increase of range in winter conditions could be reached. Since the use of thermal high-performance storage as a heating system in a battery-powered electric bus is a new approach, the requirements for such a system first need to be known to be able to proceed with further steps. To find these requirements, a review of the relevant state of the art of battery-powered electric buses, with a focus on heating systems, was done. Other relevant aspects were vehicle types, electric architecture, battery systems, and charging strategies. With the help of this review, requirements for thermal high-performance storage as a heating system for a battery-powered electric bus were produced. Categories for these requirements were the thermal capacity and performance, long-term stability, mass and volume, cost, electric connection, thermal connection, efficiency, maintenance, safety, adjustment, and ecology.

Keywords: electric buses; thermal energy storage; latent heat storage; metallic phase change material; cabin heating

---

1. Introduction

The latest reports on global warming show a significant increase in the worldwide average surface temperatures on earth compared to the pre-industrial era. Since this effect is related mainly to the rise of human CO_2 emissions from burning fossil fuels, such as coal, oil, or gas, reducing the use of them seems to be a way to restrict global warming.

In terms of public transport, buses using diesel fuel are the dominant vehicle category. Out of the 80,519 buses registered in Germany, a total of 78,472 (a relative portion of 97.5%) are diesel-fueled buses, amounting to a large majority. Toward the aim of protecting the climate and reducing local emissions like NO_x or noise, replacing diesel-fueled buses with battery-powered electric buses could have positive effects. However, the number of fully electric buses, which is 228 and therefore contributing to an amount of 0.28% to the total amount of buses, is currently insignificant [1].

Since public transport is often uneconomical and has to be subsidized, low costs for the acquisition, energy consumption, and maintenance of public transport are highly relevant. However, prices for electric buses are currently roughly double that of diesel buses. Additionally, high investments in the charging infrastructure are necessary. Besides the costs, the range of electric buses is limiting the application possibilities in public transport, since not every tour can be served if the buses cannot be readily charged. Since the interior heating capacity can exceed the necessary power for traction in low ambient temperatures, the range issue becomes even worse if interior heating is required on low ambient temperatures and heating systems using electric energy are used. A further description of these challenges is given in Section 2.4. As an example, the dependence of the required heating capacity on the ambient temperature of a 12 m city bus is shown in Figure 1. By way of comparison, the energy demand for the traction of a 12 m city bus is in the magnitude of 1.1 kWh/km. [2]

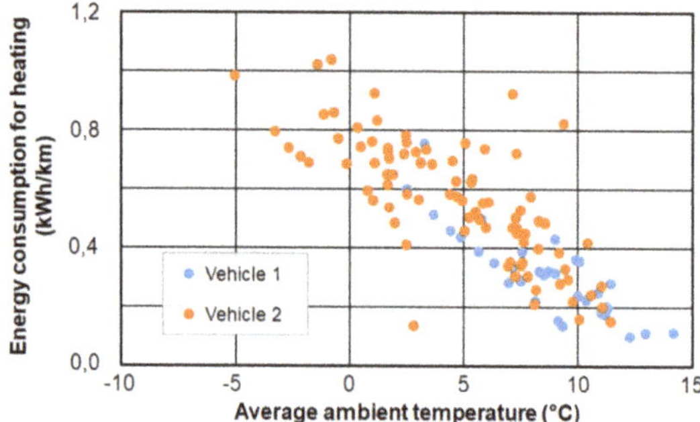

Figure 1. Dependence of the specific heating demand of a 12 m bus on the ambient temperature [2].

Since the prices for batteries are high (up to approx. 930 €/kWh, see Section 2.1.3) and the energy densities are limited (up to 160–180 Wh/kg, see Section 2.1.3), using a high battery capacity is part of the two described issues and currently low usage of electric busses. Therefore, replacing the battery capacity with a more lightweight and more affordable energy storage method could be a solution to solve this problem.

As mentioned, interior heating has a huge contribution to the energy consumption of an electric bus in winter conditions. To overcome the problem of the high cost and limited range in winter conditions, the use of thermal high-performance storage (THS) made of metallic phase change materials (mPCM) could be a potential solution, as already described in past publications. The idea is to charge the THS and the battery simultaneously, where the energy stored in the THS is provided to the cabin while driving. By doing so, no electric energy from the battery has to be used and therefore the range can be increased. Furthermore, no battery capacity needs to be available for heating and therefore can be designated only for driving. Metallic phase change materials are selected as the main choice since they offer high thermal conductivities at high energy densities and low cost. The high thermal conductivity enables the potential for fast charging, which can be very relevant. The maximum storage temperatures are intended to be 600–700 °C [3,4].

Since the application of THS as a heating unit in electric buses is a new approach, the requirements for such a system must first be known to be able to proceed with further steps, such as conceptual designing or experimental investigation. Therefore, this review aimed toward finding these requirements. To do so, first, the relevant state of the art of battery-powered electric buses is shown. The relevant aspects are vehicle types, electric architecture, battery systems, charging strategies, and especially heating systems. With the help of this knowledge, the requirements for THS are established and discussed.

2. State of the Art of Electric Buses

2.1. Electric Buses in General

2.1.1. Common Vehicle Types

Buses can be generally divided into different categories, which mainly comes from their use case or their size. In terms of their use case, categories are airport buses, city buses, regional buses, and coaches for traveling. The main differences between these kinds of buses are, e.g., the range, comfort for passengers, door sizes, engine power, and the type of construction (low floor or high floor) [5–8]. In terms of size, buses can be categorized into small buses (<6 m), mini-buses (6–8 m), midi-buses (8–10.6 m), standard buses with two axles, (10.6–13.5 m), double-decker buses (10–11 m), standard buses with three axles (13.5–15 m), articulated buses (17.5–19 m), and double-articulated buses (21–26.2 m) [9]. Since battery-powered electric buses have mainly been used as city buses, the focus will be put on this category. A detailed overview of battery-powered electric buses that are currently available on the market is given by Faltenbacher et al. [9]. As the state of the art in this category, the EvoBus eCitaro is highly regarded in the authors' view. The eCitaro comes with a maximum battery capacity of 292 kWh, offering a maximum range of about 280 km under ideal conditions. Additionally, innovative features are, e.g., the detection of occupancy by using mass sensors integrated into the axles, which is used for thermal management purposes, or the use of a CO_2 heat pump, as further described in Section 2.3.3 [10].

2.1.2. Electric Architecture

Regarding the electric architecture, a DC intermediate circle is typically used in combination with DC/DC converters and DC/AC converters for the electric consumers. The typical voltages are 650 to 750 VDC for the battery relative to the intermediate circle. The typical voltages of electric consumers, such as an air conditioner or air compressor, are 400 VAC. Additionally, there is a 24 VDC voltage level for electric consumers that have a lower electric power demand. Components with high power consumption, such as a fully electric resistance heater (see also Section 2.3.2), are typically operated with high voltages, either on the high voltage DC level or with the 400 VAC level. An example of a battery-powered bus electric architecture is shown in Figure 2 [2,11].

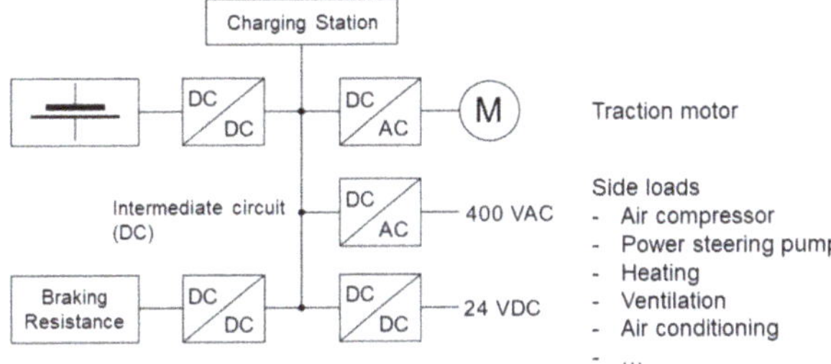

Figure 2. The basic setup of the electric architecture of a battery-powered electric bus (without a control system) [2].

2.1.3. Battery Systems Used in Electric Buses

Regarding battery systems in battery-powered electric buses, only battery systems based on lithium-ion batteries are currently used. A basic distinction is made for high-energy batteries and high-power batteries. High-energy batteries typically have higher energy at a lower power density. High-power batteries typically have higher power densities but lower energy densities. An overview of electric energy storage technologies with a focus on battery technology is shown in Figure 3.

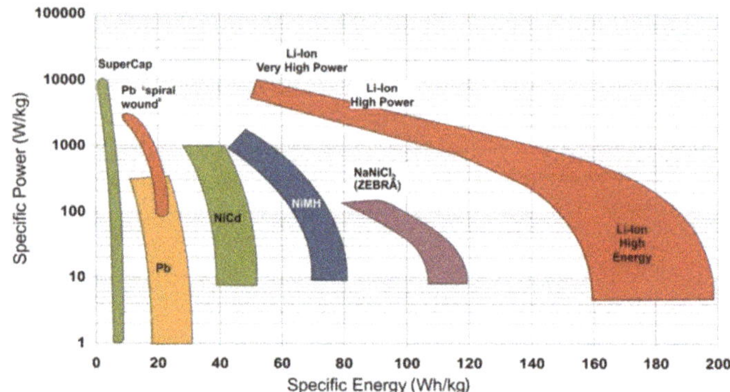

Figure 3. The energy density and power density of different electric energy storage systems with a focus on battery technology, plotted as a Ragone diagram [12].

As one state-of-the-art manufacturer of batteries for bus applications, Akasol is highly regarded in the authors' view, and is the battery supplier for the aforementioned EvoBus eCitaro, among others. Batteries from this supplier that are used in buses can use either NMC (lithium nickel manganese cobalt oxide) for high-energy batteries or LTO (lithium titanate) for high-power batteries, or a combination of both [13]. The cost for NMC is stated to be approx. 420 \$/kg (≈389 €/kWh), and for LTO, the cost is approx. 1005 \$/kg (≈931 €/kWh) [14]. Another battery type that is often used is LFP (lithium iron phosphate) [9]. The price of this type of battery is stated to be 580 \$/kWh (≈537 €/kWh).

Regarding an ultra-high-energy battery, Akasol is offering the AKM CYC battery system with a capacity of 42 kWh, a mass of 230 to 260 kg, and a volume of 17.85 L per battery pack, resulting in a gravimetric energy density of 162 to 183 Wh/kg and a volumetric energy density of 235 Wh/L. A continuous performance of 20 to 32 kW is stated, resulting in a power density of 77 W/kg to 139 W/kg. These values are in good agreement with the numbers for the high-energy batteries shown in Figure 3 [15]. Regarding an ultra-high-power battery system, Akasol is offering the AKM POC battery system. For the AKM 53 POC, a capacity of 35.3 kWh at a weight of 333 kg, a volume of 25.05 L, and a continuous performance of 60 kW are stated. This results in a gravimetric energy density of 106 Wh/kg, a volumetric energy density of 141 Wh/L, a gravimetric power density of 180.2 W/kg, and a volumetric power density of 240 W/L. However, the peak performance of discharging is stated as being 270 kW, leading to a power density of 811 W/kg (1078 W/L). For charging, the peak performance is 106 kW, offering a power density of 318 W/kg (423 W/L). The peak performance of this battery system is in agreement with the numbers given in Figure 3 [16].

Furthermore, the latest numbers regarding future estimations of battery systems for electric buses are given in the literature. For the batteries of the eCitaro, Akasol is planning to raise the capacity in 2020 by replacing the currently used battery packs with the latest battery technology, enabling a maximum capacity of 330 kWh. Another step that might be taken within the next few years is to replace the currently used NMC batteries with solid oxide batteries. This would allow for a maximum capacity of up to 400 kWh within the same space due to a higher gravimetric energy density, according to EvoBus [10].

One last thing that must be considered regarding batteries for electric buses is their lifetime. Buses are typically in operation for 12 to 14 years, as reported by Schwarzer [17]. Due to degradation, the batteries have to be changed after about half of the bus's operation time. As a criterion for changing the battery, a state of charge (SOC) of 80% is recommended. Factors to consider that come along with changing the battery are the high cost for a new battery and ecological aspects, such as high CO_2 emissions during the production process. To at least partially overcome this, batteries from electric buses are considered to be used as stationary energy storage as a second-life application [2,18].

2.2. Charging Strategies for Electric Buses

2.2.1. Overnight Charging

In an overnight charging scenario, buses are fully charged at the bus terminal. Since the charging is typically done at night, outside the service hours of city buses, it is called overnight charging. In this scenario, the energy needed for the whole service is stored within one charge. Regarding the charging infrastructure, only one location, which is normally the bus terminal, needs to be equipped. However, buses suitable for this scenario require high energy storage capacities. For this reason, the battery types used for this scenario are high-energy batteries (see Section 2.1.3). A visualization of the described charging scenario is shown in Figure 4 [2].

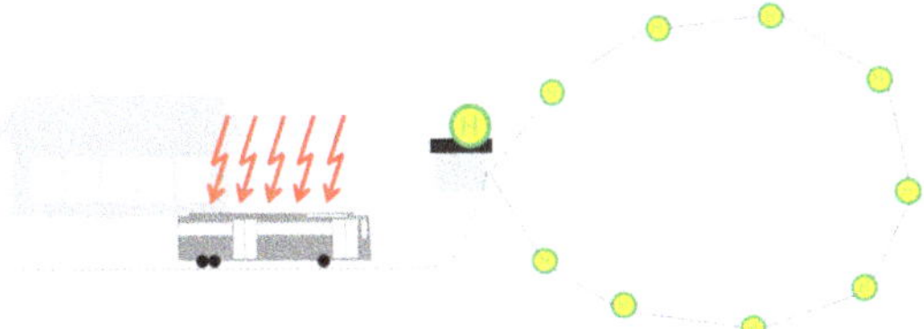

Figure 4. Illustration of the overnight charging strategy with the charging infrastructure located at the bus terminal [2].

2.2.2. Opportunity Charging at Final Stops

During opportunity charging at final stops, in addition to the charging at the bus terminal, charging is done at the final stop of a city bus´s route. For this, the regular stop at the final stop is used. Stop times at final stops can be within the range of a few minutes up to 20 or 30 min, according to Berthold [19]. Buses being operated in such a scenario require lower energy storage capacities compared to buses operated in an overnight charging scenario. However, the effort required for the installation of the charging infrastructure rises since more locations than just the bus terminal have to be equipped. For this scenario, high-power batteries are typically used (see Section 2.1.3). A visualization of the described charging scenario is shown in Figure 5 [2].

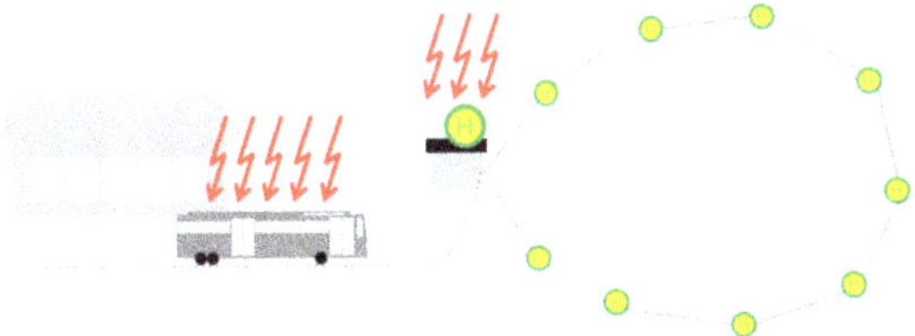

Figure 5. Illustration of the overnight charging strategy combined with opportunity charging at a final stop, with the charging infrastructure located at the bus terminal and the final stop [2].

2.2.3. Opportunity Charging at Multiple Stops

Another variation of the opportunity charging scenario is the opportunity charging scenario taking place at multiple stops, as illustrated in Figure 6. Besides the installation of the charging infrastructure at the bus terminal and the final stops, the charging infrastructure at regular bus stops is installed. Since the stop times at regular stops are pretty short (typically within the range of 20 to 90 s, according to Cundill and Watts [20]), high charging powers are required to be able to recharge significant amounts of energy. Regarding the charging infrastructure, an even higher amount of effort is necessary since more locations have to be equipped; however, the storage capacities of the buses can be lowered. For this charging scenario, ultra-high-power batteries are required (see Section 2.1.3) [2].

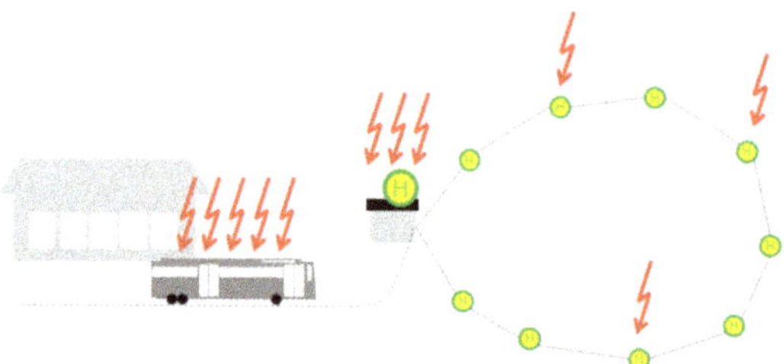

Figure 6. Illustration of the overnight charging strategy combined with opportunity charging at intermediate stops and the final stop, with charging infrastructure located at the bus terminal, at intermediate stops, and the final stop [2].

2.3. Heating Systems for Electric Buses

For providing the high thermal capacity necessary for interior heating, as shown in Figure 1, three main solutions are currently used in battery-powered electric vehicles. These solutions are fuel heaters, electric heaters, and heat pumps. Another solution found in the literature and regarded as relevant for conducting the requirements of THS is thermal energy storage proposed by Fraunhofer (Kratzing) and Konvekta (Best). However, this solution is currently in the development and field testing stage. The three state-of-the-art heating solutions and the thermal energy storage in development are further described within the next sections. Additionally, the typical architecture of the thermal management system of a battery-powered electric bus is shown.

2.3.1. Fuel Heater

Fuel heaters burn a liquid fuel and provide the released heat of the burning process to the air, or in most cases, to the vehicle's cooling fluid. Regarding the fuels used, in general, all kinds of liquid fuels, such as diesel, gas, ethanol, bio-fuel, etc., can be used. However, diesel is usually considered as the main choice. The benefit of using fuel heaters as the heater for battery-powered electric buses is the fact that no electric energy from the battery is needed for heating purposes. A negative aspect of burning fuels is local emissions, which is in contradiction with the local emission-free purpose of using an electric bus. The fuel consumption of a fuel heater for a 12 m standard bus is about 2.9 L/h, with a thermal output of 24 kW, according to the manufacturer, at an efficiency of about 76.5% to 79.5%, according to Sonnekalb et al. [21,22]. Considering a city bus's average speed of 12 km/h (SORT (Standardised on-road test cycles) heavy urban cycle [23]), this would result in fuel consumption of 24.2 L/100 km.

Typically, fuel heaters are available with different performance levels. Adjustment is mostly done using a simple ON/OFF operation. Regarding the conditions for switching on or off, the temperature of the cooling fluid is measured. According to Valeo, switching on is done when the cooling fluid temperature falls below 70 °C, while switching on is done when it goes above 85 °C. Communication with the vehicle is realized by using the CAN (Controller area network) interface, and the electric power supply is provided by connecting to the 24 V supply of the vehicle [22,24].

When it comes to maintenance, fuel heaters require at least a yearly service. However, due to the formation of soot, cleaning can be necessary more often. To keep this cleaning frequency as low as possible, a minimum burning period of 2 min is implemented during operation. Regarding maintenance and self-protection of the heater, a minimum volume flow of the cooling fluid through the heater is required to ensure a high enough heat transfer and to prevent the heater from overheating. Furthermore, a bimetal switch is used to realize a switch-off function at a temperature of 135 °C. Figure 7 shows the assembly of a fuel heater from Eberspächer (Esslingen, Germany) [24].

The dimensions of a fuel heater by Eberspächer with a thermal output of 24 kW are 600 mm × 230 mm × 222 mm, which is a volume of 30.6 L. The mass of such a system is stated to be 18 kg [22].

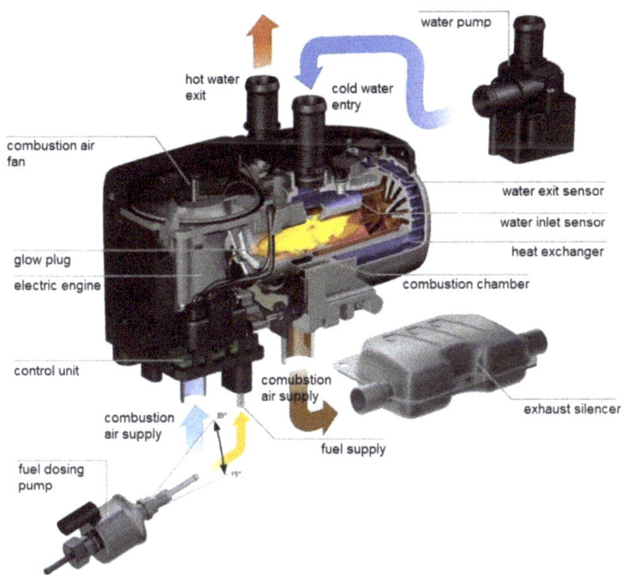

Figure 7. Detailed view of the assembly of a fuel heater from Eberspächer [22].

2.3.2. Electric Heater

Electric heaters, as shown in Figure 8, transform electric energy into thermal energy, which is transferred to the air or the vehicle's cooling fluid. Regarding heating elements, PTC (Positive Temperature Coefficient)-heaters or resistance heaters are used. The benefit of electric heaters is that it is an easy technology and there is a lack of emissions from heating. However, electric energy from the battery is used, leading to a potentially high range reduction when heating is required due to the direct transformation of electric energy to thermal energy (at an efficiency of about 98% without considering the battery's efficiency).

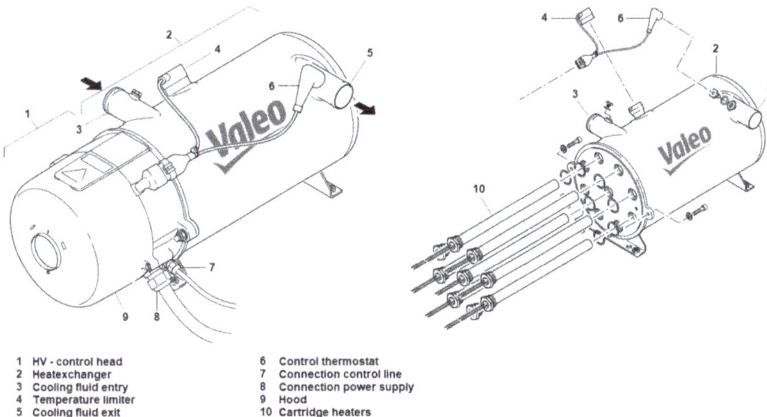

Figure 8. Detailed view on the assembly of the electric water heater Thermo AC/DC from Valeo [25].

Just as with the aforementioned fuel heaters, electric heaters are available with different performance levels. Furthermore, adjustment is typically done using ON/OFF operation with switch-on and switch-off temperature thresholds (e.g., 68 °C and 75 °C, respectively, for the Valeo Thermo AC/DC). For the electric power supply, electric heaters can either be connected to a DC with typical voltages of 600 to 750 VDC or to 400 VAC. For communication with the vehicle, the CAN interface is used, as well as the 24 V power supply of the bus [11,25].

Regarding maintenance, at least a yearly service is required. To simplify servicing, typical wearing parts, such as cartridge heaters, are designed to be exchangeable, as shown in Figure 8. Regarding the self-protection of the heating unit, a minimum volume flow of the cooling fluid, a trail of the cooling fluid pump of at least 2 min, and a switch-off threshold of 125 °C are implemented [25].

The dimensions of an electric heater from Valeo with a thermal output of 20 kW are 578 mm × 247 mm × 225 mm, which is a volume of 32.1 L. The mass of such a system is stated to be 15 kg [11].

A special version of an electric heater uses braking resistance. Every bus should have a continuous braking unit according to German traffic regulations. In terms of electric buses, this continuous braking unit is typically conducted as an electric resistance heating unit. In the case of the EvoBus eCitaro, it is connected to the vehicle's cooling fluid circle. Due to this, it can be used for heating purposes. Further description of this is given in Section 2.3.5 [26,27].

2.3.3. Heat Pump

By using heat pumps, the heating capacity can be provided with a COP (Coefficient of performance) of greater than 1 due to the use of a refrigerant circle that is mostly driven by an electric compressor [28]. Regarding the heat source, ambient air is used, while the heat sink is the air provided to the cabin. The electric energy for powering the compressor is taken from the vehicle's traction battery. Most heat pumps used within existing vehicles nowadays use R134a as a refrigerant; however, the use of R134a has been prohibited in newly registered vehicles since 2017 due to its high global warming potential (GWP). Recently introduced heat pumps use CO_2 as the refrigerant [28–33].

Using CO_2 as a refrigerant requires higher process pressures; however, this leads to benefits regarding the thermal output. The COP values stated by Konvekta are 4 for an ambient temperature of 15 °C, 2.5 at 0 °C, 2.2 at −5 °C, and 2 at −10 °C [28,34]. Figure 9 shows a comparison of the performance as a function of the ambient temperature between heat pumps designed for buses with R134 and R744 (which is CO_2) as the refrigerant. It can be seen that the thermal output that can be provided by a CO_2 heat pump is higher compared to a heat pump using R134a. Additionally, the thermal output can be provided for temperatures as low as −20 °C, while for the heat pump using R134a, a thermal output for temperatures as low as about −5 °C is stated. Referring to Basile et. al, the operation of a heat pump using R134a as refrigerant is not useful [34]. However, Lee [35] states that for electric vehicles (not buses), the operation of a heat pump using R134 could be possible, even below, even below −10 °C. Besides the positive effect of an overall reduction in energy consumption for heating using heat pumps, the range reduction is less compared to using electric heaters for heating. However, electric energy from the battery still must be used for heating. Additionally, the efficiency of a heat pump decreases with decreasing ambient temperature, which increases the heating demand. Furthermore, for very low temperatures of around −10 °C or less, an additional heating system is required, even when using a CO_2 heat pump, since the thermal output cannot fulfill the heating demand of the vehicle anymore. Another aspect that must be considered is the icing of outer heat exchangers within specific temperature ranges, which leads to a reduction in efficiency and requires a defrosting system [28,36].

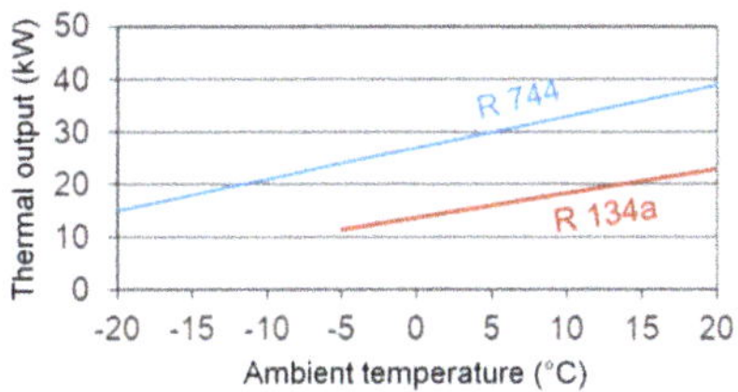

Figure 9. Comparison of the thermal outputs of heat pumps for bus heating using R134a vs. R744 as the refrigerant [37].

Regarding the adjustment of the thermal output, several performance levels can typically be set. This is done via frequency adjustment of the compressor. An optional infinitely variable control is possible, too. When the adjustment is done using defined performance levels, surplus produced heat must be extracted to the ambient air. For the electric power supply, the heat pump system is connected to the DC level. Since the compressor is working with three-phase 400 AC, a frequency converter is used in between. Communication with the vehicle is done via the CAN-interface [31,38].

Since a heat pump is a more complex system than a fuel heater or an electric heater, as seen in Figure 10, maintenance is essential. Regular maintenance is required on a six-monthly basis. For safety reasons, permanent monitoring of all sensors and checks of the plausibility of the measured values is done. In case of any irregularities, the heat pump will be switched off [38,39].

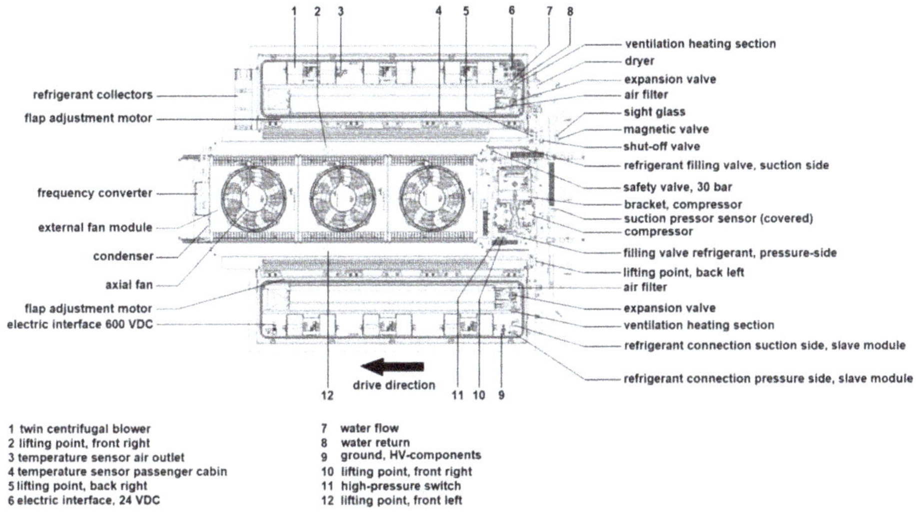

Figure 10. Detailed view of the assembly of the Revo-E heat pump from Valeo [38].

The dimensions of a REVO-E heat pump from Valeo with a thermal output of 16 kW (cooling power of 25 kW) and R134a as the refrigerant are 2800 mm × 2091 mm × 406 mm, which is a volume of 2377 L. The mass of such a system is stated to be 272 kg [29]. The dimensions of an UltraLight 500 heat pump from Konvekta with a thermal output of 18.2 kW (cooling power of 20 kW) and CO_2 as the refrigerant are 2124 mm × 2045 mm × 366 mm, which is a volume of 1590 L; the mass of this heat pump could not be found in the design proposal [33].

2.3.4. Thermal Energy Storage

Very recently, a thermal energy storage system for the interior heating of an electric city bus was proposed by Fraunhofer and Konvekta. The basic idea is the same as it is for the use of THS. The thermal energy storage and the battery can be electrically charged simultaneously, where the stored heat is used for interior heating to keep the electric energy stored in the battery for driving. For this thermal energy storage, a prototype was built and investigated in a laboratory. A field test on a real bus is planned but has not been conducted yet [40,41].

Regarding the storage material, paraffin wax is used within this storage system. The melting temperature of this material is about 69–71 °C, the storage capacity is 260 kJ/kg (sensible and latent heat within the temperature range of 62–77 °C), the densities are 0.88 kg/L (solid) and 0.77 kg/L (liquid), the heat conductivity is 0.2 W/mK, and the maximum working temperature is 100 °C [40].

The storage is designed to be in a rectangular shape. For charging, electric resistance heaters (720 VDC) are used, while for discharging, pipe–slat heat exchangers are brought directly into the storage material. Regarding the discharging medium, the cooling fluid of the vehicle is used. The conceptual design of the storage setup is shown in Figure 11, while an outer view of the installed storage is shown in Figure 11.

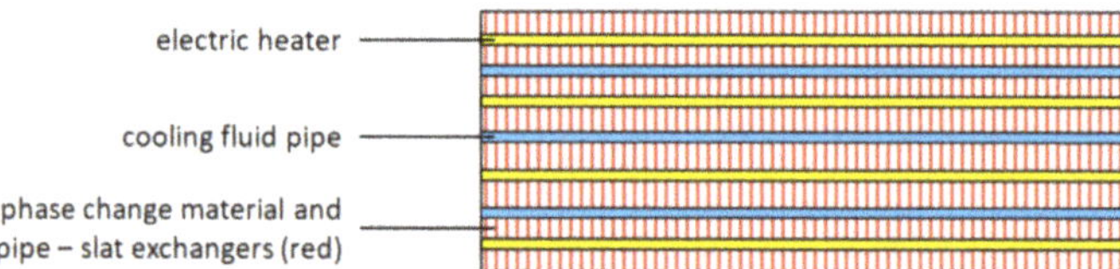

Figure 11. Conceptual design of the thermal energy storage prototype using paraffin wax for heating an electric bus from Fraunhofer and Konvekta [40].

For integration into the bus, a modular design with six modules is planned. Each module should have a storage capacity of 2.25 kWh, a charging power of 20 kW at 680 VDC, and a storage density of 39 Wh/kg. The positioning of the storage modules is intended to be spread over the interior of the vehicle, as shown in Figure 12 [40].

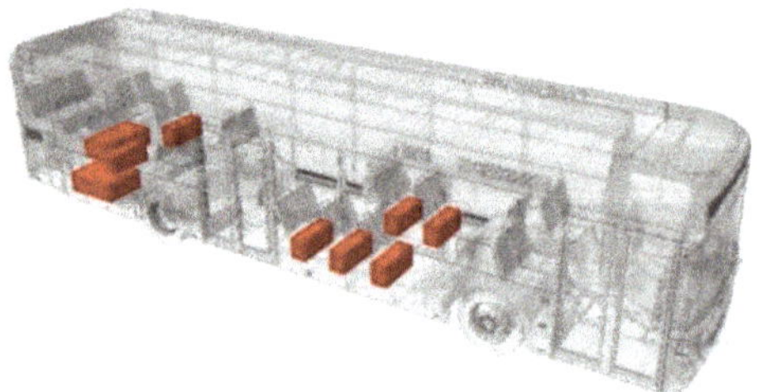

Figure 12. Intended positions of the thermal energy storage modules (red) within the interior of a bus [40].

Regarding the potential benefits of the storage system, a high lifetime with more than 16,000 full cycles, no need for an exchange due to degradation, and a much lower cost compared to batteries is stated. However, with a very low gravimetric energy density of 39 Wh/kg, the weight is a potential disadvantage of this system. Furthermore, the volumetric energy density of the storage system could be quite low since the density of the phase change material itself is low and the need for slats for improving the charging and discharging behavior requires volume as well. A specific volumetric energy density is not given within the design proposal [40].

2.3.5. Thermal Management Architecture

To give an idea of how the aforementioned components are implemented into an electric-powered city bus, the architecture of the thermal management is described within this section. As an example, the thermal management architecture of the EvoBus eCitaro is shown in Figure 13.

The relevant components for the interior heating are the CO_2 heat pump on the roof of the bus, the braking resistance in the rear, the additional heater in the rear, the floor heater, and the frontbox. The other components shown in Figure 13 are more connected to cooling functions and thermal management of the battery, and therefore are not further described. The CO_2 heat pump on the roof delivers heat to the interior through the air that flows into the cabin from the top. The braking resistance serves to heat the interior, alongside its use for energy recuperation and emergency braking. The additional heater is used as a fuel heater burning diesel fuel (or biodiesel fuel in the case of second-generation heaters). The additional heater can be used at very low temperatures when the thermal capacity of the heat pump is not sufficient and no extra electric energy should be used for heating to maintain the range of the bus. To bring the heat from the braking resistance and the additional heater into the interior of the vehicle, both are connected to a cooling fluid circle. The cooling fluid circle is connected to floor heaters within the passenger cabin and to the frontbox within the driver's cabin, which are heat exchangers used to bring the heat from the cooling fluid into the cabin [26].

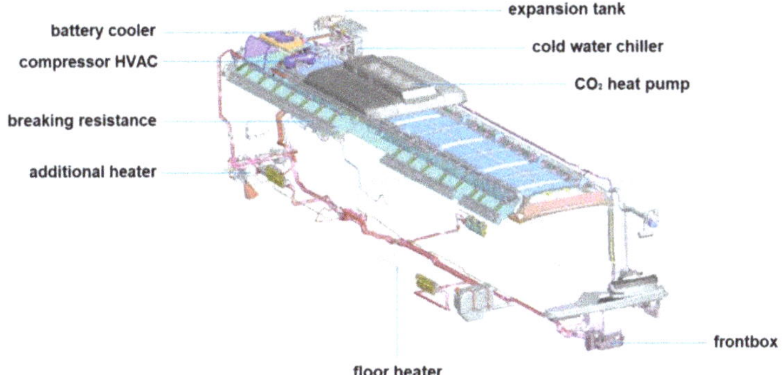

Figure 13. Thermal management architecture of the EvoBus eCitaro [26]. HVAC: heating, ventilation, and air conditioning.

2.4. Challenges Regarding the Use of Electric Buses

As reported in the introduction, the spread of electric buses is currently very limited. Reasons for this are, in the authors' opinion, two main facts: the high investment costs and the limited ranges, which are dependent on the ambient temperature, especially when heating is required. To emphasize both aspects, this section provides further information beyond that which is provided in the introduction.

The first challenge to be discussed is the high investment cost for battery-powered electric buses compared to conventional diesel buses. As reported by Knote, prices are roughly twice as high for battery-powered electric buses compared to diesel buses [2]. Since public transport is not economical for traffic enterprises in most cases, subsidization is mostly required to keep a high level of public transport services. A reason for the high investment cost of battery-powered electric buses is the high prices for the batteries, as described in Section 2.1.2. When considering a battery price of 500 €/kWh, which is at the lower end of the range given in Section 2.1.2, the battery price for the discussed example of the eCitaro would be about €146,000. Since a 12 m diesel bus costs approx. €240,000 to €350,000 [2], the battery alone accounts for about half of the cost of a whole conventional diesel bus.

The second challenge to be discussed is the insufficient range of battery-powered electric buses in non-ideal ambient conditions, especially in winter conditions when interior heating is required. A further explanation of this issue is given in Figure 14. On the x-axis of Figure 14, the distance of a city bus's daily tour relative to the range of a battery-powered electric city bus is given. On the y-axis, the number of tours relative to the tours that can be covered is given. Therefore, the blue line represents the relative number of tours covering the given distance as a function of the relative number of tours that can be covered with a city bus offering the range given on the x-axis. To show the number of tours that electric city buses can cover, grey bars are given using the EvoBus eCitaro as an example. As before mentioned, this bus is offering a range of 280 km under ideal ambient conditions, which is stated to be 20 °C. Referring to the diagram, a range of 280 km is sufficient to cover 70% of all tours within the German bus transport system. However, when interior heating is required, the range of the bus decreases if heaters consuming electric energy are used, and therefore the number of tours that could be covered with the bus decreases as well. Using an ambient temperature of about −10 °C as an example, interior heating of about 1.4 kWh/km is required, according to Figure 1. Since the eCitaro is offering an electric CO_2 heat pump, the electric energy consumption is reduced. According to the manufacturer, a COP of about 2 can be reached at an ambient temperature of −10 °C, leading to energy consumption of about 0.7 kWh/km for heating [33]. Together with an energy consumption of 1.04 kWh/km for traction (calculated based on the battery capacity and maximum range), the overall energy consumption would be approx. 1.74 kWh/km, leading to a maximum range of approx. 167 km. As can be seen in the diagram, a range of 167 km would be sufficient for only about 10% of all tours.

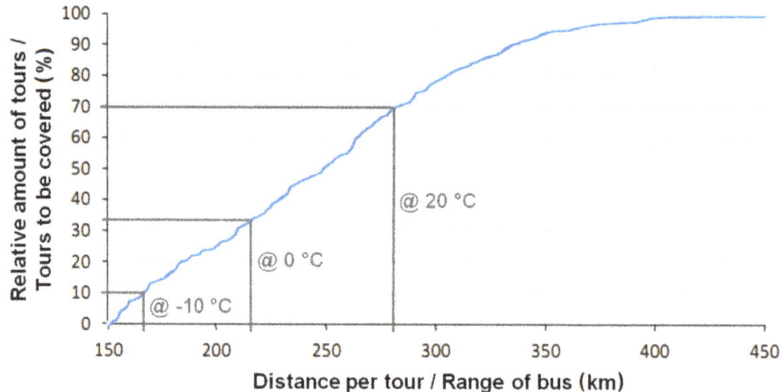

Figure 14. The distance of daily tours in the German bus transport system that need to be covered as a function of the relative number of tours that can be covered by a bus with the given range [2].

Due to the described range reduction when interior heating is required, the usability of electric city buses is strongly dependent on the ambient conditions. For traffic enterprises, this would mean that electric buses can only be used on tours with low distances if the buses are to be used throughout the year and no charging infrastructure for opportunity charging is to be built.

Based on the above discussion regarding the challenges faced, it can be concluded that providing a low-cost energy storage system for interior heating can have a major impact on the spread of battery-powered electric buses. Besides lowering the cost for the buses itself, it could reduce the necessity of installing opportunity charging infrastructure, which is currently vital if relevant ranges are to be offered independent of the ambient conditions. However, installing opportunity charging infrastructure is neither possible at any location nor affordable.

3. Requirements for Thermal High-Performance Storage

The requirements for THS are separated into eleven different categories. The derivation for each category is described within each respective section.

3.1. Thermal Capacity and Performance

Thermal capacity and performance are described together within one section since both are influenced by the same parameters, where performance refers to the charging (electric power) and discharging (thermal output) performances. Both are influenced by the purpose, vehicle size, climate conditions, and charging strategy. Figure 15 visualizes the influence of these parameters.

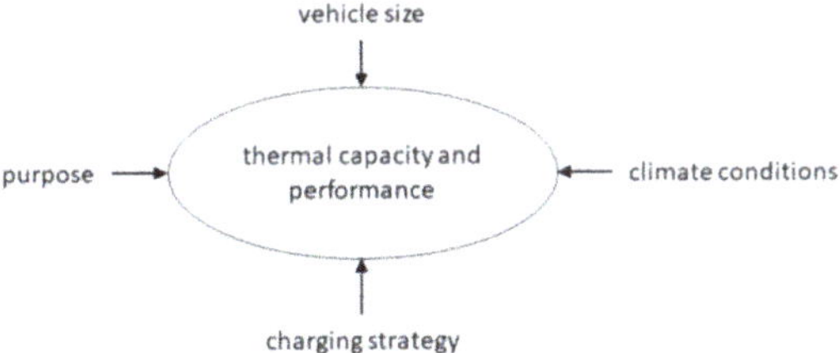

Figure 15. Influencing parameters regarding the requirements for thermal capacity and performance of THS.

The term "purpose" refers to whether the THS is used as the only heating system within the vehicle or whether it should be used as an additional heating system complementing the heat pump. If it is used as the only heating system, it has to cover any heating demand from the vehicle. If it is used as a complementary system, it only has to cover the gap between the maximum thermal output of the heat pump and the heating demand of the vehicle.

The vehicle size is highly relevant since the heating demand of a vehicle is strongly dependent on its interior volume, as described by Grossmann [42]. Since this fact is regarded as trivial, no further description is given.

Regarding climate conditions, two basic thoughts must be considered. The first one is that a bus, especially a city bus, is typically used in one fixed location. Since the climate conditions for one location are well known and do not change significantly over the typical lifetime of a bus, THS can be adopted precisely for these conditions. However, climate conditions can vary greatly between locations. For example, there would be a huge difference between the climate conditions of a city in Scandinavia, such as Oslo or Stockholm, and a city in southern Europe, such as Rome or Madrid. The second thought to be considered is that the thermal output for heating varies greatly depending on the ambient temperature, as already shown in Figure 1. Since a bus is used throughout the year, as well as typically throughout the day, the heating demand can vary greatly throughout both a year and a day. A very strong variation over the year is typically connected to areas with a very continental climate, such as in Central Northern America (e.g., Winnipeg in Canada) or central Asia (e.g., Astana in Kazakhstan). High variations of the temperature throughout a day often occur in desserts, and therefore in cities built in desserts, e.g., in Las Vegas (USA) or Tehran (Iran) [43–46].

To clarify the effect of different charging strategies on THS, some basic calculations were conducted based on the assumptions on two extreme scenarios: one extreme overnight charging scenario and one extreme opportunity charging scenario. For the extreme overnight charging scenario, it was assumed that the overall daily drive time of a bus was 17 h and the following time for charging was 7 h. Assuming a very cold winter day with a very low ambient air temperature (e.g., in Scandinavia, Canada, etc.), this led to an energy demand for heating of 425 kWh (assumed required thermal output of 25 kW, which is sufficient for −25 °C, according to Valeo; assuming an average speed of 17 km/h (SORT easy urban), this would lead to a heating demand of 1.47 kWh/km, which would be sufficient for about

−15 °C according to Figure 1). To charge THS with such a high capacity within 7 h, a charging power of 60.7 kW would be necessary. Out of these values, two indicators can be calculated that relate the storage capacity to the charging power vs. the discharging power. For charging power, it is 7 kWh/kW, while for discharging, it is 17 kWh/kW. The boundary conditions for the extreme opportunity charging scenario are a drive time of 0.47 h (28 min) and a charging time of 0.05 h (3 min), according to data regarding public transport in Amsterdam [47]. Using the same climate conditions as described for the extreme overnight charging scenario, the required storage capacity and charging power are 11.7 kWh and 234 kW, respectively. The aforementioned indicators were calculated to be 0.05 kWh/kW for charging and 0.47 kWh/kW for discharging. Comparing the indicators for both scenarios, there is a factor of about 140 for the charging indicator and 36 for the discharging indicator between both scenarios. Based on this, it can be concluded that the requirements for THS regarding storage capacity and charging/discharging performance vary greatly depending on the charging scenario.

From the given descriptions, it can be concluded that no specific requirement regarding capacity and performance can be stated. Capacity and performance have to be specifically chosen for the given scenario such that the THS is neither oversized nor undersized. Due to this, THS has to be designed to meet its specific purpose.

3.2. Long-Term Stability

The parameters that influence the long-term stability of THS are visualized in Figure 16. The main influences are the climate conditions, purpose, lifetime of the bus, and a potential substitution during the lifetime.

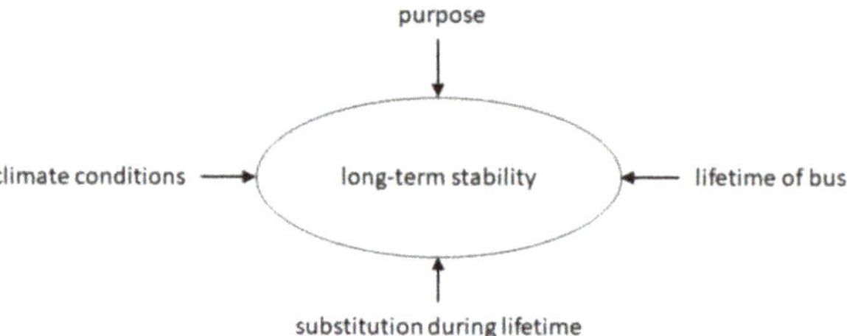

Figure 16. Influencing parameters on the requirements for the long-term stability of THS.

The first parameter is the climate conditions, which significantly influence THS since they determine whether there is a heating demand at all (heating demand only occurs below an ambient temperature of 15 °C according to Figure 1). If there is no heating demand, the THS does not need to be in operation.

The second parameter is the purpose. If the THS is used as the only heating system, it is in operation any time there is a heating demand. If it is used in addition to a heat pump, it is regularly only used if the thermal output of the heat pump is insufficient. To emphasize the influence of this, the climate data of Stuttgart is considered as an example. According to a test reference year dataset from the German Meteorological Service, temperatures of 15 °C or below occurred for 73% of that year. However, temperatures of −10 °C or below (the minimum temperature a CO_2 heat pump is sufficient as the only heating system according to Section 2.3.3) only occurred for 0.05% of that year.

The third parameter is the typical use time of a bus. With a yearly average mileage of about 57,000 km, according to Goebelt, an overall expected mileage over the lifetime of a bus (see Section 2.1.3) is 684,000 km to 798,000 km [48]. Considering an average speed of 12 km/h according to the SORT heavy urban cycle, the overall expected use time for a bus is 57,000 to 66,500 h [23].

Based on the aforementioned aspects, the expected lifetime was calculated based on use in Stuttgart for both cases. Considering a use time of the bus of 66,500 h and the probability of a heating demand of 73%, an overall long-term stability of 48,545 h would be required, which expresses an extreme maximum value for this location. Combining the lifetime of 57,000 h with the probability of 0.05%, a long-term stability of 28.5 h would be the result, which expresses an extreme minimum value for this location. A final thought on the long-term stability is the fact that THS could be substituted within the lifetime of a bus, similar to what is done with a battery, as described in Section 2.1.3. The parameter influencing this decision would probably be the cost. If it is cheaper to replace the THS after a while than manufacturing new THS that would last for the whole lifetime of a bus, a replacement would probably be preferred.

Based on the aforementioned discussion, it was concluded that the requirement for long-term stability cannot be generally described. Some particular use cases might not need much long-term stability. However, there are potentially many use cases that require long-term stabilities of several thousand or even tens of thousands of hours. Since the development of THS should be done in a way that as many use cases as possible can potentially be covered, long-term stability of several thousand or tens of thousands of hours is stated as the primary requirement in this category.

3.3. Mass and Volume

As is generally known, mass and volume are properties of high relevance in mobile applications, especially for vehicles. However, a more detailed look at both factors for buses leads to the conclusion that mass is regarded as more relevant for application in battery-powered electric buses than volume. Following reasons were found to explain why volume should not be rated as high as mass:

- The roof of a bus, which offers lots of space, is typically used for components like the air conditioner/heat-pump and for energy storage, such as batteries. Furthermore, THS could be placed on the roof.
- Components like ticket machines are sometimes installed within the interior of a bus, leading to a reduction in the number of seats. Due to this, installing extra components within the interior does not seem to matter much.
- Seats above wheel cases are often less spacious compared to the other seats within a bus and therefore are often less comfortable for passengers. Replacing these seats with other components should therefore not matter significantly.
- The space below seats often is not used at all.
- Battery electric buses typically have a higher weight compared to diesel buses, leading to a reduction of the vehicle's load capacity and therefore to a reduced number of allowed passengers. Comparing the EvoBus eCitaro (with 10 battery packs) with the regular Citaro (12 m standard-bus with two doors) as examples, the empty weight is 13,700 kg compared to 11,415 kg and the maximum number of passengers is 89 compared to 105. Therefore, mass seems very relevant for battery-powered buses [49–52].
- Mass affects energy consumption during traction. Since energy consumption correlates with the CO_2 footprint and operating cost, the mass of the vehicle should be as low as possible.

Based on the above reasons, a requirement for THS is that the mass needs to be minimized. Volume should also be kept as low as possible; however, this is regarded as a secondary concern.

Since the intention is to replace the state-of-the-art heating systems with THS, it can also be concluded that THS should be more lightweight than these systems. In the case of replacing an electric heater, the mass of the THS has to be lower than the mass of the electric heater plus the required battery capacity. Since the energy densities of batteries vary greatly between battery types (see Section 2.1.3), no general conclusion can be given. However, two examples were calculated for the extreme charging scenarios described in Section 3.1 to determine whether THS should replace an electric heater.

For the opportunity charging scenario, 11.7 kWh is required for heating. Since it is a quick charging scenario, high-power batteries are assumed to be used with a gravimetric energy density of 106 Wh/kg, leading to a battery mass of 110.4 kg. Adding the weight of the electric heater of 15 kg leads to an overall weight of 125.4 kg for the heating system, which would be a gravimetric energy density of about 93.3 Wh/kg overall. Therefore, switching to THS in this scenario would require a higher energy density than 93.3 Wh/kg. For the overnight charging scenario, 425 kWh is required for heating. In this case, high-energy batteries would be used with an assumed energy density of 170 Wh/kg, leading to a battery mass of 2500 kg. Together with the electric heater, an overall mass of 2515 kg and an overall gravimetric energy density of 169 Wh/kg would result. Therefore, switching to THS in this scenario would require a higher gravimetric energy density than 169 Wh/kg.

3.4. Cost

Regarding the cost, it is important what viewpoint is taken to determine the requirements. Possible viewpoints could be the one of a manufacturer of the heating system, a manufacturer of a bus, or the user of a bus, which is normally the transport operator. For a manufacturer, it would, e.g., be important to produce THS as cheap as possible to be able to offer attractive prices and keep the margin as high as possible. In the authors' view, the cost borne by the end user is regarded to be relevant, since the end user is the one that decides what kind of heating system they would like to have. In terms of buses, the end user would be the transport operator. For a transport operator, the total cost of ownership is relevant. The main influences on this are the cost for purchase and therefore the manufacturing cost, the energy cost while in operation, the maintenance cost, and the decomissioning cost. The influence of these four parameters is expressed in Figure 17.

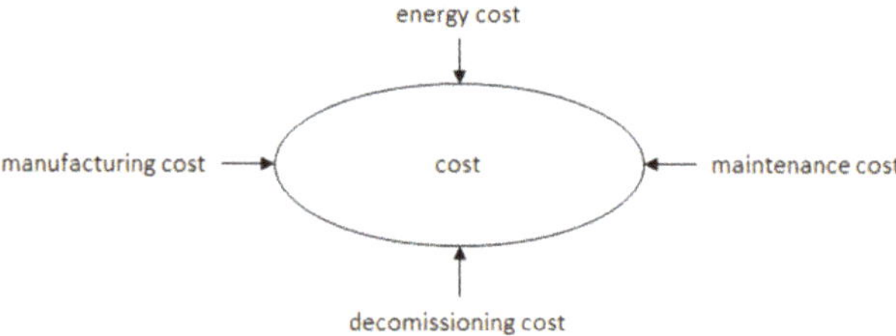

Figure 17. Influencing parameters regarding the cost requirements of THS.

The requirement for a THS is to keep the sum of these four cost categories as low as possible. However, the cost of manufacturing and the cost of maintenance are in tension with each other. A product of higher quality typically requires higher manufacturing effort and therefore a higher cost but could lead to lower demand for maintenance and therefore a lower maintenance cost. Regarding the energy cost, the primary energy cost, which is defined as the consumption of electric energy for heating, and the secondary energy cost must be considered. Regarding the secondary energy cost, a possible example is the energy consumption from traction due to weight penalties or advantages is meant. Decommissioning could either result in a cost or an income since used batteries are often used for second-life purposes.

3.5. Electric Connection

Out of the descriptions within Section 2.1.2, the heating system for charging the THS should preferably be operable at the 650 to 750 VDC level. A second option could be to use a DC/AC converter and operate at 400 VAC if having a DC-compatible heating system is inefficient or not possible. Furthermore, it should be connectable to a 24 VDC level for the energy supply of minor energy consumers, such as sensors or control electronics. Additionally, communication with the vehicle should be possible via the CAN interface.

3.6. Thermal Connection

Integration into the thermal management system can be done using two different solutions. The first one is to directly heat the air blown into the vehicle's cabin, while the second one is to transfer heat into a cooling fluid circle, which spreads the thermal energy throughout the interior of the bus. Both solutions could lead to sufficient results. Since the THS is more intended to replace a fuel heater or an electric heater in 12 m or larger buses, an integration that involves a connection to the cooling fluid of the vehicle is regarded as the preferred solution and therefore considered to be the primary requirement.

3.7. Efficiency

The efficiency of a heating system, i.e., the ratio of used thermal energy to charged electric energy, is directly connected to the consumption of electrical energy or fuel. Since lower consumption has positive ecological and economic effects, the efficiency of THS should be as high as possible. To give a specific number, the efficiency of an electric heater powered using a battery is considered. Considering a heater with an efficiency of 98% (see Section 2.3.2) and battery losses of 6.9%, according to the measurements of Berthold, an overall efficiency of about 91.2% would be the result [19]. Therefore, THS should have an efficiency of at least about 90%.

3.8. Maintenance

As discussed earlier, maintenance is conducted for all heating solutions currently used within a bus. Based on this, it was concluded that maintenance should be necessary for THS. As such, the related requirement for THS is that it should be designed in a way that means maintenance is as easy as possible, especially for wearing parts, where the wearing parts in THS could be, e.g., the electric heaters or the housing of the phase change material.

3.9. Safety

Since THS liquid metals with temperatures that are potentially higher than 600 °C are used, safety is an important criterion, in particular regarding the risk of fire or burns. Regarding the risk of fire in buses, some guidelines are given within the VDV 2303 regulations. Other relevant documents for the safety of buses are the European guidelines 2001/85/EG and 95/28/EWG. One basic conclusion drawn from these guidelines is that components with high surface temperatures or hot liquids have to be kept away from burnable parts within the interior of buses. For THS that is, e.g., positioned within the interior of the vehicle, this would become relevant and has to be considered. Other important considerations regarding safety are to prevent liquid metal from flowing out uncontrollably from the storage and to manage the isolation of electric components that are relevant for the charging of the THS [53–55].

Another safety aspect is the protection of storage from unwanted damage. Just as with the state-of-the-art heating systems, the safety mechanism that, e.g., keeps the THS from overheating needs to be implemented.

3.10. Adjustment

Looking at the adjustment of the known heating solutions, it is noticed that it is typically kept simple. Because of this, the adjustment of a THS regarding thermal output should also be kept as simple as possible; a kind of on/off solution seems sufficient.

3.11. Ecology

The last requirement for THS is that it should be as ecologically safe as possible. In particular, no critical raw materials or toxic materials should be used. Furthermore, the CO_2 footprint should be as low as possible. Besides the energy consumption in operation, the energy consumption during the manufacturing process should also be as low as possible. Additionally, recyclable materials should be used wherever possible.

4. Conclusions

The present paper provides the basics for the development of thermal energy storage using metallic phase change material for use within a heating system in battery-powered electric buses. This was done by providing a review of selected topics on battery-powered electric buses, leading to the following conclusions:

- Current battery-powered electric buses have drawbacks compared to conventional diesel buses due to their high investment costs and limited range. The investment costs are typically double that of conventional diesel buses. The range is typically limited to below 300 km and becomes even lower when interior heating is required and electric heating systems are used for heating; in such a case, the range can be cut by more than half, which can lead to ranges that prevent electric buses from a year-round use. Because of this, the spread of battery-powered electric buses is currently very low (e.g., only 0.28% of buses within Germany are battery-powered).
- The described limitations mainly result from the use of costly battery systems, which can easily cost €100,000–200,000, about half of the overall cost of a conventional diesel bus.
- The charging of a battery-powered electric bus can either be done using "overnight charging," which requires lower installation costs for the charging systems but requires high-capacity energy-storage systems. The other charging strategy is "opportunity charging," which leads to higher installation costs for the charging systems but requires low-capacity energy-storage systems. Depending on the charging scenario, either high-energy batteries or high-power batteries should be used.

By utilizing knowledge of the state of the art on battery-powered electric buses, requirements regarding thermal energy storage with metallic phase change material were produced. The main requirements are the following:

- The storage capacity and performance of a thermal energy storage system with metallic phase change material has to be easily adaptable since it is strongly dependent on the given scenario; storage capacities might vary between 11.7 kWh and 425 kWh, with the installed electric charging power varying between 60.7 kW and 235 kW.
- The long-term stability of thermal high-performance storage should allow for at least several thousand hours of use. However, maintenance is allowed to reach this, where the ease of maintenance for wearing parts should be considered.
- Mass and the associated gravimetric energy density is regarded as being more relevant than volume and the associated volumetric energy density since volume for additional components seems to be readily available in a bus; however, mass limits the maximum number of passengers and affects energy consumption for traction. The gravimetric energy density of THS should reach values higher than about 100 Wh/kg for extreme opportunity charging scenarios and values higher than about 180 Wh/kg for extreme overnight charging scenarios to be comparable with a conventional electric heater using electric energy from the traction battery.
- The integration into the thermal management architecture could either be done via integration into the airflow provided to the vehicle's interior or via integration into the liquid cooling circle; however, integration into the liquid cooling circle is regarded as the preferred solution since most of the currently used heating systems are integrated into the liquid cooling circle. Thermal energy storage using metallic phase change materials could simply replace the current heating systems if integration into the liquid cooling circle is possible.
- The efficiency of a thermal energy storage system using metallic phase change materials should be as high as possible, namely in the range of 90% or more, if possible.
- The heat supply system of the storage system should ideally be adaptable to a DC intermediate circle with voltages of about 650 to 750 VDC; alternatively, a connection to a 400 VAC level could also be possible.

- The use of thermal energy storage using metallic phase change materials should aim to minimize the overall cost, which is made up of the manufacturing cost, energy cost, maintenance cost, and decommissioning cost.

- Thermal energy storage using metallic phase change materials should be designed in such a way that no safety issues, especially regarding burning or leakage of liquid metal can occur; also, it should be equipped with safety devices to prevent the system from overheating, for example.

- Ecological aspects must be considered when designing thermal high-performance storage: if possible, recyclable raw materials, no toxic materials, and no critical raw materials should be used; additionally, the energy costs during manufacturing should be as low as possible.

Author Contributions: The contribution of the author P.V. refers to the Sections 2.1.2, 2.3.2 and 2.3.5. Main contribution was the support of the literature research in these paragraphs. The contribution of the author V.S. refers to the Sections 3.2 and 3.9. Main contrtibutions were the calculation of the lifetime and the definition of the requirements regarding safety as well as the search for relevant safety regulations. All other contributions come from the author W.K. All authors have read and agreed to the published version of the manuscript.

Funding: This research was funded by the EFRE (Europäischer Fonds für regionale Entwicklung) Leitmarktwettbewerb NRW (Nordrhein-Westfalen) Mobilität.Logistik within the project "Lathe.Go" and by Baden-Württemberg´s Department of Trade and Industry within the project "THS-Bus."

Conflicts of Interest: The authors declare no conflicts of interest.

References

1. Kraftfahrt- Bundesamt (KBA). *Fahrzeugzulassungen - Bestand an Kraftfahrzeugen nach Umwelt-Merkmalen*; Kraftfahrt- Bundesamt (KBA): Flensburg, Germany, 2019.

2. Knote, T. *Ansätze zur Standardisierung und Zielkosten für Elektrobusse*; Fraunhofer IVI: Dresden, Germany, 2017.

3. Kraft, W.; Jilg, V.; Altstedde, M.K.; Lanz, T.; Vetter, P.; Schwarz, D. Thermal High Performance Storages for use in vehicle applications. In *IAV - ETA Tagung*; IAV: Berlin, Germany, 2018.

4. Kraft, W. *Metallische Latentwärmespeicher als Heizsystem für die Anwendung in Batterieelektrischen PKWs und Bussen*; Hochschule für Technik - Seminar Erneuerbare Energien: Karlsruhe, Germany, 2019.

5. Wikipedia. Vorfeldbus. Available online: https://de.wikipedia.org/wiki/Vorfeldbus (accessed on 22 January 2020).

6. Wikipedia. Stadtbus (Fahrzeug). Available online: https://de.wikipedia.org/wiki/Stadtbus_ (accessed on 22 January 2020).

7. Wikipedia. Regionalbus. Available online: https://de.wikipedia.org/wiki/Regionalbus (accessed on 22 January 2020).

8. Wikipedia. Reisebus. Available online: https://de.wikipedia.org/wiki/Reisebus (accessed on 22 January 2020).

9. Faltenbacher, M.; Eckert, S.; Kupferschmid, S.; Klingenberg, H.; Burkhardt, J.; Schärzel, C.; Kiepsch, M.; Ramme, J.; Knote, T.; Jehle, C.; et al. *Marktübersicht Fahrzeuge und Infrastruktur - Programmbegleitforschung Bus*; BMVI: Leinfelden-Echterdingen, Germandy, 2019.

10. EvoBus GmbH. *Omnibus - Der vollelektrische Mercedes-Benz eCitaro - Special Edition*; EvoBus GmbH: Stuttgart, Germany, 2019.

11. Valeo. *Thermo AC/DC*; Valeo: Gilching, Germany, 2017.

12. Budde-Meiwes, H.; Drillkens, J.; Lunz, B.; Muennix, J.; Rothgang, S.; Kowal, J.; Sauer, D. A review of current automotive battery technology and future prospects. *J. Automob. Eng.* **2013**, *227*, 761–776. [CrossRef]

13. Bünnagel, C. *Interview mit Sven Schulz, CEO von Akasol*; Busplaner.de: Munich, Germany, 2019; pp. 28–31.

14. Batteryuniversity.com. BU-205_ Types of Lithium-ion. Available online: https://batteryuniversity.com/learn/article/types_of_lithium_ion (accessed on 19 February 2020).

15. Akasol. Akasystem AKM CYC - Ultra-Hochenergietechnologie für Langstreckenanwendungen. Available online: https://www.akasol.com/de/akasystem-akm-cyc (accessed on 19 February 2020).

16. Akasol. AKASYSTEM AKM POC. Available online: https://www.akasol.com/de/akasystem-akm-poc (accessed on 19 February 2020).

17. Schwarzer, M. Linienbusse unter Strom. ZEIT ONLINE, Bde. %1 von %2. Available online: https://www.zeit.de/auto/2010-04/linienbusse-strom (accessed on 19 February 2016).

18. Verkehrsbetriebe Hamburg-Holstein GmbH. Das zweite Leben der Elektrobus Batterien. Available online: https://vhhbus.de/second-life-energiespeicher/ (accessed on 8 April 2020).

19. Berthold, K. Techno-ökonomische Auslegungsmethodik für die Elektrifizierung urbaner. Busnetze. Dissertation, Karlsruher Schriftenreihe Fahrzeugsystemtechnik, Karlsruhe, Germany, 2019.

20. Cundill, M.; Watts, P. *Bus Boarding and Alighting Times*; Transport and Road Research Laboratory: Crowthorne, UK, 1973.

21. Sonnakalb, M.; Tegethoff, W.; Försterling, S. *CO2 basierte Air-Condition und Heizung für Stadtbusse*; Abschlussbericht über ein Entwicklungsprojekt, gefördert unter dem Aktenzeichen Az: 23864 von der Deutschen Bundesstiftung Umwelt: Braunschweig, Germany, 2008.

22. Eberspächer. Eberspächer Wasserheizung Hydronic. 2019. Available online: https://www.eberspaecher.com/produkte/fuel-operated-heaters/produktportfolio/wasserheizungen.html (accessed on 6 March 2019).

23. UITP Bus Committee. Standardised On-Road Test Cycles - SORT. In Proceedings of the 54th UITP International Congress, London, UK, 20–25 May 2001.

24. Valeo. *Werkstatthandbuch Thermo Plus*; Valeo: Gilching, Germany, 2018.

25. Valeo. *Werkstatt-Handbuch Thermo AC/DC*; Valeo: Gilching, Germany, 2018.

26. Bareiß, M.; Vorgerd, D. Thermomanagement für elektrisch angetriebene Stadtbusse. *ATZ* **2019**, *121*, 52–55. [CrossRef]

27. Bundesamt für Justiz, Straßenverkehrs-Zulassungs-Ordnung (StVZO) - § 41 Bremsen und Unterlegkeile. 2012. Available online: https://www.gesetze-im-internet.de/stvzo_2012/__41.html (accessed on 7 April 2020).

28. Jefferies, D.; Ly, T.-A.; Kunith, A.; Göhlich, A. *Energiebedarf Verschiedener Klimatisierungssysteme für Elektro-Linienbusse*; DKV-Tagung: Dresden, Germany, 2015.

29. Valeo. *RevoE*; Valeo: Gilching, Germany, 2017.

30. Konvekta, AG. *CO$_2$ Wärmepumpe für Elektrobusse*; Konvekta: Schwalmstadt, Germany, 2017.

31. Konvekta, AG. *UltraLight 500 CO$_2$ Wärmepumpe*; Konvekta: Schwalmstadt, Germany, 2018.

32. Wirtschaftswoche, Die Krux mit dem Kältemittel in Auto-Klimaanlagen. 26 Juni 2018. Available online: https://www.wiwo.de/technologie/mobilitaet/umweltbundesamt-warnt-die-krux-mit-dem-kaeltemittel-in-auto-klimaanlagen/22735826.html (accessed on 8 April 2020).

33. Konvekta, AG. *Datenblatt Ultralight 500 EM / 600 EM/700 EM - 2. Gen.*; Konvekta: Schwalmstadt, Germany, 2018.

34. Basile, R.; Scheid, H.; Tanke, D.; Moeseler, M.; Häring, R. Beheizungsstrategien für Elektrobusse. In *Transport Innovation for Sustainable Cities and Regions (POLIS)*; Valeo: Brussels, Belgium, 2017.

35. Lee, M.; Lee, H. Steady state and start-up performance charactersitics of air source heat pump for cabin heating in an electric passenger vehicle. *Int. J. Refrig.* **2016**, *69*, 232–242. [CrossRef]

36. Miller, W. Laboratory examination and seasonal analysis of frosting and defforsting for an air-to-air heat pump. *Ashrae Trans.* **1987**, *93*, 1474–1489.

37. Valeo. Heizsysteme. Available online: https://www.valeo-thermalbus.com/eu_de/Produkte/Heizsysteme (accessed on 17 April 2019).

38. Valeo. *Werkstatt-Handbuch REVO-E*; Valeo: Gilching, Germany, 2016.

39. Valeo. *Wartungs- und Serviceplan REVO-E Wärmepumpe*; Valeo: Gilching, Germany, 2017.

40. Best, P. Entwicklung ener moudlaren und schnellladefähigen Wärmespeicherheizung für vollelektrische Stadtbusse. In *3. Tagung Fahrzeugklimatisierung*; Haus der Technik e.V.: Essen, Germany, 2019.

41. Kratzing, R. *HEAT2GO - Entwicklung eines schnellladefähigen Latentwärmespeichers für die Beheizung von Elektrobussen*; Fraunhofer Institut für Verkehrs- und Infrastruktursysteme IVI: Dresden, Germany, 2017.

42. Grossmann, H. *Pkw – Klimatisierung – Physikalische Grundlagen und technische Umsetzung*; Springer: Berlin/Heidelberg, Germany, 2013.

43. Climate-Data.org. Klima Winnipeg. Available online: https://de.climate-data.org/nordamerika/kanada/manitoba/winnipeg-982/ (accessed on 5 March 2020).

44. Climate-Data.org. Klima Astana. Available online: https://de.climate-data.org/asien/kasachstan/astana/astana-491/ (accessed on 5 March 2020).

45. Climate-Data.org. Klima Las Vegas. Available online: https://de.climate-data.org/nordamerika/vereinigte-staaten-von-amerika/nevada/las-vegas-723/ (accessed on 5 March 2020).

46. Climate-Data.org. Klima Teheran. Available online: https://de.climate-data.org/asien/iran/teheran/teheran-198/ (accessed on 5 March 2020).

47. Beekman, R.; van den Hoed, R. Operational demands as determining factor for electric bus charging infrastructure. In Proceedings of the Hybrid and Electric Vehicle Conerence, London, UK, 2 November 2016.

48. Goebelt, R. *TÜV Bus-Report 2018*; Verband der TÜV e.V.: Berlin, Germany, 2018.

49. Forster, O. Mercedes-Benz eCitaro - Ausfahrt auf leisen Sohlen. *Busmagazin* **2019**, 7–10.

50. Görgler, J. Mercedes-Benz Citaro (Euro 6) - Sparsam unterwegs. *Busmagazin* **2014**, 6–10.

51. EvoBus GmbH. *Der Neue eCitaro - Technische Informationen*; EvoBus GmbH: Stuttgart, Germany, 2019.

52. EvoBus GmbH. *Die Citaro Stadtbusse*; EvoBus GmbH: Stuttgart, Germany, 2019.

53. Verband Deutscher Verkehrsunternehmen (VDV). *VDV-Mitteilung 2303 - Empfehlung zur Verhinderung von Brandschäden bei Linienbussen*; Beka Verlag: Köln, Germany, 2012.

54. Das Europäische Parlament und der Rat der Europäischen Union. *Richtlinie 2001/85/EG des europäischen Parlaments und des Rates*; Das Europäische Parlament und der Rat der Europäischen Union: Brusseles, Belgium, 2012.

55. Das europäische Parlament und der Rat der europäischen Union. *Richtlinie 95/28/EG*; Das europäische Parlament und der Rat der europäischen Union: Brusseles, Belgium, 1995.

Article

A Method for the Combined Estimation of Battery State of Charge and State of Health Based on Artificial Neural Networks

Angelo Bonfitto

Department of Mechanical and Aerospace Engineering, Politecnico di Torino, 10129 Torino, Italy; angelo.bonfitto@polito.it; Tel.: +39-011-090-6239

Received: 25 March 2020; Accepted: 17 May 2020; Published: 18 May 2020

Abstract: This paper proposes a method for the combined estimation of the state of charge (SOC) and state of health (SOH) of batteries in hybrid and full electric vehicles. The technique is based on a set of five artificial neural networks that are used to tackle a regression and a classification task. In the method, the estimation of the SOC relies on the identification of the ageing of the battery and the estimation of the SOH depends on the behavior of the SOC in a recursive closed-loop. The networks are designed by means of training datasets collected during the experimental characterizations conducted in a laboratory environment. The lithium battery pack adopted during the study is designed to supply and store energy in a mild hybrid electric vehicle. The validation of the estimation method is performed by using real driving profiles acquired on-board of a vehicle. The obtained accuracy of the combined SOC and SOH estimator is around 97%, in line with the industrial requirements in the automotive sector. The promising results in terms of accuracy encourage to deepen the experimental validation with a deployment on a vehicle battery management system.

Keywords: battery; state of charge; state of health; artificial intelligence; artificial neural networks; hybrid vehicles; electric vehicles; estimation

---

1. Introduction

The automotive industry is recently dedicating increasing attention to sustainability, with the objective of mitigating the negative effects of vehicular mobility on the environment. Carmakers cope with the always more stringent regulations about CO_2 emissions, focusing their efforts on the development of advanced powertrain architectures [1,2]. Solutions based on the adoption of full electric (battery electric vehicles (BEVs)) powertrains or on the combination of an internal combustion engine (ICE) and electric traction (hybrid/plug-in hybrid electric vehicles (HEVs/PHEVs)) are now established as reliable alternatives to conventional powertrains [3,4]. They exploit batteries as the primary energy source in BEVs or as an auxiliary source in HEVs and PHEVs [5]. In the automotive industry, the most common battery technology exploits lithium because of its remarkable advantages in terms of the energy density, fast charging, low maintenance, and long lifetime allowances. Moreover, lithium-based solutions allow for obtaining powerful, compact, and light configurations together with satisfactory levels of autonomy, which is currently settled in the order of a few hundreds of kilometers [6]. However, the reliability and performance of these type of batteries are strongly influenced by the management of the charging and discharging phases. It is indeed well known that an appropriate handling of these operations is mandatory to avoid the occurrence of overcharging or deep discharging, that would lead to permanent or hardly reversible damages of the pack. A continuous and accurate monitoring of the battery state takes on significant importance to extend the battery lifetime, effectively plan the trip route and charging stops, optimize the energy flow management of HEVs [7,8], and mitigate psychological effects, such as the range anxiety that is commonly experienced by a large number of BEV drivers [6]. The main parameters to be assessed for a correct battery monitoring are the residual available energy in the pack, known as state of charge (SOC), and the degradation suffered by the battery, indicated by the state of health (SOH) [9]. As is well known, these two states cannot be directly measured, since the technology to make a sensor that plays the equivalent role of a fuel gauge is not available. Therefore, the adoption of some estimation techniques becomes mandatory [10,11]. Typically, carmakers exploit look-up tables (LUTs), where the SOC and SOH behavior is mapped during the preliminary experimental characterizations conducted in a laboratory environment. These tests are done following the so-called direct methods, which are based on ampere-hour counting or the measurement of the internal impedance and open circuit voltage of the battery [10,12]. However, the adoption of LUTs may have a high computational cost and imposes the storage of a huge amount of data in the electronic control unit memory, particularly in the case of the SOH estimation. A further class of methods exploits model-based techniques for the real-time assessment of both the SOC and SOH [13]. The most common are the Kalman filter [14] and its derivations, namely the extended (EKF) [15] and unscented Kalman filters (UKF) [16,17], the adaptive particle filter (APF) [18], and the smooth variable structure filter (SVSF) [19]. Although these solutions can be implemented in real time on a vehicle, they may suffer problems of inaccuracies if the reference model is not completely and accurately tuned in all the possible operating conditions. An alternative and promising approach to overcome this limitation is represented by artificial intelligence (AI). In most cases, these solutions adopt artificial neural networks (ANNs) and allow getting rid of the model while obtaining satisfactory levels of accuracy and reliability, provided that the networks are properly trained. An extensive literature is dedicated to the methods for the estimation of the SOC [20–23] or SOH [24–27] with AI. Nevertheless, to the best of the author's knowledge, very few works deal with the combined estimation of the SOC and SOH and most of them describe model-based techniques [28–30].

This paper proposes a technique for the combined estimation of the SOC and SOH with a set of five ANNs: four regression networks dedicated to the SOC estimation and one classification network for the SOH identification. The method is independent by the battery model and is designed with a training phase conducted with datasets obtained from the preliminary laboratory experimental characterizations. The SOC estimation exploits four nonlinear autoregressive neural networks with exogenous input. Each of them is associated with a specific class of ageing (SOH) of the battery. The correct estimation among the four outputs is selected according to the SOH identification, which is obtained separately by a classifier that is done with a pattern recognition neural network. The SOH estimator provides a class of ageing among four possibilities, ranging from 80% to 100% with a step of 5%. A further class is associated to exhausted batteries and covers the range from 0% to 80% of the SOH, where 80% is the degradation threshold in the automotive sector. The output of the SOH classifier is used to select the correct SOC estimation among the four outputs of the regression ANNs, while the SOC estimation is used as an input for the SOH classifier in a closed-loop recursive architecture. The SOH estimator is an algorithm which is triggered only when a specific battery load condition in terms of the mean charging/discharging capacity request in a predefined time window is detected. This procedure allows reducing the training dataset of the SOH neural classifier to only one specific case. This aspect represents a relevant advantage in terms of a size reduction of the training dataset and a consequent time saving during the dataset collection and learning procedures. Additionally, the size of the network is smaller with a consequent reduction of the memory occupation when deployed on the battery management system (BMS).

The paper describes the design of the two estimators and the validation phase is conducted with the adoption of driving cycles acquired on a mild hybrid electric vehicle. The performance of the SOC estimator is evaluated by comparing the temporal evolution of the expected and estimated state of charge, whereas the SOH classifier accuracy is measured by using a confusion matrix, a common evaluation tool of classification algorithms.

The novel contributions of this work are as follows: a) the proposal of a combined estimation of the SOC and SOH with ANNs, allowing to make the method independent from the model and valid for every operating condition, provided that the network training dataset is complete and accurate; and b) the proposal of an SOH estimation method that is triggered only when a specific load condition corresponding to a predefined charging/discharging current profile is detected: this results in a compact algorithm that can be trained with a dataset that is smaller with respect to what would be needed in the case of a reproduction of the whole set of ageing conditions.

2. Method

The proposed method aims to provide a combined estimation of both the SOC and SOH of a battery. The approach is equally valid for a battery pack, module, or for the single cell.

Figure 1 illustrates the overall layout of the method that is composed of two subsystems: the SOC estimator, consisting of four regression ANNs, that is illustrated in the top left dashed box, and the SOH estimator, that exploits a neural classifier, that is reported in the bottom right dotted box. As is well known, the behavior of the two parameters is connected: the SOC of a battery is strongly influenced by the ageing, as well as the SOH estimation needing the information of the SOC variation during the charging/discharging operations. This motivates the adoption of a recursive loop architecture, where the SOC output is provided as an input to the SOH classifier and vice-versa. Both algorithms were trained on the basis of the preliminary experimental characterizations conducted in a laboratory. The two subsystems are described in detail in the following sections.

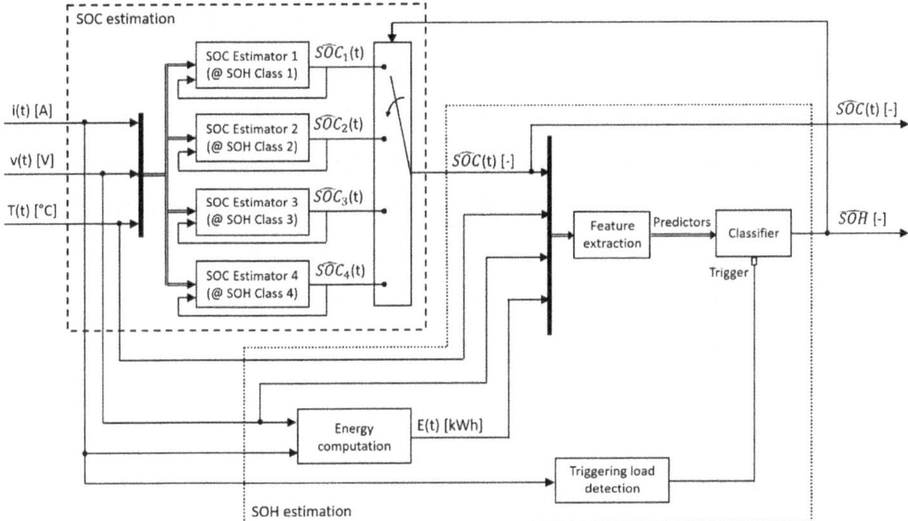

Figure 1. Overall method architecture. Dashed box: state of charge (SOC) estimation. Dotted box: state of health (SOH) estimation. $i(t)$: charging/discharging current. $v(t)$: voltage at battery terminals. $T(t)$: battery temperature. $E(t)$: energy request. SOH classes: 1: $(100 \div 95)\%$; 2: $(95 \div 90)\%$; 3: $(90 \div 85)\%$; 4: $(85 \div 80)\%$.

The battery pack considered for the study is composed of 168 cells (the cell model is Kokam SLPB 11543140H5, its characteristics are reported in Table 1) in the configuration 12p14s (p: parallel, s: series). The pack has a nominal voltage of 48 V, a nominal capacity of 60 Ah, and is designed for a mild hybrid electric vehicle with a peak electric power of around 20 kW, obtained considering a discharge rate of around 7C in nominal conditions.

Table 1. Main characteristics of the battery cell.

Typical Capacity (@0.5C, 4.2 V ÷ 2.7 V, 25 °C)		5 Ah
Nominal Voltage		3.7 V
Cut-off voltage		2.7 V
Continuous current		150 A
Peak current		250 A
Cycle life (Charge/Discharge @ 1C)		>800 cycles
Charge condition	Max. Current	10 A
	Voltage	4.2 V ± 0.03 V
Operating Temperature	Charge	0–40 °C
	Discharge	−20–60 °C
Mass		128.0 ± 4 g
Dimension	Thickness	11.5 ± 0.2 mm
	Width	42.5 ± 0.5 mm
	Length	142.0 ± 0.5 mm

2.1. SOC Estimation

The SOC estimator consists of four parallel regression ANNs (dashed box in Figure 1) working on the same inputs. Each network is associated with a specific ageing condition: SOH class 1 (from 100% to 95%), SOH class 2 (from 95% to 90%), SOH class 3 (from 90% to 85%), and SOH class 4 (from 85% to 80%). The threshold of 80% was decided considering that in the automotive sector, a battery has to be considered exhausted when the capacity or power fading is higher than 20%. The step of 5% is aligned with the typical precision that can be reached when dealing with the SOH estimation problem [31,32].

Each of the four regression ANNs receive, simultaneously, the following signals as inputs: charging/discharging current ($i(t)$ [A]), voltage at battery terminals ($v(t)$ [V]), and temperature ($T(t)$ [C]). They provide four different outputs: $S\hat{O}C_1(t)$, $S\hat{O}C_2(t)$, $S\hat{O}C_3(t)$, and $S\hat{O}C_4(t)$. The final SOC estimation ($S\hat{O}C(t)$) is obtained with a downstream selector that is operated by a signal fed back from the SOH classifier output, that is running separately, as indicated in Figure 1.

The structure of the four SOC estimators is the nonlinear autoregressive neural network with exogenous input (NARX) architecture. Typically, this layout is adopted for prediction tasks and finds an application in industrial engineering fields as well as in other sectors, namely linguistic search engines or weather forecasting. However, its effectiveness has been demonstrated also for estimation tasks and has been presented as an effective solution to estimate the SOC of lithium batteries in [21], where an additional comparison with other ANN architectures in terms of the computational cost and estimation accuracy is provided. The scheme of the NARX is reported in Figure 2, where the two adopted configurations are illustrated: an open-loop configuration (a), often indicated also as the series–parallel (SP) mode, that is adopted during the training procedure, and a closed-loop configuration (b), or equivalently the parallel (P) mode, that is the final architecture adopted for the estimation when the network is deployed on the vehicle for the real-time execution.

The output of the regression is defined as

$$y(n) = \varphi\left[y(n-1), y(n-2), \ldots, y(n-d_y); x(n-1), x(n-2), \ldots, x(n-d_x)\right] \tag{1}$$

where $y(n) \in \mathbb{R}$ and $x(n) \in \mathbb{R}$ denote the output (state of charge) and inputs (current, voltage, and temperature) of the NARX model at the discrete timestep n, respectively, d_x and d_y are the input and output memory delays used in the model, respectively, and φ is the function, generally non-linear, represented by the ANN. During the regression computation, the next value of the dependent output signal $y(n)$ is regressed on the previous d_y values of the output signal and previous d_x values of the independent (exogenous) input signal. In the open-loop configuration, the output regressor is

$$y(n) = \varphi\left[\overline{y}(n-1), \overline{y}(n-2), \ldots, \overline{y}(n-d_y); x(n-1), x(n-2), \ldots, x(n-d_x)\right] \qquad (2)$$

A supervised training procedure is conducted using the measured output as the target. This approach allows for enriching the information to be processed by the network and permits using a common static backpropagation algorithm, the Levenberg–Marquardt in this case, for the training process, since the resulting network has a purely feedforward architecture.

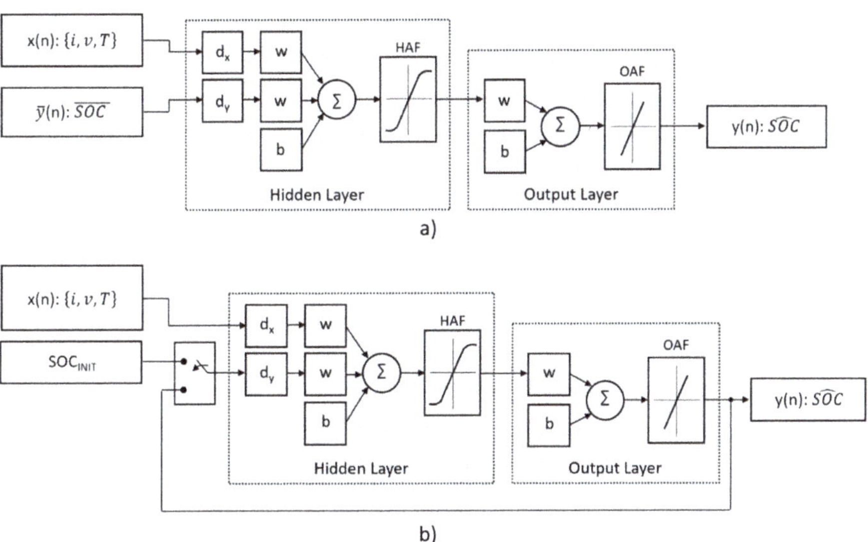

Figure 2. Nonlinear autoregressive neural network with exogenous input (NARX) architecture. (**a**) Series–parallel (SP) mode (open-loop configuration) adopted during the training. (**b**) Parallel (P) mode (closed-loop configuration) adopted for the estimation when the network is deployed. HAF: hidden activation function. OAF: output activation function. w: weight. b: bias.

In the first second of computation, the value of the algorithm output is not stable and is unpredictable. Therefore, if this value is fed back and provided as input to the ANN, it generates an estimation divergence over time. To avoid the occurrence of this irremediable condition, during the first second of estimation the feedback of the estimated SOC is replaced by the last estimation value (SOC_{INIT} in Figure 2b) recorded on a non-volatile memory at the previous shut down of the vehicle. After 1 second, when the output has become stable, the SOC input of the network switches from the previously recorded value to the real feedback of the estimation so that the regular operation of the algorithm can start.

Referring to Figure 2b and indicating with $n = n_0$ the time instant when the feedback signal switches from SOC_{INIT} to the estimated output, the characteristic equations of the model are written as

$$y(n) = \varphi[SOC_{INIT}; x(n-1, x(n-2), \ldots, x(n-d_x))], \ n < n_0 \qquad (3)$$

and

$$y(n) = \varphi\left[y(n-1), y(n-2), \ldots, y(n-d_y); x(n-1), x(n-2), \ldots, x(n-d_x)\right], \ n \geq n_0 \qquad (4)$$

The four networks have the same size in terms of layers, neurons, and delays and adopts the same activation functions. All these parameters have been designed with a trial and error approach aimed to maximize the estimation accuracy and avoid the risk of overfitting. Specifically, each network has one layer with eight neurons, the delays d_x and d_y are equal to two, the activation function in the hidden layer (*HAF*) and output layer (*OAF*) are the hyperbolic tangent and linear functions respectively, and the training function is the Levenberg–Marquardt function.

During the design phase, the training precision is evaluated by computing the mean square error (MSE) that reached a value of 1×10^{-13} as indicated in the small box embedded in Figure 3, and the estimation accuracy is measured with the maximum relative error (MRE), that is computed as

$$\text{MRE } [\%] = \max_{1<i<n}\left(\left|\frac{SOC_{exp}(i) - SOC_{est}(i)}{SOC_{exp,max} = 1}\right|\right) \times 100 \tag{5}$$

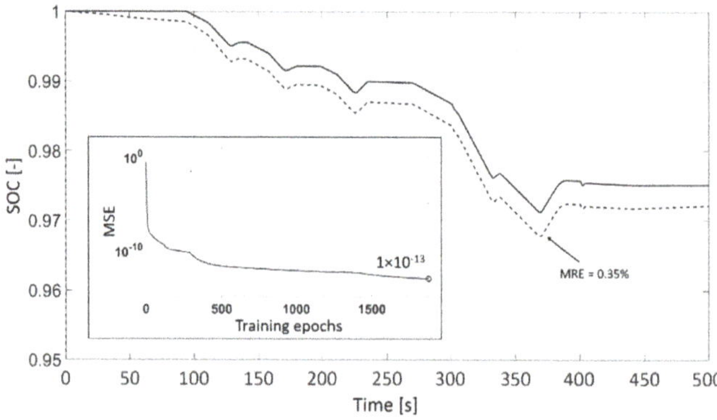

Figure 3. Comparison performance between the estimation (dashed line) and expected values (solid line) of the SOC in the case of an SOH = 100%. The obtained maximum relative error (MRE) is equal to 0.35%. The small box in the bottom left indicates the trend of the mean square error during the training phase.

This parameter reached the value of 0.35% as indicated in Figure 3, where the comparison between the estimation (dashed line) and the expected value (solid line) of the state of charge is reported in the case of a new battery (SOH = 100%). This plot wants to represent an indication of the training evaluation during the design phase.

The time length of the training dataset for the four regression ANNs is 13 h.

A more detailed description of the overall method results is reported in the final section of the paper.

2.2. SOH Estimation

The degradation of the battery is estimated with an algorithm reproducing a pattern recognition classifier with an ANN. Since the algorithm is proposed for the automotive sector, the method considers 20% as the maximum admitted capacity fading. Therefore, the considered life cycle of the battery ranges from an SOH of 100% when the battery is new to an SOH of 80% when the battery has to be considered exhausted. The proposed solution aims at quantifying the degradation suffered by the battery by identifying the five different levels of ageing which correspond to the five classes provided as an output by the classification algorithm. The first class covers the interval of ageing below the level of 80% (assumed as the threshold of the maximum degradation of the battery) of the SOH. The other four classes are equally distributed between 80% and 100% with four intervals of 5%, a percentage that is considered as consistent with the reasonable level of accuracy that can be reached when dealing with the problem of the SOH estimation.

As in the case of the SOC network design, the proposed algorithm for the SOH estimation exploits a preliminary experimental characterization phase conducted on the battery in a laboratory environment. The obtained results are used to build the training dataset to be adopted for the learning phase of the neural classifier. Specifically, the data of interest are recorded in a specific battery load condition corresponding to a mean request of 12 Ah in an interval of time of 120 s. This condition was selected because it can be detected quite frequently during a common driving cycle of an electric or hybrid vehicle. Afterwards, the network is trained with the dataset corresponding to this specific operating condition obtained at different values of temperature. Therefore, when the algorithm is deployed on the vehicle, it is called to estimate the level of ageing whenever the same condition is detected during the real driving cycle. This implies that when driving the vehicle, consecutive buffers of 120 s are analyzed back-to-back by a control logic that is implemented in the "Triggering load detection" block in Figure 1. As soon as the specific load condition of interest (mean capacity request of 12 Ah in 120 s) is detected, the classifier is triggered and provides the SOH classification as an output. Therefore, the estimation rate is not continuous over time, but it is produced in a discrete and not time deterministic way, only in correspondence with the detection of the predefined known load condition. The output of the estimator is kept equal to the last SOH estimation if the triggering condition is not occurring.

Figure 4 reports a part of the ANN training dataset obtained by the preliminary experimental characterization conducted on the battery. Subplot "a)" illustrates the behaviour of the degradation of the battery as a function of the number of discharging cycles at different values of temperature [33]. The discharging is conducted with the predefined load above-mentioned. Subplot "b)" reports the coupling effects between the SOH, capacity, SOC, and battery voltage. In this test, the temperature is set to 25 °C and the variation of the capacity is motivated by the difference in the time needed to discharge the battery at the different levels of ageing.

The time length of the training dataset covering all the considered levels of ageing is equal to 916 h obtained from 27,494 buffers with a duration of 120 s.

The SOH classifier works on discrete inputs, the so-called predictors, that are extracted in the "Feature extraction" block in Figure 1 from the time histories of the following signals: current, voltage, temperature, SOC, and energy. The latter is obtained from the "Energy computation" block in Figure 1 and is defined as

$$E = \int_{t_0}^{t_0+t_b} v(t)i(t)dt \tag{6}$$

where t_0 is the initial time of the buffer and t_b is the time length of the processed buffer that is set equal to 120 s.

The list of the extracted predictors is state of charge variation (-) (ΔSOC), voltage variation (V) (ΔV), requested energy (Wh) (E), and mean temperature (°C) (T).

The architecture of the classifier is illustrated in Figure 5. The training phase of the neural classifier is performed exploiting the scaled conjugate gradient (SCG) backpropagation training function [27]. This algorithm is designed to minimize the cost function including the difference between the estimated and expected outputs. This approach gives a good performance over a large number of pattern recognition problems that may include numerous parameters and guarantees a low performance degradation while reducing the training error. Additionally, this function is characterized by a relatively low computational cost and memory requirements [21], and its ability to provide well-separated classes in data mining and classification problems has been proven in many research works [34].

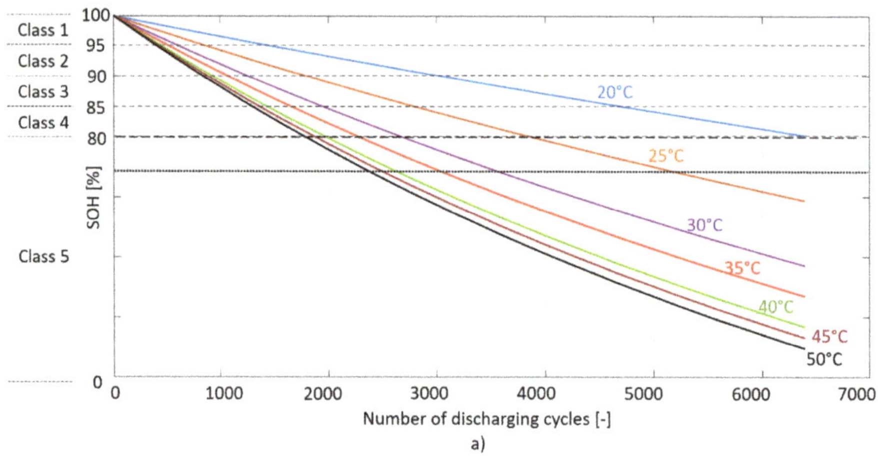

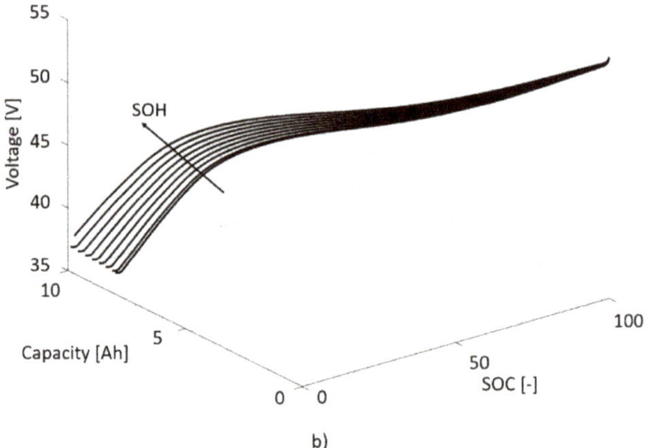

Figure 4. Battery experimental characterization for the SOH estimation. (**a**) SOH as a function of the number of discharging cycles and of the temperature. (**b**) Behavior of the SOH as a function of the voltage, capacity, and SOC. The temperature in this case is set equal to 25 °C.

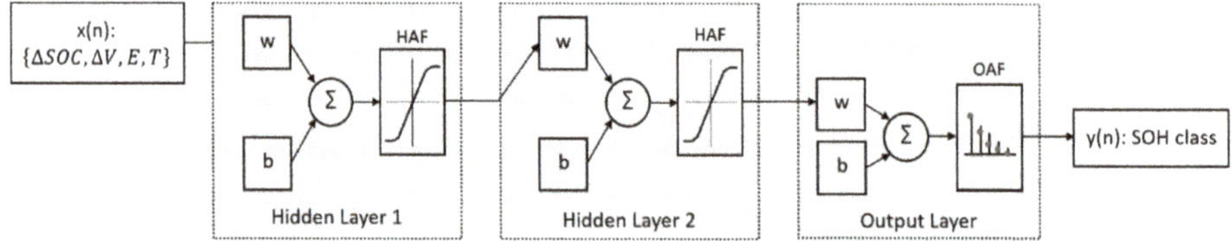

Figure 5. Pattern recognition a feed-forward artificial neural network (ANN) architecture for the SOH classification. HAF: hidden activation function. OAF: output activation function. *w*: weight. *b*: bias.

The classifier is composed of one input, two hidden and one output layer. As in the case of the SOC network design, the number and size of the hidden layers is defined heuristically, by means of a trial and error procedure. Specifically, the hidden layers consist of ten neurons each, HAF is a hyperbolic tangent sigmoid, and OAF is a normalized exponential function. The performance of the training process is evaluated by means of the cross-entropy cost function, that at the end of the training process is equal to 1×10^{-3}, after around 3000 training epochs.

3. Results and Discussion

The validation of the method is conducted in two separate phases: (a) an analysis of the performance of the SOH identifier and (b) an evaluation of the accuracy of the overall SOC estimation that includes the ageing classification.

3.1. SOH Classification

As is described above, the classification algorithm is called to identify the class of degradation only when a specific load condition is detected during the driving operations. To evaluate the effectiveness of the method, a profile corresponding to the specific charging/discharging profile was created artificially to have an exhaustive number of occurrences in the different operating conditions to test.

The profile is reported in Figure 6, where it has a duration of 5000 s and includes 42 different consecutive buffers with the time length of 120 s and a mean capacity request of 12 Ah. The profile was cycled until reaching a total duration of 50 h, to sweep the range of the SOC of the battery, corresponding to 1500 buffers of 120 s, for each class of ageing. The resulting timeseries was provided to the LUT representing the battery. This LUT was tuned after the preliminary laboratory experimental characterization and allows for extracting the predictors provided to the classifier in the five ageing conditions.

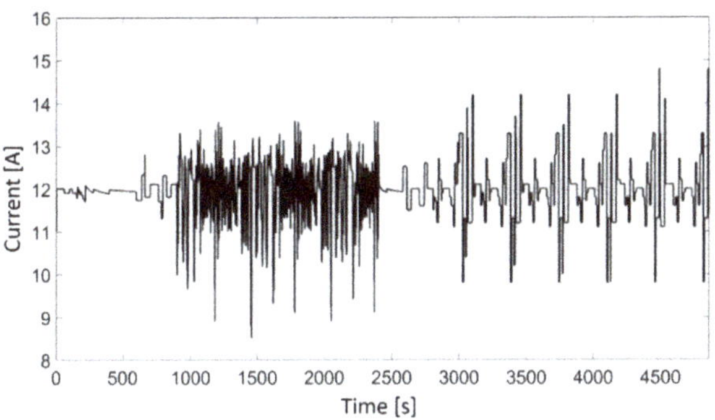

Figure 6. Current profile created to validate the SOH classifier. The profile is replicated until reaching the total duration of 50 h and a number of buffers of 1500 for each class of ageing.

The resulting validation dataset is therefore composed of 7500 different buffers with a time length of 120 s each. The resulting profile represents the different operating conditions at different degradation levels and is given as an input to the classifier.

The tool adopted to evaluate the accuracy of the SOH estimation is the confusion matrix reported in Figure 7. The classified and actual ageing condition instances are reported in the rows and columns, respectively. The values contained in the main diagonal cells indicate the correct classifications, whereas the off-diagonal cells report the number of the misclassifications. The overall obtained estimation accuracy is equal to 2.4%, which is equal to the number of misclassifications (178 buffers) over the total number of tested occurrences (7500 buffers). This result is aligned with the expected accuracy.

Classification output	Class 1 100÷95	Class 2 95÷90	Class 3 90÷85	Class 4 85÷80	Class 5 80÷0	
Class 1 (100÷95)	1469	19	2	0	0	
Class 2 (95÷90)	21	1459	15	0	0	
Class 3 (90÷85)	10	21	1467	15	14	
Class 4 (85÷80)	0	1	16	1472	21	
Class 5 (80÷0)	0	0	0	23	1465	
Error	2%	2.7%	2.2%	2.5%	2.3%	2.4%
	Class 1 100÷95	Class 2 95÷90	Class 3 90÷85	Class 4 85÷80	Class 5 80÷0	

Classification target

Figure 7. Evaluation of the SOH classification performance. Confusion matrix obtained for the ANN trained with the scaled conjugate gradient (SCG) algorithm. The cell in the grey background indicates the overall accuracy of the method.

3.2. SOC Estimation

The second part of the validation is dedicated to the evaluation of the accuracy of the SOC estimation. To this end, the profiles illustrated in Figure 8 have been adopted as validation timeseries. The subplot "a)" reports the current profile, and the subplot "b)" illustrates the behavior of the battery terminal voltage at different levels of ageing. The voltage is only an occurrence of the many possibilities that are associated to a class of degradation. The plots in the right part of the figure are zoomed-in areas with a time length of 2000 s. When providing these timeseries to the SOC estimation block (dashed box in Figure 1), the regression ANNs will provide four different outputs. The one corresponding to the correct ageing level of the battery is then selected according to the output of the SOH classifier (dotted box in Figure 1).

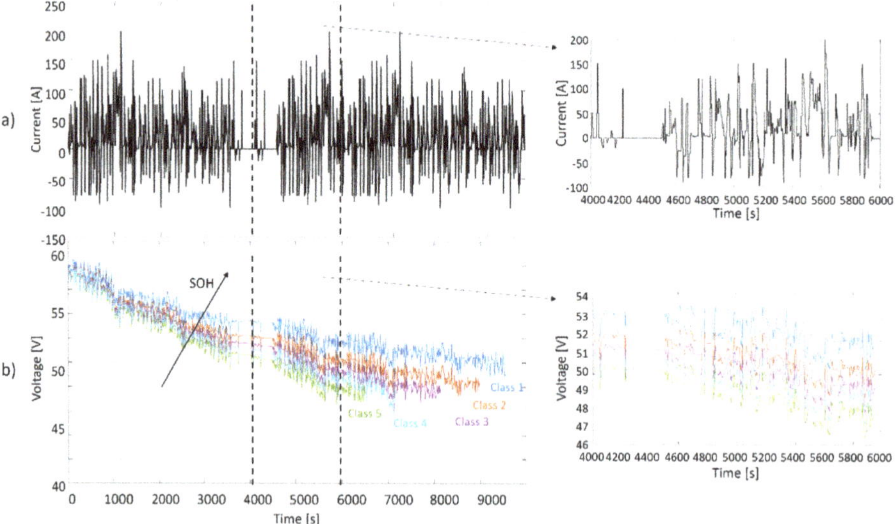

Figure 8. ANN validation datasets recorded from a real mild hybrid vehicle. (**a**) Current $i(t)$. (**b**) Voltage $v(t)$ at four different degradation levels corresponding to the four SOH classes.

The results obtained in the five ageing levels are illustrated in Figure 9, where for each SOH class, the estimated SOC, on the blue line, is compared with the expected value, on the red line. The expected value is the one obtained from the preliminary experimental characterization conducted in the laboratory. The estimation error is reported in the lower subplot for each case. The accuracy of the estimation is demonstrated by the error that is limited to a maximum value of 3%. The results obtained for the class of ageing going from 0% to 80% (subplot "e") demonstrate that the algorithm keeps being valid also under the threshold of 80%. The reported test has been conducted at a temperature of around 25 °C. A more exhaustive validation of the method should be conducted in a climatic test chamber to evaluate the accuracy of the estimation at different environmental conditions.

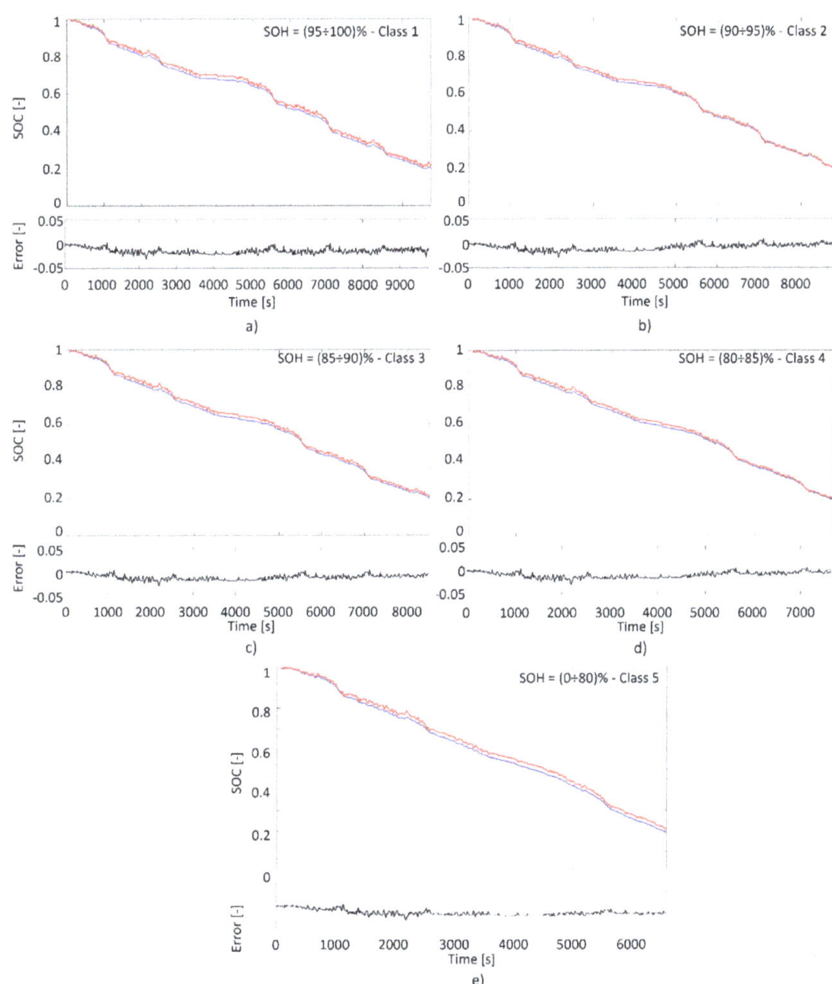

Figure 9. SOC estimation at different degradation levels. Red line: expected value. Blue line: estimation. Error indicates the difference between the estimated and expected values. (**a**): ageing class 1 (SOH: 95 ÷ 100%); (**b**): ageing class 2 (SOH: 90 ÷ 95%); (**c**): ageing class 3 (SOH: 85 ÷ 90%); (**d**): ageing class 4 (SOH: 80 ÷ 85%); (**e**): ageing class 5 (SOH: 0 ÷ 80%).

4. Conclusions

This paper presented a method for the combined estimation of the state of charge and state of health of batteries with artificial intelligence. The technique is valid at the cell, module, and pack levels and is suitable for adoption in the automotive sector in the case of hybrid and full electric vehicles. The design procedure of the algorithm and specifically the training phase of the artificial neural networks were presented. The method was demonstrated to be effective in terms of the estimation accuracy when tested on real driving cycles extracted from the acquisition on-board of an electric vehicle. The estimation error of the combined method is around 3%. The good potential and the promising results encourage the adoption of the proposed method for deployment in a vehicle battery management system for a real-time battery monitoring.

Funding: This research received no external funding.

Acknowledgments: This work was developed in the framework of the activities of the Interdepartmental Center for Automotive Research and Sustainable Mobility (CARS) of Politecnico di Torino (www.cars.polito.it).

Conflicts of Interest: The author declares no conflict of interest.

References

1. Ahman, M. Assessing the future competitiveness of alternative powertrains. *Int. J. Veh.* **2003**, *33*, 309. [CrossRef]
2. Wang, Y.; Diaz, D.F.R.; Chen, K.S.; Wang, Z.; Adroher, X.C. Materials, technological status, and fundamentals of PEM fuel cells—A review. *Mater. Today* **2020**, *32*, 178–203. [CrossRef]
3. Walther, G.; Wansart, J.; Kieckhafer, K.; Schneider, E.; Spengler, T.S. Impact assessment in the automotive industry: Mandatory market introduction of alternative powertrain technologies. *Syst. Dyn. Rev.* **2010**, *26*, 239–261. [CrossRef]
4. Bishop, J.D.K.; Martin, N.P.D.; Boies, A.M. Cost-effectiveness of alternative powertrains for reduced energy use and CO_2 emissions in passenger vehicles. *Appl. Energy* **2014**, *124*, 44–61. [CrossRef]
5. Chan, C.C.; Chau, K.T. *Modern Electric Vehicle Technology*; Oxford University Press: New York, NY, USA, 2002.
6. Onori, S.; Serrao, L.; Rizzoni, G. *Hybrid Electric Vehicles: Energy Management Strategies*; Springer: London, UK, 2016.
7. Anselma, P.G.; Huo, Y.; Roeleveld, J.; Belingardi, G.; Emadi, A. Slope-Weighted Energy-Based Rapid Control Analysis for Hybrid Electric Vehicles. *IEEE Trans. Veh. Technol.* **2019**, *68*, 4458–4466. [CrossRef]
8. Anselma, P.G.; Huo, Y.; Roeleveld, J.; Belingardi, G.; Emadi, A. Integration of On-Line Control in Optimal Design of Multimode Power-Split Hybrid Electric Vehicle Powertrains. *IEEE Trans. Veh. Technol.* **2019**, *68*, 3436–3445. [CrossRef]
9. Vetter, J.; Novák, P.; Wagner, M.; Veit, C.; Möller, K.C.; Besenhard, J.O.; Winter, M.; Wohlfahrt-Mehrens, M.; Vogler, C.; Hammouche, A. Ageing mechanisms in lithium-ion batteries. *J. Power Sources* **2005**, *147*, 269–281. [CrossRef]
10. Chang, W.Y. The State of Charge Estimating Methods for Battery: A Review. *Appl. Math.* **2013**, *2013*, 953792. [CrossRef]
11. Piller, S.; Perrin, M.; Jossen, A. Methods for state-of-charge determination and their applications. *J. Power Sources* **2001**, *96*, 113–120. [CrossRef]
12. Remmlinger, J.; Buchholz, M.; Meiler, M.; Bernreuter, P.; Dietmayer, K. State-of-health monitoring of lithium-ion batteries in electric vehicles by on-board internal resistance estimation. *J. Power Sources* **2011**, *196*, 5357–5363. [CrossRef]
13. Huang, S.; Tseng, K.; Liang, J.; Chang, C.; Pecht, M.G. An Online SOC and SOH Estimation Model for Lithium-Ion Batteries. *Energies* **2017**, *10*, 512. [CrossRef]
14. Wei, Z.; Zhao, J.; Ji, D.; Tseng, K.T. A multi-timescale estimator for battery state of charge and capacity dual estimation based on an online identified model. *Appl. Energy* **2017**, *204*, 1264–1274. [CrossRef]
15. Pérez, G.; Garmendia, M.; Reynaud, J.F.; Crego, J.; Viscarret, U. Enhanced closed loop State of Charge estimator for lithium-ion batteries based on Extended Kalman Filter. *Appl. Energy* **2015**, *155*, 834–845. [CrossRef]
16. Yu, Q.; Xiong, R.; Lin, C. Online estimation of state-of-charge based on H-infinity and unscented Kalman filters for lithium ion batteries. *Energy Procedia* **2017**, *105*, 2791–2796. [CrossRef]

17. Zeng, M.; Zhang, P.; Yang, Y.; Xie, C.; Shi, Y. SOC and SOH Joint Estimation of the Power Batteries Based on Fuzzy Unscented Kalman Filtering Algorithm. *Energies* **2019**, *12*, 3122. [CrossRef]
18. Ye, M.; Guo, H.; Xiong, R.; Yang, R. Model-based state-of-charge estimation approach of the Lithium-ion battery using an improved adaptive particle filter. *Energy Procedia* **2016**, *103*, 394–399. [CrossRef]
19. Kim, T.; Wang, Y.; Sahinoglu, Z.; Wada, T.; Hara, S.; Qiao, W. State of Charge Estimation Based on a Realtime Battery Model and Iterative Smooth Variable Structure Filter. In Proceedings of the IEEE Innovative Smart Grid Technologies—Asia, Kuala Lumpur, Malaysia, 20–23 May 2014.
20. Charkhgard, M.; Farrokhi, M. State-of-Charge Estimation for Lithium-Ion Batteries Using Neural Networks and EKF. *IEEE Trans. Ind. Electron.* **2010**, *57*, 4178–4187. [CrossRef]
21. Bonfitto, A.; Feraco, S.; Tonoli, A.; Amati, N.; Monti, F. Estimation Accuracy and Computational Cost Analysis of Artificial Neural Networks for the State of Charge Estimation in Lithium Batteries. *Batteries* **2019**, *5*, 47. [CrossRef]
22. He, W.; Williard, N.; Chen, C.; Pecht, M. State of charge estimation for Li-ion batteries using neural network modeling and unscented Kalman filter-based error cancellation. *Electr. Power Energy Syst.* **2014**, *62*, 783–791. [CrossRef]
23. Chaoui, H.; Ibe-Ekeocha, C.C. State of Charge and State of Health Estimation for Lithium Batteries Using Recurrent Neural Networks. *IEEE Trans. Veh. Technol.* **2017**, *66*, 8773–8783. [CrossRef]
24. Chang, C.; Liu, Z.; Huang, Y.; Wei, D.; Zhang, L. Estimation of Battery state of Health Using Back Propagation Neural Network. *Comput. Aided Draft. Des. Manuf.* **2014**, *24*, 60.
25. Bonfitto, A.; Feraco, S.; Ezemobi, E.; Tonoli, A.; Amati, N.; Hegde, S. State of Health Estimation of Lithium Batteries for Automotive Applications with Artificial Neural Networks, IEEE. In Proceedings of the 2019 AEIT International Conference of Electrical and Electronic Technologies for Automotive, AEIT AUTOMOTIVE, Turin, Italy, 2–4 July 2019.
26. Lin, H.; Liang, T.; Chen, S. Estimation of Battery State of Health Using Probabilistic Neural Network. *IEEE Trans. Ind. Inform.* **2013**, *9*, 679–685. [CrossRef]
27. Yang, D.; Wang, Y.; Pan, R.; Chen, R.; Chen, Z. A neural network based state-of-health estimation of lithium-ion battery in electric vehicles. In Proceedings of the 8th International Conference on Applied Energy—ICAE 2016, Beijing, China, 8–10 October 2016.
28. Huet, F. A review of impedance measurement for determination of state-of-charge or state-of-health of secondary battery. *J. Power Sources* **1998**, *70*, 59–69. [CrossRef]
29. Wassiliadis, N.; Adermann, J.; Frericks, A.; Pak, M.; Reiter, C. Revisiting the dual extended Kalman filter for battery state-of-charge and state-of-health estimation: A use-case life cycle analysis. *J. Energy Storage* **2018**, *19*, 73–87. [CrossRef]
30. Andre, D.; Appel, C.; Soczka-Guth, T.; Sauer, D.U. Advanced mathematical methods of SOC and SOH estimation for lithium-ion batteries. *J. Power Sources* **2013**, *224*, 20–27. [CrossRef]
31. Zhang, C.W.; Chen, S.R.; Gao, H.B.; Xu, K.J.; Yang, M.Y. State of Charge Estimation of Power Battery Using Improved Back Propagation Neural Network. *Batteries* **2018**, *4*, 69. [CrossRef]
32. Samolyk, M.; Sobczak, J. Development of an Algorithm for Estimating Lead-Acid Battery State of Charge and State of Health. Master's Thesis, Blekinge Institute of Technology, Karlskrona, Sweden, 2013.
33. Leng, F.; Tan, C.; Pecht, M. Effect of Temperature on the Aging rate of Li Ion Battery Operating above Room Temperature. *Sci. Rep.* **2015**, *5*, 12967. [CrossRef]
34. Bonfitto, A.; Feraco, S.; Tonoli, A.; Amati, N. Combined regression and classification artificial neural networks for sideslip angle estimation and road condition identification. *Veh. Syst. Dyn.* **2019**, 1–22. [CrossRef]

Article

Analytical Solution for Coupled Diffusion Induced Stress Model for Lithium-Ion Battery

Davide Clerici [†], Francesco Mocera [*,†] and Aurelio Somà [†]

Department of Mechanical and Aerospace Engineering, Politecnico di Torino, Corso duca degli Abruzzi 24, 10129 Torino, Italy; davide.clerici@polito.it (D.C.); aurelio.soma@polito.it (A.S.)
* Correspondence: francesco.mocera@polito.it
† These authors contributed equally to this work.

Received: 17 January 2020; Accepted: 31 March 2020; Published: 4 April 2020

Abstract: Electric cycling is one of the major damage sources in lithium-ion batteries and extensive work has been produced to understand and to slow down this phenomenon. The damage is related to the insertion and extraction of lithium ions in the active material. These processes cause mechanical stresses which in turn generate crack propagation, material loss and pulverization of the active material. In this work, the principles of diffusion induced stress theory are applied to predict concentration and stress field in the active material particles. Coupled and uncoupled models are derived, depending on whether the effect of hydrostatic stress on concentration is considered or neglected. The analytical solution of the coupled model is proposed in this work, in addition to the analytical solution of the uncoupled model already described in the literature. The analytical solution is a faster and simpler way to deal with the problem which otherwise should be solved in a numerical way with finite difference method or a finite element model. The results of the coupled and uncoupled models for three different state of charge levels are compared assuming the physical parameters of anode and cathode active material. Finally, the effects of tensile and compressive stress are analysed.

Keywords: diffusion induced stress; hydrostatic stress influence on diffusion; electrode particle model; battery mechanical aging; li-ion battery

---

1. Introduction

Lithium-ion batteries are actually one of the most widespread rechargeable energy-storage systems [1]. They have a large field of application from small electronic systems up to electric vehicles in automotive and industrial applications [2,3]. Indeed, they can span a great capacity and power range with a good energy density and long lifetime.

Safety and batteries performance have been analysed through modelling and experimental tests in the last decades [4,5]. Progressive damage with charge/discharge cycles is the major weak point of lithium ion cells, since it affects the lifetime considerably [6]. The lithium ions are stored/withdrawn in the active material of the electrodes through insertion/extraction processes. These processes must be studied at a micrometre scale, at which active material of both electrodes appears as a particulate matter, as depicted in Figure 1a. Insertion/extraction processes induce expansion/contraction of the active material particles, and in turn mechanical stress which damage the electrode structure. The electrode damage causes indirectly the increase of solid electrolyte interface (SEI) growth which in turn affects the battery performance and the available capacity [7]. Stress and strain due to insertion/extraction processes are recently investigated via in-situ measurements in battery electrodes [8–10]. However, the lack of experimental stress measurements in active material particles does not allow to validate the results derived in this work.

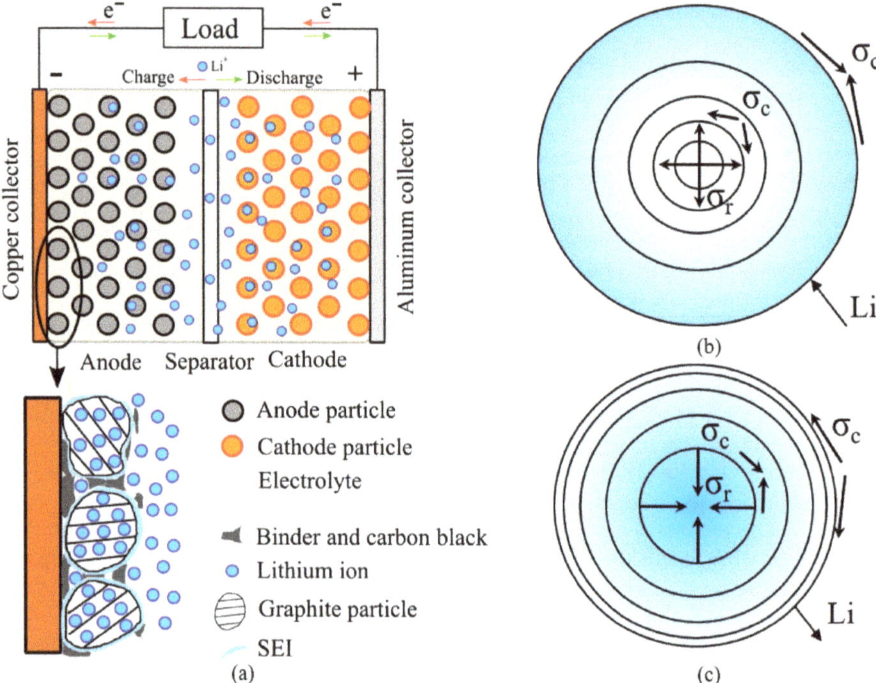

Figure 1. Battery scheme at micrometer scale and focus on the lithium intercalation mechanism in anode particles (**a**). Stress and deformation of the active material particles during lithium insertion (**b**) and extraction (**c**). Blue shadows depict the lithium concentration (blue means high concentration, white means low concentration). Concentration level affects tension (arrows pointing toward each other) and compression (arrows pointing away from each other) of radial and hoop stress. The concentric lines show expansion or shrinking of the particle: they are evenly spaced in the undeformed configuration.

These type of stresses are described originally by Prussin [11] as chemical stress or diffusion induced stress (DIS), which manages the interaction between chemical and mechanical problem. Later, Larché and Cahn [12,13] and Chu and Lee [14,15] studied the interaction between thermodynamic of diffusion and stress.

Several continuum models of diffusion induced stress were proposed in the last decade, each of them highlights some particular features. The models mainly divide in two groups: the coupled models consider the influence of the hydrostatic stress on the lithium ions diffusion, namely "pressure diffusion effect," and the uncoupled models neglect this effect.

Cheng and Verbrugge presented an uncoupled model for stress evolution in spherical particle in Reference [16] and they studied the effect of surface mechanics in nanometre particles, showing that tensile stress may be significantly reduced in magnitude or even be reverted to a state of compression with small particle radius [17].

Christensen and Newman were among the first to study DIS models applied to lithium ions cell [18–20] deriving a stand-alone electrode particle model. Meanwhile, Zhang et al. performed a numerical implementation of coupled DIS problem, and studied the influence of the aspect ratio of an ellipsoidal particle of lithium manganese oxide (LMO) on stress [21]. Later they studied the stresses which arise both from concentration gradient and heat generation in cathode particles, assumed as ellipsoidal with varying aspect ratio [22]. The basis of DIS theory applied to lithium-ion cells are clearly explained in Reference [23–25] for different geometry domain. Recently Bagheri et al. presented the numerical results of coupled DIS problem for galvanostatic and potentiostatic insertion in spherical

LMO particle [26]. Eshghinejad et al. presented a continuum model which couples diffusion and mechanics of ions intercalation for non-dilute solutions [27], according to Haftbaradaran et al. [28], and solved it with a finite elements (FE) model. Eshghinejad et al. highlighted that compressive stress results in lower lithium ions solubility and thus lower achievable capacity. Recently Wu et al. extended the study of diffusion induced stress to the interaction with other particles. Indeed, they developed a three-dimensional particle network model in a FE platform and a multiscale model which considers the stress in the particle as the superposition of the concentration gradient induced stress and the stress which comes from the interaction with other particles [29].

Tensile hoop stress due to insertion/extraction processes is the driving force for the analytical crack propagation model described by Deshpande et al. [30]. The authors described the capacity reduction rate due to electrolyte decomposition on the new free surfaces created by the cracks. Other works performed numerical analysis to predict crack propagation in active material with a stochastic approach [31] and considering the effect of current rate [32]. Grantab et al. developed a numerical method for studying lithiation-induced crack propagation in silicon nanowires that accounts for the effects of pressure-diffusion on the stress, and compared the results with an uncoupled model [33].

A continuum model of diffusion-mechanical problem in spherical geometry is presented in this paper. The model described in Section 2 gives a tool to estimate the stresses and strains in active material particles during lithiation/delithiation on the basis of particle size and mechanical properties with specific assumptions. Furthermore the model takes into account the hydrostatic stress influence on lithium diffusion, namely the "pressure diffusion effect", which couples mechanical and diffusive problem. The hydrostatic stress effect is highlighted comparing the results of the coupled and uncoupled model, which considers or neglects the stress influence, respectively. The definition of an equivalent diffusion coefficient which accounts for the hydrostatic stress effect allows to derive an analytical solution even for the coupled model, still not available in literature, as far as the authors know. The results of the analytical model are compared with numerical results in literature in Section 3. Furthermore, lithium concentration and stress are derived in galvanostatic insertion and extraction with different state of charge (SOC) for anode and cathode insertion material in Section 3. It is highlighted that tensile stresses are the driving force for crack propagation, and thus solid electrolyte interface (SEI) growth and capacity fade. On the other hand, compressive stresses induce a reduction of lithium flux which in turn affects the achievable capacity.

2. Problem Formulation

The mechanical stresses which arise in the active material particles are directly linked to lithium intercalation/de-intercalation phenomena which occur over charge/discharge cycling. For this reason, this model works just for intercalation materials, and not for conversion material, such as Silicon, since lithium ions would interact with host material differently. The particle is assumed spherical, isotropic and linear elastic, and the current density is assumed uniform all over the particle surface. These assumptions make the problem axisymmetric, leading to a one-dimensional problem in the radial coordinate. A couple of boundary conditions are given: traction-free condition is applied on the outer surface of the particle, thus neglecting the interaction with surroundings, and a fixed central point is imposed to prevent rigid body motion.

The lithium ions diffuse gradually in the particle during insertion and extraction. The ions diffusion makes the lithium concentration inhomogeneous along the particle radius, as shown in Figure 1b,c.

The concentration gradient causes a mechanical stress state in analogy to temperature gradient. Therefore, the areas with greater concentration gradient are affected by larger deformation compared to the areas where the gradient is lower. Therefore, a diffusion-elastic problem must be studied, since the stress state described by the elastic problem depends on the concentration level which is described by the diffusive problem.

The concentration level in insertion and extraction affects the sign of the stresses. Referring to insertion in Figure 1b, the greater concentration of the outer layers causes a tensile radial stress all over the particle because the surface expands more than the core. The hoop stress is compressive in the outer layers and tensile in the core because the greater deformation of the surface is prevented by the core. In extraction (Figure 1c) the particle shrinks, and the radial stress is compressive because the concentration level decreases along the radius. The hoop stress is tensile in the surface and compressive in the core because the greater expansion of the core is prevented by outer layers which are characterized by a lower concentration level.

2.1. Mechanical Problem

Strains are characterized by an elastic and a chemical contribution, which is expressed according to Prussin [11] in Equation (1).

$$\varepsilon_{ch} = \frac{\Omega}{3}C(r),\tag{1}$$

where Ω is the partial molar volume, which describes the volume variation of the solution host material and intercalated lithium. C is the lithium concentration field within the particle. The total strain is expressed as the sum of elastic strain and chemical strain [11], so the constitutive equations are expressed in spherical coordinates in Equations (2) and (3).

$$\varepsilon_r = \frac{1}{E}\left(\sigma_r - 2v\sigma_c\right) + \frac{\Omega C}{3},\tag{2}$$

$$\varepsilon_c = \frac{1}{E}\left[(1-v)\sigma_c - v\sigma_r\right] + \frac{\Omega C}{3},\tag{3}$$

where E and v are the Young modulus and Poisson ratio. The last term in Equations (2) and (3) couples the mechanical and chemical aspects. It is worth noting that this expression is written in analogy to an elastic-thermal problem, so the diffusivity part of the equation is totally equivalent to a thermal problem if the concentration field is replaced by the temperature and the partial molar volume is replaced by thermal expansion coefficient.

It is useful to rearrange the constitutive equations in terms of stresses for later purposes:

$$\sigma_r = \frac{E\left[\left(\varepsilon_r - \frac{\Omega C}{3}\right)(1-v) + 2v\left(\varepsilon_c - \frac{\Omega C}{3}\right)\right]}{(1+v)(1-2v)},\tag{4}$$

$$\sigma_c = \frac{E\left[\left(\varepsilon_c - \frac{\Omega C}{3}\right) + v\left(\varepsilon_r - \frac{\Omega C}{3}\right)\right]}{(1+v)(1-2v)}.\tag{5}$$

The congruency equations, which relate strain and displacement are:

$$\varepsilon_r = \frac{du}{dr},\tag{6}$$

$$\varepsilon_c = \frac{u}{r}.\tag{7}$$

The characteristic time of solid elastic deformation is much smaller than the diffusion of atoms in solids, for this reason the mechanical equilibrium is reached much faster than the diffusive one, and the elastic problem is treated as a quasi-static problem: the transient is not considered, and the equilibrium is assumed to be reached instantaneously.

The mechanical equilibrium equation over a spherical domain is given by Equation (8) according to References [16,17,26,34].

$$\frac{d\sigma_r}{dr} + \frac{2}{r}\left(\sigma_r - \sigma_c\right) = 0. \tag{8}$$

The mechanical stress state within a spherical particle is completely described by the set of Equations (4)–(8). The elastic problem is solved for the displacement replacing the congruency equations (Equations (6) and (7)) in the constitutive ones (Equations (4) and (5)), and the latter in the equilibrium equation (Equation (8)).

This leads to a second order differential equation (Equation (9)) depending on displacement and concentration field.

$$\frac{d^2u}{dr^2} + \frac{2}{r}\frac{du}{dr} - \frac{2u}{r^2} = \frac{1+v}{1-v}\frac{\Omega}{3}\frac{dC}{dr}. \tag{9}$$

Equation (9) is solved for the displacement integrating it twice. A first integration leads to:

$$\frac{du}{dr} + \frac{2u}{r} + \int_0^r \frac{2u}{r^2}\,dr - \int_0^r \frac{2u}{r^2}\,dr = \frac{1+v}{1-v}\frac{\Omega}{3}C(r) + C_1. \tag{10}$$

The displacement field is got in Equation (11) multiplying Equation (10) for r^2, integrating another time and rearranging the terms.

$$u(r) = \frac{1+v}{1-v}\frac{\Omega}{3}\frac{1}{r^2}\int_0^r C(r)r^2\,dr + \frac{C_1}{3}r + \frac{C_2}{r^2}. \tag{11}$$

The constants C_1 and C_2 are obtained imposing the boundary condition for $r = 0$ and $r = R$. The first boundary condition is null displacement at the center of the sphere which leads to $C_2 = 0$. This result is obtained solving $\lim_{r \to 0} u(r)$, since Equation (11) is not defined for $r = 0$.

The second boundary condition is derived from the behaviour on the surface. When free expansion is considered, the radial stress must vanish on the surface. This condition is got solving Equation (12) according to the constitutive equation (Equation (4)).

$$\left(\frac{du}{dr}\bigg|_{r=R} - \frac{1}{3}\Omega C(R)\right)(1-v) + 2v\left(\frac{u(R)}{R} - \frac{1}{3}\Omega C(R)\right) = 0. \tag{12}$$

The displacement and its derivative valued for $r = R$ are replaced in Equation (12). Then Equation (12) is solved for C_1, leading to:

$$C_1 = 2\frac{1-2v}{1-v}\frac{\Omega}{R^3}\int_0^R C(r)r^2dr. \tag{13}$$

At this stage all the boundary conditions are set, and the displacement field is expressed in Equation (14).

$$u(r) = \frac{\Omega}{3(1-v)}\left[(1+v)\frac{1}{r^2}\int_0^r C(r)r^2\,dr + 2(1-2v)\frac{r}{R^3}\int_0^R C(r)r^2\,dr\right]. \tag{14}$$

Therefore Equation (14) is replaced in the congruency equations (Equations (6) and (7)) and radial and hoop strains are obtained in Equation (15).

$$\begin{cases} \varepsilon_r(r) = \frac{1+v}{1-v}\frac{\Omega}{3}\left[-\frac{2}{r^3}\int_0^r C(r)r^2\,dr + C(r)\right] + \frac{2\Omega}{3}\frac{1-2v}{1-v}\frac{1}{R^3}\int_0^R C(r)r^2\,dr \\[2mm] \varepsilon_c(r) = \frac{1+v}{1-v}\frac{\Omega}{3}\frac{1}{r^3}\int_0^r C(r)r^2\,dr + \frac{2}{3}\Omega\frac{1-2v}{1-v}\frac{1}{R^3}\int_0^R C(r)r^2\,dr \end{cases}. \tag{15}$$

Then, the stresses are derived replacing the strain equations (Equation (15)) in the constitutive ones (Equations (4) and (5)).

$$\begin{cases} \sigma_r(r) = \frac{2\Omega}{3}\frac{E}{1-\nu}\left[\frac{1}{R^3}\int_0^R C(r)r^2\,dr - \frac{1}{r^3}\int_0^r C(r)r^2\,dr\right] \\ \sigma_c(r) = \frac{\Omega}{3}\frac{E}{1-\nu}\left[\frac{2}{R^3}\int_0^R C(r)r^2\,dr + \frac{1}{r^3}\int_0^r C(r)r^2\,dr - C(r)\right]. \end{cases} \tag{16}$$

Finally, the hydrostatic stress, defined according to Equation (17), is obtained. The value of hydrostatic stress is the coupling factor between elastic and diffusion in the diffusive problem.

$$\sigma_h(r) = \frac{\sigma_1 + \sigma_2 + \sigma_3}{3} = \frac{\sigma_r + 2\sigma_c}{3} = \frac{2\Omega E}{9(1-\nu)}\left[\frac{3}{R^3}\int_0^R C(r)r^2\,dr - C(r)\right]. \tag{17}$$

The results of the mechanical problem, namely the displacement, the strains and the stresses in Equations (14)–(17) depend on the concentration field, that is solved in the diffusion problem.

2.2. Diffusive Problem

The diffusive equation is derived with the thermodynamic approach described by Chu and Lee [15] and later adopted by most of the works concerning stress in intercalation materials [21–26]. Chemical potential gradient is the driving force for mass transport. The chemical potential of a solute in an ideal solution subjected to stress is written in Equation (18) according to Chu and Lee [15].

$$\mu = \mu_0 + R_g T ln(C) - \sigma_h \Omega, \tag{18}$$

where μ_0 is a constant, R_g the universal constant of gasses, C the concentration field, T the temperature, σ_h the hydrostatic stress experienced by the particle and Ω the partial molar volume. The expression of chemical potential in Equation (18) is formulated for dilute solution and does not take into account non ideality: namely the lithium migrates among interstitial sites and the structure of the host material is not modified by the intercalation process. The flux of lithium ions inside the particle due to the chemical potential is expressed in Equation (19) according to Reference [15].

$$J = -MC\frac{\partial \mu}{\partial r}, \tag{19}$$

where M is the mobility of the solute.

The mass conservation law of lithium ions in radial coordinate is introduced in Equation (20) [15].

$$\frac{\partial C}{\partial t} + \frac{1}{r^2}\frac{\partial (r^2 J)}{\partial r} = 0 \tag{20}$$

2.3. Uncoupled Problem

At first the uncoupled problem is solved, so the hydrostatic stress term in the chemical potential expression in Equation (18) is neglected, allowing to decouple the elastic problem from the diffusive one.

This simplified version of chemical potential is replaced in Equation (19). Therefore, the lithium ions flux is expressed as Equation (21).

$$J = -D\frac{\partial C}{\partial r} \tag{21}$$

where D is the diffusion coefficient, defined as $D = MR_g T$.

Finally Equation (21) is replaced in the mass conservation law (Equation (20)), and the diffusion equation in radial coordinate is derived in Equation (22).

$$\frac{\partial C}{\partial t} = \frac{D}{r^2}\frac{\partial}{\partial r}\left(r^2\frac{\partial C}{\partial r}\right) \tag{22}$$

Diffusion Equation (22) is written in perfect analogy with thermal diffusion, if concentration field is replaced by the temperature one.

Equation (22) is associated to a couple of boundary conditions which describe two possible cell operations: constant voltage (potentiostatic operation, Equation (23)) or constant current (galvanostatic operation, Equation (24)).

A constant lithium concentration equal to C_R is imposed at the cell boundary over time with potentiostatic operation. The diffusion Equation (22) has a singularity for $r = 0$, so the second boundary condition prescribes a finite concentration value at the centre of the sphere. Finally, a constant initial concentration value C_0 is prescribed for $t = 0$ across the domain.

$$\begin{cases} \frac{\partial C}{\partial t} = \frac{D}{r^2}\frac{\partial}{\partial r}\left(r^2\frac{\partial C}{\partial r}\right), & \\ C(r,0) = C_0, & \text{for } 0 \leq r \leq R \\ C(R,t) = C_R, & \text{for } t \geq 0 \\ C(0,t) = finite, & \text{for } t \geq 0 \end{cases} \tag{23}$$

A constant lithium flux, proportional to the current density, is applied on the external surface of the particle over time in galvanostatic operation. The lithium flux goes from the particle surface radially toward the centre, so the flux must be zero for $r = 0$. An initial concentration value C_0 is present all over the particle, as defined before.

$$\begin{cases} \frac{\partial C}{\partial t} = \frac{D}{r^2}\frac{\partial}{\partial r}\left(r^2\frac{\partial C}{\partial r}\right) & \\ C(r,0) = C_0, & \text{for } 0 \leq r \leq R \\ \left.\frac{\partial C(r,t)}{\partial r}\right|_{r=R} = \frac{I}{FD}, & \text{for } t \geq 0 \\ \left.\frac{\partial C(r,t)}{\partial r}\right|_{r=0} = 0, & \text{for } t \geq 0 \end{cases} \tag{24}$$

The solutions of the problem in Equations (23) and (24) are derived analytically by variable separation. However the solutions are well known in the literature [16] and are reported in Equation (25) for potentiostatic operation and in Equation (26) for galvanostatic operation.

$$\frac{C(r,t) - C_R}{C_0 - C_R} = -2\sum_{n=1}^{\infty}\frac{(-1)^{n+1}}{\pi n x}sin(n\pi x)\,e^{-n^2\pi^2\tau}, \tag{25}$$

$$C(r,t) = C_0 + \frac{IR}{FD}\left[3\tau + \frac{x^2}{2} - \frac{3}{10} - \frac{2}{x}\sum_{n=1}^{\infty}\left(\frac{sin(\lambda_n x)}{\lambda_n^2 sin(\lambda_n)}e^{-\lambda_n^2\tau}\right)\right]. \tag{26}$$

The solutions are normalized according to the dimensionless characteristic time expressed as $\tau = Dt/R^2$ and to the radial position $x = r/R$. λ_n are the positive roots of the transcendent equation $\lambda_n = tan(\lambda_n)$. C_R is the concentration value on the particle surface, C_0 is the initial concentration, F is the Faraday constant, I is the current density and R is the radius of the particle.

The concentration solutions in Equations (25) and (26) are replaced in Equations (14)–(17) to get displacement, strains and stress expressions for the uncoupled model.

2.4. Coupled Problem

The lithium ion flux for the coupled problem reported in Equation (27) is obtained by replacing the chemical potential (Equation (18)) in lithium flux (Equation (19)).

$$J = -D\left(\frac{\partial C}{\partial r} - \frac{\Omega C}{RT}\frac{\partial \sigma_h}{\partial r}\right).$$
(27)

Finally, the concentration equation for the coupled problem is derived in Equation (28) replacing Equation (27) in the mass conservation law (Equation (20)).

$$\frac{\partial C}{\partial t} = D\left[\frac{\partial^2 C}{\partial r^2} + \frac{2}{r}\frac{\partial C}{\partial r} - \frac{\Omega}{R_g T}\frac{\partial C}{\partial r}\frac{\partial \sigma_h}{\partial r} - \frac{\Omega C}{R_g T}\left(\frac{\partial^2 \sigma_h}{\partial r^2} + \frac{2}{r}\frac{\partial \sigma_h}{\partial r}\right)\right].$$
(28)

Equation (28) cannot be solved analytically as Equation (22), so the flux expression in Equation (27) is rearranged in order to obtain an expression similar to Equation (21) which brings back to a concentration equation analogue to Equation (22), whose analytical solution is already known.

The term $\frac{\partial \sigma_h}{\partial r}$ in Equation (27) is written according to the chain rule method as: $\frac{\partial \sigma_h}{\partial r} = \frac{\partial \sigma_h}{\partial C}\frac{\partial C}{\partial r}$. Then the hydrostatic stress in Equation (17) is differentiated with respect to the concentration: $\frac{\partial \sigma_h}{\partial C} = -\frac{2\Omega E}{9(1-\nu)}$. Finally, the lithium flux is expressed in Equation (29) factoring the term $\frac{\partial C}{\partial r}$.

$$J = -D\left(1 + \frac{2\Omega^2 EC}{9R_g T(1-\nu)}\right)\frac{\partial C}{\partial r}$$
(29)

Equation (29) is similar to the expression of lithium ions flux derived for the uncoupled problem (Equation (21)), but the diffusion coefficient is multiplied by a new factor. So, the equivalent diffusion coefficient for the coupled problem is introduced in Equation (30), in accordance with Chu and Lee [15].

$$D_{eqvn}(r) = D\left(1 + \frac{2\Omega^2 EC}{9R_g T(1-\nu)}\right) = D(1 + k \cdot C).$$
(30)

The equivalent diffusion coefficient is composed by the physical diffusion coefficient D and by an artificial contribution $k \cdot C$ due to the hydrostatic stress effect. This factor is always greater than one, both for material with positive and negative fraction molar volume. Thus, hydrostatic stress effect always enhances lithium diffusion decreasing the concentration gradient within the particle.

This result can be derived even with general boundary conditions on the surface, that is, a general form of C_1. The displacement in Equation (11) as a function of C_1 ($C_2 = 0$) is replaced in the expression of the radial and hoop stress, and the hydrostatic stress is calculated according to Equation (17). After some calculations it gives:

$$\sigma_h(r) = \frac{2\Omega E}{9(1-\nu)}\left[C(r) + C_1\frac{3(1-\nu)}{2(1-\nu)\Omega}\right]$$
(31)

Even in this condition the derivative of the hydrostatic stress with respect to the concentration has the same value computed before, leading to the same artificial diffusion contribution.

Lithium ion flux of the coupled problem in Equation (32) is formally similar to the flux of the uncoupled problem in Equation (21): namely it is equal to a coefficient multiplied for the derivative of the concentration with respect to the radius.

$$J = -D_{eqvn}\frac{\partial C}{\partial r}$$
(32)

This observation allows to derive the same diffusion equation of the uncoupled problem (Equation (22)) introducing the new diffusion coefficient D_{eqv} instead of the physical diffusion coefficient. Therefore, the same solutions derived for the uncoupled problem can be used also for the coupled one replacing the physical diffusion coefficient with the equivalent diffusivity.

The concentration solutions reported in Equations (25) and (26) become non-linear because the equivalent diffusion coefficient depends itself on concentration. Thus, an iterative calculation, described in Figure 2 must be performed as follows:

- The concentration field is calculated according to the physical diffusion coefficient.
- The equivalent diffusion coefficient is calculated for each radial position with the concentration function.
- A new concentration function is computed with the equivalent diffusion coefficient computed in the previous iteration.

The iterations go on until the maximum difference between the concentration computed in the kth iteration and $k - 1$th is below a certain threshold. This iterative computation converges in few iterations to a stable value.

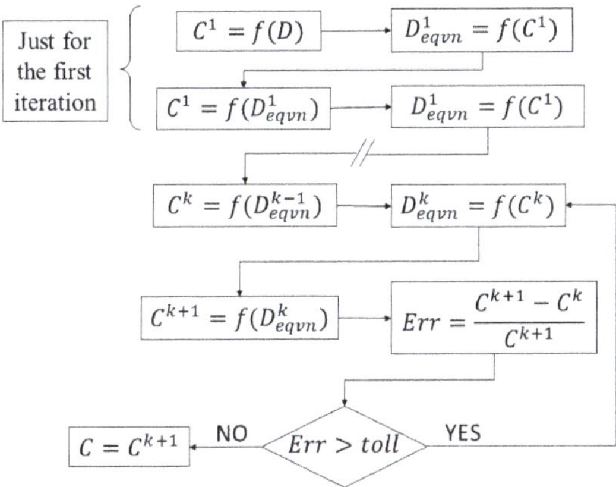

Figure 2. Iterative calculation procedure.

Once the concentration field is obtained, it is replaced in Equations (14)–(17) to get displacement, strains and stresses expressions for the coupled model.

3. Results and Discussion

In this section, the numerical results derived with the model explained in Section 2 are presented. The concentration functions for galvanostatic and potentiostatic insertion are compared with the results available in literature derived through numerical simulation. Then, concentration and stress function computed with the analytical model are presented in case of galvanostatic operation. The comparison between the coupled and uncoupled model highlights the hydrostatic stress influence on lithium diffusion. The results are derived according to the physical parameters reported in Tables 1 and 2, referring to anode and cathode materials.

Table 1. Graphite properties (anode).

Parameter	Symbol	Value	Unit	Reference
Diffusion coefficient	D	$2 \cdot 10^{-14}$	m^2/s	[35,36]
Partial molar volume	Ω	$3.42 \cdot 10^{-6}$	m^3/mol	[37]
Maximum concentration	C_{max}	$3.18 \cdot 10^4$	mol/m^3	[37]
Young modulus	E	15	GPa	[18]
Poisson ratio	v	0.3	-	[18]
Particle radius	R	$5 \cdot 10^{-6}$	m	[18]

Table 2. $Li_xMn_2O_4$ (LMO) properties (cathode).

Parameter	Symbol	Value	Unit	Reference
Diffusion coefficient	D	$7.08 \cdot 10^{-15}$	m^2/s	[21,26]
Partial molar volume	Ω	$3.497 \cdot 10^{-6}$	m^3/mol	[21,26]
Maximum concentration	C_{max}	$2.29 \cdot 10^4$	mol/m^3	[21,26]
Young modulus	E	10	GPa	[21,26]
Poisson ratio	v	0.3	-	[21,26]
Particle radius	R	$5 \cdot 10^{-6}$	m	[21,26]

3.1. Compatibility between Model Assumptions and Real Material

Graphite and lithium manganese oxide (LMO) are chosen as case study for anode and cathode intercalation materials respectively. The compatibility between the assumptions of the analytical model and real active materials is discussed in this section. The linear elastic assumption is respected for all the insertion materials, because they show slight deformation. There are two aspects to discuss: geometry and homogeneity assumptions.

About the geometry assumption, it is necessary to understand how a sphere is capable to represent the random geometry of the active material particles. For some materials, such as NMC or graphite the real particle geometry is close to a sphere, as showed by SEM images in Reference [37–39], but for other materials, such as LMO or LCO, particles shape is more elongated and irregular. However, a modelling approach must overcome the statistical variation of the samples and give reasonable and general results which can be valid for the entire samples population. For this reason, an ideal geometry which can be extended and representative of all the other particles is adopted in this work, according the following reasoning. Simulations confirm that greater particle size results in higher stress, because of longer diffusion path. Hence, the radius is chosen so that the ideal spherical particle adopted in simulation circumscribes the mean real particle, once the statistical distribution of the particle size of a powder is known. In this manner, it is possible to give a safe estimation of the stress in the powder particles based on their size, overcoming the limit of the random distribution of the particles shapes. The results of the simulation at least overestimate the real stress in the particles of a powder, since a bigger spherical equivalent particle is used to simulate all the different particles shapes present in a powder.

For what concerns different ideal geometries, Zhang et al. studied the influence of the aspect ratio of an ellipsoidal LMO particle on the stress [21,22]. The results show that the stresses computed with aspect ratios different from one (one corresponds to the sphere) are ±10% of the stress computed with spherical geometry. Hence, spherical geometry minimizes the error due to the statistical distribution of the particle shape [21].

Some works analysed the stress over a realistic particle geometry extracted from SEM images [40–43]. The results show that the main difference between realistic particle and ideal spherical particle resides in the stress concentrations in notches. In future works, it is meaningful to study a generalized notch factor, based on the statistical analysis of powder samples, which can be representative of the notching effect of different active material particles. Indeed, stress concentration due to notching effect causes cracks propagation: the main reason for electrode damage and capacity fade.

Moreover, a recent work [44] demonstrated how the assumptions of spherical geometry and isotropic and linear elastic material are accurate for LCO and graphite, since they found a good agreement between the macroscopic deformation predicted by their numerical model and the experimental measurements conducted on pouch cell. Finally, spherical assumption is widely accepted among several works [20–22,26,45].

The second aspect concerns the homogeneity assumption. This aspect can be still divided in two issues: phase transition during Li insertion and inhomogeneity of primary particles. Phase transition occurs in some materials when lithium content exceeds the equilibrium concentration. Lithium ions begin to intercalate in different type of interstices above the equilibrium concentration, modifying the crystallographic structure: this leads to a stress and concentration jump between the two phases because they are characterized by different physical parameters. Christensen et al. pointed out that LMO shows a single phase in the 4V plateau, and the transformation in cubic-tetragonal phase occurs with deep discharge [20]. In LCO, which has a negative partial molar volume, Li poor phase has a larger partial molar volume than the lithium rich phase [46]. Therefore, as the second phase starts to form during the discharge process, the magnitude of stress starts to decrease—phase change leads to a safer condition in this case.

However, a simple general approach which considers phase transition in our model is the following: it is necessary to change the physical parameters according to the concentration level with a moving boundary concept. When the concentration level exceeds the equilibrium concentration of a phase in a certain radial position, equilibrium concentration and physical parameters (Young modulus, Poisson ratio and especially partial molar volume) must be changed with the physical parameters of the new phase. During insertion the "boundary" between the two phases moves towards the core, and the physical parameters must be changed accordingly. So, phase transition can be implemented in the model in Section 2 tuning the input parameters, but the framework of the model remains unchanged.

The particles of some materials, such as NMC, are synthesized from primary particles [39]. However, it is hard to model the influence of randomly distributed primary particles which compose secondary particle, and because of the random orientation of primary particles, secondary particles are usually assumed as a continuum material [47,48]. However, Wu et al. [49] tried to consider the influence of primary particle on the stress.

3.2. Comparison with the Results of Numerical Models in Literature

The analytical solution of DIS problem was proposed by Cheng et al. [16] according the uncoupled formulation. As far as the authors know, analytical solution of the coupled problem are still not available in literature. Generally, the finite difference method [18,20,21,23,24] or FE simulations [16,21,27,33] in commercial codes were adopted in order to solve the DIS problem in the coupled formulation. Following the procedure explained in Section 2, the analytical solutions of the standard diffusion problem are used to get the solution of the DIS coupled problem via an iterative computation, which is a much simpler way compared to solve the whole coupled equation (Equation (28)) with finite difference method or building up a FE analysis.

The concentration function computed with the analytical model in galvanostatic and potentiostatic insertion are compared with the results derived via finite different method by Bagheri et al. [26] in Figure 3. A good agreement is achieved between the analytical model proposed in this work and Reference [26] derived via numerical computation. Furthermore the concentration trend with and without "stress effect" matches with the results of Zhang et al. [21] and Christensen et al. [19]. Moreover the results of the uncoupled model matches with the model derived by Cheng et al. [16]. Even the stress functions fits faithfully the numerical results available in the literature, since they depend uniquely on concentration.

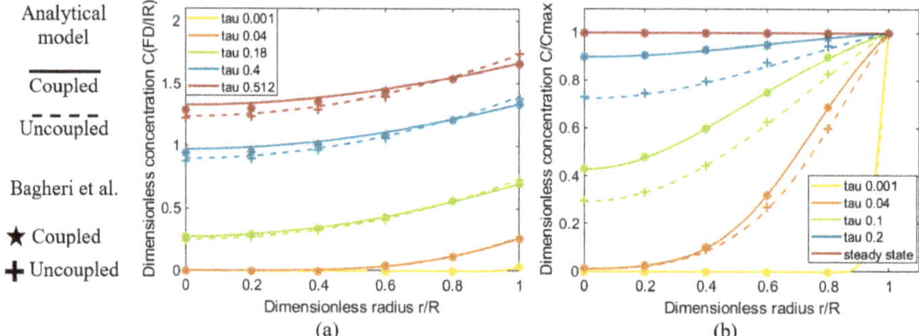

Figure 3. Concentration level as a function of dimensionless radial position for different time constants τ in galvanostatic (**a**) and potentiostatic (**b**) insertion. The analytical model is reported with solid (coupled) and dashed (uncoupled) lines. The solution of the numerical model [26] is reported in discrete radial coordinates with dots (coupled) and crosses (uncoupled).

The lack of stress measurements in insertion material particles makes the results in this work impossible to be validated experimentally, as pointed out even by other authors [26], and mathematical modelling is the only way to make these computations.

3.3. Insertion under Galvanostatic Control

Concentration level and the stress functions in galvanostatic insertion are computed with the analytical model proposed in Section 2 assuming null initial concentration. The results are derived for three different SOC levels: 25%, 50%, 75%. The SOC level is calculated according to Zhang [26,50] as:

$$SOC = \frac{\int_0^R c(r)r^2 dr}{\int_0^R c_{max} r^2 dr} \cdot 100. \tag{33}$$

The current density over the particle surface is 3 A/m^2, which corresponds roughly to 2C. The temperature is assumed to be constant at 298 K. The results for anode and cathode particles whose physical parameters are listed in Tables 1 and 2 are reported in Figures 4 and 5.

The concentration level normalized with the maximum concentration value which can be stored in the active material particles is reported in Figure 4a. So, when the normalized concentration is equal to the unity all the available sites in the particle are occupied by lithium ions. Radial, hoop and Von Mises stress are reported in Figure 4b–d for graphite particles. The same data are reported in Figure 5 for LMO particles.

The pressure diffusion dependence determines a greater equivalent diffusion coefficient which allows a faster lithium diffusion within the particle. In particular enhances mass transport from areas subjected to compression to areas subjected to tensile stress. This fact makes the lithium concentration gradient predicted by the coupled model lower if compared to the uncoupled model, as shown in Figures 4a and 5a. Remembering the thermal analogy, lower gradient means lower stress, so the coupled model predicts a lower stress state, as highlighted in Figures 4b–d and 5b–d; the differences between the two model are up to 40%. The differences between the stresses computed with the three SOC levels are small (the maximum is 15% for graphite and 7% for LMO), this suggests that the active material particles experience similar stresses during almost the whole SOC range.

The comparison between the results obtained for anode (Figure 4) and cathode (Figure 5) shows the influence of the diffusion coefficient. Indeed, the slight difference in the diffusion coefficient between LMO and graphite produces an important increase of the concentration gradient in the cathode particles, which in turn causes a serious increase of the stress. Consequently, a greater equivalent stress makes the coupling between mechanical and diffusion aspect stronger.

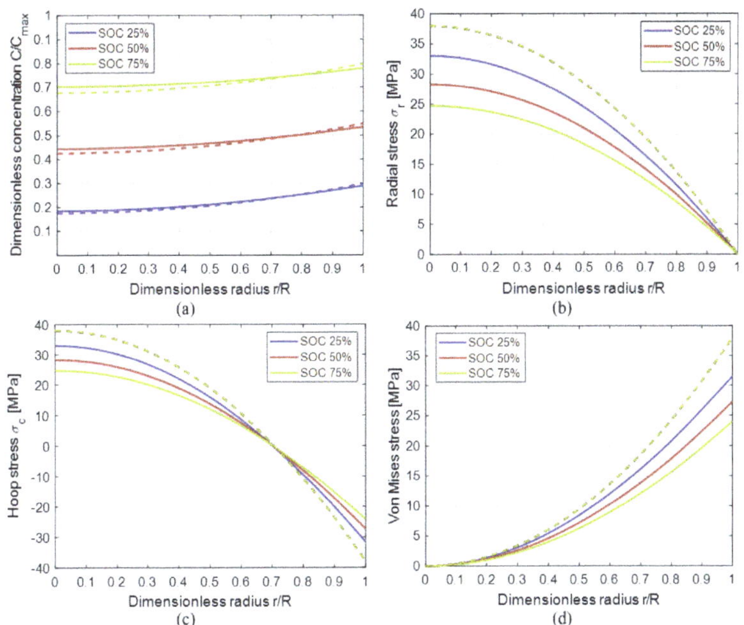

Figure 4. Lithium concentration (**a**), radial stress (**b**), hoop stress (**c**) and Von Mises stress (**d**) for different SOC levels in galvanostatic insertion in anode material. Dashed lines refer to the uncoupled model and solid lines refer to the coupled model.

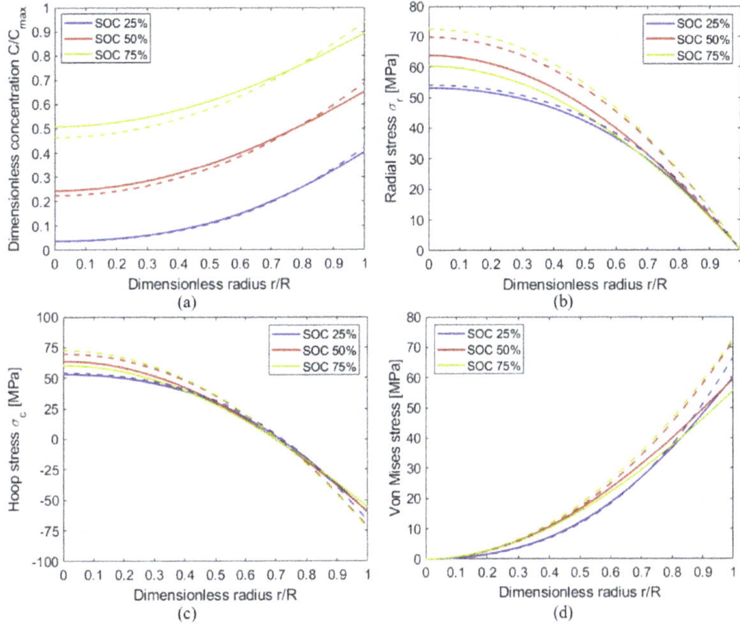

Figure 5. Lithium concentration (**a**), radial stress (**b**), hoop stress (**c**) and Von Mises stress (**d**) for different SOC levels in galvanostatic insertion in cathode material. Dashed lines refer to the uncoupled model and solid lines refer to the coupled model

This results in a greater difference between the concentration computed with the coupled model or the uncoupled one in LMO because the mechanical component in the chemical potential (Equation (18)) becomes greater.

3.4. Extraction under Galvanostatic Control

The galvanostatic extraction is modelled assuming an initial concentration within the particle equal to C_{max}, namely SOC 100%, and surface current density equal to -3 A/m^2. The extraction determines a reduction of the SOC level, and the result for SOC equal to 75%, 50% and 25% are reported in Figures 6 and 7 for anode and cathode material, respectively. Concentration is reported in Figure 6a, radial, hoop and Von Mises stress are reported in Figure 6b–d for graphite. The same data are reported in Figure 7 for LMO.

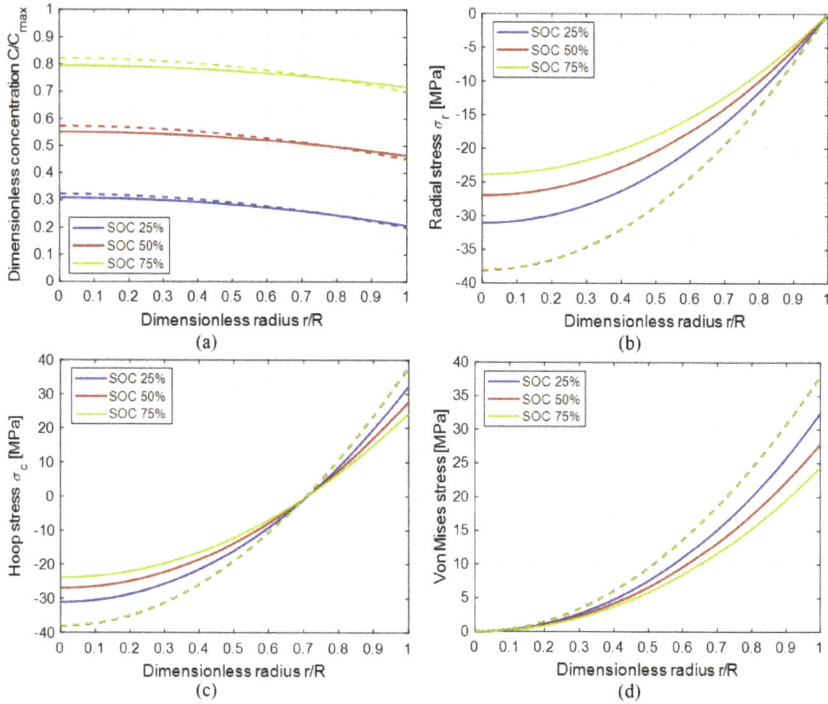

Figure 6. Lithium concentration (**a**), radial stress (**b**), hoop stress (**c**) and Von Mises stress (**d**) for different SOC levels in galvanostatic extraction in anode material. Dashed lines refer to the uncoupled model and solid lines refer to the coupled model.

The same conclusions made for insertion about the differences of concentration and stress state between the coupled and uncoupled model can be made also for extraction. The differences between the stress computed with the coupled and uncoupled model are up to 35%. In Figures 6c and 7c it is highlighted that the particle experiences tensile hoop stress on its surface during galvanostatic extraction, which is supposed to be the driving force for crack propagation in active material [30].

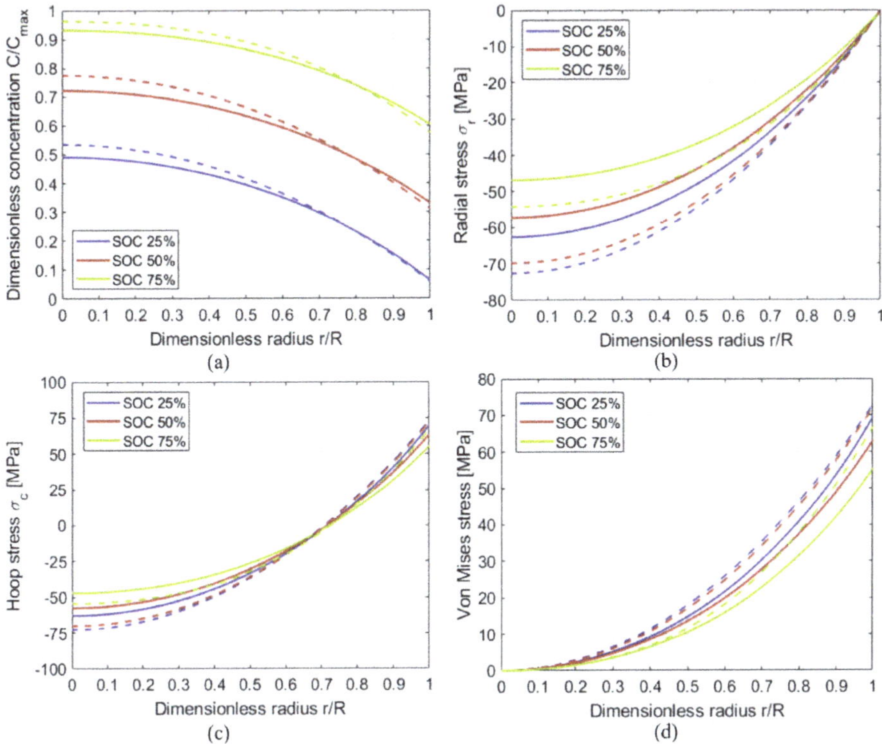

Figure 7. Lithium concentration (**a**), radial stress (**b**), hoop stress (**c**) and Von Mises stress (**d**) for different SOC levels in galvanostatic extraction in cathode cathode. Dashed lines refer to the uncoupled model and solid lines refer to the coupled model.

3.5. Evolution of Von Mises Stress in Time

The continuous evolution in time of the Von Mises stress is shown in Figure 8. The hydrostatic stress effect homogenizes the lithium concentration within the particle, so the stress values computed by the coupled model are lower than the uncoupled one, in accordance with Reference [21]. On the contrary, for high extraction time the stress predicted by the coupled model tends to the uncoupled one.

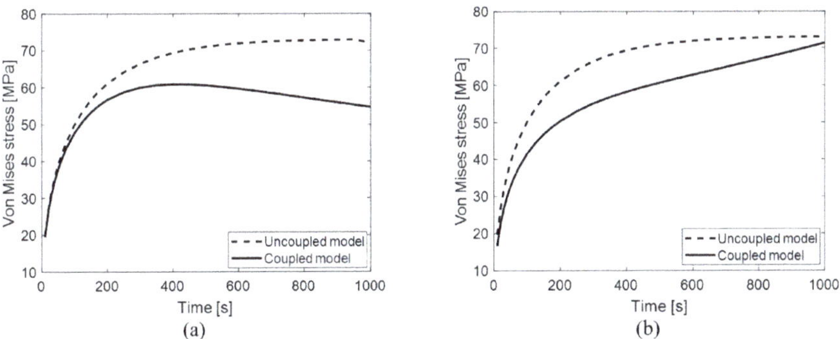

Figure 8. Von Mises stress as a function of insertion (**a**) and extraction (**b**) time. The results are derived with physical parameters of Table 2.

3.6. Influence of Hydrostatic Stress on Concentration

Hydrostatic stress influence always enhances lithium diffusion. This assertion is justified from a quantitative point of view by the artificial contribution to the equivalent diffusion coefficient in Equation (30). This contribution is always positive regardless of the insertion or extraction operation, time constant or radial position, so the coupled model is always characterized by a greater diffusion coefficient which decreases the concentration gradient within the particle, as showed in Figures 4a–7a.

This concept is qualitatively explained in Figure 9. Indeed, the chemical potential in Equation (18) can be split in the concentration contribution ($\mu_C = R_g T \ln(C)$) and the hydrostatic stress contribution ($\mu_\sigma = -\Omega \sigma_h$) as follow: $\mu = \mu_C + \mu_\sigma$. In case of insertion, referring to Figure 9a, the concentration C_2 at $r + dr$ is greater than the concentration C_1 at $r - dr$ for a generic radial coordinate r, resulting in a positive concentration contribution. In the same way the hydrostatic stress $\sigma_{h,2}$ at $r + dr$ is lower than the hydrostatic stress $\sigma_{h,1}$ at $r - dr$, resulting in a positive hydrostatic stress contribution. This analysis can be extended to all the radial coordinates from zero to R because both concentration and hydrostatic stress are monotonic. Therefore, both the contributions are concordant due to the shape of concentration and hydrostatic stress functions showed in Figure 9a, and concur to the incoming flux. This analysis is also valid for extraction, as explained in Figure 9b. In this case, concentration, hydrostatic stress and thus chemical potential contributions are the opposite, resulting in an outgoing flux.

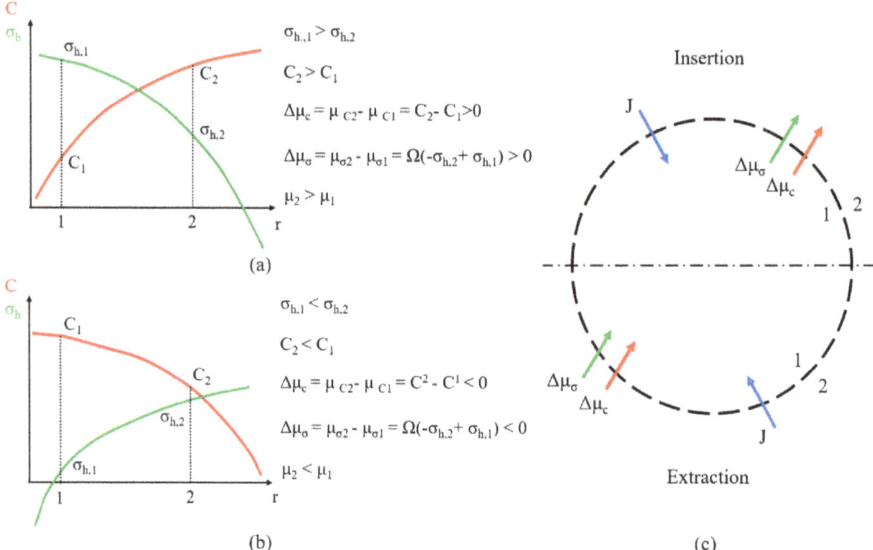

Figure 9. Hydrostatic stress influence on lithium diffusion for insertion and extraction operation. The qualitative trend of the lithium concentration (red) and the hydrostatic stress (green) for insertion (**a**) and extraction (**b**) are reported in the graphics. The chemical potential is split in concentration contribution and hydrostatic stress contribution whose increments are valued for a general radial position r. The stress and concentration contributions for a general radial coordinate are graphically reported with green and red arrows which goes from the lower to the higher value (**c**). The lithium flux due to the chemical potential difference is reported with the blue arrows (**c**).

On the other hand, hydrostatic stress influences the concentration level in the particle. Tensile hydrostatic stress allows to store a greater amount of lithium ions, on the contrary compressive hydrostatic stress determines a reduction of the storable lithium ions. These differences are highlighted comparing the results with the uncoupled model, which neglects the influence of the hydrostatic stress on the lithium concentration. Referring to Figure 10a, black arrows mark the border between positive and negative hydrostatic stress. This trend matches with the differences in concentration function computed with the two models in Figure 10b: where the hydrostatic stress is positive the concentration level of the coupled model is higher than the uncoupled one, where the hydrostatic stress is negative the coupled model predicts a lower concentration level compared to the uncoupled model. About this issue it is worth noting that the model in this work assumes free expansion of the particle surface, neglecting the interaction with its surroundings.

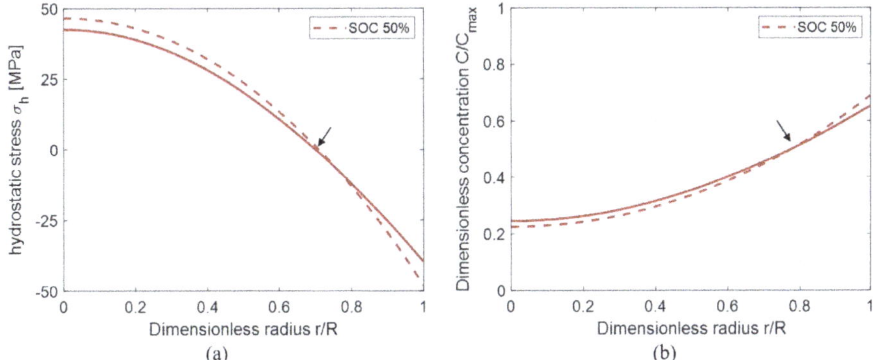

Figure 10. Hydrostatic stress (**a**) and concentration (**b**) in cathode material. Black arrows show the points where the hydrostatic stress changes sign which partially reflect the turnaround of the concentration function. Dashed lines refer to the uncoupled model and solid lines refer to the coupled model.

A more accurate model should consider the surface constraints of the particle which are supposed to generate an increase in the compressive stress and in turn lower achievable concentration values. Future works should be carried out in order to confirm this hypothesis.

4. Conclusions

The stress state within active material particle of graphitic anode and LMO cathode are computed with coupled and uncoupled model according to DIS theory, assuming no constraints on the external surface. The analytical solution of the coupled model is proposed in this work defining an equivalent diffusion coefficient composed by a physical term and by an artificial contribution connected to the hydrostatic stress. This definition allows to exploit the analytical solutions of the uncoupled model even for the coupled one with an iterative calculation, since the equivalent diffusion coefficient depends itself on concentration.

The results derived with the analytical solution of the coupled model are compared with the solutions derived with numerical methods in literature and show a good agreement. The analytical solution proposed in this work is easier and requires a lower computing time if compared to strongly non-linear FE method or finite difference method. The concentration function and the stress state within the particle are computed for three SOC levels: 25%, 50%, 75%. The differences between the stresses computed with the different SOC levels are small: this fact suggests that particles experience almost the same stress state during about the whole SOC window. The differences between the stress state in LMO and graphite is mainly due to diffusion: a smaller diffusion coefficient causes higher lithium concentration gradient, and higher stress consequently (up to 40%), according to thermal analogy concept. Thus, the coupling factor $\Omega\sigma_h$ in chemical potential becomes higher and the differences between coupled and uncoupled model are not negligible.

Finally, it is pointed out that tensile stress, in particular tensile hoop stress which occurs on the particle surface during extraction, is the driving force for crack propagation, which in turn damages the active material and accelerates the SEI growth. On the other hand, compressive hydrostatic stress influences the lithium solubility decreasing the achievable capacity, namely the lithium ions which can be stored in the host material. An increase of compressive stress is expected if surface constraints are considered, because the particle expansion is prevented. Future works should analyse this issue which can result in a not negligible achievable capacity reduction.

Author Contributions: Conceptualization, D.C., F.M. and A.S.; methodology, D.C., F.M. and A.S.; formal analysis, D.C.; investigation, D.C., F.M.; Writing—Original draft preparation, D.C.; Writing—review and editing, F.M. and A.S.; supervision, F.M. and A.S. All authors have read and agreed to the published version of the manuscript.

Funding: This research received no external funding.

Conflicts of Interest: The authors declare no conflict of interest.

Abbreviations

The following abbreviations are used in this manuscript:

C	Concentration	mol/m^3
C_0	Initial concentration	mol/m^3
C_{max}	Maximum concentration	mol/m^3
C_r	Surface concentration	mol/m^3
D	Diffusion coefficient	m^2/s
D_{eqv}	Equivalent diffusion coefficient	m^2/s
E	Young Modulus	MPa
F	Faraday constant	96485.332 As/mol
I	Current density	A/m^2
J	Lithium flux	mol/m^2s
M	Mobility	mol· s/Kg
r	Radius	m
R	Particle radius	m
R_g	Gas constant	8.3145 J/mol K
SOC	State of charge	-
T	Temperature 298	K
u	Displacement	m
x	Normalized radial coordinate	-
ϵ_c	Hoop strain	-
ϵ_{ch}	Chemical strain	-
ϵ_r	Radial strain	-
μ	Chemical potential	J/mol
μ_0	Reference chemical potential	J/mol
v	Poisson ratio	-
σ_c	Hoop stress	MPa
σ_h	Hydrostatic stress	MPa
σ_r	Radial stress	MPa
τ	Characteristic time	-
Ω	Partial molar volume	m^3/mol

References

1. Miao, Y. Current li-ion battery technologies in electric vehicles and opportunities for advancements. *Energies* **2019**, *12*, 1074. [CrossRef]
2. Soma, A. Hybridization Factor and Performance of Hybrid Electric Telehandler Vehicle. *IEEE Trans. Ind. Appl.* **2016**, *52*, 5130–5138. [CrossRef]
3. Mocera, F. Battery performance analysis for working vehicles applications. *IEEE Trans. Ind. Appl.* **2019**, *56*, 644–653. [CrossRef]
4. Vergori, E. Battery modelling and simulation using a programmable testing equipment. *Computers* **2018**, *7*, 20. [CrossRef]

5. Kleiner, J. Thermal modelling of a prismatic lithium-ion cell in a battery electric vehicle environment: Influences of the experimental validation setup. *Energies* **2019**, *13*, 62. [CrossRef]

6. Gantenbein, S. Capacity fade in lithium-ion batteries and cyclic aging over various state-of-charge ranges. *Sustainability* **2019**, *11*, 6697. [CrossRef]

7. Lee, D. Modeling the Effect of the Loss of Cyclable Lithium on the Performance Degradation of a Lithium-Ion Battery. *Energies* **2019**, *12*, 4386. [CrossRef]

8. Willenberg, L.K. High-Precision Monitoring of Volume Change of Commercial Lithium-Ion Batteries by Using Strain Gauges. *Sustainability* **2020**, *12*, 557. [CrossRef]

9. Qi, Z.; Shan, Z.; Ma, W.; Li, L.; Wang, S.; Li, C.; Wang, Z. Strain Analysis on Electrochemical Failures of Nanoscale Silicon Electrode Based on Three-Dimensional in Situ Measurement. *Appl. Sci.* **2020**, *10*, 468. [CrossRef]

10. Cheng, X. In situ stress measurement techniques on li-ion battery electrodes: A review. *Energies* **2017**, *10*, 591. [CrossRef]

11. Prussin, S. Generation and distribution of dislocations by solute diffusion. *J. Appl. Phys.* **1961**, *32*, 1876–1881. [CrossRef]

12. Larchè, J. A linear theory of thermochemical equilibrium of solids under stress. *Acta Metall.* **1973**, *8*, 1051–1063. [CrossRef]

13. Larchè, J. The effect of self-stress on diffusion in solids. *Acta Metall.* **1982**, *30*, 1835–1845. [CrossRef]

14. Chu, J.L. Chemical stresses in composite circular cylinders. *J. Appl. Phys.* **1993**, *73*, 2239–2248. [CrossRef]

15. Chu, J.L. The effect of chemical stresses on diffusion. *J. Appl. Phys.* **1994**, *75*, 2823–2829. [CrossRef]

16. Cheng, Y.T. Evolution of stress within a spherical insertion electrode particle under potentiostatic and galvanostatic operation. *J. Power Sources* **2009**, *190*, 453–460. [CrossRef]

17. Cheng, Y.T. The influence of surface mechanics on diffusion induced stresses within spherical nanoparticles. *J. Appl. Phys.* **2008**, *104*, 083521. [CrossRef]

18. Christensen, J. Modeling diffusion-induced stress in Li-ion cells with porous electrodes. *J. Electrochem. Soc.* **2010**, *157*, 336–386. [CrossRef]

19. Christensen, J. Stress generation and fracture in lithium insertion materials. *J. Solid State Electrochem.* **2006**, *10*, 293–319. [CrossRef]

20. Christensen, J. A mathematical model of stress generation and fracture in lithium manganese oxide. *J. Electrochem. Soc.* **2006**, *153*, 1019–1030. [CrossRef]

21. Zhang, X. Numerical simulation of intercalation-induced stress in Li-Ion battery electrode Particles. *J. Electrochem. Soc.* **2007**, *154*, 910–916. [CrossRef]

22. Zhang, X. Intercalation-induced stress and heat generation within single lithium-ion battery cathode particles. *J. Electrochem. Soc.* **2008**, *155*, 542–555. [CrossRef]

23. Wang, W.L. Effect of chemical stress on diffusion in a hollow cylinder. *J. Appl. Phys.* **2002**, *91*, 9584–9590. [CrossRef]

24. Wang, W.L. Interaction between diffusion and chemical stresses. *Mater. Sci. Eng. A* **2005**, *409*, 153–159.

25. Zhao, Y. A review on modeling of electro-chemo-mechanics in lithium-ion batteries. *J. Power Sources* **2018**, *413*, 259–283. [CrossRef]

26. Bagheri, A. On the effects of hydrostatic stress on Li diffusion kinetics and stresses in spherical active particles of Li-ion battery electrodes. *J. Power Sources* **2019**, *137*, 103–134. [CrossRef]

27. Eshghinejad, A. The coupled lithium ion diffusion and stress in battery electrodes. *Mech. Mater.* **2015**, *91*, 343–350. [CrossRef]

28. Haftbaradaran, H. Continuum and atomistic models of strongly coupled diffusion, stress, and solute concentration. *J. Power Sources* **2011**, *196*, 361–370. [CrossRef]

29. Wu, B. A consistently coupled multiscale mechanical–electrochemical battery model with particle interaction and its validation. *J. Mech. Phys. Solids* **2019**, *125*, 89–111. [CrossRef]

30. Deshpande, R. Battery cycle life prediction with coupled chemical degradation and fatigue mechanics. *J. Electrochem. Soc.* **2012**, *159*, 1730–1738. [CrossRef]

31. Kotak, N. Electrochemistry-mechanics coupling in intercalation electrodes. *J. Electrochem. Soc.* **2018**, *165*, 1064–1083. [CrossRef]

32. Zhang, Y. Simulation of crack behavior of secondary particles in Li-ion battery electrodes during lithiation/de-lithiation cycles. *J. Mech. Sci.* **2018**, *155*, 178–186. [CrossRef]

33. Grantab, R. Pressure-gradient dependent diffusion and crack propagation in lithiated silicon nanowires. *J. Electrochem. Soc.* **2012**, *159*, 584–591. [CrossRef]
34. Verbrugge, M.W. Stress and strain-energy distributions within diffusion-controlled insertion-electrode particles subjected to periodic potential excitations. *J. Electrochem. Soc.* **2009**, *156*, 927–937. [CrossRef]
35. Tang, Z.Y. Determination of the Lithium Ion Diffusion Coefficient in Graphite Anode Material. *Acta Phys. Chim. Sin.* **2001**, *17*, 388.
36. Jun, G.H. Diffusion coefficient of lithium in artificial graphite, mesocarbon microbeads, and disordered carbon. *New Carbon Mater.* **2007**, *22*, 7–10.
37. Barai, P. Stochastic analysis of diffusion induced damage in lithium-ion battery electrodes. *J. Electrochem. Soc.* **2013**, *160*, 955–967. [CrossRef]
38. Korsunsky, A.M. Explicit formulae for the internal stress in spherical particles of active material within lithium ion battery cathodes during charging and discharging. *Mater. Des.* **2015**, *69*, 247–252. [CrossRef]
39. Pimenta, V. Synthesis of Li-Rich NMC: A Comprehensive Study. *Mater. Des.* **2017**, *23*, 9923–9936. [CrossRef]
40. Mendoza, H. Mechanical and Electrochemical Response of a $LiCoO_2$ Cathode using Reconstructed Microstructures. *Electrochim. Acta* **2016**, *190*, 1–15. [CrossRef]
41. Malavé, V. A computational model of the mechanical behavior within reconstructed Li_xCoO_2 Li-ion battery cathode particles. *Electrochim. Acta* **2014**, *130*, 707–717. [CrossRef]
42. Korsunsky, A.M. Generation of Realistic Particle Structures and Simulations of Internal Stress: A Numerical/AFM Study of $LiMn_2O_4$ Particles. *J. Electrochem. Soc.* **2011**, *158*, 434–442.
43. Chung, M. Implementing Realistic Geometry and Measured Diffusion Coefficients into Single Particle Electrode Modeling Based on Experiments with Single $LiMn_2O_4$ Spinel Particles. *J. Electrochem. Soc.* **2011**, *158*, 371–378. [CrossRef]
44. Ai, W. Electrochemical Thermal-Mechanical Modelling of Stress Inhomogeneity in Lithium-Ion Pouch Cells. *J. Electrochem. Soc.* **2011**, *167*, 013512. [CrossRef]
45. Huang, S. Stress generation during lithiation of high-capacity electrode particles in lithium ion batteries. *Acta Mater.* **2013**, *61*, 4354–4364. [CrossRef]
46. Renganathan, S. Theoretical Analysis of Stresses in a Lithium Ion Cell. *J. Electrochem. Soc.* **2010**, *157*, 155–163. [CrossRef]
47. Xu, R. Computational analysis of chemomechanical behaviors of composite electrodes in Li-ion batteries. *Mater. Res.* **2016**, *31*, 2715–2727. [CrossRef]
48. Goldin, G. Three-dimensional particle-resolved models of Li-ion batteries to assist the evaluation of empirical parameters in one-dimensional models. *Mater. Res.* **2012**, *64*, 118–129. [CrossRef]
49. Wu, B. Mechanical-electrochemical modeling of agglomerate particles in lithium-ion battery electrodes. *Mater. Res.* **2016**, *163*, 3131–3139. [CrossRef]
50. Zhang, T. Effect of reversible electrochemical reaction on Li diffusion and stresses in cylindrical Li-ion battery electrodes. *J. Appl. Phys.* **2014**, *115*, 083504. [CrossRef]

Article

Butyronitrile-Based Electrolytes for Fast Charging of Lithium-Ion Batteries

Peter Hilbig [1], Lukas Ibing [1], Martin Winter [1,2] and Isidora Cekic-Laskovic [2,*]

[1] MEET Battery Research Center/Institute of Physical Chemistry, University of Münster, Corrensstrasse 46,
    48149 Münster, Germany
[2] Helmholtz-Institute Münster, IEK-12, Forschungszentrum Jülich GmbH, Corrensstrasse 46,
    48149 Münster, Germany
* Correspondence: i.cekic-laskovic@fz-juelich.de

Received: 19 June 2019; Accepted: 18 July 2019; Published: 25 July 2019

Abstract: After determining the optimum composition of the butyronitrile: ethylene carbonate: fluoroethylene carbonate (BN:EC:FEC) solvent/co-solvent/additive mixture, the resulting electrolyte formulation (1M $LiPF_6$ in BN:EC (9:1) + 3% FEC) was evaluated in terms of ionic conductivity and the electrochemical stability window, as well as galvanostatic cycling performance in NMC/graphite cells. This cell chemistry results in remarkable fast charging, required, for instance, for automotive applications. In addition, a good long-term cycling behavior lasts for 1000 charge/discharge cycles and improved ionic conductivity compared to the benchmark counterpart was achieved. XPS sputter depth profiling analysis proved the beneficial behavior of the tuned BN-based electrolyte on the graphite surface, by confirming the formation of an effective solid electrolyte interphase (SEI).

Keywords: lithium-ion batteries; non-aqueous electrolyte; nitrile-based solvents; butyronitrile; SEI forming additives; fast charging

---

1. Introduction

Thanks to their excellent performance characteristics, lithium ion battery (LIB) cells find application in a broad spectrum of different fields, comprising the consumer and automotive industries as well as application in small portable devices, like mobile phones or laptops [1–4]. The main reason behind the broad field of application relates, among other reasons, to the high specific energy and energy density of LIBs [5–7] and the numerous cell materials, that can be employed [8].

In standard LIBs, organic carbonate-based non aqueous aprotic electrolytes are employed. Although given as state of the art electrolytes, they display several disadvantages (e.g., moderate ionic conductivity and low flash points) [9–12]. To further advance state of the art battery electrolytes, many solvent classes were comprehensively investigated to replace organic carbonates [9,13–24]. Nitriles and other cyano-compounds display high ionic conductivity as well as low temperature cycling performance [25–34]. Nevertheless, many examples of this class of compounds are known for being incompatible with metallic lithium and not able to form an effective solid electrolyte interphase (SEI) on graphite [35–40]. For this reason, the presence of SEI forming electrolyte additive(s) is inevitably required to enable their application in graphite based LIBs [41].

In the case of organic carbonate-based electrolytes, ethylene carbonate (EC) is typically involved in the formation of the SEI on graphite in the first charge/discharge cycles [42]. In addition to EC [43,44], other SEI forming agents on graphite were reported in the literature, e.g., lithium difluoro-(oxalate)borate (LiDFOB), vinylene carbonate (VC) or fluoroethylene carbonate (FEC) [45–53] and many more. Among them, VC and FEC are preferred as SEI additives on graphite anodes for organic carbonate-based electrolytes [9,13].

In this contribution, butyronitrile (BN)-based electrolytes containing EC, FEC or both as co-solvents/functional additives are considered for fast charging application in lithium-nickel-manganese-cobalt-oxide (NMC)/graphite cells. The investigations were mainly performed in full cell setup. The optimum BN:EC:FEC solvent/co-solvent-functional additive ratio was investigated in terms of long-term cycling and C-rate performance in NMC/graphite cells. The obtained electrochemical results were correlated to the surface analysis of the graphite electrodes via XPS measurements.

2. Experimental Section

2.1. Electrolyte Formulation

All considered electrolytes were formulated using volume percent (vol.%) in an argon-filled glovebox (MBRAUN, Garching, Germany) with a water and oxygen content below 0.1 ppm. BN 99% (MERCK, Darmstadt, Germany), lithium hexafluorophosphate (LiPF$_6$, BASF, battery grade, Ludwigshafen, Germany), FEC (BASF, battery grade, Ludwigshafen, Germany) and EC 99.8% anhydrous (MERCK, Darmstadt, Germany) were used as received. As reference electrolyte, 1M LiPF$_6$ in EC:DMC (1:1 wt.%) (LP30, BASF, battery grade, Ludwigshafen, Germany) was used.

2.2. Preparation of T44 Graphite and Lithium Manganese Oxide Electrodes

The composition of graphite electrodes was as follows: 87 wt.% T44 graphite (Imerys, Paris, France) 8 wt.% polyvinylidene difluoride (PVdF, Arkema, Colombes, France) and 5 wt.% conductive additive Super C65 (Imerys, Paris, France). T44 graphite was used as the active material due to its high BET surface area, leading to a pronounced reduction of the electrolyte [54] The LiMn$_2$O$_4$ (LMO) electrodes were composed of 80 wt.% LMO (Toda, Hiroshima, Japan), 10 wt.% PVdF and 10 wt.% Super C65. In the fabrication process of the electrodes, PVdF was dissolved in *N*, *N*-dimethylformamide 99.8% anhydrous (DMF, Alfa Aesar, Haverhill, MA, USA). Subsequently, conductive additive (Super C65) and active material (T44 or LMO) were added to the solution and mixed with a dissolver. The suspension was thereafter coated with a special film applicator, on a copper foil (negative electrodes; T44; 120 μm wet thickness) and on aluminum foil (positive electrodes; LMO; 100 μm wet thickness). The coated foils were dried in an oven (Binder, Tuttlingen, Germany) at 80 °C overnight. The obtained electrodes were cut with a punching tool (Hohsen Corp. Osaka, Japan) into a diameter of 12 mm and thereafter dried at 120 °C in vacuum for 24 h in a Buchi Glass Oven 585 with a rotary vane pump vacuum (Büchi, Flawil, Switzerland). After weighting (Sartorius laboratory balance; Sartorius, Göttingen, Germany), the resulting electrodes had an active mass loading between 2.5 and 3 mg cm^{-2} [53]. For the investigations in the full-cell setup, balanced NMC111 (Litarion, Kamenz, Germany) and graphite electrodes (Litarion, Kamenz, Germany) both 12 mm diameter, were used.

2.3. Electrochemical Measurements

2.3.1. Cell Set-Up

The electrochemical measurements were performed in a two electrode, coin cell (2032) (Hohsen Corp. Osaka, Japan), setup as well as in three-electrode T-cell setup (Swagelok® Solon, OH, USA). The NMC was used as the working electrode (WE), graphite as the counter electrode (CE), whereas lithium foil (Albemarle, Charlotte, NC, USA) was taken as the reference electrode (RE).

2.3.2. Linear Sweep Voltammetry Measurements

The electrochemical stability window of the considered BN-based electrolytes was determined by means of linear sweep voltammetry (LSV) using a VMP3 potentiostat (Bio-Logic, Seyssinet-Pariset, France). A lithium manganese oxide (LMO) based electrode was used as WE, whereas lithium foil was used as the CE and RE. The measurements were performed in the potential range between the open circuit potential (OCP) and 5.0 V vs. Li/Li $^+$, using a scan rate of 100 μV s^{-1} at room temperature (20 °C).

2.3.3. Cyclic Voltammetry Measurements

Cyclic voltammetry (CV) measurements were carried out at room temperature, using a Bio-Logic VMP3 potentiostat, in the potential range from 0.02–2.00 V vs. Li/Li$^+$. The cells were cycled with a scan rate of 20 μV s^{-1}. T44 graphite was used as WE. Lithium foil was used as the CE and RE.

2.3.4. Galvanostatic Measurements

Measurements were carried out at 20 °C by means of battery cycler (MACCOR Series 4000, Tulsa, OK, USA). The cathode limited NMC/graphite cells (20% capacity-oversized anode) were cycled for five formation cycles at 0.1C in a voltage range between 3.00–4.30 V. After the formation sequence, cells were cycled with a charge and discharge rate of 372 mA g^{-1} (1C). For the C-rate evaluation, C-rate of the charge step, the discharge step and of both the charge and the discharge steps was always altered after five cycles in the following manner: five cycles with a C-rate of 1C followed by a C-rate of 0.2C, 1C, 2C, 5C, 10C, 15C, 20C followed by 30 cycles with a C-rate of 1C. During the performance assessment of the charge behavior, the C-rate of the charge step was altered, and the C-rate of the discharge step was set to 1C. The performance assessment was based on the discharge capacity. In the discharge performance evaluation, the C-rate of the discharge step was altered, and the C-rate of the charge step was set to 1C. In the charge/discharge performance rating, the C-rate of the charge step as well as the C-rate of the discharge step were altered in afore mentioned way. During the 5C performance evaluation, the C-rate was set to 1C after the formation sequence for 10 cycles followed by 95 charge/discharge cycles with 5C for each charge and discharge step. During the long-time cycling evaluation, the C-rate of the charge and discharge step was set to 1C after the formation procedure. Furthermore, a current-limited CV step of 0.05C was introduced to the galvanostatic cycling procedure, for the 1000 cycle measurement.

2.4. Conductivity Measurements

AC impedance measurements were used to determine the conductivity of the considered BN-based electrolyte formulations. All measurements were carried out on a Solartron 1260A (AMETEK, Berwyn, PA, USA) impedance gain phase analyzer, connected to a Solartron 1287A (AMETEK, Berwyn, PA, USA) potentiostat using a customized cell having two stainless steel disk-electrodes. A frequency range from 1 kHz to 1 MHz using an AC amplitude of 20 mV was applied to the cell for each temperature (−40 to 60 °C), which was regulated via a climate chamber.

2.5. X-ray Photoelectron Spectroscopy (XPS) Analysis

For the XPS measurements, an AXIS Ultra DLD (Kratos, Shimadzu Corporation, Kyoto, Japan) was used. An area of 300 μm × 700 μm was irradiated using a filament voltage of 12 kV, an emission current of 10 mA and a pass energy of 20 eV. The obtained spectra were calibrated against the adventitious carbon signal at 284.5 eV. For the XPS sputter depth profiling measurements a sputter crater diameter of 1.1 mm, an emission current of 8 mA, and a filament voltage of 0.5 kV as well as a pass energy of 40 eV and a 110 μm aperture were applied. The fitting of the resulted spectra was performed with the help of CasaXPS.

3. Results and Discussion

Nitrile-based electrolytes are known to deliver higher ionic conductivity values compared to the state of the art organic carbonate-based counterparts (Figure 1) [55]. This solvent class is particularly interesting when it comes to fast charging behavior of LIBs. Having in mind that BN is not stable against metallic lithium or graphite, a SEI-forming co-solvent was added to the BN-based electrolyte.

With this in line, 1M LiPF$_6$ in BN:EC (1:1) as well as 1M LiPF$_6$ in BN:FEC (1:1) electrolyte formulations were compared with the 1M LiPF$_6$ in EC:DMC (1:1) electrolyte, taken as reference.

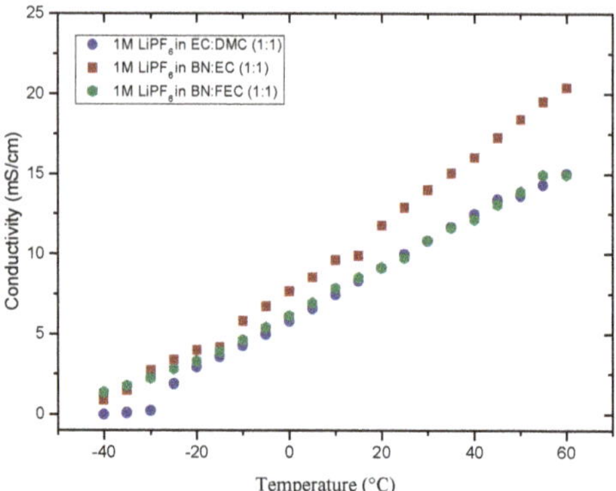

Figure 1. Temperature dependent conductivity measurements of 1M LiPF$_6$ in EC:DMC (1:1), 1M LiPF$_6$ in BN:FEC (1:1) and 1M LiPF$_6$ in BN:EC (1:1), in the temperature range from −40 to 60 °C.

When using EC as co-solvent, the BN-based electrolyte delivers higher conductivity values compared to the 1M LiPF$_6$ in EC:DMC (1:1) electrolyte (Figure 1). Especially at low temperature (0 °C), the conductivity of the considered BN:EC-based electrolyte is at least 32% higher (7.69 mS/cm) compared to the organic carbonate-based counterpart (5.83 mS/cm). Substitution of EC with FEC leads to a decreased conductivity (from 11.80 mS/cm to 9.16 mS/cm) at 20 °C. In the temperature range of 20 °C to 60 °C, the conductivity of 1M LiPF$_6$ in EC:DMC (1:1) is equal to the conductivity values of the 1M LiPF$_6$ in BN:FEC (1:1) electrolyte. In contrast to the high conductivity of the BN:EC mixture, the conductivity of the BN:FEC mixture was shown to be quite poor. The high conductivity of the BN:EC mixture-based electrolytes makes them suitable for fast charging (>1C).

The conductivity values of the investigated electrolytes can be explained by means of relevant physicochemical properties of the used solvents. The conductivity is related to the viscosity and to the relative permittivity of the electrolyte formulation. The ion mobility is linked to the viscosity whereas the salt dissociation capability is related to the relative permittivity. To obtain a high conductivity, the viscosity of the electrolyte formulation should be low, and the relative permittivity must be high enough to ensure a sufficient dissolution of the conducting salt.

To determine the oxidative stability of the BN-based electrolytes, compared to the reference electrolyte, corresponding voltammograms were recorded using LMO as WE (Figure 2). A content of 50% of FEC was chosen to overcome the instability of nitriles towards metallic lithium and to ensure the passivation of the metallic lithium [13,61]. Whereas with organic carbonate-based electrolyte Li metal is stable, [62] with an EC content of only 50%, in the mixture the degradation of the electrolyte could not be inhibited. Therefore, EC:BN mixtures could not be investigated in combination with lithium metal. Nevertheless, this mixture should display the same oxidative stability (as confirmed by later full cell experiments). With 1M LiPF$_6$ in EC:DMC (1:1) as reference electrolyte, the maxima of the de-insertion peaks of LMO are positioned at. 4.05 V vs. Li/Li$^+$ and 4.16 V vs. Li/Li$^+$ [63]. In this setup, the reference electrolyte was found to be electrochemically stable up to 4.90 V vs. Li/Li$^+$ [64].

Compared to the reference electrolyte, the voltammogram of the cell containing 1M LiPF$_6$ in BN:FEC (1:1) displays de-insertion peak maxima of LMO at 4.05 V vs. Li/Li$^+$ and 4.19 V vs. Li/Li$^+$. This electrolyte formulation shows electrochemical stability up to 4.50 V vs. Li/Li$^+$, which is much higher compared to other literature known nitriles [65]. Furthermore, this result fits well with literature showing known density functional theory (DFT) calculations [55]. This behavior makes the combination of BN-based electrolytes with cathode materials, such as lithium nickel cobalt aluminum oxide (NCA) and NMC possible.

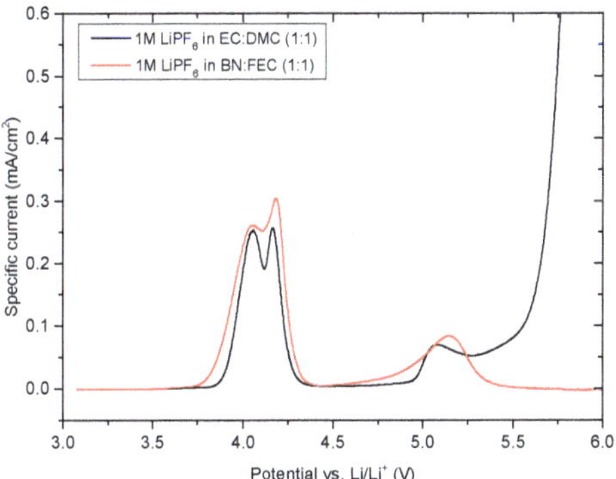

Figure 2. Linear sweep voltammograms of cells containing 1M $LiPF_6$ in BN:FEC (1:1) and 1M $LiPF_6$ in EC:DMC (1:1) (wt.%), LMO as WE, and Li as CE and RE, at scan rate of 100 μV s^{-1} at room temperature.

Cyclic voltammetry measurements in T44 graphite/lithium cells containing 1M $LiPF_6$ in various BN:FEC solvent/co-solvent ratios were performed to determine the reductive stability of the considered electrolyte formulations vs. the anode (Figure 3).

The decomposition of FEC starts at a potential of 1.60 V vs. Li/Li$^+$, reaching the peak maximum at a potential value of 1.50 V vs. Li/Li$^+$ (Figure 3a–d). Due to the SEI formation in presence of FEC, the decomposition of the 1M $LiPF_6$ in BN:FEC (1:1) (Figure 3a), 1M $LiPF_6$ in BN:FEC (6:4) (Figure 3b), 1M $LiPF_6$ in BN:FEC (7:3) (Figure 3c), 1M $LiPF_6$ in BN:FEC (8:2) (Figure 3d) electrolyte formulations are inhibited, thus leading to the reversible intercalation and deintercalation of lithium ions into the graphite host structure, as indicated by the presence of the corresponding peaks (starting at a potential of 0.30 V vs. Li/Li$^+$). Compared to the aforementioned electrolyte formulations, 1M $LiPF_6$ in BN:FEC (9:1) electrolyte (Figure 3e) is not able to form an effective SEI on graphite and results in a severe decomposition. As a consequence, no intercalation/deintercalation steps take place. The amount of FEC seems not to be enough to protect the BN against decomposition on both T44 graphite and lithium electrode. Compared to FEC, the decomposition of in the 1M $LiPF_6$ in EC:DMC (1:1) (Figure 3f) mixture starts at 0.9 V vs. Li/Li$^+$ and the peak maximum is reached at 0.80 V vs. Li/Li$^+$.

To prove the fast charging ability of the NMC/graphite cells containing afore mentioned BN-based electrolyte formulations, a C-rate evaluation up to 5C was performed, starting with five formation cycles at 0.1C. After the formation, 10 cycles at 1.0C were conducted, followed by 95 charge/discharge cycles with a C-rate of 5C. The obtained results are shown in Figure 4. As depicted in Figure 4a, the NMC/graphite cell containing 1M $LiPF_6$ in BN:EC (1:1) electrolyte, reaches a Coulombic efficiency of 87% in the first cycle (see Meister et al. for the meanings of efficiencies) [66]. The specific discharge capacity amounts to 176 mAh/g with a C-rate of 0.1 C in the first five cycles, whereas in the consecutive 10 charge/discharge cycles, a specific discharge capacity of 153 mAh/g with a Coulombic efficiency of 99% is achieved. After 15 cycles, the C-rate evaluation was started with a C-rate of 5C for each charge and discharge step for the consecutive 95 charge/discharge cycles. The specific discharge capacity displays a negligible fading and drops from 76 mAh/g in the 30th cycle to 68 mAh/g in the 110th cycle. The Coulombic efficiency drop in the 6th and 16th cycle is related to the change of the C-rate and observed in each chart in Figure 4. The cell containing 1M $LiPF_6$ in BN:EC (7:3) + 1% FEC electrolyte formulation displays a first cycle Coulombic efficiency of 87%, as illustrated in Figure 4b. The specific discharge capacity amounts to 177 mAh/g for each cycle with a C-rate of 0.1C. A specific discharge capacity of 155 mAh/g with a Coulombic efficiency of 99% is reached in the following 10 charge/discharge cycles. In the C-rate evaluation, the specific discharge capacity shows a notable fading and drops from 93 mAh/g in cycle 30 to 68 mAh/g in cycle 110. The cell chemistry outlined in Figure 4c comprises of 1M $LiPF_6$ in BN:EC (9:1) + 3% FEC electrolyte. The specific discharge capacity amounts to 176 mAh/g in the first five cycles using a C-rate of 0.1C, whereas the first cycle

Coulombic efficiency amounts to 86%. Unlike other considered electrolyte formulations displayed in Figure 4, 99% Coulombic efficiency is not reached in the second but in the third cycle. The specific discharge capacity in the consecutive 10 charge/discharge cycles amounts to 155 mAh/g. During the 5C sequence, the capacity drops down to 105 mAh/g in the 30th cycle and decreases to 100 mAh/g in the 110th cycle, without a substantial fading. The cell containing 1M $LiPF_6$ in EC:DMC (1:1) electrolyte formulation, (Figure 4d) displays a first Coulombic efficiency of 87%. The specific discharge capacity amounts to 176 mAh/g for each cycle with a C-rate of 0.1 C. The following 10 charge/discharge cycles display a specific discharge capacity of 155 mAh/g with a Coulombic efficiency of 99%. In the C-rate evaluation the specific discharge capacity drops from 67 mAh/g in the 30th cycle to 64 mAh/g in the 110th cycle. A comparison between the 1M $LiPF_6$ in BN:EC (9:1) + 3% FEC and the reference electrolyte indicates a similar cycling performance at C-rates up to 1C and a superior higher performance of BN-based electrolyte at 5C. During cycling at 5C, the specific discharge capacity is decreased by 5% from 105 mAh/g to 100 mAh/g comparable to 4% with the reference electrolyte. In addition, the average specific discharge capacity at 5C is $\approx$ 103 mAh/g compared to $\approx$ 66 mAh/g for the reference electrolyte. The deviation amounts to 37 mAh/g (56%).

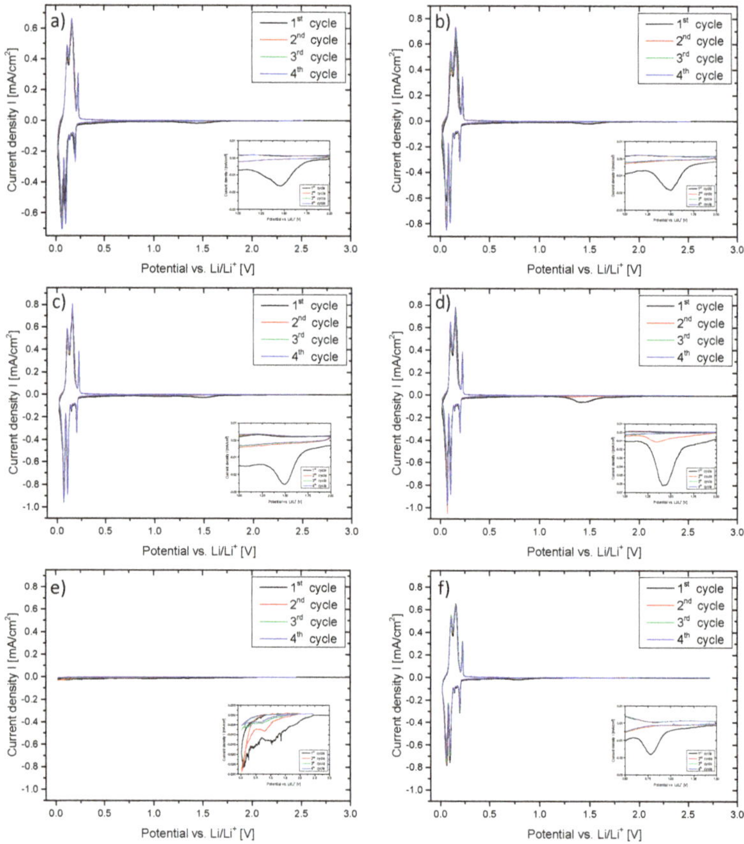

Figure 3. Cyclic voltammograms of T44 graphite/lithium cells containing 1M $LiPF_6$ in (**a**) BN:FEC (1:1), (**b**) BN:FEC (6:4), (**c**) BN:FEC (7:3), (**d**) BN:FEC (8:2), (**e**) BN:FEC (9:1) and (**f**) EC:DMC (1:1) as electrolyte formulation, in the potential range between 0.02–2.00 V vs. Li/Li$^+$, at scan rate of 20 µV s^{-1}; insert shows the magnification of the reductive decomposition peak of fluoroethylene carbonate (FEC) at 1.5 V vs. Li/Li$^+$.

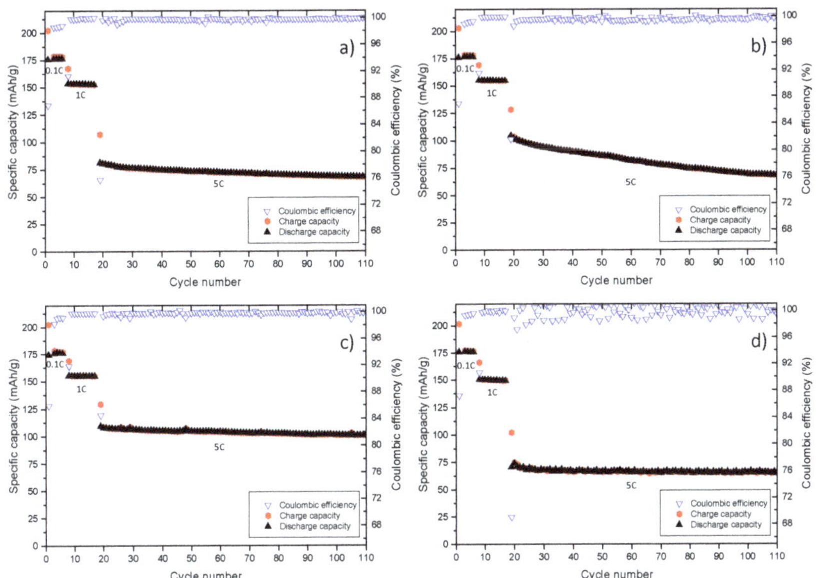

Figure 4. C-rate evaluation of the NMC/graphite cells containing (**a**) 1M LiPF$_6$ in BN:EC (1:1), (**b**) 1M LiPF$_6$ in BN:EC (7:3) + 1% FEC, (**c**) 1M LiPF$_6$ in BN:EC (9:1) + 3% FEC and (**d**) 1M LiPF$_6$ in EC:DMC (1:1) in the voltage range of 3.00–4.30 V.

The obtained results show, that NMC/graphite cells containing 1M LiPF$_6$ in BN:EC (9:1) + 3% FEC electrolyte display a remarkably stable cycling behavior at 5C. In addition, a C-rate evaluation, in NMC/graphite cells, up to 20C was carried out. Figure 5 shows the C-rate evaluation of 1M LiPF$_6$ in BN:EC (9:1) + 3% FEC compared to the reference organic carbonate-based 1M LiPF$_6$ in EC:DMC (1:1) electrolyte. Two types of C-rate evaluations were performed to determine whether the C-rate for charge (Figure 5a,b) or the C-rate for discharge (Figure 5c,d) has a more pronounced impact on the cycling stability of the NMC/graphite cells. In the first C-rate evaluation, the charge current is altered from 0.1C to 20C, whereas the C-rate of the discharge step was kept constant. In the second C-rate evaluation, the C-rate of the charge step remained constant while the C-rate of the discharge step was changed. The C-rate evaluation started with five formation cycles at 0.1C followed by five cycles at 1C.

When comparing the overall performance of the considered NMC/graphite cells with 1M LiPF$_6$ in BN:EC (9:1) + 3% FEC and the 1M LiPF$_6$ in EC:DMC (1:1) electrolytes, a better C-rate performance is achieved for the BN-based electrolyte containing cells, as depicted in in Figure 5a. Especially at a C-rate (charge step) of 5C and 10C, the specific discharge capacity is much higher for the BN-based electrolyte containing cell. At low C-rates (charge step), the specific discharge capacities of both electrolyte containing cells are quite similar. At 0.1C and 0.2C, the specific discharge capacity of the BN-based electrolyte containing cell has a value of 174 mAh/g and 160 mAh/g, respectively. On the other hand, the organic carbonate-based electrolyte containing cell delivers a specific discharge capacity of 172 mAh/g at 0.1 C and 154 mAh/g at 0.2C, which is nearly similar to the cell containing 1M LiPF$_6$ in BN:EC (9:1) + 3% FEC. At 5C and 10C, the better electrochemical performance of the BN-based electrolyte containing cell becomes clear, as a discharge capacity of 125 mAh/g is reached, compared to the 82 mAh/g for the organic carbonate-based counterpart. Even though the specific discharge capacity of the BN-based cell is not constant at 10C, the specific discharge capacity value is 62 mAh/g, is higher compared to the 21 mAh/g obtained in the cell with the organic carbonate-based electrolyte. At a C-rate (charge step) of 20C, the BN-based electrolyte containing cell delivers a specific capacity of 8 mAh/g. The decrease of C-rate to 1C results in a stable cycling performance for both considered cell chemistries. Nevertheless, the cell containing BN-based formulation has a slightly higher specific discharge capacity. The corresponding Coulombic efficiency values are depicted in Figure 5b.

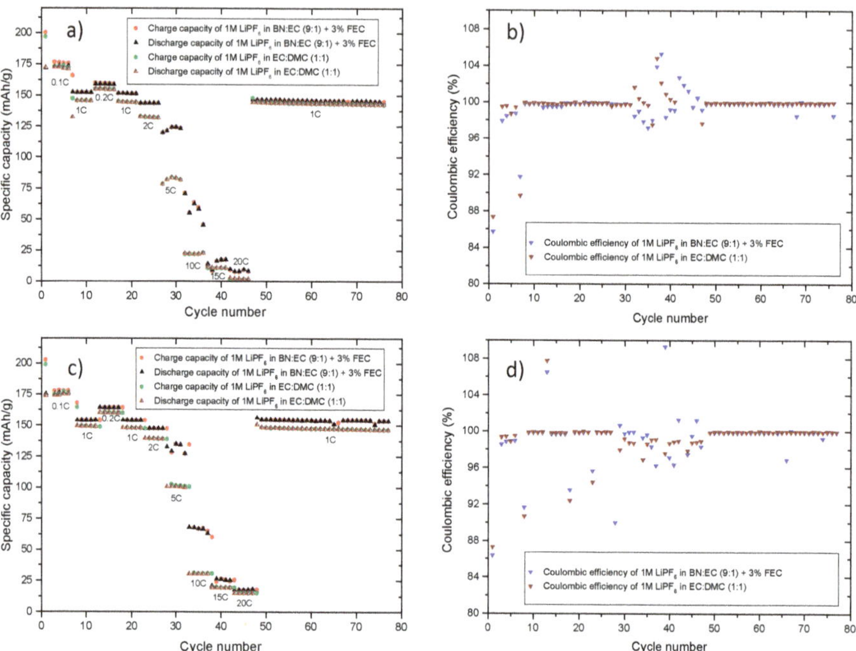

Figure 5. Cycling profiles and Coulombic efficiencies of the NMC/graphite cells containing 1M LiPF$_6$ in BN:EC (9:1) + 3% FEC and 1M LiPF$_6$ in EC:DMC (1:1) electrolyte formulations cycled in a voltage range of 3.00–4.30 V using a C-rate procedure (**a,b**) the C-rate of the charge step is increasing while the C-rate of discharge step stays constant at 1C and a C-rate procedure (**c,d**) where the C-rate of the discharge step is increasing whereas the C-rate of the charge step stays constant at 1C.

For the C-rate (of the discharge step) performance, a similar behavior can be observed (Figure 5c). At low C-rates (of the discharge step), the electrochemical performance of both cells is nearly similar, whereas with increasing C-rate (discharge step), the cell with the BN-based electrolyte shows a much better performance. At 5C, a specific discharge capacity of 135 mAh/g is achieved. Increasing the discharge rate up to 10C, a value of 67 mAh/g is reached for the BN-based electrolyte containing cell. On the other side, the specific discharge capacity is much lower (101 mAh/g and 29 mAh/g, respectively) for the organic carbonate-based electrolyte containing cell. The decrease of the C-rate (discharge step) to 1C, results in stable cycling performance for both cell chemistries. Nevertheless, the cell containing BN-based electrolyte shows a higher specific discharge capacity (155 mAh/g vs. 144 mAh/g) at a C-rate (charge and discharge step) of one 1C after the 100th cycle. The corresponding Coulombic efficiency values are depicted in Figure 5d.

For both C-rate (both the charge and the discharge step) evaluations, it was shown that the cells containing a BN-based electrolyte outperform the organic carbonate-based counterpart. This is especially observed, at 5C and 10C. Even at higher C-rates (15C and 20C), a cell containing 1M LiPF$_6$ in BN:EC (9:1) + 3% FEC electrolyte shows better electrochemical performance compared to the reference organic carbonate-based electrolyte counterpart. The simultaneous charge/discharge behavior of the considered BN-based cell and organic carbonate-based cell at different C-rates was evaluated further. As depicted in Figure 6, a similar behavior in terms of specific discharge capacity can be observed. An increase in the C-rate (of the charge and discharge step) results in higher difference between the specific discharge capacities of the cells containing BN-based electrolyte and the ones with the organic carbonate-based electrolyte. The cell containing BN as solvent shows much better cycling performance at higher C-rates (both charge and discharge), compared to the state-of-the-art electrolyte containing counterpart. At 1C, a specific discharge capacity of 154 mAh/g and 150 mAh/g for the cell containing organic carbonate-based electrolyte is achieved. By increasing the C-rate (both the charge and the discharge step) to 2C, the specific discharge capacity reach values of 140 mAh/g and 129 mAh/g, respectively, whereas an increase in the C-rate (both the charge and the discharge step to 10C results in a specific capacity value of 42 mAh/g and 22 mAh/g, respectively. After increasing the C-rate (both the charge and the discharge step) to 20C, the BN-based electrolyte containing cell reaches a specific capacity of 9 mAh/g in contrast to 1 mAh/g for the state-of-the-art electrolyte containing counterpart. After the C-rate (both the charge and the discharge step) is decreased to 1C again, both electrolyte containing cells exhibit a stable cycling behavior (for both Coulombic efficiency as well as specific discharge capacity). During cycling with a C-rate (both the charge and the discharge step) of 1C, the cells deliver specific discharge capacity of 151 mAh/g in case of the BN-based electrolyte and 148 mAh/g for the organic carbonate-based electrolyte. The corresponding Coulombic efficiency values are depicted in Figure 6b. Table 1 summarizes the results obtained from graphs presented in Figures 5 and 6 for the NMC/graphite cells cycled with 1M LiPF$_6$ in BN:EC (9:1) + 3% FEC and 1M LiPF$_6$ in EC:DMC (1:1) electrolytes, respectively.

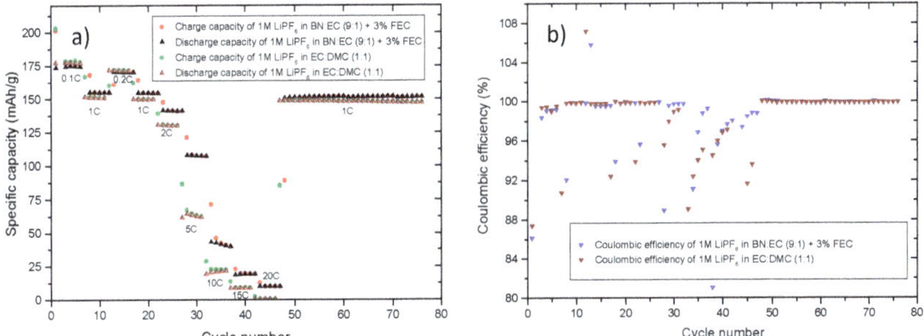

Figure 6. Cycling profiles (**a**) and Coulombic efficiencies (**b**) of NMC/graphite cells containing 1M LiPF$_6$ in BN:EC (9:1) + 3% FEC as well as 1M LiPF$_6$ in EC:DMC (1:1), respectively as electrolyte, cycled in the voltage range of 3.00–4.30 V.

Table 1. Summary of solvents used in this work. Physical properties are reported at 25 °C if not stated otherwise [55].

Structure	Compound	Abbreviation	η (cP)	ε_r
	Ethylene carbonate	EC	1.9 [a] [13]	89.8 [a] [9]
	Dimethyl carbonate	DMC	0.59 [56]	3.12 [57]
	Fluoroethylene carbonate	FEC	4.1 [58]	107 [58]
	Butyronitrile	BN	0.553 [59]	20.7 [60]

[a] Viscosity (η) and relative permittivity (ε_r) values for EC are determined at 40 °C.

As data listed in Table 2 show, the cells containing BN-based electrolytes deliver higher specific capacity values at higher C-rate compared to their state-of-the-art electrolyte counterparts. The C-rate (discharge step) evaluation setup leads to higher specific discharge capacities compared to the other two C-rate evaluations. This might be explained on the basis of the intercalation and deintercalation steps on graphite: the deintercalation process for graphite is always favored therefore, higher discharge capacities can be reached for both electrolytes [57]. Based on the obtained results, the main use for the BN-based electrolyte formulation would be in applications with high demands to power, fast charge ability or even both.

Table 2. Selected specific discharge capacities for the cells cycled with the BN-based electrolyte (cell a) and 1M $LiPF_6$ in EC:DMC (1:1) (cell b), respectively.

C-Rate	C-Rate Performance (Charge) Specific Capacity (mAh/g)		C-Rate Performance (Discharge) Specific Capacity (mAh/g)		C-Rate Performance (Charge/Discharge) Specific Capacity (mAh/g)	
	Cell a	Cell b	Cell a	Cell b	Cell a	Cell b
0.1C	176	172	177	174	175	177
1C	152	145	154	149	154	150
2C	144	133	148	139	140	129
5C	125	84	134	99	108	63
10C	62	23	68	31	42	22
20C	9	2	18	15	9	1

As BN:EC (9:1) + 3% FEC electrolyte containing cells show remarkable C-rate performance, long-time cycling experiments were conducted to enable deeper characterization of the electrochemical behavior of the considered cell chemistry. In Figure 7, two different long-term cycling measurements (1000 charge/discharge cycles) were performed for the BN-based electrolyte containing NMC/graphite cell as well as the state-of-the-art electrolyte containing counterpart.

The afore mentioned C-rate evaluation was performed without using a constant voltage (CV) step after the charge step. A CV step is typically used to enhance the capacity of the graphite slightly, making sure, that the graphite is fully lithiated [67].

The long-time cycling measurements depicted in Figure 7 show that, without CV step, the long term cycling performance of the 1M $LiPF_6$ in BN:EC (9:1) + 3% FEC containing cells (Figure 7c) is comparable to the state of the art electrolyte based on 1M $LiPF_6$ in EC:DMC (1:1) cell counterparts, as depicted in Figure 7a. The 1st cycle Coulombic efficiency of the cell containing 1M $LiPF_6$ in EC:DMC (1:1) electrolyte (87%) matches the Coulombic efficiency resulting with the 1M $LiPF_6$ in BN:EC (9:1) + 3% FEC electrolyte (87% Coulombic efficiency). Ninety-nine percent Coulombic efficiency is reached in the second cycle for the state-of-the-art electrolyte as well as for the BN-based counterpart. From this point onwards, the Coulombic efficiency values of both cells containing considered electrolytes are nearly similar, amounting to ≈99% during the long-term cycling performance. In the initial cycles, in which SEI formation takes place, a specific discharge capacity of 177 mAh/g is reached for the cell containing BN-based electrolyte, whereas the one with the EC:DMC-based electrolyte shows a specific discharge capacity of 173 mAh/g. After the initial cycles (five cycles with 0.1C), the cells were cycled with 1C until the 1000th charge/discharge cycle. For both cell chemistries, a stable long-term cycling is observed, with an absence of strong fading in capacity. In the 10th cycle, a specific discharge capacity of 155 mAh/g is reached and decreases slightly to 129 mAh/g in the last (1000th) cycle, for the cell with 1M $LiPF_6$ in BN:EC (9:1) + 3% FEC as electrolyte. For the cell containing 1M $LiPF_6$ in EC:DMC (1:1) as electrolyte, a specific discharge capacity of 145 mAh/g in the 10th cycle and 135 mAh/g in the 1000th cycle is reached. Comparing the 10th cycle with the 1000th cycle, both electrolytes reach over 80% of the initial capacity, meeting the automotive requirements (80% state of health after 1000 charge/discharge cycles).

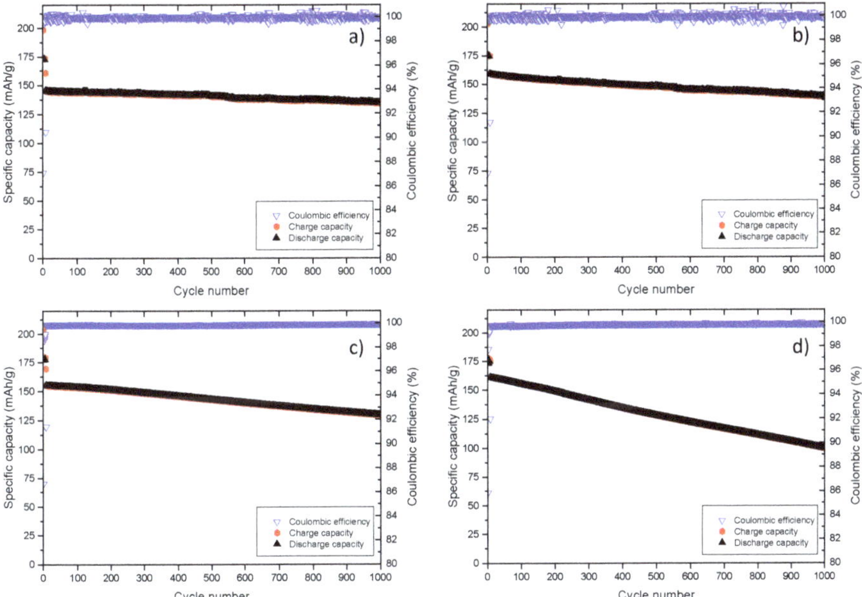

Figure 7. Long-term cycling profiles of the NMC/T44graphite cells containing electrolyte formulations: (**a**) 1M LiPF$_6$ in EC:DMC (1:1) without cyclic voltammetry (CV) step (cell a), (**b**) 1M LiPF$_6$ in EC:DMC (1:1) including CV step (cell b), (**c**) 1M LiPF$_6$ in BN:EC (9:1) + 3% FEC without CV step (cell c), (**d**) 1M LiPF$_6$ in BN:EC (9:1) + 3% FEC including CV step (cell d) in the voltage range of 3.00–4.30 V.

For the long-term evaluations comprising a current-limited constant voltage step the results are different. The long-term cycling performance of the 1M LiPF$_6$ in BN:EC (9:1) + 3% FEC (Figure 7d) is decreased compared to the state-of-the-art electrolyte-based cell depicted in Figure 7b. The 1st cycle Coulombic efficiency of the 1M LiPF$_6$ in EC:DMC (1:1) electrolyte (87%) containing cell is similar to the Coulombic efficiency obtained for the 1M LiPF$_6$ in BN:EC (9:1) + 3% FEC electrolyte (86% Coulombic efficiency) based counterpart. The Coulombic efficiency of 99% for the state-of-the-art electrolyte containing cell is reached in the second cycle. For the BN-based electrolyte containing cell, a Coulombic efficiency amounts to o 99% only in the 4th cycle. From this point onwards, the Coulombic efficiency values of both electrolyte containing cells are nearly similar, >99% prolong the long-term cycling. During the initial cycles, a specific discharge capacity of 174 mAh/g is reached for the BN-based electrolyte containing cell, whereas the one with the EC:DMC-based electrolytes displays a specific discharge capacity of 175 mAh/g. After the initial cycles (five charge/discharge cycles with 0.1C), the cells were cycled at 1C until the 1000th cycle. Both cells passed the long-term cycling procedure, thus indicating a good cycling performance. However, a slight fading of the cell with the BN-based electrolyte (Figure 7d) is noticeable. In the 10th cycle, a specific discharge capacity of 161 mAh/g is reached and decreases to 100 mAh/g in the 1000th cycle, for the cell with 1M LiPF$_6$ in BN:EC (9:1) + 3% FEC as electrolyte. For the cell containing 1M LiPF$_6$ in EC:DMC (1:1) as electrolyte, a specific discharge capacity of 159 mAh/g in the 10th cycle and 139 mAh/g in the 1000th cycle is reached. Comparing the 10th cycle with the 1000th cycle only the state of the art electrolyte has reached over 80% of the initial capacity, meeting the automotive requirements [68]. Table 3 summarizes afore mentioned cycling performance and comperes both cycling procedures (with and without CV step).

The capacity retention values show, that a CV step deteriorates the electrochemical performance of the NMC/graphite cells. The same effect is observed with the reference electrolyte containing cell however, the effect is less pronounced. This outcome is explained by the time at which the cells remain at the cut-off voltage. For the cells cycled with a CV step, this duration is much larger and is leading to a pronounced degradation (shorter lifespan) of these cells. To correlate the obtained results with the surface chemistry of electrodes containing BN-based electrolyte, XPS sputter depth profiling of graphite electrodes was performed, to prove the stability of the electrolyte towards graphite. The electrochemical decomposition of considered BN-based electrolyte formulations on the graphite surface was analyzed by means of XPS (see Figure 8).

Table 3. 1st Coulombic efficiency, discharge capacity as well as capacity retention between the 10th and 1000th cycle for the reference electrolyte (cells a/b) and the nitrile-based electrolyte formulation (cells c/d) containing cells.

Selected Parameters	Cell a (Reference Electrolyte)	Cell b (CV) (Reference Electrolyte)	Cell c (BN-Based Electrolyte)	Cell d (CV) (BN-Based Electrolyte)
1st Coulombic efficiency (%)	87	87	87	86
Discharge capacity in the10th cycle (mAh/g)	145	159	155	161
Discharge capacity in the 1000th cycle (mAh/g)	135	139	129	100
Capacity retention (%)	93	87	83	62

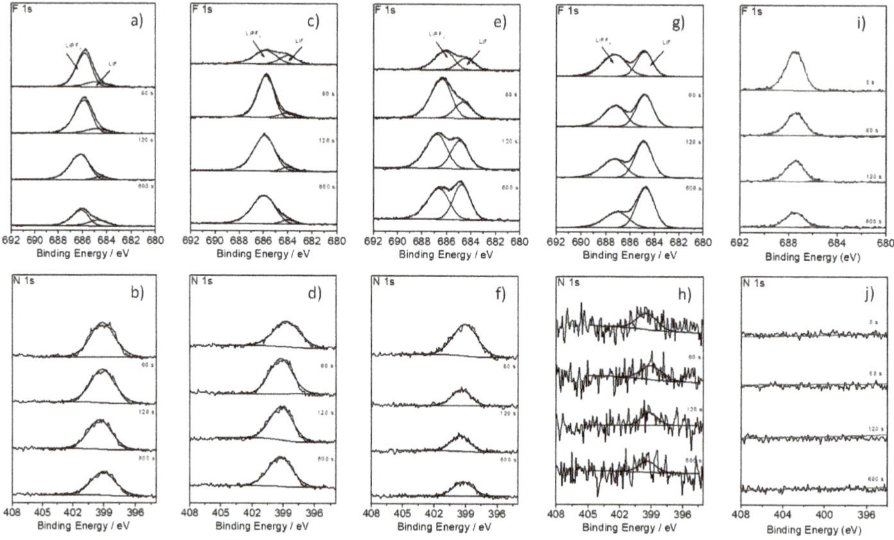

Figure 8. F 1s and N 1s core spectra of graphite electrodes, after five charge/discharge cycles at 0.1C in a NMC/graphite cells with different amounts of FEC in the electrolyte: (**a,b**) 1M LiPF$_6$ in BN, (**c,d**) 1M LiPF$_6$ in BN + 5% FEC, (**e,f**) 1M LiPF$_6$ in BN:EC (9:1) + 2% FEC, (**g,h**), 1M LiPF$_6$ in BN:EC (9:1) + 3% FEC and (**i,j**) pristine electrode.

Figure 8 depicts the XPS F 1s and N 1s core spectra of graphite electrode-based cells cycled in presence of a,b) 1M LiPF$_6$ in BN; c,d) 1M LiPF$_6$ in BN with 5% FEC; e,f) 1M LiPF$_6$ in BN:EC (9:1) + 2% FEC; as well as g,h) 1M LiPF$_6$ in BN:EC (9:1) + 3% FEC as electrolyte. As reference spectra, the XPS F 1s and N 1s core spectra of a pristine graphite electrode (Figure 8i,j) are shown. In the F 1s spectra (Figure 8i), a signal located at 687 eV is observable, which can be attributed to the polyvinylidene difluoride (PVdF) binder [69]. The decrease of the peak intensity during sputtering is related to the decomposition of the binder during XPS measurement [70]. On the other hand, no nitrogen signal was observed on the pristine electrode surface (Figure 8j). In the F 1s spectra of the cycled electrodes (Figure 8a,c,e,g), an additional signal attributed to lithium fluoride (LiF), formed due to the decomposition of the conducting salt $LiPF_6$, occurs at 685 eV. As depicted in Figure 8a,c the amount of LiF remains unaffected relatively to the intensity of the PVdF peak with increasing the sputter time. This could be explained by a limited degradation of $LiPF_6$ in the BN-based electrolytes without EC. Due to the severe decomposition of BN, the corresponding peaks overlap the peaks assigned to the decomposition of $LiPF_6$. In addition, in the N1s core spectra of the electrodes containing pure BN-based electrolyte (Figure 8b,d) a signal at 399 eV is observed, attributed to the decomposition of the nitrile during cycling. In the absence of BN decomposition, the peak of LiF increases relatively to the PVdF peak with increasing the sputter time (Figure 8e,g), as LiF is the main component of the inorganic part of the SEI [70]. The increase of the LiF peak indicates absence of decomposition, meaning that a SEI was formed on graphite surface. Nevertheless, the N 1s core spectra of the graphite electrode cycled with 1M $LiPF_6$ in BN:EC (9:1) + 2% FEC, exhibit a peak 399 eV related to BN decomposition, thus indicating that the formed SEI does not fully prevent the decomposition of the nitrile. By adding 3% FEC to the BN:EC (9:1) electrolyte formulation, the peak in the corresponding N1s core spectrum at 399 eV disappears (Figure 8h), thus indicating an effective SEI formation, which prevents BN against decomposition. Table 4 lists the corresponding surface concentration given in arbitrary units (a.u.).

Table 4. Surface concentration on graphite in arbitrary unit (a.u.) of the performed XPS measurements using different BN-based electrolytes.

Pure BN-Based Electrolyte: Surface Concentration (a.u.)			
Sputter time	**0 s**	**60 s**	**120 s**	**600 s**
LIF	2.13 (0.25)	0.99 (0.71)	1.28 (0.05)	1.84 (0.66)
N 1 s	10.74 (0.98)	10.06 (1.24)	8.69 (2.07)	9.14 (0.38)

BN + 5%FEC based electrolyte: surface concentration (a.u.)			
Sputter time	**0 s**	**60 s**	**120 s**	**600 s**
LIF	3.80 (1.02)	2.14 (2.43)	1.26 (0.45)	4.00 (0.53)
N 1 s	11.83 (0.58)	12.26 (0.99)	9.37 (0.32)	9.78 (0.73)

BN:EC (9:1) + 2%FEC based electrolyte: surface concentration (a.u.)			
Sputter time	**0 s**	**60 s**	**120 s**	**600 s**
LIF	4.37 (0.79)	7.88 (0.22)	8.79 (0.28)	9.90 (0.16)
N 1 s	3.53 (0.35)	3.71 (0.28)	3.22 (0.11)	3.28 (0.21)

BN:EC (9:1) + 3%FEC based electrolyte: surface concentration (a.u.)			
Sputter time	**0 s**	**60 s**	**120 s**	**600 s**
LIF	12.15 (0.47)	16.49 (1.03)	18.40 (1.16)	20.61 (0.52)
N 1 s	0.44 (0.11)	0.38 (0.11)	0.27 (0.30)	0.25 (0.25)

Reference electrode (pristine): surface concentration (a.u.)			
Sputter time	**0 s**	**60 s**	**120 s**	**600 s**
LIF	0.86 (0.84)	2.33 (0.73)	1.52 (1.47)	1.32 (0.93)
N 1 s	0 (0)	0 (1.12)	0 (0)	0 (0)

For the electrodes with pure BN-based electrolyte, as well as for the electrolyte formulation containing 5% FEC, only small amounts of LiF were detected. The origin of the spectra can be dedicated to the decomposition of small amounts of the conducting salt $LiPF_6$. The N 1s surface concentration for both electrolytes indicates a severe decomposition of BN. Without formation of an effective SEI, an ongoing decomposition of the solvent (BN) takes place. For the electrolyte formulations BN:EC (9:1) with addition of 2% and 3% FEC respectively, the amount of LiF increases during sputtering [70,71]. The intensity of the N 1s signal decreases for both electrolytes corresponding to a less pronounced decomposition of the BN-solvent. However, regarding the N 1s surface concentration, the addition of 2% FEC is not enough for the formation of an effective SEI on graphite. The N 1s surface concentration of the formulation containing 3% FEC is comparable to the N 1s surface concentration of the reference electrode, however, both do not show a significant signal.

4. Conclusions

With two successfully tuned BN-based electrolyte formulations (one used in half-cell and the other in full-cell configuration), the decomposition on both lithium metal and graphite, could be prevented. Both electrolytes were comparable or even better compared to the state-of-the-art organic carbonate-based electrolyte. In half cell experiments, 1M $LiPF_6$ in BN:FEC (1:1) containing cell showed the most promising results. EC was compared to FEC, due to its lower passivation capability towards metallic lithium, not suitable to protect BN against decomposition. Nevertheless, 1M $LiPF_6$ in BN:FEC formulation showed lower ion conductivity values compared to BN:EC counterparts. Especially at low temperatures around 0 °C, the conductivity of the BN:EC-based electrolytes was at least 32% higher (7.69 mS/cm) compared to the BN:FEC-based and the organic carbonate-based electrolyte (5.83 mS/cm). Since the main focus of this paper is related to the possible the automotive applicability of BN, investigations in NMC/graphite cells containing BN:EC-based electrolytes were studied in detail. In each investigation, the cell containing 1M $LiPF_6$ in BN:EC (9:1) + 3% FEC showed a superior high performance compared to the organic-carbonate-based counterpart. To match automotive requirements, a C-rate evaluation of 5C was performed. It was shown, that the average specific discharge capacity at 5C amounted to ≈103 mAh/g for the investigated BN-based electrolyte containing cell, which was nearly twice the capacity of the cell with the reference electrode (≈66 mAh/g). Further, a C-rate evaluation up to 20C was performed. The cells containing investigated BN-based electrolyte formulation showed a superior C-rate performance compared to the organic state of the art counterpart. In addition, the CV step in the CCCV measurements, typically used in case of organic carbonate-based electrolyte containing cells, was investigated. It was found out, that a CV step increases the charge/discharge capacity at the beginning of the cycling procedure so that more lithium ions can be intercalated into graphite. Nevertheless, the CV step reduces the overall cycle-life of the cell as well, due to the pronounced electrolyte degradation. In the long-term cycling experiment (Figure 7c,d) the advantage of the CV step is lost after the 150th cycle. From the 150th cycle onwards the capacity of the cell cycled without CV step is higher compared to the cell cycled with CV step.

XPS analysis of the NMC electrodes complements well to the electrochemical characterization of the BN-based electrolytes, showing that a minimum amount of 3% FEC is needed to prevent the BN-based electrolyte formulation of 1M $LiPF_6$ in BN:EC (9:1) from decomposition on graphite in NMC/graphite cell setup.

Author Contributions: P.H., I.C.-L. and M.W. conceived and designed theexperiments; P.H. and L.I. performed the experiments; P.H., L.I., I.C.-L. and M.W. analyzed the data; P.H. and I.C.-L. wrote the paper.

Funding: Financial support by the German Federal Ministry for Education and Research (BMBF) within the project Electrolyte Lab 4E (project reference 03X4632) is gratefully acknowledged.

Conflicts of Interest: The authors declare no conflict of interest.

References

1. Winter, M.; Brodd, R.J. What are batteries, fuel cells, and supercapacitors? *Chem. Rev.* **2004**, *104*, 4245–4269. [CrossRef] [PubMed]
2. Andre, D.; Kim, S.-J.; Lamp, P.; Lux, S.F.; Maglia, F.; Paschos, O.; Stiaszny, B. Future generations of cathode materials: an automotive industry perspective. *J. Mater. Chem. A* **2015**, *3*, 6709–6732. [CrossRef]
3. Schmuch, R.; Wagner, R.; Hörpel, G.; Placke, T.; Winter, M. Performance and cost of materials for lithium-based rechargeable automotive batteries. *Nat. Energy* **2018**, *3*, 267. [CrossRef]
4. Winter, M.; Barnett, B.; Xu, K. Before Li ion batteries. *Chem. Rev.* **2018**, *118*, 11433–11456. [CrossRef] [PubMed]

5.   Lu, L.; Han, X.; Li, J.; Hua, J.; Ouyang, M. A review on the key issues for lithium-ion battery management in electric vehicles. *J. Power Sources* **2013**, *226*, 272–288. [CrossRef]

6.   Placke, T.; Kloepsch, R.; Dühnen, S.; Winter, M. Lithium ion, lithium metal, and alternative rechargeable battery technologies: the odyssey for high energy density. *J. Solid State Electrochem.* **2017**, *21*, 1939–1964. [CrossRef]

7.   Betz, J.; Bieker, G.; Meister, P.; Placke, T.; Winter, M.; Schmuch, R. Theoretical versus Practical Energy: A Plea for More Transparency in the Energy Calculation of Different Rechargeable Battery Systems. *Adv. Energy Mater.* **2019**, *9*, 1803170. [CrossRef]

8.   Wagner, R.; Preschitschek, N.; Passerini, S.; Leker, J.; Winter, M. Current research trends and prospects among the various materials and designs used in lithium-based batteries. *J. Appl. Electrochem.* **2013**, *43*, 481–496. [CrossRef]

9.   Xu, K. Nonaqueous liquid electrolytes for lithium-based rechargeable batteries. *Chem. Rev.* **2004**, *104*, 4303–4417. [CrossRef]

10.   Wen, J.; Yu, Y.; Chen, C. A Review on Lithium-Ion Batteries Safety Issues: Existing Problems and Possible Solutions. *Mater. Express* **2012**, *2*, 197–212. [CrossRef]

11.   Wrodnigg, G.H.; Besenhard, J.O.; Winter, M. Ethylene sulfite as electrolyte additive for lithium-ion cells with graphitic anodes. *J. Electrochem. Soc.* **1999**, *146*, 470–472. [CrossRef]

12.   Chawla, N.; Bharti, N.; Singh, S. Recent advances in non-flammable electrolytes for safer lithium-ion batteries. *Batteries* **2019**, *5*, 19. [CrossRef]

13.   Xu, K. Electrolytes and Interphases in Li-Ion Batteries and Beyond. *Chem. Rev.* **2014**, *114*, 11503–11618. [CrossRef] [PubMed]

14.   Cekic-Laskovic, I.; von Aspern, N.; Imholt, L.; Kaymaksiz, S.; Oldiges, K.; Rad, B.R.; Winter, M. Synergistic effect of blended components in nonaqueous electrolytes for lithium ion batteries. *Top. Curr. Chem.* **2017**, *375*, 37. [CrossRef] [PubMed]

15.   Xue, L.; Lee, S.-Y.; Zhao, Z.; Angell, C.A. Sulfone-carbonate ternary electrolyte with further increased capacity retention and burn resistance for high voltage lithium ion batteries. *J. Power Sources* **2015**, *295*, 190–196. [CrossRef]

16.   Lewandowski, A.; Kurc, B.; Stepniak, I.; Swiderska-Mocek, A. Properties of Li-graphite and LiFePO4 electrodes in LiPF6–sulfolane electrolyte. *Electrochim. Acta* **2011**, *56*, 5972–5978. [CrossRef]

17.   Abu-Lebdeh, Y.; Davidson, I. High-Voltage Electrolytes Based on Adiponitrile for Li-Ion Batteries. *J. Electrochem. Soc.* **2009**, *156*, A60–A65. [CrossRef]

18.   Winter, M.; Novák, P. Chloroethylene Carbonate, a Solvent for Lithium-Ion Cells, Evolving CO_2 during Reduction. *J. Electrochem. Soc.* **1998**, *145*, L27–L30. [CrossRef]

19.   Wrodnigg, G.H.; Besenhard, J.O.; Winter, M. Cyclic and acyclic sulfites: new solvents and electrolyte additives for lithium ion batteries with graphitic anodes? *J. Power Sources* **2001**, *97*, 592–594. [CrossRef]

20.   Krämer, E.; Passerini, S.; Winter, M. Dependency of aluminum collector corrosion in lithium ion batteries on the electrolyte solvent. *ECS Electrochem. Lett.* **2012**, *1*, C9–C11. [CrossRef]

21.   Winter, M.; Moeller, K.-C.; Besenhard, J.O. *Lithium Batteries: Science and Technology*; Nazri, G.A., Pistoia, G., Eds.; Kluwer Academic Publishers: New York, NY, USA, 2004; p. 144.

22.   Safa, M.; Chamaani, A.; Chawla, N.; El-Zahab, B. Polymeric ionic liquid gel electrolyte for room temperature lithium battery applications. *Electrochim. Acta* **2016**, *213*, 587–593. [CrossRef]

23.   Safa, M.; Adelowo, E.; Chamaani, A.; Chawla, N.; Herndon, M.; Baboukani, A.; Wang, C.; El-Zahab, B. Poly (Ionic Liquid) based Composite Gel Electrolyte for Lithium Batteries. *ChemElectroChem* **2019**, *6*, 3319–3332. [CrossRef]

24.   Hilbig, P.; Ibing, L.; Wagner, R.; Winter, M.; Cekic-Laskovic, I. Ethyl methyl sulfone-based electrolytes for lithium ion battery applications. *Energies* **2017**, *10*, 1312. [CrossRef]

25.   Yamada, Y.; Furukawa, K.; Sodeyama, K.; Kikuchi, K.; Yaegashi, M.; Tateyama, Y.; Yamada, A. Unusual Stability of Acetonitrile-Based Superconcentrated Electrolytes for Fast-Charging Lithium-Ion Batteries. *J. Am.Chem. Soc.* **2014**, *136*, 5039–5046. [CrossRef]

26.   Isken, P.; Dippel, C.; Schmitz, R.; Schmitz, R.W.; Kunze, M.; Passerini, S.; Winter, M.; Lex-Balducci, A. High flash point electrolyte for use in lithium-ion batteries. *Electrochim. Acta* **2011**, *56*, 7530–7535. [CrossRef]

27. Schmitz, R.W.; Murmann, P.; Schmitz, R.; Müller, R.; Krämer, L.; Kasnatscheew, J.; Isken, P.; Niehoff, P.; Nowak, S.; Röschenthaler, G.-V.; et al. Investigations on novel electrolytes, solvents and SEI additives for use in lithium-ion batteries: Systematic electrochemical characterization and detailed analysis by spectroscopic methods. *Prog. Solid State Chem.* **2014**, *42*, 65–84. [CrossRef]

28. Korepp, C.; Santner, H.; Fujii, T.; Ue, M.; Besenhard, J.; Möller, K.-C.; Winter, M. 2-Cyanofuran—A novel vinylene electrolyte additive for PC-based electrolytes in lithium-ion batteries. *J. Power Sources* **2006**, *158*, 578–582. [CrossRef]

29. Brox, S.; Röser, S.; Streipert, B.; Hildebrand, S.; Rodehorst, U.; Qi, X.; Wagner, R.; Winter, M.; Cekic-Laskovic, I. Innovative, Non-Corrosive LiTFSI Cyanoester-Based Electrolyte for Safer 4 V Lithium-Ion Batteries. *ChemElectroChem* **2017**, *4*, 304–309. [CrossRef]

30. Brox, S.; Röser, S.; Husch, T.; Hildebrand, S.; Fromm, O.; Korth, M.; Winter, M.; Cekic-Laskovic, I. Alternative Single-Solvent Electrolytes Based on Cyanoesters for Safer Lithium-Ion Batteries. *ChemSusChem* **2016**, *9*, 1704–1711. [CrossRef]

31. Pohl, B.; Grünebaum, M.; Drews, M.; Passerini, S.; Winter, M.; Wiemhöfer, H.D. Nitrile functionalized silyl ether with dissolved LiTFSI as new electrolyte solvent for lithium-ion batteries. *Electrochim. Acta* **2015**, *180*, 795–800. [CrossRef]

32. Kirshnamoorthy, A.N.; Oldiges, K.; Winter, M.; Heuer, A.; Cekic-Laskovic, I.; Holm, C.; Smiatek, J. Electrolyte solvents for high voltage lithium ion batteries: ion correlation and specific anion effects in adiponitrile. *Chem. Chem. Phys.* **2018**, *20*, 25701–25715. [CrossRef] [PubMed]

33. Oldiges, K.; von Aspern, N.; Cekic-Laskovic, I.; Winter, M.; Brunklaus, G. Impact of Trifluoromethylation of Adiponitrile on Aluminum Dissolution Behavior in Dinitrile-Based Electrolytes. *J. Electrochem. Soc.* **2018**, *165*, A3773–A3781. [CrossRef]

34. Santner, H.J.; Möller, K.C.; Ivanco, J.; Ramsey, M.G.; Netzer, F.P.; Yamaguchi, S.; Besenhard, J.O.; Winter, M. Acrylic acid nitrile, a film-forming electrolyte component for lithium-ion batteries, which belongs to the family of additives containing vinyl groups. *J. Power Sources* **2003**, *119*, 368–372. [CrossRef]

35. Peled, E. The Electrochemical Behavior of Alkali and Alkaline Earth Metals in Nonaqueous Battery Systems—The Solid Electrolyte Interphase Model. *J. Electrochem. Soc.* **1979**, *126*, 2047–2051. [CrossRef]

36. Peled, E.; Golodnitsky, D.; Ardel, G. Advanced model for solid electrolyte interphase electrodes in liquid and polymer electrolytes. *J. Electrochem. Soc.* **1997**, *144*, L208–L210. [CrossRef]

37. Zhang, S.S.; Ding, M.S.; Xu, K.; Allen, J.; Jow, T.R. Understanding solid electrolyte interface film formation on graphite electrodes. *Electrochem. Solid State Lett.* **2001**, *4*, A206–A208. [CrossRef]

38. Winter, M.; Appel, W.K.; Evers, B.; Hodal, T.; Moller, K.C.; Schneider, I.; Wachtler, M.; Wagner, M.R.; Wrodnigg, G.H.; Besenhard, J.O. Studies on the anode/electrolyte interface in lithium ion batteries. *Chem. Mon.* **2001**, *132*, 473–486. [CrossRef]

39. Li, F.S.; Wu, Y.S.; Chou, J.; Winter, M.; Wu, N.L. A mechanically robust and highly ion-conductive polymer-blend coating for high-power and long-life lithium-ion battery anodes. *Adv. Mater* **2015**, *27*, 130–137. [CrossRef]

40. Winter, M. The solid electrolyte interphase–the most important and the least understood solid electrolyte in rechargeable Li batteries. *Z. Phys. Chem.* **2009**, *223*, 1395–1406. [CrossRef]

41. Santner, H.J.; Korepp, C.; Winter, M.; Besenhard, J.O.; Möller, K.-C. In-situ FTIR investigations on the reduction of vinylene electrolyte additives suitable for use in lithium-ion batteries. *Anal. Bioanal. Chem.* **2004**, *379*, 266–271. [CrossRef]

42. Märkle, W.; Lu, C.-Y.; Novák, P. Morphology of the Solid Electrolyte Interphase on Graphite in Dependency on the Formation Current. *J. Electrochem. Soc.* **2011**, *158*, A1478–A1482. [CrossRef]

43. Besenhard, J.O.; Wagner, M.W.; Winter, M.; Jannakoudakis, A.D.; Jannakoudakis, P.D.; Theodoridou, E. Inorganic Film-Forming Electrolyte Additives Improving the Cycling Behavior of Metallic Lithium Electrodes and the Self-Discharge of Carbon Lithium Electrodes. *J. Power Sources* **1993**, *44*, 413–420. [CrossRef]

44. Wang, K.; Xing, L.; Zhi, H.; Cai, Y.; Yan, Z.; Cai, D.; Zhou, H.; Li, W. High stability graphite/electrolyte interface created by a novel electrolyte additive: A theoretical and experimental study. *Electrochim. Acta* **2018**, *262*, 226–232. [CrossRef]

45. Wang, Y.; Nakamura, S.; Tasaki, K.; Balbuena, P.B. Theoretical Studies to Understand Surface Chemistry on Carbon Anodes for Lithium-Ion Batteries: How Does Vinylene Carbonate Play Its Role as an Electrolyte Additive? *J. Am.Chem. Soc.* **2002**, *124*, 4408–4421. [CrossRef] [PubMed]

46. Aurbach, D.; Gamolsky, K.; Markovsky, B.; Gofer, Y.; Schmidt, M.; Heider, U. On the use of vinylene carbonate (VC) as an additive to electrolyte solutions for Li-ion batteries. *Electrochim. Acta* **2002**, *47*, 1423–1439. [CrossRef]

47. Zhang, L.; Ma, Y.; Cheng, X.; Zuo, P.; Cui, Y.; Guan, T.; Du, C.; Gao, Y.; Yin, G. Enhancement of high voltage cycling performance and thermal stability of LiNi1/3Co1/3Mn1/3O2 cathode by use of boron-based additives. *Solid State Ionics* **2014**, *263*, 146–151. [CrossRef]

48. Meister, P.; Qi, X.; Kloepsch, R.; Krämer, E.; Streipert, B.; Winter, M.; Placke, T. Anodic Behavior of the Aluminum Current Collector in Imide-Based Electrolytes: Influence of Solvent, Operating Temperature, and Native Oxide-Layer Thickness. *ChemSusChem* **2017**, *10*, 804–814. [CrossRef]

49. Markevich, E.; Salitra, G.; Aurbach, D. Fluoroethylene carbonate as an important component for the formation of an effective solid electrolyte interphase on anodes and cathodes for advanced Li-ion batteries. *ACS Energy Lett.* **2017**, *2*, 1337–1345. [CrossRef]

50. Zugmann, S.; Moosbauer, D.; Amereller, M.; Schreiner, C.; Wudy, F.; Schmitz, R.; Schmitz, R.; Isken, P.; Dippel, C.; Muller, R.; et al. Electrochemical characterizaiton of electrolytes for lithium-ion batteries based on lithium difluoromono(oxalato)borate. *J. Power Sources* **2011**, *196*, 1417–1424. [CrossRef]

51. Dagger, T.; Grützke, M.; Reichert, M.; Haetge, J.; Nowak, S.; Winter, M.; Schappacher, F.M. Investigation of lithium ion battery electrolytes containing flame retardants in combination with the film forming electrolyte additives vinylene carbonate, vinyl ethylene carbonate and fluoroethylene carbonate. *J. Power Sources* **2017**, *372*, 276–285. [CrossRef]

52. Heine, J.; Hilbig, P.; Qi, X.; Niehoff, P.; Winter, M.; Bieker, P. Fluoroethylene carbonate as electrolyte additive in tetraethylene glycol dimethyl ether based electrolytes for application in lithium ion and lithium metal batteries. *J. Electrochem. Soc.* **2015**, *162*, A1094–A1101. [CrossRef]

53. Kasnatscheew, J.; Schmitz, R.W.; Wagner, R.; Winter, M.; Schmitz, R. Fluoroethylene Carbonate as an Additive for γ-Butyrolactone Based Electrolytes. *J. Electrochem. Soc.* **2013**, *160*, A1369–A1374. [CrossRef]

54. Winter, M.; Novak, P.; Monnier, A. Graphites for lithium-ion cells: The correlation of the first-cycle charge loss with the Brunauer-Emmett-Teller surface area. *J. Electrochem. Soc.* **1998**, *145*, 428–436. [CrossRef]

55. Hall, D.S.; Eldesoky, A.; Logan, E.R.; Tonita, E.M.; Ma, X.; Dahn, J.R. Exploring Classes of Co-Solvents for Fast-Charging Lithium-Ion Cells. *J. Electrochem. Soc.* **2018**, *165*, A2365–A2373. [CrossRef]

56. Pal, A.; Kumar, A. Excess Molar Volumes, Viscosities, and Refractive Indices of Diethylene Glycol Dimethyl Ether with Dimethyl Carbonate, Diethyl Carbonate, and Propylene Carbonate at (298.15, 308.15, and 318.15) K. *J. Chem. Eng. Data* **1998**, *43*, 143–147. [CrossRef]

57. Smart, M.C.; Ratnakumar, B.V.; Surampudi, S. Electrolytes for Low-Temperature Lithium Batteries Based on Ternary Mixtures of Aliphatic Carbonates. *J. Electrochem. Soc.* **1999**, *146*, 486–492. [CrossRef]

58. Ue, M.; Sasaki, Y.; Tanaka, Y.; Morita, M. Nonaqueous Electrolytes with Advances in Solvents. In *Electrolytes for Lithium and Lithium-Ion Batteries*; Jow, T.R., Xu, K., Borodin, O., Ue, M., Eds.; Springer: New York, NY, USA, 2014; pp. 93–165.

59. D'Aprano, A.; Fuoss, R.M. Conductance in isodielectric mixtures. I.n-butyronitrile with dioxane, benzene, and carbon tetrachloride. *J. Solut. Chem.* **1974**, *3*, 45–55. [CrossRef]

60. Wyman, J. Polarization and Dielectric Constant of Liquids. *J. Am.Chem. Soc.* **1936**, *58*, 1482–1486. [CrossRef]

61. Grünebaum, M.; Buchheit, A.; Lürenbaum, C.; Winter, M.; Wiemhöfer, H.-D. Ester-Based Battery Solvents in Contact with Metallic Lithium: Effect of Water and Alcohol Impurities. *J. Phys. Chem.* **2019**, *123*, 7033–7044. [CrossRef]

62. Bieker, G.; Winter, M.; Bieker, P. Electrochemical in situ investigations of SEI and dendrite formation on the lithium metal anode. *Phys. Chem. Chem. Phys.* **2015**, *17*, 8670–8679. [CrossRef] [PubMed]

63. Xu, K.; Ding, S.P.; Jow, T.R. Toward reliable values of electrochemical stability limits for electrolytes. *J. Electrochem. Soc.* **1999**, *146*, 4172–4178. [CrossRef]

64. Kasnatscheew, J.; Streipert, B.; Röser, S.; Wagner, R.; Laskovic, I.C.; Winter, M. Determining oxidative stability of battery electrolytes: validity of common electrochemical stability window (ESW) data and alternative strategies. *Phys. Chem. Chem. Phys.* **2017**, *19*, 16078–16086. [CrossRef] [PubMed]

65. Hilbig, P.; Ibing, L.; Streipert, B.; Wagner, R.; Winter, M.; Cekic-Laskovic, I. Acetonitrile-Based Electrolytes for Lithium-Ion Battery Application. *Curr. Top. Electrochem.* **2018**, *20*, 1–30.

66. Meister, P.; Jia, H.; Li, J.; Kloepsch, R.; Winter, M.; Placke, T. Best Practice: Performance and Cost Evaluation of Lithium Ion Battery Active Materials with Special Emphasis on Energy Efficiency. *Chem. Mater.* **2016**, *28*, 7203–7217. [CrossRef]

67. Besenhard, J.O.; Winter, M. Insertion reactions in advanced electrochemical energy storage. *Pure Appl. Chem.* **1998**, *70*, 603–608. [CrossRef]

68. Mller, T. Lithium ion battery automotive applications and requirements. In Proceedings of the Seventeenth Annual Battery Conference on Applications and Advances, Long Beach, CA, USA, 18 January 2002; pp. 113–118.

69. Wachtler, M.; Wagner, M.R.; Schmied, M.; Winter, M.; Besenhard, J.O. The effect of the binder morphology on the cycling stability of Li–alloy composite electrodes. *J. Electroanal. Chem.* **2001**, *510*, 12–19. [CrossRef]

70. Niehoff, P.; Passerini, S.; Winter, M. Interface Investigations of a Commercial Lithium Ion Battery Graphite Anode Material by Sputter Depth Profile X-ray Photoelectron Spectroscopy. *Langmuir* **2013**, *29*, 5806–5816. [CrossRef] [PubMed]

71. Niehoff, P.; Winter, M. Composition and Growth Behavior of the Surface and Electrolyte Decomposition Layer of/on a Commercial Lithium Ion Battery LixNi1/3Mn1/3Co1/3O2 Cathode Determined by Sputter Depth Profile X-ray Photoelectron Spectroscopy. *Langmuir* **2013**, *29*, 15813–15821. [CrossRef]

Review

Overview and Comparative Assessment of Single-Phase Power Converter Topologies of Inductive Wireless Charging Systems

Phuoc Sang Huynh *, Deepak Ronanki, Deepa Vincent and Sheldon S. Williamson *

Smart Transportation Electrification and Energy Research (STEER) Group, Department of Electrical, Faculty of Engineering and Applied Science, Computer and Software Engineering, University of Ontario Institute of Technology (Ontario Tech University), Oshawa, ON L1G 0C5, Canada; dronanki@ieee.org (D.R.); deepa.vincent@ontariotechu.net (D.V.)

* Correspondence: phuoc.huynh@ontariotechu.net (P.S.H.); sheldon.williamson@uoit.ca (S.S.W.)

Received: 22 March 2020; Accepted: 22 April 2020; Published: 1 May 2020

Abstract: The acquisition of inductive power transfer (IPT) technology in commercial electric vehicles (EVs) alleviates the inherent burdens of high cost, limited driving range, and long charging time. In EV wireless charging systems using IPT, power electronic converters play a vital role to reduce the size and cost, as well as to maximize the efficiency of the overall system. Over the past years, significant research studies have been conducted by researchers to improve the performance of power conversion systems including the power converter topologies and control schemes. This paper aims to provide an overview of the existing state-of-the-art of power converter topologies for IPT systems in EV charging applications. In this paper, the widely adopted power conversion topologies for IPT systems are selected and their performance is compared in terms of input power factor, input current distortion, current stress, voltage stress, power losses on the converter, and cost. The single-stage matrix converter based IPT systems advantageously adopt the sinusoidal ripple current (SRC) charging technique to remove the intermediate DC-link capacitors, which improves system efficiency, power density and reduces cost. Finally, technical considerations and future opportunities of power converters in EV wireless charging applications are discussed.

Keywords: AC–AC converters; battery chargers; electric vehicles; power conversion harmonics; wireless power transmission

---

1. Introduction

The electrification of transportation has been considered as a promising solution to tackle greenhouse gas emissions and fossil fuel depletion. To boost the market share of electric vehicles (EVs), their inherent issues such as limited driving range, long charging time, and costly and cumbersome energy storage systems should be resolved. Wireless charging technology can mitigate the aforementioned issues [1–10]. Wireless power transfer (WPT) enabling transferring energy from a source to a load without electrical contact has been extensively studied and successfully demonstrated using various techniques, namely, acoustic power transfer (APT) [11,12], radio frequency power transfer (RFPT) [13,14], optical power transfer (OPT) [15,16], capacitive power transfer (CPT) [17], and inductive power transfer (IPT) [18]. However, it is well demonstrated from the literature that the IPT technology is the most suitable for EV charging applications where the power requirement is form few to several kW, and the air gap varies from a few centimeters to a few meters [5]. Particularly, researchers and engineers have fitted the outcomes of the IPT to EV battery charging applications with various commercial products and standards [19]. The IPT chargers offer with several benefits such as safety, convenience, flexibility, weather immunity, and the possibility of range extension and battery volume reduction [1,8,10,20,21]. The wireless chargers can be deployed in residential garages, and office/service/shopping center parking lots for static wireless charging [22], or they can be placed at bus stops, taxi ranks, and traffic lights to implement quasi-dynamic wireless charging [23]. Moreover, dynamic wireless charging systems can be installed on the roads to constantly charge the EVs, in turn, to extend the driving range and reduce the battery volume of the vehicles [24–26].

Essentially, an IPT charging system comprises an inductive coupling coil pair, compensation networks, primary converters to generate high-frequency inputs, and a secondary rectifier to convert AC to DC current to charge the battery. In the IPT charging systems, power electronic converters make a significant contribution to the size and cost, and efficiency of the overall system. Typically, dual-stage conversion (AC–DC–AC) systems have been employed to excite the IPT systems, as shown in Figure 1a. The dual-stage converter topologies are intensively studied and widely used in industry [27–29]. The main advantage of these topologies is that each conversion stage can be separately designed and controlled to optimize specific performance indices. However, the presence of multiple conversion stages and a bulk DC-link capacitor increases the cost, size, and weight of the system. In recent years, the use of matrix converters (MCs) for feeding the IPT systems has drawn increasing attention [30–38]. MCs enable direct conversion of low-frequency AC inputs (50–60 Hz) to high-frequency outputs (up to 85 kHz) without any intermediate conversion stage; therefore, they enhance the system performance in terms of power density, reliability, and cost [32,39]. The single-phase matrix converter-based IPT systems remove the DC-link energy storage elements in the primary side to absorb double line frequency ripple, thus it appears on the battery side. Sinusoidal ripple current (SRC) charging technique reported in [40–45] allows batteries to be charged by double line frequency (100 or 120 Hz) current with minor side effects on their performance. Therefore, matrix converter-based IPT systems can use the sinusoidal charging technique advantageously and remove the intermediate DC-link capacitor. The single-stage EV IPT charging system using MCs is illustrated in Figure 1b.

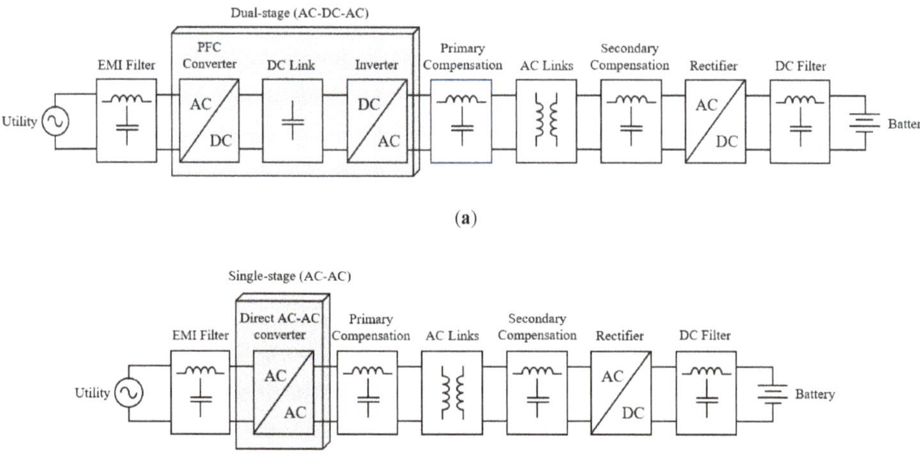

Figure 1. Configuration of electric vehicle (EV) inductive power transfer (IPT) systems with (**a**) dual-stage power conversion and (**b**) single-stage power conversion. PFC, power factor correction; EMI, electromagnetic interference.

This paper aims to provide an extensive overview of single-phase power conversion topologies employed in static wireless charging. Then, a comprehensive performance comparison between the conventional dual-stage (power factor correction (PFC) and full-bridge voltage source inverter (VSI)) and single-stage topologies including the buck-derived full-bridge (FB)MC and boost-derived FBMC in the IPT EV charging application is presented. The comparison involves the input power factor, input current distortion, power losses, switching stress, and normalized cost, while taking into account the requirements of Standard J2954 [19] established by the Society of Automotive Engineers (SAE). The Standard SAE J2954 defines acceptable criteria for interoperability, electromagnetic compatibility, electromagnetic field (EMF), minimum performance, safety, and testing for wireless charging of light-duty electric vehicles. Table 1 shows the power classes, operating frequency, and efficiency performance targets of the WPT systems defined in the SAE J2954. As can been seen, four wireless power transfer (WPT) classes are defined based on the maximum input volt-amp (VA) drawn from the grid by the primary side or ground assembly (GA) electronics. The input real power depends on the input power factor, while the output power depends on the efficiency of the system. The SAE J2954 specifies that WPT systems should be operated at a single nominal frequency of 85 kHz. However, for WPT systems using frequency control to compensate operating variations, their operating frequency must be tuned in the band of 81.38 to 90.00 kHz. Finally, the improvement opportunities for each of the IPT charging topologies are discussed in this paper.

Table 1. Wireless power transfer (WPT) power classification for light-duty electric vehicles—SAE J2954.

WPT Levels	1	2	3	4
Maximum AC input power (kVA)	3.7	7.7	11	22
Minimum target efficiency at nominal alignment (%)	>85	>85	>85	To be defined (TBD)
Minimum target efficiency at offset position (%)	>80	>80	>80	TBD
Operating frequency (kHz)			81.38–90 (typical 85)	

2. Power Converter Topologies for Inductive Wireless Charging

In this section, an overview of front-end converter topologies for WPT applications is provided. They can be classified into two groups, namely dual-stage and single-stage based on the power conversion stages. The classification of single-phase converter topologies for IPT systems is shown in Figure 2.

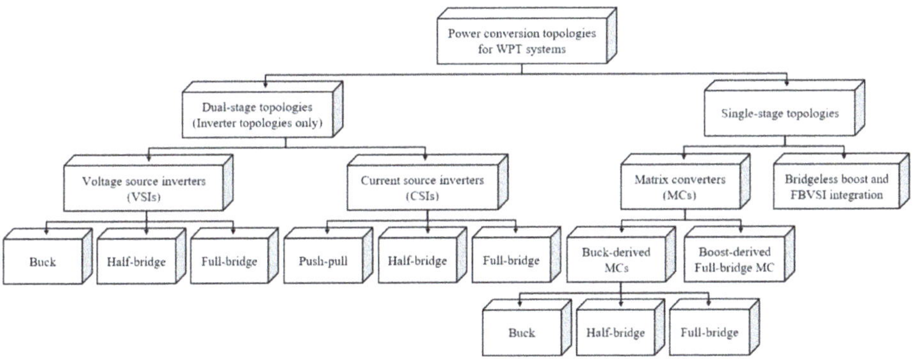

Figure 2. Classification of front-end converter topologies for IPT applications. FB, full-bridge; WPT, wireless power transfer.

2.1. Dual-Stage Power Conversion

A front-end AC–DC converter is used to convert the supply AC voltage to an intermediate DC-link voltage and to shape the input current for both PFC and harmonic reduction. A comprehensive review for the PFC rectifiers is presented in [46,47]. For the inversion stage, a current-source inverter (CSI) or a VSI can be employed.

Two CSI topologies commonly used in IPT systems are push-pull, half-bridge [48–53], and full-bridge [54,55]. Figure 3a–c shows the configuration of CSIs. The requirement of blocking diodes and bulky inductors that increases the size and cost of the whole IPT system is one of the major drawbacks of the CSIs. A single parallel compensating capacitor in the primary circuit is normally used with CSIs; however, the inverter switches suffer high voltage stress in high-power applications [48,50,52,54]. In order to overcome this drawback, a parallel-series *CC* compensation circuit is introduced in [53,55]. The CSIs combined with the parallel-series *CC* compensation circuit mitigates current and voltage stress on inverter switches and harmonic contents in primary coil current.

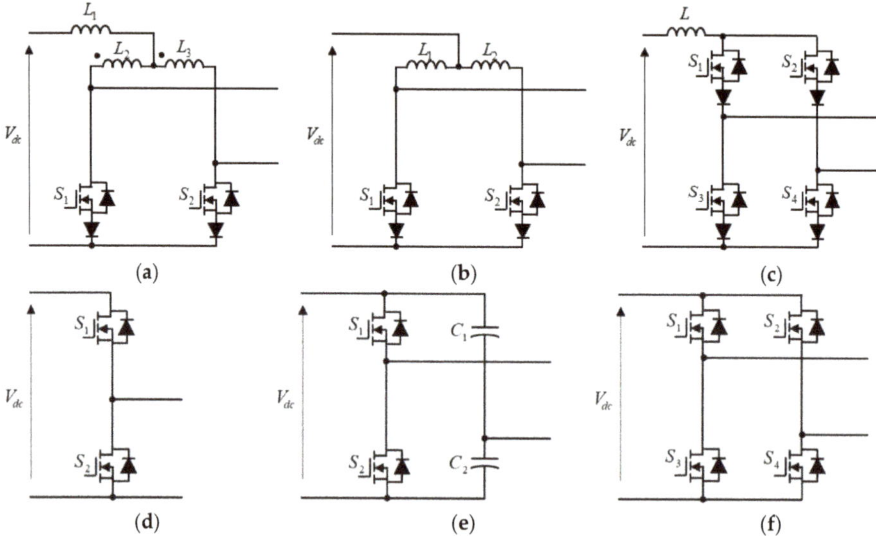

Figure 3. Inverter topologies: (**a**) push-pull, (**b**) half-bridge CSI, (**c**) full-bridge CSI, (**d**) buck, (**e**) half-bridge VSI, and (**f**) full-bridge VSI.

For VSI topologies, buck, half-bridge, and full-bridge topologies shown in Figure 3d–f can be used in the IPT systems, and they are compatible with single capacitor series, *LCL*, and *LCCL* compensation networks [1,21,56–66]. The series compensation is simple and cost-effective. However, under light load conditions or in the absence of the receiver, the system experiences severe instability [67,68]. *LCL* or *LCCL* tanks overcome these issues and have a high tolerance to coil misalignments [68]. However, a significant amount of lower-order harmonics in the output current of the VSIs connected with *LCL* and *LCCL* compensation circuits deviates zero-phase-angle operation of the inverters, increasing their switching losses [69]. Moreover, the inductors in *LCL* and *LCCL* compensation circuits must be designed precisely as the effective power transfer capability is highly sensitive to the inductance value [57,61]. Figure 4 shows the compatibility of the inverter types and primary compensation circuits of the IPT systems. Table 2 shows the comparison of the inverter topologies regarding the component requirement. It can be seen that the CSIs require more components than the VSIs.

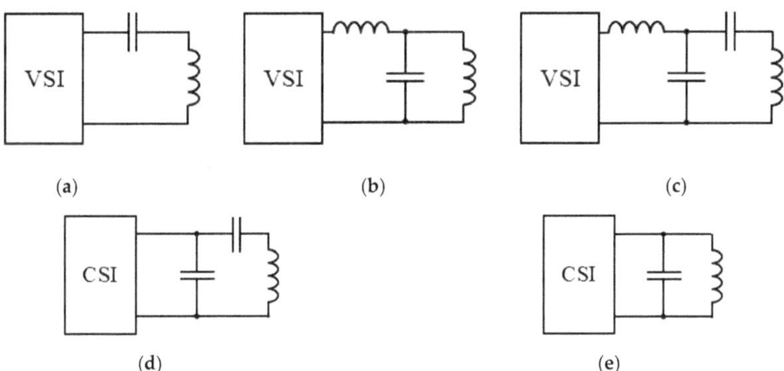

Figure 4. Compatibility between inverter types and primary compensation circuits in IPT systems. (**a**) Series *C*, (**b**) *LCL*, (**c**) *LCCL*, (**d**) parallel *C*, and (**e**) parallel-series *CC*.

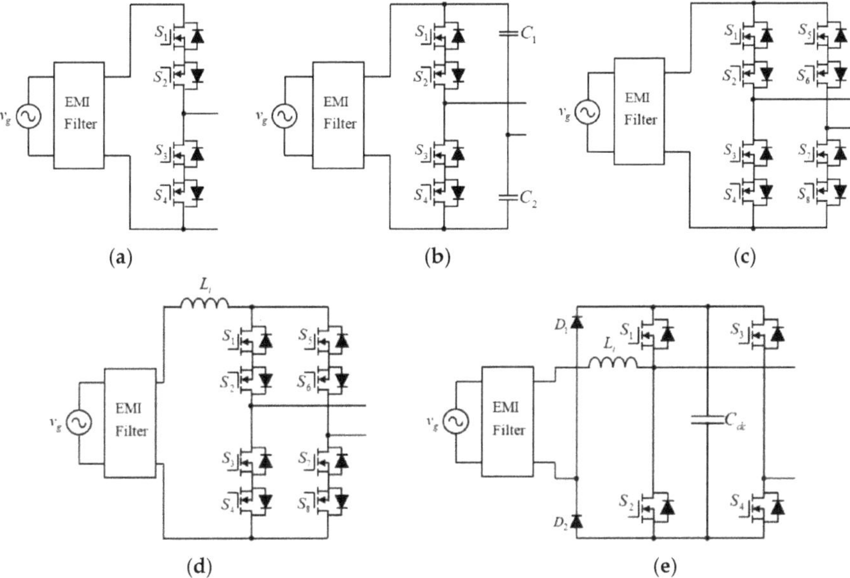

Figure 5. Single-stage conversion topologies: (**a**) buck matrix converter, (**b**) half-bridge matrix converter, (**c**) full-bridge matrix converter, (**d**) boost-derived matrix converter, and (**e**) bridgeless boost converter.

Table 2. Power conversion topologies and control schemes of the inductive power transfer (IPT) systems.

Power Conversion Topologies		Figure	Component Requirement		Control Schemes
			Switches	Passive Components	
Dual-stage (excluding front-end PFC stage)	Current-source push-pull	Figure 3a	2 reverse blocking	1 inductor, 1 phase-splitting transformer	• Variable switching frequency • DC-link voltage control • Secondary-side control • Dual-side control
	Current-source half-bridge	Figure 3b	2 reverse blocking	2 inductors	
	Current-source full-bridge	Figure 3c	4 reverse blocking	1 inductor	• Variable switching frequency • DC-link voltage control • Pulse width modulation (phase-shift control) • Secondary-side control • Dual-side control
	Voltage-source buck	Figure 3d	2 reverse conducting	None	• Discrete energy injection • Variable switching frequency • DC-link voltage control • Pulse width modulation (duty cycle control) • Secondary-side control • Dual-side control
	Voltage-source half-bridge	Figure 3e	2 reverse conducting	2 capacitors	
	Voltage-source full-bridge	Figure 3f	4 reverse conducting	None	• Discrete energy injection • Variable switching frequency • DC-link voltage control • Pulse width modulation (phase-shift control) • Secondary-side control • Dual-side control
Single-stage	Buck MC	Figure 5a	2 bidirectional	None	• Discrete energy injection • Pulse width modulation (duty cycle control) • Secondary-side control • Dual-side control
	Half-bridge MC	Figure 5b	2 bidirectional	2 capacitors	
	Full-bridge MC	Figure 5c	4 bidirectional	None	• Discrete energy injection • Pulse width modulation (phase-shift control) • Secondary-side control • Dual-side control
	Boost-derived full-bridge MC	Figure 5e	4 bidirectional	1 inductor	• Pulse width modulation • Secondary-side control • Dual-side control
	Bridgeless boost	Figure 5f	2 diodes, 4 reverse conducting	1 inductor, 1 capacitor	• Pulse width modulation • Secondary-side control • Dual-side control

2.2. Single-Stage AC–AC Conversion

Matrix converters (MCs) are considered as a prominent candidate for powering the WPT systems with only single-stage power conversion. Several MCs including buck [36,37], half-bridge [30,31], and full-bridge [35] have been introduced to IPT applications in the literature. All MCs reported in [30,31,35–37] have a buck-derived configuration, as shown in Figure 5a–c, thus line-current regulation is compromised. In EV charging application, if a highly nonlinear diode-bridge rectifier is used at the battery side, there will be severe line current distortion and power factor deterioration, as explained in [70]. In [35], a secondary active full-bridge rectifier whose phase shift angle follows the line-voltage waveform is used to shape the line current. In this topology, the primary and secondary converters must be controlled synchronously in every switching cycle, which increases the implementation complexity.

In order to overcome the above issue, a boost-derived full-bridge MC (FBMC) compatible with a primary parallel-series CC compensation network is proposed in [38]. The proposed converter topology is able to shape the line current and regulate power flow through two control loops, which are similar to those of a conventional boost converter. In [39], a single-stage topology integrating bridgeless boost PFC converter and full-bridge VSI is proposed for IPT applications. The converter is operated in discontinuous conduction mode, thereby, the line current control loop is eliminated. However, in discontinuous conduction mode (DCM), the converter incurs more current stress, losses, and electromagnetic interference (EMI) problems, which is not suitable for high-power applications. Figure 5d and e show the configuration of boost-derived FBMC and bridgeless boost PFC converter in IPT systems.

3. Power Control Schemes

Figure 6 shows the classification of power control schemes for IPT systems. Power control in IPT systems can be implemented on the primary side, secondary side, or both sides. The secondary side control is suitable for the IPT applications where multiple secondary coils are coupled to a single primary coil. In these applications, the frequency and the magnitude of primary current are fixed, and the power flow is controlled on the secondary side by an active rectifier or a back-end DC–DC converter illustrated in Figure 6 for each secondary coil [30,58,59,71–74]. These topologies are normally employed in long-power track systems where a constant track current is required to power independent secondary coils.

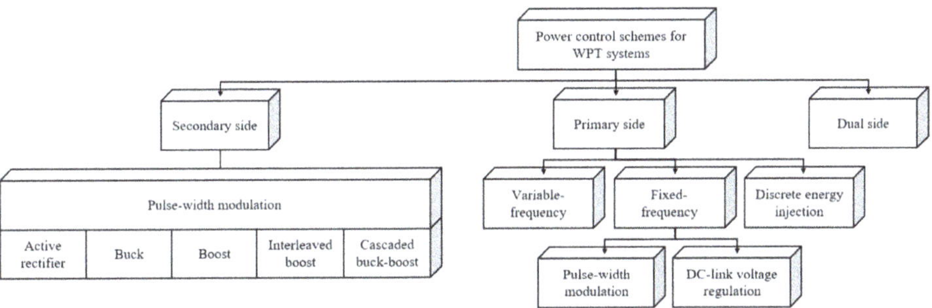

Figure 6. Classification of power control schemes for IPT applications.

However, in charging applications where only one a secondary coil is coupled to a primary coil and keeping the secondary-side configuration as simple as possible is a priority, the primary side control is selected. The primary side control can be divided into three groups: fixed frequency, variable frequency, and discrete energy injection. In fixed frequency control, the switching frequency of the inverter is kept at a constant value, which is slightly different from the primary resonant frequency to offer soft-switching operation. In order to control the power flow, the phase (phase shift control) or the duty cycle of the inverter switches is varied [75,76]. This allows the inverters to produce output voltage/current with variable pulse width. The other way to regulate the power flow with the fixed switching frequency is controlling the input DC voltage of the inverter using a front-end DC–DC converter [48]. For the variable switching frequency control scheme, the duty cycle of the gating signals is maintained constant at 50% and the switching frequency is varied to regulate the output power [49]. However, if the operating frequency is largely different from the resonant frequency, the resonant tank will incur a large circulating current, causing an efficiency drop in the overall system owing to large losses in switches and the coils. Moreover, the bifurcation phenomenon must be carefully considered in this control technique [77]. In [36], a discrete energy injection control is used for the matrix buck converter in order to control the magnitude of the primary current. The control technique reduces the switching frequency and enables soft switching. However, the zero-crossing detection of primary high-frequency current that is required to ensure the converter to be operated in zero current switching (ZCS) conditions is an implementation challenge. Moreover, current sag occurs during the zero-crossing of its single-phase input voltage, which degrades the average power transferred. The dual-side control is suitable for bidirectional IPT systems where power flow can be regulated in both directions by controlling the duty cycle of the primary and secondary converters and the phase-shift between them [56,57,78]. Table 2 shows the compatibility of power conversion topologies and control schemes of the IPT applications.

4. Performance Comparison and Discussion

In this section, the performance of a conventional dual-stage topology and two potential single-stage topologies including buck- and boost-derived FBMCs are compared regarding input power quality, current stress, voltage stress, power losses, and cost.

4.1. Design Considerations

The conventional dual-stage IPT charging system is illustrated in Figure 7a. At the front end, a conventional boost rectifier is used to shape the grid current and maintain a constant DC voltage V_{dc} across DC-link capacitor C_i. As a bulky and costly inductor is required for the CSIs, an FBVSI is the most common choice at the primary side to generate a high-frequency voltage (v_p) feeding the primary coil. A series-series (SS) compensation topology is used because it is simple and cost-effective, and its primary compensation is independent of the coupling coefficient and load. Moreover, the efficiency of SS compensation is high even at a low coupling coefficient [68,79]. In order to maximize the power transfer capabilities and minimize the VA rating of the primary inverter, the resonant circuits at both sides of the coupling are usually tuned to the same resonant frequency (ω_0) equal to the switching frequency of the inverter.

$$\omega_0 = \frac{1}{\sqrt{L_p C_p}} = \frac{1}{\sqrt{L_s C_s}}$$

(1)

where L_p and L_s are primary and secondary coil self-inductances, respectively, and C_p and C_s are primary and secondary tuning capacitors, respectively.

Power regulation is conducted using the phase-shift control at the primary inverter side. Considering an ideal IPT system operating at the resonance frequency (ω_0), power transferred from the primary to the secondary side can be given by

$$P_o = \frac{8 V_{dc} V_b}{\pi^2 \omega_0 M} \sin \pi D_p$$

(2)

where D_p is the duty cycle of the primary voltage (v_p) and M is the mutual inductance and can be calculated as

$$M = k \sqrt{L_p L_s}$$

(3)

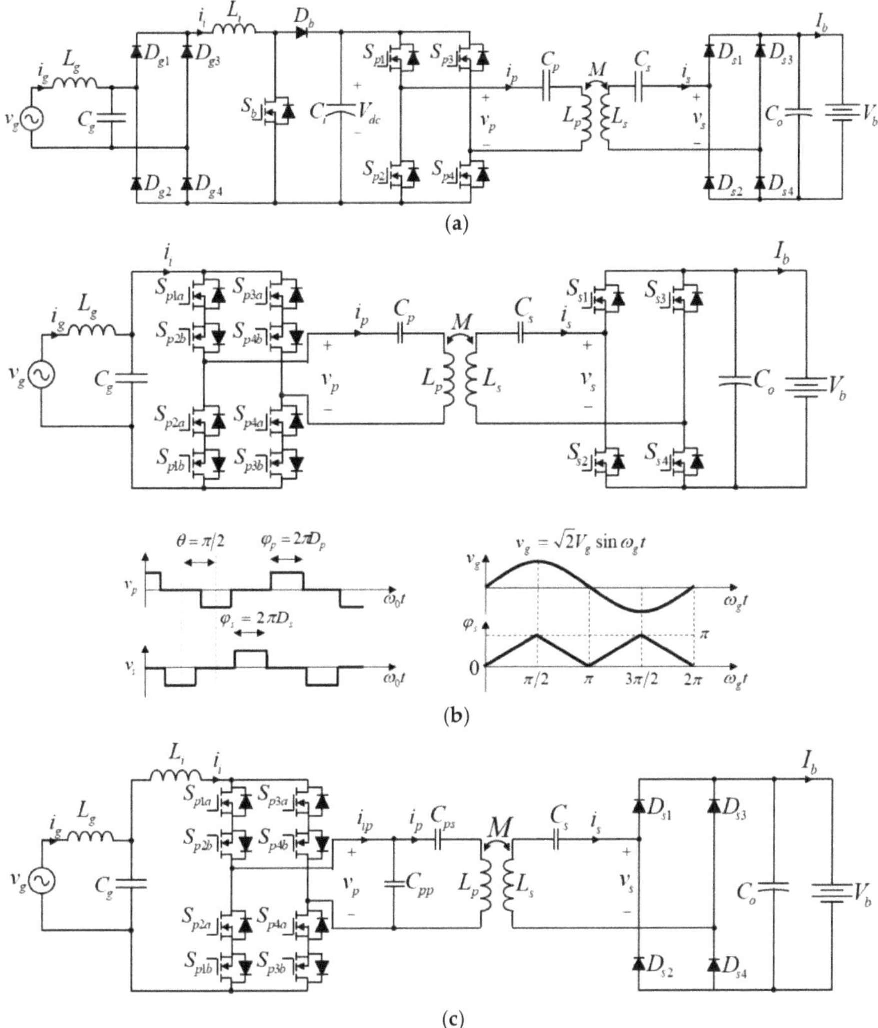

Figure 7. IPT charging system fed by (**a**) dual-stage power converter (PFC and full-bridge VSI), (**b**) buck-derived FBMC, and (**c**) boost-derived FBMC.

In EV wireless charging applications, the coupling coefficient k may be in the range of 0.1–0.3. In a dual-stage topology, the major drawback is low power density owing to multiple conversion stages and a bulky DC-link capacitor. The reduction of the number of power conversion stages can be obtained using MCs. Figure 7b shows the IPT charging system using a buck-derived FBMC. The FBMC constituted by four bidirectional switches can directly convert low frequency (50–60 Hz) grid voltage to resonant frequency (85 kHz) voltage feeding the inductive coil. During the positive half cycle of the grid voltage v_g, switches S_{pnb} (n = 1, 2, 3, 4) are turned on and switches S_{pna} are the phase-shift pulse-width modulation (PWM) strategy. Otherwise, during the negative half-cycle, the switches S_{pna} are kept on and switches S_{pnb} are controlled by the phase-shift PWM strategy.

An active rectifier is employed in the battery side for shaping the input current. The primary and secondary converters are synchronized in every switching cycle so that primary voltage v_p is 90°

lagging with secondary voltage v_s, and the duty cycle of the secondary voltage is controlled following grid voltage waveform to correct input current, as shown in Figure 7b. The power transferred is controlled by adjusting the duty cycle of the primary voltage.

$$P_o = \frac{4\sqrt{2}V_g V_b}{\pi^2 \omega_0 M}\sin \pi D_p \tag{4}$$

where V_g is the root mean square (RMS) value of the grid voltage.

Although the buck-derived FBMC-based IPT charging system removes the intermediate conversion stage, high-frequency communication is required to synchronize the PWM patterns of the primary and secondary converters in every switching cycle, which increases the control complexity. The boost-derived MC can solve the above issue. It is capable of correcting the grid current and regulating power flow through two control loops, which are similar to those of a conventional boost converter. Figure 7c shows an IPT topology fed by a boost-derived FBMC [38]. On the primary side, parallel-series CC compensation is used to reduce voltage stress on the MC switches. The tuning capacitor C_{ps} is selected so as to limit the maximum peak of v_p across the converter switches. It is desirable to restrict v_p to 0.5~0.7 of the rating voltage of the switches [48]. The switching scheme and controller design for the boost-derived full-bridge matrix converter are described in [38]. Tables 3 and 4 show the passive component design and the switching stresses of the converter components.

Table 3. Passive component design. FB, full-bridge.

Topologies	Components	Parameters	
Dual-stage [80]	Boost inductor L_i	Inductance $L_i = \frac{1}{\%\Delta I_i}\frac{V_g^2}{P_{o(max)}f_s}\left(1 - \frac{\sqrt{2}V_g}{V_{dc}}\right)$	Peak current $\hat{I}_{Li} = \sqrt{2}\frac{P_o}{V_g} + \frac{\Delta I_i}{2}$
	Boost capacitor C_i	Capacitance $C_i = \frac{1}{\%\Delta V_{dc}}\frac{P_o}{\omega_g V_{dc}^2}$	Peak voltage $\hat{V}_{Ci} = V_{dc} + \frac{\Delta V_{dc}}{2}$
	Compensation capacitors C_p and C_s	Capacitance $C_p = \frac{1}{\omega_0^2 L_p}$ $C_s = \frac{1}{\omega_0^2 L_s}$	Peak voltage $\hat{V}_{Cp} = \frac{4V_b}{\pi\omega_0^2 C_p M}$ $\hat{V}_{Cs} = \frac{4V_{dc}}{\pi\omega_0^2 C_s M}$
Buck-derived FB MC [81,82]	Compensation capacitors C_p and C_s	Capacitance $C_p = \frac{1}{\omega_0^2 L_p}$ $C_s = \frac{1}{\omega_0^2 L_s}$	Peak voltage $\hat{V}_{Cp} = \frac{4V_b}{\pi\omega_0^2 C_p M}$ $\hat{V}_{Cs} = \frac{4\sqrt{2}V_g}{\pi\omega_0^2 C_s M}$
	Boost inductor L_i	Inductance $L_i = \frac{\sqrt{2}V_g}{2\Delta I_i f_s}$	Peak current $\hat{I}_{Li} = \frac{\sqrt{2}P_o}{V_g} + \frac{\Delta I_i}{2}$
Boost-derived FB MC [38]	Compensation capacitors C_{pp}, C_{ps}, and C_s **Note:** C_{ps} is designed to limit the peak of primary voltage v_p, which is the voltage stress on MC switches.	Capacitance $C_s = \frac{1}{\omega_0^2 L_s}$ $C_{pp} = \dfrac{L_p - \frac{1}{\omega_0^2 C_{ps}}}{\frac{\omega_0^4 M^4}{R_{oeq}^2} + \omega_0^2\left(L_p - \frac{1}{\omega_0^2 C_{ps}}\right)^2}$	Peak voltage $\hat{V}_{Cs} = \frac{\pi P_o}{\omega_0 C_s V_b}$ $\hat{V}_{Cpp} = \hat{V}_p = \sqrt{\left(\frac{4V_b}{\pi M}\right)^2\left(L_p - \frac{1}{\omega_0^2 C_{ps}}\right)^2 + \left(\frac{\pi\omega_0 M P_o}{V_b}\right)^2}$ $\hat{V}_{Cps} = \frac{4V_b}{\pi\omega_0^2 C_{ps} M}$

Table 4. Stress on switching devices.

Topologies	Components	Current Stress	Voltage Stress
Dual-stage	Boost switch S_b and diode D_b	Peak current $\hat{I}_{Sb} = \hat{I}_{Db} = \frac{\sqrt{2}P_o}{V_g} + \frac{\Delta I_i}{2}$	Break down voltage $\hat{V}_{Sb} = \hat{V}_{Db} > V_{dc} + \frac{\Delta V_{dc}}{2}$
	Primary inverter switches S_{pn} ($n = 1, 2, 3, 4$)	Peak current $\hat{I}_{Spn} = \frac{4V_b}{\pi \omega_0 M}$	Break down voltage $\hat{V}_{Spn} > V_{dc} + \frac{\Delta V_{dc}}{2}$
	Secondary rectifier diodes D_{sn} ($n = 1, 2, 3, 4$)	Peak current $\hat{I}_{Dsn} = \frac{4V_{dc}}{\pi \omega_0 M}$	Break down voltage $\hat{V}_{Dsn} > V_b$
Buck-derived FB MC	Primary inverter switches S_{pnx} ($n = 1, 2, 3, 4$ and $x = a, b$)	Peak current $\hat{I}_{Spnx} = \frac{4V_b}{\pi \omega_0 M}$	Break down voltage $\hat{V}_{Spnx} > \sqrt{2}V_g$
	Secondary rectifier switches S_{sn} ($n = 1, 2, 3, 4$)	Peak current $\hat{I}_{Ssn} = \frac{4\sqrt{2}V_g}{\pi \omega_0 M}$	Break down voltage $\hat{V}_{Ssn} > V_b$
Boost-derived FB MC	Primary inverter switches S_{pnx} ($n = 1, 2, 3, 4$ and $x = a, b$)	Peak current $\hat{I}_{Spnx} > \frac{\sqrt{2}P_o}{V_g} + \frac{\Delta I_i}{2}$	Break down voltage $\hat{V}_{Spnx} > \hat{V}_p$
	Secondary rectifier diodes D_{sn} ($n = 1, 2, 3, 4$)	Peak current $\hat{I}_{Dsn} > \frac{\pi P_o}{V_b}$	Break down voltage $\hat{V}_{Dsn} > V_b$

4.2. Performance Comparison

In this section, the performance of IPT configurations is compared in terms of input power factor, input current distortion, current stress, voltage stress, power losses on converters, and normalized cost. The IPT charging systems are designed in compliance with the level 1 (WPT1) of static wireless charging standard for light-duty vehicles provided in SAE J2954 technical information report [19] with power rating P_o = 3.3 kW, operating frequency f_s = 85 kHz, grid voltage V_g = 208 V, and battery voltage V_b = 300–400 V. The parameters of the charging system with each type of power conversion configuration are shown in Table 5. All the components are designed based on Tables 3 and 4. The selection of components is based on their maximum current and voltage stresses. Note that available discrete Rohm SiC MOSFETs and Schottky diodes are considered for all power conversion topologies. Moreover, *LC* filters are used as interfaces between the grid and the charging systems to limit current harmonic injection owing to the switching power converters. The *LC* filters are designed based on the spectrum analysis of the input current waveforms (i_i). The details of the selected components for different power conversion stages are listed in Table 6.

Figure 8 shows the typical waveforms of IPT charging systems with different power supply topologies. It can be seen that the absence of DC-link energy storage in MC-based topologies causes a double line frequency fluctuation in transferred power. This results in a fluctuating charging current as shown in Figure 8b and c. As reported in [40–44,83,84], batteries can be charged by double line frequency (100 or 120 Hz) current with negligible side effects on their performance.

Table 5. Specifications of IPT charging systems.

Topologies	Parameter	Symbol	Value	Unit
Dual-stage	Primary, secondary, mutual inductance	L_p, L_s, M	356, 328, 65	μH
	Compensation capacitors	C_p, C_s	10, 11	nF
	Boost inductor	L_i	0.215	mH
	DC-bus capacitor	C_i	1540	μF
	DC-bus voltage	V_{dc}	400	V
	Grid inductor	L_g	0.215	mH
	Grid capacitor	C_g	0.78	μF
	Output capacitor	C_o	500	μF
Buck-derived FBMC	Primary, secondary, mutual inductance	L_p, L_s, M	111, 111, 24	μH
	Compensation capacitors	C_p, C_s	32, 32	nF
	Grid inductor	L_g	0.215	mH
	Grid capacitor	C_g	0.78	μF
	Output capacitor	C_o	500	μF
Boost-derived FBMC	Primary, secondary, mutual inductance	L_p, L_s, M	111, 111, 24	μH
	Compensation capacitors	C_{ps}, C_{pp}, C_s	43, 115, 32	nF
	Boost inductor	L_i	0.215	mH
	Grid inductor	L_g	0.036	mH
	Grid capacitor	C_g	0.136	μF
	Output capacitor	C_o	500	μF

Table 6. Main components of power conversion stages.

Topologies	Components	Symbol	Part Number	Quantity
Dual-stage	Front-end rectifier diodes	$D_{gn}{}^{*}$	SCS240AE2C-ND	4
	Boost diode	D_b	SCS240AE2C-ND	1
	Boost switch	S_b	SCT3060ALGC11-ND	1
	Primary inverter switches	$S_{pn}{}^{*}$	SCT3120ALHRC11-ND	4
	Secondary rectifier diodes	$D_{sn}{}^{*}$	SCS230AE2HRC-ND	4
	Boost inductor	L_i	HF5712-561M-25AH	2 parallel
	DC-bus capacitor	C_i	LGN2X221MELC50	7 parallel
	Grid inductor	L_g	HF5712-561M-25AH	2 parallel
	Grid capacitor	C_g	B32656T7394K000	2 parallel
Buck-derived FBMC	Primary MC switches	$S_{pna}, S_{pnb}{}^{*}$	SCT3030ALGC11-ND	8
	Secondary rectifier diodes	$S_{sn}{}^{*}$	SCT3060ALGC11-ND	4
	Grid inductor	L_g	HF5712-561M-25AH	2 parallel
	Grid capacitor	C_g	B32656T7394K000	2 parallel
Boost-derived FBMC	Primary MC switches	$S_{pna}, S_{pnb}{}^{*}$	SCT2080KEC-ND	8
	Secondary rectifier diodes	$D_{sn}{}^{*}$	SCS240AE2C-ND	4
	Boost inductor	L_i	HF5712-561M-25AH	2 parallel
	Grid inductor	L_g	HF467-980M-25AV	2 parallel
	Grid capacitor	C_g	B32654A1683K000	2 parallel

$^{*}n = 1, 2, 3, 4.$

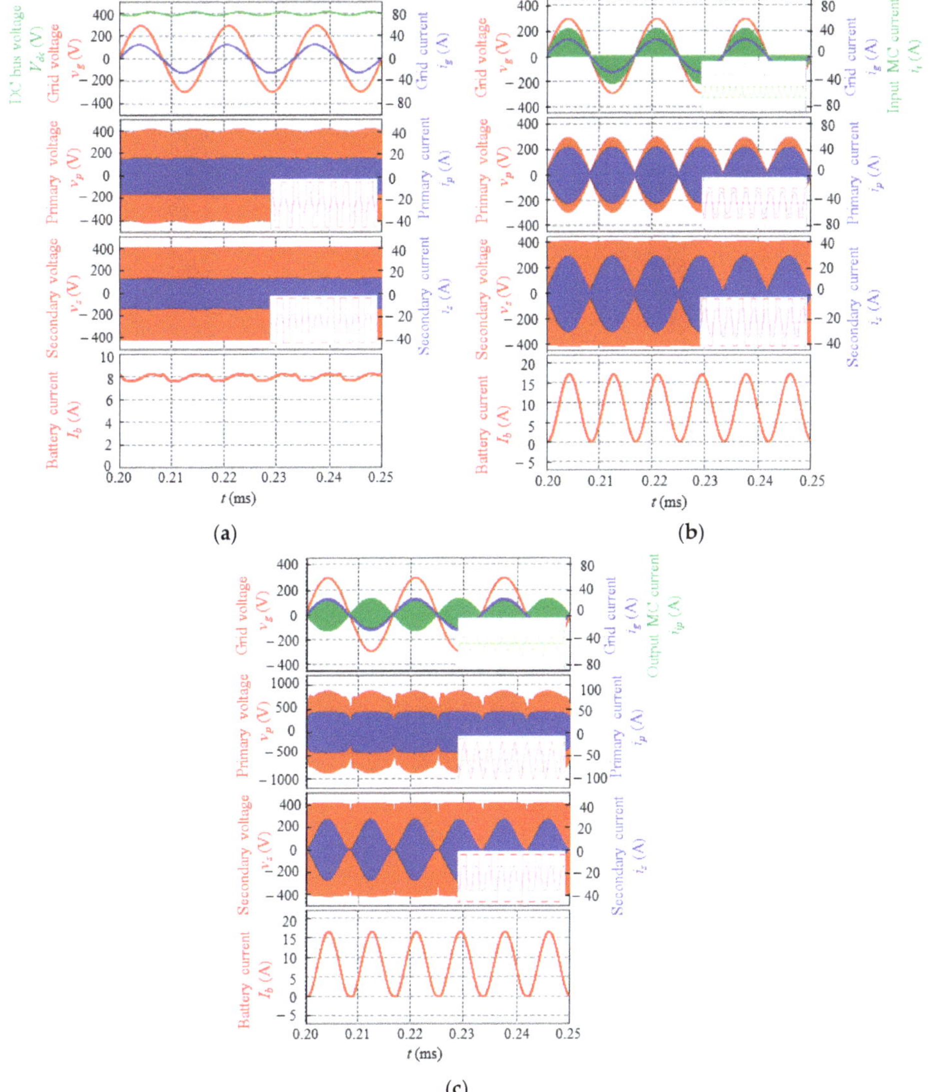

Figure 8. Simulation waveforms of IPT charging systems fed by (**a**) dual-stage power converter (PFC and full-bridge VSI), (**b**) buck-derived FBMC, and (**c**) boost-derived FBMC.

4.2.1. Input Power Factor and Input Current Distortion

An EV charger must ensure a good grid power quality with a high power factor and low current distortion. All three topologies provide sinusoidal grid currents with the power factor of 0.99. Figure 9a shows total harmonic distortion (THD) of the grid current under different load conditions (20%, 50%, and 100% of load). It can be seen that the three topologies can be preferred in order of boost-derived FBMC, dual-stage converter, and buck-derived FBMC, regarding grid current distortion. Despite having the identical input *LC* filter, the buck-derived FBMC injects higher current harmonics to the grid than dual-stage topology, because its input current is discontinuous. The boost-derived FBMC has the continuous input current with ripple frequency at a twofold switching frequency, thereby gaining the significant harmonic reduction of grid current with a smaller input filter.

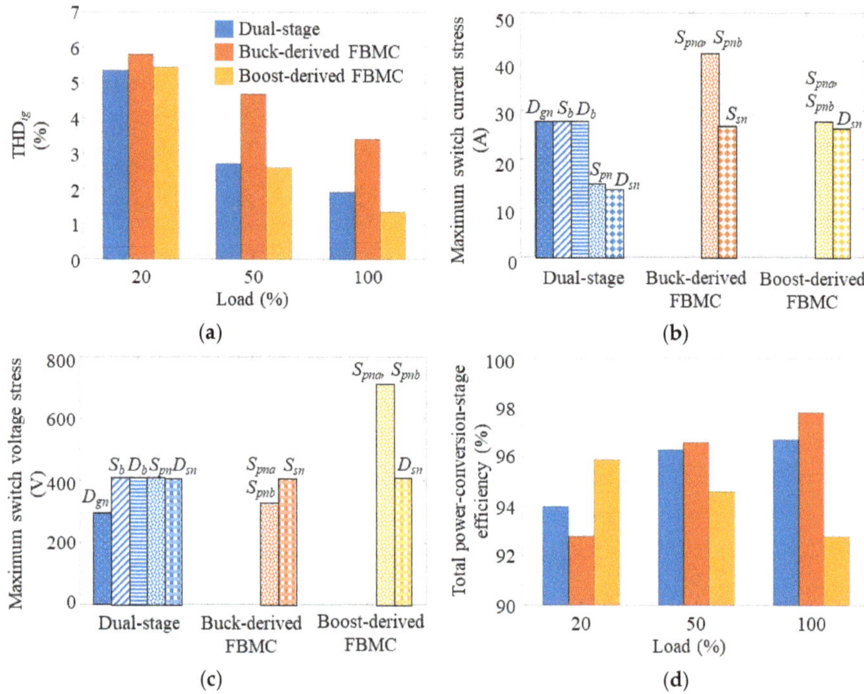

Figure 9. Performance comparison of three IPT charging systems: (**a**) grid current total harmonic distortion (THD), (**b**) switch current stress, (**c**) switch voltage stress, and (**d**) power-conversion-stage efficiency.

4.2.2. Switching Stress

Figure 9b and c show the maximum current and voltage stress on the converter switches. Although the parallel-series CC compensation is used, the switches of boost-derived FBMC still suffer from high voltage stress. The buck-derived FBMC is characterized by low switch voltage stress (grid voltage peak) and high switch current stress. The dual-stage topology exhibits the lowest switch current stress in the primary inverter and secondary rectifier.

4.2.3. Efficiency and Loss Distribution

The losses on the conversion stages of each system are simulated and analyzed using the thermal modules in PSIM simulation. The efficiency of the power conversion stages of each system versus various output power is illustrated in Figure 9d. It is clear that the efficiency of the buck-derived FBMC system is the highest (almost 98%) at full load conditions, but it decreases gradually to 93% at the light load conditions. In contrast, the efficiency of the boost-derived FBMC system steadily increases from 92.5% to 96% when the load decreases from 100% to 20%. The dual-stage system maintains fairly high efficiency (94~96.5%) in a wide load range.

The detailed loss distribution of the three systems is shown in Figure 10. It can be observed that the conduction losses of primary converters dominate the total losses of power conversion stages. In the dual-stage system, the conduction losses of the front-end rectifier and the primary inverter are the two major parts. For the single-stage systems, the conduction losses of matrix converters contribute to the largest proportions (>60%).

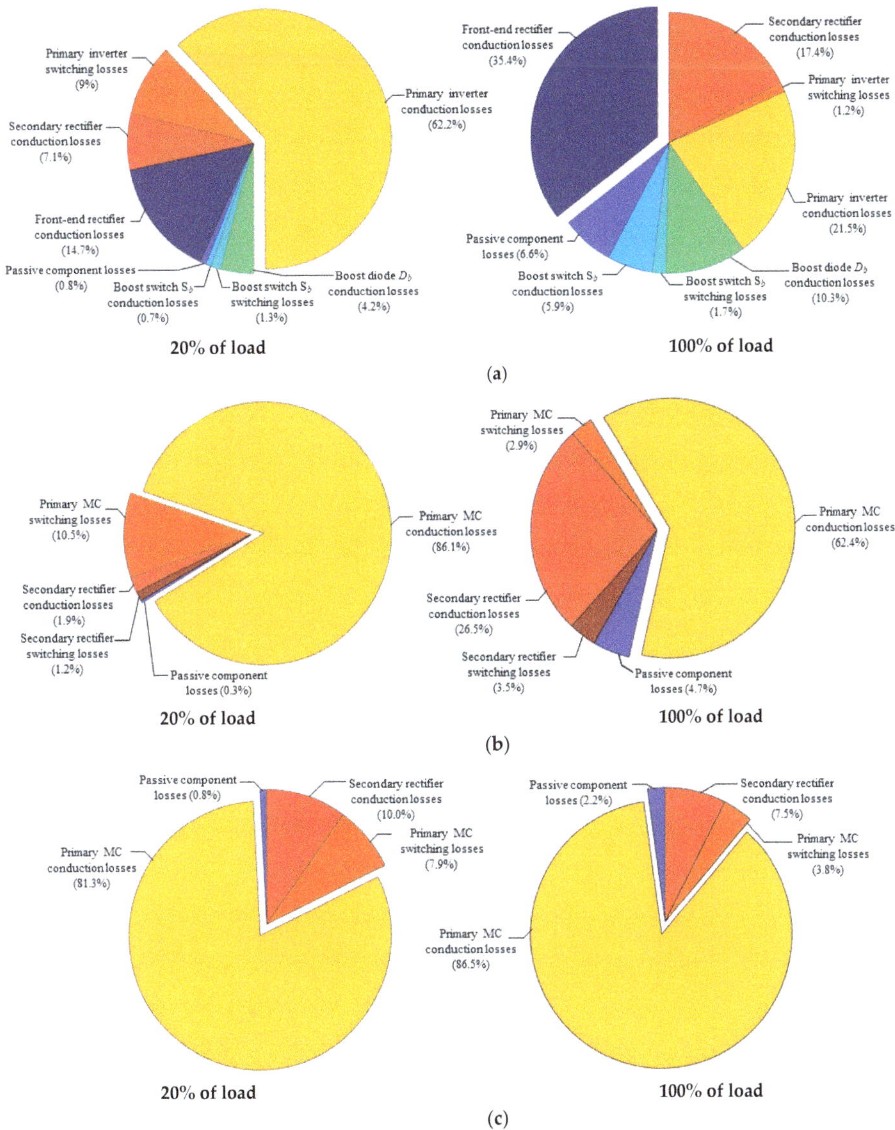

Figure 10. Loss distribution: (**a**) dual-stage topology, (**b**) buck-derived FBMC, and (**c**) boost-derived FBMC.

4.2.4. Cost

Cost is also an important quantity to evaluate the performance of a power converter. The cost structure of each charging system excluding inductive coupling coils and compensation networks is illustrated in Figure 11. The costs of the power conversion stages are calculated based on the component cost provided in Table 7. In order to simplify the cost analysis, the auxiliary cost including printed circuit board (PCB) cost, cooling system cost, and housing cost is assumed to be 10% of the power converter cost. Note that MOSFETs are driven by isolated gate drivers, and MOSFETs having common-source connection utilize a common gate driver power supply to reduce the system cost. This shows that the cost of single-stage systems is lower than that of the dual-stage counterpart. The buck-derived FBMC system is the most cost-effective solution, as it presents 8.4% less cost than the dual-stage system. It is found that the costs of the passive components dominate in the dual-stage system, whereas the semiconductor devices of matrix converters occupy the largest portions in the total cost of the single-stage systems.

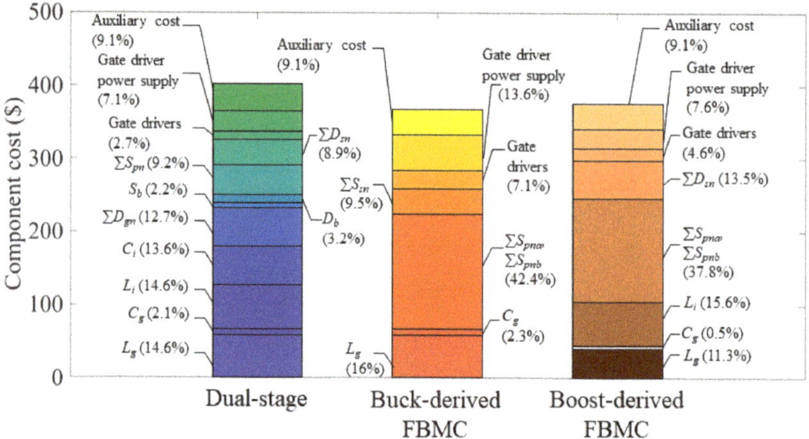

Figure 11. Component cost structure of the charging system excluding inductive coupling coils and compensation networks.

Table 7. Cost of components.

Components	Manufacturer Part Number	Rating	Unit Cost ($)
Diode	SCS230AE2HRC-ND	650 V/30 A	8.97
	SCS240AE2C-ND	650 V/40 A	12.75
MOSFET	SCT3120ALHRC11-ND	650 V/21 A	9.27
	SCT3060ALGC11-ND	650 V/39 A	8.74
	SCT3030ALGC11-ND	650 V/70 A	19.46
	SCT2080KEC-ND	1200 V/40 A	17.77
Gate driver IC	UCC5390SCD	N/A	2.16
Gate driver supply	R12P21503D	+15 V/−3 V/2 W	7.11
Inductor	HF467-980M-25AV	25 A/72 μH	21.15
	HF5712-561M-25AH	25 A/430 μH	29.25
Capacitor	LGN2X221MELC50 (Electrolytic)	600 V/220 μF	7.78
	B32656T7394K000 (Film)	500 V/0.39 μF	4.23
	B32654A1683K000	500 V/0.068 μF	1.01

4.3. Discussions

From the above analysis, it can be observed that the three IPT charging systems have their own advantages and disadvantages. A comparison summary of the three IPT charging systems is shown in Figure 12, where performance indices are presented in a scale range from 1 (worst) to 3 (best). In order to evaluate the efficiencies of the three systems, their average values under all load conditions are considered. The switching stresses are assessed based on the product of the maximum current and voltage stresses. Figure 12 shows that the buck-derived FBMC surpasses the other counterparts with the advantages of high efficiency, cost reduction and possible power density improvement due to less component count, while the boost-derived FBMC has the greatest input current quality due to the feature of the continuous input current with ripple frequency at a twice switching frequency. The conventional dual-stage topology has the lowest stress on switching devices, and its efficiency maintains a comparable high level over wide load range. Moreover, the dual-stage converter topology is highly matured in terms of manufacturability and control as it has been developed by many manufacturers and widely used in the industry. Also, this topology allows each converter stage to be separately designed and optimized.

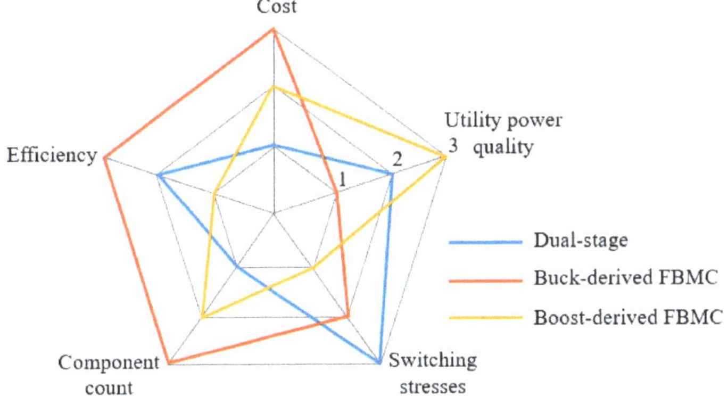

Figure 12. Comparison summary of the dual-stage and single-stage IPT charging systems.

5. Future Trends and Opportunities

Over the past decade, there have been significant developments in power converter topologies and control schemes for EV IPT charging. One of the important challenges is the design of high-frequency power converters for IPT to meet future requirements. Still, there is a lot of scope for further improvements to enhance the performance in terms of efficiency, power density, scalability, and reliability to promote the IPT-based systems for EV charging. Reducing power losses in power conversion from the source to the input of the coil is a vital factor in improving efficiency. One of such initiatives is to develop advanced soft-switching modulation techniques for the existing converter topologies and new reduced-switch-count converter topologies to reduce switching losses. This is expected to improve the thermal design and power density of the overall system. Several soft-switching control schemes have been reported in the literature. Generally, they can be divided into three groups: with auxiliary DC-DC converters [85–87], with variable resonant networks [88,89], and with active inverter/rectifier control [59,90–92]. However, the proposed control schemes require extra DC-DC converters or resonant components, or have an operation range limitation and high control complexity.

Direct power conversion topologies such as matrix converters can be one of the possible candidates with the elimination of life-limited bulky DC capacitors employing enhanced charging techniques like SRC charging. Another important performance enhancement is employing wide bandgap devices in the existing converter topologies or development of advanced topologies, which can operate at higher switching frequencies with low switching losses [93]. This can boost the performance of power converters in the wireless charging applications [94]. The application of gallium nitride (GaN) in IPT systems has opened up a new scope in improvement in power transfer and power density. These devices have a low voltage drop, ability to operate at the higher switching frequency, and comparatively lower thermal generation during operation, which allows for passive cooling to increase the converter power density and cost-effectiveness [95]. However, some challenges regarding the manufacturing process, packaging with higher current ratings, the gate driver design, device characterization, busbar layout, thermal management, and reliability need to be addressed. Therefore, much more research initiatives in the aforementioned issues could decide the power density of the converter. In addition, employing the GaN devices allows an increase in switching frequencies even at higher current standards, which improves the performance of WPT, such as transfer distance extension, higher tolerance to coil misalignment, and passive component size reduction. Furthermore, the magnetic integration can be used to integrate magnetic components of power converters, compensation networks, and coupling coils, in turn, to enhance using higher flux density material to reduce the system size and losses. One of the possible ways is the utilization of advanced materials and nanotechnology to reduce the size and weight of passive components.

In the recent days, modular power converters with fault-tolerance are demanded due to industrial requirements such as flexibility in assembly, manufacturing process, scalability and reduced mean time to failure (MTTF). Some of the possible potential candidates are multiphase parallel inverter [96], modular multilevel converter [97], parallel IPT power supply topologies [98] and extreme fast charging architectures [99] to improve the output power capability and fault tolerance for WPT systems, which can open up more research in developing advanced power topologies and fault-tolerant control schemes. Other stimulating research areas for developing technology are bidirectional power flow, integration with hybrid energy storage systems and multiple energy sources [100]. However, these areas are still under research that further attention and investigation for developing advanced multi-port converter topologies and newly advanced control schemes must comply with future charging standards to promote IPT systems for EV charging applications.

6. Conclusions

This paper presents an extensive overview of power conversion topologies and control schemes for IPT-based EV charging applications. The design considerations and performance evaluation of the conventional dual-stage topology and two potential single-stage topologies including buck- and boost-derived FBMCs were discussed. It is concluded that the conventional dual-stage topology has the lowest stress on switching devices, and the boost-derived FBMC provides the greatest input current quality. On the other hand, the superiorities of the buck-derived FBMC over the other topologies are high efficiency, low component count and cost. However, further investigation on IPT-based charging systems is needed including scalability to higher power levels, adoption of soft-switching technology, fault-tolerability technology, active power decoupling methods, magnetic integration, green energy based-IPT systems, multi-mode operation systems, wide bandgap device technology, high-performance advanced and non-invasive control schemes. With the continual improvements and the aforementioned advancements, IPT-based systems will definitely increase the availability and economic viability of the EVs in the near future.

Author Contributions: Conceptualization, P.S.H. and S.S.W.; Software, P.S.H.; Validation, P.S.H.; Formal Analysis, P.S.H.; Investigation, P.S.H.; Resources, P.S.H. and D.R.; Data Curation, P.S.H.; Writing Original Draft Preparation, P.S.H. and D.R.; Writing Review & Editing, P.S.H, D.R., D.V., and S.S.W.; Visualization, P.S.H., D.R., and D.V.; Supervision, S.S.W.; Project Administration, S.S.W.; Funding Acquisition, S.S.W. All authors have read and agreed to the published version of the manuscript.

Funding: This work was supported by the Natural Sciences and Engineering Research Council (NSERC), Canada.

Conflicts of Interest: The authors declare no conflict of interest.

References

1.   Khaligh, A.; Dusmez, S. Comprehensive topological analysis of conductive and inductive charging solutions for plug-in electric vehicles. *IEEE Trans. Veh. Technol.* **2012**, *61*, 3475–3489. [CrossRef]

2.   Wang, C.S.; Stielau, O.H.; Covic, G.A. Design considerations for acontactless electric vehicle battery charger. *IEEE Trans. Ind. Electron.* **2005**, *52*, 1308–1314. [CrossRef]

3.   Hui, S.Y.R.; Ho, W.W.C. A new generation of universal contactless battery charging platform for portable consumer electronic equipment. *IEEE Trans. Power Electron.* **2005**, *20*, 620–627. [CrossRef]

4. Daga, A.; Miller, J.M.; Long, B.R.; Kacergis, R.; Schrafel, P.; Wolgemuth, J. Electric fuel pumps for wireless power transfer: Enabling rapid growth in the electric vehicle market. *IEEE Power Electron. Mag.* **2017**, *4*, 24–35. [CrossRef]

5. Kazmierkowski, M.P.; Moradewicz, A.J. Unplugged but connected: Review of contactless energy transfer systems. *IEEE Ind. Electron. Mag.* **2012**, *6*, 47–55. [CrossRef]

6. Bosshard, R.; Kolar, J.W. Inductive power transfer for electric vehicle charging: Technical challenges and tradeoffs. *IEEE Power Electron. Mag.* **2016**, *3*, 22–30. [CrossRef]

7. Choi, S.Y.; Gu, B.W.; Jeong, S.Y.; Rim, C.T. Advances in wireless power transfer systems for roadway-powered electric vehicles. *IEEE J. Emerg. Sel. Top. Power Electron.* **2015**, *3*, 18–36. [CrossRef]

8. Ahmad, A.; Alam, M.S.; Chabaan, R. A comprehensive review of wireless charging technologies for electric vehicles. *IEEE Trans. Transp. Electrific.* **2018**, *4*, 38–63. [CrossRef]

9. Mi, C.C.; Buja, G.; Choi, S.Y.; Rim, C.T. Modern advances in wireless power transfer systems for roadway powered electric vehicles. *IEEE Trans. Ind. Electron.* **2016**, *63*, 6533–6545. [CrossRef]

10. Covic, G.A.; Boys, J.T. Modern trends in inductive power transfer for transportation applications. *IEEE J. Emerg. Sel. Top. Power Electron.* **2013**, *1*, 28–41. [CrossRef]

11. Roes, M.G.L.; Duarte, J.L.; Hendrix, M.A.M.; Lomonova, E.A. Acoustic energy transfer: A review. *IEEE Trans. Ind. Electron.* **2013**, *60*, 242–248. [CrossRef]

12. Hu, Y.; Zhang, X.; Yang, J.; Jiang, Q. Transmitting electric energy through a metal wall by acoustic waves using piezoelectric transducers. *IEEE Trans. Ultrason. Ferroelectr. Freq. Control* **2003**, *50*, 773–781. [CrossRef] [PubMed]

13. Chow, E.Y. Wireless powering and the study of RF propagation through ocular tissue for development of implantable sensors. *IEEE Trans. Antennas Propag.* **2011**, *59*, 2379–2387. [CrossRef]

14. Sasaki, S.; Tanaka, K.; Maki, K.I. Microwave power transmission technologies for solar power satellites. *Proc. IEEE* **2013**, *101*, 1438–1447. [CrossRef]

15. Wang, N.; Zhu, Y.; Wei, W.; Chen, J.; Liu, S.; Li, P.; Wen, Y. One-to-multipoint laser remote power supply system for wireless sensor networks. *IEEE Sens. J.* **2012**, *12*, 389–396. [CrossRef]

16. Sahai, A.; Graham, D. Optical wireless power transmission at long wavelengths. In Proceedings of the International Conference on Space Optical Systems and Applications, Santa Monica, CA, USA, 11–13 May 2011; pp. 164–170.

17. Kline, M.; Izyumin, I.; Boser, B.; Sanders, S. Capacitive power transfer for contactless charging. In Proceedings of the 26th Annual IEEE Applied Power Electronics Conference and Exposition, Fort Worth, TX, USA, 6–11 March 2011; pp. 1398–1404.

18. Covic, G.A.; Boys, J.T. Inductive Power Transfer. *Proc. IEEE* **2013**, *101*, 1276–1289. [CrossRef]

19. *J2954: Wireless Power Transfer for Light-Duty Plug-In/Electric Vehicles and Alignment Methodology*; SAE International: Warrendale, PA, USA, 2016.

20. Musavi, F.; Eberle, W. Overview of wireless power transfer technologies for electric vehicle battery charging. *IET Power Electron.* **2014**, *7*, 60–66. [CrossRef]

21. Siqi, L.; Mi, C.C. Wireless power transfer for electric vehicle applications. *IEEE J. Emerg. Sel. Top. Power Electron.* **2015**, *3*, 4–17. [CrossRef]

22. *J1772: SAE Electric Vehicle and Plug in Hybrid Electric Vehicle Conductive Charge Coupler*; SAE International: Warrendale, PA, USA, 2017.

23. Mohamed, A.A.S.; Lashway, C.R.; Mohammed, O. Modeling and feasibility analysis of quasi-dynamic WPT system for EV applications. *IEEE Trans. Transp. Electrif.* **2017**, *3*, 343–353. [CrossRef]

24. Cirimele, V.; Diana, M.; Bellotti, F.; Berta, R.; Sayed, N.E.; Kobeissi, A.; Guglielmi, P.; Ruffo, R.; Khalilian, M.; Ganga, A.L.; et al. The fabric ICT platform for managing wireless dynamic charging road lanes. *IEEE Trans. Veh. Technol.* **2020**, *69*, 2501–2512. [CrossRef]

25. Miller, J.M.; Onar, O.C.; White, C.; Campbell, S.; Coomer, C.; Seiber, L.; Sepe, R.; Steyerl, A. Demonstrating dynamic wireless charging of an electric vehicle: The benefit of electrochemical capacitor smoothing. *IEEE Power Electron. Mag.* **2014**, *1*, 12–24. [CrossRef]

26. Cirimele, V.; Diana, M.; Freschi, F.; Mitolo, M. Inductive power transfer for automotive applications: State-of-the-art and future trends. *IEEE Trans. Ind. Appl.* **2018**, *54*, 4069–4079. [CrossRef]

27. Everts, J.; Krismer, F.; Keybus, J.V.d.; Driesen, J.; Kolar, J.W. Optimal ZVS modulation of single-phase single-stage bidirectional DAB AC–DC converters. *IEEE Trans. Power Electron.* **2014**, *29*, 3954–3970. [CrossRef]

28. Bosshard, R.; Kolar, J.W. All-SiC 9.5 kW/dm3 On-Board Power Electronics for 50 kW/85 kHz Automotive IPT System. *IEEE J. Emerg. Sel. Topics Power Electron.* **2017**, *5*, 419–431. [CrossRef]

29. Kolar, R.B.a.J.W. Multi-objective optimization of 50 kW/85 kHz IPT system for public transport. *IEEE J. Emerg. Sel. Topics Power Electron.* **2016**, *4*, 1370–1382.

30. Koushki, B.; Jain, P.; Bakhshai, A. A bi-directional AC-DC converter for electric vehicle with no electrolytic capacitor. In Proceedings of the IEEE 7th International Symposium on Power Electronics for Distributed Generation Systems (PEDG), Vancouver, BC, Canada, 27–30 June 2016; pp. 1–8.

31. Shin, C.J.; Lee, J.Y. An electrolytic capacitor-less bi-directional EV on-board charger using harmonic modulation technique. *IEEE Trans. Power Electron.* **2014**, *29*, 5195–5203. [CrossRef]

32. Moghaddami, M.; Anzalchi, A.; Sarwat, A.I. Single-stage three-phase AC–AC matrix converter for inductive power transfer systems. *IEEE Trans. Ind. Electron.* **2016**, *63*, 6613–6622. [CrossRef]

33. Bac, N.X.; Vilathgamuwa, D.M.; Madawala, U.K. A SiC-based matrix converter topology for inductive power transfer system. *IEEE Trans. Power Electron.* **2014**, *29*, 4029–4038.

34. Ecklebe, A.; Lindemann, A.; Schulz, S. Bidirectional switch commutation for a matrix converter supplying a series resonant load. *IEEE Trans. Power Electron.* **2009**, *24*, 1173–1181. [CrossRef]

35. Weerasinghe, S.; Madawala, U.K.; Thrimawithana, D.J. A matrix converter-based bidirectional contactless grid interface. *IEEE Trans. Power Electron.* **2017**, *32*, 1755–1766. [CrossRef]

36. Li, H.L.; Hu, A.P.; Covic, G.A. A direct AC–AC converter for inductive power-transfer systems. *IEEE Trans. Power Electron.* **2012**, *27*, 661–668. [CrossRef]

37. Sulistyono, W.; Enjeti, P. A series resonant AC-to-DC rectifier with high-frequency isolation. *IEEE Trans. Power Electron.* **1995**, *10*, 784–790. [CrossRef]

38. Samanta, S.; Rathore, A.K. A new inductive power transfer topology using direct ac–ac converter with active source current waveshaping. *IEEE Trans. Power Electron.* **2018**, *33*, 5565–5577. [CrossRef]

39. Liu, J.; Chan, K.W.; Chung, C.Y.; Chan, N.H.L.; Liu, M.; Xu, W. Single-stage wireless-power-transfer resonant converter with boost bridgeless power-factor-correction rectifier. *IEEE Trans. Ind. Electron.* **2018**, *65*, 2145–2155. [CrossRef]

40. Kwon, M.; Choi, S. An electrolytic capacitorless bidirectional EV charger for V2G and V2H applications. *IEEE Trans. Power Electron.* **2017**, *32*, 6792–6799. [CrossRef]

41. Prasad, R.; Namuduri, C.; Kollmeyer, P. Onboard unidirectional automotive G2V battery charger using sine charging and its effect on Li-ion batteries. In Proceedings of the IEEE Energy Conversion Congress and Exposition, Montreal, QC, Canada, 20–24 September 2015; pp. 6299–6305.

42. Beh, H.Z.; Covic, G.A.; Boys, J.T. Effects of pulse and DC charging on lithium iron phosphate (LiFePO4) batteries. In Proceedings of the IEEE Energy Conversion Congress and Exposition, Denver, CO, USA, 15–19 September 2013; pp. 315–320.

43. Jeong, S.; Jeong, Y.; Kwon, J.; Kwon, B. A soft-switching single-stage converter with high efficiency for a 3.3-kW on-board charger. *IEEE Trans. Ind. Electron.* **2019**, *66*, 6959–6967. [CrossRef]

44. Xue, L.; Shen, Z.; Boroyevich, D.; Mattavelli, P.; Diaz, D. Dual active bridge-based battery charger for plug-in hybrid electric vehicle with charging current containing low frequency ripple. *IEEE Trans. Power Electron.* **2015**, *30*, 7299–7307. [CrossRef]

45. Chen, L.; Wu, S.; Shieh, D.; Chen, T. Sinusoidal-ripple-current charging strategy and optimal charging frequency study for Li-ion batteries. *IEEE Trans. Ind. Electron.* **2013**, *60*, 88–97. [CrossRef]

46. Singh, B.; Singh, B.N.; Chandra, A.; Al-Haddad, K.; Pandey, A.; Kothari, D.P. A review of single-phase improved power quality AC-DC converters. *IEEE Trans. Ind. Electron.* **2003**, *50*, 962–981. [CrossRef]

47. Singh, B.; Singh, B.N.; Chandra, A.; Al-Haddad, K.; Pandey, A.; Kothari, D.P. A review of three-phase improved power quality AC-DC converters. *IEEE Trans. Ind. Electron.* **2004**, *51*, 641–660. [CrossRef]

48. Kamineni, A.; Neath, M.J.; Covic, G.A.; Boys, J.T. A mistuning-tolerant and controllable power supply for roadway wireless power systems. *IEEE Trans. Power Electron.* **2017**, *32*, 6689–6699. [CrossRef]

49. Si, P.; Hu, A.P.; Malpas, S.; Budgett, D. A frequency control method for regulating wireless power to implantable devices. *IEEE Trans. Biomed. Circuits Syst.* **2008**, *2*, 22–29. [CrossRef] [PubMed]

50. Green, A.W.; Boys, J.T. 10 kHz inductively coupled power transfer-concept and control. In Proceedings of the IET 5th International Conference on Power Electronics and Variable-Speed Drives, London, UK, 26–28 October 1994; pp. 694–699.

51. Tian, J.; Hu, A.P. A DC-voltage-controlled variable capacitor for stabilizing the ZVS frequency of a resonant converter for wireless power transfer. *IEEE Trans. Power Electron.* **2017**, *32*, 2312–2318. [CrossRef]

52. Kamineni, A.; Covic, G.A.; Boys, J.T. Self-tuning power supply for inductive charging. *IEEE Trans. Power Electron.* **2017**, *32*, 3467–3479. [CrossRef]

53. Samanta, S.; Rathore, A.K.; Thrimawithana, D.J. Bidirectional current-fed half-bridge (C) (LC)–(LC) configuration for inductive wireless power transfer system. *IEEE Trans. Ind. Appl.* **2017**, *53*, 4053–4062. [CrossRef]

54. Samanta, S.; Rathore, A.K. Wireless power transfer technology using full-bridge current-fed topology for medium power applications. *IET Power Electron.* **2016**, *9*, 1903–1913. [CrossRef]

55. Samanta, S.; Rathore, A.K. A new current-fed CLC transmitter and LC receiver topology for inductive wireless power transfer application: Analysis, design, and experimental results. *IEEE Trans. Transp. Electrif.* **2015**, *1*, 357–368. [CrossRef]

56. Twiname, R.P.; Thrimawithana, D.J.; Madawala, U.K.; Baguley, C.A. A dual-active bridge topology with a tuned CLC network. *IEEE Trans. Power Electron.* **2015**, *30*, 6543–6550. [CrossRef]

57. Thrimawithana, D.J.; Madawala, U.K. A generalized steady-state model for bidirectional IPT systems. *IEEE Trans. Power Electron.* **2013**, *28*, 4681–4689. [CrossRef]

58. Asa, E.; Colak, K.; Czarkowski, D. Analysis of a CLL resonant converter with semi-bridgeless active rectifier and hybrid control. *IEEE Trans. Ind. Electron.* **2015**, *62*, 6877–6886. [CrossRef]

59. Colak, K.; Asa, E.; Bojarski, M.; Czarkowski, D.; Onar, O.C. A novel phase-shift control of semibridgeless active rectifier for wireless power transfer. *IEEE Trans. Power Electron.* **2015**, *30*, 6288–6297. [CrossRef]

60. Nam, I.I.; Dougal, R.A.; Santi, E. Novel unity-gain frequency tracking control of series–series resonant converter to improve efficiency and receiver positioning flexibility in wireless charging of portable electronics. *IEEE Trans. Ind. Appl.* **2015**, *51*, 385–397. [CrossRef]

61. Madawala, U.K.; Thrimawithana, D.J. A bidirectional inductive power interface for electric vehicles in V2G systems. *IEEE Trans. Ind. Electron.* **2011**, *58*, 4789–4796. [CrossRef]

62. Miller, J.M.; Onar, O.C.; Chinthavali, M. Primary-side power flow control of wireless power transfer for electric vehicle charging. *IEEE J. Emerg. Sel. Top. Power Electron.* **2015**, *3*, 147–162. [CrossRef]

63. Buja, G.; Bertoluzzo, M.; Mude, K.N. Design and experimentation of WPT charger for electric city car. *IEEE Trans. Ind. Electron.* **2015**, *62*, 7436–7447. [CrossRef]

64. Bavastro, D.; Canova, A.; Cirimele, V.; Freschi, F.; Giaccone, L.; Guglielmi, P.; Repetto, M. Design of wireless power transmission for a charge while driving system. *IEEE Trans. Magn.* **2014**, *50*, 965–968. [CrossRef]

65. Aditya, K.; Williamson, S.S.; Sood, V.K. Impact of zero-voltage switching on efficiency and power transfer capability of a series-series compensated IPT system. In Proceedings of the IEEE Transportation Electrification Conference (ITEC-India 2017), Pune, India, 13–15 December 2017; pp. 1–7.

66. Aditya, K.; Sood, V.K. Design of 3.3 kW wireless battery charger for electric vehicle application considering bifurcation. In Proceedings of the IEEE Electrical Power and Energy Conference (EPEC 2017), Saskatoon, SK, Canada, 22–25 October 2017; pp. 1–6.

67. Aditya, K.; Williamson, S.S. A review of optimal conditions for achieving maximum power output and maximum efficiency for a series–series resonant inductive link. *IEEE Trans. Transp. Electrif.* **2017**, *3*, 303–311. [CrossRef]

68. Patil, D.; McDonough, M.K.; Miller, J.M.; Fahimi, B.; Balsara, P.T. Wireless power transfer for vehicular applications: Overview and challenges. *IEEE Trans. Transp. Electrif.* **2018**, *4*, 3–37. [CrossRef]

69. Yao, Y.; Wang, Y.; Liu, X.; Lin, F.; Xu, D.G. A novel parameter tuning method for double-sided LCL compensated WPT system with better comprehensive performance. *IEEE Trans. Power Electron.* **2018**, *33*, 8525–8536. [CrossRef]

70. Boys, J.T.; Huang, C.Y.; Covic, G.A. Single-phase unity power-factor inductive power transfer system. In Proceedings of the IEEE Power Electronics Specialists Conference, Rhodes, Greece, 15–19 June 2008; pp. 3701–3706.

71. Keeling, N.A.; Covic, G.A.; Boys, J.T. A unity-power-factor IPT pickup for high-power applications. *IEEE Trans. Ind. Electron.* **2010**, *57*, 744–751. [CrossRef]

72. Colak, K.; Bojarski, M.; Asa, E.; Czarkowski, D. A constant resistance analysis and control of cascaded buck and boost converter for wireless EV chargers. In Proceedings of the Applied Power Electronics Conference and Exposition (APEC), Charlotte, NC, USA, 15–19 March 2015; pp. 3157–3161.

73. Colak, K.; Asa, E.; Bojarski, M.; Czarkowski, D. A novel common mode multi-phase half-wave semi-synchronous rectifier for inductive power transfer applications. In Proceedings of the IEEE Transportation Electrification Conference and Expo (ITEC), Dearborn, MI, USA, 14–17 June 2015; pp. 1–6.

74. Fu, M.; Ma, C.; Zhu, X. A cascaded boost-buck converter for high-efficiency wireless power transfer systems. *IEEE Trans. Ind. Inform.* **2014**, *10*, 1972–1980. [CrossRef]

75. Aditya, K.; Williamson, S.S. Comparative study on primary side control strategies for series-series compensated inductive power transfer system. In Proceedings of the IEEE 25th International Symposium on Industrial Electronics (ISIE), Santa Clara, CA, USA, 8–10 June 2016; pp. 811–816.

76. Li, H.L. High Frequency Power Converters Based on Energy Injection Control for IPT Systems. Ph.D. Thesis, Department of Electrical and Computer Engineering, University of Auckland, Auckland, New Zealand, 2011.

77. Wang, C.S.; Covic, G.A.; Stielau, O.H. Power transfer capability and bifurcation phenomena of loosely coupled inductive power transfer systems. *IEEE Trans. Ind. Electron.* **2004**, *51*, 148–157. [CrossRef]

78. Weearsinghe, S.; Thrimawithana, D.J.; Madawala, U.K. Modeling bidirectional contactless grid interfaces with a soft DC-link. *IEEE Trans. Power Electron.* **2015**, *30*, 3528–3541. [CrossRef]

79. Yeo, T.D.; Kwon, D.; Khang, S.T.; Yu, J.W. Design of maximum efficiency tracking control scheme for closed-loop wireless power charging system employing series resonant tank. *IEEE Trans. Power Electron.* **2017**, *32*, 471–478. [CrossRef]

80. Aditya, K.; Williamson, S.S. Design guidelines to avoid bifurcation in a series-series compensated inductive power transfer system. *IEEE Trans. Ind. Electron.* **2019**, *66*, 3973–3982. [CrossRef]

81. Mohamed, A.A.S.; Berzoy, A.; de Almeida, F.G.N.; Mohammed, O. Modeling and assessment analysis of various compensation topologies in bidirectional IWPT system for EV applications. *IEEE Trans. Ind. Appl.* **2017**, *53*, 4973–4984. [CrossRef]

82. Kalra, G.R.; Huang, C.Y.; Thirmawithana, D.J.; Madawala, U.K.; Neuburger, M. A comparative study on grid-integration techniques used in bi-directional IPT based V2G applications. In Proceedings of the IEEE 2nd Annual Southern Power Electronics Conference (SPEC), Auckland, New Zealand, 5–8 December 2016; pp. 1–6.

83. Rosekeit, M.; Broeck, C.; Doncker, R.W.D. Dynamic control of a dual active bridge for bidirectional ac charging. In Proceedings of the IEEE International Conference on Industrial Technology, Seville, Spain, 17–19 March 2015; pp. 2085–2091.

84. Bala, S.; Tengner, T.; Rosenfeld, P.; Delince, F. The effect of low frequency current ripple on the performance of a Lithium Iron Phosphate (LFP) battery energy storage system. In Proceedings of the IEEE Energy Conversion Congress and Exposition, Raleigh, NC, USA, 15–20 September 2012; pp. 3485–3492.

85. Gunji, D.; Imura, T.; Fujimoto, H. Operating point setting method for wireless power transfer with constant voltage load. In Proceedings of the IEEE IECON 2015—41st Annual Conference, Yokohama, Japan, 9–12 November 2015; pp. 881–886.

86. Yang, Y.; Zhong, W.; Kiratipongvoot, S.; Tan, S.; Hui, S.Y.R. Dynamic improvement of series–series compensated wireless power transfer systems using discrete sliding mode control. *IEEE Trans. Power Electron.* **2018**, *33*, 6351–6360. [CrossRef]

87. Li, H.; Li, J.; Wang, K.; Chen, W.; Yang, X. A maximum efficiency point tracking control scheme for wireless power transfer systems using magnetic resonant coupling. *IEEE Trans. Power Electron.* **2015**, *30*, 3998–4008. [CrossRef]

88. Qu, X.; Han, H.; Wong, S.-C.; Tse, C.K.; Chen, W. Hybrid IPT topologies with constant current or constant voltage output for battery charging applications. *IEEE Trans. Power Electron.* **2015**, *30*, 6329–6337. [CrossRef]

89. Chen, Y.; Kou, Z.; Zhang, Y.; He, Z.; Mai, R.; Cao, G. Hybrid topology with configurable charge current and charge voltage output-based WPT charger for massive electric bicycles. *IEEE J. Emerg. Sel. Top. Power Electron.* **2018**, *6*, 1581–1594. [CrossRef]

90. Jiang, Y.; Wang, L.; Wang, Y.; Liu, J.; Li, X.; Ning, G. Analysis, design, and implementation of accurate ZVS angle control for EV battery charging in wireless high-power transfer. *IEEE Trans. Ind. Electron.* **2019**, *66*, 4075–4085. [CrossRef]

91. Mishima, T.; Morita, E. High-frequency bridgeless rectifier based ZVS multiresonant converter for inductive power transfer featuring high-voltage GaN-HFET. *IEEE Trans. Ind. Electron.* **2017**, *64*, 9155–9164. [CrossRef]
92. Li, Y.; Hu, J.; Chen, F.; Li, Z.; He, Z.; Mai, R. Dual-phase-shift control scheme with current-stress and efficiency optimization for wireless power transfer systems. *IEEE Trans. Circuits Syst. I Regul. Pap.* **2018**, *65*, 3110–3121. [CrossRef]
93. Zhao, C.; Trento, B.; Jiang, L.; Jones, E.A.; Liu, B.; Zhang, Z.; Costinett, D.; Wang, F.; Tolbert, L.M.; Jansen, J.F.; et al. Design and implementation of a GaN-based, 100-kHz, 102-W/in3 single-phase inverter. *IEEE J. Emerg. Sel. Top. Power Electron.* **2016**, *4*, 824–840. [CrossRef]
94. Cai, A.Q.; Siek, L. A 2-kW, 95% efficiency inductive power transfer system using Gallium Nitride gate injection transistors. *IEEE J. Emerg. Sel. Top. Power Electron.* **2017**, *5*, 458–468. [CrossRef]
95. Jones, E.A.; Wang, F.F.; Costinett, D. Review of commercial GaN power devices and GaN-based converter design challenges. *IEEE J. Emerg. Sel. Top. Power Electron.* **2016**, *4*, 707–719. [CrossRef]
96. Deng, Q.; Liu, J.; Czarkowski, D.; Hu, W.; Zhou, H. An inductive power transfer system supplied by a multiphase parallel inverter. *IEEE Trans. Ind. Electron.* **2017**, *64*, 7039–7048. [CrossRef]
97. Ronanki, D.; Williamson, S.S. Modular multilevel converters for transportation electrification: Challenges and opportunities. *IEEE Trans. Transport. Electrific.* **2018**, *4*, 399–407. [CrossRef]
98. Hao, H.; Covic, G.A.; Boys, J.T. A parallel topology for inductive power transfer power supplies. *IEEE Trans. Power Electron.* **2014**, *29*, 1140–1151. [CrossRef]
99. Ronanki, D.; Kelkar, A.; Williamson, S.S. Extreme fast charging technology—Prospects to enhance sustainable electric transportation. *Energies* **2019**, *12*, 3721. [CrossRef]
100. Jain, P.; Jain, T. Impacts of G2V and V2G power on electricity demand profile. In Proceedings of the 2014 IEEE International Electric Vehicle Conference (IEVC), Florence, Italy, 17–19 December 2014; pp. 1–8. [CrossRef]

Article

Research on the Critical Issues for Power Battery Reusing of New Energy Vehicles in China

Zongwei Liu [1,2,3], **Xinglong Liu** [1,2], **Han Hao** [1,2,4], **Fuquan Zhao** [1,2,*], **Amer Ahmad Amer** [5] **and Hassan Babiker** [5]

[1]   State Key Laboratory of Automotive Safety and Energy, Tsinghua University, Beijing 100084, China; liuzongwei@tsinghua.edu.cn (Z.L.); lxl19@mails.tsinghua.edu.cn (X.L.); hao@tsinghua.edu.cn (H.H.)
[2]   Tsinghua Automotive Strategy Research Institute, Tsinghua University, Beijing 100084, China
[3]   Sloan Automotive Laboratory, Massachusetts Institute of Technology, Cambridge, MA 02139, USA
[4]   China Automotive Energy Research Center, Tsinghua University, Beijing 100084, China
[5]   Research and Development Center, Saudi Aramco, Dhahran 31311, Saudi Arabia; amer.amer.4@aramco.com (A.A.A.); hassan.babiker@aramco.com (H.B.)
*   Correspondence: zhaofuquan@tsinghua.edu.cn

Received: 15 March 2020; Accepted: 7 April 2020; Published: 14 April 2020

Abstract: With the rapid development of new energy vehicles (NEVs) industry in China, the reusing of retired power batteries is becoming increasingly urgent. In this paper, the critical issues for power batteries reusing in China are systematically studied. First, the strategic value of power batteries reusing, and the main modes of battery reusing are analyzed. Second, the economic benefit models of power batteries echelon utilization and recycling are constructed. Finally, the economic benefits of lithium iron phosphate (LIP) battery and ternary lithium (TL) battery under different reusing modes are analyzed based on the economic benefit models. The results show that when the industrial chain is fully coordinated, LIP battery echelon utilization is profitable based on a reasonable scenario scheme. However, the multi-level echelon utilization is only practical under an ideal scenario, and more attention should be paid to the first level echelon utilization. Besides, the performance matching of different types of batteries has a great impact on the echelon utilization income. Thus, considering the huge potentials of China's energy storage market, the design of automobile power batteries in the future should give due consideration to the performance requirements of energy storage batteries. Moreover, the TL battery could only be recycled directly, while the LIP has the feasibility of echelon utilization at present. At the same time, it will strengthen the cost advantage of the LIP battery, which deserves special attention.

Keywords: new energy vehicle; power battery; battery reusing; echelon utilization; battery recycling

---

1. Introduction

With the continuous support of the government, the number of NEVs (new energy vehicles) has been increasing rapidly in China, which has led to the rapid development of the power battery industry [1–3]. As shown in Figure 1, the installed capacity of China's traction battery is already very large. There was an increase of more than 60 GWh in 2019 and an accumulated installed capacity of more than 205 GWh, which is still growing rapidly [4,5]. At the same time, more and more power batteries are approaching the state of retirement with the passage of time. There are two reasons that the scale of batteries to be retired is further increased. First, the service life of power batteries is usually lower than that of the vehicles resulting that a large number of retired batteries will appear before the vehicles are scrapped. Second, the technical level of early NEV products is relatively low; the service life of many power batteries is far shorter than the newly developed batteries. Therefore, it can be expected that China will soon usher in the peak period of the retirement of NEV power batteries [6].

Obviously, it will bring serious environmental and security risks if these retired batteries cannot be effectively recycled and managed [7].

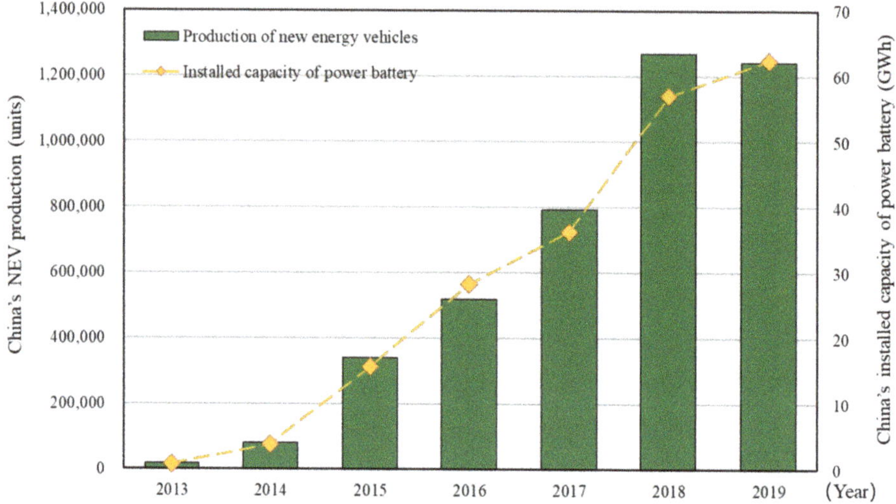

Figure 1. Production of new NEVs (new energy vehicles) and installed capacity of power batteries in China (2013–2019).

For this reason, the Chinese government is stepping up the development of relevant policies on the reusing of power batteries. As shown in Figure 2, the frequency and content of recent relevant policies are getting higher and higher [8,9]. All kinds of signs indicate that China will issue regulations and policies on the mandatory recycling of retired power batteries soon. Thus, it would require enterprises to solve the problem of retired power batteries in the form of laws so as to ensure the sustainable development of the new energy automobile industry. According to the principle of the "Extended Producer Responsibility System", power battery reusing will become the responsibility of vehicle enterprises. The vehicle enterprises will definitely decompose the responsibility along the supply chain, so the whole power battery industry will be affected [10]. Therefore, relevant enterprises need to think ahead, lay out, and prepare in advance to meet the requirements of national laws and regulations.

In fact, the power battery of NEVs contains a large number of metal materials, such as lithium, cobalt, nickel, etc. Its reusing is not only a matter of legal responsibility but also affects the supply status of these metals directly [11,12]. So, it may change the trend of the price of power battery, and then affect the development trend of the industry and the income of enterprises significantly.

For this reason, this paper systematically studies the key issues for NEV power battery reusing in China, including the strategic value, main reusing modes, echelon utilization value, recycling value, and overall value analysis of power battery reusing.

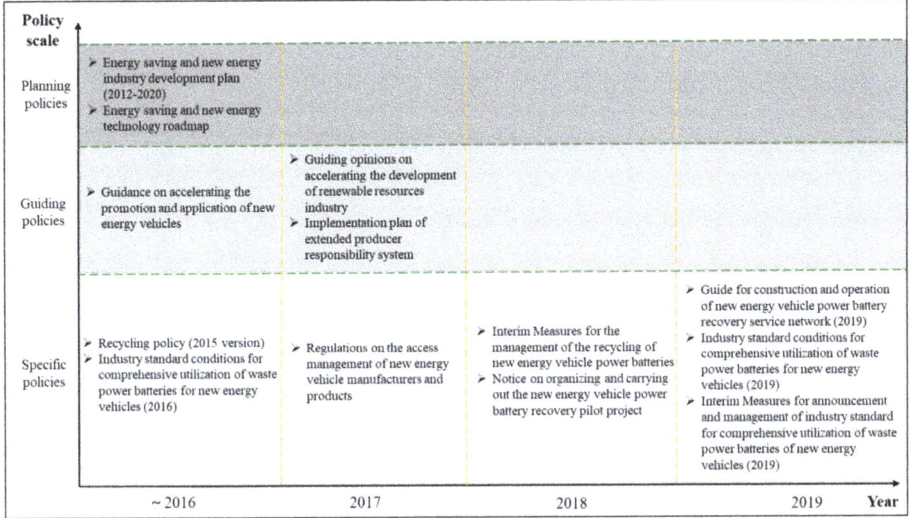

Figure 2. List of relevant policies for power battery reusing in China.

2. Strategic Value of Power Battery Reusing

As mentioned above, the retired volume of power batteries for NEVs will be huge and grow rapidly in China. Reusing these batteries has three important strategic values [13].

First, it could solve the environmental pollution and potential safety problems caused by retired power batteries [14]. Based on the historical data of China's NEVs and their power batteries, this paper makes a quantitative estimation of the scale of retired power batteries in the next six years. The evaluation method is shown in Equation (1):

$$R_M = \sum_{M-A}^{M-B} \frac{R_k}{A - B + 1} \cdot I_k \tag{1}$$

where the k is the year; R_M indicates the retirement volume in year M; A/B indicates the up/down year of retired power battery life; I_k represents the conversion coefficient of influencing factors in the year k, which is related to market size, battery type, battery life and installed capacity of batteries [15].

The results of this estimation are shown in Figure 3. It is estimated that the retired volume of power batteries in China will reach about 27 GWh in 2020 and 146 GWh in 2025. It will become a serious threat to the ecological environment. To be specific, it will result in water, soil, and air pollution, as well as public security risks such as short-circuit combustion if the retired power battery with such a huge scale cannot be reused and recycled effectively [16].

Second, it could realize the recycle and reuse of metal resources and reduce the supply risk of power battery raw materials [17]. On the one hand, the demand for lithium, cobalt, nickel, etc. for NEV power batteries is growing. For example, China's power battery industry consumed about 11,000 tons of lithium, 41,000 tons of nickel, and 17,000 tons of cobalt in 2018 [15]. At the same time, China's external dependence on these three metal materials is more than 80%, especially for cobalt, with the overseas import volume as high as 97% [15]. In this case, the recycling of these metal materials must be realized through the reusing of retired power batteries so as to avoid resource bottlenecks and ensure the safety and control of the industry [18].

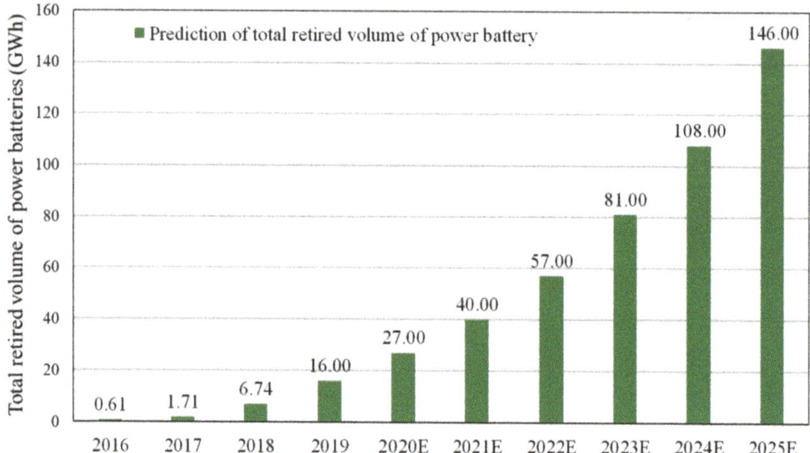

Figure 3. The total retired volume of the new energy vehicle power battery in China.

On the other hand, the current price of power battery is still relatively expensive, which is the fundamental reason why NEVs are at a price disadvantage compared with traditional internal combustion engine vehicles (ICEVs) [19]. As China's subsidy policy for NEVs will be completely phased out after 2020, the NEVs will be mostly market-driven. Therefore, to improve the competitiveness of the NEVs, their cost-performance needs to approach the level of ICEVs without subsidies. The expectation of this industry is that the scale effect will make the price of power battery drop rapidly with the increase of NEVs. However, if a large number of retired power batteries cannot be reused, it is likely that the supply of metal materials would be a bottleneck. It will lead to the price of power batteries rising instead of falling, resulting in a higher cost of the NEVs [14]. In this regard, the reusing of power battery is one of the important factors affecting the sustainable development of China's NEVs industry in the future.

Third, it has the potential to reduce the cost of power battery in automobile enterprises and generate additional social value [16]. The retired power battery should not only simply be dismantled and recycled and the reusing for raw materials. If its potential value can be fully exploited before dismantling and recycling, the retired power battery can be utilized to the maximum extent. Thus, the use cost of the power battery will be effectively shared at different utilization stages [20]. In this case, the recovery and utilization of power battery may not be a burden for the automotive enterprise, and may even bring some profits to the enterprise instead. As a result, it would promote the cost reduction of new NEV products further, and thus constitute a competitive advantage for the enterprise. At the same time, the field of application of retired power batteries is also expected to benefit from waste utilization [21,22]. For the whole society, this means the upgrading of the efficiency of social resources utilization. Therefore, it would also produce certain social values.

It can be seen that China's subsidies for NEVs are about to be phased out entirely. Hence, the industry must attach great importance and study how to maximize the benefits of the reusing of retired power batteries seriously under the situation. This strategically important for the sustainable development of China's NEV industry as well as the construction and improvement of a green energy-saving society in the future.

3. Main Modes of Power Battery Reusing

Firstly, the concept of power battery reusing is defined, as shown in Figure 4.

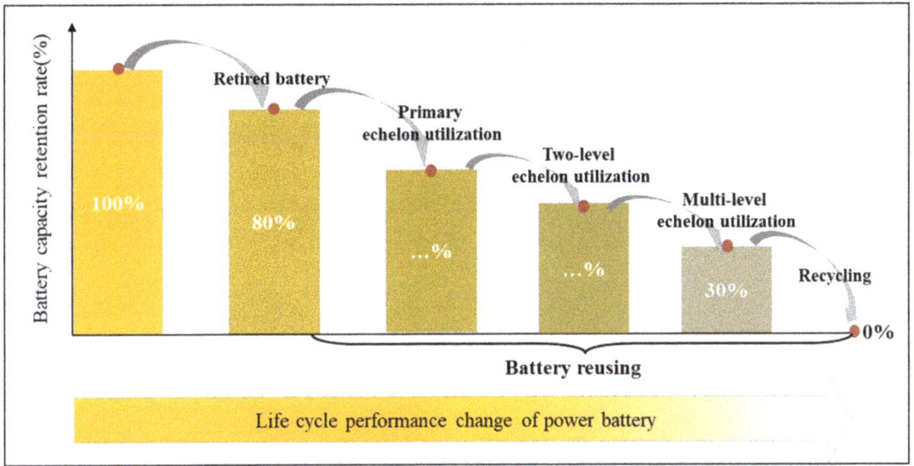

Figure 4. Main ways of power battery reusing of NEVs.

Retired battery: when the performance of the power battery has declined to a certain extent, it needs to be retired from use and then enter the reusing process. For NEVs, the indicator is that the battery capacity retention rate decays below 80% [23].

Echelon utilization: the batteries retired from the original products can be used in other applications with relatively low requirements on battery performance [24–27]. Theoretically, this kind of secondary utilization can be performed many times, which is the so-called multi-level echelon utilization. NEVs have high requirements for batteries, but the power batteries could be used in micro electric vehicles, communication base stations, energy storage [28], and other fields until the battery capacity retention rate is reduced to 30% [28].

Recycling: the power battery is dismantled and recycled, which is the last link of reusing [29]. From the legal point of view, it will meet the requirements if the retired battery is recycled. However, the retired battery cannot achieve the maximum benefit when dismantled and recycled directly from the perspective of enterprise management.

Reusing: This is actually a big concept, including two aspects of echelon utilization and recycling. It covers the whole process of processing and processing retired batteries with the basic goal of environmental protection and resource reuse [13,30].

According to the above definition, the reusing of retired power batteries of NEVs could be divided into two main modes, as shown in Figure 5. The first mode is to recycle retired batteries directly, that is, to realize resource recovery and meet regulatory requirements [7]. Mode II is to select a suitable scenario for the echelon utilization of retired batteries to give full play to the residual capacity and then implement the recycling [23]. Obviously, Mode II can obtain greater returns than Mode I in theory, but the actual situation is far more complex than the theory [31]. There are many factors influencing the benefit of echelon utilization, including appropriate echelon utilization scenarios; therefore, there are key technologies, and costs of detecting battery residual energy and regrouping, cooperation modes of all parties involved in echelon utilization, etc. Thus, the specific value of echelon utilization needs to be systematically evaluated to obtain a greater return from the reusing of retired power batteries.

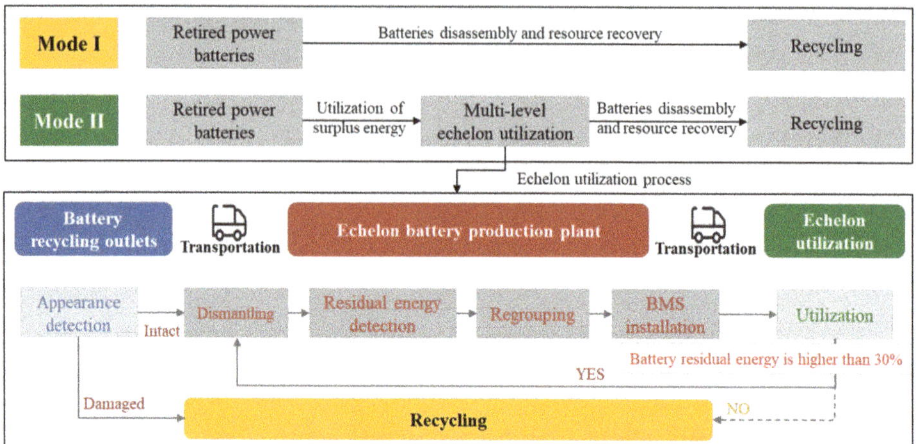

Figure 5. Main reusing modes and the echelon utilization process of retired power battery.

4. Research on the Key Issues of Power Battery Echelon Utilization

At present, the mainstream power batteries are TL batteries and LIP batteries. The mass-energy density of the TL battery is higher, which plays a leading role in passenger cars, but its cost is higher than that of the LIP battery [32]. However, the LIP battery exhibits better thermal stability, lower cost, and lower volume energy density. It is more widely used in commercial vehicles [33]. Overall, they have their own advantages and disadvantages and are widely used in NEVs. By the end of 2018, LIP battery accounted for 54%, and TL battery accounted for 42% of the total installed of NEV power batteries in China [34].

However, there are technical bottlenecks in TL battery when facing echelon utilization. On the one hand, the safety of the TL battery is poor due to the low thermal runaway temperature (180 °C), meaning it is more prone to cause fire and explosion. On the other hand, more nickel is used in the cathode material of the TL battery in order to pursue higher energy density, which results in shorter cycle life of TL batteries [35]. The solution to these problems needs a corresponding technological breakthrough and high costs. So, it is difficult to implement the TL battery echelon utilization at the current technical level. Instead, direct recycling for TL battery is a more reasonable choice. Therefore, this paper only studies the echelon utilization of LIP battery [36].

4.1. Scenario Selection of Power Battery Echelon Utilization

There are three important factors to consider when selecting echelon utilization scenarios, including demand matching, utilization cost, and scale capacity. First of all, the performance of retired power batteries must meet the basic needs of echelon utilization scenarios. It includes service life, charging and discharging speed, internal resistance, and working voltage, etc. Secondly, the retired power battery can be used for echelon utilization only after being tested and regrouped. The cost must be low enough to ensure the economy of echelon utilization. Therefore, the technical requirements for batteries in relevant scenarios could not be too high. Finally, the ideal echelon utilization scenario should have a large and stable battery demand due to the number of retired power batteries increasing year by year. Otherwise, it cannot play a great role even if the demand matches, and the cost is controllable.

Based on the above three factors, this study conducts a comprehensive analysis of various possible echelon utilization scenarios, and finally extracts the six most likely scenarios [23,25,37]. They are divided into two categories: power battery and energy storage battery. The former includes a microelectric vehicle (MEV), electric special vehicle (ESV) (such as sightseeing vehicle, ferry car, etc.), and electric bike (E-Bike). While the latter includes communication base station (CBS), renewable energy power station (REPS), and parallel micro-grid (PMG). Each of these scenarios has different battery use characteristics, presented in detail in Table 1.

Table 1. The most likely echelon utilization scenarios and their battery use characteristics.

Echelon Utilization Scenarios	Power Battery			Energy Storage Battery		
	MEV	ESV	E-Bike	CBS	REPS	PMS
Depth of discharge	80%	80%	100%	80%	80%	80%
Cycle	450	450	180	600	800	650
Capacity (kWh)	13.5	13.5	1	-	-	-
Current price (€/kWh)	65	65	90	31	58	39

China's low-speed road vehicles, including MEV, ESV, and E-Bike, etc. listed in Table 1, mostly use lead-acid (LA) batteries nowadays. LA batteries have serious environmental pollution problems. At the same time, the energy density, cycle life, and other main indicators are far lower than LIP batteries, as shown in Table 2. But its main advantage is the low cost [38]. With the continuous growth of the power battery scale of NEVs, the retired LIP batteries can be completely used to replace the LA battery. The retired LIP batteries are not only with better performance and lower cost, but also meet the performance requirement is due to the similar power battery application scenario. Therefore, the replacement of LA batteries in low-speed road vehicles should be one of the best scenarios for the echelon utilization of retired power batteries.

Table 2. Performance comparison for different batteries.

Performance	LA	LIP	TL
Specific energy (mAh/g)	40–70	130–165	150–210
Cycle life (cycle)	400–800	2000–6000	800–2000

In addition, China currently has a wide range of demand for energy storage batteries. For example, the renewable energy power stations (photovoltaic power generation, wind power generation, etc.) in Table 1, the microgrid of the distributed independent power sources and communication base stations all have considerable demand for energy storage batteries. With the optimization of China's energy structure and the upgrading of information infrastructure, the demand scale of energy storage batteries will continue to expand. The new production of the energy storage battery to meet the demand will inevitably consume a lot of resources and bear high costs. Therefore, if the retired power battery of NEVs can be used for energy storage, it can not only fully accommodate the increasing number of retired power batteries, but also meet the growing demand for energy storage batteries. This will produce huge social benefits [39]. In fact, the electric vehicle itself is also regarded as a movable energy storage device. The power batteries have already been and will play the role of energy storage to a certain extent before retired.

As mentioned above, the retired battery can be used in a multi-level echelon in theory. That is to say, until the last remaining energy is exhausted, it can be repurposed from one echelon utilization scenario to the next echelon utilization scenario and so on. However, the switching of each echelon utilization scenario will produce certain costs, including battery purchase, detection, regrouping, and transportation, etc. Especially at present, echelon utilization technology is still in development, leading to the high cost of battery detection and regrouping. Thus, too many levels of echelon utilization will lead to high costs and reduce expected profits. So, the multi-level echelon utilization has no realistic possibility under the existing technology level.

For this reason, this paper only studies the echelon utilization scheme of one or two levels. In the two-level echelon utilization scheme, the power battery with a higher matching degree of battery demand is taken as the first level, and the energy storage battery is taken as the second level. Note that the use of E-Bike as the first link is not considered because it is relatively scattered, and the single-car battery load is small. Therefore, six primary application schemes and six two-level application schemes are determined as the research objects, as shown in Table 3.

Table 3. Echelon utilization schemes of retired power battery.

Echelon Utilization Schemes	Primary Echelon Utilization Scheme						Two-level Echelon Utilization Scheme					
	F-1	F-2	F-3	F-4	F-5	F-6	S-1	S-2	S-3	S-4	S-5	S-6
MEV	X						X	X	X			
ESV		X								X	X	X
E-Bike			X									
CBS				X			XX			XX		
REPS					X			XX			XX	
PMG						X			XX			XX

Note: "X" represents the first-level echelon utilization; "XX" represents the second-level echelon utilization.

4.2. Economic Benefit Analysis of Power Battery Echelon Utilization

4.2.1. Economic Benefit Evaluation Model of Echelon Utilization

In order to evaluate the economic benefits and influencing factors of different echelon utilization schemes, this paper constructs an evaluation model of echelon utilization economic benefits, as shown in Figure 6. There are two key assumptions in this model. First, this study assumes that there are enterprises with enough technology and capital to complete the corresponding work in each link of echelon utilization. This paper only studies the overall cost and benefit of the whole process from the perspective of the industrial chain. However, as for which specific enterprises complete which link of work, and the benefit distribution of enterprises with different roles is not in the scope of this study. Second, in order to simplify the calculation, the battery capacity retention rate is used to measure the overall performance of the battery and the only judgment condition of battery state of health, service life, and whether it should be transferred to the next link [40,41].

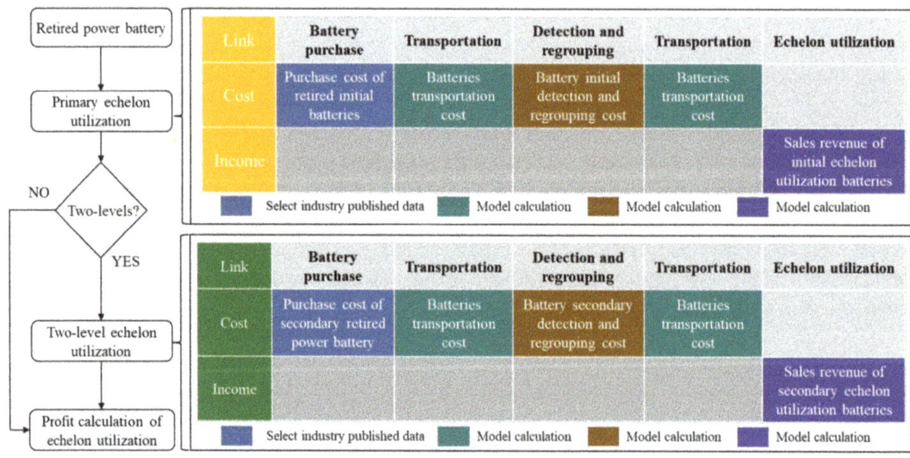

Figure 6. Block diagram of economic benefit evaluation model for the echelon utilization of retired power battery.

The time point of decommissioning of the battery from the whole vehicle and the judgment condition of the ultimate scrapping are all expressed by battery performance; therefore, all expressions related to the battery service life are also unified to the form of capacity retention rate in the process of model calculation. The specific definition of the capacity retention rate is shown in Equation (2).

$$Battery\ capacity\ retention\ rate = \frac{C_l}{C_{new}} \tag{2}$$

where the C_l represents the battery capacity after l cycles, and C_{new} indicates the nominal capacity of the new power battery.

With the use of the battery, the capacity retention rate will continue to decline. The capacity decay rate (η) of each cycle is defined as the impact of the new charge-discharge cycle on the battery performance. The specific calculation is as follows:

$$\eta = \frac{C_{l+1}}{C_l} \tag{3}$$

where C_{l+1} indicates the battery capacity after $l + 1$ cycle.

According to the study conducted by Schuster, the battery capacity retention rate can be approximately considered constant before the battery performance drop to the capacity diving point [41]. The capacity retention rate can be calculated using the capacity decay rate (η) and cycles as follows:

$$Battery\ capacity\ retention\ rate = \eta^l \tag{4}$$

And the battery capacity retention rate would be different due to the depth of discharge (DOD) under the use condition. According to the three DOD corresponding to the six application scenarios as abovementioned, different η values are used to calculate the life of the echelon utilization battery, as shown in Table 4.

Table 4. Capacity decay rates under different depth of discharge (DOD).

DOD	Capacity Decay Rate η
60%	99.98%
80%	99.97%
100%	99.96%

As shown in Figure 6, the costs incurred in the process of echelon utilization mainly include the purchase and transportation costs of retired batteries as well, as the costs of echelon utilization battery detection and regrouping. The revenue comes from battery sales revenue [42]. It should be noted that the cost of each link is different due to the difference in battery performance in primary and secondary echelon utilization. As abovementioned, the echelon utilization profit can be calculated by the total revenue and expenditure, as shown in Equation (5). In order to make the results as broadly applicable as possible, the cost, revenue, and profit are all in term of €/kWh in this study. Note that the study converts RMB to Euro at the 2019 exchange rate.

$$P_c = \sum_{i=1}^{N}\left(SR_e - C_{p1} - C_{tr} - C_r\right) \tag{5}$$

where the P_c is the profit of echelon utilization; SR_e represents the sales revenue of echelon utilization batteries; C_{p1} indicates the purchase cost of retired batteries; C_{tr} is the batteries transportation cost; C_r indicates the cost of battery detection and regrouping.

For the sales revenue of the echelon utilization battery: there are two calculation methods. The first method is based on the residual capacity of the retired LIP battery. It calculates profits according to the discount coefficient on the basis of the new battery price [37]. The specific calculation is Equation (6) as follows:

$$SR_e = \frac{P_{new} \cdot TO_{retired}}{TO_{new} \cdot DR} \tag{6}$$

where the P_{new} is the price of new LIP battery; TO_{new} represents a nominal capacity of the LIP battery; $TO_{retired}$ indicates the capacity of LIP battery after decommissioning; DR is the discount coefficient [37].

The second method is based on the price of the new LA battery and the different cycle life of the two batteries. The specific calculation is Equation (7) as follows:

$$SR_e = \frac{P_{qs} \cdot L_{qs}}{L_{Li}} \tag{7}$$

where the P_{qs} is the price of the replaced LA battery in the echelon utilization scenario; L_{qs} represents LA battery cycle life; L_{Li} indicates the cycle life of retired LIP power battery.

Finally, the lower one in the calculation results of Equations (6) and (7) is taken as the battery sales price, which is more likely to be accepted by the demander of the echelon utilization scenario, as shown in Equation (8):

$$SR_b = min\left\{ \frac{P_{new} \cdot TO_{retired}}{TO_{new} \cdot DR}, \frac{P_{qs} \cdot L_{qs}}{L_{Li}} \right\} \tag{8}$$

Purchase cost of retired battery: when the LIP battery is retired from the vehicle, the capacity retention rate is 80%. Its value is taken from the industry average data, about 13.4 €/kWh [10].

Battery transportation cost: in this study, only the use cost of freight cars is included, and the cost of vehicle purchase and depreciation is not considered. In order to minimize the cost, trucks with different carrying capacity will be selected according to different transportation needs. In addition, the transportation process is divided into two types according to the distance, including intercity transportation and trans provincial transportation. In this study, the average distances of the two types of transportation are 50 km and 500 km, respectively. Most of the echelon utilization scenarios are short-distance intercity transportation. Only when retired batteries are applied to renewable energy power stations and large-scale communication base stations, can long-distance trans provincial transportation be considered [43]. According to the current use of freight in China, the cost of intercity transportation and trans provincial transportation can be calculated, as shown in Equation (9):

$$\begin{cases} C_t^{intercity} = 0.00052 \cdot D \\ C_t^{trans\ provincial} = 0.000077 \cdot D \end{cases} \tag{9}$$

where the $C_t^{intercity}$ is the intercity transportation cost; $C_t^{trans\ provincial}$ represents trans provincial transportation cost; D indicates the transportation distance.

Detection and regrouping cost: as echelon utilization is still in the exploration and preliminary stage, there is no open direct cost data available. Based on the operation cost of the existing battery treatment plant, this study estimates the detection and regrouping cost of the battery through cost apportionment methods. Considering the different technical difficulties of the different echelon utilization scenarios, the scenario correction coefficient is added to modify the echelon utilization calculation model developed by the National Renewable Energy Laboratory (NREL) [43], as shown in Equation (10):

$$C_r = C_{factory} \cdot CF/Q \tag{10}$$

where the $C_{factory}$ is the enterprise detection and regrouping cost of echelon utilization, including fixed asset cost and labor cost; CF represents the correction coefficient of battery detection and group cost in different echelon utilization scenarios which is related to the processing difficulty and time [37]; Q indicates the annual production of echelon battery.

4.2.2. Economic Benefit Analysis of Echelon Utilization

Based on the evaluation model, the benefits of the echelon utilization of various schemes are calculated, and the results are shown in Figure 7. Firstly, it can be seen that most of the schemes of primary and secondary echelon utilization can achieve positive benefits. It shows that if the relevant industrial chain is coordinated, the echelon utilization of retired power battery is profitable. Then the cost of using power battery for BEVs is leveraged by echelon utilization [20]. With the development of battery detecting and regrouping technology, the benefits of echelon utilization are expected to expand in the future further. In addition, only LIP battery has the technical feasibility of echelon utilization, which also means that echelon utilization is expected to further increase the cost advantage of LIP battery comparing to TL battery.

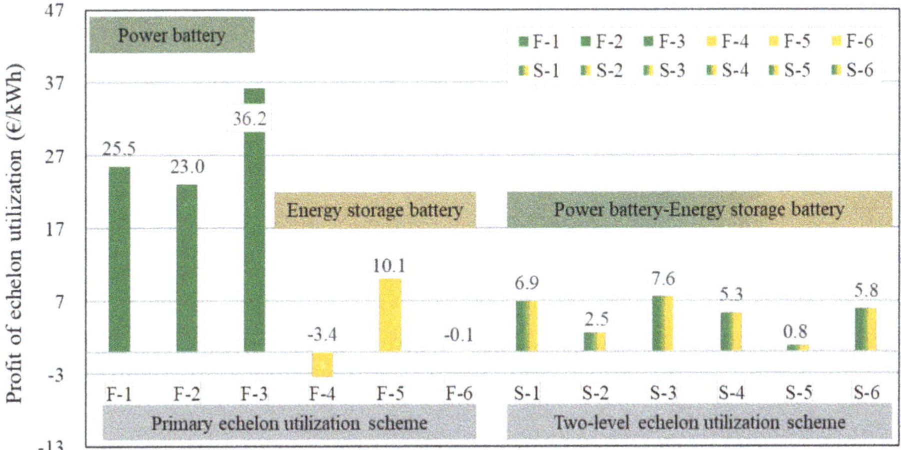

Figure 7. Comparison of economic benefits of the different echelon utilization schemes.

Second, compared with the three scenarios of power battery replacement in the primary scheme, the profits of all the secondary schemes are lower. It can be explained that the increase in echelon utilization levels will lead to the decrease of echelon utilization income. Meanwhile, this shows that it is reasonable to study only the primary and secondary schemes in this paper. In a word, multi-level echelon utilization is only a theoretical concept at present. Hence, more attention should be paid to the development of the first level echelon utilization scheme in industrial practice in the future. Moreover, the two-level scheme should be properly considered according to the urgency of the scene demand.

Third, the performance matching of different batteries has a significant impact on the benefits of echelon utilization. From the analysis of the primary echelon utilization scheme, it can be seen that the echelon utilization scenario based on the power battery for transportation purposes can achieve greater benefits, with the maximum profit of 30.21 €/kWh. While there are echelon utilization scenarios for energy storage, only the renewable energy power station can achieve positive benefits. Its profit level is far lower than the power battery scheme. This is mainly due to the difference in performance requirements between the power battery and the energy storage battery. That is to say that the retired batteries of NEVs that were originally used as power batteries are more suitable to continue to be used as low-performance power batteries. If it is to be used as the energy storage battery, the processing cost of the battery will be much higher, making it difficult to make profits. Of course, the battery design of NEVs is only based on the requirement of the power system itself at present without considering the potential echelon utilization in the future, especially for energy storage scenarios [44–46]. Thus, if the performance requirements of the energy storage battery are properly considered at the beginning of the battery design, the evaluation results may be significantly different. This is also an important decision-making factor for relevant enterprises to consider seriously.

Generally speaking, E-Bike is the application scenario with the highest profit from echelon utilization at present. However, the demand for retired batteries of E-Bike is limited, and urban regulatory policies are being tightened continuously. It is expected that the follow-up market scale will not be able to grow significantly and cannot accommodate the increasing number of retired power batteries of NEVs. At the same time, microelectric vehicles, commonly known as low-speed electric vehicles, have not been officially recognized by the Chinese government. Therefore, the electric-special vehicle may be the most feasible entry point for automobile enterprises to try to use battery echelons in the near future. In the long run, although the benefits of energy storage scenarios such as renewable energy power stations are relatively low, there is a very broad further market. Especially, China has made the strategic goal of building a low-carbon society and low-carbon industry. China needs a large number of energy storage batteries, which will play an important role in reducing clean power waste and optimizing the balance of power grid urgently. In this sense, the echelon utilization of energy storage scenarios may have more market potential in the future. [37].

5. Research on the Recycling Value of Power Battery

The recycling of power battery mainly includes the process of dismantling, crushing, repairing, and smelting. Its purpose is to realize the recycling of resources on the basis of eliminating environmental pollution and potential safety hazards [47]. The recovery and utilization of battery materials, especially metals such as lithium, nickel, and cobalt, could also generate certain economic benefits [48]. Considering that $Ni_{0.5}Co_{0.2}Mn_{0.3}$ (NCM523) lithium battery is the most recently retired TL battery, this paper analyzes the recycling value of LIP power battery and TL battery NCM523.

In this paper, the economic benefit evaluation model of power battery recycling is built. It can be seen in Figure 8 for details.

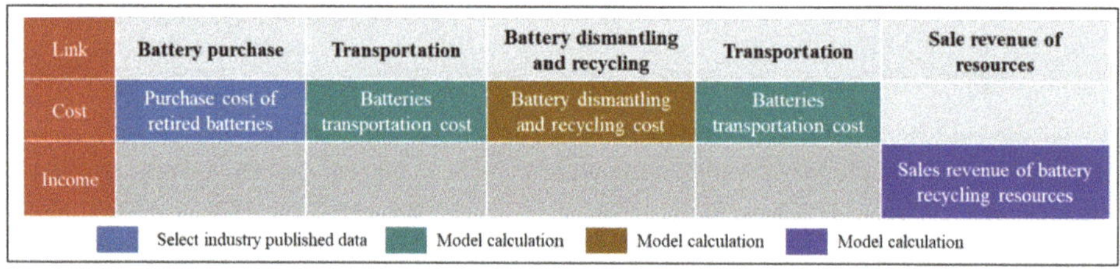

Figure 8. Block diagram of the economic benefit evaluation model for battery recycling.

As shown in Figure 8, the cost of the recycling process mainly includes purchase, transportation, and dismantling cost of the retired batteries (including the two situations of direct retirement from the NEVs or retirement after echelon utilization), and the revenue mainly comes from the sale of the recovered materials [42]. Therefore, the economic benefits of recycling can be calculated by Equation (11):

$$P_r = SR_r - C_{p2} - C_{dr} - C_{tr} \tag{11}$$

where the P_r is the profit of recycling; SR_r represents the sales revenue of battery recycling resources; C_{p2} indicates the battery purchase cost; C_{dr} represents battery dismantling and recovery cost; C_{tr} indicates battery transportation cost.

The sales revenue of battery recycling resources is mainly from the sales of lithium, cobalt, nickel, and other metal materials. It is assumed that all recovered metal materials can be fully reused. The calculation equation is as follows:

$$C_{p2} = \sum_j \left(P_j \cdot Q_j \cdot RR_j \right) \qquad (12)$$

where the P_j is the price of the metal j, note that the price of various metals in 2018 is taken in this paper, as shown in Table 5 [10]; Q_j represents the recyclable amount of the metal j in retired batteries per kWh; RR_j represents the recovery rate of the metal j, which is determined by the current recovery technology, as shown in Table 5 [10].

Table 5. Average market price and recovery rate of raw metal materials for power battery (2018).

Element Type	Lithium	Cobalt	Nickel	Manganese	Aluminum
Average price (€/t)	117,828	771,769	3600	1858	1845
Recovery rate	85%	98%	98%	98%	90%

Retired battery purchase cost: the initial purchase cost of the retired LIP battery is the same as that of the echelon utilization analysis, referring to the average industry price. As for the battery after echelon utilization (LIP battery), this paper assumes that the repurchase cost will be greatly reduced when recycling due to the large performance degradation. This paper takes the average price of the industry, about 3.87 €/kWh [10]. However, for the TL battery with high energy density and metal resources, this paper assumes that the purchase cost of TL battery to LIP battery is proportional to the price ratio of the new batteries.

Cost of battery dismantling and recycling: the wet recovery method is widely used in the industry at present. This method extracts the required metal materials after soaking the cathode materials with chemical reagents. According to the current technical level, the cost of dismantling and recovering the LIP battery is about 7.55 €/kWh, and the cost of recovery of the TL battery (NCM523) is about 12.58 €/kWh [15].

Battery transportation cost: only intercity transportation is considered because dismantling and recycling enterprises would be near the power battery recovery areas generally. At the same time, its calculation method is the same as that in the echelon utilization model.

Based on the above model, the TL (NCM523) battery and the LIP battery are calculated. Among them, the TL (NCM523) battery is analyzed in terms of the abovementioned mode I (recycling directly after retirement), while the LIP battery is analyzed in terms of mode I and mode II (recycling after echelon utilization). The evaluation results of battery recycling are shown in Figure 9.

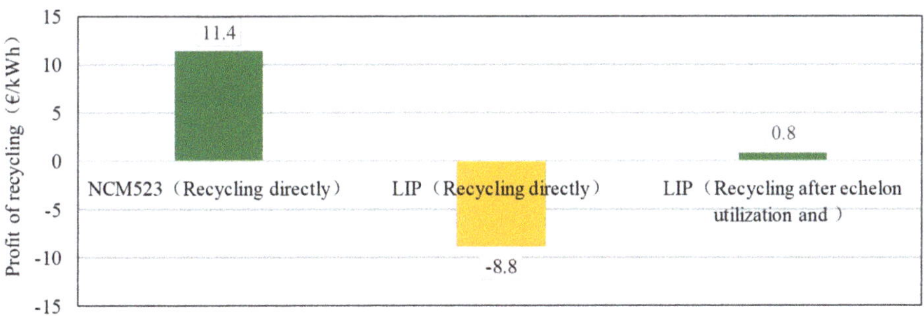

Figure 9. Analysis of the economic benefits of recycling of the retired power battery.

It can be seen from the figure that the TL battery (NCM523) has a high recycling income with a profit of about 11.41 €/kWh, which is mainly due to the large number of metal materials contained in the TL (NCM523) battery. However, due to the low value of reusable materials, if the LIP battery is recycled directly, it will cause a loss of 8.77 €/kWh. If the government begins to implement the mandatory recycling of retired power batteries, the loss will become an additional cost for enterprises. However, if the LIP battery is first used in echelon utilization and then disassembled for recycling, the profit of 0.79 €/kWh from battery recycling can be realized. It is equivalent to reduce the use cost of LIP battery when recycling after echelon utilization.

6. Overall Analysis of Retired Power Battery Reusing

The following is an analysis of the overall economic benefits of the power battery reusing of NEVs. Meanwhile, a comparative analysis of the use economy of TL (NCM523) battery and LIP battery from the perspective of industrial concerns is conducted.

According to the industry average data, the selling prices of TL battery (NCM523) and LIP battery are 155 €/kWh and 129 €/kWh at present, respectively [34]. Combined with the above economic benefit evaluation results of echelon utilization and recycling, the use costs of two power batteries in NEVs can be calculated. Note that the use cost represents that enterprises need to pay for the use of the power battery. The specific calculation equation is as follows:

$$P_u = P_p - P_c - P_r \tag{13}$$

where the P_u indicates the cost of using the battery; P_p represents the original battery price; P_c indicates the profit of the echelon utilization of retired power battery. Note that the profit is zero if echelon utilization is not carried out; P_r represents the profit of recycling of retired power batteries.

The specific calculation results are shown in Figure 10. For the LIP battery, if it is recycled directly, its cost will increase. However, if the echelon utilization with the appropriate scenario is implemented (Figure 10 shows the minimum benefit in the low-performance power battery scenario, which is the electric special vehicle scenario), the use cost of the LIP battery will be significantly reduced to 105.3 €/kWh. Thus, the best reusing mode is to carry out primary echelon utilization and then recycle for the LIP power battery.

The list in Figure 10 shows the change of the use cost of the TL (NCM523) battery and the LIP battery in different reusing modes relative to the original price of the TL (NCM523) battery. It can be found that the use cost of the LIP battery recycling directly is 5.5 €/kWh, or 3.9% lower than that of the TL (NCM523) battery. Nevertheless, it weakens the cost advantage of the original LIP battery compared with the TL (NCM523) battery. Instead, the use cost of the LIP battery after echelon utilization and recycling could be 38.2 €/kWh lower than the TL (NCM523) battery, or about 27%. There is a 23.1% cost reduction potential for the echelon utilization of the LIP battery if regulations mandate battery recycling. This will improve the cost competitiveness of the LIP battery further. However, there is no feasible technical scheme for the echelon utilization of the TL battery at present. If we can make a breakthrough in this aspect in the future, it may improve the cost-effectiveness of the TL battery significantly.

According to the technological roadmap for energy-saving and new energy vehicles in China, the industry-wide unit cost of a battery pack will be reduced to 104 €/kWh in 2025 and 73 €/kWh in 2030 [9]. In this paper, we assume that the price of the LIP battery is the same as the technological roadmap. The price of the TL battery is assumed that the degree of reduction is the same as that of LIP battery, as the technological roadmap suggested. Moreover, there is no available cost data for the battery reusing technological progress. The dismantling and recycling techniques for recycling and detection and regrouping techniques for echelon utilization are assumed to keep constant. The transportation cost in reusing is assumed to be constant. The other costs are assumed to be proportional to the battery prices.

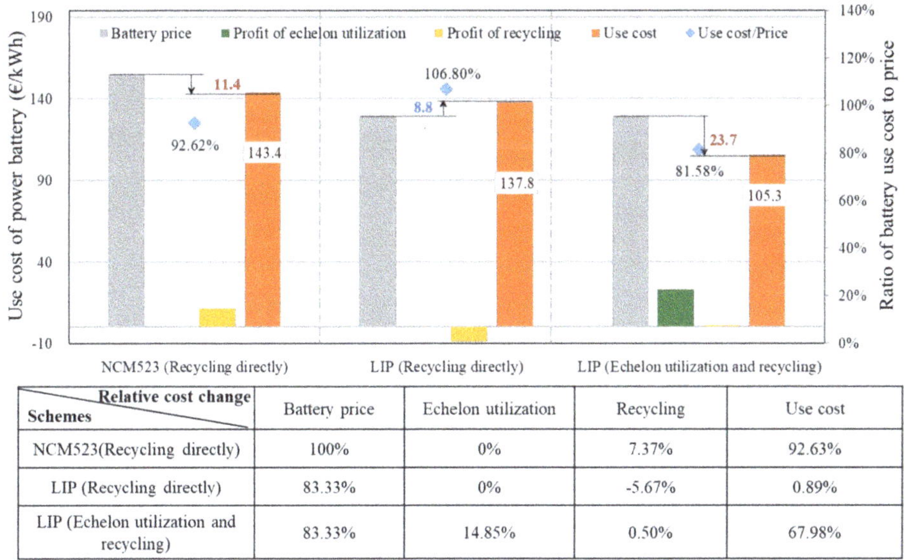

Schemes / Relative cost change	Battery price	Echelon utilization	Recycling	Use cost
NCM523(Recycling directly)	100%	0%	7.37%	92.63%
LIP (Recycling directly)	83.33%	0%	-5.67%	0.89%
LIP (Echelon utilization and recycling)	83.33%	14.85%	0.50%	67.98%

Figure 10. Comparison of use costs of NCM523 battery and LIP battery.

Therefore, according to the abovementioned assumptions, the use costs of the TL and LIP batteries could be predicted, as shown in Figure 11. It could be seen that the use costs of power batteries are decreased with the battery prices decreased regardless of the varieties of batteries. The TL battery could reduce to 89.1 €/kWh in 2030 by recycling directly. The LIP battery could reduce to 72.5 €/kWh in 2030 by recycling after echelon utilization. It should be noted that the extent of reduction of the use cost of TL batteries is more than that of the LIP battery due to the assumption that the reusing technical progress cost is constant. In fact, the cost of recycling technology reduces with the development of technology gradually, so the use cost of the LIP battery will be relatively lower than the TL battery.

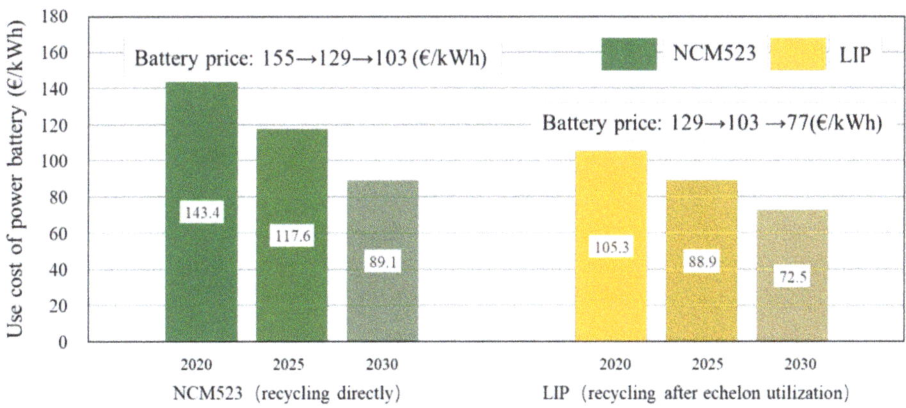

Figure 11. The use costs of different power batteries with the battery prices decreased.

7. Policy Suggestion

Based on the research and evaluation of the critical issues in the reusing of retired power batteries, this paper proposes some policy suggestions for the government and enterprises. First, China is about to usher in the peak period of retired power batteries, and mandatory recycling is imminent by the government. The relevant enterprises should make arrangements and preparations in advance so as to meet the regulations. Second, while multi-level echelon utilization is practical in theory, it is not economically feasible at present. In the industrial practice, more attention should be paid to the first level of echelon utilization in the future. At the same time, the two-level echelon utilization scheme should be considered properly according to the urgency of the scene demand. Third, the feasibility and profit of the echelon utilization are highly dependent on the requirements of battery performance at each utilization stage, implying the importance of performance matching. Thus, special consideration should be given to the huge market potential of energy storage in China. Besides, the future design of NEV power batteries may need to give due consideration to the performance requirements of the energy storage battery. Finally, the TL battery can only be recycled directly, while the LIP battery is suitable for echelon utilization and recycling at present. This would further improve the cost advantage of the lithium iron phosphate battery. Therefore, automotive enterprises should manage the power battery from the perspective of the whole life cycle to achieve the lowest use cost.

Looking forward to the future, it is imperative to reuse the retired power battery of NEVs. With the decrease of the battery price and the maturity of the reusing technology, the revenue from the reuse of retired power battery will be further improved. The government and related enterprises should increase the research of battery materials and recycling technology so as to reduce the price of batteries and the cost of recycling. Besides, the use cost of the LIP after reusing is lower than TL batteries; the enterprise should make it a key factor in determining the power battery technology route in the future.

8. Conclusions and Limitation

This paper aims to reveal the critical issues for power battery reusing of NEVs in China by the strategic value analysis and building the economic benefit model. Based on the results and discussion, some conclusions could be drawn:

1. Power battery reusing has three aspects strategic values such as protecting the environment and eliminating potential safety problems of retired power batteries, realizing resource recovery and reducing the risk of battery material supply and reducing the use cost of power battery and then improving the competitiveness of NEVs.
2. The echelon utilization of retired LIP batteries would be profitable if the relevant industry chain could be coordinated. The primary echelon utilization of retired LIP could obtain the maximum profit in the E-Bike application scenario of 30.21 €/kWh. While the profits of two-level echelon utilization are relatively lower. Thus, more attention should be paid to the first level echelon utilization in the future.
3. The TL battery (NCM523) has a high recycling income with a profit of about 11.41 €/kWh, while the LIP battery is recycled directly, causing a loss of 8.77 €/kWh for LIP recycling directly. Instead, if the LIP battery is first used in echelon utilization and then disassembled for recycling, the profit of 0.79 €/kWh from battery recycling can be realized.
4. The use cost of the LIP battery after echelon utilization and recycling could be 38.2 €/kWh lower than the TL (NCM523) battery recycling directly. This would further improve the cost advantage of the lithium iron phosphate battery. Hence, the best reusing mode for the TL battery is recycling directly, while the LIP battery is suitable for echelon utilization and recycling at present.
5. The use costs of power batteries are decreased, with the battery prices decreased regardless of the varieties of batteries. The TL battery could reduce to 89.1 €/kWh in 2030 by recycling directly. The LIP battery could reduce to 72.5 €/kWh in 2030 by recycling after echelon utilization.

However, it should be noted that this study is based on the ideal assumption that every link of recycling has been fully coordinated. Meanwhile, the model assumes the cost of dismantling and regrouping at the first-level and second-level echelon utilization is the same, while the cost of dismantling and regrouping at the second-level echelon utilization is higher. Therefore, we will refine the model by taking these limitations into consideration and improve the accuracy of the model in future work.

Author Contributions: Conceptualization, Z.L. and F.Z.; methodology, Z.L. and X.L.; validation, Z.L., X.L.; formal analysis, X.L.; investigation, X.L.; resources, X.L.; data curation, X.L.; writing—original draft preparation, Z.L. and X.L.; writing—review and editing, H.H. and Z.L.; visualization, X.L.; supervision, A.A.A. and H.B.; project administration, F.Z. and Z.L.; funding acquisition, Z.L., A.A.A. and H.B. All authors have read and agreed to the published version of the manuscript.

Funding: This study is supported by the National Natural Science Foundation of China (U1764265) and the Chinese Academy of Engineering (2019-XZ-55-01-01).

Acknowledgments: The authors would like to thank Zaidan Chen for her help collect the industrial data of power batteries. Additionally, the authors would like to thank the anonymous reviewers for their reviews and comments.

Conflicts of Interest: The authors declare no conflict of interest.

Abbreviations

NEV	New Energy Vehicle
ICEV	Internal Combustion Engine Vehicles
MEV	Micro Electric Vehicle
ESV	Electric Special Vehicle
E-Bike	Electric Bike
CBS	Communication Base Station
REPS	Renewable Energy Power Station
PMG	Parallel Micro-grid
LIP	Lithium Iron Phosphate batteries
TL	Ternary Lithium battery
LA	Lead-acid battery
DOD	Depth of Discharge
NCM523	$Ni_{0.5}Co_{0.2}Mn_{0.3}$ ternary lithium battery
NREL	National Renewable Energy Laboratory

References

1. Chen, K.; Zhao, F.; Hao, H.; Liu, Z. Synergistic impacts of China's subsidy policy and new energy vehicle credit regulation on the technological development of battery electric vehicles. *Energies* **2018**, *11*, 3193. [CrossRef]
2. Hao, H.; Cheng, X.; Liu, Z.; Zhao, F. China's traction battery technology roadmap: Targets, impacts and concerns. *Energy Policy* **2017**, *108*, 355–358. [CrossRef]
3. Liu, F.; Zhao, F.; Liu, Z.; Hao, H. China's electric vehicle deployment: Energy and greenhouse gas emission impacts. *Energies* **2018**, *11*, 3353. [CrossRef]
4. CAAM. *China's Automotive Industry Satus in 2019*; China Association of Automobile Manufacturers: Beijing, China, 2020. [CrossRef]
5. China Battery Enterprise Alliance. Available online: http://pba.cbea.com/ (accessed on 15 January 2020).
6. Tang, Y.; Zhang, Q.; Li, Y.; Wang, G.; Li, Y. Recycling mechanisms and policy suggestions for spent electric vehicles' power battery—A case of Beijing. *J. Clean. Prod.* **2018**, *186*, 388–406. [CrossRef]
7. Huang, B.; Pan, Z.; Su, X.; An, L. Recycling of lithium-ion batteries: Recent advances and perspectives. *J. Power Sources* **2018**, *399*, 274–286. [CrossRef]
8. MIIT. Policy Document. Available online: http://www.miit.gov.cn/n1146295/n1652858/index.html (accessed on 16 December 2019).
9. SAE-China. *Technology Roadmap for Energy-Saving and New Energy Vehicles*; Chinese Society of Automotive Engineers: Beijing, China, 2016. [CrossRef]
10. Orient Securities. Power Battery Recycling. Available online: http://www.nxny.com/stype_54_p153/ (accessed on 23 October 2018).
11. Mayyas, A.; Steward, D.; Mann, M. The case for recycling: Overview and challenges in the material supply chain for automotive li-ion batteries. *Sustain. Mater. Technol.* **2019**, *19*. [CrossRef]

12. Sonoc, A.; Jeswiet, J.; Soo, V.K. Opportunities to improve recycling of automotive lithium ion batteries. *Procedia Cirp* **2015**, *29*, 752–757. [CrossRef]

13. Pagliaro, M.; Meneguzzo, F. Lithium battery reusing and recycling: A circular economy insight. *Heliyon* **2019**, *5*. [CrossRef]

14. Wang, T.; Song, D.; Li, G.; Huang, J.; Zhu, H.; He, W. Recycling Potential for Waste Electric Vehicle Lithium-ion Batteries in China. Available online: https://pdfs.semanticscholar.org (accessed on 10 October 2018).

15. China Automotive Technology & Research Center. *China New Energy Vehicle Power Battery Recycling Industry Development Report*; Publishing House of Electronics Industry: Beijing, China, 2019. [CrossRef]

16. Ahmadi, L.; Yip, A.; Fowler, M.; Young, S.B.; Fraser, R.A. Environmental feasibility of re-use of electric vehicle batteries. *Sustain. Energy Technol. Assess.* **2014**, *6*, 64–74. [CrossRef]

17. Steward, D.M.; Mayyas, A.T.; Mann, M.K. Economics and challenges of Li-ion battery recycling from end-of-life vehicles. *Procedia Manuf.* **2019**, *33*. [CrossRef]

18. Unterreiner, L.; Jülch, V.; Reith, S. Recycling of battery technologies—Ecological impact analysis using life cycle assessment (LCA). *Energy Procedia* **2016**, *99*, 229–234. [CrossRef]

19. Liu, Z.; Hao, H.; Cheng, X.; Zhao, F. Critical issues of energy efficient and new energy vehicles development in China. *Energy Policy* **2018**, *115*, 92–97. [CrossRef]

20. Reinhardt, R.; Christodoulou, I.; Gassó-Domingo, S.; García, B.A. Towards sustainable business models for electric vehicle battery second use: A critical review. *J. Environ. Manag.* **2019**, *245*, 432–446. [CrossRef] [PubMed]

21. Qiao, Q.; Zhao, F.; Liu, Z.; Hao, H. Recycling-based reduction of energy consumption and carbon emission of China's electric vehicles: Overview and policy analysis. *WCX World Cong. Exp.* **2018**, *1*. [CrossRef]

22. Qiao, Q.; Zhao, F.; Liu, Z.; Hao, H. Electric vehicle recycling in China: Economic and environmental benefits. *Resour. Conserv. Recycl.* **2019**, *140*, 45–53. [CrossRef]

23. Casals, L.C.; Amante Garcia, B.; Canal, C. Second life batteries lifespan: Rest of useful life and environmental analysis. *J. Environ. Manag.* **2019**, *232*, 354–363. [CrossRef]

24. Madlener, R.; Kirmas, A. Economic viability of second use electric vehicle batteries for energy storage in residential applications. *Energy Procedia* **2017**, *105*, 3806–3815. [CrossRef]

25. Cusenza, M.A.; Guarino, F.; Longo, S.; Mistretta, M.; Cellura, M. Reuse of electric vehicle batteries in buildings: An integrated load match analysis and life cycle assessment approach. *Energy Build.* **2019**, *186*, 339–354. [CrossRef]

26. Diouf, B.; Pode, R.; Osei, R. Recycling mobile phone batteries for lighting. *Renew. Energy* **2015**, *78*, 509–515. [CrossRef]

27. Neubauer, J.; Pesaran, A. The ability of battery second use strategies to impact plug-in electric vehicle prices and serve utility energy storage applications. *J. Power Sources* **2011**, *196*, 10351–10358. [CrossRef]

28. Song, Z.; Feng, S.; Zhang, L.; Hu, Z.; Hu, X.; Yao, R. Economy analysis of second-life battery in wind power systems considering battery degradation in dynamic processes: Real case scenarios. *Appl. Energy* **2019**, *251*. [CrossRef]

29. Boyden, A.; Soo, V.K.; Doolan, M. The environmental impacts of recycling portable lithium-ion batteries. *Procedia Cirp* **2016**, *48*, 188–193. [CrossRef]

30. Harper, G.; Sommerville, R.; Kendrick, E.; Driscoll, L.; Slater, P.; Stolkin, R.; Walton, A.; Christensen, P.; Heidrich, O.; Lambert, S. Recycling lithium-ion batteries from electric vehicles. *Nature* **2019**, *575*, 75–86. [CrossRef] [PubMed]

31. Martinez-Laserna, E.; Gandiaga, I.; Sarasketa-Zabala, E.; Badeda, J.; Stroe, D.I.; Swierczynski, M.; Goikoetxea, A. Battery second life: Hype, hope or reality? A critical review of the state of the art. *Renew. Sustain. Energy Rev.* **2018**, *93*, 701–718. [CrossRef]

32. Guo, Y.; Luo, M.; Zou, J.; Liu, Y.; Kang, J. *Temperature Characteristics of Ternary-Material Lithium-Ion Battery for Vehicle Applications*; 0148-7191; SAE Technical Paper: Warrendale, PA, USA, 2016.

33. Tredeau, F.P.; Salameh, Z.M. Evaluation of lithium iron phosphate batteries for electric vehicles application. In Proceedings of the 2009 IEEE Vehicle Power and Propulsion Conference, Dearborn, MI, USA, 7–10 September 2009; pp. 1266–1270.

34. China Automotive Technology & Research Center. *Annual Report on the Development of New Energy Vehicle Power Battery Industry in China*; Social Sciences Academic Press: Beijing, China, 2019.

35. Gilbert, J.A.; Bareño, J.; Spila, T.; Trask, S.E.; Miller, D.J.; Polzin, B.J.; Jansen, A.N.; Abraham, D.P. Cycling behavior of NCM523/graphite lithium-ion cells in the 3–4.4 V range: Diagnostic studies of full cells and harvested electrodes. *J. Electrochem. Soc.* **2017**, *164*, A6054–A6065. [CrossRef]

36. Wang, W.; Wu, Y. An overview of recycling and treatment of spent LiFePO 4 batteries in China. *Resour. Conserv. Recycl.* **2017**, *127*, 233–243. [CrossRef]

37. Liu, J. Second use potential of retired EV batteries in power system and associated cost analysis. *Energy Storage Sci. Technol.* **2017**, *6*, 243–249.

38. Li, M.; Yang, J.; Liang, S.; Hou, H.; Hu, J.; Liu, B.; Kumar, R.V. Review on clean recovery of discarded/spent lead-acid battery and trends of recycled products. *J. Power Sources* **2019**, *436*. [CrossRef]

39. Yang, Y.; Zhu, W.; Xie, C.; Shi, Y.; Liu, F.; Li, W.; Tang, Z. A layered bidirectional active equalization method for retired power lithium-ion batteries for energy storage applications. *Energies* **2020**, *13*, 832. [CrossRef]

40. Liu, X.; Deng, X.; He, Y.; Zheng, X.; Zeng, G. A dynamic state-of-charge estimation method for electric vehicle lithium-ion batteries. *Energies* **2019**, *13*, 121. [CrossRef]

41. Schuster, F.S.; Bach, T.; Fleder, E. Nonlinear aging characteristics of lithium-ion cells under different operational conditions. *J. Energy Storage* **2015**, *1*, 44–53. [CrossRef]

42. Li, L.; Dababneh, F.; Zhao, J. Cost-effective supply chain for electric vehicle battery remanufacturing. *Appl. Energy* **2018**, *226*, 277–286. [CrossRef]

43. Newbauer, J.; Pesaran, A. *NREL's PHEV/EV Li-Ion Battery Secondary-Use Project*; National Renewable Energy Lab.(NREL): Golden, CO, USA, 2010.

44. Manzetti, S.; Mariasiu, F. Electric vehicle battery technologies: From present state to future systems. *Renew. Sustain. Energy Rev.* **2015**, *51*, 1004–1012. [CrossRef]

45. Baron, C.; Al-Sumaiti, A.S.; Rivera, S. Impact of energy storage useful life on intelligent microgrid scheduling. *Energies* **2020**, *13*, 957. [CrossRef]

46. Gaines, L. The future of automotive lithium-ion battery recycling: Charting a sustainable course. *Sustain. Mater. Technol.* **2014**, *1–2*, 2–7. [CrossRef]

47. Ordoñez, J.; Gago, E.J.; Girard, A. Processes and technologies for the recycling and recovery of spent lithium-ion batteries. *Renew. Sustain. Energy Rev.* **2016**, *60*, 195–205. [CrossRef]

48. Wang, X.; Gaustad, G.; Babbitt, C.W.; Richa, K. Economies of scale for future lithium-ion battery recycling infrastructure. *Resour. Conserv. Recycl.* **2014**, *83*, 53–62. [CrossRef]

Article

A Layered Bidirectional Active Equalization Method for Retired Power Lithium-Ion Batteries for Energy Storage Applications

Yang Yang [1,2], Wenchao Zhu [1,2], Changjun Xie [1,2,*], Ying Shi [1,*], Furong Liu [1], Weibo Li [1] and Zebo Tang [3]

[1] School of Automation, Wuhan University of Technology, Wuhan 430070, Hubei, China; whutyangyang@whut.edu.cn (Y.Y.); zhuwenchao@whut.edu.cn (W.Z.); lfr@whut.edu.cn (F.L.); liweibo@whut.edu.cn (W.L.)
[2] Hubei Key Laboratory of Advanced Technology for Automotive Components, Wuhan University of Technology, Wuhan 430070, China
[3] DongFeng Motor Corporation Technical Center, Wuhan 430058, China; tangzeb@dfmc.com.cn
[*] Correspondence: jackxie@whut.edu.cn (C.X.); a_laly@163.com (Y.S.); Tel.: +86-185-0717-4180 (C.X.); +86-186-7295-2831 (Y.S.)

Received: 17 January 2020; Accepted: 11 February 2020; Published: 14 February 2020

Abstract: The power from lithium-ion batteries can be retired from electric vehicles (EVs) and can be used for energy storage applications when the residual capacity is up to 70% of their initial capacity. The retired batteries have characteristics of serious inconsistency. In order to solve this problem, a layered bidirectional active equalization topology is proposed in this paper. Specifically, a bridge-type equalization topology based on an inductor is adopted in the bottom layer, and the distributed equalization topological structure based on the bidirectional BUCK-BOOST circuit is adopted in the top layer. State of charge (SOC) is used as the equalization target variable, and the bottom layer equalization algorithm based on a "partition" idea and route optimization is proposed. The static equalization experiments and charge equalization experiments are performed by the 12 retired batteries selected from an electric sanitation vehicle. The results show that the proposed equalization method can reduce the SOC difference between retired batteries and can effectively improve the inconsistency of the retired battery pack with a faster equalization speed.

Keywords: retired batteries; energy storage applications; layered bidirectional equalization; equalization algorithm

---

1. Introduction

The number of electric vehicles (EVs) has grown rapidly due to increasing anxieties of environment deterioration and shortages of fossil fuels in recent years. Lithium-ion (Li-ion) power batteries are the main types of power batteries used in EVs because they offer a low self-discharge rate, high energy density, long cycle life, reasonable usage cost, and non-pollution [1–4]. Nowadays, battery manufacturers and governments are facing great pressure to recycle and dispose of batteries, because a large number of lithium-ion batteries are retired from EVs for safe operation and longer driving range after being employed in EVs for a few years [5]. However, the retired lithium-ion batteries' capacity is 70% to 80% of their initial capacity, and the certain residual capacity can be used for energy storage in an electricity grid after they are tested, selected, and classified [6,7]. The accumulatively installed capacity of electrochemical energy storage projects reached 1072.7 MW in China by the end of 2018, and out of all electrochemical energy storage projects in China, the quotient of lithium-ion batteries was maximal and achieved 70.7% [8]. However, of all the global energy storage projects, the quotient of electrochemical energy storage projects is very small. One of the key factors is exceptionally costly. Low-cost retired batteries bring opportunities and the potential for energy storage. Retired batteries, after being classified and regrouped, have no sharp capacity attenuation and still have good discharge ability. They can be employed for energy storage systems, which could help to reduce the initial cost of EVs' owners, reduce the cost of an energy storage system, and avoid energy waste and secondary pollution to the environment [9]. Several studies have shown, by establishing reasonable retired battery cycle life models, by using cost models, as well as by estimating retired battery cycle life accurately and controlling the depth of discharge, that the cycle life of second-use retired batteries can be prolonged, and the economic benefits of retired batteries used for energy storage are considerable [10–13]. Due to the limited voltage and capacity of a single battery cell, retired Li-ion power batteries need to be connected in series or in parallel to be used for an energy storage system [14]. However, because of the factory differences of batteries, environmental factors, and long-term use, the inconsistency between the retired batteries is more serious than that of the new batteries, which is the bottleneck of second-use retired batteries, and may result in low available capacity of a battery pack, a short cycle life, and security risks caused by over-charge and over-discharge [15,16]. Therefore, executing equalization is necessary before a second-use of retired batteries, and appropriate and better equalizers are critical.

Battery equalization is generally classified into two categories, passive equalization and active equalization. Passive equalization is an energy dissipative equalization method, which is achieved by using the resistors added at both ends of the battery to consume the excessive energy of the battery [17]. This is a simple and low-cost method, but it has some disadvantages, such as energy waste, serious resistance temperature rise, and slow equalization speed [18].

Active equalization transfers energy between batteries, and has advantages with high energy transfer efficiency and a shorter equalization time in comparison to passive equalization [19]. The studies focus on equalization topologies and equalization algorithms. Equalization topologies mainly include switched capacitor type, inductor type, and transformer type. The topology using a capacitor to transfer energy from a higher voltage battery to a lower voltage battery has the advantage of higher equalization efficiency, but the equalization time is long and the cost is high [20]. A bidirectional cell-to-cell active equalization method using a multiwinding transformer is proposed in [21]. It gives a short balancing path and guarantees a fast equalization speed. Improved transformer type topologies have a fast equalization speed and high equalization efficiency, such as a bidirectional transformer topology based on a DC/DC converter, which is proposed for active equalization in [22], and a bidirectional flyback DC/DC converter, which is introduced for charge equalization in [23]. However, they suffer from magnetic saturation, complicated switch sequence control, and high-voltage stress of rectifier diodes. Many equalizers based on an inductor are designed in the literature. Dual-inductor based charge equalizer energy is proposed for transfer energy from high-voltage cells to the battery module directly in [24]. The advantages are that it is easy to implement, has a simple structure, and a simple control, but it is not suitable when the batteries that need to be balanced are numerous. A multi-switched inductor equalization circuit is designed in [25], and an inductor based equalization circuit is proposed in [26] for constant-voltage/current charging and discharging, their topologies are simple and easy to control, but could only transfer energy between adjacent batteries. An interleaved equalization architecture based on a resonant *LC* converter and buck converter is proposed in [15], which can exchange energy from a battery module to the next adjacent battery module and from a battery module to a cell, and achieve high equalization efficiency and a large equalization current. However, when the consistency of the batteries in the same module is poor, a more effective equalizer is needed. Energy-bus battery equalization topology based on the Cuk circuit is proposed in [27], which allows the cell-pack equalization. Numerous equalization algorithms have been researched, which mainly include strategy based on open-circuit voltage (OCV) [24], a method based on voltage difference [16], a method based on the state of charge (SOC) [23,26,28–30], and a method based on residual available energy [31].

Due to the large number and serious inconsistency of retired batteries, in this paper, a novel layered bidirectional active equalization topology based on inductor for retired batteries for energy storage is proposed. SOC is used as the equalization target variable. A bottom layer equalization algorithm based on the "partition" idea and route optimization is employed. This paper is organized as follows: In Section 2, the detailed retired battery sorting process is introduced and a retired battery pack connected with 12 retired batteries in series is selected. In Section 3, a layered bidirectional equalization topology based on an inductor is first introduced. We subsequently propose a SOC-based bottom layer equalization algorithm based on the "partition" idea and route optimization. In Section 4, static equalization and charge equalization experiments with 12 selected retired batteries are implemented in order to verify the proposed equalization method and show the equalization performance. The numerical results have shown that the proposed method is suitable for the equalization of retired lithium-ion batteries for energy storage applications.

2. Selection of Retired Battery Pack for Equalization

2.1. Retiring Standards for Lithium-Ion Batteries

In practical applications, there are several situations in which electric vehicle power batteries are retired: (1) the residual capacity of the power battery is reduced to less than 80% of the rated value, and the running mileage is reduced seriously; (2) the power battery has been used for more than five years, the performance of the battery has fallen seriously, especially in safety performance; (3) the driving mileage of the EV is over 80 thousand kilometers; (4) the performance degradation of one or a few cells in a battery pack, the difference with other batteries in the battery pack is obvious, result in serious inconsistency of the battery pack, and the capacity of the battery pack declines rapidly during use. In most cases, an individual battery cannot be replaced in the electric vehicle power battery pack, so the whole battery pack must be retired.

2.2. Selecting Process of the Retired Battery Pack

After the power from lithium-ion battery packs is retired from EVs, the dump energies can be used for the energy storage field. The retired battery pack studied in this paper retired from an electric sanitation vehicle made by Dongfeng Automobile Co., Ltd in Wuhan, China. The factory parameters of the power from lithium-ion battery are as follows: the rated capacity is 40 Ah, the nominal voltage is 3.2 V, the internal resistance is 1 mΩ, the charging cut-off voltage is 3.6 V, the discharging cut-off voltage is 2.5 V, and the standard charge and discharge current is 0.3 C. The whole retired battery pack has 106 battery cells and consists of two layers. The echelon use of retired batteries for the energy storage system is considered in this paper. Safety testing and consistency sorting are required before the retired batteries are balanced. Firstly, through appearance screening, short circuit test, acupuncture test, high-temperature test, and over-charge and over-discharge test, a total of 50 retired batteries with good appearance and safety were selected. Secondly, OCV of 50 retired batteries was measured and the voltage distribution of 50 retired batteries is shown in Figure 1. After removing 14 retired batteries with a voltage less than 2.4 V, a total of 36 retired batteries with a voltage of between 2.4 V and 3.2 V were obtained. Then, in order to prevent over-charge, the 36 retired batteries were charged with 10 A constant current for 2 h, and then we measured the voltage of each battery. As shown in Figure 2, there were 30 retired batteries with a voltage greater than 3 V after charging, which has an echelon use value. Thirdly, we determined the capacity sorting for the remaining 30 retired batteries. Because the capacity of the retired battery is much lower than that of the factory battery, it was necessary to recalibrate the rated capacity of retired batteries before balancing them. The capacity test was based on national standards GB/T 34015-2017 "Recycling of traction battery used in electric vehicle—Test of residual capacity", and the capacity test steps of the retired battery are as follows: (1) under normal temperature, discharge the residual energy of the retired batteries and let it stand for 15 min. (2) Charge the retired batteries at a constant current of 1 C to the cut-off voltage of 3.6 V, and then switch to constant-voltage charging until the current drops to 0.05 C. (3) Let them stand for 0.5 h. (4) Discharge the battery at a constant current of 1 C to the discharge cut-off voltage of 2.5 V, record the discharge capacity. (5) Repeat Steps (1) to (4) five times in total, use the average value of the discharge capacity of five times as the new rated capacities of the retired batteries. When the range of three continuous test results is less than 3% of the factory rated capacity, the test can be terminated in advance, and the average value of the last three test results is used as the new rated capacities of the retired batteries. The new rated capacity distribution of 30 retired batteries is shown in Figure 3. It can be seen from Figure 3 that the capacities of most retired batteries are between 24 Ah and 32 Ah. As shown in Table 1, a total of 12 retired batteries with a larger rated capacity were selected from 30 retired batteries and were connected in series, numbered as B_1, B_2, ... , B_{12}, as the object of simulation and experiment in this paper.

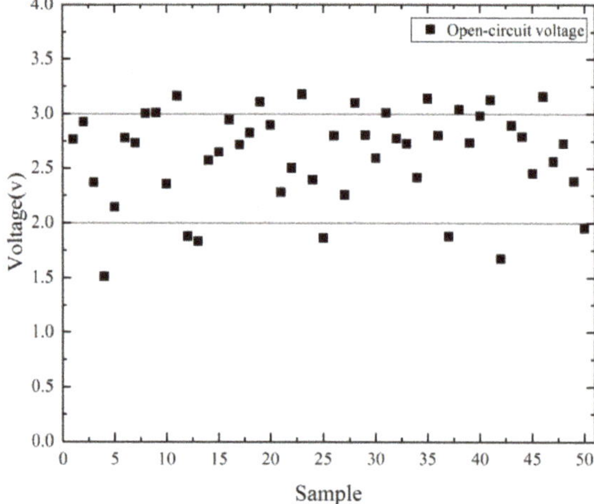

Figure 1. Voltage distribution of 50 retired batteries.

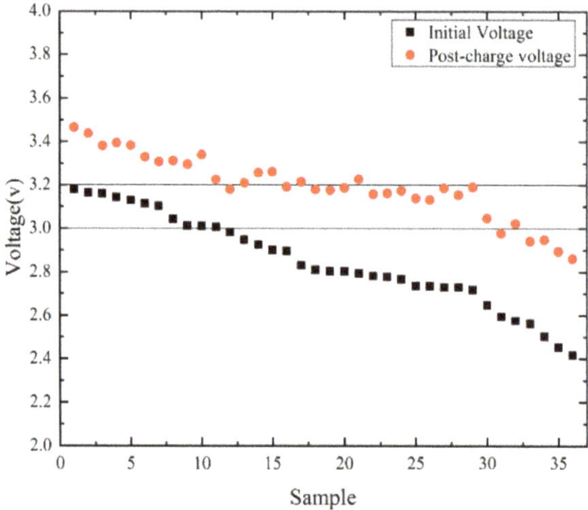

Figure 2. Post-charge voltage of 36 retired batteries.

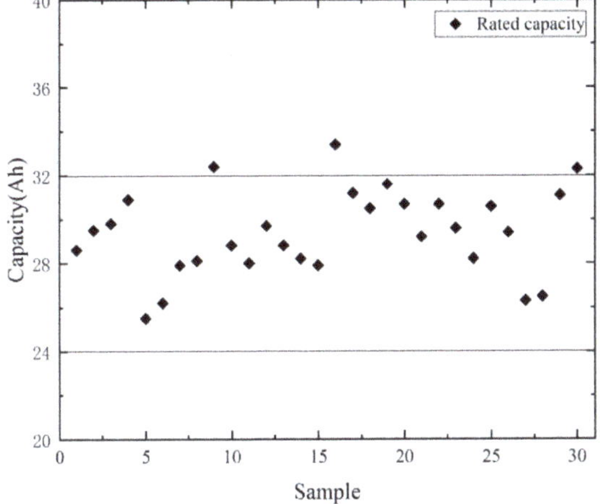

Figure 3. New rated capacity distribution of 30 retired batteries.

Table 1. New rated capacities of the selected 12 retired cells.

Number	16	9	30	19	17	29	4	22	20	25	18	3
Cell	B_1	B_2	B_3	B_4	B_5	B_6	B_7	B_8	B_9	B_{10}	B_{11}	B_{12}
Capacity (Ah)	33.4	32.4	32.3	31.6	31.2	31.1	30.9	30.7	30.7	30.6	30.5	29.8

3. Proposed Active Equalization Strategy

3.1. Equalization Topology

3.1.1. Proposed Equalization Topology

In order to improve the equalization efficiency, simplify the structure of the equalization system, and improve the expansibility of the equalization structure, as shown in Figure 4a, a novel layered bidirectional active equalization topology is proposed. n retired battery cells are grouped into m groups with p cells in each group. In this paper, 12 retired battery cells are grouped into three groups with four cells in each group. The three groups are denoted P_1, P_2, and P_3.

As shown in Figure 4b, a bridge-type equalization topology based on an inductor is adopted in the bottom layer. Each battery group is equipped with a bottom circuit m, consisting of p retired battery cells, 2(p + 1) MOSFET switch tubes, 2 p diodes, and inductor L, can implement energy transfer between any single cell in the battery group and has the advantages of high energy transmission efficiency, a simple circuit structure, and is easy to control.

As shown in Figure 4c, the distributed equalization topological structure based on the bidirectional BUCK-BOOST circuit is adopted in the top layer. Each two adjacent battery groups equipped with one distributed bidirectional balancing circuit, battery group m-1 and battery group m, are equipped with a top circuit named top circuit m − 1. They can implement an energy flow between adjacent battery groups, which makes the equalization system have strong expansibility features and makes them easy to modularize.

The balancing master controller collects retired battery data in real-time and executes the corresponding equalization algorithm, decides in turn whether bottom layer equalization or top layer equalization is needed, and opens or closes the corresponding switches to realize the equalization control of the entire retired battery pack.

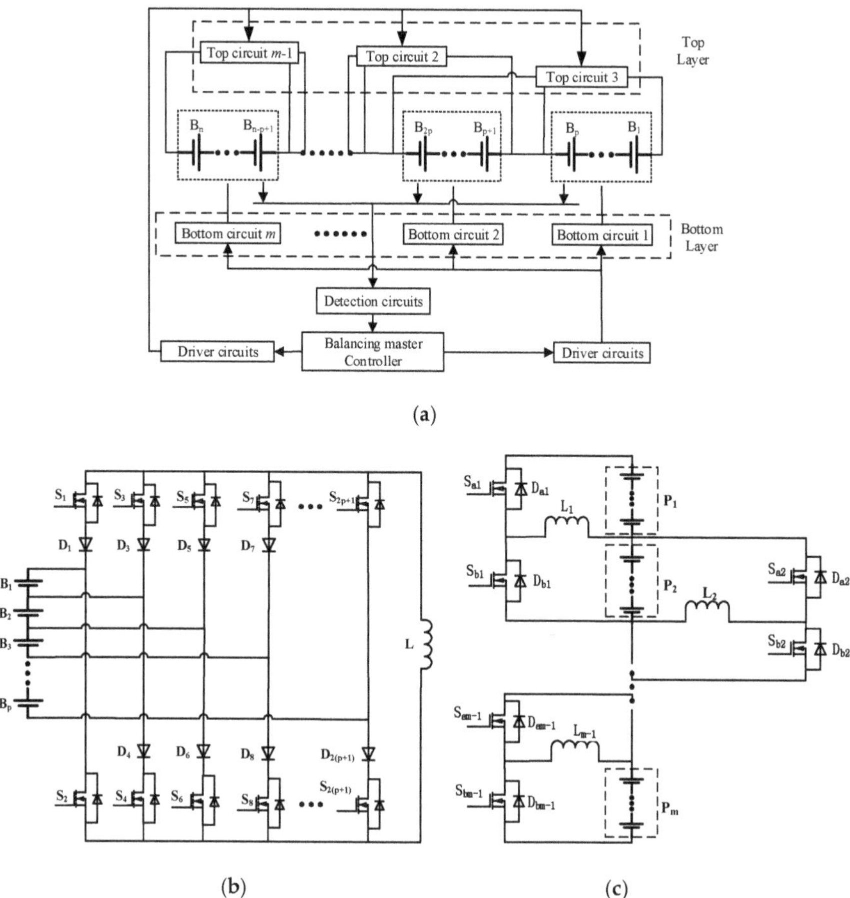

Figure 4. (**a**) Structure of the proposed bidirectional equalizer, (**b**) structure of bottom layer balancing circuit, (**c**) structure of top layer balancing circuit.

3.1.2. Equalization Principle of the Bottom Layer Circuits

Four retired battery cells connected in series were taken as an example to elaborate the equalization principle. It consisted of 4 retired battery cells, 10 MOSFET switch tubes, 8 diodes, and inductor L, and was able to implement energy transfer between any single cell in the group. Inductor L was used as an energy transmission medium between retired batteries in the group. The intragroup energy transfer process can be divided into two stages. Suppose that cell B_2 has the highest SOC value and cell B_4 has the lowest SOC value, the equalization process with two stages is shown in Figure 5.

Stage 1: charge L. As shown in Figure 5a, the switches S_4 and S_5 are turned ON, other switches are turned off. Cell B_2 charges the inductor L. The inductor L stores energy with the current i_L.

Stage 2: discharge L. As shown in Figure 5b, the switches S_7 and S_{10} are turned ON, other switches are turned off. The inductor L charges cell B_4, the energy transmission is from inductor L to cell B_4.

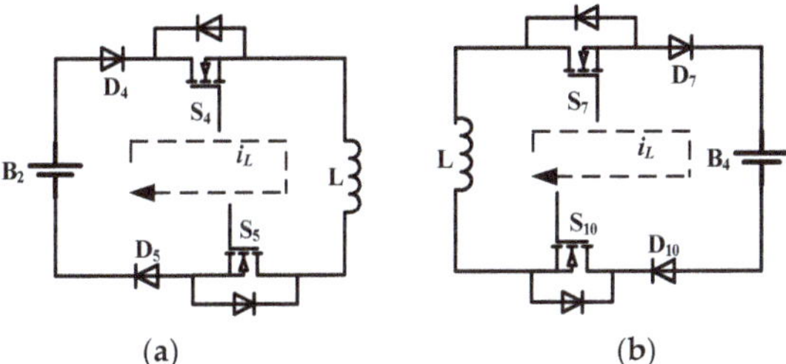

(a)   (b)

Figure 5. The equalization process of a bottom layer equalization circuit. (**a**) Charge L, (**b**) discharge L.

3.1.3. Equalization Principle of the Top Layer Circuits

Figure 6 shows the equalization process of the top layer circuits, P_1, P_2, and P_3, represent three battery groups. L_1 and L_2 are energy storage inductors, which can achieve bidirectional energy flow between adjacent retired battery groups. Suppose that the SOC mean value of battery group P_1 is higher than battery group P_2, the equalization process with two stages is shown in Figure 6.

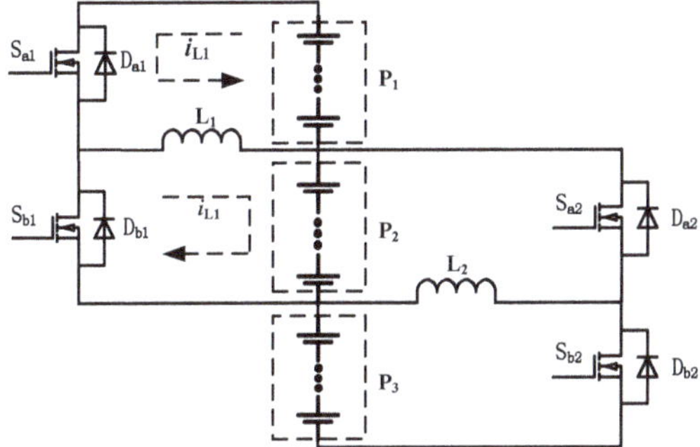

Figure 6. The equalization process of top layer equalization circuit.

Stage 1: The switch S_{a1} is turned ON, the battery group P_1 charges the inductor L_1, the energy is stored in inductor L_1, the inductor current i_{L1} is gradually increased from 0 to the maximum.

Stage 2: The switch S_{a1} is turned OFF and S_{b1} is turned ON, the inductor L_1 charges the battery group P_2, the energy transmission is from inductor L_1 to battery group P_2, the inductor current i_{L1} is gradually reduced from maximum to 0.

3.2. Equalization Algorithm

The performances of the retired batteries have degraded. In order to make a reasonable equalization control strategy, first of all, selecting a parameter that can accurately characterize the charged and discharged conditions of the cells and determine the consistency of the battery pack is needed. In this paper, the SOC is used as the equalization target variable, the battery pack consistency can be judged well, and the differences of the rated capacities of the battery cells are not needed to be considered. The capacity of the battery pack can be fully utilized. A fuzzy unscented Kalman filtering algorithm proposed by our research group was used to estimate the SOCs of retired batteries [32].

Assume that p denotes the battery number of the mth retired battery group, SOC_i is the SOC of the ith retired battery, $\overline{SOC_m}$ is the average SOC of the mth retired battery group, r_m is the maximum deviation from the average of the mth retired battery group, SOC_{avg} is the average SOC of the whole retired battery pack, and r denotes the maximum difference between SOC_{avg} and $\overline{SOC_m}$:

$$r_m = \max(SOC_i - \overline{SOC_m}), i = 1, 2, \ldots, p \tag{1}$$

$$r = \max(\overline{SOC_m} - SOC_{avg}) \tag{2}$$

Because the consistency between the battery cells in the retired battery pack is poor, if the top layer equalization is carried out firstly, the degree of inconsistency between batteries will be aggravated, and the switching loss during the equalization process will be increased. Therefore, in order to reduce the difficulty of battery equalization and improve the energy utilization efficiency, the strategy of carrying out bottom layer equalization first and top layer equalization second is adopted, which can first ensure the retired batteries in a group are consistent. Figure 7 shows the flowchart of the proposed equalization algorithm. The following steps are used for the proposed equalization algorithm.

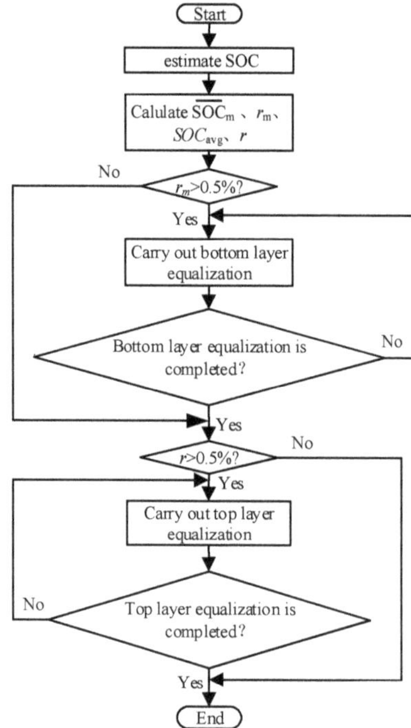

Figure 7. Flowchart of the proposed equalization algorithm.

(1) Estimate the SOC of each single retired battery;

(2) Calculate the $\overline{SOC}_m$, r_m, SOC_{avg} and r;

(3) Assume that the set bottom layer equalization threshold value Δ_m is 0.5%, check the r_m, if $r_m > \Delta_m$, go to Step 4, and execute bottom layer equalization algorithm. If $r_m \leq \Delta_m$, then go to Step 6;

(4) Carry out bottom layer equalization algorithm. In order to improve the equalization efficiency, the "partition" idea is introduced and equalization paths are optimized. As shown in Figure 8a, according to SOC from low to high, batteries in the same retired battery group are sorted. The SOC values of retired batteries in area a are lower and need to be charged. Area b is divided into area b_1 and area b_2, the difference between the SOC values of batteries in area b and $\overline{SOC}_m$ is within the threshold range. The SOC values of retired batteries in area c are higher and need to be discharged.

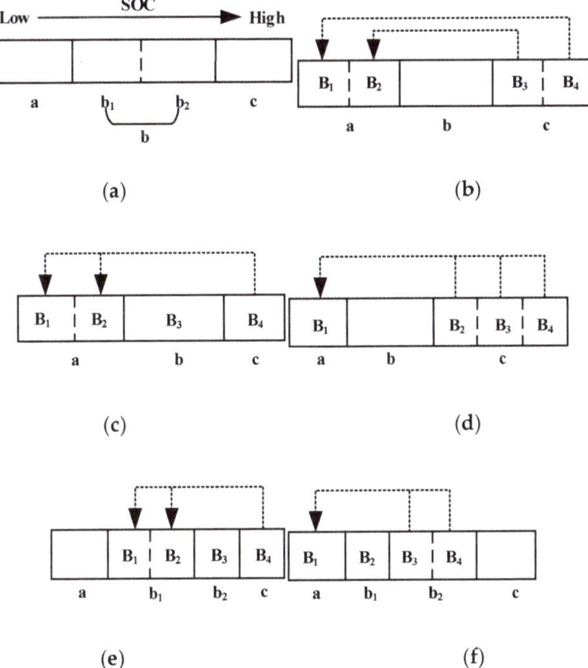

Figure 8. Partition and equalization path sketch maps. (**a**) Batteries partition sketch map, (**b**) equalization path of Case 1, (**c**) equalization path of Case 2, (**d**) equalization path of Case 3, (**e**) equalization path of Case 4, (**f**) equalization path of Case 5.

Which areas the batteries in the mth retired battery group belong to can be expressed by a piecewise function, where i represents the ith retired battery in the mth retired battery group.

$$i \in \begin{cases} a, SOC_i < \overline{SOC_m} - \Delta_m \\ b_1, \overline{SOC_m} \geq SOC_i \geq \overline{SOC_m} - \Delta_m \\ b_2, \overline{SOC_m} < SOC_i \leq \overline{SOC_m} + \Delta_m \\ c, SOC_i > \overline{SOC_m} + \Delta_m \end{cases}, i = 1,2,\ldots,p \tag{3}$$

The numbers of retired batteries in areas a, b, and c are denoted N_a, N_b, and N_c, respectively, because there are four retired batteries in one battery group. Therefore, according to the different N_a, N_b, and N_c, there are the following five cases:

① Case 1: $N_a = N_c$, and $N_a \neq 0$, $N_c \neq 0$. Energy complementary pairs are formed by the retired batteries in area a and area c, batteries in area c discharged, and batteries in area a charged. For example, as shown in Figure 8b, battery B_4 and battery B_1 exchange energy, battery B_3 and battery B_2 exchange energy;

② Case 2: $N_a > N_c$, and $N_a \neq 0$, $N_c \neq 0$. Energy complementary pairs are formed by the retired batteries in area a and area c, batteries in area c discharged, and batteries in area a charged. For example, as shown in Figure 8c, the battery with the highest SOC is complementary to two or more than two batteries with lower SOC. The SOC of B_4 is highest, B_4 transfers energy to B_1 and B_2;

③ Case 3: $N_a < N_c$, and $N_a \neq 0$, $N_c \neq 0$. Energy complementary pairs are formed by the retired batteries in area a and area c, batteries in area c discharged, and batteries in area a charged. For example, as shown in Figure 8d, the SOC of B_1 is lowest, B_4, B_3, and B_2 transfer energy to B_1;

④ Case 4: $N_a = 0$, $N_c > 0$. Energy complementary pairs are formed by the retired batteries in area b_1 and area c, batteries in area c discharged, and batteries in area b_1 charged. For example, as shown in Figure 8e, retired battery B_4 in area c is discharged, retired batteries B_1 and B_2 in area b_1 are charged;

⑤ Case 5: $N_a > 0$, $N_c = 0$. Energy complementary pairs are formed by the retired batteries in area a and area b_2, batteries in area b_2 discharged, and batteries in area a charged. For example, as shown in Figure 8f, retired battery B_1 in area a is charged, retired batteries B_3 and B_4 in area b_2 are discharged.

(5) According to the energy complementary pairs established in Step 4, set switching frequency and duty cycle, control the switching of the corresponding MOSFET switch tubes. If $r_m \leq \Delta_m$, bottom layer equalization has completed and go to Step 6; if $r_m > \Delta_m$, go to Step 4 to continue the bottom layer equalization;

(6) Assume that the set top layer equalization threshold value Δ is 0.5%, check the r, if $r > \Delta$, go to Step 7 and execute the top layer equalization algorithm; if $r \leq \Delta$, then go to Step 9;

(7) Carry out top layer equalization algorithm. The three retired battery groups are sorted according to $\overline{SOC_m}$ ($m = 1, 2, 3$) from low to high, and then top layer equalization paths are determined. There are the following six sorting cases: ① $\overline{SOC_1} > \overline{SOC_2} > \overline{SOC_3}$, ② $\overline{SOC_1} > \overline{SOC_3} > \overline{SOC_2}$, ③ $\overline{SOC_2} > \overline{SOC_1} > \overline{SOC_3}$, ④ $\overline{SOC_2} > \overline{SOC_3} > \overline{SOC_1}$, ⑤ $\overline{SOC_3} > \overline{SOC_1} > \overline{SOC_2}$, ⑥ $\overline{SOC_3} > \overline{SOC_2} > \overline{SOC_1}$. It can be divided into two categories: (1) cases ②, ③, ④, and ⑤ belong to direct equalization and transfer energy between adjacent battery groups; case ① and case ⑥ belong to indirect equalization, the battery group P_1 and the battery group P_3 are not adjacent, the battery group P_2 is taken as the energy transmission medium to achieve the equalization between them;

(8) Check the r. If $r > \Delta$, the retired battery pack is unbalanced, go to Step 7 to continue top layer equalization; if $r \leq \Delta$, then go to Step 9;

(9) End of the whole retired battery pack equalization.

4. Simulation and Experimental Verification

In order to verify the proposed equalization method and show the equalization performance, an equalization experiment with selected 12 retired batteries was implemented. The balancing experiment test bench and battery test system are illustrated in Figure 9a,b, respectively. The balancing experiment test bench included battery management system (BMS), master control board, voltage and current detection board, MOSFET switch array, bidirectional BUCK-BOOST circuit, voltage inspection board, retired battery pack, and power. The ITS5300 battery charge and discharge test system was used for battery performance testing. The inductors worked in discontinuous current mode (DCM).

4.1. Static Equalization Experiment

In order to verify the effectiveness of the bottom layer equalization algorithm based on the idea of "partitioning" and the optimization of equalization path, the selected initial states of charge (SOCs) are needed to involve different cases of the bottom layer partitioning equalization algorithm. Table 2 shows the initial SOCs of 12 retired cells for static equalization, three retired battery groups belonging to three different cases of bottom equalization algorithm, and $\overline{SOC_1} > \overline{SOC_2} > \overline{SOC_3}$. At the beginning of the equalization, the initial SOC range of 12 retired cells was 10.6%, the average of SOCs was 75.458%, r_1 was 1.225%, r_2 was 2%, r_3 was 2.1%, r was 4.058%, maximum deviation from average was 5.442%, and standard deviation was 3.596%. Bottom layer equalization threshold value Δ_m and top layer equalization threshold value Δ both were set to 0.5%.

The static equalization process lasted for two hours. Figure 10a illustrates the SOC response. It was observed that the differences between the retired cells' SOCs gradually converged, bottom layer equalization time was about 1100 s, equalization times of battery group P_1, battery group P_2 and battery group P_3 were 960 s, 1080 s, and 1100 s, respectively, top layer equalization phase was 1100 s–6060 s, equalization time of the whole retired battery pack was about 6060 s; as shown in Table 4, at the end of the equalization, the SOC range was decreased to 1.6%, the average of SOCs was 74.841%, maximum deviation from average was decreased to 0.92%, and standard deviation was decreased to 0.531%, r was decreased to 0.464%. The charge transfer efficiency of static equalization was 63.83%.

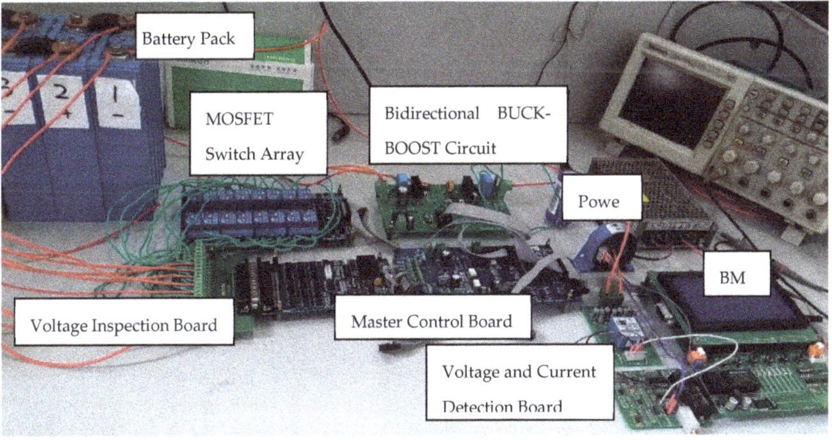

(a)

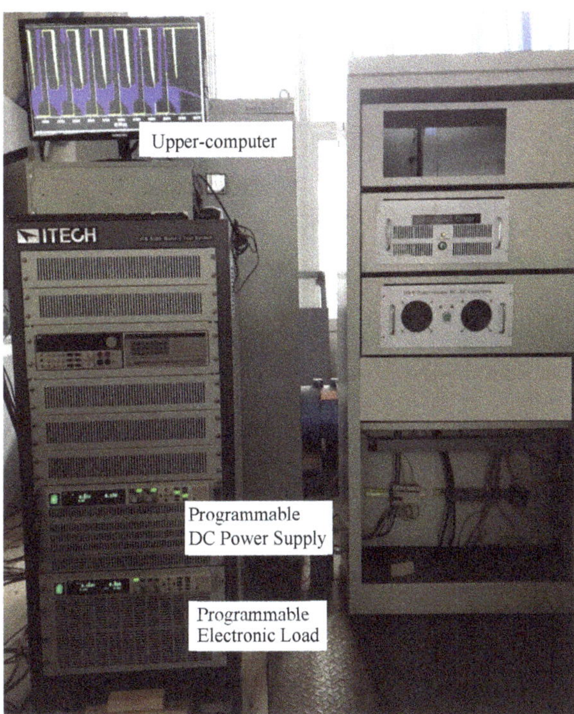

(b)

Figure 9. (a) Balancing experiment test bench, (b) battery test system.

Table 2. The initial SOCs of 12 retired cells for static equalization.

Groups	P_1						P_2				P_3	
Cells	B_1	B_2	B_3	B_4	B_5	B_6	B_7	B_8	B_9	B_{10}	B_{11}	B_{12}
SOC (%)	80.9	80.6	78.5	78.7	77.3	74.9	73.6	75.4	73.5	71.2	70.3	70.6

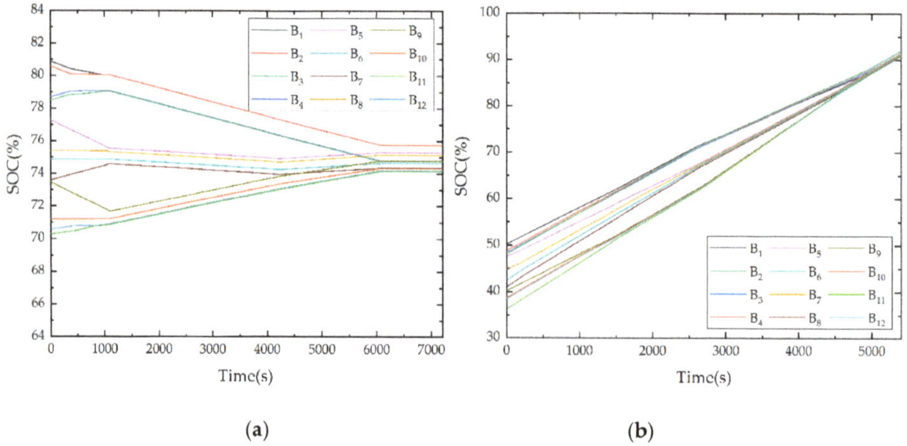

Figure 10. (**a**) SOC response with 12 retired cells of static equalization, (**b**) SOC response with 12 retired cells of charging equalization.

4.2. Charging Equalization Experiment

Table 3 shows the initial SOCs of 12 retired cells for charging equalization, three retired battery groups belonging to three different cases of bottom equalization algorithm, and $\overline{SOC_1} > \overline{SOC_2} > \overline{SOC_3}$. At the beginning of equalization, the initial SOC range of 12 retired cells was 13.9%, the average of SOCs was 43.77%, r_1 was 0.775%, r_2 was 3.6%, r_3 was 2.125%, r was 5.345%, maximum deviation from average was 7.47%, and standard deviation was 4.6%. Bottom layer equalization threshold value Δ_m and top layer equalization threshold value Δ both were set to 0.5%.

Table 3. The initial SOCs of 12 retired cells for charging equalization.

Groups	P_1						P_2				P_3	
Cells	B_1	B_2	B_3	B_4	B_5	B_6	B_7	B_8	B_9	B_{10}	B_{11}	B_{12}
SOC (%)	50.2	48.1	48.3	48.8	47.6	42.5	44.9	41	40.2	38.7	36.3	38.5

The charging equalization process lasted for 1.5 h. Figure 10b illustrates the SOC response. It was observed that the differences between the retired cells' SOCs gradually converged during the charging process. Bottom layer equalization time was about 2625 s, equalization time of the whole retired battery pack was about 4925 s; as shown in Table 4, at the end of the equalization, the SOC range was decreased to 1.41%, the average of SOCs was 86.91%, maximum deviation from average was decreased to 0.87%, and standard deviation was decreased to 0.42%, r was decreased to 0.467%. The charge transfer efficiency of charging equalization was 63.18%.

Table 4. Consistency comparison of retired battery pack before and after equalization.

	SOC Range (%)	SOC$_{avg}$ (%)	Maximum Deviation from Average (%)	Standard Deviation (%)	Equalization Time (s)	Charge Transfer Efficiency (%)
Static equalization					6060	63.83
Before equalization	10.6	75.458	5.442	3.596		
After equalization	1.6	74.841	0.92	0.531		
Charging equalization					4925	63.18
Before equalization	13.9	43.77	7.47	4.6		
After equalization	1.41	86.91	0.87	0.42		

5. Conclusions

This paper proposes a layered bidirectional active equalization method based on the SOC of batteries. The main concluding remarks can be made below:

(1) Power lithium-ion batteries retired from EVs can be used for energy storage applications. Moreover, a layered bidirectional active equalization topology is introduced for retired battery equalization, which can be used for the equalization of a large number of retired batteries. Equalization is based on the SOC of cells, and a bottom layer equalization algorithm based on the "partition" idea and route optimization is proposed;

(2) The balancing experiment test bench is developed and an ITS5300 battery test system is used for static equalization of 12 retired batteries and a charge equalization experiment. The experiment results have verified the proposed equalization method is effective, although the selected initial SOCs for static and charge equalization have a large range and involve different cases of the bottom layer partitioning equalization algorithm. The equalization time is short and the consistency of the retired battery pack is greatly improved. The SOC range of 12 retired cells is decreased from 10.6% to 1.6% and from 13.9% to 1.41% after static equalization and charge equalization, respectively;

(3) Although the equalization charge transfer efficiency is moderate, improved consistency is much more important for retired batteries, and the retired battery pack has a much larger available capacity after equalization, which is more critical for a second-use retired battery pack for energy storage.

Author Contributions: Conceptualization, Y.S. and W.L.; data curation, Y.Y.; formal analysis, Y.Y.; funding acquisition, C.X. and Z.T.; investigation, W.Z.; methodology, W.Z.; project administration, F.L.; software, Y.Y.; supervision, C.X.; validation, Y.S. and W.L.; writing—original draft preparation, Y.Y.; writing—review and editing, C.X. and Y.S. All authors have read and agreed to the published version of the manuscript.

Funding: This research was funded by the National Natural Science Foundation of China, grant number 51977164; the Hubei Science Fund for Distinguished Young Scholars, grant number 2017CFA049; and the Hubei province technological innovation major project, grant number 2018AAA059.

Conflicts of Interest: The authors declare no conflicts of interest.

References

1. Xie, C.; Xu, X.; Bujlo, P.; Shen, D.; Zhao, H.; Quan, S. Fuel cell and lithium iron phosphate battery hybrid powertrain with an ultracapacitor bank using direct parallel structure. *J. Power Sources* **2015**, *279*, 487–494. [CrossRef]
2. Farmann, A.; Sauer, D.U. A comprehensive review of on-board State-of-Available-Power prediction techniques for lithium-ion batteries in electric vehicles. *J. Power Sources* **2016**, *329*, 123–137. [CrossRef]
3. Li, X.; Xie, C.; Quan, S.; Huang, L.; Fang, W. Energy management strategy of thermoelectric generation for localized air conditioners in commercial vehicles based on 48 V electrical system. *Appl. Energy* **2018**, *231*, 887–900. [CrossRef]
4. Wang, Y.; Zhang, C.; Chen, Z. An adaptive remaining energy prediction approach for lithium-ion batteries in electric vehicles. *J. Power Sources* **2016**, *305*, 80–88. [CrossRef]
5. Jiang, Y.; Jiang, J.; Zhang, C.; Zhang, W.; Gao, Y.; Guo, Q. Recognition of battery aging variations for LiFePO4 batteries in 2nd use applications combining incremental capacity analysis and statistical approaches. *J. Power Sources* **2017**, *360*, 180–188. [CrossRef]

6. Liao, Q.; Mu, M.; Zhao, S.; Zhang, L.; Jiang, T.; Ye, J.; Shen, X.; Zhou, G. Performance assessment and classification of retired lithium ion battery from electric vehicles for energy storage. *Int. J. Hydrog. Energy* **2017**, *42*, 18817–18823. [CrossRef]

7. Lai, X.; Qiao, D.; Zheng, Y.; Ouyang, M.; Han, X.; Zhou, L. A rapid screening and regrouping approach based on neural networks for large-scale retired lithium-ion cells in second-use applications. *J. Clean Prod.* **2019**, *213*, 776–791. [CrossRef]

8. China Energy Storage Alliance. Energy Storage Industry Research White Paper 2019. Available online: http://www.cnesa.org/index/inform_detail?cid=5cf0ef98b1fd3772048b4567 (accessed on 18 May 2019).

9. Zhang, Y.; Li, Y.; Tao, Y.; Ye, J.; Pan, A.; Li, X.; Liao, Q.; Wang, Z. Performance assessment of retired EV battery modules for echelon use. *Energy* **2020**, *193*, 116555. [CrossRef]

10. Debnath, U.K.; Ahmad, I.; Habibi, D. Quantifying economic benefits of second life batteries of gridable vehicles in the smart grid. *Int. J. Electr. Power Energy Syst.* **2014**, *63*, 577–587. [CrossRef]

11. Neubauer, J.; Pesaran, A. The ability of battery second use strategies to impact plug-in electric vehicle prices and serve utility energy storage applications. *J. Power Sourcesces* **2011**, *196*, 10351–10358. [CrossRef]

12. Debnath, U.K.; Ahmad, I.; Habibi, D. Gridable vehicles and second life batteries for generation side asset management in the Smart Grid. *Int. J. Electr. Power Energy Syst.* **2016**, *82*, 114–123. [CrossRef]

13. Madlener, R.; Kirmas, A. Economic Viability of Second Use Electric Vehicle Batteries for Energy Storage in Residential Applications. *Energy Procedia* **2017**, *105*, 3806–3815. [CrossRef]

14. Huang, W.; Abu Qahouq, J.A. Energy Sharing Control Scheme for State-of-Charge Balancing of Distributed Battery Energy Storage System. *IEEE Trans. Ind. Electron.* **2015**, *62*, 2764–2776. [CrossRef]

15. Zhang, C.; Shang, Y.; Li, Z.; Cui, N. An Interleaved Equalization Architecture with Self-Learning Fuzzy Logic Control for Series-Connected Battery Strings. *IEEE Trans. Veh. Technol.* **2017**, *66*, 10923–10934. [CrossRef]

16. Gallardo-Lozano, J.; Romero-Cadaval, E.; Milanes-Montero, M.I.; Guerrero-Martinez, M.A. A novel active battery equalization control with on-line unhealthy cell detection and cell change decision. *J. Power Sourcesces* **2015**, *299*, 356–370. [CrossRef]

17. Liu, X.; Wan, Z.; He, Y.; Zheng, X.; Zeng, G.; Zhang, J. A Unified Control Strategy for Inductor-Based Active Battery Equalisation Schemes. *Energies* **2018**, *11*, 405. [CrossRef]

18. Gallardo-Lozano, J.; Romero-Cadaval, E.; Milanes-Montero, M.I.; Guerrero-Martinez, M.A. Battery equalization active methods. *J. Power Sourcesces* **2014**, *246*, 934–949. [CrossRef]

19. Bouchhima, N.; Schnierle, M.; Schulte, S.; Birke, K.P. Active model-based balancing strategy for self-reconfigurable batteries. *J. Power Sourcesces* **2016**, *322*, 129–137. [CrossRef]

20. Daowd, M.; Antoine, M.; Omar, N.; van den Bossche, P.; van Mierlo, J. Single Switched Capacitor Battery Balancing System Enhancements. *Energies* **2013**, *6*, 2149–2174. [CrossRef]

21. Chen, Y.; Liu, X.; Cui, Y.; Zou, J.; Yang, S. A Multi-Winding Transformer Cell-to-Cell Active Equalization Method for Lithium-Ion Batteries with Reduced Number of Driving Circuits. *IEEE Trans. Power Electron.* **2016**, *31*, 4916–4929. [CrossRef]

22. Wang, Y.; Zhang, C.; Chen, Z.; Xie, J.; Zhang, X. A novel active equalization method for lithium-ion batteries in electric vehicles. *Appl. Energy* **2015**, *145*, 36–42. [CrossRef]

23. Hannan, M.A.; Hoque, M.M.; Peng, S.E.; Uddin, M.N. Lithium-Ion Battery Charge Equalization Algorithm for Electric Vehicle Applications. *IEEE Trans. Ind. Appl.* **2017**, *53*, 2541–2549. [CrossRef]

24. Dai, H.; Wei, X.; Sun, Z.; Wang, D. A novel dual-inductor based charge equalizer for traction battery cells of electric vehicles. *Int. J. Electr. Power Energy Syst.* **2015**, *67*, 627–638. [CrossRef]

25. Cui, X.; Shen, W.; Zhang, Y.; Hu, C. A Fast Multi-Switched Inductor Balancing System Based on a Fuzzy Logic Controller for Lithium-Ion Battery Packs in Electric Vehicles. *Energies* **2017**, *10*, 1034. [CrossRef]

26. Zheng, X.; Liu, X.; Yao, H.; Zeng, G. Active vehicle battery balancing scheme in the condition of constant-voltage/current charging and discharging. *IEEE Trans. Veh. Technol.* **2017**, *66*, 3714–3723. [CrossRef]

27. Wu, Z.; Ling, R.; Tang, R. Dynamic battery equalization with energy and time efficiency for electric vehicles. *Energy* **2017**, *141*, 937–948. [CrossRef]

28. Zhang, Z.; Gui, H.; Gu, D.; Yang, Y.; Ren, X. A Hierarchical Active Balancing Architecture for Lithium-Ion Batteries. *IEEE Trans. Power Electron.* **2017**, *32*, 2757–2768. [CrossRef]

29. Ouyang, Q.; Chen, J.; Zheng, J.; Hong, Y. SOC Estimation-Based Quasi-Sliding Mode Control for Cell Balancing in Lithium-Ion Battery Packs. *IEEE Trans. Ind. Electron.* **2018**, *65*, 3427–3436. [CrossRef]

30. Ma, Y.; Duan, P.; Sun, Y.; Chen, H. Equalization of Lithium-Ion Battery Pack Based on Fuzzy Logic Control in Electric Vehicle. *IEEE Trans. Ind. Electron.* **2018**, *65*, 6762–6771. [CrossRef]
31. Diao, W.; Xue, N.; Bhattacharjee, V.; Jiang, J.; Karabasoglu, O.; Pecht, M. Active battery cell equalization based on residual available energy maximization. *Appl. Energy* **2018**, *210*, 690–698. [CrossRef]
32. Zeng, M.; Zhang, P.; Yang, Y.; Xie, C.; Shi, Y. SOC and SOH Joint Estimation of the Power Batteries Based on Fuzzy Unscented Kalman Filtering Algorithm. *Energies* **2019**, *12*, 3122. [CrossRef]

Article

An Optimal Fast-Charging Strategy for Lithium-Ion Batteries via an Electrochemical–Thermal Model with Intercalation-Induced Stresses and Film Growth

Guangwei Chen, Zhitao Liu * and Hongye Su

State Key Laboratory of Industrial Control Technology, Institute of Cyber-Systems and Control,
Zhejiang University, Hangzhou 310027, China; gwchen@zju.edu.cn (G.C.); hysu@iipc.zju.edu.cn (H.S.)
* Correspondence: ztliu@zju.edu.cn; Tel.: +86-0571-87952233 (ext. 8239)

Received: 15 March 2020; Accepted: 7 May 2020; Published: 11 May 2020

Abstract: Optimal fast charging is an important factor in battery management systems (BMS). Traditional charging strategies for lithium-ion batteries, such as the constant current–constant voltage (CC–CV) pattern, do not take capacity aging mechanisms into account, which are not only disadvantageous in the life-time usage of the batteries, but also unsafe. In this paper, we employ the dynamic optimization (DP) method to achieve the optimal charging current curve for a lithium-ion battery by introducing limits on the intercalation-induced stresses and the solid–liquid interface film growth based on an electrochemical–thermal model. Furthermore, the backstepping technique is utilized to control the temperature to avoid overheating. This paper concentrates on solving the issue of minimizing charging time in a given target State of Charge (SoC), while limiting the capacity loss caused by intercalation-induced stresses and film formation. The results indicate that the proposed optimal charging method in this paper offers a good compromise between the charging time and battery aging.

Keywords: electrochemical–thermal model; lithium-ion battery; fast charging

---

1. Introduction

Lithium-ion batteries have been used in many electronic products due to their high cell voltage, high energy density, high power density, convenient operating temperature range, lack of memory property, and high cycle life [1]. When operating a lithium-ion battery efficiently and safely during charging, long charging time, capacity degradation, capacity wastage, and overheating are the main difficulties that need to be overcome.

In recent years, many researchers have made efforts to optimize lithium battery charging. Many results are based on traditional charging patterns, such as the constant-current (CC) pattern and the constant current–constant voltage (CC–CV) pattern, without considering the aging process. For example, Liu and Luo [2] proposed a Taguchi-based algorithm and adopted orthogonal arrays to obtain the optimal rapid-charging strategy for a piecewise CC charging approach, which can charge a battery cell from 0% to 75% within 40 min and increase the cycle life by more than 60%. On the basis of the open circuit voltage (OCV)-resistance equivalent circuit model, Abdollahi et al. [3] presented a closed-form solution for optimally charging a lithium-ion battery; the target function is established through a combination of two consumption functions: time-to-charge (TTC) and energy losses (EL). Here, the CC–CV pattern is selected as the optimal charging scheme, where the current in the CC stage is a function of the ratio of weighting on TTC and EL. Then, Abdollahi et al. [4] proposed the objective function consisting of TTC, EL, and a temperature rise index (TRI). Then, the value of the current in the CC stage is also considered as a function of the ratio of weighting on TTC and EL, and finally the analytical solution for the optimal problem is derived. Monem et al. [5] studied the influence of three charging strategies including CC, CC–CV, and constant current–constant voltage with negative pulse (CC–CVNP) on the battery's cycle life. The results show that the CC–CVNP pattern with low amplitude and less negative pulses is more efficient than the CC and CC–CV patterns. Liu et al. [6] firstly put forward a triple-objective function for optimal battery charging on the basis of a coupled thermoelectric model. Then, the CC–CV charge strategy is optimized, which offers the best compromise among three significant performance indexes consisting of charging time, energy loss, and temperature rise. Fang et al. [7] permit users to specify charging objectives and reach them by dynamic optimal control for the first time and proposed two charging methods based on the linear quadratic control theory without real-time constrained optimization computation. Compared to the conventional open-loop regulation of fast charging, the close-loop optimal method can be used to accurately control the specific parameters, such as temperature, current, and voltage. Patnaik et al. [8] came up with a constant temperature–constant voltage (CT-CV) charging algorithm that considers battery temperature as a key feedback variable. Then, a simple and easy-to-realize proportional-integral-derivative (PID) controller is employed to implement this close-loop method and the results indicate that the proposed approach achieved a 20% faster charging rate with an identical total temperature increase as compared to the constant current–constant voltage (CC–CV) technique. Klein et al. [9] paid attention to minimum charging time and proposed a simple one-step predictive control algorithm that is capable of solving the time-optimal solution and satisfying the real-time requirement. On the basis of the electrochemical battery model, Pramanik et al. [10] introduced a novel method for optimally charging the lithium-ion battery cell, which establishes the objective function aiming to minimize the charging cost. The result indicates that the optimal charging method presented in the paper [10] can decrease the charging time of a lithium-ion cell, meanwhile guaranteeing the temperature limit when compared with the traditional constant current charging. Considering the electrolyte and thermal dynamics based on a single particle model, the Legendre–Gauss–Radau (LGR) pseudo-spectral approach is used to solve the problem of nonlinear multistate optimal control, and the minimum time charge strategies are analyzed minutely while taking the solid and electrolyte phase concentration limits and temperature constraint into account [11].

Generally speaking, fast charging can accelerate the battery aging processes. For reducing the aging rate, some researchers consider the aging process when optimizing the fast-charging strategy and some good results have been obtained. For example, through coupling incremental capacity (IC) and IC peak area analysis with the mechanistic model, Ansean et al. [12] quantified the mechanism of degradation that leads to the aging of the battery cell. In addition, the results show that aging is caused by a loss of lithium ions and a lower level of loss of active material on the negative electrode. On the basis of cycle-life testing (up to 4500 cycles), Ansean et al. [13] proposed a multistage fast-charging algorithm which allows a full recharging of the cell (0% to 100% SoC) within 20 min (indeed after 4500 cycles are reached) and does not cause any remarkable degradation to the battery cell. Considering the influence of intercalation-induced stress on aging, Suthar et al. [14] used dynamic optimization to achieve the optimal current profile to fast charge a lithium-ion battery through a single-particle model while coupling this with the intercalation-induced stress generation model. In addition, this was the first time protocols for optimally charging batteries while ensuring a minimal mechanical cost to the electrode particles during intercalation were demonstrated. Torchio et al. [15] used the first-principles pseudo-two-dimensional (P2D) model together with the capacity fade mechanisms that work when the battery is operating. Then, the model predictive control (MPC) method based on a linearized model of the P2D model was proposed to approach a target value of the state of charge (SoC) while considering the degradation process of the system as well as the thermal and voltage limits. To analyze the effect of static and dynamic fast-charging current strategies on the degradation properties of lithium-ion batteries, Monem et al. [16] applied both the static and dynamic fast-charging current profiles to a lithium-ion battery cell. After 1700 cycles, the result shows that the dynamic fast-charging current profile had a more outstanding role in reducing the aging rate and the charging time of cells than the static fast-charging profile. To reduce the influence of fast charging on battery aging,

Ali et al. [17] proposed a temperature control method based on fuzzy logic that protects the batteries from overvoltage and overheating. The result illustrates that the proposed fast-charging pattern spends 9.76% less time during full battery recharging than the conventional CC–CV method, and the approach does not bring significant degradation. As a compromise between the three objectives of high safety, longer lifetime, and a lower charging time, Zou et al. [18] proposed a fast-charging method on the basis of the electrochemical model and MPC theory. Here, the battery optimal charging issue is described in a linear time-varying model for the implementation of the MPC algorithm. Similarly, Lina et al. [19] presented an electrolyte enhanced single particle model with aging mechanisms which considers the effect of electrolyte dynamics, then the dynamic programming DP approach is adopted to obtain the optimal charging profiles to reduce the charge time and battery aging. On the basis of the electrochemical–thermal capacity fade model, Xu et al. [20] used the DP optimization algorithm to minimize capacity fade, temperature rise, and charging time. Although there are many papers considering the aging process in terms of optimizing the fast charge curve, the aging models used in the above-mentioned papers seldom involve intrinsic aging mechanisms and never consider the effects of intercalation-induced stress and film growth together on the aging process.

In this paper, the electrochemical–thermal model is employed to obtain the optimal charging profile and control the temperature of cell. Controlling the temperature of the hottest point inside the cell based on the backstepping method can help avoid overheating. During fast charging, intercalation-induced stress will cause particle fracture which can accelerate the aging process. Hence, restricting intercalation-induced stress to a given range is significant to reduce the degradation of cells when seeking the optimal fast-charging profile. Furthermore, the main aging cause is the growth of film on the surface of particles, and this film is a compound containing lithium which cannot be reused. Therefore, confining the growth rate of surficial film to a proper range can maximize the available lithium, which is another contribution of this paper. To sum up, on the basis of maintaining a constant temperature using the backstepping control method and minimizing the charging time while limiting intercalation-induced stress and the film growth to an appropriate range are the main contributions of this paper.

2. Electrochemical–Thermal Model with Intercalation-Induced Stress and Film Growth

2.1. SPM-Electrolyte-T Model

The SPM-Electrolyte-T Model is a simplified version of the first principle electrochemical model of batteries composed of a single particle model, electrolyte dynamics, and thermal dynamics. Considering the effects of electrolyte and thermal dynamics can help us investigate the fast-charging scheme to reduce battery aging to prolong the lifetime of lithium-ion batteries. The single particle model can be described by the following partial differential equation [21]:

$$\frac{\partial c_s^{\pm}}{\partial t} = D_s^{\pm}(T_{avg}) \times \left(\frac{\partial^2 c_s^{\pm}}{\partial r^2} + \frac{2}{r} \frac{\partial c_s^{\pm}}{\partial r} \right) \tag{1}$$

with the following conditions:

$$\frac{\partial c_s^{\pm}}{\partial r}(0, t) = 0, \ \frac{\partial c_s^{\pm}}{\partial r}(R_s^{\pm}, t) = \frac{\pm I(t)}{Fa^{\pm}AL^{\pm}} \tag{2}$$

The definitions of the relevant electrochemical parameters are listed in the attached table.

Most studies relevant to the electrochemical model of lithium-ion batteries ignore the influence of electrolyte dynamics for inexpensive computation. However, when the charging or discharging current is high, there would be a remarkable electrolyte concentration potential between the cathode and the anode, which has a significant impact on the prediction of the terminal voltage. Specifically, the process of fast charging requires more than a 2 C current, which has enough power to produce an obvious concentration potential. Hence, during the operation of fast charging, it is wise to consider electrolyte dynamics.

The electrolyte dynamics are governed by

$$\varepsilon_e^\pm \frac{\partial c_e^\pm}{\partial t}(x,t) = \frac{\partial}{\partial x}\left[D_e^{eff}\left(c_e^\pm, T_{avg}\right)\frac{\partial c_e^\pm}{\partial x}(x,t)\right] \mp \frac{\left(1-t_c^0\right)I(t)}{FAL^\pm} \tag{3}$$

$$\varepsilon_e^{sep} \frac{\partial c_e^{sep}}{\partial t}(x,t) = \frac{\partial}{\partial x}\left[D_e^{eff}\left(c_e^{sep}, T_{avg}\right)\frac{\partial c_e^{sep}}{\partial x}(x,t)\right] \tag{4}$$

with the boundary conditions

$$\frac{\partial c_e^-}{\partial x}(0^-,t) = \frac{\partial c_e^+}{\partial x}(0^+,t) = 0 \tag{5}$$

$$D_e^{eff}\left(L^-, T_{avg}\right)\frac{\partial c_e^-}{\partial x}(L^-,t) = D_e^{eff}\left(0^{sep}, T_{avg}\right)\times\frac{\partial c_e^{sep}}{\partial x}(0^{sep},t) \tag{6}$$

$$D_e^{eff}\left(L^{sep}, T_{avg}\right)\frac{\partial c_e^{sep}}{\partial x}(L^{sep},t) = D_e^{eff}\left(L^+, T_{avg}\right)\times\frac{\partial c_e^+}{\partial x}(L^+,t) \tag{7}$$

$$c_e(L^-,t) = c_e(0^{sep},t), \quad c_e(L^{sep},t) = c_e\left(L^+,t\right) \tag{8}$$

where $D_e^{eff}(c_e, T_{avg}) = D_e(c_e, T_{avg})(c_e)\cdot(\varepsilon_e)^{brug}$ is the effective electrolyte diffusivity, which means that the diffusion rate of lithium ions is related to its concentration and temperature; t_c^0 is the transference number, which is deemed to be constant; $\varepsilon_e^+, \varepsilon_e^-, \varepsilon_e^{sep}$ are the volume fractions of the electrolyte in the cathode, anode, and separator, respectively, and L^+, L^-, L^{sep} are the lengths of the cathode, anode, and separator. The effects of electrolyte dynamics on the terminal voltage equation is presented as follows, considering:

$$\begin{aligned} V(t) = \quad & \frac{RT_{avg}(t)}{\alpha F}\sinh^{-1}\left(\frac{I(t)}{2a^+AL^+i_0^+(t)}\right) - \frac{RT_{avg}(t)}{\alpha F}\sinh^{-1}\left(\frac{-I(t)}{2a^-AL^-i_0^-(t)}\right) \\ & +U^+\left(c_{ss}^+(t)\right) - U^-\left(c_{ss}^-(t)\right) + \left(\frac{R_f^+}{a^+AL^+} + \frac{R_f^-}{a^-AL^-} + \frac{R_{ce}\left(T_{avg}(t)\right)}{A}\right)I(t) \\ & + \left(\frac{L^+ + 2L^{sep} + L^-}{2A\overline{\kappa}^{eff}\left(T_{avg}\right)}\right)I(t) + k_{conc}(t)\left[\ln c_e(0^+,t) - \ln c_e(0^-,t)\right] \end{aligned} \tag{9}$$

where

$$i_0^\pm(t) = k^\pm\left(T_{avg}\right)[c_{ss}^\pm(t)]^{\alpha_c}\left[c_e^\pm(x,t)\left(c_{s,max}^\pm - c_{ss}^\pm(t)\right)\right]^{\alpha_a} \tag{10}$$

Most model parameters of battery cells are relative to the temperature and are expressed by the following equation:

$$P\left(T_{avg}\right) = P_0\exp\left(\frac{E_{ap}}{R}\left(\frac{1}{T_0} - \frac{1}{T_{avg}}\right)\right) \tag{11}$$

where P_0 is the parameter when the temperature is equal to T_0; and E_{ap} is the activation energy of parameter P. The average temperature T_{avg} can be obtained by the following equations:

$$\frac{\partial T(r_1,t)}{\partial t}\frac{1}{\alpha} = \frac{\partial T^2(r_1,t)}{\partial r_1^2} + \frac{1}{r_1}\frac{\partial T(r_1,t)}{\partial r_1} + \frac{\dot{q}}{k} \tag{12}$$

$$T_{avg}(t) = \frac{1}{r_1}\int_0^{r_1} T(s,t)ds \tag{13}$$

with the boundary conditions

$$\frac{\partial T(0,t)}{\partial r_1} = 0, \frac{\partial T(R_b,t)}{\partial r_1} = U = \frac{h}{k}(T_a - T(R_b,t)) \tag{14}$$

where $r \in [0, R_b]$ is the radial coordinate of a cylindrical battery; $\alpha = k/(\rho C_p)$ is proportional to the average thermal conductivity k; ρ is the average mass density and C_p is the average specific thermal capacity; $\dot{q}$ is the volumetric heat generated rate; and T_a is the ambient temperature. $\dot{q}$ can be computed by the following equation:

$$\dot{q} = I(t) \times \left(U_o(t) - U_{ter} - \overline{T}(t)\frac{\partial U_o(t)}{\partial \overline{T}(t)}\right) \tag{15}$$

where $U_o(t)$ is the open circuit voltage, which is a function of the boundary concentration; $U_{ter} = V(t)$ is the terminal voltage; and $\partial U_o(t)/\partial \overline{T}(t)$ is the entropic heat generation which is too small to neglect. The open circuit voltage $U_o(t)$ can be given by the following equation:

$$U_o(t) = U^+(c_{ss}^+(t)) - U^-(c_{ss}^-(t)) \tag{16}$$

Then, the volumetric heat generated rate $\dot{q}$ can be rewritten as follows:

$$\begin{aligned}
\dot{q} &= I(t) \times (U_o(t) - V(t)) = I(t) \times \left\{ \frac{RT_{avg}(t)}{\alpha F}\sinh^{-1}\left(\frac{I(t)}{2a^+AL^+i_0^+(t)}\right) \right. \\
&- \frac{RT_{avg}(t)}{\alpha F}\sinh^{-1}\left(\frac{-I(t)}{2a^-AL^-i_0^-(t)}\right) + \left(\frac{R_f^+}{a^+AL^+} + \frac{R_f^-}{a^-AL^-} + \frac{R_{ce}(T_{avg}(t))}{A}\right)I(t) \\
&+ \left. \left(\frac{L^+ + 2L^{sep} + L^-}{2A\overline{\kappa}^{eff}(T_{avg})}\right)I(t) + k_{conc}(t)[\ln c_e(0^+, t) - \ln c_e(0^-, t)] \right\}
\end{aligned} \tag{17}$$

2.2. Intercalation-Induced Stress

The intercalation-induced stress generated by lithium-ion intercalation and deintercalation influences the diffusion of lithium and even causes electrode particle fracture. This electrode particle fracture will accelerate aging. Considering the effects of intercalation-induced stress helps us to investigate the fast-charging scheme to reduce the electrode particle fracture and thus prolong the lifetime of the lithium-ion battery. The intercalation-induced stresses are composed of radial stress σ_r and tangential stress σ_t, which are dependent of the lithium concentration [22]:

$$\begin{aligned}
\sigma_r^{\pm}(r, t) &= 2\beta^{\pm}\left[\frac{1}{\left(R_s^j\right)^3}\int_0^{R_s^{\pm}}\widetilde{c}_s^{\pm}r^2 dr - \frac{1}{r^3}\int_0^r \widetilde{c}_s^{\pm}\rho^2 d\rho\right] \\
\sigma_t^{\pm}(r, t) &= \beta^{\pm}\left[\frac{2}{\left(R_s^j\right)^3}\int_0^{R_s^{\pm}}\widetilde{c}_s^{\pm}r^2 dr + \frac{1}{r^3}\int_0^r \widetilde{c}_s^{\pm}\rho^2 d\rho - \widetilde{c}_s^{\pm}\right]
\end{aligned} \tag{18}$$

where, $\beta^{\pm} = \Omega^{\pm}E^{\pm}/3(1 - v^{\pm})$. $\Omega^{\pm}$, $E^{\pm}$, and $v^{\pm}$ are the partial molar volume, Young's Modulus, and Poisson's ratio of the electrode material in the cathode and anode, respectively, which have different sensitivities to temperature. $\widetilde{c}^{\pm}$ is the concentration change from the stress-free value. The resultant stress is a weighted sum of σ_r and σ_t:

$$\sigma_{res}^{\pm} = \frac{\sigma_r^{\pm} + 2\sigma_t^{\pm}}{3} = \frac{2}{3}\beta^{\pm}\left[\frac{3}{(R_s^{\pm})^3}\int_0^{R_s^{\pm}}\widetilde{c}_s^{\pm}r^2 dr - \widetilde{c}_s^{\pm}\right] \tag{19}$$

From [23], we know that the maximum absolute values of radial and tangential stress are located at the center and the surface of the particle, respectively. Taking the anode as an example, we have

$$\begin{aligned}
\sigma_{r,max}^-(t) &= 2\beta^-\left[\frac{1}{(R_s^-)^3}\int_0^{R_s^-}\widetilde{c}_s^-r^2 dr - \frac{1}{3}\widetilde{c}_s^-(0, t)\right] \\
\sigma_{t,max}^-(t) &= \beta^-\left[\frac{3}{(R_s^-)^3}\int_0^{R_s^-}\widetilde{c}_s^-r^2 dr - \widetilde{c}_s^-(R_s^-, t)\right]
\end{aligned} \tag{20}$$

According to Equation (19), we find that the maximum value of the resultant stress is located at the point where the lithium concentration is the minimum value of the electrode particle. That is

$$\bar{\sigma}_{res,max} = \frac{2}{3}\beta^{-}\left[\frac{3}{(R_s^{-})^3}\int_0^{R_s^{-}} \widetilde{c}_s^{-} r^2 dr - \widetilde{c}_{s,min}^{-}\right] \tag{21}$$

Figure 1 shows the input current under Urban Dynamometer Driving Schedule (UDDS), which is applied to a battery cell. Then, the maximum absolute values of radial stress, tangential stress, and the resultant stresses are investigated. According to Equation (21), the intercalation-induced stresses are caused by the lithium-ion concentration nonuniformity in space, which is related to the values of the charging and discharging current. When the current is positive, the lithium-ion concentration in the surface of the particle is smaller than in the center of the particle, that is, the maximum radical stress is positive and the maximum tangential stress is negative, and vice versa. From the input current under UDDS, we know that the current is rapidly changing, which can cause the oscillation of lithium-ion concentration and enlarge the imbalance of lithium-ion concentration in space. Hence, the maximum absolute values of radial stress, tangential stress, and the resultant stresses change rapidly over time, as shown in Figures 2–4. From the three figures, we can see that when the input current is negative, the absolute values of the three stresses will increase gradually, and vice versa. This is because the negative current increases the lithium-ion concentration nonuniformity in space and the positive current decreases this nonuniformity. Besides that, we can find that the absolute values of the three stresses have the same changing curve, which agrees with Equations (20) and (21). Another finding is that the value of the maximum tangential stress is larger than the maximum resultant stress, and the maximum resultant stress is always larger than the maximum radial stress. Hence, by adjusting the input current, the radial stress, tangential stress, and the resultant stress can be controlled to decrease the aging of a battery cell.

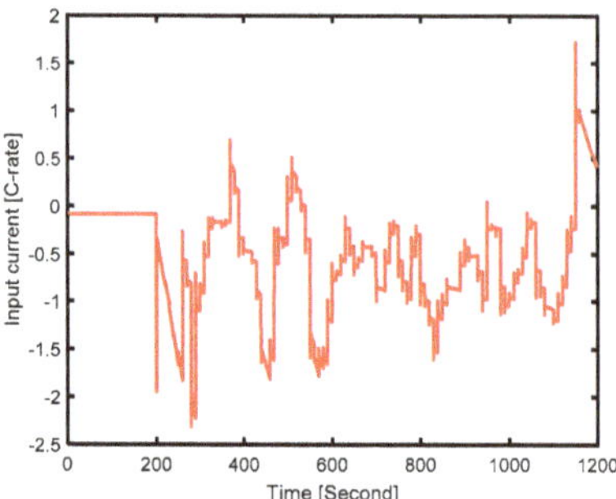

Figure 1. Input current under UDDS.

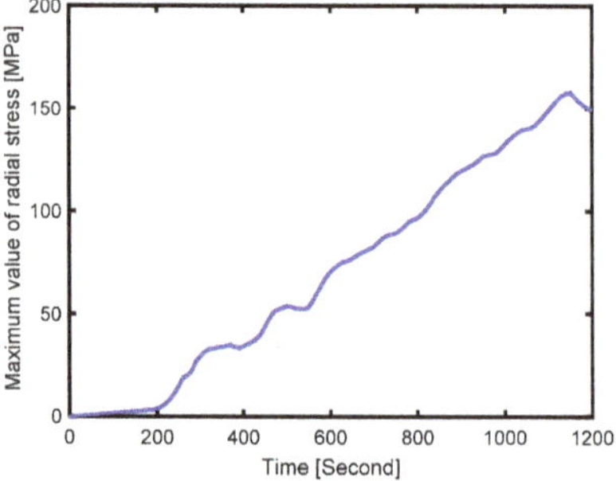

Figure 2. Maximum radial stress.

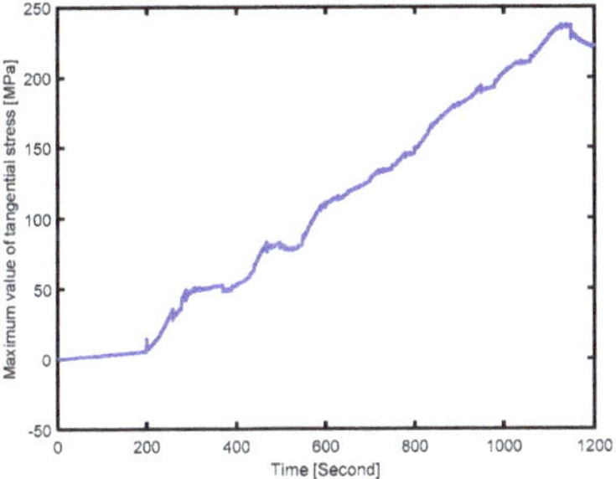

Figure 3. Maximum tangential stress.

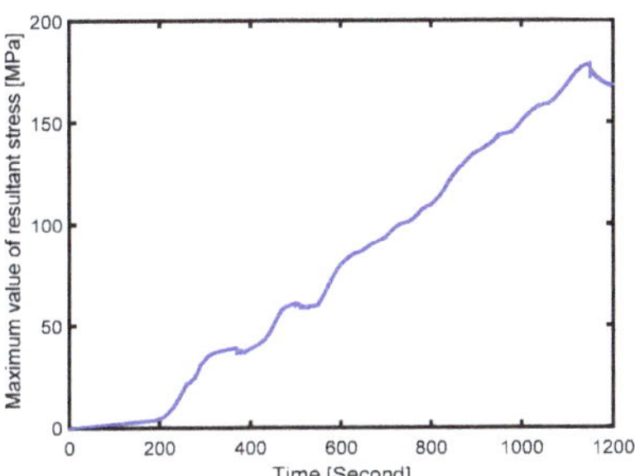

Figure 4. Maximum resultant stress.

2.3. Film Growth Model

Ramadass et al. [24] studied a resistive film formed on the anode electrode/electrolyte interphase, which is the main cause of capacity loss. The veracious chemical side reaction generating the resistive film lies on the chemical component of the anode electrode and electrolyte. Furthermore, a simplified and universal model for capacity reduction based on film growth originated from a single particle model is presented here:

$$S + Li^+ + e^- \rightarrow P \tag{22}$$

where S is the type of solvent, and P is the reaction product.

On account of this nonreversible side reaction, the reaction product builds up a film at the interface of electrode/electrolyte with a thickness varying over time $\delta_{film}(t)$. The irreversibly formed membrane together with the initial solid interphase resistance R_{SEI} forms the total resistance of the electrode/electrolyte interface as follows:

$$R_{film}(t) = R_{SEI} + \frac{\delta_{film}(t)}{\kappa_P} \tag{23}$$

where κ_p is the conductivity of the formed film.

The key state related to the increasing rate of interface film thickness, because of the unwanted lithium loss, is described by

$$\frac{\partial \delta_{film}(t)}{\partial t} = -\frac{M_P}{a_n \rho_P F} J_s(t) \tag{24}$$

where, a_n, M_p, ρ_p, and F are the formed film's specific surface area, molecular weight, mass density, and Faraday's constant, respectively. The variable $J_s(t)$ is the local volume current density used for the side reaction and described by Butler–Volmer kinetics. Assuming that the lithium reduction reaction is nonreversible and the concentration of lithium ions in the solution changes little, it is possible to approximate $J_s(t)$ through the following equation:

$$J_s(t) = -i_{0,s}a_n e^{\left(\frac{-0.5F}{RT_{avg}}\eta_s(t)\right)}$$

(25)

where $i_{0,s}$ is current density occupied by the lithium-loss side reaction; R is the universal gas constant; and T_{avg} is the average temperature of the battery cell. The component $\eta_s(t)$ denotes the overpotential generated by the side reaction, which is expressed by the following equation:

$$\eta_s(t) = \Delta\phi(t) - U_{s,ref} - \frac{J_{tot}(t)}{a_n}R_{film}(t)$$

(26)

$$J_{tot} = J_1 + J_s$$

(27)

where $\Delta\phi(t)$ denotes the difference between the solid phase potential and electrolyte potential. The variable $U_{s,ref}$ is the reference equilibrium potential, which is considered as a constant value. The total intercalation current J_{tot} is used for the anode-side solution, which is presented by the sum of the current between the solid and solution J_1, and the solution decreasing reaction and solution J_s.

Under three different charging currents (1 C, 2 C, and 3 C), the resistance film growth rates are computed according to Equations (23)–(27) and are presented in Figure 5. From this figure, we see that the film growth rate increases as the charging current or the time increases, owing to the augmentation of the reduction reaction rate and the rise in $\Delta\phi(t)$ according to Equation (6). This means that a continuous high current and a high SoC is detrimental to reducing the lithium loss.

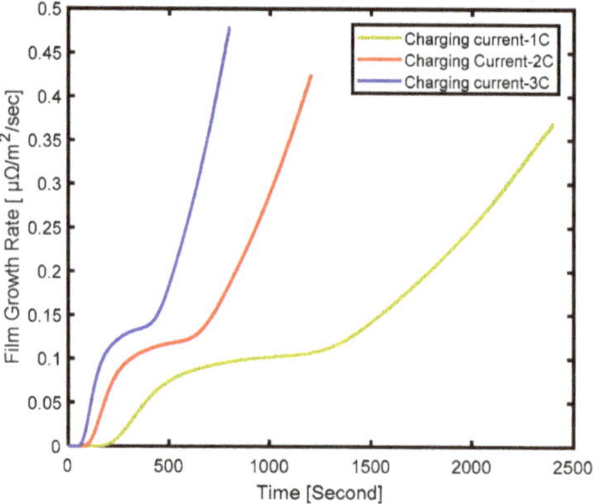

Figure 5. Film growth under three charging current.

3. Temperature Control Based on the Backstepping Technique

The core temperature of a battery cell is higher than its surface temperature, hence, the control of the core temperature being more significant. Confining the maximum temperature of the battery to a certain value can guarantee both the safe operation and efficient charging. However, the core temperature of the battery is not accessible and cannot be measured by sensor directly. Therefore, we have to estimate the core temperature of a battery cell to make sure that the control goal is achieved. H. E. Perez et al. utilized the thermal model based on the equivalent circuit model to obtain the core temperature of a battery cell, which consists of the heat conduction resistance R_c, convection resistance R_u, core heat capacity C_c, and surface heat capacity C_s [25]. The thermal model used in this paper is described by the partial differential equation, which can accurately present the evolution of spatial temperature and is more accurate than the model employed by H. E. Perez et al.

To maintain the temperature of the battery cell where we want it to be, we can adjust the temperature or flow velocity of the cooling fluid. Herein, a backstepping method is utilized to design the control law for a prespecified time-invariant reference temperature $T_{ref}(r_1)$.

The following error system is thus introduced:

$$T_e(r_1,t) = T(r_1,t) - T_{ref}(r_1), \quad T_{ref}(r_1) = \frac{T_{sref} - T_{cref}}{R_b^2}r_1^2 + T_{cref} \tag{28}$$

where the expected temperature distribution $T_{ref}(r_1)$ is a quadratic parabola, which is in accordance with the true case; T_{sref} and T_{cref} are the reference surface temperature and core temperature, respectively.

The time derivative of $T_e(r_1,t)$ can be obtained according to Equations (12) and (28).

$$\begin{aligned}
\frac{\partial T_e(r_1,t)}{\partial t} &= \frac{\partial T(r_1,t)}{\partial t} - \frac{\partial T_{ref}(r_1)}{\partial t} = \frac{\partial T^2(r_1,t)}{\partial r_1^2} + \frac{1}{r_1}\frac{\partial T(r_1,t)}{\partial r_1} + \frac{\dot{q}}{k} \\
&= \frac{\partial T_e^2(r_1,t)}{\partial r_1^2} + \frac{1}{r_1}\frac{\partial T_e(r_1,t)}{\partial r_1} + 2\frac{T_{sref} - T_{cref}}{R_b^2} + \frac{\dot{q}}{k}
\end{aligned} \tag{29}$$

with the conditions

$$\frac{\partial T_e(0,t)}{\partial r_1} = 0, \frac{\partial T_e(R_b,t)}{\partial r_1} = U - 2\frac{T_{sref} - T_{cref}}{R_b}$$

where U is the boundary control law, which is applied to the boundary condition of Equation (12) for restricting $T(r_1,t)$ to around $T_{ref}(r_1)$.

Here, the following invertible backstepping transformation is adopted:

$$w(r_1,t) = T_e(r_1,t) - \int_0^{r_1} K(r_1,\rho)T_e(\rho,t)d\rho \tag{30}$$

which maps (29) into the stable system:

$$w_t = w_{r_1 r_1} + \frac{w_{r_1}}{r_1} - \frac{cw}{r_1^2} \tag{31}$$

with boundary conditions

$$w_{r_1}(0,t) = 0, w_{r_1}(R_b,t) = 0 \tag{32}$$

where $c > 0.5$. $K(r_1,\rho)$ can be solved easily by the following kernel function:

$$K_{r_1 r_1} + \frac{K_{r_1}}{r_1} - K_{\rho\rho} + \frac{K_\rho}{\rho} - \frac{K}{\rho^2} = \frac{c}{r_1^2}K \tag{33}$$

with the following conditions:

$$K(r_1,0) = 0, \ K_\rho(r_1,0) = 0, \ K(r_1,r_1) = -\int_0^r \frac{c/r_1^2}{2}d\rho \tag{34}$$

The detailed solution process can be seen in [26].

According to the boundary condition $w_{r_1}(R_b,t) = 0$ and Equation (29), we have

$$\frac{\partial T_e(R_b,t)}{\partial r_1} = K(R_b,R_b)T_e(R_b,t) + \int_0^{R_b} K_{r_1}(R_b,\rho)T_e(\rho,t)d\rho \tag{35}$$

Then, the control law is obtained:

$$U = K(R_b,R_b)T_e(R_b,t) + \int_0^{R_b} K_{r_1}(R_b,\rho)\big[T(\rho,t) - T_{ref}(\rho)\big]d\rho + 2\frac{T_{sref} - T_{cref}}{R_b} \tag{36}$$

The control law requires access to the full states $T(\rho, t)$. However, the core temperature cannot be measured by sensor directly. For that, the following boundary temperature observer is proposed:

$$\frac{\partial \hat{T}(r_1, t)}{\partial t} \frac{1}{\alpha} = \frac{\partial \hat{T}^2(r_1, t)}{\partial r_1^2} + \frac{1}{r_1} \frac{\partial \hat{T}(r_1, t)}{\partial r_1} + \frac{\dot{q}}{k} + P(r_1)(T(R_b, t) - \hat{T}(R_b, t)) \tag{37}$$

with boundary conditions

$$\frac{\partial \hat{T}(0, t)}{\partial r_1} = 0, \quad \frac{\partial \hat{T}(R_b, t)}{\partial r_1} = U = \frac{h}{k}(T_a - T(R_b, t))$$

The estimation error $\widetilde{T}(r_1, t) = T(r_1, t) - \hat{T}(r_1, t)$ is introduced, then substituting Equation (37) for Equation (12), one has

$$\frac{\partial \widetilde{T}(r_1, t)}{\partial t} \frac{1}{\alpha} = \frac{\partial \widetilde{T}^2(r_1, t)}{\partial r_1^2} + \frac{1}{r_1} \frac{\partial \widetilde{T}(r_1, t)}{\partial r_1} - P(r_1)(\widetilde{T}(R_b, t)) \tag{38}$$

with boundary conditions

$$\frac{\partial \widetilde{T}(0, t)}{\partial r_1} = 0, \quad \frac{\partial \widetilde{T}(R_b, t)}{\partial r_1} = 0$$

Using the following backstepping transform, which is similar with (30), we have

$$w(r_1, t) = \widetilde{T}(r_1, t) - \int_0^{r_1} M(r_1, \rho)\widetilde{T}(\rho, t)d\rho \tag{39}$$

where $w(r_1, t)$ satisfies the following stable system:

$$w_t = w_{r_1 r_1} + \frac{w_{r_1}}{r_1} - \frac{dw}{r_1^2} \tag{40}$$

with boundary conditions

$$w_{r_1}(0, t) = 0, w_{r_1}(R_b, t) = 0$$

where $d > 0.5$ as well. $M(r_1, \rho)$ satisfies the following kernel function:

$$M_{r_1 r_1} + \frac{M_{r_1}}{r_1} - M_{\rho\rho} + \frac{M_\rho}{\rho} - \frac{M}{\rho^2} = -\frac{c}{r_1^2}M \tag{41}$$

with the boundary conditions:

$$M_\rho(0, \rho) = 0, \quad M(r_1, r_1) = -\int_0^{r_1} \frac{c/r_1^2}{2} d\rho$$

The solution procedure of $M(r_1, \rho)$ is similar to $K(r_1, \rho)$.

To verify the performance of the backstepping-based observer and controller, a constant current of 2 C is applied to a battery cell; the curve of the temperature rising is presented in Figure 6. $r = 0$ and $r = 1$ represent the center and the surface of a battery cell, respectively. The two horizontal lines are the target temperature of the surface and the center of a battery cell. From Figure 6, we can see that the temperature estimation of both the surface and the center is capable of tracking the true temperature and approaching the target temperature gradually. That is to say that we can control the temperature to a safe range using the proposed temperature controller.

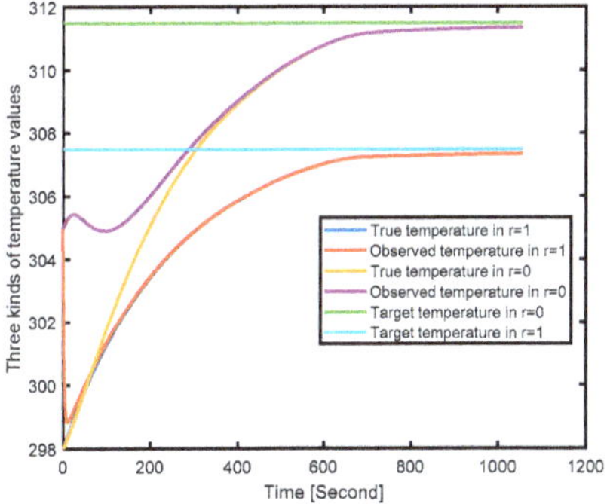

Figure 6. Estimation and control of temperature.

4. Problem Formulation

This paper concentrates on the minimization of the charging time in a given target SoC with limits involving current, voltage, intercalation-induced stress, and film formation. The simultaneous nonlinear programming approach is used in this paper. Considering the optimal charging profile with predetermined final SoC under the objective of minimization of charging time, the objective function J is given by

$$J = \min_{I(t),s(t),t_f} \int_{t_0}^{t_f} 1 \cdot dt \tag{42}$$

with bounds

$$I_{\min} \leq I(t) \leq I_{\max} \tag{43}$$

$$t_0 \leq t_f \leq t_{\max} \tag{44}$$

$$V_{\min}(t) \leq V(t) \leq V_{\max}(t) \tag{45}$$

$$\sigma_{r,\max}^-(t) \leq \sigma_{r,\text{upper}}^- \tag{46}$$

$$-\sigma_{t,\max}^-(t) \leq \sigma_{t,\text{upper}}^- \tag{47}$$

$$\partial \dot{\delta}_{film}(t) \leq \partial \dot{\delta}_{film\_\max}(t) \tag{48}$$

where $I(t)$ is the input current (A); t_f is the final time (s); $V(t)$ is the terminal voltage (V); $\sigma_{r,\text{upper}}^-(t)$ and $\sigma_{t,\text{upper}}^-$ are the radial and tangential stress upper bounds, respectively; and $\partial \dot{\delta}_{film\_\max}(t)$ is the upper bound of the film growth rate.

5. Simulation and Results

In order to verify the performance of the proposed optimal fast-charging method, CC–CV, the most common pattern of fast charging, was used for the comparative study, including four aspects: charging time, maximum tangential stress, maximum radial stress, and film growth rate.

In the process of fast charging a lithium battery, the temperature of the battery cell grows quickly and the aging rate increases. To avoid overheating, the backstepping method is used to keep the battery cell at a constant temperature, as shown in Figure 7. A constant temperature in the battery cell not only helps avoid overheating, but also helps compare the performance of the proposed fast-charging method with the CC–CV method. In order to find a compromise between the charging time and battery aging, the proposed optimal fast-charging method in this paper considers two aging factors: intercalation-induced stress and film growth. In the comparative study, the CC–CV pattern adopts three kinds of maximum charging current: 1 C, 2 C, and 3 C. This means that the effect of the choice of the maximum charging current on the comparative results is avoided.

In this paper, the upper limits of the radial and tangential stress are predetermined and optimal charging curves were obtained using the dynamic optimization approach. As we know, when charging occurs, the maximum tangential stress is located at the surface of the particle, and the maximum radial stress is located at the center of the particle. For an anode made of graphite, neither the maximum tangential stress nor the maximum radial stress is not allowed to exceed the yielding stress for any length of time to reduce the risk of anode fracture. The following case is discussed: charging a battery from 0% SoC to 100% SoC.

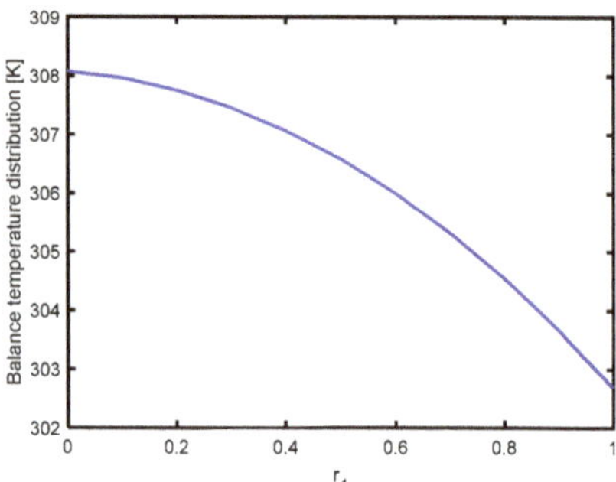

Figure 7. The balance temperature distribution.

Figure 7 presents the balance of temperature distribution. If this balance is broken, the temperature controller would work to drag the temperature to the target position using various methods, such as changing the fluid temperature and adjusting the fluid flow velocity. Thus, a stable cell temperature can be achieved for safe operation. Figures 8 and 9 show the current and voltage curves of charging a battery from 0% SoC to 100% SoC by utilizing the CC–CV pattern under three maximum charging currents. At the initial charging stage, the input current maintains at the maximum and the terminal voltage climbs fast. When the output voltage reaches the upper value, this voltage is held until the battery cell is charged fully. With the increase in the maximum charging current, the time cost for full charging becomes less and less. Figure 10 indicates the optimal charging current obtained by the dynamic optimization method. The upper value and lower values of the constraints are listed in Table 1.

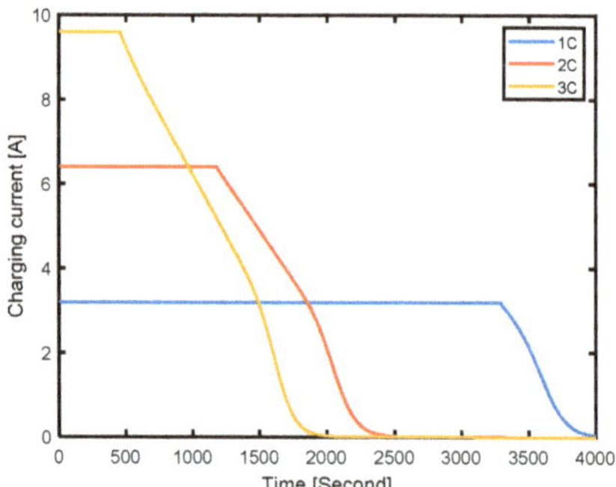

Figure 8. Three Current profiles under CC–CV.

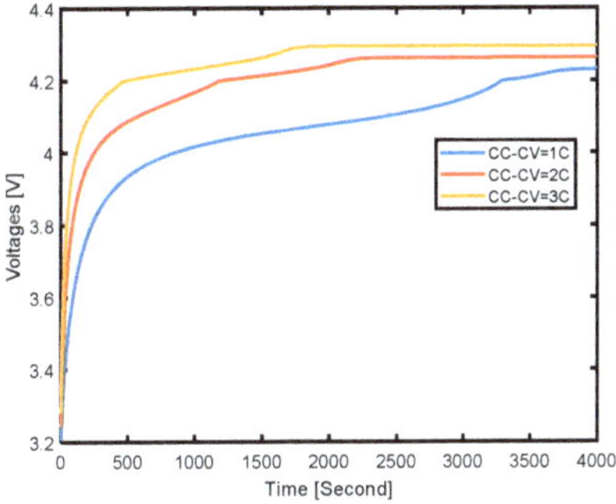

Figure 9. Corresponding voltages under CC–CV.

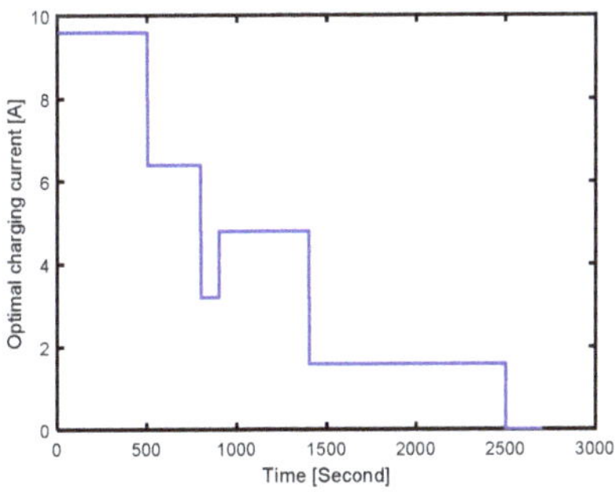

Figure 10. Optimal charging profile.

Table 1. The upper value or lower values of the constraints.

Range	$I(t)$ (C)	t_f (min)	$V(t)$ (V)	$\sigma^-_{r,\max}(t)$ (MPa)	$-\sigma^-_{t,\max}(t)$ (MPa)	$\partial\dot{\delta}_{film}(t)$ ($\mu\Omega/\text{m}^2/\text{s}$)
Upper	3	80	4.3	160	220	0.5
Lower	0	15	2	0	0	0

From Figure 11, we see that the area under the film growth curve is smaller than those of the other three curves, which means the least lithium loss during full charging. At the same time, the charging time of the optimal charging profile just takes 200 s more than the time cost under CC–CV with a 3 C current, which is fast enough and acceptable.

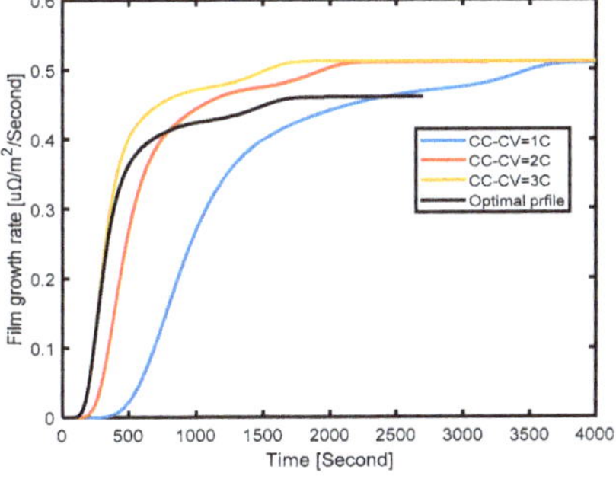

Figure 11. Film growth rate under four profiles.

The yielding stress of the radial stress and tangential stress in a graphite-based anode are both around 30 MPa. If the radial stress or tangential stress is more than 30 Mpa for a long time, it is possible to cause fatigue failure and accelerate the aging process. From Figure 12, the tangential stress generated by optimal charging profile becomes greater than its yielding stress within 1400 s, while the tangential stress under CC–CV with 1 C, 2 C, and 3 C is larger than 30 MPa for more than 1700 s, 2200 s, and 3700 s, respectively. From the maximum values of radial stress as shown in Figure 13, we can see that the optimal charging profile causes a shorter period of overstress, which represents less electrode fatigue damage. Hence, the result shows that the optimal charging profile obtained by the dynamic optimization method has less probability of causing stress than the CC–CV profiles and so can slow the aging process and extend the battery life.

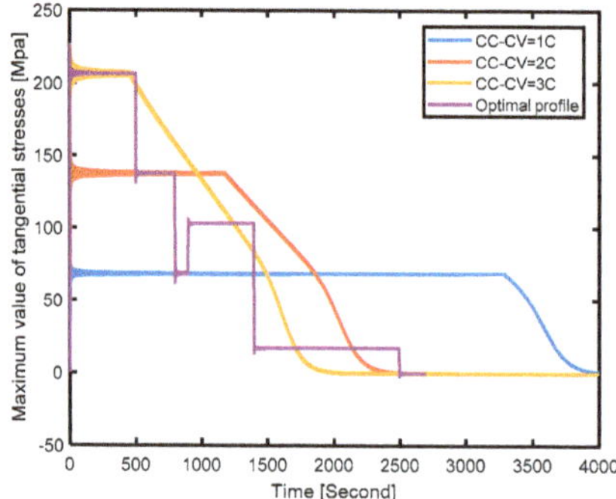

Figure 12. Maximum value of tangential stress.

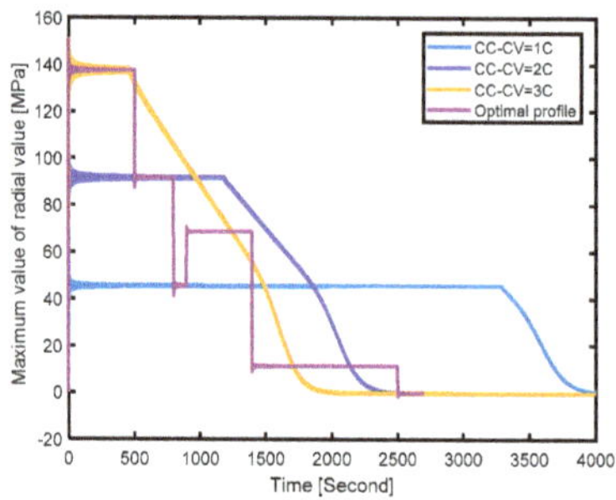

Figure 13. Maximum value of radial stress.

Therefore, while optimizing the charging speed, the proposed optimal charging method offers a good compromise between charging time and battery aging.

6. Conclusions

In this paper, the electrochemical–thermal model was employed to obtain the optimal charging profile and control the temperature of the cell. A temperature controller-based backstepping method was proposed to keep a relatively constant cell temperature to avoid overheating. Then, the effects of intercalation-induced stress were considered, because these cause particle fracture, which can accelerate the aging process. Furthermore, the growth of film on the surface of particles was also taken into account due to its ability to reduce the reused lithium ions. After that, the optimization objective was established, which minimizes the time cost during charging, while confining intercalation-induced stress and the growth rate of surficial film to a given range. Finally, the simulation was implemented and the results show that the film growth of the optimal charging curve is smaller than that of the other three charging curves, thus demonstrating the least lithium loss during full charging. Besides that, the tangential stress and the radial stress generated by the optimal charging profile were both greater than their yielding stresses for less time during charging as compared with the CC–CV pattern, which means there is less risk of electrode fatigue fracture. Furthermore, the charging time of the optimal charging profile just takes 200 s more than the time cost under CC–CV with a 3 C current, which is fast enough and acceptable. Hence, the proposed optimal charging method offers a good compromise between charging time and battery aging.

Author Contributions: Conceptualization, G.C. and Z.L.; methodology, H.S.; software, G.C.; validation, Z.L., G.C. and H.S.; formal analysis, Z.L.; investigation, G.C.; resources, H.S.; data curation, Z.L.; writing—original draft preparation, G.C.; writing—review and editing, G.C.; visualization, Z.L.; supervision, H.S.; project administration, Z.L.; funding acquisition, H.S. All authors have read and agreed to the published version of the manuscript.

Funding: This work was supported in part by the National Key R &D Program of China (Grant NO.2018YFA0703800), Science Fund for Creative Research Group of the National Natural Science Foundation of China (Grant NO.61621002), Nation Natural Science Foundation of China (NSFC:61873233, 61633019), Fundamental Research Funds for the Central Universities.

Conflicts of Interest: The authors declare no conflict of interest.

Abbreviations

The following abbreviations are used in this manuscript:

BMS	Batteries management system
DP	Dynamic optimization
SoC	State of Charge
CC-CV	Constant current-constant voltage
CC	Constant current
IC	Incremental capacity
OCV	Open circuit voltage
TTC	Time-to-charge
EL	Energy losses
TRI	Temperature rise index
P2D	Pseudo-two dimensional
MPC	Model predictive control
CT-CV	Constant temperature-constant voltage
LGR	Legendre-Gauss-Radau
PID	Proportional-integral-derivative
SPM	Single particle model
PDE	Partial differential equation
ODE	Ordinary differential equation
UDDS	Urban Dynamometer Driving Schedule

References

1. Dehghani-Sanij, A.; Tharumalingam, E.; Dusseault, M.; Fraser, R. Study of energy storage systems and environmental challenges of batteries. *Renew. Sustain. Energy Rev.* **2019**, *104*, 192–208. [CrossRef]
2. Liu, Y.; Luo, Y. Search for an Optimal Rapid-Charging Pattern for Li-Ion Batteries Using the Taguchi Approach. *IEEE Trans. Ind. Electron.* **2010**, *57*, 3963–3971. [CrossRef]
3. Abdollahi, A.; Raghunathan, N.; Han, X.; Avvari, G.V.; Balasingam, B.; Pattipati, K.R.; Bar-Shalom, Y. Battery Charging Optimization for OCV-Resistance Equivalent Circuit Model. In Proceedings of the 2015 American Control Conference Palmer House Hilton, Chicago, IL, USA, 1–3 July 2015.
4. Abdollahi, A.; Han, X. Optimal battery charging, Part I: Minimizing time-to-charge, energy loss, and temperature rise for OCV-resistance battery model. *J. Power Sources* **2016**, *303*, 388–398. [CrossRef]
5. Monem, M.A.; Trad, K.; Omar, N.; Hegazy, O.; Mantels, B.; Mulder, G.; Bossche, P.V.D.; Van Mierlo, J. Lithium-ion batteries: Evaluation study of different charging methodologies based on aging process. *Appl. Energy* **2015**, *152*, 143–155. [CrossRef]

6. Liu, K.; Li, K.; Yang, Z.; Zhang, C.; Deng, J. An advanced Lithium-ion battery optimal charging strategy based on a coupled thermoelectric model. *Electrochim. Acta* **2017**, *225*, 330–344. [CrossRef]

7. Fang, H.Z.; Wang, Y.B.; Chen, J. Health-Aware and User-Involved Battery Charging Management for Electric Vehicles: Linear Quadratic Strategies. *IEEE Trans. Control Syst. Technol.* **2017**, *25*, 911–923. [CrossRef]

8. Patnaik, L.; Praneeth, A.V.J.S.; Williamson, S.S. A Closed-Loop Constant-Temperature Constant-Voltage Charging Technique to Reduce Charge Time of Lithium-Ion Batteries. *IEEE Trans. Ind. Electron.* **2019**, *66*, 1059–1067. [CrossRef]

9. Klein, R.; Chaturvedi, N.A.; Christensen, J.; Ahmed, J.; Findeisen, R.; Kojic, A. Optimal charging strategies in lithium-ion battery. In Proceedings of the 2011 American Control Conference, San Francisco, CA, USA, 29 June–1 July 2011. [CrossRef]

10. Pramanik, S.; Anwar, S. Electrochemical model based charge optimization for lithium-ion batteries. *J. Power Sources* **2016**, *313*, 164–177. [CrossRef]

11. Perez, H.E.; Dey, S.; Hu, X.; Moura, S.J. Optimal Charging of Li-Ion Batteries via a Single Particle Model with Electrolyte and Thermal Dynamics. *J. Electrochem. Soc.* **2017**, *164*, A1–A9. [CrossRef]

12. Anseán, D.; Gonzalez, M.; Viera, J.C.; García, V.; Blanco, C.; Valledor, M. Fast charging technique for high power lithium iron phosphate batteries: A cycle life analysis. *J. Power Sources* **2013**, *239*, 9–15. [CrossRef]

13. Anseán, D.; Dubarry, M.; Devie, A.; Liaw, B.Y.; García, V.; Viera, J.; Gonzalez, M. Fast charging technique for high power LiFePO4 batteries: A mechanistic analysis of aging. *J. Power Sources* **2016**, *321*, 201–209. [CrossRef]

14. Suthar, B.; Ramadesigan, V.; De, S.; Braatz, R.D.; Subramanian, V.R. Optimal charging profiles for mechanically constrained lithium-ion batteries. *Phys. Chem. Chem. Phys.* **2014**, *16*, 277–287. [CrossRef] [PubMed]

15. Torchio, M.; Magni, L.; Braatz, R.; Raimondo, D.M. Optimal Health-aware Charging Protocol for Lithium-ion Batteries: A Fast Model Predictive Control Approach. In Proceedings of the International Federation of Automatic Control, Sherbrooke, QC, Canada, 21–25 August 2016.

16. Abdel-Monem, M.; Trad, K.; Omar, N.; Hegazy, O.; Bossche, P.V.D.; Van Mierlo, J. Influence analysis of static and dynamic fast-charging current profiles on ageing performance of commercial lithium-ion batteries. *Energy* **2017**, *120*, 179–191. [CrossRef]

17. Umair Ali, M.; Hussain Nengroo, S.; Adil Khan, M.; Zeb, K.; Ahmad Kamran, M.; Kim, H.J. A Real-Time Simulink Interfaced Fast-Charging Methodology of Lithium-Ion Batteries under Temperature Feedback with Fuzzy Logic Control. *Energies* **2018**, *11*, 1122. [CrossRef]

18. Wang, Y.; Hu, X.; Wei, Z.; Wik, T.; Egardt, B. Electrochemical Estimation and Control for Lithium-Ion Battery Health-Aware Fast Charging. *IEEE Trans. Ind. Electron.* **2018**, *65*, 6635–6645.

19. Lin, X.; Hao, X.; Liu, Z.; Jia, W. Health conscious fast charging of Li-ion batteries via a single particle model with aging mechanisms. *J. Power Sources* **2018**, *400*, 305–316. [CrossRef]

20. Xu, M.; Wang, R.; Zhao, P.; Wang, X. Fast charging optimization for lithium-ion batteries based on dynamic programming algorithm and electrochemical-thermal-capacity fade coupled model. *J. Power Sources* **2019**, *438*, 15–27. [CrossRef]

21. Moura, S.J.; Argomedo, F.B.; Klein, R.; Mirtabatabaei, A.; Krstic, M. Battery State Estimation for a Single Particle Model with Electrolyte Dynamics. *EEE Trans. Control Syst. Technol.* **2016**, *25*, 453–468. [CrossRef]

22. Zhang, D.; Dey, S.; Couto, L.D.; Moura, S.J. Battery Adaptive Observer for a Single Particle Model with Intercalation-Induced Stress. *IEEE Trans. Control. Syst. Technol.* **2019**, 1–15. [CrossRef]

23. Zhang, X.C.; Shyy, W.; Sastrya, A.M. Numerical Simulation of Intercalation-Induced Stress in Li-Ion Battery Electrode Particles. *J. Electrochem. Soc.* **2007**, *154*, A910–A916. [CrossRef]

24. Ramadass, P.; Haran, B.; Gomadam, P.M.; White, R.; Popov, B.N. Development of first principles capacity fade model for Li-ion cells. *J. Electrochem. Soc.* **2004**, *151*, 196–203. [CrossRef]

25. Perez, H.E.; Hu, X.; Dey, S.; Moura, S. Optimal Charging of Li-Ion Batteries with Coupled Electro-Thermal-Aging Dynamics. *IEEE Trans. Veh. Technol.* **2017**, *66*, 7761–7770. [CrossRef]

26. Vazquez, R.; Krstic, M. Boundary control and estimation of reaction–diffusion equations on the sphere under revolution symmetry conditions. *Int. J. Control* **2019**, *92*, 2–11. [CrossRef]

MDPI
St. Alban-Anlage 66
4052 Basel
Switzerland
Tel. +41 61 683 77 34
Fax +41 61 302 89 18
www.mdpi.com

Energies Editorial Office
E-mail: energies@mdpi.com
www.mdpi.com/journal/energies